Springer Proceedings in Physics

Volume 246

Indexed by Scopus

The series Springer Proceedings in Physics, founded in 1984, is devoted to timely reports of state-of-the-art developments in physics and related sciences. Typically based on material presented at conferences, workshops and similar scientific meetings, volumes published in this series will constitute a comprehensive up-to-date source of reference on a field or subfield of relevance in contemporary physics. Proposals must include the following:

- name, place and date of the scientific meeting
- a link to the committees (local organization, international advisors etc.)
- scientific description of the meeting
- list of invited/plenary speakers
- an estimate of the planned proceedings book parameters (number of pages/articles, requested number of bulk copies, submission deadline).

More information about this series at http://www.springer.com/series/361

Olena Fesenko · Leonid Yatsenko
Editors

Nanomaterials and Nanocomposites, Nanostructure Surfaces, and Their Applications

Selected Proceedings of the 7th International Conference Nanotechnology and Nanomaterials (NANO2019), 27–30 August 2019, Lviv, Ukraine

Springer

Editors
Olena Fesenko
National Academy of Sciences of Ukraine
Institute of Physics
Kyiv, Ukraine

Leonid Yatsenko
National Academy of Sciences of Ukraine
Institute of Physics
Kyiv, Ukraine

ISSN 0930-8989 ISSN 1867-4941 (electronic)
Springer Proceedings in Physics
ISBN 978-3-030-51904-9 ISBN 978-3-030-51905-6 (eBook)
https://doi.org/10.1007/978-3-030-51905-6

This Springer imprint is published by the registered company Springer Nature Switzerland AG
The registered company address is: Gewerbestrasse 11, 6330 Cham, Switzerland

Preface

This book highlights the most recent advances in nanoscience from leading researchers in Ukraine, Europe, and beyond. It features contributions from participants of the 7th International Research and Practice Conference "Nanotechnology and Nanomaterials" (NANO-2019), held in Lviv, Ukraine, on August 27–30, 2019. This event was organized jointly by the Institute of Physics of the National Academy of Sciences of Ukraine, Lviv Polytechnic National University (Ukraine), University of Tartu (Estonia), University of Turin (Italy), and Pierre and Marie Curie University (France). Internationally recognized experts from a wide range of universities and research institutes shared their knowledge and key results in the areas of nanooptics and nanophotonics, nanoplasmonics, surface-enhanced spectroscopy, nanocomposites and nanomaterials, nanostructured surfaces, microscopy of nano-objects, nanochemistry, nanobiotechnology, and nanobiotechnology for health care.

Today, nanotechnology is becoming one of the most actively developing and promising fields of science. Numerous nanotechnology investigations are already producing practical results that can be applied in various areas of human life from science and technology to medicine and pharmacology. The aim of this book is to highlight the latest investigations from different areas of nanoscience and to stimulate new interest in this field. Volume II of this two-volume work covers important topics such as structural, morphological, magnetic, and photovoltaic properties of nanostructures, achievements in the field of nanomaterials and nanocomposites, and a new effect in nanoscale physics and its applications.

This book is divided into three sections: Part I—Nanoscale Physics, Part II—Nanostructured Interfaces and Surfaces, and Part III—Nanocomposites and Nanomaterials. Sections covering Nanochemistry and Biotechnology, and Nanooptics and Photonics can be found in Volume II: **Nanooptics and Photonics, Nanochemistry and Nanobiotechnology, and Their Applications**.

The papers published in these five sections fall under the broad categories of nanomaterial preparation and characterization, nanobiotechnology, nanodevices and quantum structures, and spectroscopy and nanooptics. The book will help readers to familiarize with current research and practical applications in nanoscience and thus to promote further implementation of nanotechnologies into innovations according to public needs. We hope that both volumes will be equally useful and interesting for young scientists or Ph.D. students and mature scientists alike.

Kyiv, Ukraine

Olena Fesenko
Leonid Yatsenko

Contents

Nanostructured Interfaces and Surfaces

Contributors

O. Aksimentyeva Ivan Franko National University of Lviv, Lviv, Ukraine

Vitalina Alisova Igor Sikorsky Kyiv Polytechnic Institute, Kyiv, Ukraine

A. E. Artyukhov Processes and Equipment of Chemical and Petroleum-Refineries Department, Sumy State University, Sumy, Ukraine

T. Avramenko Physics Faculty, Research Laboratory of Physics of Metals and Ceramics, Taras Shevchenko National University of Kyiv, Kyiv, Ukraine

V. F. Bashev Oles Honchar Dnipro National University, Dnipro, Ukraine

A. A. Baturin National Technical University "Kharkiv Polytechnic Institute", Kharkiv, Ukraine

O. R. Bediukh Taras Shevchenko National University of Kyiv, Kiev, Ukraine

N. I. Berezovska Physics Faculty, Taras Shevchenko National University of Kyiv, Kiev, Ukraine

Nataliia B. Bielousova Departments of International Information, Taras Shevchenko National University of Kyiv, Kiev, Ukraine

Alexander Bildyukevich Institute of Physical Organic Chemistry, National Academy of Science of Belarus, Minsk, Belarus

Zinoviy Blikharskyy Institute of Building and Environmental Engineering, Lviv Polytechnic National University, Lviv, Ukraine

I. V. Blonskiy Photon Processes Department, Institute of Physics of the NAS of Ukraine, Kiev, Ukraine

Ya. V. Bobitski Department of Photonics, Lviv Polytechnic National University, Lviv, Ukraine;
Faculty of Mathematics and Natural Sciences, University of Rzeszow, Rzeszow, Poland

V. L. Borblik Department of Electrical and Galvanomagnetic Properties of Semiconductors, V.Lashkarev Institute of Semiconductor Physics, Kiev, Ukraine

M. C. Bouleklab Laboratoire de thermodynamique et traitement de surfaces des Matériaux, Université Frères Mentouri Constantine1, Constantine, Algeria

A. Brik M.P. Semenenko Institute of Geochemistry, Mineralogy and Ore Formation, NAS of Ukraine, Kiev, Ukraine

Mary O. Chaban VI Vernadskii Institute of General and Inorganic Chemistry of the National Academy of Science of Ukraine, Kiev, Ukraine

A. P. Chebanenko Odessa Mechnikov National University, Odessa, Ukraine

Vasyl Chumachenko Taras Shevchenko National University of Kyiv, Kiev, Ukraine

Zbigniew Czapla Department of Physics, Opole University of Technology, Opole, Poland

V. V. Datsyuk Kiev, Ukraine

Irina Davidenko Taras Shevchenko National University of Kyiv, Kiev, Ukraine

Nikolay Davidenko Taras Shevchenko National University of Kyiv, Kiev, Ukraine

P. Demchenko Ivan Franko National University of Lviv, Lviv, Ukraine

L. Didukh Ternopil Ivan Puluj National Technical University, Ternopil, Ukraine

I. M. Dmitruk Physics Faculty, Taras Shevchenko National University of Kyiv, Kiev, Ukraine

A. M. Dmytruk Photon Processes Department, Institute of Physics of the NAS of Ukraine, Kiev, Ukraine

Tetiana Dontsova Igor Sikorsky Kyiv Polytechnic Institute, Kyiv, Ukraine

Yu. Dovhopyaty Ternopil Ivan Puluj National Technical University, Ternopil, Ukraine

O. Dubikovskyi Lashkaryov Institute of Semiconductor Physics, Kiev, Ukraine

N. Dudchenko M.P. Semenenko Institute of Geochemistry, Mineralogy and Ore Formation, NAS of Ukraine, Kiev, Ukraine

Alina V. Dvornichenko Sumy State University, Sumy, Ukraine

Yuliya S. Dzyazko VI Vernadskii Institute of General and Inorganic Chemistry of the National Academy of Science of Ukraine, Kiev, Ukraine

O. M. Fesenko Institute of Physics of National Academy Sciences of Ukraine, Kyiv, Ukraine

Gracja Fijałkowska Faculty of Chemistry, Department of Radiochemistry and Environmental Chemistry, Institute of Chemical Sciences, Maria Curie-Sklodowska University in Lublin, Lublin, Poland

L. M. Filevska Odessa Mechnikov National University, Odessa, Ukraine

V. M. Fitio Department of Photonics, Lviv Polytechnic National University, Lviv, Ukraine

Angelina Gab Assoc. Prof., Open International University of Human Development "Ukraine", Kyiv, Ukraine

I. I. Gab Frantsevich Institute for Problems of Materials Science of National Academy Sciences of Ukraine, Kyiv, Ukraine

I. E. Garkusha National Science Center "Kharkiv Institute of Physics and Technology" (NSC KIPT), Institute of Plasma Physics, Kharkiv, Ukraine;
National Technical University of Agriculture, Kharkiv, Ukraine;
National Technical University "Kharkiv Polytechnic Institute", Kharkiv, Ukraine;
V. N. Karazin Kharkiv National University, Kharkiv, Ukraine

Marcelle Gaune-Escard Directeur de Recherche, Ecole Polytechnique, CNRS, UMR 6595, Technopôle deChateau Gombert, Marseille Cedex 13, France

Z. Gburski Silesian Centre for Education and Interdisciplinary Research, Institute of Physics, University of Silesia in Katowice, Chorzów, Poland;
Katowice Institute of Information Technologies, Katowice, Poland

E. I. Get'man Vasyl' Stus Donetsk National University, Vinnytsia, Ukraine

JU. Gomeniuk Lashkaryov Institute of Semiconductor Physics, Kiev, Ukraine

O. Gomenyuk Oleksandr Dovzhenko Hlukhiv National Pedagogical University, Hlukhiv, Ukraine

Alla M. Gorb Departments of Physics, Taras Shevchenko National University of Kyiv, Kyiv, Ukraine

Borys M. Gorelov Chuiko Institute of Surface Chemistry of the National Academy of Sciences of Ukraine, Kyiv, Ukraine

V. S. Grinevych Odessa Mechnikov National University, Odessa, Ukraine

Iurii Gudyma Department of General Physics, Institute of Physical, Technical and Computer Sciences, Yuriy Fedkovych Chernivtsi National University, Chernivtsi, Ukraine

O. Gudymenko Lashkaryov Institute of Semiconductor Physics, Kiev, Ukraine

S. Hamamda Laboratoire de thermodynamique et traitement de surfaces des Matériaux, Université Frères Mentouri Constantine1, Constantine, Algeria

Anna O. Herus Department of Spectroscopy of Molecular Systems and Nanostructured Materials, B. Verkin Institute for Low Temperature Physics and Engineering of the National Academy of Sciences of Ukraine, Kharkiv, Ukraine

L. A. Honcharova Institute of Macromolecular Chemistry, National Academy of Sciences of Ukraine of, Kyiv, Ukraine

Yu. Horbenko Ivan Franko National University of Lviv, Lviv, Ukraine

Adam Ingram Department of Physics, Opole University of Technology, Opole, Poland

K. Ivanenko Physics Faculty, Research Laboratory of Physics of Metals and Ceramics, Taras Shevchenko National University of Kyiv, Kyiv, Ukraine

A. V. Ivaniia Processes and Equipment of Chemical and Petroleum-Refineries Department, Sumy State University, Sumy, Ukraine

O. D. Kachkovsky Physics Faculty of Kyiv National Taras Shevchenko University, Kiev, Ukraine

Olexandr O. Kalenyk Departments of Physics, Taras Shevchenko National University of Kyiv, Kiev, Ukraine

Waldemar Kaliszczak Faculty of Chemistry, Department of Analytical Chemistry, Institute of Chemical Sciences, Maria Curie Skłodowska University, Lublin, Poland

Andriy Kalynovskyy Department of Foreign and Custom Activities, Lviv Polytechnic National University, Lviv, Ukraine

Gennadii V. Kamarchuk Department of Spectroscopy of Molecular Systems and Nanostructured Materials, B. Verkin Institute for Low Temperature Physics and Engineering of the National Academy of Sciences of Ukraine, Kharkiv, Ukraine

L. V. Karabanova Institute of Macromolecular Chemistry, National Academy of Sciences of Ukraine of, Kyiv, Ukraine

Hanna Karakurkchi National Technical University, Kharkiv Polytechnic Institute, Kharkiv, Ukraine

M. V. Karpets Institute for Superhard Materials of the NAS of Ukraine, Kyiv, Ukraine

M. T. Kartel Department of Physics and Chemistry of Carbon Nanomaterials, O. Chuiko Institute of Surface Chemistry National Academy of Sciences of Ukraine, Kyiv, Ukraine

T. A. Khalyavka Institute for Sorption and Problems of Endoecology, NAS of Ukraine, Kyiv, Ukraine

Yevhen Kharchenko Institute of Building and Environmental Engineering, Lviv Polytechnic National University, Lviv, Ukraine

O. Y. Khyzhun Institute for Problems of Materials Science, NAS of Ukraine, Kyiv, Ukraine

V. Kladko Lashkaryov Institute of Semiconductor Physics, Kiev, Ukraine

V. Koda G. V. Kurdyumov Institute for Metal Physics of the N.A.S. of Ukraine, Kiev, Ukraine

R. S. Kolodka Physics Faculty, Taras Shevchenko National University of Kyiv, Kiev, Ukraine

Vitaliy Korendiy Department of Mechanics and Automation Engineering, Lviv Polytechnic National University, Lviv, Ukraine

A. M. Korol Laboratory on Quantum Theory in Linkoping, ISIR, Linkoping, Sweden;
National University for Food Technologies, Kiev, Ukraine

A. M. Korostil Institute of Magnetism (IMAG) NASU and MESU, Kiev, Ukraine

O. Korotchenkov Faculty of Physics, Taras Shevchenko Kyiv National University, Kiev, Ukraine

G. V. Korzhak L.V. Pisarzhevskii Institute of Physical Chemistry, NAS of Ukraine, Kyiv, Ukraine

Marek Kostrzewa Department of Physics, Opole University of Technology, Opole, Poland

B. D. Kostyuk Frantsevich Institute for Problems of Materials Science of National Academy Sciences of Ukraine, Kyiv, Ukraine

O. Kosulya Lashkaryov Institute of Semiconductor Physics, Kiev, Ukraine

Yu. Kotsiuba Karpenko Physico-Mechanical Institute of the NAS of Ukraine, Lviv, Ukraine

B. V. Kovalchuk G. V. Kurdyumov Institute for Metal Physics of the N.A.S. of Ukraine, Kiev, Ukraine

I. G. Kovzun F.D, Ovcharenko Institute of Biocolloid Chemistry of NAS of Ukraine, Kiev, Ukraine

O. Kramar Ternopil Ivan Puluj National Technical University, Ternopil, Ukraine

O. Yu. Kravchenko Taras Shevchenko National University of Kyiv, Kyiv, Ukraine

J. Krmela Department of Numerical Methods and Computational Modeling, Alexander Dubcek University of Trencin, Puchov, Slovak Republic

V. Krmelova Department of Materials Technologies and Environment, Alexander Dubcek University of Trencin, Puchov, Slovak Republic

M. M. Krupa Institute of Magnetism (IMAG) NASU and MESU, Kiev, Ukraine

Kateryna Kudelko V.I. Vernadskii Institute of General and Inorganic Chemistry, National Academy of Science of Ukraine, Kiev, Ukraine

Vasyl V. Kuryliuk Departments of Physics, Taras Shevchenko National University of Kyiv, Kyiv, Ukraine

O. I. Kushnerov Oles Honchar Dnipro National University, Dnipro, Ukraine

Nataliya Kutsevol Taras Shevchenko National University of Kyiv, Kiev, Ukraine

Yuliia Kuziv Taras Shevchenko National University of Kyiv, Kiev, Ukraine

Andrii Lapinskyi Igor Sikorsky Kyiv Polytechnic Institute, Kyiv, Ukraine

O. Lavrynenko Frantsevich Institute for Problems of Material Science, NAS of Ukraine, Kiev, Ukraine

O. Liubchenko Lashkaryov Institute of Semiconductor Physics, Kiev, Ukraine

Anatoliy V. Luzanov SSI "Institute of Single Crystals", NAS of Ukraine, Kharkiv, Ukraine

A. V. Lytvynenko Processes and Equipment of Chemical and Petroleum-Refineries Department, Sumy State University, Sumy, Ukraine

D. Makieła Silesian Centre for Education and Interdisciplinary Research, Institute of Physics, University of Silesia in Katowice, Chorzów, Poland

Artur Maksymov Marian Smoluchowski Institute of Physics, Jagiellonian University, Kraków, Poland

T. V. Maltsev National Technical University of Agriculture, Kharkiv, Ukraine

Viktor Malyshev Director of Institute of Engineering & Technology, Open International University of Human Development "Ukraine", Kyiv, Ukraine

E. V. Manuilov Institute of Physics, NAS of Ukraine, Kyiv, Ukraine

Oleg D. Marinin Departments of Physics, Taras Shevchenko National University of Kyiv, Kiev, Ukraine

Aleksandr Stepanovich Martseniuk Department of Science and Education of Ukraine, National University of Food Technology, Kiev, Ukraine

Ludmila Stepanovna Martseniuk Institute of Nuclear Researches of the National Academy of Science of Ukraine, Kiev, Ukraine

I. S. Maruschak Taras Shevchenko National University of Kyiv, Kyiv, Ukraine

Luidmila Yu. Matzui Departments of Physics, Taras Shevchenko National University of Kyiv, Kiev, Ukraine

N. V. Medvid' National University for Food Technologies, Kiev, Ukraine

V. Melnik Lashkaryov Institute of Semiconductor Physics, Kiev, Ukraine

O. M. Mishchenko Medical Faculty no. 3, Zaporizhzhia State Medical University. Maiakovskyi Prospect, Zaporizhzhia, Ukraine

Elena Mokrinskaya Taras Shevchenko National University of Kyiv, Kiev, Ukraine

R. M. Muratov National Science Center "Kharkiv Institute of Physics and Technology" (NSC KIPT), Institute of Plasma Physics, Kharkiv, Ukraine

H. Yu. Mykhailova G. V. Kurdyumov Institute for Metal Physics of the N.A.S. of Ukraine, Kiev, Ukraine

A. Nadtochiy Departments of Physics, Taras Shevchenko National University of Kyiv, Kyiv, Ukraine

Svitlana Nahirniak Igor Sikorsky Kyiv Polytechnic Institute, Kyiv, Ukraine

Y. Naoui Laboratoire de thermodynamique et traitement de surfaces des Matériaux, Université Frères Mentouri Constantine1, Constantine, Algeria

S. Nedilko Physics Faculty, Taras Shevchenko National University of Kyiv, Kyiv, Ukraine

T. A. Nenastina National Technical University "Kharkiv Polytechnic Institute", Kharkiv, Ukraine

O. M. Nikipelova SA «Ukrainian Research Institute of Medical Rehabilitation and Balneology, Ministry of Health of Ukraine», Odessa, Ukraine

A. Nikolenko V. Lashkaryov Institute of Semiconductor Physics, the National Academy of Science of Ukraine, Kyiv, Ukraine

Agnieszka Nosal-Wiercińska Faculty of Chemistry, Department of Analytical Chemistry, Institute of Chemical Sciences, Maria Curie Skłodowska University, Lublin, Poland

I. E. Novikova Taras Shevchenko National University of Kyiv, Kiev, Ukraine

O. Oberemok Lashkaryov Institute of Semiconductor Physics, Kiev, Ukraine

Vladimir Ogenko V.I. Vernadskii Institute of General and Inorganic Chemistry, National Academy of Science of Ukraine, Kiev, Ukraine

V. O. Oliinyk F.D, Ovcharenko Institute of Biocolloid Chemistry of NAS of Ukraine, Kiev, Ukraine

A. I. Olkhovyk Processes and Equipment of Chemical and Petroleum-Refineries Department, Sumy State University, Sumy, Ukraine

O. P. Ostash Karpenko Physico-Mechanical Institute of the NAS of Ukraine, Lviv, Ukraine

Iryna V. Ovsiienko Departments of Physics, Taras Shevchenko National University of Kyiv, Kiev, Ukraine

Olexii Palchik V.I. Vernadskii Institute of General and Inorganic Chemistry, National Academy of Science of Ukraine, Kiev, Ukraine

A. V. Panko F.D, Ovcharenko Institute of Biocolloid Chemistry of NAS of Ukraine, Kiev, Ukraine

O. Pavlenko Frantsevich Institute for Problems of Material Science, NAS of Ukraine, Kiev, Ukraine

Valeriy Pavlov Taras Shevchenko National University of Kyiv, Kiev, Ukraine

V. V. Permyakov Institute of Geological Sciences, NAS of Ukraine, Kyiv, Ukraine

H. Petrovska Department of Photonics, Lviv Polytechnic National University, Lviv, Ukraine

Tatiana Plisko Institute of Physical Organic Chemistry, National Academy of Science of Belarus, Minsk, Belarus

Inna V. Plyushchay Departments of Physics, Taras Shevchenko National University of Kyiv, Kiev, Ukraine

Olexandr I. Plyushchay Departments of Physics, Taras Shevchenko National University of Kyiv, Kiev, Ukraine

V. Ya. Podhurska Karpenko Physico-Mechanical Institute of the NAS of Ukraine, Lviv, Ukraine

A. Podolian Faculty of Physics, Taras Shevchenko Kyiv National University, Kiev, Ukraine

M. V. Pogorielov Department of Public Health, Medical Institute of Sumy State University, Sumy, Ukraine

Sergey I. Pokutnyi Chuiko Institute of Surface Chemistry of National Academy of Sciences of Ukraine, Kyiv, UA, Ukraine

Oleksiy I. Polovina Departments of Physics, Taras Shevchenko National University of Kyiv, Kyiv, Ukraine

V. Popov Lashkaryov Institute of Semiconductor Physics, Kiev, Ukraine

Alexander P. Pospelov Department of Physical Chemistry, National Technical University "Kharkiv Polytechnic Institute", Kharkiv, Ukraine

D. R. Potapov Processes and Equipment of Chemical and Petroleum-Refineries Department, Sumy State University, Sumy, Ukraine

T. O. Prikhna Institute for Superhard Materials of the NAS of Ukraine, Kyiv, Ukraine

V. A. Prokopenko F.D, Ovcharenko Institute of Biocolloid Chemistry of NAS of Ukraine, Kiev, Ukraine;
National Technical University of Ukraine «KPI», Kiev, Ukraine

S. V. Radio Vasyl' Stus Donetsk National University, Vinnytsia, Ukraine

Tatyana Rakitskaya Odessa II Mechnikov National University, Odessa, Ukraine

S. Revo Physics Faculty, Research Laboratory of Physics of Metals and Ceramics, Taras Shevchenko National University of Kyiv, Kyiv, Ukraine

S. P. Romaniuk National Technical University of Agriculture, Kharkiv, Ukraine

B. Romanyuk Lashkaryov Institute of Semiconductor Physics, Kiev, Ukraine

Ludmila Rozhdestvenska V.I. Vernadskii Institute of General and Inorganic Chemistry, National Academy of Science of Ukraine, Kiev, Ukraine

S. I. Ryabtsev Oles Honchar Dnipro National University, Dnipro, Ukraine

T. Sabov Lashkaryov Institute of Semiconductor Physics, Kiev, Ukraine

Yu. I. Sachanova National Technical University "Kharkiv Polytechnic Institute", Kharkiv, Ukraine

Galina V. Saenko Departments of Physics, Taras Shevchenko National University of Kyiv, Kiev, Ukraine

N. D. Sakhnenko Department of Physical Chemistry, National Technical University "Kharkiv Polytechnic Institute", Kharkiv, Ukraine

Yu. I. Sementsov Department of Physics and Chemistry of Carbon Nanomaterials, O. Chuiko Institute of Surface Chemistry National Academy of Sciences of Ukraine, Kyiv, Ukraine

T. B. Serbeniuk Institute for Superhard Materials of the NAS of Ukraine, Kyiv, Ukraine

Alla Serhiienko Igor Sikorsky Kyiv Polytechnic Institute, Kyiv, Ukraine

D. B. Shakhnin Assoc. Prof., Open International University of Human Development "Ukraine", Kyiv, Ukraine

N. D. Shcherban L.V. Pisarzhevskii Institute of Physical Chemistry, NAS of Ukraine, Kyiv, Ukraine

Olga M. Shchokotova Institute of Applied Physics, National Academy of Sciences of Ukraine, Sumy, Ukraine

V. Sheludko Oleksandr Dovzhenko Hlukhiv National Pedagogical University, Hlukhiv, Ukraine

O. Yu. Shevchenko National University for Food Technologies, Kiev, Ukraine

L. V. Shmeleva Physics Faculty of Kyiv National Taras Shevchenko University, Kiev, Ukraine

V. Shmid Faculty of Physics, Taras Shevchenko Kyiv National University, Kiev, Ukraine

V. I. Shtompel Institute of Macromolecular Chemistry, National Academy of Sciences of Ukraine of, Kyiv, Ukraine

V. V. Shymanovska Institute of Physics, NAS of Ukraine, Kyiv, Ukraine

O. I. Sidashenko National Technical University of Agriculture, Kharkiv, Ukraine

Nadia V. Sigareva Chuiko Institute of Surface Chemistry of the National Academy of Sciences of Ukraine, Kyiv, Ukraine

T. S. Skoblo National Technical University of Agriculture, Kharkiv, Ukraine

Yu. Skorenkyy Ternopil Ivan Puluj National Technical University, Ternopil, Ukraine

Vitalii Smokal Taras Shevchenko National University of Kyiv, Kiev, Ukraine

V. A. Smyntyna Odessa Mechnikov National University, Odessa, Ukraine

A. I. Sokolenko National University for Food Technologies, Kiev, Ukraine

Andrew E. Stetsko Lviv, Ukraine

Yaryna T. Stetsko Lviv, Ukraine

T. V. Stetsyuk Frantsevich Institute for Problems of Materials Science of National Academy Sciences of Ukraine, Kyiv, Ukraine

V. Strelchuk V. Lashkaryov Institute of Semiconductor Physics, the National Academy of Science of Ukraine, Kyiv, Ukraine

A. D. Suprun Physics Faculty of Kyiv National Taras Shevchenko University, Kiev, Ukraine

V. B. Sverdun Institute for Superhard Materials of the NAS of Ukraine, Kyiv, Ukraine

I. Sydorchenko G. V. Kurdyumov Institute for Metal Physics of the N.A.S. of Ukraine, Kiev, Ukraine

Katarzyna Szewczuk-Karpisz Institute of Agrophysics, Polish Academy of Sciences, Lublin, Poland

A. V. Taran National Science Center "Kharkiv Institute of Physics and Technology" (NSC KIPT), Institute of Plasma Physics, Kharkiv, Ukraine

V. S. Taran National Science Center "Kharkiv Institute of Physics and Technology" (NSC KIPT), Institute of Plasma Physics, Kharkiv, Ukraine

Alla Truba Odessa II Mechnikov National University, Odessa, Ukraine

Tatiana L. Tsaregradskaya Departments of Physics, Taras Shevchenko National University of Kyiv, Kiev, Ukraine

O. A. Tsyganovich F.D, Ovcharenko Institute of Biocolloid Chemistry of NAS of Ukraine, Kiev, Ukraine;
National Technical University of Ukraine «KPI», Kiev, Ukraine

L. M. Ushakova Department of Physics and Chemistry of Carbon Nanomaterials, O. Chuiko Institute of Surface Chemistry National Academy of Sciences of Ukraine, Kyiv, Ukraine

Volodymyr L. Vakula Department of Spectroscopy of Molecular Systems and Nanostructured Materials, B. Verkin Institute for Low Temperature Physics and Engineering of the National Academy of Sciences of Ukraine, Kharkiv, Ukraine

B. D. Vasyliv Karpenko Physico-Mechanical Institute, NAS of Ukraine, Lviv, Ukraine;
Department of Mechanics and Automation Engineering, Lviv Polytechnic National University, Lviv, Ukraine

S. V. Vasylyuk Physics Faculty of Kyiv National Taras Shevchenko University, Kiev, Ukraine

Maryna Ved' Department of Physical Chemistry, National Technical University, Kharkiv Polytechnic Institute, Kharkiv, Ukraine

A. O. Vikulova Taras Shevchenko National University of Kyiv, Kiev, Ukraine

Volodymyr Vira Institute of Building and Environmental Engineering, Lviv Polytechnic National University, Lviv, Ukraine

Olexii Vishnevskii M.P. Semenenko Institute of Geochemistry, Mineralogy and Ore Formation, National Academy of Science of Ukraine, Kiev, Ukraine

Zoriana Voitko Igor Sikorsky Kyiv Polytechnic Institute, Kyiv, Ukraine

Yurii M. Volfkovich A.N. Frumkin Institute of Physical Chemistry and Electrochemistry of the RAS, Moscow, Russia

Vitaliya Volkova Odessa II Mechnikov National University, Odessa, Ukraine

Luidmila L. Vovchenko Departments of Physics, Taras Shevchenko National University of Kyiv, Kiev, Ukraine

Sylwester Wacke Department of Physics, Opole University of Technology, Opole, Poland

Małgorzata Wiśniewska Faculty of Chemistry, Department of Radiochemistry and Environmental Chemistry, Institute of Chemical Sciences, Maria Curie-Sklodowska University in Lublin, Lublin, Poland

Olena Yanushevska Igor Sikorsky Kyiv Polytechnic Institute, Kyiv, Ukraine

Pavel Yaremov Institute of Physical Chemistry NAS of Ukraine, Kyiv, Ukraine

Oleksandr Yasko Student, Open International University of Human Development "Ukraine", Kyiv, Ukraine

I. Yu. Yermolenko National Technical University "Kharkiv Polytechnic Institute", Kharkiv, Ukraine

S. M. Yezhov Kiev, Ukraine

Vladimir Zakharov National University of Food Technologies, Kiev, Ukraine

Yurii Zmievskii National University of Food Technologies, Kiev, Ukraine

Nanoscale Physics

Anomalously Large Absorption of Electromagnetic Radiation by Nanometer Gold Film in the Band 26÷37.5 GHz

O. R. Bediukh, I. E. Novikova, and A. O. Vikulova

1 Introduction

The reducing impact of electromagnetic radiation (EMR) on technical devices and biological objects, especially due to the sharp increase in the number of sources of electromagnetic radiation (communication, consumer electronics, information systems, etc.) is an important task in a wide frequency band.

The development of new improved radar absorbing materials (RAM) is one of the main components of the task of protection from EMR (along with screening, the use of design elements, etc.).

The resonant absorbers are one type of RAM [1]. They use semitransparent metal films. If such film is situated at a distance of λ/4 (λ—wavelength of electromagnetic radiation) before conductive surface which has to be protected, it is provided the conditions for optimal matching at this wavelength. When several films are used, it is possible, with the help of modern computational methods, to optimize not only the absorption band but number of layers and the total thickness and weight of the RAM.

Additional studies were conducted for studying of the absorbing properties of metal films on the assumption that the film thickness was much smaller than the

O. R. Bediukh (✉) · I. E. Novikova · A. O. Vikulova
Taras Shevchenko National University of Kyiv, Volodymirska, 64, Kiev 01033, Ukraine
e-mail: bediukh@gmail.com

I. E. Novikova
e-mail: ie_novikova@meta.ua

A. O. Vikulova
e-mail: vikulova.alyona@gmail.com

O. Fesenko and L. Yatsenko (eds.), *Nanomaterials and Nanocomposites, Nanostructure Surfaces, and Their Applications*, Springer Proceedings in Physics 246, https://doi.org/10.1007/978-3-030-51905-6_1

thickness of the skin layer in the selected band of wavelengths [2, 3]. At a film thickness of 10÷100 nm, this assumption holds for the microwave range.

In this paper, an anomalously large absorption of electromagnetic radiation by a gold film of 10 nm thickness in the band 26÷37.5 GHz was studied.

2 Measurement Method

A gold film of 10 nm thickness obtained by thermal vacuum deposition on a dielectric polymer substrate of 0.1 mm thickness was used in research. Controlled film thickness was implemented by the known method of quartz oscillator.

The measurements were made in the band 26÷37.5 GHz with using of *VSWR* (voltage standing wave ratio) meter panoramic P2-65 (similar to a scalar network analyzer). The sample—the gold film on polymer substrate—was placed in a segment of the rectangular waveguide with cross section 3.4 × 7.2 mm^2. Measurements were carried out for two cases: the segment of the waveguide with the sample was connected or to the shorting in the form of metal plate (see Fig. 1a) and to the matched load (see Fig. 1b).

The sample was cut with size slightly smaller than the cross section of the waveguide so that it was easily placed in it. The sample was placed between two layers of foam as shown in Fig. 1. The gold film had no galvanic contact with the walls of the waveguide.

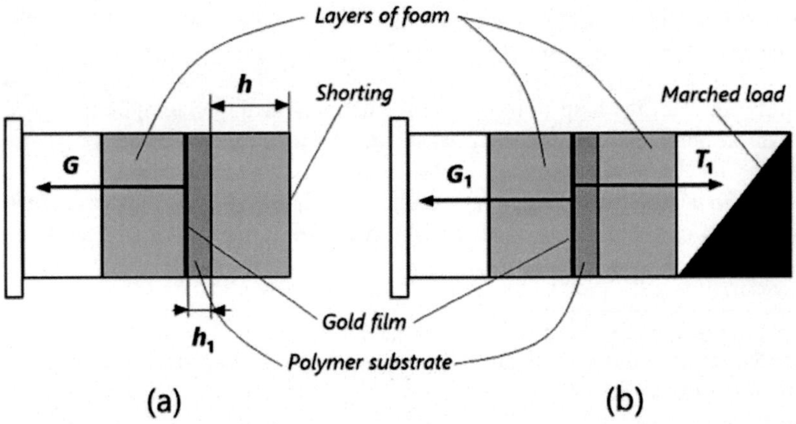

Fig. 1 Placement of the sample in the wave guideline: **a** a segment of the waveguide is connected to shorting, **b** a segment of the waveguide is connected to the matched load; *h*—the distance from the film to the shorting, h_1—the thickness of the polymer substrate, *G*—the reflection coefficient of the field in case of shorting, G_1 and T_1—the reflection and transmission coefficients of the field in case of matched load

The complex reflection coefficient of the field G from the gold film in case of shorting and the complex reflection G_1 and transmission T_1 coefficients of the field in case of matched load are also indicated in Fig. 1.

The measurements of *VSWR* were made in the waveguide line which was connected to the waveguide section with the sample as shown in Fig. 1.

Distance h was chosen to be 1 mm. The substrate thickness h_1 was 0.1 mm.

The power reflection coefficient R was calculated by the known formula

$$R = \frac{(VSWR - 1)^2}{(VSWR + 1)^2}. \tag{1}$$

The absorption coefficient A was calculated by

$$A = 1 - R. \tag{2}$$

3 Experimental Results and Discussion

The power reflection coefficients R obtained by means of (1) from the measured *VSWR*, in two cases, shorting (black curve) and matched load (red curve), are shown in Fig. 2.

As shown in Fig. 2, experimental results of power reflection coefficient for the case of the shorting (black curve) were almost independent of frequency and were at level of 0.1 or slightly less.

It means that the power absorption was more than 90% in the whole frequency band of measurements.

The experimental results of power reflection coefficient for the case of the matched load (red curve) were varied in the range of 0.32–0.42 from frequency.

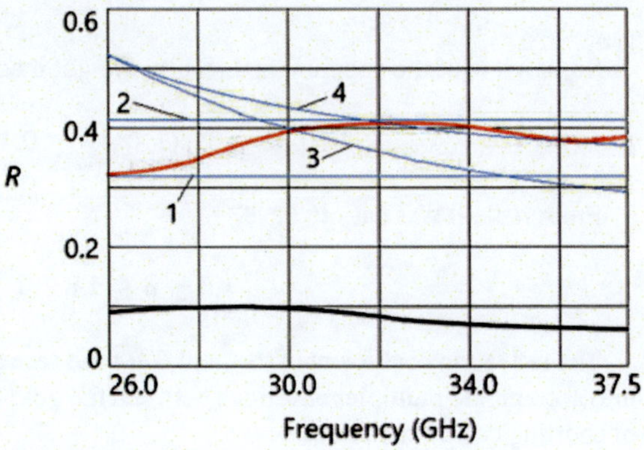

Fig. 2 Experimental (red and black curves) and theoretical (lines 1, 2 and curves 3 and 4) measurement results of power reflection coefficient R as a function of frequency for the sample with the gold film of 10 nm thickness. Lines 1, 2 and red curve—for the case of connection with matched load; curves 3, 4 and black curve—for the case of connection with shorting

Relative dielectric constant of the foam was about 1.1 and of polymer used for the substrate was about 2. The dielectric loss tangents of both of these materials were less than 10-3.

In view of the fact that the thickness of the polymer substrate was 0.1 mm, which was about 1/100 of EMR wavelength in all measuring range, in calculation, we neglected reflections associated with the presence of the foam and the polymer substrate.

Therefore, the calculations were carried out in the approximation that the gold film was in an empty waveguide.

Skin layer thickness in the selected frequency band for gold was more than an order greater than the gold film thickness. In this case when EMR penetrated the entire thickness of the film, the reflection coefficient G_1 for the film having a thickness d and a conductivity σ can be written as [3]

$$G_1 = -\beta/(1+\beta), \tag{3}$$

where $\beta = \sigma d/(\varepsilon_0 c)$, ε_0—the permittivity of vacuum, and c—the speed of light in vacuum.

Accordingly, the transmission coefficient is equal to

$$T_1 = 1 + G_1 = 1/(1+\beta). \tag{4}$$

For the 10-nm-thick gold film, the conductivity σ depends both on the thickness and on the method of preparation of the film.

On the other hand in the case of matched load, G_1^2 corresponds to the power reflection coefficient that is given by the formula (1).

Therefore, to determine β value, we had the opportunity to use experimental measurements of the power reflection coefficient for the matched load case.

Power reflection coefficient is equal to square of modulus of complex reflection coefficient of the field

$$R = |G|^2. \tag{5}$$

So, in view of the experimental results, we got inequality

$$0.32 \le \beta^2/(1+\beta)^2 \le 0.42. \tag{6}$$

From here, it was easy to have

$$1.3 \le \beta \le 1.8. \tag{7}$$

The reflection coefficient of the field G for the case of shorting connection, taking into account the multiple reflections between the gold film and the reflecting surface of shorting can be written as

$$G = G_1 - \frac{T_1^2 e^{-i4\pi h/\lambda_0}}{1 + G_1 e^{-i4\pi h/\lambda_0}}, \tag{8}$$

where $\lambda_0 = \lambda/\sqrt{1-(\lambda/2a)^2}$—the wavelength in the waveguide, and $a = 7.2$ mm—the size of the wide wall of the waveguide.

Knowing G and using (5), we calculated R for the shorting case.

Lines 1 and 2 in Fig. 2 correspond to the two extreme values of the power reflection coefficient (with $\beta = 1.3$ and $\beta = 1.8$) for the case of the matched load.

Curves 3 and 4 correspond to the two theoretical values of R as a function of frequency, calculated by (5) and (8) for the two extreme values of β (curve 3 for $\beta = 1.3$, curve 4 for $\beta = 1.8$) for the case of shorting.

As shown in Fig. 2, the calculations for the case of shorting did not give satisfactory agreement with the experimental results. The theoretical power reflection coefficients (curves 3 and 4) are in 3÷5 times higher than obtained from the experiment (black curve).

Moreover, the theoretical reflection coefficient in the shorting case decreases with increasing frequency, which corresponds to the expected reduction of reflection with frequency increasing.

In this case, $\lambda_0/4$ increases and approaches the value of h, and equality with h corresponds to a theoretical minimum of reflection.

4 Conclusion

Experimental data of the power reflection coefficient obtained for the gold film thickness of 10 nm at a distance 1 mm from the shorting reflecting surface were significantly less than that were calculated according to traditional theory in the frequency band 26÷37.5 GHz.

Such a discrepancy was probably due to the absence of galvanic contact between the gold film and the walls of the waveguide. This leads to the effect of an islet film when there is no continuous conductivity over its entire surface. In this case, in the calculations, it is necessary to take into account the reactive resistance introduced by such kind of film.

The absorption was fixed on a level of 90% and almost not depended on frequency that can significantly expand the ability to use such films in practice [4].

References

1. Saville P (2005) Review of radar absorbing materials. Tehnical Memorandum, DRDC Atlantic, TM 2005-003, January 2005
2. Andreev VG, Vdovin VA, Voronov PS (2003) An experimental study of millimeter wave absorption in thin metal films. Techn Phys Lett 29:953–955

3. Antonets IV, Kotov LN, Nekipelov SV, Karpushov EN (2004) Conductive and reflective properties of thin metal films. Techn Phys 49:1496–1500
4. Bedjukh O (2010) Radio wave absorbing covering. WO/2010/147568, Publication Date: 23.12.2010

Electric Field-Controlled Magnetic Anisotropy in Magnetic Nanoheterostructures

A. M. Korostil and M. M. Krupa

1 Introduction

Bilayers containing an interface between a thin layer of a heavy metal and a magnetic one are important hybrid materials in spintronics, as they combine magnetic order, strong spin–orbit interaction, and broken inversion symmetry. This together with exchange interactions of the magnetic layer can lead to perpendicular magnetic anisotropy (PMA) [1], Dzyaloshinskii–Moriya interactions [2, 3], spin–orbit torques, Rashba–Edelstein effects, and more [4]. Spin–orbit interactions near the interface provide a handle to alter these properties by tuning chemical composition, interface structure, or gate voltages, as demonstrated most extensively for magnetic anisotropy [5, 6].

Magnetic anisotropy energy (MAE) refers to the dependence of the total energy of a magnetic system on the real-space orientation of its magnetization. The MAE is responsible for the orientational stability of magnetic domains, and hence lies at the heart of both magnetic hard disk drives and magnetic random access memories. There are two main contributions to the MAE: the magnetocrystalline anisotropy which arises from electronic spin–orbit interactions, and shape anisotropy which arises from the magnetostatic dipolar interaction. For a thin ferromagnetic film, the magnetostatic energy is minimized when the magnetization is in the plane of the film, leading to in-plane magnetic anisotropy (IMA).

To stabilize perpendicular magnetic anisotropy (PMA), the magnetocrystalline anisotropy energy must overcome the shape anisotropy. From the technological point of view, PMA is very important, since it enables an increased bit storage density, through a reduced size of the magnetic domains that store each bit of information. Solving this problem involves description of the magnetocrystalline anisotropy of

A. M. Korostil (✉) · M. M. Krupa
Institute of Magnetism (IMAG) NASU and MESU, Kiev, Ukraine
e-mail: korostilandrii@gmail.com

O. Fesenko and L. Yatsenko (eds.), *Nanomaterials and Nanocomposites, Nanostructure Surfaces, and Their Applications*, Springer Proceedings in Physics 246, https://doi.org/10.1007/978-3-030-51905-6_2

a ferromagnet/heavy metal bilayer, driven by interfacial Rashba SOC, highlighting different physical regimes [7]. The corresponding description is based on an electron tight-binding model, which allows us to analyze the MAE with respect to the three competing energy scales: the non-relativistic kinetic energy, t', the Rashba SOC strength t'', and the strength of the exchange coupling, J to the ferromagnetic order parameter. This leads to conditions for the realization of the PMA and its dependence on the electron structure of the system.

2 Model of Two-Layer Magnetic Nanostructure with the Spin–Orbit Rashba Interaction

The properties of itinerant electrons with broken inversion symmetry and SOC in a two-dimensional lattice with one orbital per site, nearest-neighbor hopping, and Rashba-like spin–momentum is described by the Hamiltonian [7]:

$$H_e = -\frac{1}{2}\sum_{<i,j>}\sum_{s,s'} c_{is}^{\dagger}\left(t'\sigma_{ss'}^{0} - it''\left[z \times \boldsymbol{R}_{ij}\right]\cdot \boldsymbol{\sigma}_{ss'}\right)c_{js'}. \tag{1}$$

Here, the sum is over near-neighbor links, $c_{is}^{\dagger}$ and c_{js} are the creation and annihilation operators for an electron with spin s at a lattice site $\boldsymbol{R}_i$, $\sigma_{ss'}^{0}$ is the unit 2×2 spin matrix, and $\sigma = (\sigma^x, \sigma^y, \sigma^z)$ is the vector of Pauli matrices. The vector connecting site i to site j is $\boldsymbol{R}_{ij} = \boldsymbol{R}_i - \boldsymbol{R}_j$, and the cross-product favors spin orientations perpendicular to the bond direction, $\hat{\boldsymbol{R}}_{ij} = \boldsymbol{R}_{ij}/|\boldsymbol{R}_{ij}|$ and the normal to the lattice plane, z. The hopping strength is given t, and the angle ϕ_R characterizes the relative strength of conventional spin-independent hopping, $t' = 2t\cos\phi_R$ and chiral Rashba hopping $t'' = 2t\sin\phi_R$. We impose Born–von Karman periodic boundary conditions and introduce the lattice Fourier transforms of the operators,$c_{is} = \sum_k e^{i\boldsymbol{k}\boldsymbol{R}_i}$, where N is the number of lattice sites and $\boldsymbol{k}$ is the electron pulse.

This transforms the Hamiltonian (1) to the form

$$H_e(\boldsymbol{k}) = H_0(\boldsymbol{k}) + H_R(\boldsymbol{k}), \tag{2}$$

where

$$H_0(\boldsymbol{k}) = -t'\left(\cos k_x + \cos k_y\right)\sigma^0, \tag{2a}$$

$$H_R(\boldsymbol{k}) = -t''\left(\sin k_x\sigma^y - \sin k_y\sigma^x\right). \tag{2b}$$

For small k, the Hamiltonian (2) describes a Rashba electron gas.

The total Hamiltonian also involves the ferromagnetic interaction between quasiparticles and magnetic condensate, $H_B = -\boldsymbol{B}\cdot\boldsymbol{\sigma}$, i.e.,

$$H(\boldsymbol{k}) = H_e(\boldsymbol{k}) + H_B = E_0(\boldsymbol{k})\sigma^0 + \boldsymbol{b}(\boldsymbol{k})\cdot\boldsymbol{\sigma}, \tag{3}$$

where

$$E_0(\boldsymbol{k}) = -t'\left(\cos k_x + \cos k_y\right), \tag{4}$$

$$\boldsymbol{b}(\boldsymbol{k}) = \boldsymbol{b}_R(\boldsymbol{k}) + \boldsymbol{B}, \tag{4a}$$

$$\boldsymbol{b}_R(\boldsymbol{k}) = t''\left(\sin k_y\,\boldsymbol{x} - \sin k_x\,\boldsymbol{y}\right), \tag{4b}$$

$$\boldsymbol{B} = J(\sin\theta(\cos\varphi\boldsymbol{x} + \sin\varphi\boldsymbol{y}) + \cos\theta)\boldsymbol{z} \tag{4c}$$

Here $\boldsymbol{b}_R(\boldsymbol{k})$ is the Rashba spin–orbit field, and the coupling to the ferromagnetic background is given by $\boldsymbol{B}$, where the spherical angles θ and φ specify the magnetization orientation and J is the strength of the coupling.

In the diagonalized form, the Hamiltonian

$$H(\boldsymbol{k}) = E_+(\boldsymbol{k})P_+(\boldsymbol{k}) + E_-(\boldsymbol{k})P_-(\boldsymbol{k}), \tag{5}$$

where the band energies

$$E_\pm = E_0(\boldsymbol{k}) \mp |\boldsymbol{b}(\boldsymbol{k})| \tag{5a}$$

and the eigenvector projectors,

$$P_\pm(\boldsymbol{k}) = \frac{1}{2}\left(\sigma^0 \pm \hat{\boldsymbol{b}}(\boldsymbol{k})\cdot\boldsymbol{\sigma}\right),\quad \hat{\boldsymbol{b}}(\boldsymbol{k}) = \frac{\boldsymbol{b}(\boldsymbol{k})}{|\boldsymbol{b}(\boldsymbol{k})|} \tag{5b}$$

The plus sign corresponds to the lower energy majority band and the minus sign to the higher energy minority band. Band dispersions are plotted in Fig. 1 for some representative cases.

The electronic density of states (DOS) is given as

$$\rho(E) = \sum_{n=\pm} \frac{d\boldsymbol{k}}{(2\pi)^2}\delta(E - E_n(\boldsymbol{k})), \tag{6}$$

which leads to the number of electrons per lattice site,

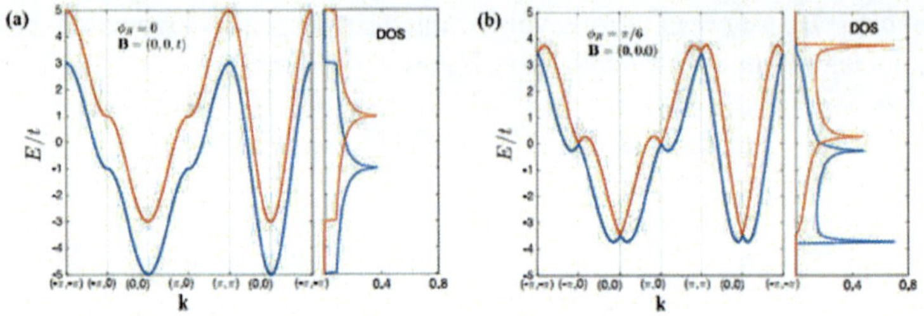

Fig. 1 Band and dispersions and the respective densities of states for representative cases: **a** $\phi_R = 0 (t' = 2t,\ t'' = 0), J = t$; **b** $\phi_R = \pi/6\ (t' = \sqrt{3}t,\ t'' = t)$

$$N(E) = \sum_{n=\pm} \frac{d\boldsymbol{k}}{(2\pi)^2} \delta(f_+(\boldsymbol{k}) - f_+(\boldsymbol{k})) = \int_{-\infty}^{E_F} dE\rho(E). \tag{7}$$

The integral in (7) is over the first Brillouin zone, and the function $f_n(\boldsymbol{k}) = \theta(E_F - E_n(\boldsymbol{k}))$ describes the occupation of the corresponding eigenstate, $E_n(\boldsymbol{k})$. The coupling to the ferromagnetic background induces a net spin moment on the itinerant electrons, given as

$$\boldsymbol{M} = \int \frac{d\boldsymbol{k}}{(2\pi)^2} (f_+(\boldsymbol{k}) - f_-(\boldsymbol{k}))\hat{\boldsymbol{b}}(\boldsymbol{k}) = \int_{\infty}^{E_F} dE\, \boldsymbol{m}(E), \tag{8}$$

where $\boldsymbol{m}(E)$ is the spin-polarized density of electron states (DOS). The energetics of the itinerant electrons can be obtained from the internal energy, which at zero temperature is defined as

$$U = \sum_{n=\pm} \int \frac{d\boldsymbol{k}}{(2\pi)^2} f_n(\boldsymbol{k}) E_n(\boldsymbol{k}) = \int_{-\infty}^{E_F} dE\rho(E)E. \tag{9}$$

Due to additivity contributions of the bare band, Rashba, and exchange interactions to the internal energy,

$$U = \sum_{n=\pm} \int \frac{d\boldsymbol{k}}{(2\pi)^2} f_n(\boldsymbol{k}) Tr\ P_n(\boldsymbol{k}) H(\boldsymbol{k}) = U_0 + U_R + U_B \tag{10}$$

Solving the Hamiltonian (5) is based on the Green function,

$$G(\boldsymbol{k}, E) = (E - H(\boldsymbol{k}))^{-1} = \sum_n \frac{P_n(\boldsymbol{k})}{E - E_n(\boldsymbol{k})}, \tag{11}$$

which specifically is related to the internal energy and its derivatives. For instance, the DOS and the spin-polarized DOS are given by

$$\rho(E) = -\frac{1}{\pi}\mathrm{ImTr}\int \frac{d\boldsymbol{k}}{(2\pi)^2} G(\boldsymbol{k}, E) \tag{12}$$

and

$$\boldsymbol{m}(E) = -\frac{1}{\pi}\mathrm{ImTr}\int \frac{d\boldsymbol{k}}{(2\pi)^2} \sigma(\boldsymbol{k}, E) G(\boldsymbol{k}, E), \tag{13}$$

respectively, where the traces are over the spin components. The uniform static spin susceptibility for a fixed number of electrons is determined as

$$\chi^{\alpha\beta} = -\frac{1}{\pi}\mathrm{ImTr}\int_{-\infty}^{E_F} dE \int d\boldsymbol{k} \frac{d\boldsymbol{k}}{(2\pi)^2} \prod_{\varsigma=\alpha,\beta} \sigma^{\varsigma} G(\boldsymbol{k}, E) - R_{\alpha\beta} \tag{14}$$

where $R_{\alpha\beta} = m^{\alpha}(E_F) m^{\beta}(E_F)/\rho(E_F)$ comes from ensuring that $\partial N_e/\partial B^{\beta} = 0$. Substituting spectral representation of the Green function (11) into (14) results in the following expression:

$$\chi^{\alpha\beta} = -\frac{1}{\pi}\mathrm{ImTr}\int \frac{d\boldsymbol{k}}{(2\pi)^2} \sum_{n,n'} \int_{-\infty}^{E_F} dE \frac{\sigma^{\alpha} P_{n'}(\boldsymbol{k})}{E - E_{n'}(\boldsymbol{k})} \frac{\sigma^{\beta} P_n(\boldsymbol{k})}{E - E_n(\boldsymbol{k})}, \tag{15}$$

where summation over $n = n'$ corresponds to intraband electron transitions and, consequently, to the intraband part ($\chi^{\alpha\beta}_{\mathrm{intra}}$) of the magnetic susceptibility. Summation over $n \neq n'$ corresponds to interband electron transition and, consequently, to the interband part ($\chi^{\alpha\beta}_{\mathrm{inter}}$) of the magnetic susceptibility. Therefore, the total magnetic susceptibility is sum, $\chi^{\alpha\beta} = \chi^{\alpha\beta}_{\mathrm{intra}} + \chi^{\alpha\beta}_{\mathrm{inter}}$. The interband magnetic susceptibility, $\chi^{\alpha\beta}_{\mathrm{inter}}$, contains the integrand, the partial fraction decompositions of which, in accordance with the Sokhotsky formula, leads to the relation

$$-\frac{1}{\pi}\mathrm{ImTr}\int_{-\infty}^{E_F} dE \frac{1}{(E - E_n(\boldsymbol{k}))} = f_n(\boldsymbol{k}). \tag{16}$$

The intraband magnetic susceptibility, $\chi^{\alpha\beta}_{\mathrm{intra}}$, contains the integrand which is the partial fraction with a second-order pole that results in the relation

$$-\frac{1}{\pi}\mathrm{ImTr}\int_{-\infty}^{E_F} dE\frac{1}{(E-E_n(\boldsymbol{k}))^2}=\frac{\partial f_n(\boldsymbol{k})}{\partial E_n(\boldsymbol{k})}, \tag{17}$$

which is the partial derivative of (16) with respect to $E_n(\boldsymbol{k})$.

Taking into account (16) and (17), the intra- and interband magnetization susceptibilities can be represented as

$$\chi_{\text{intra}}^{\alpha\beta}=\int\frac{d\boldsymbol{k}}{(2\pi)^2}\hat{b}_\alpha(\boldsymbol{k})\hat{b}_\beta(\boldsymbol{k})\sum_{n=\pm}\delta(E_F-E_n(\boldsymbol{k})) \tag{18}$$

and

$$\chi_{\text{inter}}^{\alpha\beta}=\int\frac{d\boldsymbol{k}}{(2\pi)^2}\left(\hat{b}_\alpha(\boldsymbol{k})\hat{b}_\beta(\boldsymbol{k})-\delta_{\alpha\beta}\right)\frac{f_-(\boldsymbol{k})-f_+(\boldsymbol{k})}{E_-(\boldsymbol{k})-E_+(\boldsymbol{k})}, \tag{19}$$

respectively. The intraband term collects the contributions from the Fermi surface, while the interband term collects those from the Fermi sea.

Band dispersions given by (5a) and respective densities of states essentially depend on the parameters t'' and J of the Rashba spin–orbit coupling strength and the exchange coupling to the ferromagnetic order parameter, respectively, as it is represented in Fig. 1 [1].

None of the interfacial Rashbs SOC ($t''=0$) and finite exchange coupling (J) to the background magnetization leads to a constant vertical splitting of the bands (Fig. 1a). Finite Rashba SOC ($t''\neq 0$) and no background magnetization ($J=0$) lead to $\boldsymbol{k}$-dependent horizontal splitting of the bands (Fig. 1b). The finite Rashba SOC and background magnetization lead to dependence of the dispersion on the orientation of the magnetization with respect to the lattice. When the magnetization is normal to the plane ($\boldsymbol{B}||z$), the system has fourfold rotational symmetry. When the magnetization is along a nearest-neighbor direction ($\boldsymbol{B}||\boldsymbol{x}$), the bands have a unidirectional shift in the perpendicular direction ($\boldsymbol{y}$).

3 Magnetic Anisotropy Energy

The MAE is due to the variation of the internal energy of the itinerant electrons as the ferromagnetic background orientation rotates. The MAE vanishes if there is no spin–orbit coupling, i.e., in the given model if there is no Rashba coupling ($\phi_R=t''=0$). Phenomenologically, the MAE is expanded in angular functions with respect to the symmetry of the system (7). For the square lattice (effectively tetragonal symmetry),

$$U_{\text{MAE}}(\theta,\varphi)\approx K_2\sin^2\theta+(K_4+K_4'\cos 4\varphi)\sin^4\theta, \tag{20}$$

with θ and φ the spherical angles describing the orientation of the ferromagnetic background. It follows from perturbation theory arguments that $K_{2n} \propto t''(t''/J)^{2n-1}$. Higher order anisotropy constants should decline rapidly in magnitude, as they are proportional to higher powers of the ratio between the spin–orbit interaction strength and the spin splitting, which is often small. The anisotropy constants can then be determined by fitting the angular dependence of the internal energy. Keeping all other parameters fixed, the internal energy given by (9) is an explicit function of the angles describing the ferromagnetic orientation,$U(\theta, \varphi)$. Assuming that the model form in (20) holds, evaluating the internal energy for three orientations is sufficient to fix the anisotropy. The system will have PMA provided that both of the following inequalities are satisfied:

$$\left.\begin{aligned} U(\pi/2, 0) - U(0,0) &= K_2 + K_4 + K_4', \\ U(\pi/2, \pi/4) - U(0,0) &= K_2 + K_4 - K_4'. \end{aligned}\right\} > 0 \tag{21}$$

Here, the magnetic anisotropy coefficients K_i as functions of the macroscopic parameters are determined by derivatives of the relation, $U = U_{MAE}$, with respect to the spherical angles, θ and φ. It can be shown that the anisotropy energy goes from IMA $\to$ PMA $\to$ IMA as function of the band filling.

The first and second derivatives of the internal energy with respect to the ferromagnetic moment orientation are described by the expressions

$$\frac{\partial U}{\partial \theta} = \boldsymbol{M} \cdot \frac{\partial \boldsymbol{B}}{\partial \theta} = \left(K_2 + 2\left(K_4 + K_4' \cos 4\varphi\right) \sin^2 \theta\right) \sin 2\theta, \tag{22}$$

$$\frac{\partial U}{\partial \varphi} = -\boldsymbol{M} \cdot \frac{\partial \boldsymbol{B}}{\partial \varphi} = -K_4' \sin 4\varphi \sin^4 \theta, \tag{23}$$

where $\boldsymbol{M}$ is the spin magnetic moment of the electrons defined in (8), and $\boldsymbol{B}$ is the effective magnetic field produced by the local moments defined in (4c). From the phenomenological expression for $U_{MAE}(\theta, \varphi)$, it follows that the magnetic torque $[\boldsymbol{M} \times \boldsymbol{B}]$ vanishes for the high-symmetry nearest- and next-nearest-neighbor directions ($\theta = \pi/2,\ \varphi = n\pi/4$) with $n = \{0, 1, \ldots, 7\}$, and for magnetization normal to the lattice plane ($\theta = 0,\ \pi$). The second derivatives of the internal energy are particularly simple to evaluate for these high-symmetry directions, since cross derivatives involving both polar and azimuthal angles vanish. Therefore, it is required only the Cartesian component of the spin susceptibility tensor for the plane perpendicular to a chosen magnetization direction (i.e., only the transverse spin susceptibility is needed). For the high-symmetry directions, the net spin moment of the itinerant electrons is aligned with the ferromagnetic background, $\boldsymbol{M}||\boldsymbol{B}$. For the in-plane high-symmetry directions,

$$\frac{1}{2}\frac{\partial^2 U}{\partial \theta^2}\bigg|_{\boldsymbol{M}||x} = \frac{J^2}{2}\left(\frac{M}{J} - \chi^{zz}\right) = -K_2 + 2\left(K_4 + K_4'\right), \tag{24}$$

$$\frac{1}{2}\frac{\partial^2 U}{\partial \varphi^2}\bigg|_{M\|x} = \frac{J^2}{2}\left(\frac{M}{J} - \chi^{yy}\right) = -8K_4', \tag{25}$$

and for the polar magnetization orientation

$$\frac{1}{2}\frac{\partial^2 U}{\partial \varphi^2}\bigg|_{M\|z} = \frac{J^2}{2}\left(\frac{M}{J} - \chi^{zz}\right) = K_2. \tag{26}$$

These equations determine the magnetic anisotropy constants via transverse components of the magnetic anisotropy susceptibility depending on the symmetry of the system. Consequently, the type of magnetic anisotropy (PMA or IMA) related to symmetry of the system is determined MAE which depend on the symmetry of the system. When $\boldsymbol{M}\|z$, the system has fourfold rotational symmetry from which it follows that $\chi^{xx} = \chi^{yy}$ and $\chi^{xy} = \chi^{yx} = 0$.

Equations (24) and (26) contain a common term

$$\bar{\chi} = \int \frac{d\boldsymbol{k}}{(2\pi)^2}\frac{f_+(\boldsymbol{k}) - f_-(\boldsymbol{k})}{|\boldsymbol{b}(\boldsymbol{k})|}, \tag{27}$$

which enters the expression,

$$\chi^0 = \int \frac{d\boldsymbol{k}}{(2\pi)^2}\frac{\boldsymbol{B}\cdot\boldsymbol{b}_R(\boldsymbol{k})}{J^2}\frac{f_-(\boldsymbol{k}) - f_+(\boldsymbol{k})}{|\boldsymbol{b}_R(\boldsymbol{k})|} \tag{28}$$

describing the volume susceptibility.

The expression for the intraband part of the spin susceptibility follows from (18) and does not contain $\bar{\chi}$:

$$\chi_{\text{intra}}^{\alpha\alpha} = \int \frac{d\boldsymbol{k}}{(2\pi)^2}(\hat{b}(\boldsymbol{k}))^2 \sum_{n=\pm} \delta(E_F - E_n(\boldsymbol{k})) \tag{29}$$

Subtracting this term from the interband part of the spin susceptibility in (19) results in the equation

$$\bar{\chi}_{\text{inter}}^{\alpha\alpha} = \chi_{\text{inter}}^{\alpha\alpha} - \bar{\chi} = -\int \frac{d\boldsymbol{k}}{(2\pi)^2}(\hat{b}_\alpha(\boldsymbol{k}))^2 \frac{f_+(\boldsymbol{k}) - f_-(\boldsymbol{k})}{|\boldsymbol{b}(\boldsymbol{k})|}. \tag{30}$$

In (29) and (30), $\alpha = x, y, z$, and $\hat{b}_\alpha(\boldsymbol{k})$ is the Cartesian component of the unit vector defining the spin quantization axis for each $\boldsymbol{k}$. As it is seen, $\chi_{\text{intra}}^{\alpha\alpha}$ is positive definite, while $\bar{\chi}_{\text{inter}}^{\alpha\alpha}$ is negative definite.

The magnetic anisotropy type, i.e., PMA or IMA, can be established in two ways. When $\boldsymbol{M}||\boldsymbol{x}$ the longitudinal component, $\chi^{zz} = 0$, and the sign of the MAE is determined by χ^0. On the other hand, for $\boldsymbol{M}||z$ the volume susceptibility, $\chi^0 = 0$, so the sign of the MAE is decided by the competition between the intraband and interband contributions to the spin susceptibility χ^{xx}.

Conditions favoring PMA are determined by the magnetic anisotropy constants. When $\boldsymbol{M}||z$ and $t'' << J$, from (29) it follows that

$$\chi^{xx}_{\text{intra}} \approx \left\langle (\hat{b}(\boldsymbol{k}))^2 \right\rangle \int \frac{d\boldsymbol{k}}{(2\pi)^2} \sum_{n=\pm} \delta(E_F - E_n(\boldsymbol{k})) = \frac{(t'')^2}{2J^2} \rho(E_F), \tag{31}$$

where $\rho(E_F)$ is the total density of states at the Fermi energy. The average of the matrix element was simplified by assuming that the exchange fields are much stronger than the spin–orbit fields, so that $|\boldsymbol{b}(\boldsymbol{k})| \approx -M(t'')^2/2J^3$. Combining these expressions leads to an approximate form, $K_2 \approx (M - J\rho(E_F))$ for the uniaxial anisotropy constant. The scale of anisotropy constant $K_{\text{ref}} = (t'')^2/J$ is a useful figure of merit for MAE. PMA is likely to be stable when the density of states at the Fermi level is small: since $0 \leq M \leq 1$, it can expect PMA if $J\rho(E_F) \leq 1$. The states at the Fermi level are the ones affected by SOC in the most important way, in energetic terms. A large DOS at the Fermi level then translates to a large number of single-particle states that gain the most energy from SOC once the magnetization is tilted away from the perpendicular direction, which explains why this contribution favors IMA.

4 PMA for the Gapped Half-Filled Band Spectrum

In the case, in which the ferromagnetic exchange splitting is large enough to produce a gape for the half-filled ferromagnetic insulator, the Fermi level lies in this gape. The majority band is full, $f_+(\boldsymbol{k}) = 1$, and the minority band is empty, $f_-(\boldsymbol{k}) = 0$. Starting from (8), (20), and (27), the internal energy for this case is

$$U = \int \frac{d\boldsymbol{k}}{(2\pi)^2} (E_0(\boldsymbol{k}) - |\boldsymbol{b}(\boldsymbol{k})|), \tag{32}$$

where only the second term in the integrand contains information about the orientation of the ferromagnetic background, given by the angles θ and φ.

The expansion of the spin splitting $|\boldsymbol{b}(\boldsymbol{k})|$ gives its explicit dependence on the θ and φ, which have the form

$$|\boldsymbol{b}(\boldsymbol{k})| = b_0(\boldsymbol{k}) \sum_{n=0}^{\infty} \binom{1/2}{n} (\cos\gamma\,(\boldsymbol{k}))^n, \tag{33}$$

where

$$b_0(\boldsymbol{k}) = \sqrt{|\boldsymbol{b}_R(\boldsymbol{k})|^2 + J^2}, \quad \cos\gamma\,(\boldsymbol{k}) = \frac{2\boldsymbol{B}\cdot\boldsymbol{b}_R(\boldsymbol{k})}{|\boldsymbol{b}_R(\boldsymbol{k})|^2 + J^2}. \tag{33a}$$

The expansion (33) can be written more explicitly in the form

$$|\boldsymbol{b}(\boldsymbol{k})| = \sum_{n=0}^{\infty}\sum_{p=0}^{n} B_n^p(\boldsymbol{k})(\sin\theta)^n(\cos\varphi)^p(\sin\theta)^{n-p} \tag{34}$$

with expansion coefficients,

$$B_n^p(\boldsymbol{k}) = (-1)^p 2^n \binom{\frac{1}{2}}{n}\binom{n}{p}\frac{(Jt'')^n(\sin k_y)^p(\sin k_x)^{n-p}}{\left(|\boldsymbol{b}_R(\boldsymbol{k})|^2 + J^2\right)^{n-1/2}}. \tag{35}$$

The internal energy then has the corresponding expansion,

$$U(\theta,\varphi) = \sum_{n=0}^{\infty}\sum_{p=0}^{n} U_n^p(\sin\theta)^n(\cos\varphi)^p(\sin\varphi)^{n-p}, \quad U_n^p = -\int\frac{d\boldsymbol{k}}{(2\pi)^2}B_n^p(\boldsymbol{k}). \tag{36}$$

Because the integrand is odd under $k_x \to -k_x$ and $k_y \to -k_y$, coefficients $U_n^{2p+1} = 0$, and $U_{2p+1}^k = 0$, i.e., only terms even in both p and n survive. In combination with the symmetry of the binomial coefficients, this also leads to the equality, $U_{2n}^{2n-2p} = U_{2n}^{2p}$.

Then, the first terms in the expansion are

$$U(\theta,\varphi) \approx U_0^0 + U_2^0\sin^2\theta + \frac{1}{8}\left(U_4^0 + U_4^2 + \left(2U_4^0 - U_4^2\right)\cos 4\varphi\right)\sin^4\varphi \tag{37}$$

as in the phenomenological representation (20).

For the gapped half-filled case, it is consistent to expand the integrand in the $|t''| \ll J$ limit,

$$B_0^0(\boldsymbol{k}) \approx J + r(\boldsymbol{k})\left(2J - \frac{r(\boldsymbol{k})}{J}\right), \quad B_2^0(\boldsymbol{k}) \approx -\frac{(t'')^2}{2J}\sin^2 k_x(1 - 3r(\boldsymbol{k})),$$
$$B_4^0(\boldsymbol{k}) \approx -\frac{(t'')^4}{8J^3}\sin^4 k_x, \quad B_4^2(\boldsymbol{k}) \approx -\frac{(t'')^4}{4J^3}\sin^2 k_x\sin^2 k_y, \tag{38}$$

where $r(\boldsymbol{k}) = |\boldsymbol{b}_R(\boldsymbol{k})|^2/4J^2$.

Substituting (38) into (36) results in the expressions

$$U_2^0 = \frac{1}{4}\frac{(t'')^2}{J} - \frac{15}{64}\frac{(t'')^4}{J^3}, \quad U_4^0 = -\frac{15}{64}\frac{(t'')^4}{J^3}, \quad U_4^2 = \frac{1}{4}U_4^0, \tag{39}$$

which describe the coefficients in the expansion of internal energy (37).

Then, comparing (37) with the phenomenological expression (20) results in the expressions

$$K_2 = \frac{1}{4}\frac{(t'')^2}{J} - \Delta, \quad K_4 = -\frac{5}{8}\Delta, \quad K_4' = -\frac{K_4}{5} \tag{40}$$

for the magnetic anisotropy coefficients. These anisotropy coefficients satisfy the PMA condition (21).

Thus, PMA is realized in the two-layered magnetic nanostructure composing of an insulating magnetic layer and heavy metal with the Rashba spin splitting effect in the case of the half-filled bands. In so doing, MAE depends directly on the exchange interaction between the spin-polarized iterant electrons and magnetic states of the ferromagnetic layer.

5 Physical Regimes of the Magnetic Anisotropy Formation

The magnetic anisotropy formation in the two-layered magnetic nanostructure, ferromagnetic/heavy metal, depends on the relations between parameters of the exchange interaction (J), kinetic energy (t'), and the Rashba SOC (t''). Different interfacial magnetic states are determined by conditions of strong exchange ($J \gg t' \gg t''$), intermediate exchange ($J \sim t' \gg t''$), and weak exchange ($t' \gg J \sim t''$). It is assumed that the SOC strength is smaller than the non-relativistic bandwidth. These three cases are defined by how the exchange energy due to the ferromagnetic coupling compares these two energy scales. The magnetic anisotropy is determined by the comparison of the local characterization of the MAE via the susceptibility with the global characterization via internal energy differences. For the present model, the contribution to the MAE from the volume susceptibility (28) vanishes when $\boldsymbol{M}||\boldsymbol{z}$, while it is the only non-vanishing contribution for $\boldsymbol{M}||\boldsymbol{x}$.

In the first case, the domination of the exchange energy leads to two well-separated bands. The dependence of the MAE on the number of electrons per lattice site (N_e), obtaining from the spin susceptibility, for two stable orientations of the ferromagnetic background, is characterized by PMA only in narrow range around $N_e = 1$ (Fig. 2a) and for most values Ne the IMA is realized.

When $\boldsymbol{M}||\boldsymbol{z}$, the interband contribution to the susceptibility (30) favors PMA, while the intraband contribution (29) favors IMA. The amplitude of the intraband contribution is larger than the interband one and is maximized when the Fermi level is at the Van Hove singularity in the DOS of each band. When $N_e = 1$ and $\boldsymbol{M}||\boldsymbol{z}$, the intraband contribution must vanish because the system is gapped. Only the interband

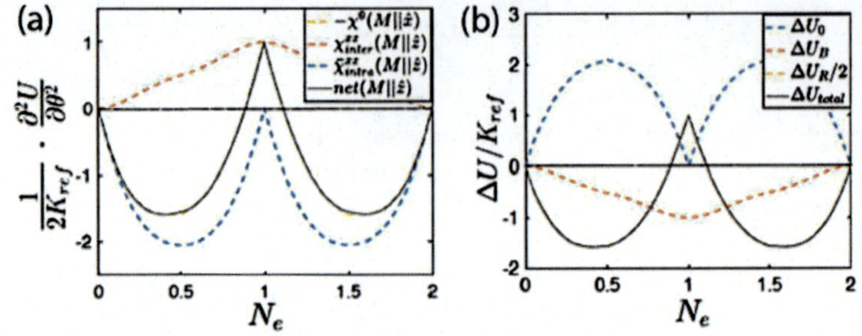

Fig. 2 MAE as the function of N_e for the strong exchange case, $J >> t' >> t''$. a The MAE from the second derivatives of the band energy, from the connection to its phenomenological form. b Internal energy differences decomposed using (10)

term remains finite and it favors PMA. When $M||x$, the volume susceptibility (28) is the only nonzero contribution and reproduces essentially the same MAE as found for $M||z$. This agreement shows that the higher order anisotropy constants (K_4 and K_4') are very small when compared with K_2. The MAE from the band energy difference between $M||x$ and $M||z$ is in perfect agreement with the one extracted from the susceptibility (Fig. 2b). The PMA is realized near half-filling. Moving from electron per site $N_e = 1$ to $N_e = 0$, the interband contribution is accurately proportional to M, which decreases monotonically to zero. The intraband contribution qualitatively follows $\rho(E_F)$, which increases up to the Van Hove singularity and then decreases again.

In the second case (Fig. 3), where the exchange energy is comparable to the non-relativistic bandwidth, by setting $J = t'$ the two bands overlap with minority band occupation beginning for $N_e > 0.5$, and the lower band being completely full for $N_e > 1.5$. This intermediate exchange coupling strength case is applicable to many ferromagnetic metals. The MAE from the spin susceptibility is characterized

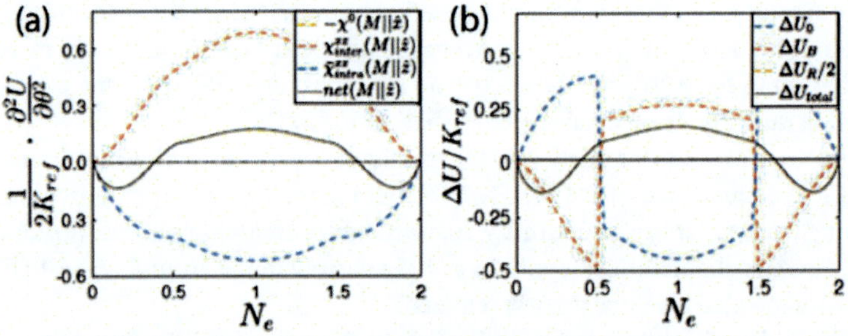

Fig. 3 MAE as the function of N_e for the intermediate exchange case, J ~ t' > > t". a The MAE from the second derivatives of the band energy, from the connection to its phenomenological form. It is shown the dependence of the spin susceptibility contributions on the magnetization direction. b Internal energy differences, decomposed using (10), which agrees with the results of the susceptibility calculations

by PMA in a much wider range of N_e than in the strong exchange interaction case (Fig. 3a).

Comparing (29) and (30), it appears that the Fermi sea term can be enhanced by reducing the $\boldsymbol{k}$-dependent spin splitting, $|\boldsymbol{b}(\boldsymbol{k})|$, which follows from weakening J, so that the interband contribution has a larger amplitude than the intraband one. However, near $N_e = 0$ (likewise near $N_e = 2$), the intraband contribution is linear in N_e while the interband one is quadratic, so that the former can overtake the latter, and thus favors IMA. The MAE from the band energy difference between $\boldsymbol{M}||x$ and $\boldsymbol{M}||z$ and its decomposition (Fig. 3b) exhibits half of the difference in the SOC energy that overlaps with the net internal energy difference. The energetic competition between the Rashba SOC and the coupling to the ferromagnetic background is settled differently when either only one or when both bands are partially filled, due to an allowed transfer of electronic occupation between the two bands at the Fermi sea.

In the third case where the exchange energy is compared to the SOC strength ($J = t''$), the splitting between the two bands is small. The intraband and interband contributions to the susceptibility are almost identical leading to a small net value of the MAE.

6 Conclusions

The pulse-field manipulation by magnetic properties of multilayer magnetic nanostructures is based on an excitation of the coherent electron spin polarization coupled by the exchange interaction with localized magnetic states and heat-induced demagnetization. The impact of the pulse laser field on magnetic states can occur directly via the effective bias field of the inverse Faraday magneto-optic effect or indirectly via the laser-induced heat demagnetization with subsequent action of the effective bias field. The controlled impact of the strong enough laser field on the magnetic nanostructures with an antiferromagnetic exchange interaction occur due to the effective bias fields caused by the different time of the heat demagnetization of magnetic sublattices and the relaxation of the antiferromagnetic exchange interaction.

The electric field control can occur via the spin polarization of an electric current under the ferromagnetic exchange interaction in ferromagnetic layers and its exchange interaction with the localized magnetic states. In multilayer magnetic nanostructures with the spin Hall effect, the electric field control occurs via a spin–orbit-induced conversion of the electric current to the spin current coupled with the localized magnetic states by the exchange interaction.

The pure electric field can control the magnetic anisotropy in the magnetic nanostructures, with the Rashba spin–orbit interaction, by the external gate voltage applied across a dielectric layer. In this case, the connection of the electric field with the localized magnetic states occurs via the exchange interaction of polarized itinerant electrons with the localized magnetic states. The corresponding mechanism of the magnetic anisotropy formation and its dependence on the interface electron structure are described in the framework of a tight-binding model of spin–orbit-coupled electrons exchange-coupled to the localized magnetic states.

Acknowledgements This work is part of a project that has received funding from the European Union's Horizon 2020 research and innovation program under the Marie Skłodowska-Curie grant agreement No 778308.

References

1. Johnson MT, Bloemen PJH, den Broeder FJA, de Vries JJ (1996) Magnetic anisotropy in metallic multilayers. Rep Prog Phys 59:1409
2. Dzyaloshinksky EI (1958) Thermodynamic theory of "weak" ferromagne-tism of antiferromagnetics. J Phys Chem Solids 4:241
3. Moriya T (1960) Anisotropic superexchange interaction and weak ferro-magnetism. Phys Rev Lett 120:91
4. Manchon A, Koo HC, Nitta J, Frolov SM, Duine RA (2015) New perspectives for Rashba spin-orbit coupling. Nat Mater 14:871
5. Chiba D, Sawicki M, Nishitani Y, Nakatani Y, Matsukura F, Ohno H (2008) Magnetization vector manipulation by electric fields. Nature 455:515
6. Maruyama T, Shiota Y, Nozaki T, Ohta K, Toda N, Mizuguchi M, Tulapurkar AA, Shinjo T, Shiraishi M, Mizukami S et al (2009) Large voltage-induced magnetic anisotropy change in a few atomic layers of iron. Nat Nanotechnol 4:158
7. Chaudhary G, Dias MS, MacDonald AH, Lounis S (2018) Anatomy of magnetic anisotropy induced by Rashba spin-orbit interaction. Phys Rev B 98:134404

Heating and Coagulation of Nanoparticles in a Plasma Jet

O. Yu. Kravchenko and I. S. Maruschak

1 Introduction

Plasma-assisted technologies represent important tools for deposition of nanostructured films on substrates. The growth of thin and ultra-thin films may be achieved using a large variety of techniques such as chemical vapour deposition, RF sputtering, pulsed laser deposition or plasma enhanced chemical vapour deposition [1–3]. Recently, a new process, which uses a plasma torch operating at low pressure has been developed with the aim of depositing uniform thin layers on large surfaces [4, 5]. In this plasma spraying process plasma jets are used as a heat sources to melt and accelerate the injected nanoparticles which subsequently impinge and solidify on a substrate. Modelling the nanoparticles, which create and assemble the film it is possible to enhance the physical properties of thin films. As is known, nanoparticles have the ability to coagulate, resulting in a change in their size. This process can be significant in plasma and it must be taken into account when transporting nanoparticles to a substrate in a plasma jet. It is important to be able to control the size of the nanoparticles, their kinetic energy, the temperature and the magnitude of the flow on the substrate. In order to control efficiently the technological parameters of the plasma's action, it is necessary to calculate the plasma flow characteristics by means of the proper mathematical model.

For the simulation of the supersonic flow of an ordinary temperature gas with dust particles, there have already been many papers in the literatures [6–8]. However, for the supersonic low pressure plasma flow, only relatively limited modeling results can be found. At plasma pressures $P = 1 - 100\,\text{Torr}$, the temperature of the electron component is substantially higher than the temperature of ions and atoms, and the plasma is thermodynamically nonequilibrium. To model processes in such a plasma, a multi-component and multi-temperature hydrodynamic model should be used. The aim of this work is to simulate the dynamics, heating and coagulation of nanoparticles in a low pressure plasma jet expanding through a round hole into a dilute gas.

O. Yu. Kravchenko (✉) · I. S. Maruschak
Taras Shevchenko National University of Kyiv, 64/13, Volodymyrska Street, Kyiv 01601, Ukraine
e-mail: kay@univ.kiev.ua

O. Fesenko and L. Yatsenko (eds.), *Nanomaterials and Nanocomposites, Nanostructure Surfaces, and Their Applications*, Springer Proceedings in Physics 246, https://doi.org/10.1007/978-3-030-51905-6_3

2 Model Description

In this paper, we simulate the outflow of the plasma jet with dust particles through a round hole of radius R_0 in the rarefied neutral gas. It is assumed that the plasma velocity V_0 and its density ρ_0 are constant at the inlet during the plasma jet expansion. The plasma considered in this article consists of four species, namely electrons, neutral argon atoms, singly ionized argon ions and nanoparticles. They will be denoted by the subscript e, a, i and d, respectively. We use hydrodynamic model to describe the expansion of the plasma jet with dust particles. In the model ions, electrons and neutral atoms have the same drift velocity $\mathbf{w} = (u, v)$ due to effective momentum exchange, and dust particles have drift velocity $\mathbf{w_d} = (u_d, v_d)$. Here u, u_d are radial velocity components and v, v_d are axial velocity components.

The ions temperature equals to neutral atoms temperature T, but electrons temperature T_e can differ from them.

The continuity equation for heavy plasma component (ions and neutral atoms) is equal to

$$\frac{\partial n}{\partial t} + div(n\mathbf{w}) = 0 \tag{1}$$

Here n is sum of ion density n_i and neutral atom density n_a.

The continuity equation for ions is equal to

$$\frac{\partial n_i}{\partial t} + div(n_i\mathbf{w}) = -\frac{I_i n_d}{e}. \tag{2}$$

The right hand side describes the ion destruction due to the recombination at the interaction with dust particles.

The ion current on dust particle I_i is described by OLM theory [9] and is equal to

$$I_i = \pi r_d^2 n_i e|\mathbf{w}| \left(1 - \frac{eq_d}{4\pi\epsilon_0 r_d \left(m|\mathbf{w}|^2/2 + kT\right)}\right) \tag{3}$$

Here r_d is dust particle radius, T is ion temperature, q_d is dust particle charge, e is proton charge, m is the ion and neutral atom mass.

The continuity equation for dust particles is equal to

$$\frac{\partial n_d}{\partial t} + div(n_d\mathbf{w_d}) = 0, \tag{4}$$

where n_d is dust particles density.

The momentum equations for heavy plasma particles (ions and atoms) are given by

$$\frac{\partial (nu)}{\partial t} + div(nu\mathbf{w}) = -\frac{1}{m}\frac{\partial P}{\partial r} - \frac{n_d f_r}{m} + \frac{e}{m} n_i E_r, \tag{5}$$

$$\frac{\partial(nv)}{\partial t} + div(nv\mathbf{w}) = -\frac{1}{m}\frac{\partial P}{\partial z} - \frac{n_d f_z}{m} + \frac{e}{m}n_i E_z, \tag{6}$$

where $P = nkT$ is the plasma pressure, E_r and E_z are radial and axial electric field components, $\mathbf{f}$ is force of the aerodynamic interaction between plasma and dust particle (f_r and f_z are its components along axis r and z). It consists of a friction force between dust particles and neutral particles $\mathbf{f}_{dn}$, as well as between the ions and the dust particles $\mathbf{f}_{di}$.

According [9] neutral drag force can be approximate as

$$\mathbf{f}_{dn} = \frac{8}{3}\sqrt{2\pi}r_d^2 n_a m V_{tn}(\mathbf{w} - \mathbf{w}_d), \tag{7}$$

where $V_{tn} = \left(w^2 + 8kT/\pi m\right)^{1/2}$ is total atom speed (a combination of directed and thermal speeds). The ion drag force can be expressed as

$$\mathbf{f}_{di} = n_i m V_{tn} \sigma^{coul} \mathbf{w}, \tag{8}$$

where $\sigma^{col}(\sigma^{coul})$ is the momentum collision cross section corresponding to the collection of ions by direct ion impacts (electrostatic Coulomb collisions).

The momentum equations for dust particles are given by

$$\frac{\partial(n_d u_d)}{\partial t} + div(n_d u_d \mathbf{w}_d) = -\frac{\alpha_d}{m_d}\frac{\partial P}{\partial r} + \frac{n_d f_r}{m_d} + \frac{q_d}{m_d}n_d E_r, \tag{9}$$

$$\frac{\partial(n_d v_d)}{\partial t} + div(n_d v_d \mathbf{w}_d) = -\frac{\alpha_d}{m_d}\frac{\partial P}{\partial z} + \frac{n_d f_z}{m_d} + \frac{q_d}{m_d}n_d E_z, \tag{10}$$

where α_d is volume fraction of dust particles.

Equations for internal energies ions and atoms ϵ, electrons ϵ_e ana dust particles ϵ_d are given by

$$\frac{\partial \rho\epsilon}{\partial t} + div(\rho\epsilon\mathbf{w}) + P div(\mathbf{w}) = Q_{ie} + Q_{en} - Q_{id} - Q_{nd}, \tag{11}$$

$$\frac{\partial \rho\epsilon_e}{\partial t} + div(\rho\epsilon_e\mathbf{w}) + P_e div(\mathbf{w}) + div(q_e) = -Q_{ie} - Q_{en} - Q_{ed}, \tag{12}$$

$$\frac{\partial \rho_d\epsilon_d}{\partial t} + div(\rho_d\epsilon_d\mathbf{w}_d) = Q_{ed} - Q_{nd} + Q_{rec} + Q_{id} - Q_{\mathrm{rad}}. \tag{13}$$

Here the heat flux is given by $\mathbf{q_e} = -\kappa(T_e)\nabla T_e$, where $\kappa(T_e)$ is the coefficient of electron thermal conductivity. Energies, which are transferred from ions to dust particles Q_{id}, from electrons to dust particles Q_{ed}, due to the recombination of ions on the surface of dust particles Q_{rec}, due to collisions of neutral atoms with dust particles Q_{nd} are defined according to [10, 11].

Let us dwell on the process of energy exchange between dust particles and plasma. As is known, the electric current from plasma flows into a dust particle, the value of which is determined by the parameters of the plasma and the geometric dimensions of the dust. This process is accompanied not only by the transfer of the electric charge from the plasma to the dust surface, but also by the transfer of energy. The heat fluxes between the plasma and the dust particle can be divided into the following types: radiation, kinetic energy transfer of electrons and ions upon contact with the surface of the dust particle, recombination of ions on the surface of the particle with the release of the corresponding energy.

Consider the process of energy exchange in kinetic processes. We assume that the fluxes to the dust particle can be described using the orbit-limited (OLM) approach [9], considering that the plasma is sufficiently rarefied. Then the electron energy flow to the dust particle can be written:

$$Q_{ed} = \int_{v_{min}}^{\infty} 4\pi\sigma_{ed} v_e^3 \left(\frac{m_e v_e^2}{2} + e\phi_d\right) f_e(v_e) dv_e. \quad (14)$$

Here v_e and f_e are the velocity and distribution function of electrons, ϕ_d is dust potential. The value of v_{min} is determined by the law of conservation of energy $m_e v_{min}^2/2 = -e\phi_d$. The cross section of electron deposition on the dust particle is determined by the expression

$$\sigma_{ed} = \pi r_d^2 \left(1 + \frac{e\Phi_d}{k_B T_e}\right).$$

A similar approach can be applied to determine the heat flux of ions to the surface of a dust particle. It is only necessary to bear in mind that the ions in the particle field acquire additional energy $e\phi_d$. Then the ion energy flow to the dust particle is determined by the expression

$$Q_{id} = \int_{0}^{\infty} 4\pi\sigma_{id} v_i^3 \left(\frac{m v_i^2}{2} - e\phi_d\right) f_i(v_i) dv_i, \quad (15)$$

where v_i and f_i are the velocity and distribution function of ions.

Assuming the Maxwell distributions of electrons and ions, after integration into the (14) and (15), we obtain the following expressions for the energy flows per dust particle

$$Q_{ed} = 8\sqrt{2\pi} r_d^2 n_e v_{te} exp\left(\frac{e\phi_d}{T_e}\right) T_e, \quad (16)$$

$$Q_{id} = 4\sqrt{\pi} r_d^2 n_i v_{ti} \frac{e\phi_d}{T} T. \quad (17)$$

Here v_{te} and v_{ti} are thermal velocities of electrons and ions.

The energy flow to the dust particle due to the recombination of ions on its surface is determined

$$Q_{rec} = \frac{I_i}{e}\epsilon_{ion}, \tag{18}$$

where ϵ_{ion} is the energy of ionization.

When the microparticles collide with neutral atoms, the dust particles will be cooled, thereby heating the neutral component of the plasma. The exchange of energy between the dust particle and the neutral component of the plasma per unit of time is determined

$$Q_n = 2\pi r_d^2 n_a v_{tn} k_B \Delta T, \tag{19}$$

where $\Delta T = T_d - T$ is temperature difference between dust and gas.

Significant contribution to the thermal balance also contributes the thermal radiation. The heat losses of such a process from the dust surface per unit time can be described using the Stefan-Boltzmann law, which has the form

$$Q_{\rm rad} = 4\pi r_d^2 \theta \sigma T_d^4, \tag{20}$$

where θ is the coefficient of grayness of the microparticle substance, σ is Stephan-Boltzman's constant.

To determine the distribution of nanoparticles by charge, we use the model proposed in [12, 13]. This model takes into account the stochastic nature of the charging of dust particles associated with the chaos of the thermal motion of electrons and ions. As a result, dust particles with different charges are present in each elemental volume of plasma. Nanoparticles in the plasma are charged because of collisions with electrons and ions. The electron and ion currents collected by a dust particle in the nanometer regime can be described by the orbital-motion-limited (OML) probe theory [9]. A particle with radius r_d which carries a charge $Z_k = ke$ (with e the elementary charge and k an integer) is charged to a surface potential of $\Phi_k = Z_k/4\pi\epsilon_0 r_d$, with ϵ_0 the vacuum dielectric constant. Using OML theory, expressions for the frequency with which a particle with charge Z_k is hit by electrons and ions, respectively, can be derived

$$\nu_{e,i} = n_{e,i} S v_{te,ti} exp\left(-\frac{q_{e,i}\Phi_k}{k_B T_{e,i}}\right), q_{e,i}\Phi_k \geq 0 \tag{21}$$

$$\nu_{e,i} = n_{e,i} S v_{e,i}\left(1 - \frac{q_{e,i}\Phi_k}{k_B T_{e,i}}\right), q_{e,i}\Phi_k < 0, \tag{22}$$

$S = 4\pi r_d^2$ is the particle surface area, $v_{te,ti} = (k_B T_{e,i}/2\pi m_{e,i})^{1/2}$, k_B is Boltzmann constant.

The charge distribution of particles of a given radius r_d is described by the fraction of particles F_k carrying a charge ke. It is normalized by $\sum F_k = 1$. The rate equation for a charge state k can then be written as

$$\frac{dF_k}{dt} = \nu_e^{k+1} F_{k+1} - \nu_e^k F_k - \nu_i^k F_k + \nu_i^{k-1} F_{k-1}. \tag{23}$$

It is assumed that the charging of particles is much faster than coagulation so the charge distribution can be considered in steady state [15]. This assumption enables the use of recursive relations for the charge distribution

$$F_{k+1} = \frac{\nu_i^k}{\nu_e^{k+1}} F_k.$$

In addition, in the presented model, coagulation of dust particles is considered, which is described by the model proposed in [14, 15]. The volume distribution function of dust particles is described by the general dynamic equation

$$\frac{\partial n(v)}{\partial t} = \frac{1}{2}\int_0^v \beta(v', v - v')n(v')n(v - v')dv' - \int_0^\infty \beta(v, v')n(v)n(v')dv', \tag{24}$$

where v is the volume of the dust particle, $n(v)dv$ denotes the particle number density in a volume range $[v, v + dv]$. Coefficient $\beta(v, v')$ is the frequency for coagulation between two particles with a volume v and v'. According to [15], $\beta(v, v')$ is given

$$\beta(v, v') = \alpha(v, v') \left(\frac{3}{4\pi}\right)^{1/6} \left(\frac{6k_B T}{\rho_p}\right)^{1/2} \left(\frac{1}{v} + \frac{1}{v'}\right)^{1/2} (v^{1/3} + v'^{1/3})^2$$

where v and v' are the volumes of the particles interacting, ρ_p is the density of the particles, and T is the temperature of the particles, $\alpha(v, v')$ is a coefficient which describes that the effective cross section for coagulation depends on the charge of both particles

$$\alpha(v, v') = \sum_{k=-\infty}^{\infty} \sum_{k'=-\infty}^{\infty} F_k(v) F_{k'}(v') Q(k, k', v, v')$$

with

$$Q(k, k', v, v') = exp\left(-\frac{kk'e^2}{4\pi\epsilon_0 R_s k_B T}\right), kk' > 0 \tag{25}$$

$$= 1 - \frac{kk'e^2}{4\pi\epsilon_0 R_s k_B T}, kk' \leq 0. \tag{26}$$

The above system of hydrodynamic equations is solved numerically by the method of large particles [16], and the distribution of nanoparticles by volume is determined by the sectional modeling [14].

3 Numerical Results and Discussion

The plasma jet with nanoparticles is simulated for different plasma pressures and nanoparticles densities at the inlet. Simulations continued until a steady-state flow of plasma. As results, spatial distributions of the plasma parameters and disperse phase parameters (densities, drift velocities, temperatures and the plasma pressure) were obtained in various times after the start of injection of the plasma jet into the space filled with gas.

This section presents the results of the heating of nanoparticles and their coagulation in the plasma jet, as well as their influence on the temperature profile of the plasma and the velocity of the dust particles.

Figure 1 shows the spatial distributions of the nanoparticle temperature at different their sizes along the axial coordinate. Figure 1 a corresponds to the plasma pressure at the inlet $P_0 = 4$ Torr and (b)—$P_0 = 40$ Torr. It is seen that the temperature of the nanoparticles in the plasma stream first grows rapidly, and then its fixed value is set. Heating of nanoparticles is caused mainly by recombination of ions when they are collide with nanoparticles. At this, ion ionization energy is transmitted to the heating of nanoparticles. At low plasma pressure (Fig. 1a), the temperature of nanoparticles increases with a decrease in their radius. This result can be explained on a simplified model describing the temperature change of nanoparticles:

$$Cm_d \frac{dT_d}{dt} = \sum_j P_j.$$

Here m_d, C are mass and heat capacity of dust particle. In the right part of the equation are energy streams on the surface of the dust particles. Approximate can be assumed that these streams are proportional to the surface area of the dust, that is proportional to its radius in the square. At the same time, the mass of the dust is proportional to its radius in Cuba. It follows that the temperature of nanoparticle is inversely proportional to its radius. This dependence of the dust temperature on their radius corresponds to the results obtained for low pressure plasma ($P_0 = 4$ Torr), but contradicts the results obtained at the pressure of $P_0 = 40$ Torr. The reason for this discrepancy is the cooling of dust particles upon contact with a neutral gas, which becomes significant at the pressure $P_0 = 40$ Torr. It should be noted that the cooling of dust particles with a smaller radius occurs more efficiently than the dust of a larger radius (provided that they have the same mass). This is due to the fact that in the first case the total surface of the dust component in contact with the plasma will be larger. Significant cooling of small dust particles in the case of $P_0 = 40$ Torr and leads to a decrease in their temperature compared to the temperature of the larger dust particles.

The above confirms Fig. 2 where the spatial distributions of the temperature of the heavy plasma component (ions and atoms) along the axial axis are presented for the case $r_d = 50$ nm (a) and $r_d = 200$ nm (b). The solid curves correspond to the plasma pressure at the inlet $P_0 = 4$ Torr, and the dotted ones corresponds to

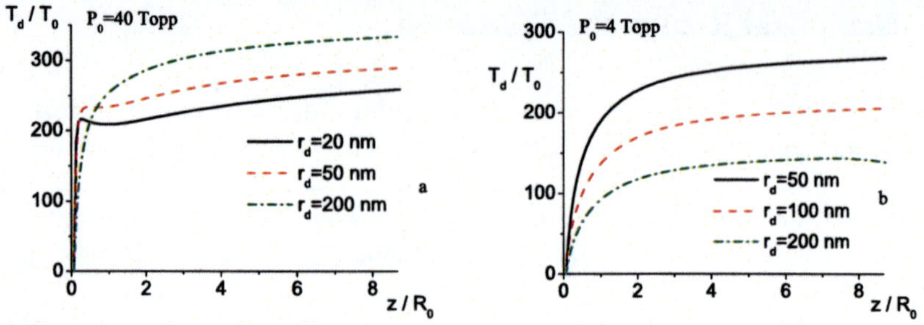

Fig. 1 Axial profiles on jet axis of the nanoparticles temperature for different plasma pressures at the inlet and nanoparticles radii

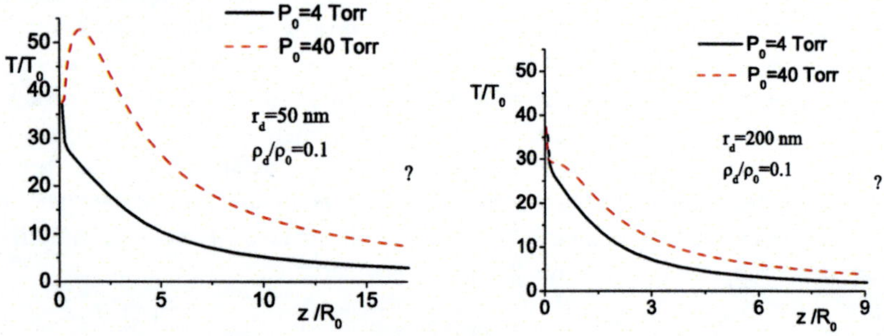

Fig. 2 Spatial distributions of the temperature of the heavy plasma component along the jet axis for the cases $r_d = 50$ nm (**a**) and $r_d = 200$ nm (**b**)

$P_0 = 40$ Torr. The ratio of the dust component density to the plasma density at the inlet in these modes was $\rho_d/\rho_0 = 0.1$. It is seen that the plasma temperature decreases with increasing z coordinate, which is caused by the conversion of thermal energy into energy of the directional motion of the plasma as it expands. It should be noted that with increasing pressure the plasma temperature in the jet increases significantly due to a more efficient exchange of energy between hot dust and plasma. Moreover, such energy exchange is larger in the case of smaller dust particles, which leads to a decrease in their temperature and an increase in the plasma temperature. The energy exchange process between the dust particles and the plasma explains the above-mentioned increase in the plasma temperature as the coordinate increases: first, the dust particles are heated by the interaction with the plasma, and then heated the plasma.

Let us now consider the spatial distributions of the temperature of dust particles at their different densities at the inlet (Fig. 3a) for the case $r_d = 20$ nm, $P_0 = 40$ Torr. The figure shows that at a lower density of the dust component ($\rho_{d0}/\rho_0 = 0.02$) the temperature of the dust particles is higher, and it increases monotonically with increasing z. At $\rho_{d0}/\rho_0 = 0.2$, dust particles are heated only at the initial stage of jet expansion (at $z < 0.2R_0$. To explain this result, let us consider Fig. 3b, which shows

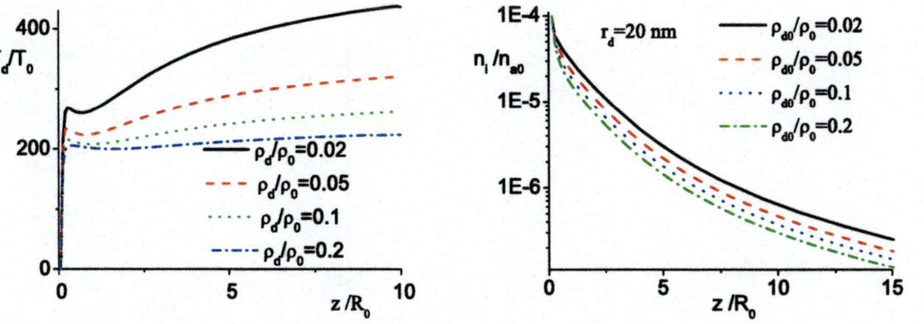

Fig. 3 Spatial distributions of dust particle temperature (**a**) and spatial distributions of ion density (**b**) along the jet axis at different dust particles densities

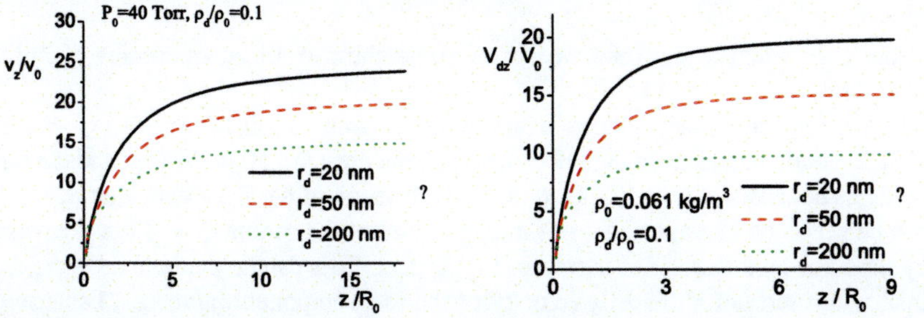

Fig. 4 Spatial distributions of plasma velocity (**a**) and dust particles velocity (**b**) along the jet axis at different dust particle radii

the spatial distributions of the ion concentration along the z axis for the same modes as Fig. 3.

In all modes, a decrease in n_i is observed as the coordinate is increased, but at a lower concentration of dust particles in the stream, the ion concentration remains higher. This causes higher energy flows to the dust surface and, as a consequence, increases their temperature.

Let us now consider the effect of dust particles on the dynamics of plasma flow, in particular on plasma velocity and dust components. Figure 4 shows the spatial distributions of the velocity of the plasma (a) and the dust component (b) along the jet axis at different radii of the particles, but equal to their mass at the inlet. It is seen that as the dust radius decreases, the velocity of the plasma jet increases, which can be explained by the increase in the plasma temperature (Fig. 2) and, accordingly, its pressure. Also from Fig. 4b it follows that as the dust radius decreases, their velocity increases. However, in all modes presented, the dust velocity is less than the plasma flow rate.

As noted, our model takes into account the coagulation of nanoparticles in plasma. As a result, nanoparticles of different sizes appear in the plasma stream. Figure 5a depicts the spatial distributions of nanoparticles of several sizes along the jet axis

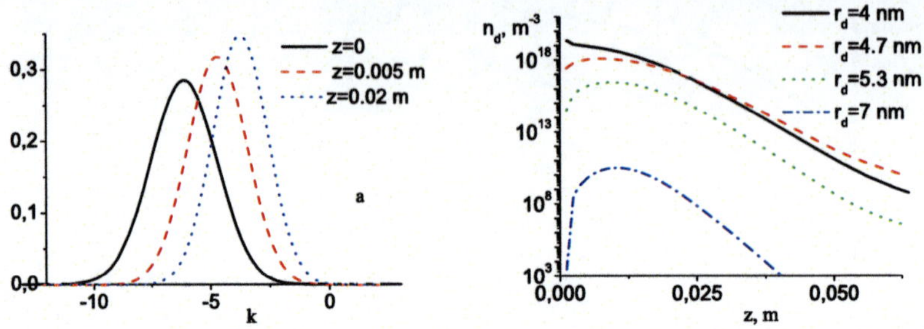

Fig. 5 Spatial profiles on jet axis of the dust densities for different their radii (**a**) and distributions by charge of nanoparticles at different distances from the inlet of plasma jet (**b**)

in the case when at the inlet we specify only the flux of nanoparticles of radius $r_d = 4\,\text{nm}$.

Figure 5b shows distributions on the charge of nanoparticles of a radius $r_d = 4\,\text{nm}$ at different distances from the inlet for he plasma pressure $P_0 = 40\,\text{Torr}$ and the dust particles density $\rho_{d0} = 0.1\rho_0$. Here F_k is the fraction of particles with charge ke. As can be seen, the average charge of nanoparticles of the radius $r_d = 4$ nm decreases by modulus as the distance increases to the inlet. This can be explained by the fact that, with increasing the drift velocity of the ion component increases, because of which the ion current onto the dust particle increases and, consequently, decreases its charge. We also note the decrease of the width of the charge particle distribution at the distance from the inlet.

4 Conclusion

The heating and coagulation of dust particles in a plasma stream expanding into a rarefied neutral gas has been investigated. The main findings can be summarized as follows.

In the low-pressure plasma stream ($P_0 = 4\,\text{Torr}$), the temperature of the dust particles decreases as their radius increases, and at higher plasma pressures ($P_0 = 40\,\text{Torr}$) the opposite dependence occurs. This result is a consequence of the dust particles being cooled by a neutral gas, which is more efficient for smaller dust particles.

At $P_0 = 40\,\text{Torr}$, the plasma temperature increases with the presence of dust particles in the flow and depends on their radius: at smaller dust radii, provided that the same dust mass in the plasma, it increases and leads to an increase in the flow velocity of the plasma and dust particles.

As the mass of the dust component increases, a decrease in the dust particle temperature and an increase in the plasma temperature are observed in the plasma stream.

Dust particles of larger radii appear in the jet due to coagulation. The maximum concentrations of these particles are at some distance from the inlet. We found that with the increase of the distance from the inlet the average charge of dust particles per module and the width of their distribution by charge decreases.

References

1. Roca i Cabarrocas P, Chaabane N, Kharchenko AV, Tchakarov S (2004) Polymorphous silicon thin films produced in dusty plasmas: application to solar cells, Plasma Phys Controlled Fusion 46:235
2. Hwang NM, Lee DK (2010) Charged nanoparticles in thin film and nanostructure growth by chemical vapour deposition. J Phys D 43:483001
3. Chaabane N, Suendo V, Vach H, Roca i Cabarrocas P (2006) Soft landing of silicon nanocrystals in plasma enhanced chemical vapor deposition. Appl Phys Lett 88:2031111
4. Biganzoli I, Fumagalli F, Di Fonzo F, Barni R, Riccardi C (2012) A supersonic plasma jet source for controlled and efficient thin film deposition. J Mod Phys 3:1626–1638
5. Trifiletti V, Ruffo R, Turrini C, Tassetti D, Brescia R, Di Fonzo F, Riccardi C, Abbotto A (2013) J Mater Chem A(1):11665
6. Ozturk A, Cetegen BM (2005) Modeling of axially and transversely. injected precursor droplets into a plasma environment. Int J Heat Mass Transf 48:4367–4383
7. Marchand C, Chazelas C, Mariaux G, Vardelle A (2007) Liquid. Precursor plasma spraying: modeling the interactions between the transient plasma jet and the droplets. J Therm Spray Technol 16(5–6):705–711
8. Jabbari F, Jadidi M, Wuthrich R, Dolatabadi A (2014) A numerical study of suspension injection in plasma-spraying process 23:3–13
9. Shukla PK, Mamun AA (2002) Introduction to dusty plasma physics, 270. IoP Publishing Ltd., Bristol and Philadelphia
10. Maurer HR, Kersten H (2011) On the heating of nano- and microparticles in process plasmas. J Phys D Appl Phys 44(7):174029
11. Bizyukov AA, Chibisov AD, Romashchenko EV (2012) Effect of the parameters of a gas-discharge plasma on the equilibrium temperature and floating potential of microparticle. Probl At Sci Technol 18(6):175–177
12. Sodha MS, Mishra SK, Misra S, Srivastava S (2010) Fluctuation of charge on dust particles in a complex plasma. Phys Plasmas 17:073705–7
13. Matsoukas T, Russell M, Smith M (1996) Stochastic charge fluctuations in dusty plasmas. J Vac Sci Technol A14:624–630
14. Agarwal P, Girshick SL (2012) Sectional modeling of nanoparticle size and charge distributions in dusty plasmas. Plasma Sources Sci Technol 21(12):055023
15. Kortshagen U, Bhandarkar U (1999) Modeling of particulate coagulation in low pressure plasmas. Phys. Rev. 60:887–898
16. Belotserkovskii OM (1994) Numercal Simulation in Continuum Mechanics, 448. Fiz Mat Lit, Moskow

The Effects of Interactions of the Extremely Low-Level Radiation with Quantum Coherent Nanosystems

Ludmila Stepanovna Martseniuk and Aleksandr Stepanovich Martseniuk

1 Introduction

The display of effects of small doses of radiation is most typically found out in the coherent environment of alive organisms. However, the similar effects can be found out and in some not biological objects. Such effects represent a theoretical interest as discovered by the features of functioning of the coherent systems, and, in particular, the alive organisms. Practical interest will be that the disclosing of this mechanism, possibly, will allow us to find another way to decision of the very important task—creation of working installations in which the guided cold nuclear fusion can be carried out.

The phenomena of the low-energy nuclear reactions were opened by L. Alvarets. Exactly for these openings, he was awarded the Nobel bonus on physics as early as in 1968 [1].

Now many laboratories of the various countries conduct researches on studying an opportunity of reception of superfluous energy due to low-energy reactions of nuclear synthesis commercially. Significant successes in this way are received. Among them it is possible to note, for example, creation of functioning installation of reception of superfluous heat—warm generator Rossi (nickel lithium aqueous reactor) which work was accompanied by the change of isotope structure of reaction components at the real absence of excretions of radioactive waste products and without radioactive radiation.

L. S. Martseniuk (✉)
Institute of Nuclear Researches of the National Academy of Science of Ukraine, 03680 Prospect Nauky 47, Kiev, Ukraine
e-mail: prolisok77@gmail.com

A. S. Martseniuk
Department of Science and Education of Ukraine, National University of Food Technology, Vladimirskaya 68, 601 Kiev, Ukraine

O. Fesenko and L. Yatsenko (eds.), *Nanomaterials and Nanocomposites, Nanostructure Surfaces, and Their Applications*, Springer Proceedings in Physics 246, https://doi.org/10.1007/978-3-030-51905-6_4

The similar working installation has started recently in Ukraine in the Institute of nuclear researches where the fact of an opportunity of reception of superfluous energy also has been established due to low-energy nuclear reactions (LENR), working without excretions of radioactive radiation [2].

Annual scientific conferences and numerous publications on cold nuclear fusion (CNF) and low-energy nuclear reactions (LENR) give hope that the extremely important for mankind a problem of industrial reception of superfluous heat due to low-energy nuclear reactions in the near future will be successfully solved. Especially, as follows from the researches which have been carried out still L. Kervran and his predecessors, alive organisms find their own ways for realization of such reactions for satisfactions of the vital tasks.

It specifies on actuality of researches of the low-energy phenomena in alive objects, as systems in which cold nuclear synthesis really occurs, with the purpose of creation of theoretical bases of this phenomenon and its possible reproduction in practical tasks.

Despite the significant successes achieved in this way, the uniform theory of this phenomenon till now is not created; experimental results are not steadily reproduced. It causes nonacceptance of the received results by some conservative-minded scientists. However, in the objections, they based on the theory advanced for high-temperature nuclear synthesis which *is not acceptable* for the description of the phenomena of cold synthesis in the coherent environment or *biological organisms, **the nanostructural elements of which** are the quantum coherent macrostructures.*

For the further description of such processes in coherent systems, we will take the definition of coherentness, presented by W. Nerst still in 1916 [3].

W. Nerst has come out with the assumption, that quantum fluctuations of elementary components of physical systems can be adjusted synchronously, leading to the occurrence of *collective, with a uniform phase,* oscillations. Oscillations of many components thus occur in unison. Particles behave as a unit; therefore, there is a loss of their individuality.

The dynamic regime corresponding to such possibility is named coherence, and it is really found out in superfluid helium, crystals, magnets, superconductors, etc. systems where arise a similar picture of ordering. However, for occurrence of coherence (according to W. Nersta's definition) in such macrosystems, it is necessary that for them the certain conditions have been executed (sufficient density, the temperature conditions providing certain structural transformations, etc.) which are achieved by macrosystems at temperatures T_{crit}. In nanostructural of alive organisms, such conditions are carried out at room temperatures [4].

According to [5], the state of quantum system of a macrosystem is defined as coherent if it is characterized by such size as a phase.

The quantum field is characterized by amplitude ψ_0 in which square is proportional to number of quantums, and a phase ϕ:

$$\psi = \psi_0 e^{i\phi} = a\sqrt{N}e^{i\phi}. \quad (1)$$

In this case, water system N is the number of molecules in the allocated volume.

E. Del Guidice describes dynamics of processes in alive organisms, specified in [6] that «Living systems cannot be thermal systems, but coherent systems» .

It is necessary to note three important factors at the characteristic of quantum coherent systems.

1. Character of processes in coherent quantum macrosystems has essential differences from incoherent one. For these systems, the phenomena of superradiation (effect of R. Dicke) [7], displays of effect of Josephson, [8], collective effects of a response on the external impulses of influence (like the Mossbauer effect), phenomena of trigger effect on the signals of an extremely low size and others are possible, which are not described by the laws of classical physics.

For example, according to the principle of Dicke [7], radiation in coherent quantum systems is *collective* process and intensity of radiation is proportional to a square of number of particles, i.e., ~N^2, instead of number of particles N, as in incoherent environments.

There can be also the phenomena of screening of Coulomb barrier of particles due to plasma vibrations and other quantum effects, characteristics only for the coherent systems, substantially reducing a factor of influencing potential barrier, impedimental of approaching particles on distances in which a nuclear reaction can begin on.

2. Management of processes in coherent quantum systems is possible at very deep level where processes of transfer of energy are not involved [9]. Such management can be carried out only through vector magnetic potential.

Circumstance that these two differences have not been taken into account in theoretical interpretations of events, which is taking place in coherent and not coherent environments, is one of the reasons of the absence of understanding the essence of reactions CNF and LENR. Exactly, this fact was marked repeatedly by the creator of new theory of water, built in accordance with principles of quantum theory of the field, J. Preparata and by his colleagues, researches of which have led to prominent results not only in area of physics of water [6] but also in the theory of cold nuclear fusion [10].

3. For biological systems, except for the mentioned features, there is one more factor on which alive cardinally differs from lifeless: the presence of biological fields. The most essential contribution to research of a biological field has brought in Gurvich [11].

These three factors, practically, determine all dynamics of functioning of an alive organism. We shall note that they are not independent; they should be considered in unity and the interrelation providing existence of an organism as a single whole.

So, in [12], it is underlined that partial loss of coherence can be considered as disease of an organism. In turn, renewal of an organism supposes renewal of the energy and information interrelations, which occur with the participation of biofield interactions, an essential role which is executed by processes, guided through the influence of vector magnetic potential which is carrying out the information functions.

The abovementioned three factors for biosystems cause specificity that the processes in an organism can be managed by signals having an ultralow intensity. As many researchers lead to the studying of physics alive (including F. Popp, L. Montagnier, M. Bischof) [13–15], consider that intercellular and intracellular interactions at an ultralow energy level determine the functioning of all alive.

2 An Alive Organism—Quantum Coherent System

In work [5], the new theory of water was for the first time presented by a prominent physicist-theorist J. Preparata, which gave exhaustive interpretation of the phenomena of water environment, including the phenomena in near-surface water, not interpretive from the position of traditional classic concepts.

According to this theory, water is the two-componential system including coherent component—coherent domains (CDs)—which have a uniform coherent field and not coherent component. Sizes of CD for distilled water are about 0, 1мкм. Not coherent component will consist of not coherent water molecules surrounding CDs.

The volume of coherent water makes room temperature of about 40% from total amount. Usual pure water under normal conditions stays in "flickering" regime, as the molecules of water, because of thermal fluctuations, continuously pass from regime of coherency to not coherent regime (and vice versa).

The situation essentially changes for near-surface water (EZ water) and for the water coupled by the nanostructural elements of an alive organism (*interfacial water*), where the sizes of CD are defined by sizes of the nanostructures or the cellular elements of alive organisms. Time of a life of such domains reaches weeks and months, because these formations are protected from the destroying influence of thermal fluctuations by surfaces of the molecular nanostructures.

CD can easily reserve energy (be excited) due to interaction with an environment.

In [16], the process of origin of the excited states is described as follows. In CD, every molecule of water oscillates between the basic and excited states; thus, the excited level corresponds to energy of 12.06 eV, while energy of ionization of water molecule makes up 12.60 eV. The oscillations with energy of 12.06 eV correspond to the size of CD ~ 0.1 мкм. Coherent oscillations lead to the occurrence of electrons which are easily excited. Every excitation corresponds to *the cold vortex* of electrons. Actually, every quasi-free electron belongs to the coherent state; therefore, the external indignation, less than the energy of gap of E_g (which separate the coherent state from incoherent), cannot be passed to any other individual molecule, and is saved in CD, giving the beginning to the collective excited state, which still coherently. The excited states in water cannot disintegrate in the thermal way, as motion of cold vortex of quasi-free electrons occurs without a friction just as it takes place in superconductors.

There is an enormous amount of the excited states (characterized by the angular moment of L) with energy of a few tens of KHz.

Individual excitations of quasi-free electrons are summarized; therefore, CD can reserve plenty of energy and time of the existence of the excited states is determined by time of the existence of CD. The spectrum of the excited states on every molecule in CD is limited from above by energy of gap (equal 0.26 eV/molecule); as in CD about 6 million molecules, it appears that the spectrum of the isolated water does not have a limit practically [16].

Actually, as specified in [17], energy, reserved in CD, can achieve a value, sufficient for realization of reaction of nuclear fusion!

The water structures, formed on the surfaces or on nanostructures, are necessary to examine as extensive CD, which possesses the high degree of coherence. EZ water has similar properties.

G. Pollak, studying properties of EZ water, marked that *some forces of long-range action are* responsible *for its* formation, and that molecules in it are *in* excited *state* [18].

Indeed, if to proceed from a position of traditional conceptions, the attraction of water molecules to the hydrophilic surface is possible to explain from its dipole properties, by virtue of which the water molecules must be attracted to surface, which have the superficial charges. In that case, the number of layers attracted to a surface should not exceed a few layers. Actually, near-surface water forms the layers occupying the much greater volume—sometimes and up to several hundreds of micron on thickness.

Other discrepancy is that near to the surface carrying a certain charge, layers of water have the same charge, as on surfaces (Fig. 1, [4]), that it is impossible to explain from the standard positions as the same charges must push off. It means that properties of near-surface water should be described by laws of quantum electrodynamics according to theory of J. Preparata.

As shown in Fig. 1, the schematic arrangement of water system close to a hydrophilic surface is submitted.

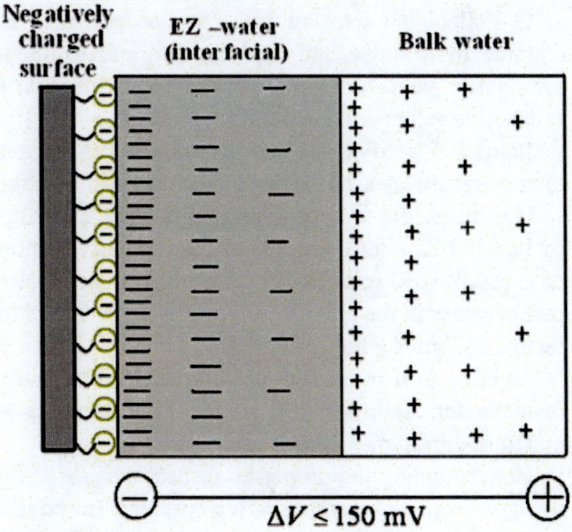

Fig. 1 The schematic image of near-surface water adjoining to negatively charged hydrophilic surface (according to work [4])

For visualization of a near-surface layer of water, G. Pollack has suggested to place in water the dyeing substances—suspensions with a diameter from 0.5 up to 2 μm. It was revealed that such substances do not remain inside a near-surface layer and a near-surface water is found out visually as undyed.

Taking into account that a near-surface water pushes out from its own structure a micro-impurity of the most various nature, G. Pollack has designated this zone—exclusion zone water (EZ water) [18].

Water, in this thick enough layer, differs from usual volumetric water practically, for all known parameters.

We list the most essential differences.

1. EZ-вода is charged negatively (potential is achieved at 150 mV) in relation to contacting with it a volume water. Along conductor, which connects an electrode, placed in EZ water, with an electrode, placed in volumetric water, passes a weak but continuous electric current.
2. At illumination of EZ water by infrared radiation with $\lambda = 3.100$ nm, a magnitude of a current increases. At the same time, the thickness of a layer of EZ water grows also (the four multiples increasing in its thickness are observed).
3. Water protons concentrate on the border between EZ water and volumetric water. Thus, the water system in which the EZ water and volumetric water coexist represents a system with division of charges, something like the condenser in which a negative plate is displayed by EZ water and positive by volumetric water with a lot of protons—a radiant energy essentially increases the capacity of this "condenser."
4. EZ water absorbs ultraviolet radiation with a maximum at $\lambda = 270$ nm. At illumination of EZ water ultraviolet radiation with $\lambda = 270$ nm, it fluoresces.
5. Viscosity of EZ water is essentially higher than usual volumetric water.

G. Pollack has selected three more properties of EZ water: its molecules—more coupled, more stable, and displaced as oriented. And it, in the opinion of G. Pollack, means that we have a liquid crystal, i.e., a situation when molecules are very well oriented in relation to each other.

Items 1–5 testify that electrons in EZ water are at much higher level of excitation than in usual water, i.e., it can be the donor of electrons.

G. Pollack did not see a possibility to explain all unusual properties observable by him in EZ water by means of traditional conceptions and specified on existence of a very distant ordering in water that is unexpected. If it is valid, then the most part of water in the universe is not the usual volumetric water which is described in textbooks, but is a liquid crystal.

G. Pollack also marked the following: «I suppose that the real attraction exists inside water. And, possibly, it other understanding will become basis for the new united understanding of nature» .

Understanding the properties of near-surface water is important in such relation that greater part of water of living organism (interfacial water) also have the characteristics of near-surface water. As indicated in [19], «the properties of such coherent interfacial wa-ter are those of EZ-water».

In [19], also it is marked: «The presence of a trapped e.m. field within the CD produces a strong field gradient on the boundary» .

The force influencing the charged particle in such field is estimated under the formula:

$$F = -Q^2/M \, \mathrm{grad} A^2, \tag{2}$$

where A is the magnetic vector potential of the field, Q is the particle charge, and M is the particle mass.

The size of the force acting on a particle, according to (2), is inversely proportional to mass of a particle. Therefore, electrons are repelled under influence of this force much more strongly than nucleus. In result on border of CD arise a jump of electric potential value from 50 up to 150 мВ.

Authors [19] point out that «The appearance of this electric potential jump on the outer surface of water CDs, whose value is in striking agreement with the generally accepted cell membrane potential, is naturally present and is not the consequence of any external supply of energy» .

Theory, developed by J. Preparata, allows to give interpretation to many phenomena in water and of water solutions, in particular, to such phenomena, as Zhadin effect, effects, opened by J. Benveniste and L. Montagnier, F. Popp, phenomena of homeopathy and other, which model the phenomena of interactions of extremely low-intensive radiation with the nanostructures of living organisms.

3 The Zhadin Effect

The Zhadin effect describes the phenomena of interaction the extremely low-intensive and very low-frequency magnetic fields with water solutions of glutamic acid and with some other water solutions.

These solutions *are not the biological objects*, but such effect illustrates the processes which can take place in living organisms. In Zhadin effect, it was shown that influences of the extremely low-frequency variable magnetic field on water solutions can result in the origin of very narrow peaks in the spectrum of dependence of conductivity of solutions from frequency of the attached external magnetic field.

The static magnetic field, the size of which was near to the natural magnetic field of Earth and the variable sinusoidal magnetic field, which was parallel to the permanent field was used.

The scheme of setting is presented in Fig. 2 [20].

The weak electric current transmitted through a solution was measured in experiment; for this purpose, two electrodes 2, 2^1 were used, between which the potential difference of 80 mV was created.

The size of a static magnetic field made up 0.04 mt and a variable magnetic field of 25 nT (i.e., in 1000 times is less than a static magnetic field) frequency changed within the limits of $1 \div 10$ Hz.

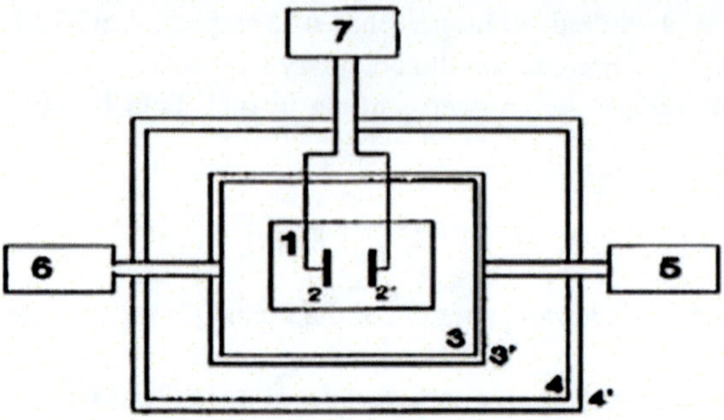

Fig. 2 From [20]. The scheme of the experimental setting. 1—Corvette with solution. 2—Electrodes. 3—Solenoid coils. 4—Magnetic screen of Permalloy. 5—Direct voltage source. 6—Sine-wave generator. 7—Measuring block: stabilizer of electrode voltage, current meter, recorder

With the position of interpretation of this effect, suggested in (19), responsible for this process are the CDs of water, to the surface of which the ions of glutamic acid are attracted. These ions can be rotated on the cyclotron orbits with gyromagnetic frequency, determined by the magnetic field of the excited vortex of quasi-free electrons of CD:

$$\nu_c = \frac{1}{2\pi}\frac{q}{m}B, \tag{3}$$

where ν_c—the cyclotron frequency; q, m—charge and mass of particle; and B—magnetic field.

Although in ordinary water (water solution) because of thermal fluctuations CDs cannot exist in the stationary regime, on various hydrophilic surfaces, they can form EZ structures in which they can exist during time, sufficient for the display of this effect. In [21], it was supposed that the surfaces of electrodes, used in an experiment, can be such surfaces.

Only at the presence of such surfaces, in accordance with interpretation, presented in [21], it is possible to observe the Zhadin effect.

The ions of impurity are attracted to the surface of CD due to forces of dispersions. In [21], it is possible to explain as follows.

CDs of water contain the captured electromagnetic field which produces a magnetic vector potential $\boldsymbol{A}$.

Hamiltonian of particle with the impulse of p, being in the field of vectorial potential A, *is* described by the next formula:

$$H = (p + qA)^2/(2\,\mathrm{m}), \tag{4}$$

from which it is possible to get the distribution of energy of the field which looks as

$$U = q^2/(2\,\mathrm{m})|A|^2. \tag{5}$$

From here, we get *ponderomotive* force: as gradient of energy:

$$F_P = -q^2/(2\,\mathrm{m})\mathrm{grad}|A|^2 = qV_P. \tag{6}$$

Vector potential *A* which arises up because of the presence of captured electromagnetic field forms a tail which extends far outside CD. [16] specified that rotor of magnetic potential outside of CD is equal to zero; therefore, the magnetic field out of area of CD must be equal to zero, by virtue of the relation: H = rot A. At least, ***A*** spreads on far greater distances, what H. It means that the dynamics of processes in the coherent systems is carried out, mainly, due to influence of vector magnetic potential.

Because of saving of the rotary moment, the extraction of ions from the cyclotron orbits cause rotary motion of quasi-free electrons of water CDs, which, thus, become energetically excited. Thus, the voltage of electric field КД changes according to the formula:

$$V = -\frac{\hbar}{e}\frac{d\phi}{dt}. \tag{7}$$

A phase ϕ is added to the primary phase of undisturbed CD. A new phase corresponds to new area in which other CDs are included and thus arise a new area with the certain gradient of phases that in turn generates not disappearing magnetic potential in accordance with a formula:

$$A = \frac{\hbar c}{e} grad\phi. \tag{8}$$

In an eventual result, the coherency is set in area far greater than primary (0.1мкм).

Thus, realization of Zhadin effect in the water environment additionally arises a field of vector magnetic potential.

In fact, any particles attracted to surface of CD can, due to an exchange of energy with CD, participate in the formation of super coherency among CDs in near-surface nanostructures.

The effect similar to the Zhadin effect has been earlier registered for ions Na^+ K^+ in the researches which have been carried out by A.R. Liboff *in biological objects* [22].

In [21] specified: «As a matter of fact, by applying to a living system the same combination of magnetic fields as that used by Zhadin, Liboff observed the selective entrance of ions within cells» .

The origin of narrow peaks at the inclusion of the external variable magnetic field corresponds to the extraction of impurity ions, rotated on the cyclotrons orbits of CDs surface. Such rotation, by virtue of regime of coherency, occurs without friction. Because the surface of CDs can be attracted, only those ions which are near-by an EZ surface, for the arising of new peak, are necessary for certain time, required for the new filling of the devastated orbits.

The surprise as it marked in [21] was the message [23] that the Zhadin effect is observed and without placement in a vessel with a solution of the electrodes. The external electric field at carrying out such experiments was created due to the use of condenser. It means that for used solutions the effect has volumetric character. Its realization can take part not only CDs, forming on the surfaces of electrodes, but also CDs distributed on all volume of a solution.

It is possible to explain such phenomenon if we assume that the solution contains certain areas where the water agglomerates are formed similar to what takes place in the dilute homeopathic preparations [24]. We will mark that in such preparations there is the special structure responsible for the occurrence of memory, which can be saved for many years. In addition, the homeopathic preparations are characterized by the high degree of coherence. CDs can exist in the regime of super coherence in such environment.

In such case, all processes, including the emission of molecules of admixtures, attracted to surfaces of domains (under influence of an external variable magnetic field), occur synchronously.

In addition, it is possible to assume that at close enough arrangement of domains in water agglomerates, the effect similar to the effect Dicke of super radiation is registered. Indeed, the great number of separate domains which can be disposed of densely enough in the volume of these formations (agglomerates) can participate in the organization of structure of the dilute solutions of glutamic acid [24].

Really, superradiation in systems which were considered by R. Dicke occurs owing to dipole interactions of the identical particles making the coherent oscillations. In [25], it is underlined that «at superradiation there is an in-phase addition of the dipole moments of radiating atoms and arises the macroscopical dipole moment proportional to number of atoms» . As is known, intensity of radiation proportional to a square of the dipole moment, therefore, appears that is proportional to N^2.

Similar reasonings can be carried out and for CDs oscillating in a regime of super coherency synchronously, with a single phase. Because their magnetic moments line up parallel to the magnetic field of Earth, they are also summarized, like the dipole's moments of atoms in the Dicke system. At a radiation, corresponded to the simultaneous extractions of impurity ions, attracted to the surface of CDs, there is a change of dipole moments of CDs because of the change of their field characteristics. This change is also summarized, and consequently, we come to the same formula of square-law dependence of intensity of radiation (extractions of ions of impurities) from the amount of CDs as well as in an effect Dicke [7].

Result can observe enough strong response signal on the ultra-weak external influence that really are displayed even then, when there are no electrodes inside a vessel with the solution of glutamic acid [23] (Fig. 3).

We will mark that a superradiation is possible only in the inverted systems, i.e., such, in which the number of atoms in the excited state exceeds the number of atoms in the basic state. The second condition is that the region, in which atoms are situated, must not exceed a wavelength, in which a radiation takes place.

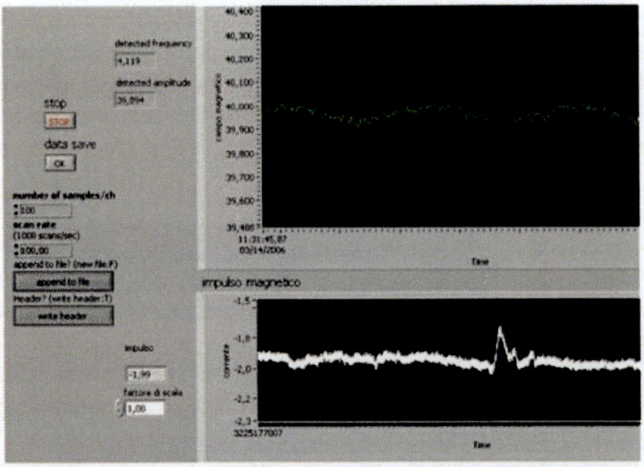

Fig. 3 From [23]. The resonance curve in the experiment with electrodes outside the cell. Cell with external electrodes and polarization 80 mV, Temperature: 22 °C, L-glutamic acid 38 mg/ml, pH 2.89. B: 40 μT, Bo: 40 nT, Sweep time: 100 s, Total data current recorded: 100 s, Scan rate: 100 point/s, Peak resonance frequency 4,119 Hz. The resonance peak with maximum near the cyclotron frequency of about 4.0 Hz. The abscissa: the scanning of frequency of alternating magnetic field. The ordinate: the current in the coil detector

It corresponds to that case when the external surfaces of CDs under influencing the ponderomotive forces have already attracted the ions of admixtures from solution of glutamic acid, and system is under influence of a variable magnetic field; CDs are located with a sufficient density.

Processes, discovered in water solutions of glutamic acid, have a direct relation to the phenomena of L. Kervran, as not only show that super weak influences of the fields can result in the very substantial response signals but also expose the mechanism of the similar phenomena in the nanostructures of living organisms. M. Zhadin considered that the effect opened by him can be the basis for the creation of installations in which the realizations of processes of cold nuclear fusion are possible.

CDs of the water systems of alive organisms formed on nanostructures of cellular systems stably exist long time; therefore, there are not the destructive effects, caused by thermal fluctuations, as in distilled water.

It means that in biological systems the display of such a phenomenon (Zhadin effect) is most complete.

Probably, such phenomenon determines the possibility of realization of reactions of cold nuclear fusion in the alive organisms described by Kervran [26].

Such processes can arise only in the coherent water environment; therefore, it is important that an external signal of electromagnetic fields should not exceed a level which collapses a coherency of the water environment. It determines the top limit of energy of the external fields corresponding to nonthermal influence on the coherent macrosystems, especially on the macrosystems of alive organisms.

As L. Kervran and other researchers showed, the low-intensive interactions prevail in a living organism. Exactly existence of such interactions is the necessary condition of realization in the living organism of LENR, including CNF. The reactions of LENR, which are opened in unbiological coherent systems, have other initial conditions for the occurrence, different from those which are incident to the biosystems. Possibly, it is related to the absence in these systems of such important element, as a biofield.

The second reason can be that some signals of influence on biostructures have a trigger character.

The similar conclusion has made F. Popp concerning a role of very weak radiation of biophotons in an organism.

Biophoton radiation, as it is underlined in [13] «is a general phenomenon of living systems. It concerns low luminescence from a few up to some hundred photons-per-second per square-centimeter surface area». The author () has noted the following: «Whatever the detailed mechanism may be, a single photon may suffice to trigger about 10^9 reactions per second» .

It is necessary also to mention the known researches of by J. Benveniste and L. Montagnier, investigating the methods of transmission of electromagnetic radiation from water solutions of high cultivation (containing originally the molecules of some substances or DNA molecules) on other objects and the influence of this low-intensive radiation on alive organisms and received very essential results.

To the not thermal methods, influences belong to also the affecting water and the water environment of living organisms of the low-intensive electromagnetic radiation in an EHF range, which, in spite of his very small intensity, can substantially influence on the vital parameters of the biosystems.

The listed kinds of low-power influence have common property that these methods do not lead to disturbance of a thermal regime of object of influence, can have essential influence only on coherent systems and, presumably, by the informative methods of influence. It means that a basic role in the transmission of such signals is taken by vector magnetic potential.

The processes guided through vector magnetic potential are not connected with the transfer of energy, because the influence occurs directly on a phase of quantum macro-object. The field of vectorial magnetic potential can exist even in the absence of electromagnetic fields (the Bohm–Aaronov effect); therefore, some authors are supposed that the fields of vector potential, in turn, can arise up as a consequence of the existence of some more than vacuum fields or to be a bridge between electromagnetic fields and thin vacuum fields. We will mark that as functions—informative—as in the transfer of energy he does not participate.

Essential difference between alive and lifeless is the presence of a biofield.

E. Del Giudice finds conceivable [27] that vectorial magnetic potential «could coincide, as suggested by Ho [28], with the morphogenetic field which is produced by each growing organism and govern its further development» . He specifies also that the biological effectiveness of very weak electromagnetic and magnetic fields can be explained: «just by assuming that the agent at work in the interaction is not the field but the potential and the mechanism of interaction is the phase-sharing» .

A biological organism is the dynamic coherent system in which all actions are interrelated. In organism, continuously there are oscillatory processes, which can reduce essentially the factor of influencing the coulomb field on various vitally important phenomena, including possible low-temperature nuclear reactions, on the existence of which specified yet by L. Kervran. Indeed, the processes related to the plasmatic coherent vibrations in a crystalline palladium [10] were the deciding factor of possibility of origin of nuclear fusion reactions in this matter.

Till now, the real reasons for the occurrence of CNF in alive organisms are not established, but the features of alive systems described above, apparently, and are exactly those basic circumstances due to which such reactions become possible.

4 Results and Conclusions

Quantum coherent systems have the cardinal differences from not coherent.

A row of the phenomena, including the reactions of low-energy nuclear fusion, can occur only in the quantum coherent macrosystems.

The origin of Zhadin effect can be responsible not only for the separate CDs but also CDs, distributed in the system of agglomerates formed in the dilute solutions of glutamic acid. Therefore, to the synchronous signal from single CDs being in solution, a simultaneous signal can be added from CDs, structured in agglomerates. Character of the answer signal from CDs in conglomerates can have square-law dependence in a maximum of the radiation corresponding to a maximum of peak in dependence of the current, measured in an experiment, from size of the extremely low-frequency to extremely low-intensive external variable magnetic field.

For quantum coherent systems, such value as the phase has essential significance; therefore, the big role in dynamics of these systems plays the vector magnetic potential. The processes controlled through vector magnetic potential are not connected to carry energy as influence is carried out directly upon the phase of quantum macro-object. The field of vector magnetic potential can exist even at the absence of electromagnetic fields (Aharonov–Bohm effect); therefore, as some authors supposed, it, in turn, can arise as a consequence of the existence of some more subtle vacuum fields or to be the bridge between electromagnetic fields and the subtle vacuum field which can be the biofield in an alive organism. Let's note that the functions of vector magnetic potential are informative, since it does not participate in carry of energy.

Essential difference of an alive from lifeless is the presence of a biofield. Only at the presence of coherency in an alive organism, the interaction between cellular nanostructure and a biofield can be carried out, in which the vector magnetic potential accepts the direct participation.

The listed features of an alive organism probably are by exactly such factors by virtue of which an alive organism occurs the reactions of cold nuclear fusion, which have been described still by L. Kervran.

References

1. Ratis JL (2009) Controlled "termonuc" or cold fusion? A drama of ideas. Samara, publishing House of the Samara scientific center of Russion Academy of Science, p 92 (in Russion)
2. Pshenichniy VA, Gritsay OO, Pavlovich VM (2019) The first researches of reactions of nikel-litium-water's termogenerator in Kyiv. Nucl Phys Atomic Energy 20(2):196–204 (2019) (in Ukraine)
3. Nernst W (1916) Uber einen Versuch, von quantentheoretischen Betrachtungen zur Annahme stetiger Energieanderungen zuruckzukehren. Verh. Deutsche Physikalische Gesellschaft. 18:83–116
4. Del Giudice E, Voeikov V, Tedeschi A, Vitiello G (2015) The origin and the special role of coherent water in living systems. In: Fels D, Cifra M, Scholkmann F (eds) Fields of the cell. Chapter 5, pp 95–111. ISBN: 978-81-308-0544-3
5. Preparata G (1995) QED, coherence in matter. World Scientific Singapore, p 236
6. Del Giudice E, De Ninno A, Fleischmann M et al (2005) Coherent quantum electrodynamics in living matter. Electromagn Biol Med 24:199–210
7. Dicke RH (1954) Coherence of in spontaneous radiation processes. Phys Rev 93:99–111
8. Josephson BD (1962) Possible new effects in superconductive tunneling. Phys Lett 1:251–253
9. Srinivasan TM (1993) Coherence and Pattern: scientific and aesthetic. Subtle Energies 3(3):i–iv
10. Preparata G (1994) Cold Fusion "93": some theoretical ideas. In: EPRI proceedings: fourth international conference on cold fusion volume 1: plenary session papers, TR-104188-V1. Electric Power Research Institute, Lahaina, Maui, Hawaii, pp 12-1–12-23
11. Gurwitsch A (1922) Über den Begriff des Embryonalen feldes. Archiv fËur Entwicklungsmechanik der Organismen 51(1):383–415
12. Kurik MV, Martseniuk LS (2012) The physical bases of life. LAP LAMBERT Academic Pabliching (Germany), p 174. IBSN: 978-3-8473-7903-4. (in Russian)
13. Popp F-A (1999) About the coherence of biophotons. Macroscopic quantum coherence. In: Proceedings of an international conference on the Boston University, edited by Boston University and MIT, World Scientific, pp 1 –12
14. Montagnier L, Aissa J, Ferris S at al (2009) Electromagnetic signals are produced by aqueous nanostructures derived from bacterial DNA sequences. Interdiscip Sci Comput Life Sci 1(2):81–90
15. Bischof M (2003) Introduction to integrative biophisics. Chapter 1. Published In: Popp F-A, Beloussov LV (eds) Integrative biophysics. Kluwer Academic Publishers, Dordrecht, pp 1–115. ISBN 1-4020-1139-3
16. Del Giudice E, Spinetti P, Tedeschi A (2010) Water dynamics at the root of metamorphosis in living organisms, №2, pp 566–586 (Water 2010). www.mdpi.com/journal/water
17. Del Giudice E (2014) Coherent quantum-electrodynamic organization of bio-chemical processes. J Formed Dir Sci 2(4):92–100. (In Russian)
18. Pollack GH, Clegg J (2008) Unexpected linkage between unstirred layers, exclusion zones, and water. In: Pollack GH, Chin WC (eds) Phase transitions in cell biology. Springer Science & Business Media, Berlin. Germany, pp 143–152
19. Voeikov VL, Del Giudice E (2009) Water respiration-the basis of the living state. Water № 1, pp 51–75
20. Zhadin MN, Novikov VV, Barnes FS, Pergola NF (1998) Combined action of static and alternating magnetic fields on ionic current in a Glutammic Acid solution, bioelectromagnetics, №19, pp 41–45
21. Del Giudice E, Giuliani L (2010) Coherence in water and the kT problem in living matter. In: Giuliani L, Soffritti MF (eds) Non-thermal effects and mechanisms of interaction between electromagnetic fields and living matter, pp 7–23
22. Liboff AR (1985) Geomagnetic cyclotron resonance in living cells. J Biol Phys 9:99–102
23. Giuliani L, Grimaldi S, Antonella L, D'Emilia E, Bobko N, Zhadin M (2008) Action of combined magnetic fields on aqueous solution of glutamic acid: the further development of investigations. BioMagnetic Res Technol 6(1):1–7

24. Yinnon TA, Liu ZQ (2015) Domains formation mediated by electromagnetic fields in very dilute aqueous solutions: 1. Quantum electrodynamic aspects. Water №7, pp 33–47
25. Trifonov ED (1996) Superradiance-spontaneous emissions of multiatomic system. Sorovs' Educ J 12:75–80. (In Russian)
26. Kervran L (2017). The proofs in biology of transmutation at weak energies. Part 1-2. J Formed Dir Sci 5(17–18):102–146. (In Rassian)
27. Del Giudice E, Stefanini P, Tedeschi A, Vitiello G (2011) The interplay of biomolecules and water at the origin of the active behavior of living organisms. In: 9th international Fr.ohlich's symposium. IOP Publishing Journal of Physics: ConferenceSeries. 012001, p 329
28. Ho MW, French A, Haffegee J, Saunders PT (1994) Can weak magnetic fields (or potentials) affect pattern formation? In: Ho MW, Popp F-A, Warnke U (eds) Bioelectrodynamics and biocommunication. World Scientific, New Jersey, London, Singapore, p 204

Structure and Properties of B_4C Coatings Obtained by RF Sputtering with External Magnetic Field

A. V. Taran, I. E. Garkusha, V. S. Taran, R. M. Muratov, T. S. Skoblo, O. I. Sidashenko, S. P. Romaniuk, T. V. Maltsev, and A. A. Baturin

In this research, the method of synthesizing B_4C coatings in Bulat-type device by using an RF plasma source with an external magnetic field is considered, which is based on the sputtering of the cathode material by high-energy argon ions when creating plasma in a weak magnetic field. The presence of an external magnetic field provided a drift and oscillating motion of the electrons and allowed the creation of plasma in the range of pressures $10 \div 10^{-2}$ Pa. The structure and chemical compositions of the obtained coatings have been investigated by SEM with EDX and XRF methods. Mechanical and tribological properties of thin B_4C coatings were investigated using nanoindentation and ball-on-disk methods.

1 Introduction

Boron carbide (B_4C) ceramics have excellent physical and mechanical properties: high melting point and hardness, good abrasion resistance, and excellent resistance to chemical agents [1–6]. Therefore, boron carbide is used in many industrial applications including cutting tools, drills, aerospace and machinery components, and

A. V. Taran (✉) · I. E. Garkusha · V. S. Taran · R. M. Muratov
National Science Center "Kharkiv Institute of Physics and Technology" (NSC KIPT), Institute of Plasma Physics, Kharkiv, Ukraine
e-mail: avtaran@ukr.net

I. E. Garkusha · T. S. Skoblo · O. I. Sidashenko · S. P. Romaniuk (✉) · T. V. Maltsev
National Technical University of Agriculture, Kharkiv, Ukraine
e-mail: romaniuk.khntusg@gmail.com

I. E. Garkusha · A. A. Baturin
National Technical University "Kharkiv Polytechnic Institute", Kharkiv, Ukraine

I. E. Garkusha
V. N. Karazin Kharkiv National University, Kharkiv, Ukraine

O. Fesenko and L. Yatsenko (eds.), *Nanomaterials and Nanocomposites, Nanostructure Surfaces, and Their Applications*, Springer Proceedings in Physics 246, https://doi.org/10.1007/978-3-030-51905-6_5

others. B_4C has low mass density of 2.52 g/cm^3 (lightweight material), low thermal expansion coefficient of 4.3×10^{-6} K^{-1}, high thermal and chemical stability, and good wear resistance. But it was reported that B_4C ceramics are hardly sinterable and relatively brittle and thus face serious obstacles for any structural material because of their low flexural strength (200–300 MPa) [7–9].

Various CVD and PVD methods are being applied to produce boron carbide coatings [10]. Two types of microstructures are reported by different authors, columnar structure by DC magnetron sputtering, reactive sputtering, pulsed ion-beam evaporation and non-columnar, featureless grown films by DC magnetron sputtering, and plasma jet CVD by PLD [10]. Hu et al. studied the effect of bias voltages on the microstructure and found that boron carbide coatings deposited at floating potential had coarse columnar microstructure, and at 200 V bias voltages it transformed to a denser zone T-type microstructure. Knotek et al. studied the effect of argon pressure on the microstructure of boron carbide coatings and found that the columnar structure for the films deposited at 4 Pa changed to no structural growth at 1 Pa [10].

Considering mechanical properties, the softer boron carbide films were reported by Chen et al. with 13 GPa for the films deposited without external heating [11] and by Ahn et al. with 18 GPa was also reported [12]. Tribological studies on RF-deposited boron carbide thin films demonstrated a quite different characteristic, especially for friction coefficient evolution. Lower friction coefficients are compared to DC-sputtered boron carbide films; about 0.4 were observed at the beginning of "pin-on-disk" tests instead of high values about 1 obtained for DC-sputtered films. Friction coefficients are stabilized around 0.5, which is believed in the friction coefficient value of RF-sputtered boron carbide films [10].

The development of vacuum-arc plasma coating methods for surface hardening is a promising direction [13, 14], and therefore, the optimization of elemental and phase composition and mechanical properties is an actual task. The aim of the research was to study the possibility of using boron carbide coating to increase the wear resistance of the working surfaces of parts made of low- and high-carbon steels.

2 Experimental Setup

Boron carbide coatings of thickness values 0.5 and 3.0 μm, were deposited onto high-carbon AISI D3 tool steel and low-carbon steel AISI 430, respectively. The chemical composition of AISI D3 and AISI 430 steels was evaluated using optical-emission spectrometer Metavision-1008i (Table 1).

Table 1 Chemical composition of AISI 430 and AISI D3 steel

Substrate	wt%							
	C	Si	Mn	Cr	Fe	P	Mo	Cu	Other < 0.05
AISI D3	2.22	0.4	0.32	12.12	84.8	0.015	0.12	–	Ti, Al
AISI 430	0.04	0.43	0.32	16.2	82.8	0.015	0.09	0.1	Ti, Ni, V

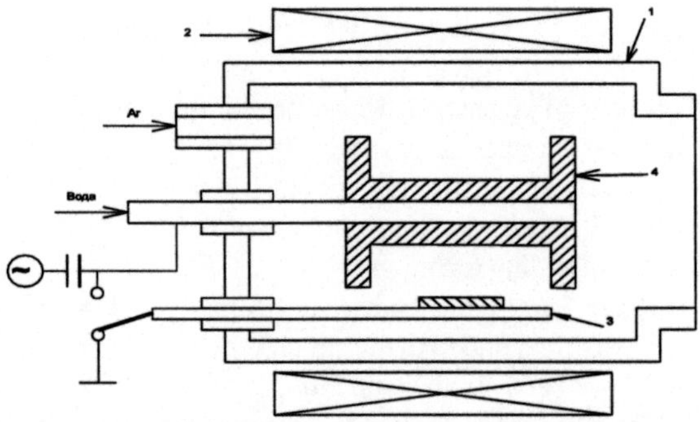

Fig. 1 Experimental scheme, 1—vacuum chamber, 2—solenoid used for external magnetic field, 3—samples, 4—RF electrode

The experimental scheme is shown in Fig. 1. The experimental setup consists of a vacuum chamber (1), and the external magnetic was created using magnetic coils (3). Samples (2) were located inside the substrate.

Bombardment with Ar ions in the RF plasma (1 kV) was used for surface cleaning and degreasing at a pressure of $P = 8 \times 10^{-3}$ Torr. The cleaning time was 10 min. For the application of B_4C coating, an RF electrode from graphite is connected to an RF generator with a closed input, and its sputtering was performed in Ar medium at a pressure of $P_{Ar} = 8 \times 10^{-3}$ Torr. The negative biasing on the RF electrode was $U_{bias} = -700$ V. The distance between the RF electrode and sample was 42 cm, and the external magnetic field comprised 30 Oersted. The thickness of the carbon coatings evaluated by XRF was 0.5 and 3 μm, respectively.

The surface topography of multilayered coating was studied using JEOL JSM-6390LV scanning electron microscope (SEM) with an accelerating voltage of 20 kV; chemical composition was examined using energy-dispersive X-ray analysis (EDX).

Energy-dispersive spectrometer SPRUT-K (AO Ukrrentgen, Ukraine) was used for X-ray fluorescent analysis, and it was equipped with Si (Li) X-100 detector (Amptek, USA) in the arrangement with Sc and KCl secondary targets. Film thickness was determined by XRF examinations and comprised of 0.5 and 3 μm, respectively.

The microhardness of the samples was measured by the Vickers method using a stationary automated device UIT HVmicro-1.

The nanohardness was measured by Nanoindenter G200 (USA). The loading and unloading rates of the nanoindentation were 10 mN/min. Samples were tested to a depth of 500 nm. Seven prints were made for each sample, and the distance between prints was 15 μm.

Dry friction tests were carried out using block-on-disk method on SMT-1 type machine. Sample loading was performed at 10 N, and the duration of the stage was 5 min. The weight and Vickers microhardness of the samples were measured before and after the tests. At the same time, the coefficient of friction was fixed during the friction tests. The rotation frequency of the rollers (counter-specimen material) was 50 min^{-1}. Material of the roller was Cr15 stainless steel. Before and after the tests, the weight of each sample was fixed on a WA-200 laboratory balance. The working surface of the samples was 10 × 8 mm.

3 Results and Discussion

Figure 2a–d presents SEM surface morphology images of B_4C coating on AISI 430 and AISI D3 steel, respectively. In Fig. 2 b, almost spherical nodules forming clusters or so-called "cauliflower" structure indicating columnar growth of B_4C coating was revealed. The surface morphology of B_4C on AISI D3 was free of droplet phase due to RF sputtering technique.

X-ray energy-dispersive spectrometry (EDX) revealed that the coatings were mostly corresponded to the stoichiometric composition of boron carbide B—75 at.% and C—20 at.% with some amount of Fe and Cr from the substrate material.

The tribological tests of the coatings have been carried out. The wear tests showed that the maximum weight loss was observed for boron carbide coating deposited on AISI 430 steel (Table 2). The gain of the mating part was revealed in these samples associated with the setting of the contacting counter-specimen material and the destruction of the coating due to poor adhesion to the substrate. The cracks were formed in B_4C coating on 3.0-μm-thick AISI 430 substrate at a load of 0.098 N (Fig. 2a). The cracks were spreading from the vertices of the pyramidal indenter tip. When the load is increased to 0.49 N, the peeling of the coating was also observed (Fig. 3b). So the surface was quite brittle due to the formation of blisters. According

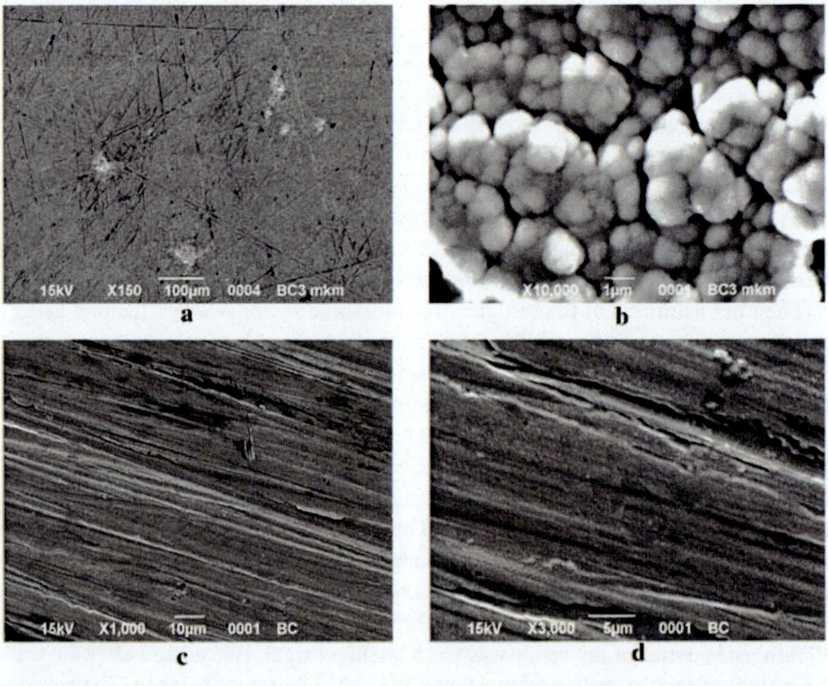

Fig. 2 SEM surface images of B_4C coating on AISI 430 (**a**, **b**) and AISI D3 (**c**, **d**) substrates

Table 2 Weight loss rate

№	Samples	Weight loss of, g	
		Coating	Roller
1	AISI 430	−0.0054	+0.0016
2	B_4C 0.5 μm	−0.0049	+0.0011
3	B_4C 3.0 μm	−0.0053	+0.0014
4	AISI D3	−0.0017	−0.0004
5	B_4C 0.5 μm	−0.0002	+0.0002
6	B_4C 3.0 μm	−0.0012	−0.0011

Fig. 3 Images of the Vickers indenter tips of B_4C/AISI 430 at various loads: **a** 0.098 N, **b** 0.49 N

to the theory of blister formation [15], which is a generalization of a large number of experimental data, blisters develop when lateral stresses of various nature appear in the surface layer. The causes of lateral stresses can be gas accumulation, the difference between the linear expansion coefficients of the substrate and the deposited layer, as well as their recrystallization process. Surface stresses, as well as surface tension forces, could determine the further modification of blisters and the destruction of their covers. The result of these processes was the destruction of the coating.

As a result of tribological tests, it was found that the use of B4C coating with a thickness of 3.0 μm on AISI D3 steel provided the highest friction characteristics and the lowest COF = 0.31 (Table 3).

The nanoindentation values of the coated low- and high-carbon steels were compared comparatively. The results are summarized in Table 4.

As a result of nanohardness measurements, it was found that the hardness of B_4C on AISI 430 is increased by 3.89 times than that of uncoated steel and increased by 1.73 times in comparison with uncoated AISI D3 steel. At the same time, the elastic modulus is lower than that of a matrix of high-carbon steel, and H/E value of 0.071 indicated the increased ability of the coating material to resist plastic deformation and affect its enhanced wear resistance, which is confirmed by the tribological tests.

Table 3 Tribological properties

№	Sample	Fiction moment, N/m	COF	Width of wear track, mm
1	Initial AISI 430	8.5	0.76	3.75
2	B_4C 0.5 μm	7	0.62	3.27
3	B_4C 3.0 μm	8	0.71	3.77
4	Initial AISI D3	6	0.53	1.78
5	B_4C 0.5 μm	7	0.62	1.64
6	B_4C 3.0 μm	3.5	0.31	0.45

Table 4 Nanoidentation results

№	AISI 430			AISI D3			B_4C		
	E, GPa	H, GPa	H/E	E, GPa	H, GPa	H/E	E, GPa	H, GPa	H/E
1	204.496	3.727	0.018	248.499	9.414	0.038	211.176	14.378	0.068
2	184.049	4.084	0.021	242.637	9.371	0.039	225.678	16.616	0.074
3	203.588	3.872	0.019	259.652	9.436	0.036	233.211	16.643	0.071
4	205.773	3.99	0.019	235.89	8.796	0.037	215.366	15.99	0.074
5	198.272	3.61	0.018	254.216	9.045	0.036	222.788	15.476	0.069
6	202.134	4.559	0.022	238.713	8.919	0.037	218.522	15.837	0.072
7	209.751	3.8	0.018	245.074	9.345	0.038	227.821	16.386	0.072
Average	**201.151**	**4.092**	**0.02**	**246.383**	**9.189**	**0.037**	**222.08**	**15.904**	**0.071**

4 Conclusions

1. A method is proposed for hardening the surface of low- and high-carbon steel products by B4C coating deposited by the ion-plasma method using a high-frequency plasma source with an external magnetic field.
2. Tribological tests of the coating of boron carbide with a thickness of 0.5 and 3.0 μm on substrates of AISI 430 and AISI D3 steels were carried out. The influence of the base on the adhesion to the coating is revealed.
3. The optimal parameters were obtained using a B4C coating with a thickness of 3.0 μm deposited on AISI D3 high-carbon steel, which provides higher friction characteristics, significantly increases the wear resistance of the working surface, and reduces the wear of the mating part.
4. The physical–mechanical characteristics of the substrate and the hardening coating are comparatively studied. The nanohardness values were in the range of 14–16.6 GPa. It was found that the B4C coating has 3.89 times higher hardness than uncoated AISI 430 steel and 1.73 times higher in comparison with bare AISI D3 steel.

References

1. Zhao H et al (2003) Eur Ceram Soc 23:1485–1490
2. Tang F et al (2008) Wear 264:555–561
3. Jiang Q et al (2005) J Alloy Compd 386:177–181
4. Yamada S et al (2003) J Eur Ceram Soc 23:1123–1130
5. Hu HM et al (2001) Mater Sci Eng A 297:94–104
6. Aizenshtein M et al (2005) J Mater Sci 40:2325–2327
7. Deng J et al (2002) Ceram Int 28:425–430
8. Hayun S et al (2008) Mater Sci Eng A 487:405–409
9. Liza S et al (2015) Science and technology of advanced materials, vol 16, Art 035007, pp 13
10. Tavsanoglu T (2009) Deposition and characterization of single and multilayered boron carbide and boron carbonitride thin films by different sputtering configurations. Engineering Sciences, École Nationale Supérieure des Mines de Paris, NNT:2009ENMP1641, pastel-00005772
11. Chen HY et al (2000) Synthesis of boron carbide films by ion beam sputtering. Surf Coat Tech 128–129:329
12. Ahn HS et al (2005) Tribological behavior of sputtered boron carbide coatings and the influence of processing gas. Wear 259:807
13. Taran A et al (2019) Nanostructured ZrO_2 ceramic PVD coatings on Nd-Fe-B permanent magnets. Nanotechnol Percept 15(1):13–20. (04/2019)
14. Taran A et al (2019) Synthesis and characterization of nanocrystalline ZrN PVD coatings on AISI 430 stainless steel. Probl Atomic Sci Technol 25(1):243–247. (02/2019)
15. Martynenko Yu V (1979) The theory of blister effect. Radiat Eff 45:93–102

Structural, Morphological, and Catalytic Properties of Cryptomelane

Tatyana Rakitskaya, Alla Truba, Vitaliya Volkova, and Pavel Yaremov

1 Introduction

Ozone belongs to extraordinarily toxic gaseous compounds; however, its maximum permissible concentration in the air of working area, MPC_{O_3}, according to existing standards, is different. For instance, MPC_{O_3} is 0.1 ppm (0.21 mg/m^3) in USA (OSHA), 0.08 ppm (0.171 mg/m^3) in China, and 0.047 ppm (0.1 mg/m^3) in Ukraine. Ozone is emitted by welding fabrication and office equipment. It is used in organic synthesis and some industries for wastewater treatment [1, 2]. Ozone decomposition can be achieved by chemical reactions between ozone and certain compounds, thermal and catalytic methods [2]. Judging by the recent publications, the most used methods are the latter. Multicomponent compositions [2–4] containing palladium, silver, and metal oxides [3–6] are the most active catalysts. Catalysts based on manganese dioxide are of particular interest due to their highest activity in the reaction of ozone decomposition [6]. In recent years, one can see a noticeable progress in understanding the most important properties determining the catalytic activity of manganese oxide forms and methods for its control. The structure [7–13] and morphology [7, 14–17] of manganese oxides are of primary importance. The most probable mechanism of ozone decomposition by manganese oxide forms involves a participation of oxygen vacancies [6, 12, 15, 18–22] and Lewis acid sites [23, 24] in this process. Catalytic activity of the manganese oxides due to their oxygen vacancies can be increased by K^+ [20] and transition metal doping [25, 26], by evacuation at different evacuation time and temperatures [19], by surface protonation [12], and

T. Rakitskaya · A. Truba (✉) · V. Volkova
Odessa II Mechnikov National University, Odessa, Ukraine
e-mail: truba@onu.edu.ua

P. Yaremov
Institute of Physical Chemistry NAS of Ukraine,
[Pleaseinsert\PrerenderUnicode{Ðš}intopreamble]yiv, Ukraine

O. Fesenko and L. Yatsenko (eds.), *Nanomaterials and Nanocomposites, Nanostructure Surfaces, and Their Applications*, Springer Proceedings in Physics 246, https://doi.org/10.1007/978-3-030-51905-6_6

by using some compositions complicated by MnO_2 anchoring on porous supports [27, 28]. However, as was reported [19, 23], the number of oxygen vacancies must be optimal. Deactivation of these catalysts is a result of a competitive adsorption of water molecules [19] and a participation of surface OH groups in the reaction with ozone with the formation of water molecules blocking surface active sites [22]. Poisoning of MnO_2 catalysts by water vapor necessitates a systematic study of water vapor adsorption by them to find out the conditions required for their application in devices for air purification from ozone [29].

As was reported [8–11], among polymorphous forms of manganese dioxide under the same conditions, α-MnO_2 (OMS-2, cryptomelane) shows the highest activity in the ozone decomposition. It is known that the selected method and conditions of OMS-2 synthesis considerably affect the product purity, its structural, morphological, and textural parameters as well as physicochemical properties. Ozone decomposition over OMS-2 was intensively studied [11, 19, 20, 22, 26, 30, 31]. OMS-2 samples were mainly synthesized by hydrothermal method; however, the precursors [15, 19, 30], synthesis temperature and duration, methods of precipitate treatment, and drying temperature [11, 19, 20, 22, 26] were different. The hydrothermal method of OMS-2 synthesis is considered as hardly realized in industrial scales [32], and another method based on thermal decomposition of manganese carbonate at 300 °C for 6 h has been proposed. A suspension method of OMS-2 synthesis from $KMnO_4$ and $Mn(Ac)_2$ lasted for 48 h was also reported [22]. Unfortunately, it was impossible to compare the catalytic activity of cryptomelane samples even through they were synthesized by the same, for instance, hydrothermal, method because, in the cited works, they were tested under varied conditions: initial ozone concentrations, $C^{in}_{O_3}$, from 14 ppm (30 mg/m^3) [8, 26] to 120 ppm (257 mg/m^3) [23, 32], temperatures from 0 °C [25] to RT, specific volume flow rates, W_{sp}, from 540 L/g × h [20, 22, 32] to 1320 L/g × h [26], and RH of the ozone–air mixture from 5 to 90% [19, 23, 26, 30, 32]. In many cases, the latter parameter was not shown despite the fact that water vapor adsorption, as stated above, resulted in catalyst deactivation. Time dependences of the ozone decomposition degree, η, %, and assertions that $\eta \rightarrow 100\%$ [20, 30–32] do not allow to evaluate ozone final (outlet) concentrations $C^{f}_{O_3}$, and their conformance to MPC_{O_3} in the air of working areas. In all cases considered, a gradual decrease in OMS-2 activity leads to $C^{f}_{O_3} >> MPC_{O_3}$. Using amorphous MnO_2-synthesized by the $KMnO_4$ reduction with formic acid as an example, we first have studied the protective capability of this catalyst at $C^{in}_{O_3} = 1.5$ mg/m^3 (0.7 ppm) and have shown that the protective time (τ_{MPC}) (a period of time during which the air purification from ozone below MPC_{O3} is provided) depends on the precipitate treatment methods [33]. Thus, in spite of numerous reported data concerning catalytic properties of polymorphic forms of manganese dioxide, an influence of the synthesis methods for manganese oxide (especially OMS-2) obtaining on their catalytic behavior in the reaction of ozone decomposition must be systematically studied.

The aim of the work was to synthesize cryptomelane samples by using four methods, i.e., solid-state (SSt), reflux (Ref), sol–gel (SG), and melting (MS), to ascertain their structural, morphologic, and textural properties, and effect of these properties on the OMS-2 catalytic activity in the reaction of ozone decomposition.

2 Experimental

2.1 Cryptomelane Synthesis

Manganese oxide samples under study were synthesized by potassium permanganate reduction: with manganese acetate (solid-state reaction), OMS-2-SSt, with manganese sulfate (reflux method), OMS-2-Ref, with maleic acid (sol–gel method), OMS-2-SG, and via KNO_3 and $MnSO_4$ melting together, and OMS-2-MS. Detailed procedures for their synthesis are presented below.

OMS-2-SSt. $KMnO_4$ and $Mn(CH_3COO)_2{\cdot}4H_2O$ were mechanically mixed in the molar ratio of 2:3 in an agate mortar, and an appropriate amount of water was added to the mixture to obtain a rheological phase. The rheological phase mixture was heated at 100 °C for 12 h. The mixture was cooled down to room temperature and washed with distilled water several times. Thus, a precursor was obtained. The precursor was heated in air at 400 °C for 4 h [34].

OMS-2-Ref. A typical synthesis [35] was as follows: 5.89 g of $KMnO_4$ in 100 mL of water was added to a solution of 8.8 g of $MnSO_4$-H_2O in 30 mL of water and 3 mL of concentrated HNO_3. The solution was refluxed at 100 °C for 24 h, and the product was filtered, washed, and dried at 120 °C.

OMS-2-SG. In a typical sol–gel synthesis [17], maleic acid (0.78 g (6.7 mmol)) and $KMnO_4$ (3.16 g (20 mmol)) were dissolved in deionized water (200 mL) at the predetermined temperature of 40 °C. After stirring for 30 min, the mixture was allowed to settle for 60 min. The sample obtained after this step was then filtered, washed with deionized water, and dried at 120 °C for 12 h. Then, calcination at 450 °C for 2 h was carried out.

OMS-2-MS. The sample was synthesized via KNO_3 and $MnSO_4$ melting together [36]. KNO_3 was put into a crucible and heated to 380 °C under stirring to form a molten solution; $MnSO_4$ was added into it, and the weight ratio of KNO_3 to $MnSO_4$ was 15. After being maintained at 380 °C for 3 h, the crucible was cooled down to RT under ambient conditions, and the product was obtained by washing with deionized water, and drying at 90 °C for 12 h.

2.2 Characterization

The samples were investigated on a Siemens D500 powder diffractometer (CuK_α radiation, $\lambda = 1.54178$ Å) with a secondary beam graphite monochromator (Siemens AG, Munich, Germany). After thorough grinding with a pestle, each sample was placed into a glass cell with an enclosed volume of $2 \times 1 \times 0.1$ cm^3 for XRD pattern recording in the 2θ range from 0° to 90° with a step of 0.03° and an accumulation time at every position of 60 s.

OMS-2 samples were also characterized by scanning electron microscopy using a JEOL-JSM5410 scanning microscope with an AZtech Energy X-max^n 50 energy-dispersive spectrometer. A resolving power and an accelerating voltage of the scanning microscope were 10 nm and 15 kV, respectively.

Infrared analysis of the OSM-2 samples was carried out using a Perkin Elmer FT-IR Spectrometer with resolution of 4 cm^{-1}; pellets consisting of 1 mg of the material under study and 200 mg of KBr were compressed under pressure of 7 tons/cm^2 for 30 s.

Nitrogen ad/desorption measurements were performed by a volumetric method on a Sorptomatic adsorption instrument (Thermo Electron Corporation) at liquid nitrogen temperature (~196 °C). Before the measurements, samples were outgassed at 200 °C for 5 h. A specific surface area was calculated using the BET method in the P/Ps ranging from 0.04 to 0.34. A pore size distribution was calculated using Barrett–Joyner–Halenda (BJH) method based on a desorption branch in the P/P_S ranging from 0.1 to 0.996. A micropore volume was determined using the t-plot method proposed by Gregg and Sing [37].

Equilibrium pH values measured for samples (0.2 g) suspended in distilled water (20 mL) were used to characterize protolytic properties of their surface functional groups. pH values were measured by a pH-340 instrument with an ESL 43-07 glass electrode and an EVL 1M3 chlorsilver electrode at continuous stirring of the suspension at 20 °C.

2.3 *Catalytic Studies*

The samples were tested in the reaction of ozone decomposition as follows. The ozone air mixture, OAM, with a required ozone concentration was obtained with the help of an IG-1Sh ozonizer by a silent electric discharge action on air oxygen. An OAM feed was controlled based on rheometer readings. The corresponding sample (0.5 g) was placed into a gas flow fixed bed reactor and the OAM with RH of 65% at 20 °C passed through the sample at a linear velocity (*u*) of 6.2 cm/s and a specific volume flow rate of 120 L/g × h. Ozone decomposition was monitored by measuring the final (outlet) ozone concentration, $C^f_{O_3}$. An initial ozone concentration, $C^{in}_{O_3}$, equal to 100 mg/m^3 was controlled by a Tsyclon-Reverse optical analyzer, whereas $C^f_{O_3}$ values were measured either by the Tsyclon-Reverse optical analyzer (detection limit of 1 mg/m^3) or by a 652EKh04 electrochemical gas analyzer (detection limit of 0.01 mg/m^3).

An activity of the manganese dioxide samples was evaluated based on the following parameters:

- τ_{MPC} is a time of protective action, i.e., a period of time from experiment beginning during which $C^{f}_{O_3} < MPC_{O_3}$,
- $\tau_{1/2}$ is a half-conversion time, i.e., a period of time from experiment beginning up to the moment when $C^{f}_{O_3}$ becomes equal to 0.5 $C^{in}_{O_3}$,
- Q_{exp} is an experimental amount of O_3 mol, i.e., the amount of ozone entered into the reaction up to a moment of experiment termination and calculated as an area under the corresponding ozonogram plotted as a ΔC_{O_3} versus τ.

3 Results and Discussion

3.1 XRD Analysis

The XRD analysis shows that all samples synthesized by us are crystalline (Fig. 1). All XRD patterns were treated by the Rietveld method. The data concerning phase identification, phase contents, crystallite sizes, and unit cell volumes are summarized in Table 1. As can be seen, OMS-2(SSt) and OMS-2(Ref), along with the dominating cryptomelane phase (h k l: 110, 200, 220; 310, 211, 301, 411, 600, 215, 002, and 541) (85.3% in OMS-2(SSt) and 90.5% in OMS-2(Ref)), contain impurity phases, i.e., pyrolusite (β-MnO_2) and bixbyite (Mn_2O_3). As opposed to them, OMS-2(SG) and OMS-2(MS) are monophase, i.e., contain only the cryptomelane phase. The samples

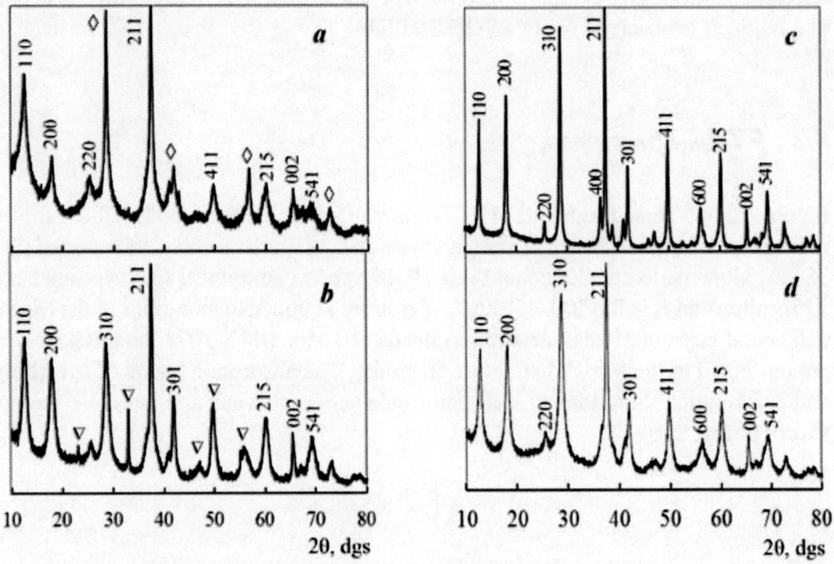

Fig. 1 X-ray diffraction patterns for OMS-2 synthesized by different methods. **a** OMS-2(SSt); **b** OMS-2(Ref); **c** OMS-2(SG); **d** OMS-2(MS) (◇ β-MnO_2 (pyrolusite); ∇ Mn_2O_3 (bixbyite))

Table 1 Phase compositions and phase parameters for the manganese oxide samples under study

Sample	Phase	Phase content, wt%	Lattice parameters (Å)	Crystallite size (nm)/V (Å^3)
OMS-2(SSt)	$K_xMn_8O_{16}$ (cryptomelane)	85.3 (4)	$a = 9.8552$	10/277.03
			$c = 2.85231$	
	β-MnO_2 (pyrolusite)	14.72 (4)	$a = 4.3915$	28/55.360
			$c = 2.8706$	
OMS-2(Ref)	$K_xMn_8O_{16}$ (cryptomelane)	90.5 (3)	$a = 9.8306$	15/275.095
			$c = 2.84655$	
	Mn_2O_3-Ht (bixbyite)	9.49 (2)	$a = 9.38922$	66/827.729
OMS-2(SG)	$K_xMn_8O_{16}$ (cryptomelane)	100.0 (4)	$a = 9.82216$	36/275.476
			$c = 2.85542$	
OMS-2(MS)	$K_xMn_8O_{16}$ (cryptomelane)	100.0 (1)	$a = 9.8389$	14/276.527
			$c = 2.8532$	

differ in their crystallite sizes. OMS-2(SG) is characterized by the largest crystallite size of 36 nm, whereas the crystallite sizes for other OMS-2 samples vary from 10 to 15 nm.

Among the impurity phases, bixbyite has the largest crystallite size of 66 nm. Different parameters and unit cell volumes of the OMS-2 samples synthesized by different methods give evidence of some structural differences. The largest expansion of a unit cell is observed for OMS-2(SSt) (Table 1).

3.2 *FT-IR Spectroscopy*

Figure 2 shows some fragments of FT-IR spectra for the synthesized OMS-2 samples in the range of 1100-400 cm^{-1}. Their analysis and comparison with data reported [36, 38–42] allow us to conclude that these FT-IR spectra are typical for cryptomelane. Absorption bands in the 700–459 cm^{-1} region are assigned to vibrations of the MnO_6 octahedral structure; weak absorption bands at 1046, 1047, 1073, and 1045 cm^{-1} are attributed to the Mn-OH structure fragment. The absorption bands of stretching and deformation vibrations of water molecules are very weak and therefore are not shown in Fig. 2.

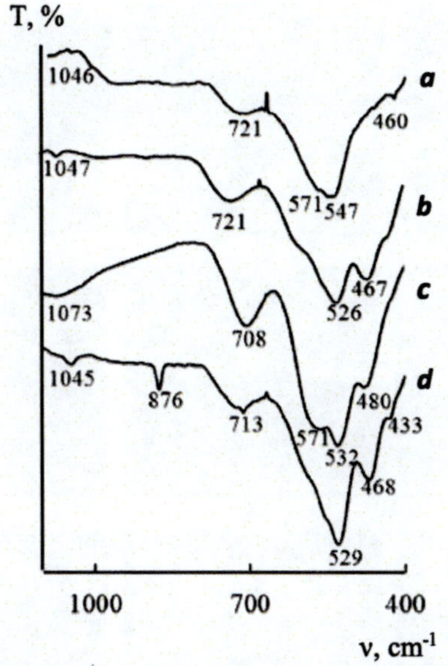

Fig. 2 IR spectra of cryptomelane synthesized by different methods: **a** OMS-2(SSt); **b** OMS-2(Ref); **c** OMS-2(SG); **d** OMS-2(MS)

3.3 Morphology

Depending on a synthesis method, various morphotypes of cryptomelane can be obtained: tubes, rods, wire, sheets, hollow spheres, spindles, urchin-like, nestle-like, flower-like, needle-like, blade-like, fibrous-like, thread-like, filiform, etc. [15, 42–45].

SEM images of surfaces of the synthesized OMS-2 (Fig. 3) indicate some morphological differences. The morphology of OMS-2(SSt) is typical for cryptomelane, i.e., spheres composed of fibbers and remaining a broken urchin-like form (Fig. 3a). Figure 3b and c explicitly shows the existence of some other phases. Staked fluffy (loosened) sheets are formed from needle-like crystallites. Enlargement of certain surface parts allows us to distinguish nanorods' characteristic of the β-MnO_2 (pyrolusite) phase [37]. As the XRD data (Table 1) indicate, the pyrolusite content in OMS-2(SSt) is 14.72%. The morphology of OMS-2(Ref) demonstrates urchin-like microspheres packed closely together (Fig. 3d, e, f). The similar SEM images for OMS-2 samples were reported earlier [43, 46].

The morphology of OMS-2(SG) indicates its non-uniform surface: in addition to urchin-like spheres packed closely together, plate- and tube-like particles can be seen (Fig. 3h, i). The morphology of OMS-2(MS) (Fig. 3j, k, l) is uniform. Even in the case of 5000 magnification (Fig. 3j), fluffy (loosened) microspheres composed of nanofibers and linked with one another by a fibrous web are clearly seen. The images

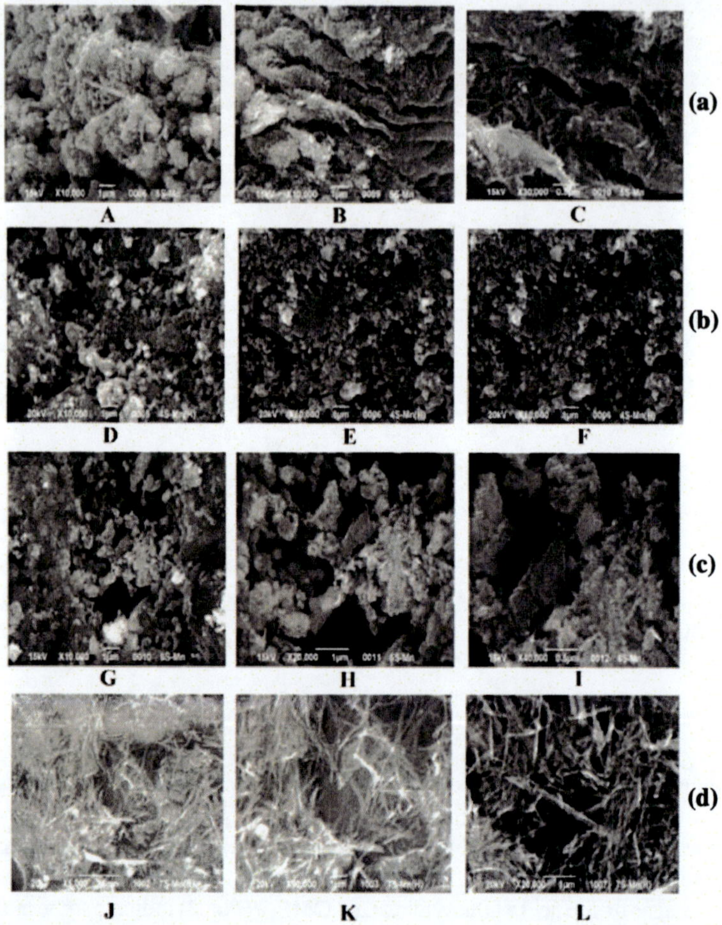

Fig. 3 SEM images of surfaces of the OMS-2 samples. **a** OMS-2(SSt); **b** OMS-2(Ref); **c** OMS-2(SG); **d** OMS-2(MS)

at 10000 and 2000 magnifications (Fig. 3k, l) clearly demonstrate single-crystal fiber characteristic of cryptomelane that is typical for OMS-2 [37, 47].

3.4 Specific Surface Area and Porosity Measurements

Figure 4 shows the isotherms of nitrogen ad/desorption by the four OMS-2 samples. Their profiles are similar and identical to type IV with H3-type hysteresis loops indicating mesoporous structure of these samples.

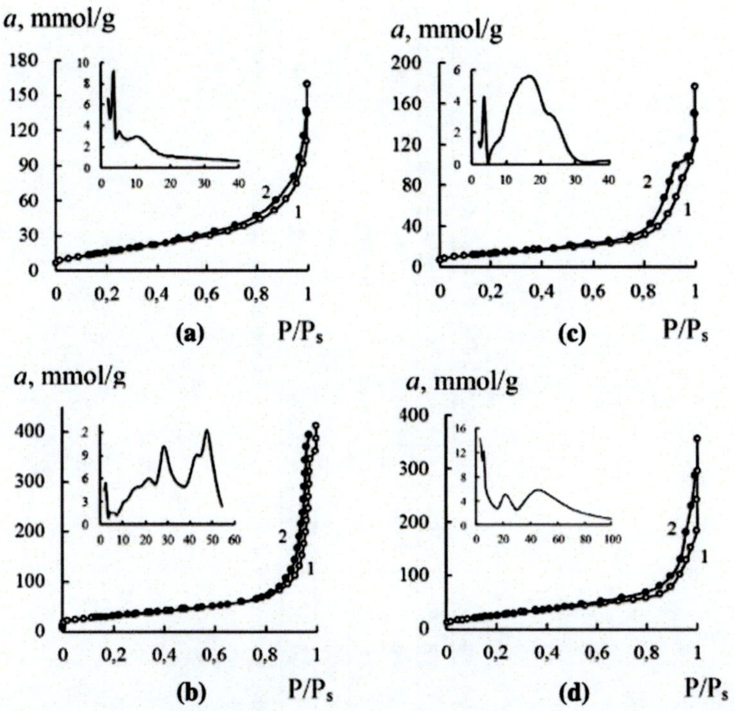

Fig. 4 Isotherms of N_2 adsorption–desorption by the OMS-2 samples: **a** OMS-2(SSt), **b** OMS-2(Ref), **c** OMS-2(SG), **d** OMS-2(MS)

Table 2 summarized data concerning textural and structural parameters of the four IMS-2 samples calculated by the BET, BJH, and t-plot methods. As can be seen, S_{BET} and S_c are somewhat different; however, in allowable limits of ± 20% [48], they can be considered as similar.

Values of an average pore diameter indicate that the OMS-2 samples are mesoporous and the pore diameter distribution shown as V × 10^3, cm^3/g versus d, nm plots (see insertions) is evidence of their non-uniform mesoporosity. The fact that values of S_t and S_{me+ext} are equal shows that N_2 is adsorbed on external surfaces of the OMS-2 samples being in agreement with the reported data [49]. The micropore volumes, V_{mi}, were calculated using the t-plot method.

Despite the fact that the OMS-2 samples under study are mesoporous, their micropore volumes form from 0.27% (OMS-2(SSt)) to 5.98% (OMS-2(MS)) of cumulative pore volume. This does not contradict to the earlier reported data concerning cryptomelane porous structure [24]. S_{sp} (BET) decreases in the order OMS-2(Ref) > OMS-2(MS) > OMS-2(SSt) > OMS-2(SG).

Table 2 Structural and adsorption parameters for the OSM-2 samples obtained by different methods

Sample	BET			BJH (D 4641-87 standard)			t-plot (D 4641-87 standard)		
	a_m, mmol/g	C	S_{sp}, m^2/g	d_m, nm	V_c, cm^3/g	S_c, m^2/g	V_{mi}, cm^3/g	S_{me+ext}, m^2/g	S_t, м^2/г
OMS-2(SSt)	0.652	25.395	63.606	17.849	0.1866	79.481	0.0005	62.484	62.484
OMS-2(Ref)	1.148	118.59	112.00	34.509	0.6099	110.27	0.011	91.365	91.365
OMS-2(SG)	0.4879	68.468	47.594	16.915	0.1698	53.121	0.0055	36.353	36.353
OMS-2(MS)	1.0762	21.247	105.00	34.914	0.4615	131.30	0.0276	61.661	61.660

d_m is a median pore diameter; V_c is a cumulative pore volume; S_c is a cumulative pore area; V_{mi} is a volume of microspores; S_{me+ext} is a mesopores+external surface area; S_t is a total surface area

3.5 Protolytic Properties

Mechanisms of water molecule protolysis depend on the difference in strength of acid (L) and basic (B) sites situated on the OMS-2 surfaces [23, 24] that becomes apparent by comparison of time dependences of pH in OMS-2 suspensions [50]. Figure 5 shows the time dependences of pH in suspensions of the initial OMS-2 samples and the OMS-2 samples after their reaction with ozone.

As can be seen, profiles of the kinetic curves and values of an equilibrium pH, pH_{eq}, depend on the methods of OMS-2 obtaining. For OMS-2(SSt) and OMS-2(MS), pH of the suspension rapidly increases and a positive value of a suspension effect, $\Delta pH_s = pH_{eq} - pH_0$ (Table 3), indicates that water molecules interact with basic sites:

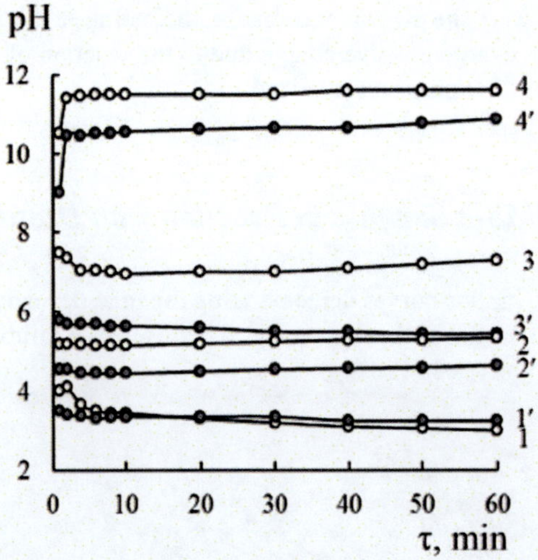

Fig. 5 Time dependences of pH in suspensions of the OMS-2 samples before (o) and after (•) their reaction with ozone: 1, 1′ OMS-2(Ref); 2, 2′ OMS-2(MS); 3, 3′ OMS-2(SG); 4, 4′ OMS-2(SSt)

Table 3 Values of pH_0, pH_{eq}, and ΔpH_s for the OMS-2 samples

Sample	Before reaction with ozone			After reaction with ozone		
	pH_0	pH_{eq}	ΔpH_s	pH_0	pH_{eq}	ΔpH_s
OMS-2(SSt)	10.5	11.6	1.1	9.0	11.0	2.0
OMS-2(Ref)	4.0	3.0	−1.0	3.5	3.2	−0.3
OMS-2(SG)	7.5	7.3	−0.2	5.8	5.5	−0.3
OMS-2(MS)	5.2	5.4	0.2	4.55	4.65	0.1

pH_0 and pH_{eq} are pH values measured in 15 s and in the equilibrium state, respectively; ΔpH_s is a suspension effect

$$B + HOH \leftrightarrow BH^{+} + OH^{-}. \tag{1}$$

In the case of OMS-2(Ref) and OMS-2(SG) samples, vice versa, suspension pH values rapidly decrease, and $\Delta pH_s < 0$ indicates the establishment of the following equilibrium:

$$L + HOH \leftrightarrow L - OH + H^{+}. \tag{2}$$

Thus, either the basic sites (OMS-2(SSt), OMS-2(MS)) or acidic sites (OMS-2(Ref), OMS-2(SG)) predominate in the synthesized samples. However, judging by the pH_{eq} values, the strength of these sites is significantly different. After the reaction of the OMS-2 samples with ozone, the character of the pH-τ curves and, consequently, the sign of the suspension effect are retained; however, in all cases, except for OMS-2(Ref), the pH values decrease, indicating an increase in the number of Lewis sites with a large positive charge due to the reaction of Mn^{3+} with ozone, i.e., the $Mn^{3+} \rightarrow Mn^{4+}$ transition is realized.

3.6 *Testing OMS-2 Samples in Reaction with Ozone*

Figure 6 shows the kinetic curves demonstrating the time dependences of the ozone concentration at the outlet of the reactor $C^{f}_{O_3}$ for the OMS-2 samples synthesized by different methods (curves 1–4).

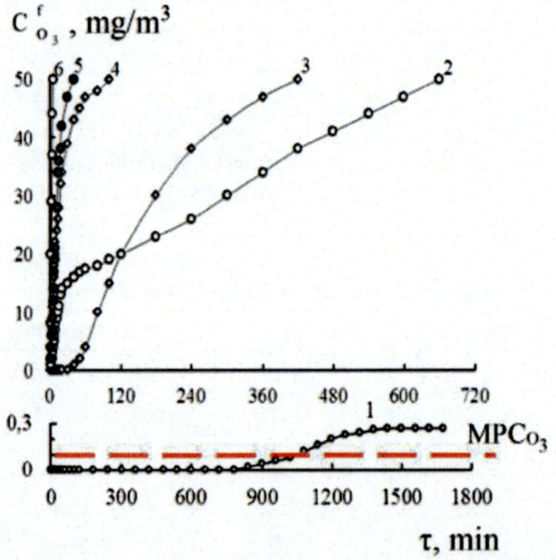

Fig. 6 The time dependences of $C^{f}_{O_3}$ for ozone decomposition by samples: 1—OMS-2(MS); 2—OMS-2(SG); 3—OMS-2(Ref); 4—OMS-2(SSt); 5—Mn_2O_3; 6— Mn_2O_3+β-MnO_2 ($C^{in}_{O_3}$ = 100 mg/m^3; m_s = 0, 5 g)

The contribution of the Mn_2O_3 and β-MnO_2 impurity phases was estimated using the reported data [51] on the kinetics of ozone decomposition by a sample containing 100% Mn_2O_3 (curve 5) and a sample containing a mixture (30% Mn_2O_3 and 48.7% β-MnO_2) [29] (curve 6). It is arguable that the contribution of these phases to the reaction of ozone decomposition by the synthesized OMS-2 cryptomelane samples can be neglected and cryptomelane, as it was earlier reported [8], makes a decisive contribution to the OMS-2 activity.

It is obvious that the kinetic curve profiles for different cryptomelane samples are different. Curves 2 and 4 for OMS-2(SG) and OMS-2(SSt) samples are characterized by a sharp increase in the final concentration of ozone in the initial section, while the time required to reach 50% decomposition of ozone ($\tau_{1/2}$) is 6 times longer in the case of OMS-2(SG). Curves 1 and 3 for OMS-2(MS) and OMS-2(Ref) samples are characterized by a section in which no ozone concentration is detected at the reactor outlet. The duration (τ_0) of this section is the longest in the case of OMS-2(MS). In addition, from curves 1 and 3, it is possible to determine the catalyst protective time (τ_{MPC}). The longest τ_{MPC} (1050 min) is observed for OMS-2(MS). Then, $C^f_{O_3}$ increases to 0.28 mg/m^3, and this value is maintained for a long time.

In Table 4, the most important characteristics of cryptomelane samples are systematized, as well as the parameters characterizing their catalytic activity in the ozone decomposition.

X-ray diffraction analysis has shown that only two products, namely, OMS-2(MS) and OMS-2(SG), are pure; the other two samples, except the cryptomelane predominant phase, contain impurities, i.e., Mn_2O_3 and β-MnO_2. Although the contribution of these impurity phases to the kinetics of ozone decomposition is insignificant (Fig. 6), it can be concluded that with an increase in the content of the cryptomelane phase, the amount of decomposed ozone, Q_{exp} (O_3), increases. At the same cryptomelane content (100%), the OMS-2(MS) sample is more active than the OMS-2(SG) sample and its crystallite size is smaller. The OMS-2 samples obtained by different methods have some structural differences, namely, the unit cell volume changes and the largest expansion has been found for OMS-2(SSt). The samples have typical cryptomelane morphology. They consist of fibrous crystals forming microspheres of different packing densities: loosely (OMS-2(MS)) and tightly packed (OMS-2(Ref), OMS-2(SG), and OMS-2(SSt)) urchin-like spheres. The sample with loose packing of nanofibers is the most active in the ozone decomposition. Two types of kinetic curves of ozone decomposition (Fig. 6) are due to structural and adsorption differences. OMS-2(MS) and OMS-2(Ref) samples have approximately the same values of S_{sp} and average pore diameter (d_m), and these parameters are greater than those for the second pair of samples, OMS-2(SG) and OMS-2(SSt).

The synthesized OMS-2 samples have demonstrated different mechanisms of water molecules protolysis and change in pH of the suspension in a wide range. The low-active OMS-2(SSt) sample is characterized by the highest pH level and the presence of basic sites inactive in the ozone decomposition [32, 53]. Luo et al. [24] characterized cryptomelane as having hydrophobic properties with a maximum adsorption value of 1,1 mmol/g according to water vapor adsorption. Our studies [29]

Table 4 Generalized data on the properties of OMS-2 synthesized by different methods and their activity in the ozone decomposition

Parameter	Sample			
	OMS-2(MS)	OMS-2(SG)	OMS-2(Ref)	OMS-2(SSt)
Phase composition, %	$K_xMn_8O_{16}$ – 100	$K_xMn_8O_{16}$ – 100	$K_xMn_8O_{16}$-90.5 Mn_2O_3-9.5	$K_xMn_8O_{16}$-85.3 β-MnO_2 – 14.7
D, nm (cryptomelane)	14	36	15	10
V, Å^3 (cryptomelane)	276.527	275.476	275.095	277.03
S_{sp} (BET), m^2/g	105.0	47.6	112.0	63.6
d_m, nm	34.9	16.9	34.0	17.8
pH_{eq}	5.4	7.3	3.0	11.6
a, mmol/g at RH = 65%	1.08	0.84	1.72	1.59
τ_{MPC}, min	1050	–	30	–
$\tau_{1/2}$, min	– [a]	660	420	100
$Q_{exp}\cdot 10^5$, moles of O_3	343.0[a]	86.6	54.4	10.8

[a]—the experiment was stopped at $C^f_{O_3} = 0.28$ mg/m^3

showed that the OMS-2 samples synthesized by different methods were characterized by a high affinity to water molecules and, at RH of 65% (conditions of catalytic activity testing used by us), they adsorbed different amounts of water (Table 4). More active OMS-2(MS) and OMS-2(SG) samples adsorbed much less water than OMS-2(Ref) and OMS-2(SSt) samples. For the latter, competing adsorption of water vapor reduces their activity in the ozone decomposition.

It is seen that the establishment of correct correlations between any OMS-2 property and catalytic activity in the ozone decomposition is difficult because of the fact that other parameters do not remain constant. The same conclusion can be drawn from the analysis of some earlier reported data [8–11, 15, 19, 20, 22, 26, 30–32]. For instance, the increase in the activity of polymorphic forms of manganese dioxide in the order $\beta\text{-}MnO_2 < \gamma\text{-}MnO_2 < \alpha\text{-}MnO_2$ was associated with the structure, and, in this case, S_{sp} increased in the same order from 13.8 to 80.7 m^2/g [8]. With the same structure, the decrease in activity in the order $\alpha\text{-}MnO_2$ (nanofibers NFs) > $\alpha\text{-}MnO_2$ (nanorods NRs) > $\alpha\text{-}MnO_2$ (nanotubes NTs) was associated with different morphologies, but, in addition, a number of other parameters changed significantly: S_{sp} and the number of oxygen vacancies decreased from 80.7 to 33.8 m^2/g and from 11 to 3.1%, respectively, whereas the average oxidation state (AOS) increased from 3.63 to 3.83 [15]. With an increase in K/Mn ratio for K-α-MnO_2 samples (AOS decreased from 3.76 to 3.37), the catalytic activity increased significantly, but, at the same time, such an important parameter as S_{sp} decreased from 61.0 to 51.94 m^2/g.

Based on both our results, and some reported data [7, 16], we can conclude that the catalytic activity of cryptomelane in the ozone decomposition and other redox processes [14, 17, 52] is determined by a combination of properties, including structure, morphology, crystallite size, and product purity; the content of oxygen vacancies and AOS; specific surface area and pore volume; the presence of acidic and basic sites, their ratio, and strength; affinity to water molecules, hydrophilicity, or hydrophobicity; and pH of the suspension.

4 Conclusions

The structural, morphological, textural, and protolytic properties of OMS-2 samples obtained by the solid-phase reaction method (OMS-2(SSt)), reflux method (OMS-2(Ref)), sol–gel method (OMS-2(SG)), and the method based on KNO_3 and $MnSO_4$ melting together (OMS-2(MS)) have been studied. The method of OMS-2 obtaining affects the product phase composition and morphology. OMS-2(SG) and OMS-2(MS) samples contain only the cryptomelane phase; their crystallite sizes are 36 and 14 nm, respectively. The cryptomelane crystallite sizes in OMS-2(Ref) and OMS-2(SSt) samples, containing impurity Mn_2O_3 and $\beta\text{-}MnO_2$, are 15 and 10 nm, respectively. For OMS-2(SSt), the largest crystal cell expansion has been found. The morphology of the obtained samples is typical for cryptomelane, i.e., agglomerates of various shapes and packing densities formed of interconnected fibrous crystals.

The most fluffy nanofibrous microspheres are typical for OMS-2(MS). The synthesized OMS-2 samples are heterogeneous micro-mesoporous materials for which the specific surface area is in the range of 48-112 m^2/g. pH values in suspensions of the OMS-2 samples change in the wide range from 3.0 to 11.6 indicating the presence of acidic and basic sites of different strengths on the OMS-2 surfaces. The method of OMS-2 obtaining also affects the hydrophilic–hydrophobic properties of the sample surfaces. At RH of 65%, the OMS-2(SG) sample adsorbs the least amount of water ($a = 0.84$ mmol/g).

The OMS-2 samples obtained by different methods demonstrate some differences in kinetics of ozone decomposition at the initial O_3 concentration of 100 mg/m^3, and their activity decreases in the order OMS-2(MS) > OMS-2(SG) > OMS-2(Ref) > OMS-2(SSt). Based on both our own results and the analysis of literature data concerning the kinetics of ozone decomposition by cryptomelane, it has been concluded that the OMS-2 catalytic activity is determined by a certain combination of their structural, morphological, textural, and physicochemical properties.

References

1. Rakitskaya TL, Bandurko AYu, Ennan AA et al (2000) Low-temperature catalytic decomposition of ozone microconcentrations by carbon fibrous materials. Adv Environ Res 3:472–487
2. Rakitskaya TL, Truba AS, Ennan AA et al (2019) Aerosols containing nanostructured polyphase magnetite: physicochemical and catalytic properties. In: Nanostructured materials: synthesis, properties and applications. Nova Science Publishers Inc., New York, pp 327–375. (ISBN: 978-1-53615-013-1)
3. Rakitskaya TL, Bandurko AYu, Raskola LA (2002) Katalizatory nizkotemperaturnogo razlozheniya ozona: sostoyaniye i perspektivy razrabotki. Visnik ONU. Khimiya. 6:13–22. (Rakitskaya TL, Bandurko AYu, Raskola LA (2002) Catalysts for low-temperature decomposition of ozone: state and development prospects. Herald ONU. Chemistry. 6:13–22)
4. Ren C, Zhou L, Shang H, Chen Y (2014) Effect of preparation method on the performance of Pd-MnO_x/γ-Al_2O_3 monolithic catalysts for ground-level O_3 decomposition. Chin J Catal 35:1883–1890. http://www.sciencedirect.com/science/journal/18722067
5. Schwab GM, Hartmann G (1956) Der katalytische Ozonzerfall. Z Phys Chem 6:72–82. https://doi.org/10.1524/zpch.1956.6.1_2.072
6. Oyama ST (2000) Chemical and catalytic properties of ozone. Catal Rev 42:279–322. https://doi.org/10.1081/CR-100100263
7. Almquist C, Krekeler M, Jiang L (2014) An investigation on the structure and catalytic activity of cryptomelane-type manganese oxide materials prepared by different synthesis routes. Chem Eng J 252:249–262. https://doi.org/10.1016/j.cej.2014.04.102
8. Jia J, Zhang P, Chen L (2016) Catalytic decomposition of gaseous ozone over manganese dioxides with different crystal structures. Appl Catal B Environ 189:210–218. https://doi.org/10.1016/j.apcatb.2016.02.055
9. Liu Y, Zhang P (2017) Catalytic decomposition of gaseous ozone over todorokite-type manganese dioxides at room temperature: effects of cerium modification. Appl Catal A Gen 530:102–110. https://doi.org/10.1016/j.apcata.2016.11.028
10. Tatibouet JM, Valange S, Touati H (2019) Near-ambient temperature ozone decomposition kinetics on manganese oxide-based catalysts. Appl Catal A Gen 569:126–133. https://doi.org/10.1016/j.apcata.2018.10.026

11. Li X, Ma J, Yang L et al (2018) Oxygen vacancies induced by transition metal doping in γ-MnO_2 for highly efficient ozone decomposition. Environ Sci Technol 52:12685−12696. https://www.ncbi.nlm.nih.gov/pubmed/30346750
12. Gopi T, Swetha G, Shekar SC et al (2017) Catalytic decomposition of ozone on nanostructured potassium and proton containing δ-MnO_2 catalysts. Catal Commun 92:51–55. https://doi.org/10.1016/j.catcom.2017.01.002
13. Liu Y, Yang W, Zhang P et al (2018) Nitric acid-treated birnessite-type MnO_2: an efficient and hydrophobic material for humid ozone decomposition. Appl Surf Sci 442:640–649. https://doi.org/10.1016/j.apsusc.2018.02.204
14. Zhang J, Li Y, Wang L et al (2015) Catalytic oxidation of formaldehyde over manganese oxides with different crystal structures. Catal Sci Technol 5:2305–2313. https://doi.org/10.1039/c4cy01461h
15. Jia J, Zhang P, Chen L (2016) The effect of morphology of α-MnO_2 on catalytic decomposition of gaseous ozone. Catal Sci Technol 6:5841–5847. https://doi.org/10.1039/c6cy00301j
16. Suib SL (2008) Structure, porosity, and redox in porous manganese oxide octahedral layer and molecular sieve materials. J Mater Chem 18:1623–1631. https://doi.org/10.1039/b714966m
17. Tian H, He J, Liu L et al (2013) Effects of textural parameters and noble metal loading on the catalytic activity of cryptomelane-type manganese oxides for formaldehyde oxidation. Ceramics Internsh 39:315–321. https://doi.org/10.1016/j.ceramint.2012.06.027
18. Dhandapani B, Oyama ST (1997) Gas phase ozone decomposition catalysts. Appl Catal B Environ 11:129–166. https://doi.org/10.1016/S0926-3373(96)00044-6
19. Zhu G, Zhu J, Jiang W et al (2017) Surface oxygen vacancy induced α-MnO_2 nanofiber for highly efficient ozone elimination. Appl Catal B Environ 209:729−737. http://dx.doi.org/doi:10.1016/j.apcatb.2017.02.068
20. Zhu G, Zhu J, Li W (2018) Tuning the K^+ concentration in the tunnels of α-MnO_2 to increase the content of oxygen vacancy for ozone elimination. Environ Sci Technol 52:8684–8692. https://doi.org/10.1021/acs.est.8b01594
21. Spasova I, Nikolov P, Mehandjiev D (2007) Ozone decomposition over alumina-supported copper, manganese and copper-manganese catalysts. Ozone Sci Eng 29:41−45. http://dx.doi.org/10.1080/01919510601111665
22. Liu Y, Zhang P (2017) Removing surface hydroxyl groups of Ce-modified MnO_2 to significantly improve its stability for gaseous ozone decomposition. J Phys Chem C 121:23488–23497. https://doi.org/10.1021/acs.jpcc.7b07931
23. Yang Y, Jia J, Liu Y et al (2018) The effect of tungsten doping on the catalytic activity of α-MnO2 nanomaterial for ozone decomposition under humid condition. Appl Catal A Gen 562:132–141. https://doi.org/10.1016/j.apcata.2018.06.006
24. Luo J, Zhang Q, Garcia-Martinez J et al (2008) Adsorptive and acidic properties, reversible lattice oxygen evolution, and catalytic mechanism of cryptomelane-type manganese oxides as oxidation catalysts. J Am Chem Soc 130:3198–3207. https://doi.org/10.1021/ja077706e
25. Özacar M, Poyraz AS, Genuino HC et al (2013) Influence of silver on the catalytic properties of the cryptomelane and Ag-hollandite types manganese oxides OMS-2 in the low-temperature CO oxidation. Appl Catal A Gen 462–463:64–74. https://doi.org/10.1016/j.apcata.2013.04.027
26. Jia J, Yang W, Zhang P et al (2017) Facile synthesis of Fe-modified manganese oxide with high content of oxygen vacancies for efficient airborne ozone destruction. Appl Catal A Gen 546:79−86. http://dx.doi.org/doi:10.1016/j.apcata.2017.08.013
27. Jiang C, Zhang P, Zhang B (2013) Facile synthesis of activated carbon-supported porous manganese oxide via in situ reduction of permanganate for ozone decomposition. Ozone Sci Eng 35:308–315. https://doi.org/10.1080/01919512.2013.795854
28. Rao Y, Zeng D, Cao X et al (2019) Synthesis of doped MnO_x/diatomite composites for catalyzing ozone decomposition. Ceramics Internsh 45:6966–6971. https://doi.org/10.1016/j.ceramint.2018.12.195
29. Rakitskaya T, Truba A, Dzhyga G et al (2018) Water vapor adsorption by some manganese oxide forms. Colloids Interfaces 2:61. https://doi.org/10.3390/colloids2040061

30. Wang C, Ma J, Liu F et al (2015) The effects of Mn^{2+} precursors on the structure and ozone decomposition activity of cryptomelane-type manganese oxide (OMS-2) catalysts. J Phys Chem C 119:23119–23126. https://doi.org/10.1021/acs.jpcc.5b08095
31. Ma J, Wang C, He H (2017) Transition metal doped cryptomelane-type manganese oxidecatalysts for ozone decomposition. Appl Catal B Environ 201:503–510. https://doi.org/10.1016/j.apcatb.2016.08.050
32. Liu Y, Zhang P, Zhan J et al (2018) Heat treatment of $MnCO_3$: an easy way to obtain efficient and stable MnO_2 for humid O_3 decomposition. Appl Surf Sci 463:374–385. https://doi.org/10.1016/j.apsusc.2018.08.226
33. Rakitskaya TL, Khitrich VF, Raskola LA et al (2004) Razlozheniye mikrokontsentratsiy ozona melkodispersnym MnO_2-katalizatorom. Visn. Odes'k. nats. un-tu. 9:117–124 (Rakitskaya TL, Khitrich VF, Raskola LA et al (2004) Decomposition of ozone microconcentrations by fine-dispersed MnO_2 catalyst. Herald ONU. Chemistry. 9:117–124)
34. Zheng H, Feng C, Kim SJ et al (2013) Synthesis and electrochemical properties of KMn_8O_{16} nanorods for Lithium ion batteries. Electrochim Acta 88:225–230. https://doi.org/10.1016/j.electacta.2012.09.119
35. DeGuzman RN, Shen YF, Neth EJ et al (1994) Synthesis and characterization of octahedral molecular sieves (OMS-2) having the hollandite structure. Chem Mater 6:815−821. https://doi.org/10.1021/cm00042a019
36. Sui N, Duan Y, Jiao X et al (2009) Large-scale preparation and catalytic properties of one-dimensional α/β-MnO_2 nanostructures. J Phys Chem 113:8560−8565. https://doi.org/10.1021/jp810452k
37. Gregg SJ, Sing KSW (1982) Adsorption, surface area, and porosity. Academic Press, London, New York. (ISBN: 0-12-300956-1)
38. Kolta GA, Kerim FMA, Azim AAA (1971) Infrared absorption spectra of some manganese dioxide modifications and their thermal products. Z Anorg Allg Chem 384:260–266. https://doi.org/10.1002/zaac.19713840311
39. Julien CM, Massot M, Poinsignon C (2004) Lattice vibrations of manganese oxides. Part I. Periodic structures. Spectrochim Acta Part A 60:689–700. https://doi.org/10.1016/S1386-1425(03)00279-8
40. Yang R, Wang Z, Dai L et al (2005) Synthesis and characterization of single-crystalline nanorods of α-MnO_2 and γ-MnOOH. Mater Chem Phys 93:149–153. https://doi.org/10.1016/j.matchemphys.2005.03.006
41. Tian H, He J, Zhang X et al (2011) Facile synthesis of porous manganese oxide K-OMS-2 materials and their catalytic activity for formaldehyde oxidation. Micropor Mesopor Mater 138:118–122. https://doi.org/10.1016/j.micromeso.2010.09.022
42. Li Z, Xu J (2016) Facile hydrothermal synthesis of flowerlike MnO_2 constructed by ultrathin nanosheets for supercapacitors. Biointerface Res Appl Chem 6:1070–1074
43. Chen Y, Hong Y, Ma Y et al (2010) Synthesis and formation mechanism of urchin-like nano/micro-hybrid α-MnO_2. J Alloys Compd 490:331–335. https://doi.org/10.1016/j.jallcom.2009.10.004
44. Musil M, Choi B, Tsutsumi A (2015) Morphology and electrochemical properties of α-, β-, γ-, and δ-MnO_2 synthesized by redox method. J Electrochem Soc 162:A2058–A2065. https://iopscience.iop.org/article/10.1149/2.0201510jes
45. Wang X, Li Y (2003) Synthesis and formation mechanism of manganese dioxide nanowires/nanorods. Chem Eur J 9:300–306. https://doi.org/10.1002/chem.200390024
46. Li D, Shen G, Tang W et al (2014) Large-scale synthesis of hierarchical MnO_2 for benzene catalytic oxidation. Particuology 14:71–75. https://doi.org/10.1016/j.partic.2013.06.010
47. Santos VP, Soares OSGP, Bakker JJW et al (2012) Structural and chemical disorder of cryptomelane promoted by alkali doping: Influence on catalytic properties. J Catal 293:165–174. https://doi.org/10.1016/j.jcat.2012.06.020
48. Jaroniec M, Kruk M, Olivier JP (1999) Standard nitrogen adsorption data for characterization of nanoporous silicas. Langmuir 15:5410–5413. https://doi.org/10.1021/la990136e

49. Wang ZM, Tezuka S, Kanoh H (2000) Gaseous molecular sieving property of a microporous hollandite-type hydrous manganese oxide. Chem Lett 29:560–561. https://doi.org/10.1246/cl.2000.560
50. Nechiporenko AP, Kudryashova AI (1987) Kislotno-osnovnoy spektr poverkhnosti α- i γ-Al_2O_3. Zhurn. Obsh. khim. 57:752–758. (Nechiporenko AP, Kudryashova AI (2017) The acid-base spectrum of the surface of α- and γ-Al_2O_3. Rus J Gen Chem 57:752–758)
51. Rakyts′ka TL, Truba AS, Nahayevs′ka AV (2017) Syntez ta katalitychna aktyvnist′ dyspersnykh oksydiv manhanu(IV) v reaktsiyi rozkladannya ozonu. Visn. Odes′k. nats. un-tu. Khimiya. 22:6–14. (Rakitskaya TL, Truba AS, Nagaevs'ka AV (2017) Synthesis and catalytic activity of dispersed manganese(IV) oxides in the reaction of ozone decomposition. Herald ONU. Chemistry. 22:6–14) http://dx.doi.org/10.18524/2304-0947.2017.4(64).115916
52. Shin JY, Cheney MA (2004) Abiotic transformation of atrazine in aqueous suspension of four synthetic manganese oxides. Coll Surf A Physicochem Eng Aspects 242:85–92. https://doi.org/10.1016/j.colsurfa.2004.04.061
53. Valdesa H, Alejandrob S, Zaror CA (2012) Natural zeolite reactivity towards ozone: the role of compensating cations. J Hazardous Mater 227–228:34–40. https://doi.org/10.1016/j.jhazmat.2012.04.067

Electrostatics of the Nanowires with Radial *p-n* or *p-i-n* Junctions

V. L. Borblik

1 Introduction

In due time, it has been noted that such important characteristic of a conducting medium as its near-surface depletion depth (which is usually minor part of the sample volume) can compete with its dimension when considering the nanoobjects. Then accurate assessment of this depth becomes especially important [1]. Furthermore, it has been proved that the boundary curvature mysteriously influences this depth. In particular, the depletion depth increases (in comparison with the case of plain boundary) with decreasing in a radius of cylindrical interface (the case of a nanowire) and reaches limit value of $\sqrt{2}W_p$ (W_p is the depletion depth for planar boundary) when it amounts to the nanowire radius, i.e., when the depletion encompasses full volume of the nanowire [1]. In the case of spherical form of the boundary (a quantum dot), increase in the depletion depth with decreasing in the quantum dot radius becomes even more significant, and the full depletion of the dot volume is reached already at the depletion depth value of $\sqrt{3}W_p$ [2, 3].

In recent time, great interest of the investigators is attracted to semiconductor nanowires, especially to the multilayer ones whose layers are doped in a different way. On the base of such objects, principally new constructions of the core–shell devices are created which use transverse (radial) transport of the current carriers (radial solar cells [4], radial photodiodes [5], and radial light-emitting devices [6]).

A *p-n* junction entering into the composition of these devices is such an interface where depletion occurs on both sides. This problem was considered in [7] where the authors have developed general electrostatic theory for radial *p-n* junction diode and have marked out four types of its electric structure depending on a set of parameters.

V. L. Borblik (✉)
Department of Electrical and Galvanomagnetic Properties of Semiconductors, V.Lashkarev Institute of Semiconductor Physics, 41, Prospect Nauky, 03680 Kiev, Ukraine
e-mail: borblik@isp.kiev.ua

O. Fesenko and L. Yatsenko (eds.), *Nanomaterials and Nanocomposites, Nanostructure Surfaces, and Their Applications*, Springer Proceedings in Physics 246, https://doi.org/10.1007/978-3-030-51905-6_7

However, explicit dependence of the depletion widths for both sides of the *p-n* junction on its radius has been not established. Partially, this has been made in paper [8] but only for the depletion width of the core which (likely to the case of semi-limited structures [1–3]) increases with decrease in the radius of metallurgical boundary of the *p-n* junction. Behavior of the depletion width of the nanowire shell as well as of the whole depletion width remained not studied.

In experiment, in order to enhance a performance of such devices, *p*- and *n*-layers are made often of different materials that allow enlargement of the built-in electric field. The radial core–shell solar cells and photodetectors have been fabricated using the hetero *p-n* junctions such as Ge/CdS [9], CdS/Cu_2S [10], ZnO/CuS [11], Si/CdS [12], GaAs/InGaAs [13], Si/ZnO [14], and others.

At the same time, the radial nanowire structures use often not *p-n* but *p-i-n* junctions [15–19]. In particular, this makes it possible to broaden the region of strong electric field in the junction that is additional advantageous in materials with the short minority carrier diffusion lengths [20].

In this paper, electrostatics of all such radial structures is analyzed theoretically.

2 Nanowire Radial *p-n* Junction

2.1 Theory

Let us consider the case of partially depleted *p*-core and *n*-shell (Fig. 1). We will proceed from the known system of two equations [7, 8] which allow us the determination of the depletion widths in the core $w_p = r_0 - r_p$ and in the shell $w_n = r_n - r_0$ where r_0 is the core radius, r_p is the depletion region boundary in the core, r_n is the depletion region boundary in the shell, and r_d is the external radius of the nanowire.

From matching of the electric fields in point r_0, we have

$$N_A\left(r_0^2 - r_p^2\right) = N_D\left(r_n^2 - r_0^2\right) \tag{1}$$

Fig. 1 Schematic view of the radial *p-n* structure

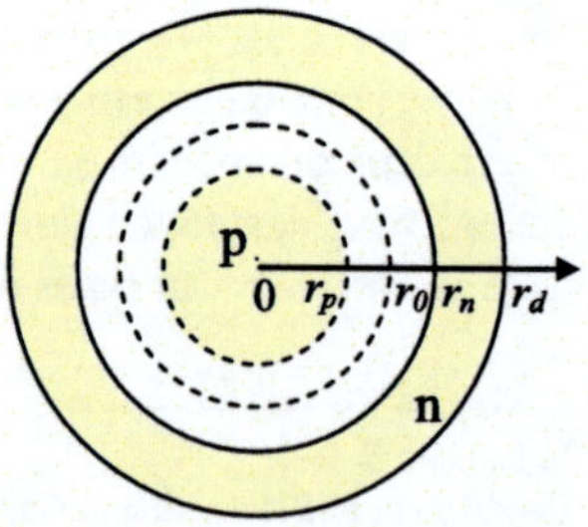

(N_A and N_D are the concentrations of acceptors and donors, respectively), and matching of the potentials in the same point gives

$$\frac{q}{2\varepsilon_S}\left(N_A r_p^2 \ln\frac{r_p}{r_0} + N_D r_n^2 \ln\frac{r_n}{r_0}\right) = V_{bi}, \tag{2}$$

where q is the electron charge, ε_S is the dielectric constant of the semiconductor, and V_{bi} is the built-in potential of the p-n junction [21]

$$V_{bi} = \frac{kT}{q}\ln\frac{N_A N_D}{n_i^2}, \tag{3}$$

k is the Boltzmann constant, T is temperature, and n_i is the intrinsic carrier concentration. Expressing r_n in terms of r_p

$$r_n = \sqrt{r_0^2 + (r_0^2 - r_p^2)N_D/N_A}, \tag{4}$$

we obtain transcendental equation in r_p

$$\left(1+\frac{N_A}{N_D}\left(1-\frac{r_p^2}{r_0^2}\right)\right)\ln\sqrt{1+\frac{N_A}{N_D}\left(1-\frac{r_p^2}{r_0^2}\right)} + \frac{N_A}{N_D}\frac{r_p^2}{r_0^2}\ln\frac{r_p}{r_0} - \frac{V_{bi}2\varepsilon_S}{qN_D r_0^2} = 0. \tag{5}$$

2.2 Numerical Results for Silicon P-N Junction

For numerical solution of (5), the parameters of silicon at room temperature have been chosen: $\varepsilon_S = 12\,\varepsilon_0$ (ε_0 is the permittivity of free space), $n_i = 6.3{\cdot}10^9$ cm^{-3}. The calculation results are presented in Fig. 2. Figure 2a corresponds to the case when the core is doped higher than the shell is done, Fig. 2b concerns with the opposite case, and Fig. 2c represents the results for the case of equal doping levels.

As it follows from these figures, in all three cases the depletion width of the core increases with decreasing its radius; meanwhile, the depletion width of the shell, on the contrary, decreases. As for the whole depletion width of the p-n junction $w = w_p + w_n$, it can both increase and decrease and even be nearly independent of the p-n junction radius at equal and high enough doping levels of both sides.

Opposite character of dependencies on the p-n junction radius of the depletion widths for the core and the shell is consequence of radial falling which is a characteristic feature for solutions of differential equations in cylindrical (as well as spherical) coordinate system. In the given case, the built-in electric field of the p-n junction which is maximal at its metallurgical boundary decreases in direction of the

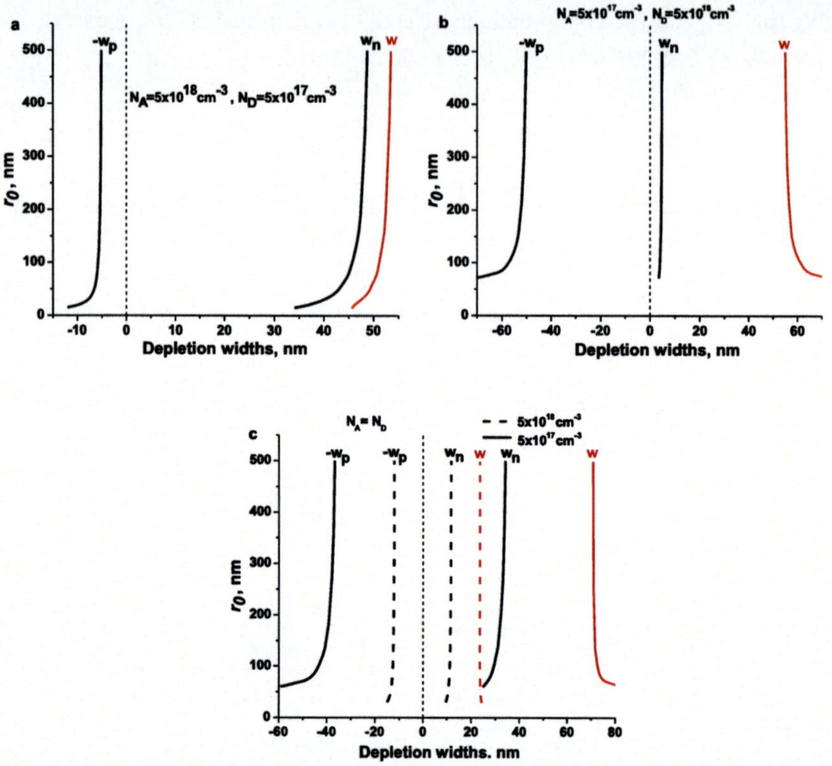

Fig. 2 Dependence of the depletion widths in the core w_p, in the shell w_n, and of the total depletion width w on radius of the *p-n* junction at $N_A >> N_D$(**a**), $N_A << N_D$ (**b**) and $N_A = N_D$ (**c**) (an abscissa equal to zero corresponds to metallurgical boundary of the *p-n* junction)

shell faster than in direction of the core. Note also that the "mysterious" increase in depletion depth in semi-limited semiconductor nanostructures with cylindrical and spherical geometry (as in [1–3]) has the same nature.

3 Nanowire Radial *p-n* Heterojunction

3.1 Theory

Let us assign number 1 to the *p*-core and number 2 to the *n*-shell. Then, in the depletion approximation, profile of the potential V is determined by next Poisson equations for two regions

$$\frac{1}{r}\frac{d}{dr}\left(r\frac{dV_1}{dr}\right) = \frac{qN_{A1}}{\varepsilon_1},\ r_p \le r \le r_0, \tag{6}$$

$$\frac{1}{r}\frac{d}{dr}\left(r\frac{dV_2}{dr}\right) = -\frac{qN_{D2}}{\varepsilon_2},\ r_0 \le r \le r_n, \tag{7}$$

where N_{A1} and N_{D2} are the acceptor and donor concentrations, respectively, and ε_1 and ε_2 are the dielectric constants of two materials.

Matching the electric inductions at the interface r_0 gives the equation

$$N_{A1}\left(r_0^2 - r_p^2\right) = N_{D2}\left(r_n^2 - r_0^2\right) \tag{8}$$

and matching the potentials results in equation

$$\frac{qN_{A1}}{2\varepsilon_1}\left(\frac{r_0^2 - r_p^2}{2} + r_p^2 \ln\frac{r_p}{r_0}\right) + \frac{qN_{D2}}{2\varepsilon_2}\left(\frac{r_0^2 - r_n^2}{2} + r_n^2 \ln\frac{r_n}{r_0}\right) = V_{bi}, \tag{9}$$

where V_{bi} is the built-in potential. Expressing r_n in terms of r_p on the basis of (8)

$$r_n = \sqrt{r_0^2 + \left(r_0^2 - r_p^2\right)N_{D2}/N_{A1}}, \tag{10}$$

one obtains the transcendental equation in r_p

$$\begin{aligned}&\frac{N_{A1}}{2N_{D2}}\left(\frac{\varepsilon_2}{\varepsilon_1} - 1\right)\left(1 - \frac{r_p^2}{r_0^2}\right) + \frac{N_{A1}\varepsilon_2}{N_{D2}\varepsilon_1}\frac{r_p^2}{r_0^2}\ln\frac{r_p}{r_0} + \\ &+ \left[1 + \frac{N_{A1}}{N_{D2}}\left(1 - \frac{r_p^2}{r_0^2}\right)\right]\ln\sqrt{1 + \frac{N_{A1}}{N_{D2}}\left(1 - \frac{r_p^2}{r_0^2}\right)} - \frac{V_{bi}2\varepsilon_2}{qN_{D2}r_0^2} = 0.\end{aligned} \tag{11}$$

At $\varepsilon_1 = \varepsilon_2$, this equation reduces to (5). Under biasing conditions, V_{bi} has to be replaced by $V_{bi} - U$ where U is the external voltage.

To solve (11), it is necessary to know the value of V_{bi}. Here, we consider the abrupt anisotype *p-n* heterojunction of the type I, for which the energy band scheme (before bringing two materials into contact) has the form shown in Fig. 3.

In the case of the non-generated semiconductors, we have

$$E_{F1} - E_{v1} = kT \ln\frac{N_{v1}}{N_{A1}}, \tag{12}$$

$$E_{c2} - E_{F2} = kT \ln\frac{N_{c2}}{N_{D2}}, \tag{13}$$

where N_{v1} and N_{c2} are the effective densities of states in the valence and conduction bands, respectively. Adding (12) and (13), we obtain

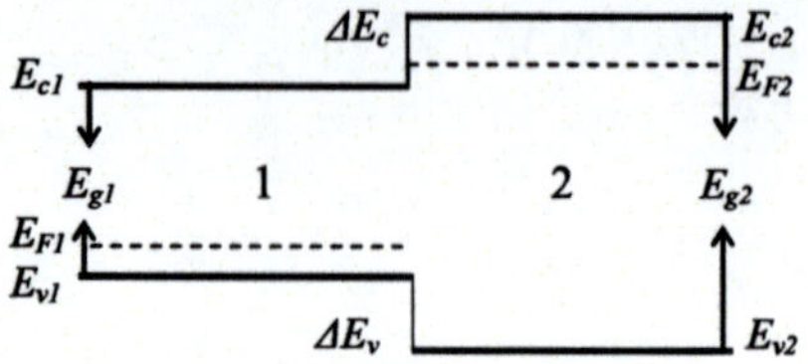

Fig. 3 Energy band arrangement of type I for *p-n* heterojunction; here, E_{c1} and E_{c2} are bottoms of the conductivity bands, E_{v1} and E_{v2} are tops of the valence bands, ΔE_c and ΔE_v are the energy bands discontinuities, E_{g1} and E_{g2} are the energy gaps, E_{F1} and E_{F2} are the Fermi levels

$$qV_{bi} \equiv E_{F2} - E_{F1} = E_{c2} - E_{v1} - kT\ln\left(\frac{N_{v1}N_{c2}}{N_{A1}N_{D2}}\right) = E_{g1} + \Delta E_c + kT\ln\left(\frac{N_{A1}N_{D2}}{N_{v1}N_{c2}}\right) \equiv$$
$$\equiv E_{g2} - \Delta E_v + kT\ln\left(\frac{N_{A1}N_{D2}}{N_{v1}N_{c2}}\right). \tag{14}$$

In concordance with Anderson's "electron affinity rule" [22], $\Delta E_c = \chi_1 - \chi_2$, where χ_1 and χ_2 are the electron affinities of two materials.

The barrier capacitance $C = \frac{dQ_p}{dU}$ where Q_p is the electron charge concentrated in depleted *p*-region of the heterojunction. This charge is given by

$$Q_p = qN_{A1}\pi\left(r_0^2 - r_p^2\right)L, \tag{15}$$

where r_p is voltage-dependent and L is length of the nanowire. Inasmuch as

$$\frac{dr_p}{dU} = \frac{\varepsilon_2}{qN_{A1}r_p}\left[\frac{\varepsilon_2}{\varepsilon_1}\ln\frac{r_p}{r_0} - \ln\sqrt{1 + \frac{N_{A1}}{N_{D2}}\left(1 - \frac{r_p^2}{r_0^2}\right)}\right]^{-1}, \tag{16}$$

the capacitance per unit area of the *p-n* junction is

$$C = \frac{2\varepsilon_1\varepsilon_2}{r_0}\left[\varepsilon_1\ln\left(\frac{r_n}{r_0}\right)^2 - \varepsilon_2\ln\left(\frac{r_p}{r_0}\right)^2\right]^{-1}. \tag{17}$$

At $\varepsilon_1 = \varepsilon_2$, (17) reduces to the corresponding expression for the radial homojunction [8].

3.2 Numerical Results for Ge/GaAs p-n Junction

The numerical calculations are performed for radial heterodiode *p*-Ge/*n*-GaAs because this heteropair has good lattice matching, i.e., no appreciable density of interface states can be associated with this heterojunction.

In accordance with data from [23], in this case $\Delta E_c = 4.13\ \text{eV}–4.07\ \text{eV} = 0.06\ \text{eV}$. Corresponding value of $\Delta E_v = E_{g2} - E_{g1} - \Delta E_c = (1.42–0.66)\ \text{eV} - 0.06\ \text{eV} = 0.7\ \text{eV}$. The densities of states for Ge and GaAs (as well as values of the energy gaps) are taken from [21].

Figure 4 represents the calculated depletion widths w_p and w_n of the core and shell, respectively, as a function of the interface radius r_0 at $N_{A1} = N_{D2} = 10^{18}\text{cm}^{-3}$(solid curves). As well as in the case of *p-n* homojunction, here depletion width of the core increases with the interface radius decreasing and depletion width of the shell, on the contrary, decreases. For comparison, the corresponding results for radial *p-n* homojunctions in Ge (dash-dot) and GaAs (dash-dot-dot) at the same impurity concentrations are shown in Fig. 4 as well.

Above values of ΔE_c and ΔE_v are not commonly accepted. Theoretical values of the energy band discontinuities from different authors as well as experimentally measured ones, as a rule, differ appreciably. In the scientific literature, the valence band discontinuities are given usually [24, 25]. For Ge/GaAs heteropair, experimental and theoretical values of this quantity are scattered in the interval of 0.25-0.71 eV. Therefore, in Fig. 4, we present also the calculation results corresponding to case of minimal $\Delta E_v = 0.25$ eV, i.e., $\Delta E_c = 0.51$ eV (dash curves). It is seen that at any value of the band discontinuity, the depletion widths of the heterojunction are situated between the corresponding values for radial homojunctions made of the constituent materials (shaded areas between solid and dash curves).

Figure 5 presents the dependence of $1/C^2$ versus applied voltage at different values of the *p-n* junction radius. At the large radius (Fig. 5a), this dependence is close to linear one but more and more deviates from it with decreasing in r_0 (Fig. 5b, c). It is seen that at any value of r_0, voltage dependences of $1/C^2$ for the heterojunction are situated between the corresponding dependences for radial homojunctions made of the constituent materials taken at the same doping levels.

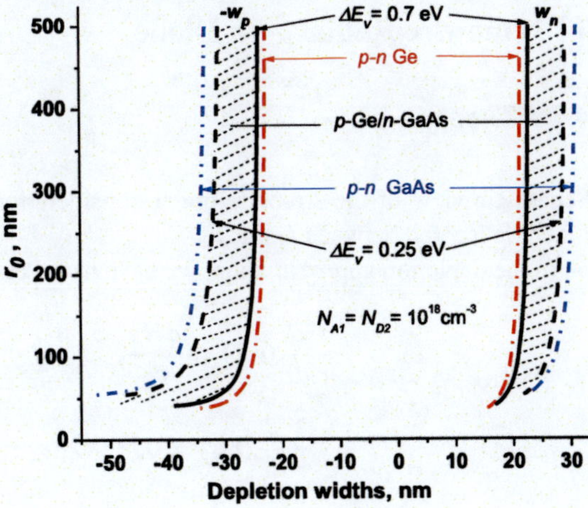

Fig. 4 Dependence of the depletion widths in the core w_p and in the shell w_n on radius of the core–shell interface at $N_{A1} = N_{D2} = 10^{18}\ \text{cm}^{-3}$ (an abscissa equal to zero corresponds to the interface); solid and dash curves—for the heterojunction at $\Delta E_v = 0.7$ and 0.25 eV, respectively, dash–dot and dash–dot–dot curves—for Ge and GaAs radial *p-n* homojunctions, respectively

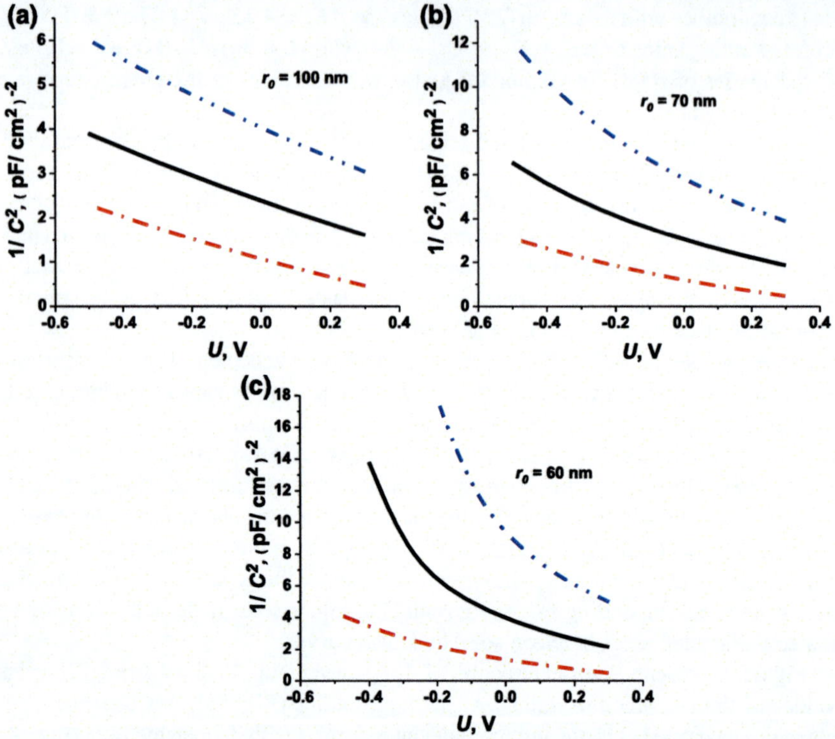

Fig. 5 Dependence of $1/C^2$ on applied voltage in *p*-Ge/*n*-GaAs heterojunction (solid curves), in Ge *p*-*n* homojunction (dash-dot), and in GaAs *p*-*n* homojunction (dash-dot-dot) at interface radius of 100 nm (**a**), 70 nm (**b**), and 60 nm, (**c**) and the same impurity concentrations $N_{A1} = N_{D2} = 10^{18}$ cm^{-3}

4 Nanowire Radial *p-i-n* Diode

4.1 Theory

Schematic view of the structure under consideration is presented in Fig. 6. Here, *i*-layer is located between r_1 and r_2.

In the depletion approximation, we have next Poisson's equations

$$\frac{1}{r}\frac{d}{dr}(rE) = \frac{qN_A}{\varepsilon_S}, \qquad r_p \le r \le r_1, \tag{18a}$$

$$\frac{1}{r}\frac{d}{dr}(rE) = 0, \qquad r_1 \le r \le r_2, \tag{18b}$$

Fig. 6 Schematic view of the nanowire *p-i-n* structure

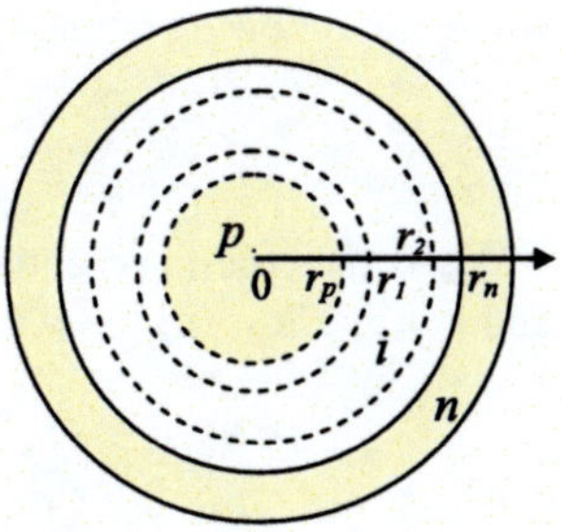

$$\frac{1}{r}\frac{d}{dr}(rE) = -\frac{qN_D}{\varepsilon_S}, \qquad r_2 \le r \le r_n. \tag{18c}$$

Solution of these equations gives the electric field distribution in the structure

$$E = -\frac{qN_A}{2\varepsilon_S}\frac{r^2 - r_p^2}{r}, \qquad r_p \le r \le r_1, \tag{19a}$$

$$E = \frac{A}{r}, \qquad r_1 \le r \le r_2, \tag{19b}$$

$$E = \frac{qN_D}{2\varepsilon_S}\frac{r^2 - r_n^2}{r}, \qquad r_2 \le r \le r_n, \tag{19c}$$

where A is the integration constant.

Matching the electric fields at r_1 and r_2, we obtain

$$A = -\frac{qN_A}{2\varepsilon_S}(r_1^2 - r_p^2) = \frac{qN_D}{2\varepsilon_S}(r_2^2 - r_n^2) \tag{20}$$

whence it follows

$$N_A(r_1^2 - r_p^2) = N_D(r_n^2 - r_2^2). \tag{21}$$

The second integration of (19a, 19b, 19c) gives the potentials

$$V(r) = \frac{qN_A}{2\varepsilon_S}\left(\frac{r^2 - r_p^2}{2} + r_p^2 \ln\left(\frac{r_p}{r}\right)\right), \qquad r_p \le r \le r_1, \tag{22a}$$

$$V(r) = -A(\ln(r) + \text{const}), \qquad r_1 \le r \le r_2, \tag{22b}$$

$$V(r) = -\frac{qN_D}{2\varepsilon_S}\left(\frac{r^2 - r_n^2}{2} + r_n^2 \ln\left(\frac{r_n}{r}\right)\right) + V_{bi}, \; r_2 \le r \le r_n, \tag{22c}$$

where the following boundary conditions are used:

$$V(r_p) = 0, \; V(r_n) = V_{bi}, \tag{23}$$

V_{bi} is the built-in potential of the junction. Matching of the potentials at $r = r_1$ and $r = r_2$ allows us to exclude *const* and obtain the equation

$$\frac{qN_A}{2\varepsilon_S} r_p^2 \ln\left(\frac{r_p}{r_1}\right) + \frac{qN_D}{2\varepsilon_S} r_n^2 \ln\left(\frac{r_n}{r_2}\right) - A \ln\left(\frac{r_2}{r_1}\right) = V_{bi}. \tag{24}$$

Equations (21) and (24) have to be solved jointly in order to obtain r_p and r_n. All the rest quantities are expressed through them.

The barrier capacitance $C = \frac{dQ_p}{dU}$, where Q_p is given by

$$Q_p = qN_A\pi\left(r_1^2 - r_p^2\right)L. \tag{25}$$

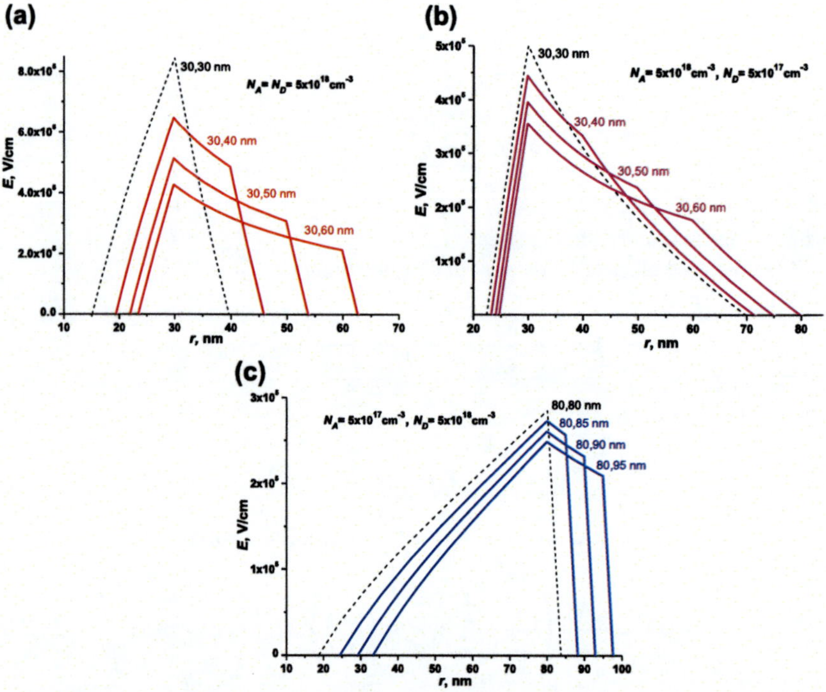

Fig. 7 Electric field distribution in the nanowire *p-i-n* diode at **a** $N_A = N_D$, **b** $N_A >> N_D$ and **c** $N_A << N_D$; numbers near the curves are radial coordinates of the *i*-layer showing its extent, dashed lines correspond to the *i*-layer of zero extent (*p-n* diode)

Inasmuch as

$$\frac{dr_p}{dU} = \frac{\varepsilon_S}{qN_A r_p} \frac{1}{\ln(r_n/r_p)}, \tag{26}$$

the capacitance per unit area of the *p-i-n* diode is

$$C = \frac{\varepsilon_S}{r_1} \frac{1}{\ln(r_n/r_p)}. \tag{27}$$

4.2 Numerical Results for Silicon p-i-n Diode

For numerical solution of (21) and (24), parameters of silicon at room temperature have been chosen. Three doping situations have been considered: $N_A = N_D$, $N_A >> N_D$, and $N_A << N_D$. The calculation results for the electric field distribution in

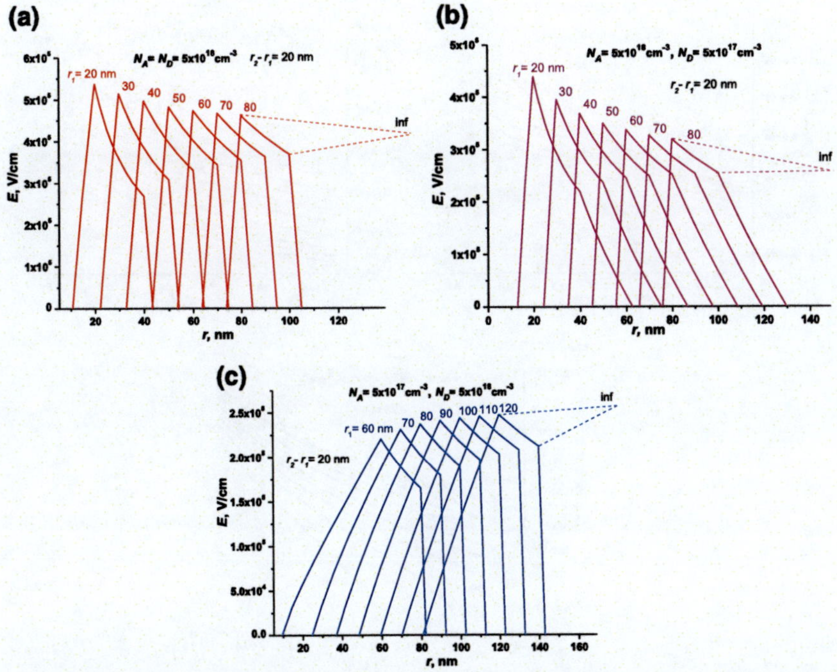

Fig. 8 Electric field distribution in the nanowire *p-i-n* diode depending on radial position of the *i*-layer with thickness of 20 nm at **a** $N_A = N_D$, **b** $N_A >> N_D$ and **c** $N_A << N_D$; dashed lines show to what magnitude both edge values of the field in the *i*-layer go when the nanowire *p-i-n* diode becomes planar one

the structure are presented in Fig. 7. Characteristic feature of these distributions is inhomogeneity of the field in the *i*-layer that sharply differs from the case of planar *p-i-n* diode where electric field in the *i*-layer is homogeneous [21]. The field inhomogeneity is especially strong when $N_A = N_D$ or $N_A >> N_D$ and diminishes with thickening of the *i*-layer. In any case, the electric field is maximal near the nanowire core.

It is of interest to study dependence of the electric field distributions on radial position of the *i*-layer in the nanowire. Figure 8 represents such dependences for three doping situations at the same thickness of the *i*-layer equal to 20 nm. It is seen that as the *i*-layer moves away from a center of the nanowire, inhomogeneity of the electric field distribution becomes more and more weak, i.e., the field goes to homogeneous one inherent to planar *p-i-n* diode.

The dash lines in these figures demonstrate asymptotical confluence of both edge values of the field in the *i*-layer when r_1 goes to infinity, i.e., the nanowire curvature becomes ignorable. It is seen also that maximum electric field in the *i*-layer of the nanowire proves to be higher than that in analogous planar diode at $N_A = N_D$ and $N_A >> N_D$ and, on the contrary, is lower than that at $N_A << N_D$.

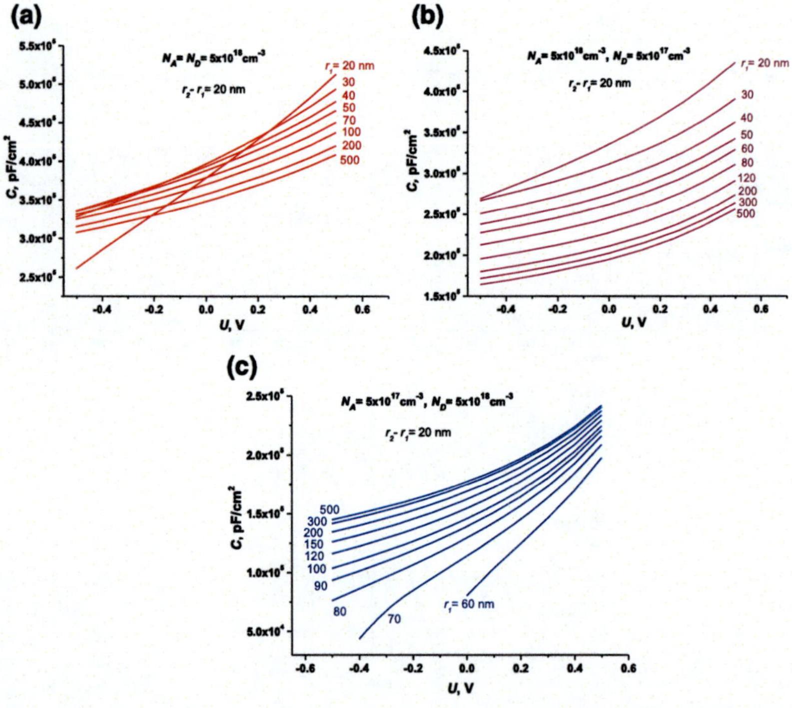

Fig. 9 Voltage dependences of the nanowire *p-i-n* diode capacitance at **a** $N_A = N_D$, **b** $N_A >> N_D$ and **c** $N_A << N_D$ as a function of the distance between the *i*-layer (with thickness of 20 nm) and the nanowire center

Figure 9 represents the voltage dependences of the nanowire *p-i-n* diode capacitance given by formula (27) for three doping combinations as a function of the distance between the *i*-layer and center of the nanowire at the same value of the *i*-layer thickness equal to 20 nm.

As it follows from these figures, capacitance of the nanowire *p-i-n* diode decreases when the *i*-layer is moving away from the nanowire center at $N_A = N_D$ and $N_A >> N_D$ and, on the contrary, increases at $N_A << N_D$. In any case, the voltage dependence of the capacitance diminishes as it has to be in planar *p-i-n* diode [21].

5 Conclusion

Thus, distinction of nanowire radial *p-n* junction electrostatics from electrostatics of analogous planar structures is conditioned by overlaying specific falling of the built-in electric field with radial coordinate (due to cylindrical geometry of the structure) on usual planar situation. As a result, electric field of the radial *p-n* junction proves to be nonsymmetrical even at symmetrical doping levels of both sides. And electric field in *i*-layer of a *p-i-n* diode acquires falling character.

Increase in the core depletion width with decrease in the *p-n* junction radius and simultaneous decrease in the shell depletion width can be important in such devices as heterostructure solar cells because this results in change of the relative contribution to the performance characteristics from the different constituent materials. Namely, the lesser the radius of the heterostructure *p-n* junction, the larger is contribution from the core material (under equal other conditions).

It should be noted that at the small *p-n* junction radiuses, dependence of $1/C^2$ on applied voltage, being nonlinear, cannot be used for determination of the built-in potential and doping level of the junction as it made usually in the case of planar structures.

Acknowledgements This work was supported by the National Academy of Sciences of Ukraine [project 2.2.6.34].

References

1. Luscombe JH, Frenzen CL (2002) Solid-State Electron 46(6):885–889
2. Nersesyan SR, Petrosyan SG (2012) Semicond Sci Technol 27(12):125009
3. Borblik VL (2015) Solid-State Electron 114:171–173
4. Tian B, Zheng X, Kempa TJ, Fang Y, Yu N, Yu G, Huang J, Lieber CM (2007) Nature 449:885–890
5. Soci C, Zhang A, Bao X-Y, Kim H, Lo Y, Wang D (2010) J Nanosci Nanotechnol 10(3):1430–1449
6. Hua B, Motohisa J, Kobayashi Y, Hara S, Fukui T (2009) Nano Lett 9(1):112–116
7. Chia ACE, LaPierre RR (2013) J Appl Phys 114(7):074317
8. Petrosyan S, Yesayan A, Nersesyan S (2012) World Acad Sci Eng Technol 71:1065–1070

9. Mondal SP, Ray SK (2009) Appl Phys Lett 94:223119
10. Tang J, Huo Z, Brittman S, Gao H, Yang P (2011) Nat Nanotechnol 6:568
11. Panigrahi S, Basa D (2012) RSC Adv 2:11963
12. Manna, S., Das, S., Mondal, S.P., Singha, R., & Ray, S.K. (2012). *J. Phys. Chem.*, C 116, 7126
13. Moratis K, Tan SL, Germanis S, Katsidis C, Androulidaki M, Tsagaraki K, Hatzopoulos Z, Donatini F, Cibert J, Niquet Y-M, Mariette H, Pelekanos NT (2016) Nanoscale Res Lett 11:176
14. Yoon IT, Cho HD, Cho HY, Kwak DW, Lee S (2017) J Electron Mater 46:4119
15. Tian B, Kempa TJ, Lieber CM (2009) Chem Soc Rev 38(1):16–24
16. Colombo C, Heiβ M, Grätzel M, Fontcuberta i Morral A (2009) Appl Phys Lett 94(17):173108
17. Yoo J, Dayeh SA, Tang W, Picraux ST (2013) Appl Phys Lett 102(9):093113
18. Zhang Y, Sanchez AM, Aagesen M, Huo S, Fonseka HA. Gott JA, Kim D, Yu X, Chen X, Xu J, Li T, Zeng H, Boras G, Liu H (2018) Small 1803684
19. Qian F, Li Y, Gradečak S, Wang D, Barrelet CJ, Lieber CM (2004) Nano Lett 4(10):1975–1979
20. Abdellatif S, Kirah K (2013) Energy Procedia 36:488–491
21. Sze SM (1981) Physics of semiconductor devices, 2nd edn. Wiley, New York, Chichester, Brisbane, Toronto, Singapore
22. Anderson RL (1962) Solid-State Electron 5:341
23. Sharma BL, Purohit RK (1974) Semiconductor heterojunction. Pergamon Press, Oxford, New York, Toronto, Sydney
24. Pollmann J, Mazur A (1983) Thin Solid Films 104:257
25. Pollard W (1991) J Appl Phys 69:3154

Composites Containing Inorganic Ion Exchangers and Graphene Oxide: Hydrophilic–Hydrophobic and Sorption Properties (Review)

Yuliya S. Dzyazko, Yurii M. Volfkovich, and Mary O. Chaban

1 Introduction

At the present time, two-dimensional materials are in a focus of attention due to a wide spectrum of their functional properties and unique potential possibilities. A number of inorganic compounds of this type are known: boron nitride, dichalcogenides of molybdenum and tungsten, chalcogenides of metals of III and IV groups, silicenes, and germanenes [1]. Two-dimensional carbides, carbonitrides, and nitrides of transition metals belong to a special group, that is, so-called MXenes. They are prospective for the development of supercapacitors, anodes for lithium batteries, and antibacterial materials [2].

Among compounds of two-dimensional structure, graphenes and graphene-like materials are most studied. Graphene is an allotrope modification of carbon, which occupies a special position in the family of carbon materials: zero-dimensional fullerenes, one-dimensional carbon nanotubes, three-dimensional graphite, and diamond. Graphenes are divided into two main types: graphene oxide (GO) and reduced graphene oxide (rGO), which are flakes of graphite. GO is characterized by one-layer structure similar to rGO, i.e., the thickness of graphite-like flakes is one atom (0.54 nm) [3, 4]. The bonds between carbon atoms are in a state of sp^2-hybridization. Carbon atoms form hexagonal two-dimensional honeycomb crystal lattice by means of σ- and π-bonds; the lattice constant is 0.246 nm. Carbon materials give sharp and narrow characteristic reflexes of ZRD patterns: at ≈26.4° (this reflex corresponds to the interlayer distance of 0.337 nm) for graphite, and at ≈11.3°

Y. S. Dzyazko · M. O. Chaban (✉)
VI Vernadskii Institute of General and Inorganic Chemistry of the National Academy of Science of Ukraine, Kiev, Ukraine
e-mail: Mary.chaban@gmail.com

Y. M. Volfkovich
A.N. Frumkin Institute of Physical Chemistry and Electrochemistry of the RAS, Moscow, Russia

O. Fesenko and L. Yatsenko (eds.), *Nanomaterials and Nanocomposites, Nanostructure Surfaces, and Their Applications*, Springer Proceedings in Physics 246, https://doi.org/10.1007/978-3-030-51905-6_8

(0.768 nm, this distance corresponds to graphene) [5]. Graphene-like materials, which consist of several graphite layers, are also known: a thickness of their flakes is several nanometers [6]. They show a diffuse peak between 11.3° and 26.4°. No electrical conductivity is attributed to GO. At the same time, rGO is an electric conductor. This gives a possibility to use it for the manufacture of electrodes for different electrochemical devices.

GO is obtained from graphite by means of Brodie [7], Staudenmaier [8], or Hummer methods [9]. These techniques provide oxidation of graphite down to various levels. Brodie and Staudenmaier used KClO3 and HNO3. KMnO4 and H2SO4 are applied to oxidation by means of Hummer's method. Other initial materials for graphene obtaining are carbon nanotubes [10, 11]. Oxidation causes the formation of phenolic, carboxyl, and epoxy groups, which are attached to the graphite layers. Carboxyl groups and a part of phenolic groups are located along the perimeter of GO flakes. Epoxy groups and other parts of hydroxyl groups are located on their basal plane [12]. rGO is synthesized by means of GO reduction.

Carboxylic groups of GO provide its adsorption properties toward inorganic cations, for instance, UO_2^{2+} [13], Cu^{2+} [14], Pb^{2+} [15], and other toxic ionic components of aqueous solutions. However, GO demonstrates practically no anion exchange properties [15]. At the same time, the adsorption of anionic azo dyes has been reported [16]. Adsorption of anionic species is evidently due to their interaction with hydrophobic regions of GO (benzene rings). These regions of GO flakes are responsible for the adsorption of organic molecules, such as aromatic [17] and nitroaromatic compounds [18]. In the case of deoxyribonucleic acid [19] or hormones [20], hydrophobic regions are adsorption centers in neutral media, when isoelectric point of proteins is achieved.

It should be stressed that GO is more attractive than rGO, since it can be used for adsorption of both inorganic ions and organic molecules. Thus, GO can be considered as a multifunctional adsorbent. However, it is impossible to fill adsorption columns with this material due to small size of its particles, and good dispersibility in water or other polar solvents.

The solution to the problem is to develop GO-containing composites, which contain the component possessing adsorption ability. In this chapter, the mutual effect of GO and inorganic particles on their structure on nano- and microlevels is considered. Hydrophilicity–hydrophobicity of different types of graphene (especially GO) and their composites are also in a focus of attention. These properties determine the adsorption ability of the composite materials toward both inorganic and organic species. Adsorption properties of the composites are also reviewed.

2 Morphology of Composites

Among inorganic adsorbents, materials based on hydrated oxides of multivalent metals are attractive as a support of GO. These adsorbents are amphoteric: depending on the solution pH, they are able to sorb either cations or anions [21]. It should be stressed that oxide materials are related to the exclusive class of inorganic materials, which possess anion exchange ability. Moreover, nanoparticles can be formed during precipitation from the sol of insoluble metal hydroxocomplexes [22]. Salts of multivalent metal can be also used for synthesis of the support [23, 24]. Due to mentioned remarkable properties, hydrated oxides are used for modifying some functional materials: inorganic [22, 25, 26] or polymer [27] membranes for electrodialysis, polymer membranes for processing of wastes of beverage industry [28–30], and ion exchange resins for sorption processes [31–33]. As found for ion exchange polymers, less soluble hydrated zirconium oxide forms larger particles compared with more soluble one, when deposition occurs under similar conditions [33].

Let us consider precipitation from sol of zirconium hydroxocomplexes [34], where nanoparticles dominate [22]. The nanoparticles are consolidated to large formations. A shape of primary nanoparticles is close to globular or elliptic (Fig. 1a).

Synthesis of GO-containing composites usually involves activation of GO suspension with ultrasound followed by adding the solution of metal salt or sol. Two-component material containing low amount of GO (about 2% relatively anhydrous ZrO_2) shows more friable structure of the inorganic support compared with pure

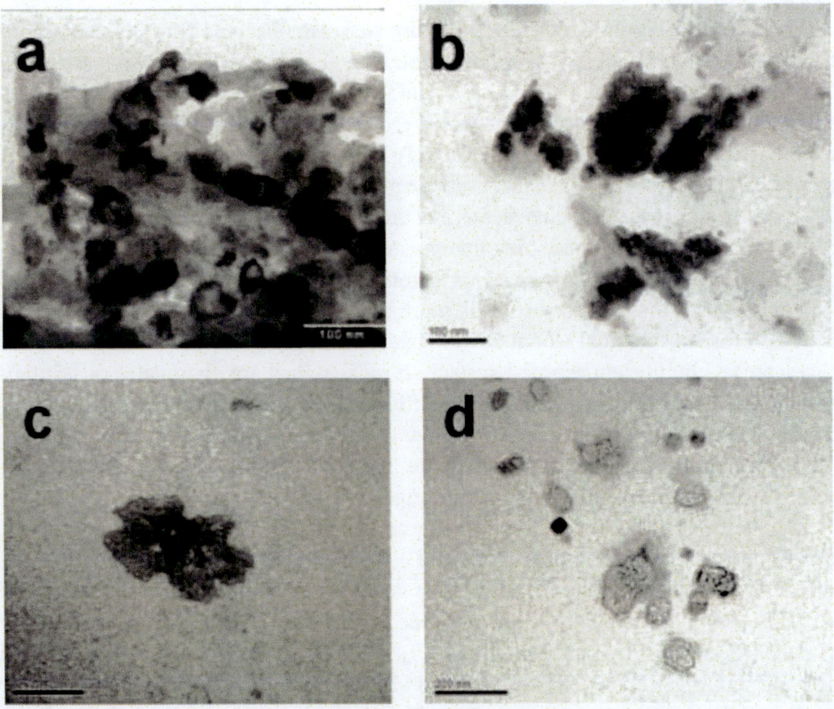

Fig. 1 TEM images of hydrated zirconium dioxide (**a**) and its composite with GO (**b**–**d**). GO nanosheets coat inorganic nanoparticles (**b**, **c**), or in aggregated form (**d**). Images (**c**, **d**) are adapted from [15]

hydrated zirconium oxide (Fig. 1b, c). Thus, isolated aggregates are observed in TEM images. Since the GO nanosheets are distributed on the particle surface, oxide nanoparticles cannot be recognized clearly. However, it is possible to suggest that their size is smaller in comparison with pure zirconium oxide. When the content of GO is rather small, the composite has been found to contain some regions, which are free from GO. Since GO tends to aggregation, its aggregates are also present in the composite. They are observed as traces. Carbon is chemically transformed under the action of accelerated electrons of microscope. As a result, degradation of GO structure occurs. Then carbon is reprecipitated on copper substrate. Nanosized bubbles are formed in this manner (they are visible as white spots).

Heterogeneity of the composite can be also caused by neutrality of zirconium-containing complexes ($[Zr_4(OH)_{16}(H_2O)_8]_n$ [35]). Moreover, sol is strongly acidic medium, where functional groups of GO are not dissociated. It means there is no electrostatic interaction of the complexes with GO surface. Regarding weakly charged particles, such as quartz sand, their interaction with GO is sufficient at $pH > 4$ [36]. However, the support nanoparticles can be easily obtained, namely, from sol.

The GO-containing composite based on hydrated zirconium dioxide is effective for adsorption of toxic Pb^{2+} ions as well as organic compounds, such as phenol and lactose.

The effect of loosening of oxide structure was also observed for rGO-containing composite that was used for adsorption of elemental mercury [37]. Manganese oxide (Mn_xO_y), which contained Mn_2O_3 and MnO_2 phases, was used as a support. It was obtained by precipitation from an $Mn(NO_3)_2$ solution. During synthesis, Mn^{2+} ions were bound with O atoms of the negatively charged oxygen-containing functional groups of the GO sheets via electrostatic forces. Further Mn_xO_y nanoparticles were deposited in situ on the GO nanosheets, and then GO was reduced with a hydrothermal method. Thus, the rGO-containing composite was obtained via GO.

TEM image shows that pure Mn_xO_y consists of elongated rod-like formations (the diameter of their cross section is more than 100 nm) and particles, a shape of which is close to spherical (about 500 nm and even larger). The globular particles are characterized by uneven serrated surface. It means, they are aggregates of primary particles, which cannot be recognized exactly. Smaller Mn_xO_y particles (10–20 nm) have been found for the composite; they are clearly visible due to weak aggregation. As suggested, most of the rGO sheets are embedded in the Mn_xO_y. However, no confirmation of this suggestion is given (Fig. 2).

Magnetic properties of GO-containing composite adsorbents are achieved by a combination of the carbon materials with Fe_3O_4 [38–44]. The materials are obtained using mainly hydrothermal synthesis. This is necessary to provide transformation of Fe(II)/Fe(III) hydroxides to magnetic oxide and to avoid GO degradation. In order to enhance adsorption properties of Fe_3O_4, additional functionalization is carried out: particularly EDTA [38] or amines [44] are used for this purpose. Different types of functionalizing agents are considered in [40]. For obtaining rather large beads, magnetic GO nanoparticles are supported, for instance, by β-cyclodextrin/poly (L-glutamic acid) [41] or chitosan [43].

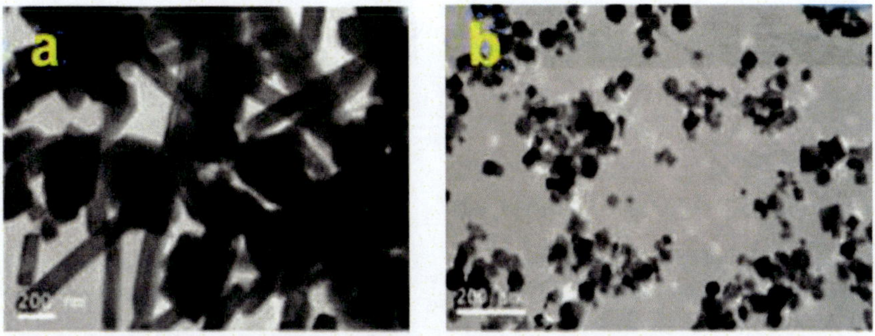

Fig. 2 TEM images of Mn_xO_y (**a**) and its composite with rGO (**b**). Adapted from [37]

As opposed to non-magnetic oxides, GO affects no sufficient disaggregation of primary magnetic particles of Fe_3O_4 (Fig. 3) evidently due to their attraction. Compared with pure Fe_3O_4, the image of the composite is less contrast due to the adsorption of nonconducting nanosheets on its surface.

GO-containing composites based on hydrated oxides of titanium [45] and aluminum [46], double iron–aluminum oxide [47], and silica [48, 49] are known as effective adsorbents. Silica is also a constituent of multicomponent composites [43]. Natural oxide materials, such as bentonite [50] or kaolin [51], are also used as a support. Except oxides, metal–organic frameworks [52], heteropolyacids [53] as well as amorphous [54] or crystalline [55] zirconium hydrophosphate are also the components of composite adsorbents. In general, phosphates of multivalent metals are applied to modifying polymer ion exchange membranes for electrodialysis [27] and low-temperature fuel cells [56]. They can be also used as fillers of filtration polymer membranes [28, 29, 57].

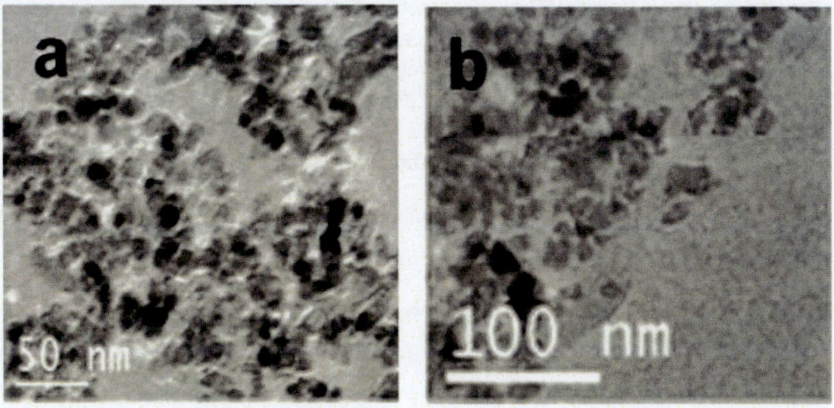

Fig. 3 TEM image of Fe_3O_4 and Fe_3O_4-GO composite (Adapted from [44])

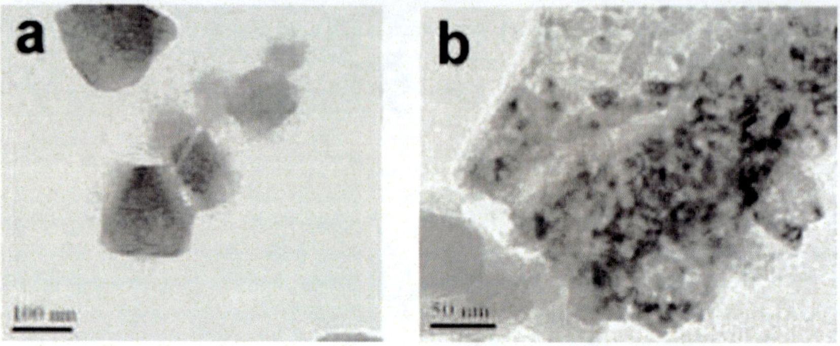

Fig. 4 TEM image of metal–organic framework (**a**) and its GO-containing composites (**b**). Adapted from [52]

In the case of composites based on metal–organic framework, their primary particles are larger compared with hydrated zirconium oxide [52] (Fig. 4a, see also Fig. 1a). In general, a size of these particles exceeds 100 nm; they cannot be related to nanoobjects. Adsorption of GO nanosheets on the surface of supporting particles has been found (Fig. 4b). The formation, a size of which is much larger compared with primary particles, is observed. Probably, it is possible to say about bonding effect of GO due to the interaction of its hydrophobic regions with organic constituents of the metal–organic framework.

When GO dominates in the composite containing inorganic ion exchanger, the last component can be considered as a modifier. In this case, GO performs a function of support. Figure 5a shows an SEM image, which shows agglomerated, overlapped, and curled sheets of GO [54]. The nanoparticles of double oxide of iron–aluminum are shown in Fig. 5b [47]. At the same time, GO forms large agglomerate with inorganic nanoparticles, when the content of this carbon constituent is 1 g per 0.2 g of metal. As found from XRD measurements of the composite, characteristic peaks for both carbon and inorganic phase disappear. When zirconium hydrophosphate is incorporated to GO, its surface becomes more wrinkly (Fig. 5d) [54]. The XRD pattern demonstrates a new broad peak at $2\theta \approx 34°$ indicating distortion of the crystal structure of GO due to the formation of $Zr - P$ moieties.

Structure transformation of the constituents undoubtedly affects their porosity and hydrophilic–hydrophobic properties. This problem is considered further.

3 Hydrophilic and Hydrophobic Pores of Graphene and Its Composites

As a rule, hydrophilicity of materials is estimated from the measurements of wetting angle. The method is suitable only for investigations of outer surface of monoliths. In order to determine hydrophilicity–hydrophobicity of pores, the method of standard

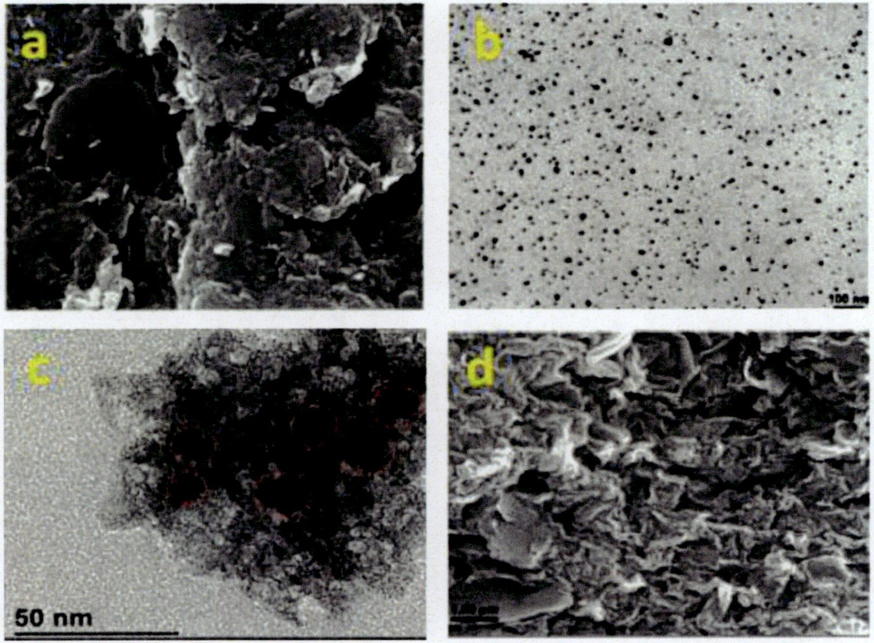

Fig. 5 SEM (a, d, adapted from [54]) and TEM (b, c, adapted from [47]) images of GO **a**, double iron–aluminum oxide **b**, composite based on GO (**c**, **d**) containing double oxide (**c**), and zirconium hydrophosphate (**d**). The regions, where the particles of double oxide are located, are pointed (**c**)

contact porosimetry (MSCP) is recommended. This technique is recognized by the IUPAC [58]. The MSCP gives a possibility to obtain pore size distributions in a very wide interval ($r = 1$ nm-100 μm).

Gravimetric measurements of the content of working liquid (water as well as octane, which is ideally wetting medium) in the sample are performed. Similar procedure is provided for porous standards, which are attached to the tested sample. Only the data for the state of capillary equilibrium are considered. The liquid amount in the system of standards and tested sample is varied by impregnation and drying. After the achievement of thermodynamic equilibrium, the liquid in contacting porous samples is characterized by the same chemical potential. The pore size distributions for the tested sample are plotted taking the known distribution for the standards into consideration. Theoretical approaches and practical details of the MSCP are given in [59–62].

Both GO and rGO are characterized by high specific surface area: the data of 560$-$900 $m^2\ g^{-1}$ have been reported [63, 64]. These results were obtained using BET measurements, which involve adsorption and desorption of nitrogen. It should be stressed that the mentioned data are lower than the theoretical value for completely isolated graphene sheets ($\approx$2600 $m^2\ g^{-1}$ [12]). Under low temperature, nitrogen molecules cannot penetrate between agglomerated, curled, and overlapped graphene sheets. The MSCP gives the values of 2000$-$2400 $m^2\ g^{-1}$ due to high disjoining pressure of octane [65], though its molecules are larger compared with nitrogen. Octane completely wets graphene penetrating between sheets. Similar behavior has been found for graphene in water medium.

Different behaviors of GO and rGO have been found as shown from the integral pore size distributions (Fig. 6) [66]. Here, rGO was obtained by the reduction of GO in a microwave oven. Each build-up of the curves corresponds to one or other type of pores. The curves are attributed to dependencies of pore volume (V) on logarithm of effective pore radius (r^*) [60–62]. The r^* term is applied to the materials containing both hydrophilic and hydrophobic pores. This magnitude is calculated via

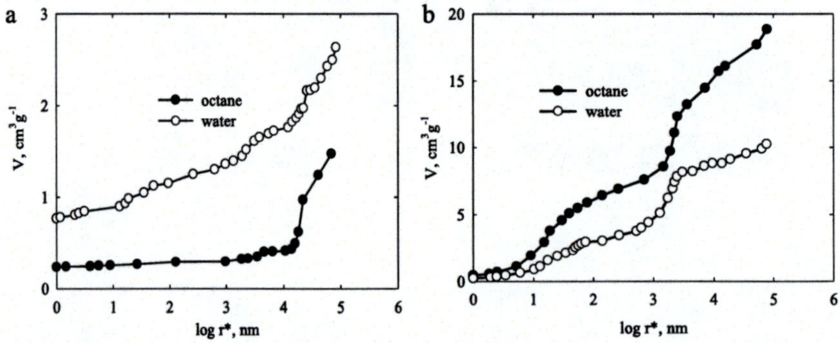

Fig. 6 Typical integral pore size distributions for GO (**a**) and rGO (**b**). Adapted from [66]

$$r^* = \frac{r}{\cos\theta}, \tag{1}$$

where θ is the wetting angle. Both hydrophilic and hydrophobic pores are determined with ideally wetting octane ($\cos\theta = 1, r^* = r$, where r is the true pore radius). Water is used for the recognition of hydrophilic pores. For completely hydrophilic materials, $r^* = r$. In the case of hydrophilic–hydrophobic solid, the $V - r^*$ distribution is shifted toward larger r^* values relatively to the curve obtained in octane. Indeed, this shift is observed for rGO. This is typical for most materials.

As opposed to rGO, higher porosity in water medium has been found for GO. This "superhydrophilicity" is caused for the hydration of its functional groups. From the formal point of view, $\cos\theta < 0$. It is similar for polymer ion exchange membranes [67] or carbon paper for gas diffusion layer of fuel cells [68]. The same results were obtained in [15] for GO obtained according to modified Hummer's method. The volume of hydrophilic pores (determined with water) is larger compared with voids, which are recognized with octane, only in the region of $\log r^* = 0-3.5$ (nm). Aerogel of rGO, which was obtained by reduction of GO using a mixture of H_3PO_2 and I_2, possesses higher hydrophilic porosity at $\log r^* = 0-2$ (nm) [69]. It means incomplete reduction of oxygen-containing functional groups of GO, when the chemical method is applied. Microwave processing allows one to reach more complete reduction (see Fig. 6b).

In order to estimate hydrophilicity–hydrophobicity of different pores, the distribution of wetting angle has been plotted for rGO (Fig. 7). Hydrophobicity is enhanced with increase in pore size. Macropores make main contributions to hydrophobic porosity.

Hydrophilic–hydrophobic properties of graphene materials were investigated also in [70, 71].

Even small amount of GO affects porous structure of the composite based on hydrated zirconium dioxide (Fig. 8) [15]. The inorganic support is characterized by developed micro- and macroporosity. The pores, which correspond to $\log r^* = 0-3$ (nm), are practically absent. The composite is characterized by porosity at $r^* = 3$ nm-1.5 μm. The pore size distributions, which were measured in water and octane, are close to each other due to the effect of hydrophilic inorganic adsorbent.

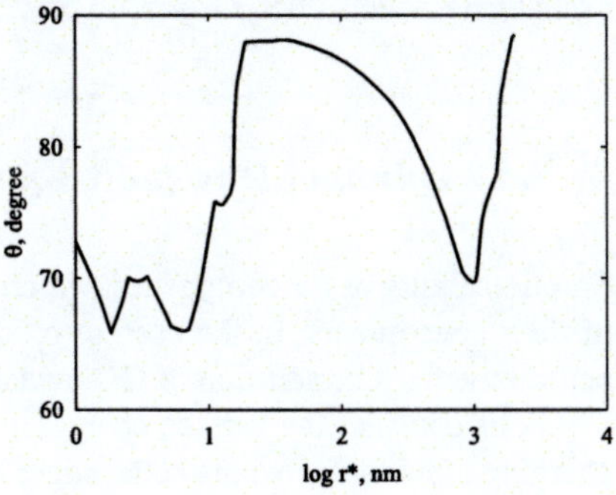

Fig. 7 Typical wetting angle distribution for rGO. Adapted from [66]

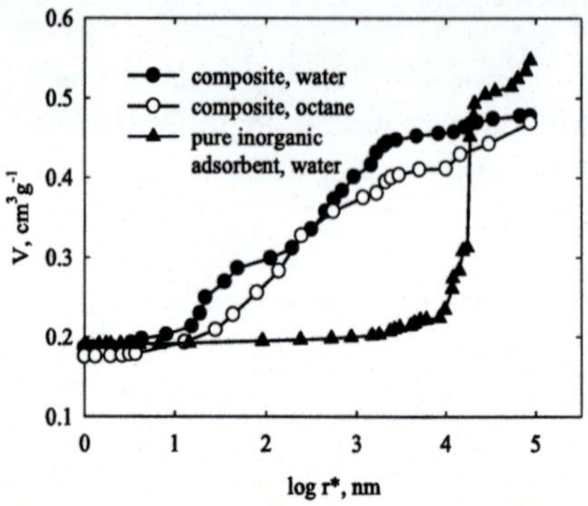

Fig. 8 Integral pore size distribution for hydrated zirconium oxide and its GO-containing composite (Adapted from [15])

Fig. 9 Schematic image of GO-containing composite based on hydrated zirconium oxide (Adapted from [15])

The volume of hydrophilic pores is larger compared with the value obtained in octane. Fast build-up of the integral curve at 20 nm (water) and slow increase at 50 nm (octane) nm are due to GO. The region of $r^* > 100$ nm corresponds to the aggregates of the nanoparticles of hydrated zirconium oxide. However, the volume of "water" pores is larger compared with "octane" voids. This shows coverage of the particle surface with GO sheets. The adsorbed sheets of the carbon material prevent compaction of the inorganic particles through Zr–O–Zr bonds (Fig. 9).

Hydrophilicity–hydrophobicity of the composites based on inorganic ion exchangers affects their adsorption properties. This problem is considered further.

4 Adsorption of Inorganic Ions

Sorption ability of the composites, which include GO, depends on the surface charge of their constituents. In the case of oxides, their point of zero charge (PZC) is in neutral media. The addition of GO evidently affects the PZC.

Let us consider the surface charge of GO. When pure GO is dispersed in water, charging of its sheets occurs at the edges by any surface sites or chemical groups that can undergo dissociation or are capable of ion adsorption [72]. Possible charging mechanisms of the GO surface are as follows:

(i) deprotonation of carboxylic groups:

$$-COOH + H_20 \rightarrow -COO^- + H_3O^+, \quad (2)$$

(ii) deprotonation of phenolic groups

$$-OH + H_20 \rightarrow -O^- + H_3O^+, \quad (3)$$

(iii) proton complexation of the π-electron system of graphene planes acting as Lewis basic sites:

$$C_\pi + 2H_2O \rightarrow C_\pi H_3O^+ + OH^-. \quad (4)$$

At last, (iv) protonation of various Bronsted basic oxygen species (epoxy and carbonyl groups) is also possible. Dissociation of functional groups (schemes (2) and (3)) provides cation adsorption. Protonation of them and π-electron system results in anion adsorption. As suggested in [72], namely, dissociation determines the surface charge of GO. Indeed, zeta potential of GO particles is negative in a wide range of the solution pH indicating the negative surface charge (Fig. 10) [73]. At the same time, protonation causes positive charge of rGO at low pH. Increase in the solution pH causes a shift of the zeta potential to the negative region due to enhancing dissociation of functional groups. The observed increase in ζ-potential in alkaline region is due to the compression of the double layer under high ionic strengths. Water dispersion of graphene is instable under these conditions. GO suspension is characterized by better stability compared with rGO, namely, due to higher zeta potential. The effect of different ions on the stability of GO suspension is reported in [74].

O has been found to shift the PZC of hydrated zirconium oxide from 7.5 to 7 [75], and GO provides a shift from 7 to 5 [23]. The shift evidently depends on the amount of this carbon material, which affects the adsorption properties of the composite.

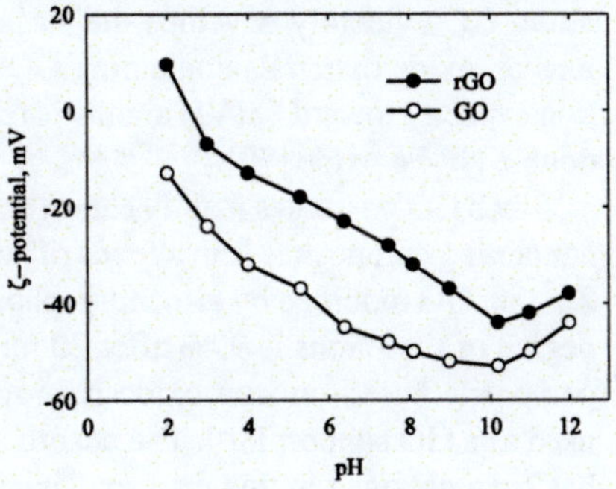

Fig. 10 Zeta potential of GO and rGO as a function of the solution pH (Adapted from [73])

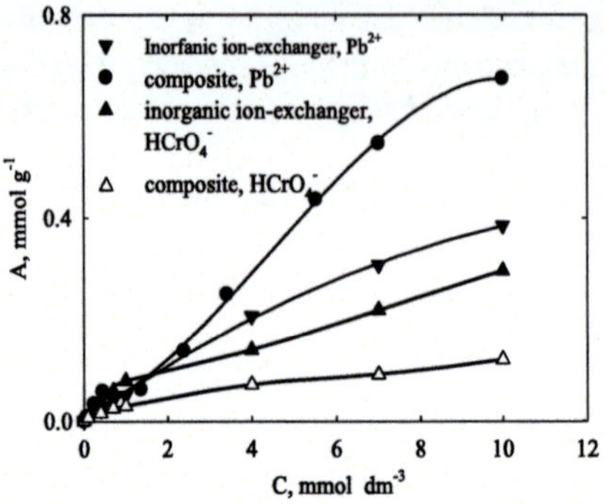

Fig. 11 Adsorption isotherms of toxic ions. The composite was based on hydrated zirconium oxide. Adapted from [15]

For example, the isotherm of Pb^{2+} adsorption on hydrated zirconium oxide is characterized by a build-up of adsorption capacity (*A*) in the region of low equilibrium concentrations (*C*) (Fig. 11) [15]. Regarding the GO-containing composite, the isotherm shows a faster growth. The curve tends plateau formation under high concentration. Adsorption capacity of the composite is higher in 1.7 times compared with pure oxide evidently due to GO contribution. Nevertheless, GO in the composite depresses $HCrO_4^-$ adsorption; this is evidently a result of screening adsorption centers on the HZD surface. No adsorption of these anions on GO has been found, since anion exchange ability is inhibited in neutral medium.

However, adsorption of H_2S [23] and SO_2 [24] on the composite based on hydrated zirconium oxide has been reported. At the same time, the material consisting of this inorganic ion exchanger and rGO gives a possibility of simultaneous adsorption of As(III) and As(V) species [75], despite no dissociation of H_3AsO_3 in a wide interval of the solution pH. The composite also contained carbon nanotubes. This three-component material allowed one to achieve high removal degree at the pH of 5–12 (As(III)) and 3–10 (As(V)). Adsorption ability of the composite based on hydrated iron oxide toward arsenate anions [76, 77] and arsenous (III) acid [47] is also suggested. Simultaneous adsorption of these compounds using magnetic materials, which contain GO or rGO, was investigated in [78]. The authors explain high adsorption of anions by the formation of inner sphere complexes with iron or zirconium. GO evidently develops the surface of the oxide support. Adsorption of F^- ions on oxide materials containing GO was considered in [46]. Significant adsorption capacity toward Cr(VI) anions is reached in the case of the composite based on double Ni–Fe oxide [79].

Since GO possesses mainly cation exchange ability, adsorption of cations on GO-containing composites is in a focus of attention. Adsorption of Pb^{2+}, Cd^{2+}, Cu^{2+}, and Zn^{2+} on GO modified by zirconium phosphate was investigated in [54]. The removal degree of these ions is 99% after 20 min. Effective removal of Hg^{2+} ions using the composite based on iron oxide is proved in [38, 39]. Manganese oxide was also used as a GO support [80]. The adsorbent, which is selective toward UO_2^{2+} cations, has been obtained in this manner. Titanium oxide [45] and bentonite [50] were also used as a support to prepare the composite for removal of U(VI) compounds from water. Silica-grafted GO demonstrates selectivity toward Pb^{2+} [48]. The materials containing heteropolyacids can be recommended for recovery of NH_4^+ ions [53].

The information regarding adsorption of heavy metal ions on GO and its composites is summarized in [40, 74].

5 Adsorption of Organic Species

Hydrophobic regions of GO sheets provide adsorption of organic compounds on the composites containing this carbon material. Despite depressed anion exchange properties, the composite adsorbs phenol that is a weak acids [15]. When the initial concentration of this compound is 5 mg dm^{-3}, the composite based on hydrated zirconium oxide removes it down to maximal allowable concentration for tap water (1 μg dm^{-3}). In the case of pure inorganic ion exchanger, adsorption of phenol is inconsiderable. Adsorption of p-cresol and p-tert-butylphenol on the composite containing manganese oxide is accompanied by their oxidation [81]. The support, which contains variable valence metal, is more effective compared with MnO_2.

A number of works are devoted to adsorption of cationic dyes on composites, for instance, Methylene Blue [39, 51, 82–84] or Malachite Green [44]. Adsorption of anionic dyes, such as Congo Red [79, 85], Methyl Orange [79], and Eriochrome Black T [44] on oxide-based materials, was also studied. It is stressed that electrostatic attraction is dominating adsorption mechanism, which is strongly dependent on the solution pH. When photocatalytic materials are used as a support, the dye adsorption is accompanied by degradation [82]. Good photocatalytic performance of the TiO_2–GO composite could be attributed to a synergy effect including the increase in specific surface area with GO amount as well as the formation of both π–π conjugations between dye molecules and aromatic rings, and the interactions between the molecule and functional groups of GO. This carbon material works as the adsorbent, electron acceptor, and photosensitizer to efficiently enhance the dye photodecomposition.

The composites are also effective toward molecular substances: lactose [15], aromatic compounds [49], such antibiotic as tetracycline [42, 43], and hormones [41]. In the case of lactose removal from water using the composite based on hydrated zirconium oxide, it is possible to reach the COD value that is lower than the maximal allowable concentration for tap water (5 mg dm^{-3}) [15]. Conversely, pure zirconium oxide shows no lactose adsorption. The material based on metal–organic framework has been proposed for acetone adsorption [52].

The GO-Fe_3O_4 nanocomposite has been tested against Gram-positive and Gram-negative bacterial strains including *Staphylococcus* aureus, *Bacillus* subtilis, *Pseudomonas* aerugino, and *Salmonella* typhimuriumupon [44]. As shown, the composite demonstrates promising antimicrobial activity.

6 Conclusions

The composites containing inorganic ion exchangers and GO are promising multifunctional materials: they are able to adsorb both inorganic ionic species and organic ions and molecules. This is due to the combination of hydrophilic and hydrophobic properties of GO. As found by means of MSCP, GO demonstrates "superhydrophilicity" similar to ion exchange polymers: larger porosity is achieved in water than that in ideally wetting octane. Except ion exchange ability, the inorganic support provides large size of granule. Thus, it is possible to use the composites for the filling of adsorption columns. A number of GO-containing adsorbents possess magnetic properties. It means that they can be easily separated from liquids and solids.

References

1. Xu M, Liang T, Shi M (2013) Graphene-like two-dimensional materials. Chem Rev 113:3766–3798
2. Anasori B, Lukatskaya MR, Gogotsi Y (2017) 2D metal carbides and nitrides (MXenes) for energy storage. Nat Rev Mater 2:16098. https://doi.org/10.1038/natrevmats.2016.98
3. Geim AK, Novoselov KS (2007) The rise of graphene. Nat Mater 6(3):183–191
4. Geim AK (2009) Graphene: status and prospects. Science 324(5934):1530–1534
5. Zhao J, Liu L, Li F (2015) Graphene oxide: physics and applications. Springer, Heidelberg
6. Stobinski L, Lesiaka B. Malolepszy A et al (2014) Graphene oxide and reduced graphene oxide studied by the XRD, TEM and electron spectroscopy methods. J Electr Spectr Rel Phen 195:145–154
7. Brodie BC (1860) Sur le poids atomique du graphite. Ann Chim Phys 59:466–472
8. Staudenmaier L (1898) Verfahren zur Darstellung der Graphitsaure. Ber Deut Chem Ges 31:1481–1487
9. Hummers WS, Offeman RE (1958) Preparation of graphitic oxide. J Am Chem Soc 80:1339–1339
10. Dimiev AM, Khannanov A, Vakhitov I et al (2018) Revisiting the mechanism of oxidative unzipping of multiwall carbon nanotubes to graphene nanoribbons. ACS Nano 12(4):3985–3993
11. Higginbotham AL, Kosynkin DV, Sinitskii A (2010) Lower-defect graphene oxide nanoribbons from multiwalled carbon nanotubes. ACS Nano 4(4):2059–2069
12. Dreyer DR, Park S, Bielawski CW et al (2010) The chemistry of graphene oxide. Chem Soc Rev 39(1):228–240
13. Li Z, Chen F, Yuan L et al (2012) Uranium(VI) adsorption on graphene oxide nanosheets from aqueous solutions. Chem Eng J 210:539–546
14. Mi X, Huang G, Xie W et al (2012) Preparation of graphene oxide aerogel and its adsorption for Cu^{2+} ions. Carbon 50(13):4856–4864
15. Dzyazko YS, Ogenko VM, Volfkovich YM et al (2018) Composite consisting of hydrated zirconium dioxide and grapheme oxide for removal of organic and inorganic components from water. Chem Phys Technol Surf 9(4):417–431
16. Konicki W, Aleksandrzak M, Moszyński D et al (2017) Adsorption of anionic azo-dyes from aqueous solutions onto graphene oxide: Equilibrium, kinetic and thermodynamic studies. J Colloid Interface Sci 496:188–200
17. Wang H, Chen P (2015) Adsorption and coadsorption of organic pollutants and a heavy metal by graphene oxide and reduced graphene materials. Chem Eng J 281:379–388

18. Chen X, Chen B (2015) Microscopic and spectroscopic investigations of the adsorption of nitroaromatic compounds on graphene oxide, reduced graphene oxide, and graphene nanosheets. Environ Sci Technol 49(10):6181–6189
19. Lu C, Huang P-JJ, Liu B et al (2016) Comparison of graphene oxide and reduced graphene oxide for DNA adsorption and sensing. Langmuir 32(41):10776–10783
20. Jiang L-h, Liu Y-g, Zeng G-m et al (2016) Removal of 17β-estradiol by few-layered graphene oxide nanosheets from aqueous solutions: External influence and adsorption mechanism. Chem Eng J 284:93–102
21. Amphlett CB (1964) Inorganic ion exchangers. Elsevier, Amsterdam
22. Dzyazko YS, Volfkovich YM, Sosenkin VE et al (2014) Composite inorganic membranes containing nanoparticles of hydrated zirconium dioxide for electrodialytic separation. Nanoscale Res Let 9(1):271. https://doi.org/10.1186/1556-276X-9-27
23. Seredych M, Bandosz TJ (2011) Reactive adsorption of hydrogen sulfide on graphite oxide/Zr(OH)4 composites. Chem Eng J 166:1032–1038
24. Seredych M, Bandosz TJ (2010) Effects of surface features on adsorption of SO_2 on graphite oxide/$Zr(OH)_4$ Composites. J Phys Chem 114(34):14552–14560
25. Dzyazko Y, Rozhdestveskaya L, Zmievskii Y et al (2019) Composite inorganic anion exchange membrane for electrodialytic desalination of milky whey. Mater Today Proc 6(2):250–259
26. Myronchuk V, Zmievskii Y, Dzyazko Y et al (2018) Whey desalination using polymer and inorganic membranes: Operation conditions. Acta Periodica Technologica 49:103–116
27. Dzyazko Y, Rozhdestveskaya L, Zmievskii Y (2015) Heterogeneous membranes modified with nanoparticles of inorganic ion-exchangers for whey demineralization. Mater Today: Proc 2(6):3864–3873
28. Myronchuk VG, Dzyazko YS, Zmievskii YG et al (2016) Organic-inorganic membranes for filtration of corn distillery. Acta Periodica Technologica 47:153–165
29. Zmievskii Y, Rozhdestvenska L, Dzyazko Y et al (2017) Organic-inorganic materials for baromembrane separation. Springer Proc Phys 195:675
30. Dzyazko Y, Rozhdestveskaya L, Zmievskii Y et al (2019) Composite inorganic anion exchange membrane for electrodialytic desalination of milky whey. Mater Today Proc 6(2):250–259
31. Dzyazko Y, Kolomyets E, Borysenko Y et al (2019) Organic-inorganic sorbents containing hydrated zirconium dioxide for removal of chromate anions from diluted solutions. Mater Today Proc 6(2):260–269
32. Perlova O, Dzyazko Yu, Halutska I et al (2018) Anion exchange resin modified with nanoparticles of hydrated zirconium dioxide for sorption of soluble U(VI) compounds. Springer Proc Phys 210:3–15
33. Maltseva TV, Kolomiets EO, Dzyazko YS et al (2019) Composite anion-exchangers modified with nanoparticles of hydrated oxides of multivalent metals. Appl Nanosci 9(5):997–1004
34. Cumbal L, SenGupta AK (2005) Arsenic removal using polymer-supported hydrated iron (III) oxide nanoparticles: role of Donnan membrane effect. Environ Sci Technol 39(17):6508–6515
35. Avvakumov GV, Senna M, Kosova NV (2001) Soft mechanochemical synthesis: a basis for new chemical technologies. Kluwer, Amsterdam
36. Sotirelis NP, Chrysikopoulos CV (2015) Interaction between graphene oxide nanoparticles and quartz sand. Environ Sci nTechnol 49:13413–13421
37. Xu H, Qu Z, Zong Ch et al (2015) MnO_x/graphene for the catalytic oxidation and adsorption of elemental mercury. Environ Sci Technol 49:6823–6830
38. Cui L, Wang Y, Gao L et al (2015) EDTA functionalized magnetic graphene oxide for removal of Pb(II), Hg(II) and Cu(II) in water treatment: Adsorption mechanism and separation property. Chem Eng J 281:1–10
39. Guo Y, Deng J, Zhu J et al (2016) Removal of mercury(II) and methylene blue from a wastewater environment with magnetic graphene oxide: adsorption kinetics, isotherms and mechanism. RSC Adv 6(86):82523–82536

40. Sherlala AIA, Raman AAA, Bello MM et al (2018) A review of the applications of organo-functionalized magnetic graphene oxide nanocomposites for heavy metal adsorption. Chemosphere 193:1004–1017
41. Jiang L, Liu Y, Liu S et al (2017) Fabrication of β-cyclodextrin/poly (L-glutamic acid) supported magnetic graphene oxide and its adsorption behavior for 17β-estradiol. Chem Eng J 308:597–605
42. Yu B, Bai Y, Ming Z et al (2017) Adsorption behaviors of tetracycline on magnetic graphene oxide sponge. Mater Chem Phys 198:283–290
43. Huang B, Liu Y, Li B et al (2017) Effect of Cu(II) ions on the enhancement of tetracycline adsorption by Fe_3O_4@SiO_2-Chitosan/graphene oxide nanocomposite. Carbohydr Polym 157:576–585
44. Raghu MS, Kumar KY, Prashant MK et al (2017) Adsorption and antimicrobial studies of chemically bonded magnetic graphene oxide-Fe_3O_4 nanocomposite for water purification. J Water Proc Eng 17:22–31
45. Yu S, Wei D. Shi L et al (2019) Three-dimensional graphene/titanium dioxide composite for enhanced U(VI) capture: Insights from batch experiments, XPS spectroscopy and DFT calculation. Environ Pollut 251:975–983
46. Barathi M, Kumar ASK, Kumar CU et al (2014) Graphene oxide–aluminium oxyhydroxide interaction and its application for the effective adsorption of fluoride. RSC Adv 4:53711–53721
47. Majia S, Ghosha A, Gupta K (2018) Efficiency evaluation of arsenic(III) adsorption of novel graphene oxide@ iron-aluminium oxide composite for the contaminated water purification. Separ Purif Technol 197: 388–400
48. Li X, Wang Z, Li Q et al (2015) Preparation, characterization, and application of mesoporous silica-grafted graphene oxide for highly selective lead adsorption. Chem Eng J 273:630–637
49. Yang K, Chen B, Zhu L (2015) Graphene-coated materials using silica particles as a framework for highly efficient removal of aromatic pollutants in water. Sci Reports 5:11641. https://doi.org/10.1038/srep11641
50. Liu H, Xie S, Liao J et al (2018) Novel graphene oxide/bentonite composite for uranium(VI) adsorption from aqueous solution. J Radioanal Nucl Chem 317(3):1349–1360
51. Kai H, Chen G, Zeng G et al (2018) Enhanced removal performance for methylene blue by kaolin with graphene oxide modification. J Taiwan Inst Chem Eng 89:77–85
52. Zhou X, Huang W, Shi J et al (2014) A novel MOF/graphene oxide composite GrO@MIL-101 with high adsorption capacity for acetone. J Mater Chem A 2(13):4722–4730
53. Petit C, Bandosz TJ (2009) Graphite oxide/polyoxometalate nanocomposites as adsorbents of ammonia. J Phys Chem C 113(9):3800–3809
54. Pourbeyram S (2016) Effective removal of heavy metals from aqueous solutions by graphene oxide–zirconium phosphate (GO–Zr-P) nanocomposite. Ind Eng Chem Res 55(19):5608–5617
55. Zhimin WuZ, Zhang L, Guan Q (2014) Preparation of a -zirconium phosphate-pillared reduced graphene oxide. Chem Eng J 258:77–84
56. Dzyazko Y, Volfkovich Y, Perlova O et al (2019) Effect of porosity on ion transport through polymers and polymer-based composites containing inorganic nanoparticles (review). Springer Proc Phys 222:235–253
57. Dzyazko YS, Rozhdestvenskaya LM, Zmievskii YG et al (2015) Organic-inorganic materials containing nanoparticles of zirconium hydrophosphate for baromembrane separation. Nanoscale Res Lett 10:64. https://doi.org/10.1186/s11671-015-0758-x
58. Rouquerol J, Baron G, Denoyel R et al (2012) 2012) Liquid intrusion and alternative methods for the characterization of macroporous materials (IUPAC Technical Report. Pure Appl Chem 84(1):107–136
59. Kononenko N, Nikonenko V, Grande D et al (2017) Porous structure of ion exchange membranes investigated by various technique. Adv Colloid Interface Sci 246:196–216
60. Volfkovich YM, Filippov AN, Bagotsky VS (2014) Structural properties of different materials and powders used in different fields of science and technology. Springer, Heidelberg

61. Volfkovich YM, Sosenkin VE (2012) Porous structure and wetting of fuel cell components as the factors determining their electrochemical characteristics. Russ Chem Rev 81(10):936–959
62. Volfkovich YM, Sakars AV, Volinsky AA (2005) Application of the standard porosimetry method for nanomaterials. Int J Nanotechnol 2(3):292–302
63. Seresht RJ, Jahanshahi M, Rashidi A et al (2013) Synthesize and characterization of graphene nanosheets with high surface area and nano-porous structure. Appl Surf Sci 276:672–681
64. McAllister MJ, Li J-L, Adamson DH et al (2007) Single sheet functionalized graphene by oxidation and thermal expansion of graphite. Chem Mater 19(18):4396–4404
65. Volfkovich YM, Rychagov AY, Sosenkin VE (2014) Measuring the specific surface area of carbon nanomaterials by different methods. Russ J Electrochem 50(11):1099–1101
66. Shulga YM, Baskakov SA, Baskakova YV et al (2015) Supercapacitors with graphene oxide separators and reduced graphite oxide electrodes. J Power Sources 279:722–730
67. Kononenko NA, Berezina NP, Vol'fkovich YM et al (1985) Investigation of ion-exchange materials structure by standard porosimetry method. J Appl Chem USSR 58(10):2029–2033
68. Volfkovich YM, Sosenkin VE, Nikolskaya NF et al (2008) Porous structure and hydrophilic-hydrophobic properties of gas diffusion layers of the electrodes in proton-exchange membrane fuel cells. Russ J Electrochem 44(3):278–285
69. Volfkovich YM, Lobach AS, Spitsyna NG et al (2019) Hydrophilic and hydrophobic pores in reduced graphene oxide aerogel. J Porous Mat 26(4):1111–1119
70. Shulga YM, Baskakov SA, Baskakova YV et al (2017) Hybrid porous carbon materials derived from composite of humic acid and graphene oxide. Micropor Mesopor Mater 245:24–30
71. Shulga YM, Baskakov SA, Baskakova YV et al (2018) Preparation of graphene oxide-humic acid composite-based ink for printing thin film electrodes for micro-supercapacitors. J Alloy Compd 730:88–95
72. Szabo T, Tombacz E, Illes E et al (2018) Enhanced acidity and pH-dependent surface charge characterization of successively oxidized graphite oxides. Carbon 44:537–545
73. Konkena B, Vasudevan S (2012) Understanding aqueous dispersibility of graphene oxide and reduced graphene oxide through pKa measurements. J Phys Chem Lett 3(7):867–872
74. Peng W, Li H, Liu Y et al (2017) A review on heavy metal ions adsorption from water by graphene oxide and its composites. J Molec Liq 230:496–504
75. Luo X, Wang C, Wang L (2013) Nanocomposites of graphene oxide-hydrated zirconium oxide for simultaneous removal of As(III) and As(V) from water. Chem Eng J 220:98–106
76. Fu D, He Z, Su S et al (2017) Fabrication of a-FeOOH decorated graphene oxide-carbon nanotubes aerogel and its application in adsorption of arsenic species. J Colloid Interface Sci 505:105–114
77. Su H, Ye Z, Hmidi N ((2017) High-performance iron oxide–graphene oxide nanocompositeadsorbents for arsenic removal. Colloids Surf A: Physicochem Eng Aspects 522:161–172
78. Yoon Y, Park WK, Hwang T-M et al (2016) Comparative evaluation of magnetite–graphene oxide and magnetite-reduced graphene oxide composite for As(III) and As(V)removal. J Hazard Mater 304:196–204
79. Zheng Y, Cheng B, You W et al (2019) 3D hierarchical graphene oxide-NiFe LDH composite with enhanced adsorption affinity to Congo red, methyl orange and Cr(VI) ions. J Hazard Mater 369:214–225
80. Yang A, Zhu Y, Huang CP (2018) Facile preparation and adsorption performance of graphene oxide-manganese oxide composite for uranium. Sci Rep 8:90584. https://doi.org/10.1038/s41598-018-27111-y
81. Zhang B, Zhao R, Sun D et al (2019) Sustainable fabrication of graphene oxide/manganese oxide composites for removing phenolic compounds by adsorption-oxidation process. J Cleaner Prod 215:165–174
82. Nguyen-Phan T-D, Pham VH, Shin FW et al (2011) The role of graphene oxide content on the adsorption-enhanced photocatalysis of titanium dioxide/graphene oxide composites. Chem Eng J 170(1):226–232
83. Fan L, Luo C, Li X et al (2012) Fabrication of novel magnetic chitosan grafted with graphene oxide to enhance adsorption properties for methyl blue. J Hazard Mater 215–215:272

84. Haque S, Gain S, Gupta K et al (2019) Methylene blue (a cationic dye) adsorption performance of graphene oxide fabricated Fe-Al bimetal oxide composite from water. Water Qual Res J (2018) 54 (1):57–69
85. Xu J, Xu D, Zhu B et al (2018) Adsorptive removal of an anionic dye Congo red by flower-like hierarchical magnesium oxide (MgO)-graphene oxide composite microspheres. Appl Surf Sci 435:1136–1142

Polyacrylamide Soil Conditioners: The Impact on Nanostructured Clay Minerals' Aggregation and Heavy Metals' Circulation in the Soil Environment

Gracja Fijałkowska, Katarzyna Szewczuk-Karpisz, and Małgorzata Wiśniewska

1 Polyacrylamide Structure and Preparation

Polyacrylamides (PAMs) are group of water-soluble, synthetic polyelectrolytes derived from acrylamide monomer in polymerization process [1–3]. They are substances which dissolve, disperse, or swell in water and thus can modify physical properties of aqueous system due to thickening, gelation, emulsification, or stabilization processes. Polyacrylamide chains consist of repeating units or blocks of units that form the structure containing hydrophilic groups acting as substituents or incorporated into the macromolecule backbone. Polyacrylamides can be classified into nonionic, anionic, and cationic ones [4, 5]. Each of these types due to their properties is used in many industries, mainly in water treatment, mineral processing, or soil conditioning [5]. Application areas of abovementioned polyacrylamide types depend on the specific function of the polymer needed. The adsorption process of macromolecular compounds such as PAMs on the suspended matter particles governs the individual polymer performance. These polyelectrolytes can interact through specific or non-specific type forces with various charged substrates. The type of interaction between the polymer macromolecule and a substrate depends on the nature and polarity of the polymer itself and the surface on which it adsorbs.

Nonionic polyacrylamide has a strong hydrophilic character, greater than other nonionic, water-soluble polymers. Despite being described as nonionic synthetic

G. Fijałkowska (✉) · M. Wiśniewska
Faculty of Chemistry, Department of Radiochemistry and Environmental Chemistry, Institute of Chemical Sciences, Maria Curie-Sklodowska University in Lublin, M. Curie-Sklodowska Sq. 3, 20-031 Lublin, Poland
e-mail: gracja.fijalkowska@poczta.umcs.lublin.pl

K. Szewczuk-Karpisz
Institute of Agrophysics, Polish Academy of Sciences, Doświadczalna 4, 20-290 Lublin, Poland

O. Fesenko and L. Yatsenko (eds.), *Nanomaterials and Nanocomposites, Nanostructure Surfaces, and Their Applications*, Springer Proceedings in Physics 246, https://doi.org/10.1007/978-3-030-51905-6_9

polyacrylamides obtained as a result of polymerization or copolymerization of acrylamides, often contain in their macromolecules about 1–3% of anionic groups [4]. These groups are formed as a result of monomer amide groups hydrolysis. In order to obtain a polyacrylamide with a lower ionic group content, factors such as temperature, pH, monomer concentration, and type of polymerization initiator should be manipulated. Nonionic polyacrylamide can be produced from acrylamide monomer by free radical polymerization using various initiation methods. This monomer is produced by the catalytic hydration of acrylonitrile. It polymerizes with a wide range of free radical initiators but often potassium persulfate or hydrogen peroxide must be used because they allow reaction in a convenient temperature range (from 40 to 67 °C). As a result, a high-molecular-weight polymer without contaminations is obtained, which can be reduced by incorporating methanol into the reaction medium (acts as a so-called chain transfer agent):

$$\underset{\text{Acrylonitrile}}{CH_2 = CH - CN} \xrightarrow[\text{catalyst}]{H_2O} \underset{\text{Acrylamide}}{CH_2 = CH - CONH_2} \tag{1}$$

$$\underset{\text{Acrylamide}}{CH_2 = CH - CONH_2} \xrightarrow{\text{initiator}} \underset{\text{Nonionic polyacrylamide}}{-(-CH_2 - \underset{CONH_2}{\underset{|}{CH_2}} -)-} \tag{2}$$

Anionic polyacrylamide is polymer with a large number of negatively charged moieties in the long chains of the PAM macromolecules. It can be obtained by partial hydrolysis of polyacrylamide neutral amide groups, which is shown in (3). This process occurs under moderate temperature by adding sodium hydroxide to the PAM solution [4, 6].

$$\underset{\text{Polyacrylamide (PAM)}}{-(-CH_2 - \underset{CONH_2}{\underset{|}{CH_2}} -)-} \xrightarrow{\text{NaOH}} \underset{\text{Anionic PAM}}{\left(\left(-(-CH_2 - \underset{CONH_2}{\underset{|}{CH}} -)-\right)_n \left(-(-CH_2 - \underset{COO^-}{\underset{|}{CH}} -)-\right)\right)} \tag{3}$$

In the case of cationic polyacrylamide, in which macromolecular chains contain positively charged groups, the basic method of preparation is the copolymerization reaction presented below:

$$\underset{\text{Polyacrylamide (PAM)}}{\left(-(-CH_2 - \underset{CONH_2}{\underset{|}{CH}} -)-\right)} + C_xH_yNCl \xrightarrow{\text{initiator}} \underset{\text{Cationic PAM}}{\left(\left(-(-CH_2 - \underset{CONH_2}{\underset{|}{CH}} -)-\right)_n \left(-(-CH_2 - \underset{COO(CH_2)_2N^+(CH_3)_3}{\underset{|}{CH}} -)-\right)_m\right)} \tag{4}$$

There are four main kinds of cationic polyacrylamide synthesis technologies: in aqueous solution, in dispersion, by inverse emulsion, or through photoinitiated polymerization [7]. These methods had experienced numerous developments and modifications such as free radical copolymerization or grafting processes [8]. Due to these methods, application of PAM with relatively high molecular weight can be obtained but characterized by short duration, high prize, and poor stability. In industry, mostly free radical polymerization method including abovementioned technologies is used. The technological process of aqueous solution polymerization is presented in Fig. 1.

The aqueous solution of cationic polyacrylamide monomer is mixed with N_2, initiator, i.e., 4,4′-azobis-4-cyanovaleric acid, 2,2′-azobis(2-amidinopropane)-dihydrochloride and by adjusting conditions (pH, temperature, time, additives such as stabilizers, e.g., poly(acryloethyl trimethyl ammonium chloride)), the polymerization process and PAM colloids' formation are induced. This technology has many advantages (short time, safety, simplicity of method and equipment, minimal environment pollution) and for this reason is widely used in many branches.

Dispersion polymerization technology (Fig. 2) relies on acrylamide and double bond of quaternary ammonium dissolving in the saline solution such as $(NH_4)_2SO_4$ or NaCl and addition of initiator and stabilizer (i.e., poly(acryloylxyethyltrimethyl ammonium chloride), poly(dimethylaminoethylmethacrylate methyl chloride)) to the system. The result of this process was a discrete form of small polymer particle precipitation. These particles show unique features such as side chain structures with high positive charge density but the obtained product had low concentration and poor stability. For this reason, this method is not usually used for large technological scale in the industrial production [9].

By water solution of cationic monomer and organic solvent mixing which results in the formation of water-in-oil emulsion and addition of initiator, the inverse emulsion polymerization occurs (Fig. 3). However, besides the fact that this method allows getting high-molecular-weight polymers, it has disadvantages, i.e., the emulsification and phase separation are difficult to control [10].

In the photoinitated polymerization method (Fig. 4), the cationic polyacrylamide is obtained by photoactivation of monomer solution. This synthesis technique is characterized by many advantages: operation simplicity, easy control, low cost, and obtained product has high purity [11, 12].

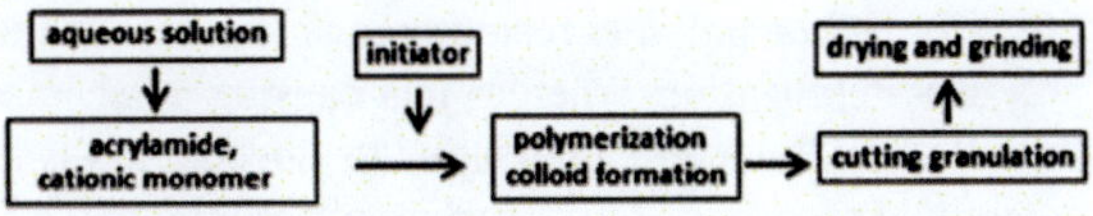

Fig. 1 Technological process of aqueous solution polymerization [7]

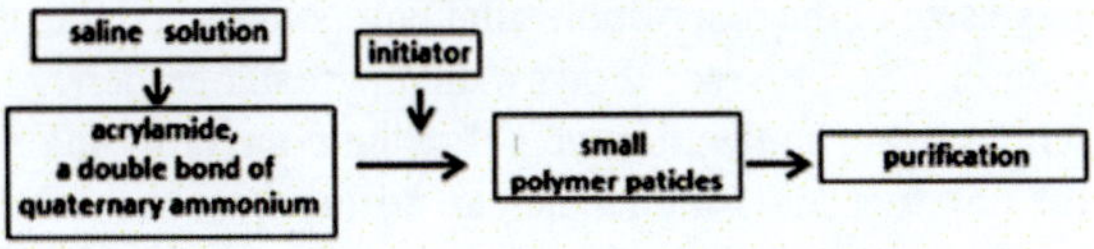

Fig. 2 Technological process of dispersion polymerization [7]

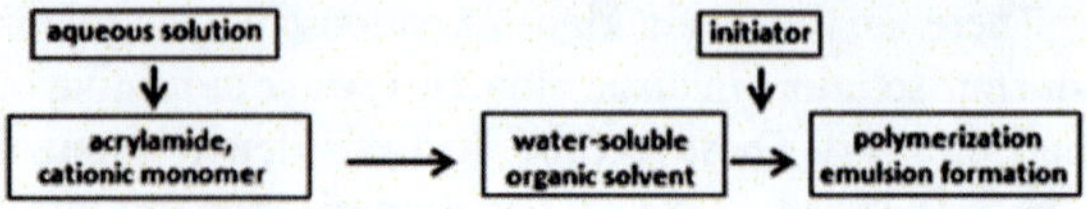

Fig. 3 Technological process of inverse emulsion polymerization [7]

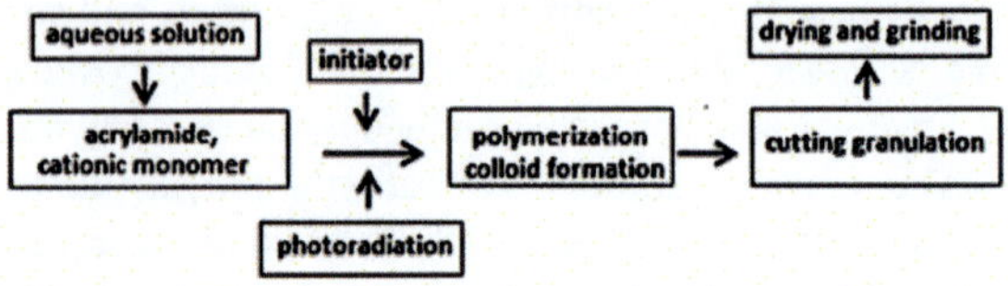

Fig. 4 Technological process of photoinitiated polymerization technological process [7]

Polyacrylamides can also be formed by acrylamide and bis-acrylamide copolymerization initiated by free radicals. This process is vinyl addition polymerization activated by the addition of initiators. The most commonly used radical initiators are ammonium persulfate and N,N,N′,N′-tetramethylethylenediamine (TEMED) or TEMED with riboflavin-5’-phosphate. Addition of TEMED catalyzes and accelerates the free radical formation rate, whereas radicals obtained from ammonium persulfate catalyze polymerization process by reaction with monomers and conversion to acrylamide free radicals. As a result of free radicals with polymer monomers reaction, they are transformed into further free radicals reacting with inactive monomers which eventually starts the polymerization reaction [13–15].

2 PAM Impact on Soil Aggregation

Particulate suspensions, including clay minerals, are used in many industrial processes that result in high-quality products (e.g., ceramics, paper coatings, cosmetics, paints, rubbers, and other plastics). These suspensions are usually highly concentrated and for this reason the maintenance of a system where solid particles are well dispersed is significantly difficult. Thus, the big importance is control of colloidal properties and dispersion stability [15, 16].

The addition of a high molecular compound affects the stability of colloidal suspensions and its rheological properties. Depending on the conditions, the presence of polymer in the system may increase or decrease aggregation stability. The protective effect of the polymer resulting in an increase in the suspension stability is called polymer stabilization, whereas that causing a decrease in this parameter—polymer flocculation. Polymer stabilization or flocculation is the result of a macromolecular compound adsorption on the surface of particles of the dispersed phase or the presence of nonadsorbed polymer macromolecules in the dispersive phase [17]. Due to the large size of the macromolecular compounds and the presence of many active sites in their chains, polymer shows a much greater tendency to adsorption than compounds with a lower molecular weight when contacting the surface of a solid. In addition, the polymer macromolecule can adsorb onto the solid surface in a variety of ways, resulting in a number of conformations such as trains, loops, and tails (Fig. 5) [6, 13, 17–20].

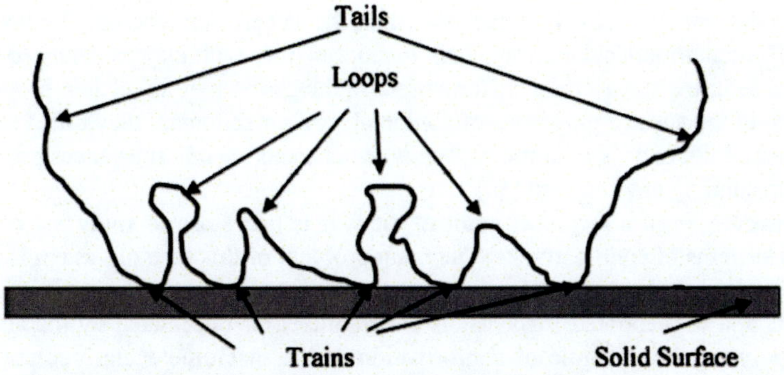

Fig. 5 Conformations of adsorbed polymer on the solid surface [13]

The concentration of the adsorbing polymer affects its ability to stabilize the colloidal system. At high concentrations, the solid surface is completely covered by the polymeric layer. When the polymer-coated particles approach each other, the interpenetration of the polymer layers occurs. Consequently, the conformational entropy of the polymer segments decreases, whereas the free enthalpy increases. As a result, the repulsion (steric stabilization) between the particles with the adsorbed polymer is observed. In the case when the adsorbing polymer is a polyelectrolyte, steric stabilization can be combined with electrostatic stabilization and electrosteric stabilization takes place in the system (also as a result of repulsion of adsorption layers with an identical charge) [6].

The polymer adsorbed on the solid surface can form loops and tails, which means that only a few fragments of polymer chains have direct contact with the solid surface. The addition of some amount of polymer to the system that will not provide complete coverage of the surface contributes to the formation of polymer bridges. This is only possible if the range of electrostatic interaction between the particles is smaller than the length of the polymer loops and tails. As a result of the bridging flocculation process occurs, i.e., formed aggregates size until they are separated from the solution in the form of a precipitate or a turbid suspension. Flocculation can also be a result of the solid charge neutralization by adsorbed polyelectrolyte macromolecules with an opposite charge [21–23].

The phenomenon of stabilization and flocculation of the dispersed systems through the addition of a polymer is widely applied in many industries [6, 21]. Adsorption of high molecular compounds is used mainly in the wastewater treatment [24, 25], in agriculture [26–28], chemical industry (production of paints, inks, varnishes, plastics) [29, 30], paper industry [31, 32], in the production of cosmetic preparations [33], food industry [34–36], ceramics [37, 38], and medicine and pharmacy [39–41]. Polymer adsorption is also used in soil stabilization and conditioning [28, 42–46]. Due to the high intermolecular bond strength, the polymer ensures an additive controlling of soil erosion, which is safe for health and the environment. It strengthens soil cohesion by binding loose mineral particles [47]. In addition, it affects the reduction of sludge transport; thus, the soil erosion process is limited [47–49]. The macromolecular compounds interact mainly with clay minerals found in soils or sediments. Adsorbing on the surface of the grains, they affect their properties. Changing the surface properties of a mineral, soil, or sediment, they affect the fate of other compounds, i.e., nutrients, heavy metals, and organic substances present in the surrounding environment [50].

However, from a technical point of view, it is necessary to study the concentrated systems of solid particles. The main problem of this type of system is maintaining adequate dispersion or achieving adequate rheological properties. It is already known that concentrated suspensions are stabilized or flocculated by the addition of a polymer, but the role of conformation of the macromolecular compound in these phenomena remains unclear. This is mainly due to difficulties in measuring adsorption and determining the conformation of the polymer substance in high-density systems. However, there are several techniques that allow such research. These are Fourier-transformed infrared spectroscopy (FT-IR), nuclear magnetic resonance (NMR), electron spin resonance (ESR), and fluorescence spectroscopy. Using the fluorescence spectroscopy and ESR technique, the conformation of poly(acrylic acid) on the solid surface and its effect on the flocculation process was determined. Based on the obtained results, it was found that the higher the molecular weight of the polymer is, the greater the degree of coiling of its macromolecules is. The more developed conformation of the high molecular compound is synonymous with the formation of a larger number of tails and loops on the surface of the solid, which in turn increases the efficiency of the flocculation process (creation of more numerous bridges) [13, 51]. It is also worth mentioning that the research on the impact of the polyacrylamide molecular weight on the adsorption, and flocculation of suspension process is also important. The higher the molecular weight of the polymer is, the greater the adsorption capacity is observed. A polymeric substance with a higher molecular weight is a much better flocculant, because a larger amount of it is adsorbed on the solid surface. In addition, the conformation of polymer chains has also been shown to depend on their molecular weight, which is manifested in an increase in chain length. This promotes the development of polymer macromolecules conformation and, as a consequence, flocculation efficiency by bridging increases.

Adsorption of polymers such as polyacrylamide and their role in the process of stabilization and flocculation of dispersed systems has a great scientific interest [52–61]. The type of action of the macromolecular compound, whether it acts as a stabilizer or flocculant, is influenced by many factors such as structure, conformation, solubility, and degree of polymerization [58, 60], as well as the energy of interaction of polymer chains with a solid surface [62]. Understanding the mechanism of polymer adsorption is important considering the production of increasingly effective dispersants. So far, many studies on adsorption and the influence of the high molecular compound presence on the interaction between solid particles have been presented. These studies concerned: the influence of pH, ionic strength, and the polymer charge on the adsorbed amount of the macromolecular compound [46, 52, 63–68]; kinetics of the flocculation process [54]; and the role of polymer conformational changes.

Currently, many studies refer to the assessment of mineral particle stability in colloidal systems, the nature of interactions between them, as well as the impact of high molecular compound presence on the suspension stability. These studies mainly refer to polyacrylamide (PAM) which is a non-toxic, environmentally friendly polymer that has the ability to stabilize and improve soil structure [69, 70]. Its beneficial effect is related to maintaining or increasing soil aggregation and pore continuity [71–73], increasing aggregate stability, and thus soil structural stability [74, 75]. The potential benefits of using PAM are influenced by a number of relationships between the properties of the polymer (its molecular weight, type, amount of surface charge, conformation of macromolecules, and density of the charge) and the properties of the soil (its type, texture, organic matter and clay mineral content, ionic strength, pH value, composition, and ion concentration in soil solution) [1, 76–79]. The type of polymer (and thus charge present in macromolecules) affects the amount of its adsorption on solid surface. The polymer adsorption increases with the following order: anionic PAM < nonionic PAM < cationic PAM [1]. Nonionic polyacrylamide adsorption on clay minerals is mainly caused by van der Waals forces [80], whereas cationic or anionic forms adsorb and bind soil particles through electrostatic forces or bridge formation between PAM charged moieties and solid surface group [1, 80]. Malik and Letey [81] concluded that due to differences in polymer chain conformations influenced by ionic PAM hydrolysis degree the stabilization of soil by polyacrylamide can vary. Adsorbed polymeric chain conformation results from solution pH value and surface charge. The higher the percent of hydrolysis (greater amount of positively or negatively charged groups) was, the more linear, well-developed macromolecular chain and more extended conformation was obtained, and thus greater adsorption was observed [81, 82]. Adsorption of PAM characterized by 2 and 20% hydrolysis degree was studied [83]. The adsorption of polymer with greater amount of hydrolyzed groups was higher, whereas the more significant stability of soil aggregates with PAM 2% was observed. Several papers related to the effect of electrolyte present on polymer adsorption and its ability to soil stabilizing. The electrolyte presence in adsorption system enhances the effectiveness of soil aggregate stabilization by polymer and its adsorption on mineral surface [83].

The addition of polymer to the soil stabilizes existing aggregates and strengthens the mutual binding of adjacent particles. The stabilizing efficiency of polyacrylamide is mainly determined by the adsorption of its macromolecules on the surface of soil particles [84]. The aggregate formation contributes to several beneficial phenomena such as stability increase, reinforcement of soil structure, water infiltration promotion as well as erosion limitation [86]. Shainberg et al. [85] and Ben-Hur et al. [86] conducted infiltration studies with polyacrylamide and soils rich in clay mineral fractions. Application of PAM flocculant promotes surface sealing prevention which results in greater infiltration rate. Fox and Bryan [87], Stern et al. [68], and Smith et al. [88] determined the influence of PAM presence on soil erosion and field runoff reduction. Soils treated with polyacrylamide maintain their original physical state in water erosion conditions. By stabilization of clay mineral aggregates by polymer adsorption, the water infiltration rate improvement and soil erosion reduction are observed.

The PAM adsorbed amount may also depend on the soil and clay mineralogy [1, 76, 82]. The amount of adsorbed polymer on the surface of soil particles is up to three orders of magnitude smaller than the amount adsorbed on clay mineral particles. Under neutral or acidic pH, the adsorption of anionic polyacrylamide on the surface of clay materials such as montmorillonite, kaolinite, and illite is comparable. However, under basic conditions, the amount of adsorbed polymer decreases in order: illite, kaolinite, and montmorillonite [46, 76]. What is more, under alkaline pH conditions greater adsorption of anionic PAM on the illite surface compared to montmorillonite one was observed. The presence of polyacrylamide with high molecular weight contributes to flocculation process. However, the structure of flocs depends on clay mineral type. These micro-aggregates of soil particles with polyelectrolyte can be formed in both neutral and acidic solutions. Other studies focused on the stability of smectic soil aggregates in the presence of PAM [74]. It turned out that the macromolecules that entered the pores of the aggregates did not have a significant effect on the stability of their structure. On the other hand, this polymer increased the percentage of stable aggregates in kaolinite soils with different structures [89] and particle size [90]. The formation of aggregates, stability of macro- and micro-aggregates, their degradation mechanism, and chemical dispersion depend not only on the amount of clay minerals and organic matter but also on soil mineralogy [91]. The influence of the mineralogical composition is difficult to determine because soils usually contain a mixture of clay materials and their properties are modified by binding with other minerals. Under unstable conditions (such as high sodium content or low electrolyte concentrations), soils with high montmorillonite content are unstable, soils with high kaolinite content are relatively stable, and soils rich in 2:1 clay minerals or with low content of montmorillonite show intermediate stability [92, 93]. The effectiveness of PAM as a soil aggregate stabilizing factor has also been studied. The addition of polymer had a positive effect on increasing the moisture content of soil aggregates, thanks to which their stability improved (in ascending order: kaolin < illitic < smectite soils).

The content of clay in the soil and their mineralogy significantly affects the stabilization of aggregates by PAM. The effectiveness of stabilization may result from the natural stability of aggregates and soil properties affecting polymer adsorption on the solid surface. The optimal pH value and the presence of easily soluble minerals, which provide divalent cations to the soil solution, can contribute to increasing PAM adsorption on the surface of soil particles and the ability to increase aggregate stability.

Polyacrylamide treatments has significant impact on crop growth and yield as well as nutrient effect [28, 94–97]. Wallace [98] studied the interaction of polyacrylamide with the following nutrients: P, Zn, Cu, Fe, Mn, and Mo. The concentration of these elements in the soil solution containing PAM is significantly different than in systems without macromolecular additive. Moreover, these differences in microelements concentration affect the plant growth—the studies shown that plant growth is improved in the nutrient solutions with polyacrylamide soil flocculant [98].

3 PAM Effect on Heavy Metals' Accumulation in the Soil Environment

Clays and modified clays have been found particularly useful for adsorption of heavy metals' ions and are the objects of several experimental works [99–114]. Clays mineral due to their internal structure can adsorb metals such as As, Cd, Cr, Co, Cu, Fe, Pb, Mn, Ni, and Zn from aqueous medium. However, their adsorption capacity depends on the type of clay used and also differs for various metals [99]. In many papers, the limiting or enhancing conditions' effect on the adsorbent efficiency of the clay materials was studied. The influence of pH value, temperature, quantitative efficiency of the individual mineral, modifications of the clay, and its composites in removing various contaminants were examined.

Investigations of pH, contact time, temperature, and initial metal cation concentration influence on uranium(VI) ions' adsorption on bentonite clay were carried out [103]. Aytas et al. [103] showed that the mineral adsorption properties can change due to temperature and time increase—the highest capacity was observed in the case of bentonite calcinated under 400 °C and at the beginning of adsorption process.

Mishra and Patel [104] studied lead and zinc ions' removal from water by various adsorbents, i.e., kaolin and bentonite depending on the time, pH, adsorbent dosage, and contact time. The obtained results showed no significant differences in heavy metals adsorption in various pH values. The highest Pb and Zn removal from solution was observed when an increase in adsorbent dosage occurred. The increase of ionic strength of electrolyte results in decrease of metal ion adsorption.

The removal of Pb, Cd, Ni, and Cu ions from aqueous solution by kaolinite was studied by Jiang et al. [111]. The adsorption was carried out in different conditions differing with initial metal ion concentration, pH, ionic strength of electrolyte, and contact time. Obtained results showed that the pH value of solution had most significant impact on heavy metal adsorption on clay surface, and adsorption equilibrium was obtained after 30 min. With the increasing metal concentration, its adsorption also increases (due to stronger driving forces to mineral surface).

Bhattachryya and Sen Gupta [100] reviewed the removal of toxic metal ions by natural kaolinite, montmorillonite, and their modified forms by pillaring with polyoxy cations such as Zr^{4+}, Al^{3+}, Si^{4+}, Ti^{4+}, Fe^{3+}, Cr^{3+}, and Ga^{3+}. Comparison of adsorption capacity of these adsorbents indicated that the natural and modified montmorillonite may adsorb much more heavy metals than kaolinite sorbents. Various metal ions can be captured from solution by clay mineral adsorbent and their modified forms [99]. The sorption affinity to As anions improvement by pre-treatment of kaolin or bentonite minerals with Fe(II), Fe(III), Al(III), and Mn(II) ions was shown [107]. Na et al. [109] demonstrated that by pillaring Ti to montmorillonite, the arsenate or arsenite removal efficiency from aqueous solution can be increased. The adsorption of As ions as a function of pH value, contact time, temperature, coexisting ions' presence, and ionic strength was studied. Comparing all examined ions (phosphate, nitrate, and sulfate), only phosphate one had noticeable effect on heavy metal adsorption. The decrease in temperature caused an increase in As-adsorbed amount on clay surface.

Oliveira et al. [101] compared nickel, cadmium, zinc, and copper cations' adsorption on bentonite clay and iron clay minerals with magnetite composite from aqueous solution. It was shown that the presence of iron oxide in adsorbent structure enhanced the adsorption capacity of bentonite. Yuan et al. [105] investigated the removal of Cr(VI) by montmorillonite supported by magnetite nanoparticles. The clay mineral modified by nanoparticles of magnetite showed greater adsorption capacity per unit mass. Moreover, the chromium adsorption process was highly pH-dependent.

Several papers related to the adsorption of heavy metals on mineral soil surface modified by polymeric substances. Gecol et al. [102] compared tungsten removal efficiency by natural and chitosan-coated montmorillonite. The effects of W concentration as well as pH value were studied. The W-adsorbed amount increased with decrease in solution pH value and metal concentration. However, clay mineral coated with biopolymer was found to be much more effective adsorbent compared to natural montmorillonite.

Wiśniewska et al. [115, 116] and Fijałkowska et al. [117, 118] examined the anionic and cationic PAM adsorption on clay minerals, i.e., montmorillonite and kaolinite and its impact on the heavy metal ions' accumulation. Both polyacrylamides may strengthen the Pb(II) and Cr(VI) ions' adsorption on the aluminosilicate surface.

Pandley and Mishra [106] used chitosan/clay nanocomposite for Cr(VI) ions' removal from aqueous solution. The adsorption of chromium(VI) ions on montmorillonite surface modified with biopolymer was carried out in different pH values. However, the highest metal ions' adsorbed amount was observed at pH 3.

Another study concerned selenium adsorption on chitosan–montmorillonite composite [108]. The adsorption process on the composite material was pH-independent. It has been shown that clay mineral modified by chitosan is a high-performance adsorbent which allows removal of the metal ions with a high concentration from solution. The Cu(II) ions' adsorption process on bentonite surface with polyacrylamide gel in the function of pH, ionic strength, adsorbent content, metal ion concentration, and temperature was studied [110]. Obtained results indicated high pH, ionic strength, and temperature dependency of copper ions' sorption process. With decrease in temperature and ionic strength, the adsorbed amount of Cu(II) ions increased.

The effects of contact time, adsorbent dosage, and pH of the initial solution on the Hg^{2+} removal by polyacrylamide/attapulgite were studied [113]. The mercury ions' adsorption on PAM–attapulgite surface increases with increasing contact time and pH of the initial suspension but decreases with clay mineral composite dosage. Zhou et al. [112] studied Hg(II), Pb(II), and Co(II) ions' adsorption on polyacrylamide/attapulgite system. The obtained results showed that the clay mineral adsorbent modified by PAM exhibits the highest sorption capacity and selectivity in relation to mercury ions, which is manifested in their greatest adsorption compared to other ions.

Gu et al. [114] studied selective heavy metal adsorption on unmodified mineral and polyacrylamide–vermiculite composite surfaces. The adsorbent modified by polymer showed much greater efficiency in Pb(II) adsorption at different pH values than unmodified one and better selectivity for Pb(II) ion in comparison to Zn(II), Cd(II), and Cu(II) ions.

References

1. Seybold CA (1994) Polyacrylamide review: soil conditioning and environmental fate. Commun Soil Sci Plan 25(11–12):2171–2185. https://doi.org/10.1080/00103629409369180
2. Kadajji VG, Betageri GV (2011) Water soluble polymers for pharmaceutical applications. Polymers 3:1972–2009. https://doi.org/10.3390/polym3041972
3. Zhang J, Huguenard C, Scarnecchia C, Menghetti R, Buffle J (1999) Stabilization and destabilization of hematite suspensions by neutral and anionic polyacrylamide. Colloids Surf A Physicochem Eng Asp 151:49–63. https://doi.org/10.1016/S0927-7757(98)00700-6
4. Shatat RS, Niazi SK, Batati FSA (2018) Synthetic Polyelectrolytes based on polyacrylamide: non-ionic, anionic and cationic Polyacrylamides and their applications in water and wastewater treatment: literature review. CSIJ 25(4):1–8. https://doi.org/10.9734/CSJI/2018/46483
5. Craciun G, Ighigeanu D, Manaila E, Stelescu MD (2015) Synthesis and characterization of poly(Acrylamide-Co-Acrylic Acid) flocculant obtained by electron beam irradiation. Mater Res 18(5):984–993. https://doi.org/10.1590/1516-1439.008715
6. Guezennec AG, Michel C, Bru K, Touze S, Desroche N, Mnif I, Motelica-Heino M (2015) Transfer and degradation of polyacrylamide-based flocculants in hydrosystems: a review. Environ Sci Pollut Res 22:6390–6406. https://doi.org/10.1007/s11356-014-3556-6
7. Jiang Z, Zhu J (2014) Cationic Polyacrylamide: Synthesis and Application in Sludge dewatering Treatment. Asian J Chem 26(3):629–633. https://doi.org/10.14233/ajchem.2014.16055
8. Ma J, Shi J, Ding H, Zhu G, Fu K, Fu X (2017) Synthesis of cationic polyacrylamide by low-pressure UV initiation for turbidity water flocculation. Chem Ang J 312:20–29. https://doi.org/10.1016/j.cej.2016.11.114
9. Chen D, Liua X, Yue Y, W. Zhang W, Wang P (2006) Dispersion copolymerization of acrylamide with quaternary ammonium cationic monomer in aqueous salts solution Eur Polym J 42:1284–1297. https://doi.org/10.1016/j.eurpolymj.2005.12.007
10. Ge X, Ye W, Xu M, Zhang Z (1998) Radiation copolymerization of acrylamide and cationic monomer in an inverse emulsion. Polymer 39:1917–1920. https://doi.org/10.1016/S0032-3861(97)00482-5
11. Yang ZL, Gao BY, Li CX, Yue QY, Liu B (2010) Photoinitiated polymerization of cationic acrylamide in aqueous solution: synthesis, characterization, and sludge dewatering performance. Chem Eng J 161:27–38. https://doi.org/10.1155/2014/465151
12. Hong KH, Liu N, Sun G (2009) UV-induced graft polymerization of acrylamide on cellulose by using immobilized benzophenone as a photo-initiator. Eur Polym J 45:2443–2449. https://doi.org/10.1016/j.eurpolymj.2009.04.026
13. Al-Hashmi AR, Luckman PF (2010) Characterization of the adsorption of high molecular weight non-ionic and cationic polyacrylamide on glass from aqueous solutions using modified atomic force microscopy. Colloids Surf A Physicochem Eng Asp 358:142–148. https://doi.org/10.1016/j.colsurfa.2010.01.049
14. Krul LP, Nareiko EI, Matusevich YI, Yakimtsova LB, Matusevich V, Seeber W (2000) Water super absorbents based on copolymers of acrylamide with sodium acrylate. Polym Bull 45:159–165. https://doi.org/10.1007/PL00006832

15. Jamshidi H, Rabiee A (2014) Synthesis and characterization of acrylamide-based anionic copolymer and investigation of solution properties. Adv Mater Sci Eng. https://doi.org/10.1155/2014/728675
16. Pan Z, Campbell A, Sonasundaran P (1991) Polyacrylic acid adsorption and conformation in concentrated alumina suspensions. Colloids Surf A Physicochem Eng Asp 191:71–78. https://doi.org/10.1016/S0927-7757(01)00765-8
17. Nowicki W, Nowicka G (2001) Effect of polymers on properties of dispersions of nanoparticles. Wiadomości Chemiczne 55:549–567
18. Bajpai AK (1997) Interface behaviour of ionic polymers. Prog Polym Sci 22:523–564. https://doi.org/10.1016/S0079-6700(96)00003-2
19. Pefferkorn E (1999) Polyacrylamide at Solid/Liquid Interfaces. J Colloid Interf Sci 216:197–220. https://doi.org/10.1006/jcis.1999.6312
20. Stuart MAC (1991) Adsorbed polymers in colloidal systems: from statics to dynamics. Polym J 23(5):669–682. https://doi.org/10.1295/polymj.23.669
21. Myagchenkov VA, Kurenkov VF (1991) Applications of acrylamide polymers and copolymers: a review. Polym-Plast Technol 30(2–3):109–135. https://doi.org/10.1080/03602559108020132
22. Adachi Y (2019) Aspects of colloid and interface in the engineering science of soil and water with emphasis on the flocculation behavior of model particles. Paddy Water Environ 17:203–210. https://doi.org/10.1007/s10333-019-00712-7
23. Brostow W, Lobland HEH, Pal S, Singh RP (2009) Polymeric flocculants for wastewater and industrial effluent treatment. J Mater Educ 31(3–4):157–166
24. Gan LM, Yeoh KW, Chew CH, Koh LL, Tan TL (1991) Poly(sodium acrylamidoalkanoate)s: Syntheses and solution properties in relation to flocculation study in water treatment. J Appl Polym Sci 42:225–232. https://doi.org/10.1002/app.1991.070420127
25. Bolto BA (1995) Soluble polymers in water purification. Prog Polym Sci Prog Polym Sci 20:987–1041. https://doi.org/10.1016/0079-6700(95)00010-D
26. Henriksen K, Berthelsen L, Matzen R (1998) Separation of liquid pig manure by flocculation and ion exchange Part 1: laboratory experiments. J Agric Eng Res 69(2):115–125. https://doi.org/10.1006/jaer.1997.0229
27. Aidarova S, Bekturganova N, Kerimkulova M, Musabekov M, Sharipova A (2012) Structure formation of the surface layer of soil as a way to prevent a wind and water erosion. Eurasian ChemTech J 14:321–325. https://doi.org/10.18321/ectj129
28. Azzam RAI (1980) Agricultural polymers polyacrylamide preparation, application and prospects in soil conditioning. Commun Soil Sci Plan 11(8):767–834. https://doi.org/10.1080/00103628009367081
29. Janardhan R, Gedam PH, Sampathkumaran PS (1990) The effect of polymer molecular weight in the adsorption process. J Colloid Interf Sci 140:391–400. https://doi.org/10.1016/0021-9797(90)90359-V
30. Spinelli JH (1999) Polymeric dispersants in ink jet technology. Adv Mater 10:1215–1218. https://doi.org/10.1002/(SICI)1521-4095(199810)10:15%3c1215:AID-ADMA1215%3e3.0.CO;2-0
31. Wagberg L, Nordgvist T (1999) Detection of polymer induced flocculation of cellulosic fibres by image analysis. Nord Pulp Pap Res J 14:247–255. https://doi.org/10.3183/npprj-1999-14-03-p247-255
32. Miller P, Wiener EM, Turowski A (1999) O/W emulsions for cosmetics products stabilized by alkyl phosphates—rheology and storage tests. Colloids Surf A Physicochem Eng Asp 152:155–160. https://doi.org/10.1016/S0927-7757(98)00630-X
33. Swerin A, Ödberg L, Wagberg L (1996) An extended model for the estimation of flocculation efficiency factors in multicomponent flocculant systems. Colloids Surf A Physicochem Eng Asp 113:25–38. https://doi.org/10.1016/0927-7757(95)03506-0
34. Kholkin IY, Viglazov VV, Kind VB, Metlee HD (1999) Purification of carbohydrate-containing substrates from plant biomass hydrolysates. Appl Biochem Biotechnol 82:135–140. https://doi.org/10.1385/ABAB:82:2:135

35. Dickinson E (1999) Caseins in emulsions: interfacial properties and interactions. Inter Dairy J 9:305–312. https://doi.org/10.1016/S0958-6946(99)00079-5
36. Crees L, Senogles E, Whayman E (1991) The flocculation of cane sugar muds with acrylamide–sodium acrylate copolymers. J Appl Polym Sci 42(3):837–844. https://doi.org/10.1002/app.1991.070420329
37. Bergstrom L, Sjostrom E (1999) Temperature induced gelation of concentrated ceramic suspensions: rheological properties. J Eur Ceram Soc 19:2117–2123. https://doi.org/10.1016/S0955-2219(99)00021-7
38. Koltay JA, Feke DL (1999) Preparation of continuous fiber ceramic composites using a combination of steric-stabilization and depletion-flocculation phenomena. Compos Part A-Appl S 30:231–237. https://doi.org/10.1016/S1359-835X(98)00165-1
39. Schmidt C, Bodmeier R (1999) Incorporation of polymeric nanoparticles into solid dosage forms. J Control Release 57(2):115–125. https://doi.org/10.1016/s0168-3659(98)00108-4
40. Ha HK, Shin JH, Rha SE, Lee YS, Park KB, Lee MG, Kim PN, Auh YH (1999) Modified small-bowel follow-through: use of methylcellulose to improve bowel transradiance and prepare barium suspension. Radiology 211(1):197–201. https://doi.org/10.1148/radiology.211.1.r99ap02197
41. Duro R, Souto C, Gomez-Amoza JL, Martinez-Pacheco R, Concheiro A (1999) Interfacial adsorption of polymers and surfactants: implications for the properties of disperse systems of pharmaceutical interest. Drug Dev Ind Pharm 25(7):817–829. https://doi.org/10.1081/DDC-100102244
42. Bouranis DL, Theodoropoulos AG, Drossopoulos JB (1995) Designing synthetic polymers as soil conditioners. Commun Soil Sci Plan 26(9–10):1455–1480. https://doi.org/10.1080/00103629509369384
43. Graveling GJ, Ragnarsdottir KV, Allen GC, Eastman J, Brady PV, Balsley SD, Skuse DR (1997) Controls on polyacrylamide adsorption to quartz, kaolinite, and feldspar. Geochim Cosmochim Acta 61:3515–3523. https://doi.org/10.1016/S0016-7037(97)00175-0
44. Lee BJ, Schlautman MA (2015) Effects of polymer molecular weight on adsorption and flocculation in aqueous kaolinite suspensions dosed with nonionic polyacrylamides. Water 7(11):5896–5909. https://doi.org/10.3390/w7115896
45. Liu J, Shi B, Lu Y, Jiang H, Huang H, Wang G, Kamai T (2012) Effectiveness of a new organic polymer sand-fixing agent on sand fixation. Environ Earth Sci 65:589–595. https://doi.org/10.1007/s12665-011-1106-9
46. Deng Y, Dixon JB, White GN (2006) Adsorption of polyacrylamide on smectite, illite, and Kaolinite. Soil Sci Soc Am J 70:297–304. https://doi.org/10.2136/sssaj2005.0200
47. Sojka RE, Lentz RD (1997) Reducing furrow irrigation erosion with polyacrylamide (PAM). J Prod Agric 10(1):47–52. https://doi.org/10.2134/jpa1997.0047
48. Sojka RE, Bjorneberg DL, Entry JA, Lentz RD, Orts WJ (2007) Polyacrylamide in agriculture and environmental land management. Adv Agr 92:75–162. https://doi.org/10.1016/S0065-2113(04)92002-0
49. Sepaskhah AR, Shahabizad V (2010) Effects of water quality and PAM application rate on the control of soil erosion, water infiltration and runoff for different soil textures measured in a rainfall simulator. Biosyst Engin 106:513–520. https://doi.org/10.1016/j.biosystemseng.2010.05.019
50. Lu S, Chen F, Ngo HN, Guo W, Feng C, Wu J, Zheng B (2016) Effect of straw and polyacrylamide on the stability of land/waterecotone soil and the field implementation. Ecol Eng 94:12–21. https://doi.org/10.1016/j.ecoleng.2016.05.076
51. Jiang T, Teng L, Wei S, Deng L, Luo Z, Chen Y, Flanagan DC (2010) Application of polyacrylamide to reduce phosphorus losses from a Chinese purple soil: A laboratory and field investigation. J Environ Manage 91:1437–1445. https://doi.org/10.1016/j.jenvman.2010.02.006
52. Shubin V (1997) Adsorption of cationic polyacrylamide onto monodisperse colloidal silica from aqueous electrolyte solutions. J Colloid Interf Sci 191:372–377. https://doi.org/10.1006/jcis.1997.4934

53. Kawaguchi M, Takashi AA (1992) Polymer adsorption at solid-liquid interfaces. Adv Colloid Interface Sci 37:219–317. https://doi.org/10.1016/0001-8686(92)80085-C
54. Hocking MB, Klimchuk KA, Lowen S (1999) Polymeric flocculants and flocculation. J Macromol Sci Polymer Rev 39:177–203. https://doi.org/10.1081/MC-100101419
55. Otsubo Y (1996) Flocculation of colloids by soluble polymers and its effect on rheology. Heterogen Chem Rev 3:327–349. https://doi.org/10.1002/(sici)1234-985x(199612)3:4%3c327:aid-hcr66%3e3.0.co;2-s
56. Adachi Y (1995) Dynamic aspects of coagulation and flocculation. Adv Colloid Interface Sci 56:1–31. https://doi.org/10.1016/0001-8686(94)00229-6
57. Yu X, Somasundaran P (1996) Kinetics of polymer conformational changes and its role in flocculation. J Colloid Interface Sci 178:770–774. https://doi.org/10.1006/jcis.1996.0176
58. Mamedov AI, Huang C, Aliev FA, Levy GJ (2016) Aggregate stability and water retention near saturation characteristics as affected by soil texture, aggregate size and polyacrylamide application. Land Degrad Dev 28(2). https://doi.org/10.1002/ldr.2509
59. Larson A, Walldal C, Wall S (1999) Flocculation of cationic polymers and nanosized particles. Colloids Surf A Physicochem Eng Asp 159:65–67. https://doi.org/10.1016/S0927-7757(99)00163-6
60. Durand-Piana G, Lafuma F, Audebert R (1987) Flocculation and adsorption properties of cationic polyelectrolytes towards Na-montmorillonite dilute suspensions. J Colloids Interface Sci 119(2):474–479. https://doi.org/10.1016/0021-9797(87)90293-1
61. Nakamura A, Murakami K (2019) The effect of cationic polymer as flocculant on bentonite aggregation under different pH and the study of aggregation mechanism. Clay Science 23:7–14. https://doi.org/10.11362/jcssjclayscience.23.1_7
62. Holmberg M, Wigren R, Erlandsson R, Claesson PM (1997) Interactions between cellulose and colloidal silica in the presence of polyelectrolytes. Colloids Surf A Physicochem Eng Asp 130:175–183. https://doi.org/10.1016/S0927-7757(97)00036-8
63. Terry RE, Nelson SD (1986) Effects of polyacrylamide and irrigation method on soil physical properties. Soil Sci 141:317–320
64. Wallace GA, Wallace A (1986) Control of soil erosion by polymeric soil conditioners. Soil Sci 141(5):363–367
65. Wallace A, Wallace GA, Abouzamzam AM (1986) Amelioration of sodic soils with polymers. Soil Sci 141:359–362
66. Levy GJ, Levin J, Gal M, Ben-Hur M, Shainberg I (1992) Polymers' effects on infiltration and soil erosion during consecutive simulated sprinkler irrigations. Soil Sci Soc Amer J 56:902–907. https://doi.org/10.2136/sssaj1992.03615995005600030037x
67. Shainberg I, Levy GJ, Rengasamy PI, Frenkel F (1992) Aggregate stability and seal formation as affected by drops' impact energy and soil amendments. Soil Sci 154:113–119
68. Stern R, Van Der Merwe AJ, Laker MC, Shainberg I (1992) Effect of soil surface treatments on runoff and wheat yields under irrigation. Agron J 84:114–119. https://doi.org/10.2134/agronj1992.00021962008400010022
69. Roshaznizarmehri M, Fotovat A, Emami H, Kehl M, Hirmas DR, Hosseinalizadeh M, Ramezanian N (2018) Combined effects of polyacrylamide and nanomagnetite amendment on soil and water quality, Khorasan Razavi Iran. J Environ Manage 223:703–712. https://doi.org/10.1016/j.jenvman.2018.06.061
70. Amiri E, Emami H, Mosaddeghi MR, Astraei AR (2019) Shear strength of an unsaturated loam soil as affected by vetiver and polyacrylamide. Soil Tillage Res 194:104331. https://doi.org/10.1016/j.still.2019.104331
71. Green VS, Stott DE (2001) Polyacrylamide: a review of the use, effectiveness, and cost of a soil erosion control amendment. Sustain Global Farm, 384–389
72. Ben-Hur M, Keren R (1997) Polymer effects on water infiltration and soil aggregation. Soil Sci Soc Am J 61(2):565–570. https://doi.org/10.2136/sssaj1997.03615995006100020028x
73. Ajwa HA, Trout TJ (2006) Polyacrylamide and water quality effects on infiltration in sandy loam soils. Soil Sci Soc Am J 70(2):643–650. https://doi.org/10.2136/sssaj2005.0079

74. Mamedov AI, Beckmann S, Huang CWM, Levy GJ (2007) Aggregate stability as affected by polyacrylamide molecular weight, soil texture, and water quality. Soil Sci Soc Am J 71(6):1909–1918. https://doi.org/10.2136/sssaj2007.0096
75. Hudek S, Stanchi S, D'Amico M, Freppaz M (2017) Quantifying the contribution of the root system of alpine vegetation in the soil aggregate stability of moraine. Int Soil Water Conservation Res 5(1):36–42. https://doi.org/10.1016/j.iswcr.2017.02.001
76. Ben-Hur M, Malik M, Letey J, Mingelgrin U (1992) Adsorption of polymers on clays as affected by clay charge and structure, polymer properties, and water quality. Soil Sci Soc Am J 153(5):349–356
77. Lu JH, Wu L, Letey J (2002) Effects of soil and water properties on anionic polyacrylamide sorption. Soil Sci Soc Am J 66:578–584. https://doi.org/10.2136/sssaj2002.5780
78. Letey J (1994) Adsorption and desorption of polymers on soil. Soil Sci Soc Am J 158:244–248. https://doi.org/10.1097/00010694-199410000-00003
79. Krauth DM, Bouldin JL, Green VS, Wren PS, Baker WH (2008) Evaluation of a polyacrylamide soil additive to reduce agricultural-associated contamination. Bull Environ Contam Toxicol 81:116–123. https://doi.org/10.1007/s00128-008-9448-z
80. Theng BKG (1982) Clay-polymer interactions: summary and perspectives. Clays Clay Miner 30:1–9. https://doi.org/10.1346/CCMN.1982.0300101
81. Malik M, Letey M (1991) Adsorption of polyacrylamide and polysaccharide polymers on soil materials. Soil Sci Soc Am J 55:380–386. https://doi.org/10.2136/sssaj1991.03615995005500020014x
82. Michaels AS, Morelos O (1955) Polyelectrolyte adsorption by kaolinite. Ind Eng Chem 47:1801–1809. https://doi.org/10.1021/ie50549a029
83. Nadler A, Letey J (1989) Adsorption isotherms of polyanions on soils using tritium labeled compounds. Soil Sci Soc Amer J 53:1375–1378. https://doi.org/10.2136/sssaj1989.03615995005300050012x
84. Stutzmann T, Siffert B (1977) Contribution to the adsorption mechanism of acetamide and polyacrylamide on to clays. Clay Clay Miner 25:392–406. https://doi.org/10.1346/CCMN.1977.0250604
85. Shainberg I, Warrington DN, Rengasamy P (1990) Water quality and PAM interactions in reducing surface sealing. Soil Sci 149:301–307. https://doi.org/10.1097/00010694-199005000-00007
86. Ben-Hur M, Faris F, Malik M, Letey J (1989) Polymers as soil conditioners under consecutive irrigations and rainfall. Soil Sci Soc Amer J 53:1173–1177. https://doi.org/10.2136/sssaj1989.03615995005300040030
87. Fox D, Bryan RB (1992) Influence of a polyacrylamide soil conditioner on runoff generation and soil erosion: field tests in Baringo District, Kenya. Soil Techn 5:101–119. https://doi.org/10.1016/0933-3630(92)90012-P
88. Smith HJC, Levy GJ, Shainberg I (1990) Water-droplet energy and soil amendments: effect on infiltration and erosion. Soil Sci Soc Amer J 54:1084–1087. https://doi.org/10.2136/sssaj1990.03615995005400040026x
89. Miller WP, Willis RL, Levy GJ (1998) Aggregate stabilization in kaolinitic soils by low rates of anionic polyacrylamide. Soil Use Manage 14(2):101–105. https://doi.org/10.1111/j.1475-2743.1998.tb00623.x
90. Levy GJ, Miller WP (1999) Polyacrylamide adsorption and aggregate stability. Soil Till Res 51:121–128. https://doi.org/10.1016/S0167-1987(99)00048-3
91. Amezketa E (1999) Soil aggregate stability: a review. J Sustain Agric 14:83–151. https://doi.org/10.1300/J064v14n02_08
92. McNeal BL, Coleman NT (1966) Effect of solution composition on soil hydraulic conductivity. Soil Sci Soc Am J 30:308–312. https://doi.org/10.2136/sssaj1966.03615995003000030007x
93. Le Bissonnais Y (1996) Aggregate stability and assessment of crustability and erodibility: 1. Theory and methodology. Eur J Soil Sci 47:425–437. https://doi.org/10.1111/ejss.2_12311

94. Wallace A, Wallace GA, Abouzamzam AM (1986) Effects of excess of levels of a polymer as a soil conditioner on yields and mineral nutrition of plants. Soil Sci 141:377–380
95. Wallace A (1987) Anionic polyacrylamide treatment of soil improves seedling emergence and growth. HortScience 22:951
96. Wallace A, Wallace GA (1986) Effects of soil conditioners on emergence and growth of tomato, cotton, and lettuce seedlings. Soil Sci 141:313–316. https://doi.org/10.1097/00010694-198605000-00002
97. Wallace A, Wallace GA, Abouzamzam AM, Cha JW (1986) Effects of polyacrylamide soil conditioner on the iron status of soybean plants. Soil Sci 141:368–370
98. Wallace A (1986) Effect of polymers in solution culture on growth and mineral composition of tomatoes. Soil Sci 141:395–396. https://doi.org/10.1097/00010694-198605000-00020
99. Srinivasan R (2011) Advances in Application of Natural Clay and Its Composites in Removal of Biological, Organic, and Inorganic Contaminants from DrinkingWater. Adv Mater Sci Eng (1687–8434). https://doi.org/10.1155/2011/872531
100. Bhattacharyya KG (2008) Sen Gupta S (2008) Adsorption of a few heavy metals on natural and modified kaolinite and montmorillonite: a review. Adv Colloid Interf Sci 140(2):114–131. https://doi.org/10.1016/j.cis.2007.12.008
101. Oliveira LCA, Rios RVRA, Fabris JD, Sapag K, Garg VK, Lago RM (2003) Clay-iron oxide magnetic composites for the adsorption of contaminants in water. Appl Clay Sci 22(4):169–177. https://doi.org/10.1016/S0169-1317(02)00156-4
102. Gecol H, Miakatsindila P, Ergican E, Sage RH (2006) Biopolymer coated clay particles for the adsorption of tungsten from water. Desalination 197(1–3):165–178. https://doi.org/10.1016/j.desal.2006.01.016
103. Aytas S, Yurtlu M, Donat R (2009) Adsorption characteristic of U(VI) ion onto thermally activated bentonite. J Hazard Mater 172(2–3):667–674. https://doi.org/10.1016/j.jhazmat.2009.07.049
104. Mishra PC, Patel RK (2009) Removal of lead and zinc ions from water by low cost adsorbents. J Hazard Mater 168(1):319–325. https://doi.org/10.1016/j.jhazmat.2009.02.026
105. Yuan P, Fan M, Yang D, He H, Liu D, Yuan A, Zhu J (2009) Montmorillonite-supported magnetite nanoparticles for the removal of hexavalent chromium [Cr(VI)] from aqueous solutions. J Hazard Mater 166(2–3):821–829. https://doi.org/10.1016/j.jhazmat.2008.11.083
106. Pandey S, Mishra SB (2011) Organic-inorganic hybrid of chitosan/organoclay bionanocomposites for hexavalent chromium uptake. J Colloid Interf Sci 361(2):509–520. https://doi.org/10.1016/j.jcis.2011.05.031
107. Dousova B, Fuitova L, Grygar T, Machovic V, Kolousek D, Herzgova L, Lhotka M (2009) Modified aluminosilicates as low-cost sorbents of As(III) from anoxic groundwater. J Hazard Mater 165(1–3):134–140. https://doi.org/10.1016/j.jhazmat.2008.09.088
108. Bleiman N, Mishael YG (2010) Selenium removal from drinking water by adsorption to chitosan-clay composites and oxides: Batch and columns tests. J Hazard Mater 183(1–3):590–595. https://doi.org/10.1016/j.jhazmat.2010.07.065
109. Na P, Jia X, Yuan B, Li Y, Na J, Chen Y, Wang L (2010) Arsenic adsorption on Ti-pillared montmorillonite. J Chem Technol Biotechnol 85(5):708–714. https://doi.org/10.1002/jctb.2360
110. Zhao G, Zhang H, Fan Q, Ren X, Li J, Chen Y, Wang X (2010) Sorption of copper(II) onto super-adsorbent of bentonite-polyacrylamide composites. J Hazard Mater 173(1–3):661–668. https://doi.org/10.1016/j.jhazmat.2009.08.135
111. Jiang M, Jin X, Lu X-Q, Chen Z (2010) Adsorption of Pnb(II), Cd(II), Ni(II) and Cu(II) onto natural kaolinite clay. Deslination 252:33–39. https://doi.org/10.1016/j.deal.2009.11.005
112. Zhou S, Xue A, Zhao Y, Wang Q, Chen Y, Li M, Xing W (2011) Competitive adsorption of Hg^{2+}, Pb^{2+} and Co^{2+} ions on polyacrylamide/attapulgite. Desalination 270:269–274. https://doi.org/10.1016/j.desal.2010.11.055
113. Zhao Y, Chen Y, Li M, Zhou S, Xue A, Ximg W (2009) Adsorption of Hg^{2+} from aqueous solution onto polyacrylamide/attapulgite. J Hazard Mater 171:640–646. https://doi.org/10.1016/j.jhazmat.2009.06.048

114. Gu S, Wang L, Mao X, Yang L, Wang C (2018) Selective Adsorption of Pb(II) from Aqueous Solution by Triethylenetetramine-Grafted Polyacrylamide/Vermiculite. Materials 11(4):514. https://doi.org/10.3390/ma11040514
115. Wiśniewska M, Fijałkowska G, Szewczuk-Karpisz K (2018) The mechanism of anionic polyacrylamide adsorption on the montmorillonite surface in the presence of Cr(VI) ions. Chemosphere 211:524–534. https://doi.org/10.1016/j.chemosphere.2018.07.198
116. Wiśniewska M, Fijałkowska G, Szewczuk-Karpisz K, Urban T, Nosal-Wiercińska A, Wójcik G (2019) Comparison of adsorption affinity of anionic and cationic polyacrylamides for montmorillonite surface in the presence of chromium(VI) ions. Adsorption 25:41–50. https://doi.org/10.1007/s10450-018-9990-x
117. Fijałkowska G, Szewczuk-Karpisz K, Wiśniewska M (2019) Anionic polyacrylamide as a substance strengthening the Pb(II) immobilization on the kaolinite surface. Int J Environ Sci Technol. https://doi.org/10.1007/s13762-019-02546-6
118. Fijałkowska G, Szewczuk-Karpisz K, Wiśniewska M (2019) Chromium(VI) and lead(II) accumulation at the montmorillonite/aqueous solution interface in the presence of polyacrylamide containing quaternary amine groups. J Mol Liq 293:111514. https://doi.org/10.1016/j.molliq.2019.111514

Modified Two-Pole Approximation for Systems with Strong Electron Correlations: Peculiarities of Spectrum and DOS

L. Didukh, O. Kramar, Yu. Dovhopyaty, and Yu. Skorenkyy

In a new variant of Hartree–Fock approximation for the calculation of electrical and magnetic properties of strongly correlated electron systems in nanoscale material, the quasiparticle energy spectrum is obtained. The spectrum is temperature- and concentration-dependent, exact in atomic and band limits and gives a consistent description of the metal–insulator transition at temperature change or under the external pressure application. Refinement of the initial approach has allowed us to take into account the energy states widening which caused an essential transformation of the electron density of energy states. The single-particle Green function is calculated analytically, and model DOS features are discussed in detail for various values of energy parameters of the model and applied for investigation of low-temperature antiferromagnetic phase and possible phase transitions under the external influences. The electron–hole asymmetry in the model reflects a peculiar behavior of specific nanomaterials, generalizing the Hubbard model and t–J model. Hopping integrals are renormalized by electron correlations and appear to be concentration-dependent. As a consequence, at change of conductance type, the bandwidth, activation energy, and conductivity are found to undergo sharp changes.

1 Introduction

For nanoscale systems' theoretical description, the improvement of both analytic and numerical methods of treatment for strong electron correlations in a wide range of electron concentration and energy parameters is instrumental.

L. Didukh · O. Kramar (✉) · Yu. Dovhopyaty · Yu. Skorenkyy
Ternopil Ivan Puluj National Technical University, 56, Rus'ka St., 46001 Ternopil, Ukraine
e-mail: okramar18@gmail.com

O. Fesenko and L. Yatsenko (eds.), *Nanomaterials and Nanocomposites, Nanostructure Surfaces, and Their Applications*, Springer Proceedings in Physics 246, https://doi.org/10.1007/978-3-030-51905-6_10

Physics of disordered low-dimensional structures is one of the most topical and fast developing branches of the modern condensed matter theory. Interest in this field is related to both new fundamental problems and phenomena and, from the other side, prospects of developing new quantum devices and systems with unmatched capacities for optoelectronics and nanoelectronics, metrology, novel IT technologies, communication techniques, etc. [1]. Studies of the low-dimensional systems resulted in discoveries of new and already widely used phenomena as integer and fractional quantum Hall effect, in two-dimensional electron gas, Wigner crystallization of quasi-two-dimensional electrons and holes, new composite quasiparticles and electron excitations with fractional charges, high-frequency Bloch oscillation, and many other effects. Operation of the modern semiconducting lasers on heterostructures is based on low-dimensional systems (quantum wells, self-organizing quantum dots, and quantum threads) as well.

In various nanoscopic structures (nanomaterials), one can observe metal–insulator transitions (MIT) similar to MIT in bulk materials. However, the noted transitions from metallic to insulator state in nanomaterials have some differences, namely, the transition takes place for a range of nanoparticle sizes and dimensions of the quasi-two-dimensional material, and there are compounds in which MIT exists only in nanomaterial. One should note that nanomaterials have simpler structure than bulk materials, and such complex phenomena as MIT can be more transparent. In recent years, MIT in VO_2 is studied intensively in nanothreads [2], by femtosecond spectroscopy [3] (ultra-fast spectroscopy has been applied recently [4] to characterize peculiarities of MIT in "canonical" Mott–Hubbard system $(VCr)_2O_3$)), in thin films [5]. These studies prove that the transition is caused by electron–electron interactions but not by the lattice changes, which take place later. Electron-interaction-driven MIT is observed also in $SrRuO_3$ nanofilms [6], Fe_3O_4 nanocrystals [7], and conjugated conducting polymer-based nanothreads [8]. All the noted studies show that further experimental and theoretical studies of electron-interaction-driven MIT in nanomaterials are of importance.

Despite the recent progress in the field, the development of the analytical methods is needed for nanoscale systems with strong electron correlations at arbitrary w to U ratios and for arbitrary electron concentrations. In this context, we would like to emphasize the approaches developed in papers [9, 10] (hereafter, approximation I) and in papers [11, 12] (approximation II).

Within the approximation I, the energy spectrum at electron concentration $n = 1$ in the paramagnetic state has the form

$$E\left(\vec{k}\right) = -\mu + (1-2d)t\left(\vec{k}\right) + \frac{U}{2} \mp \frac{1}{2}\sqrt{U^2 + \left(4dt\left(\vec{k}\right)\right)^2}. \quad (1)$$

Respective result for the approximation II is

$$E\left(\vec{k}\right) = -\mu + (1-2d)t\left(\vec{k}\right) + \frac{U}{2} \mp \frac{1}{2}\sqrt{U^2 + t^2\left(\vec{k}\right)}, \quad (2)$$

where μ is the chemical potential, $t\left(\vec{k}\right)$ is the Fourier transform of the hopping integral, U is the repulsion parameter for two electrons of opposite spin projections on the same site, and d stands for the concentration of sites with two electrons (doublons).

Spectra (1) and (2) are exact in band, and atomic limits yet these equations do not reproduce the spectrum [10] obtained on the basis of effective t–J-model Hamiltonian in the limit of small w/U; moreover, these spectra do not contain a momentum-independent term ~w^2/U, which is responsible for the kinetic exchange between the nearest neighbors, key to antiferromagnetic ordering. For small U/w, if the system is spin-polarized ($n_\uparrow \neq n_\downarrow$), spectra (1) and (2) are modified; however, Hartree–Fock terms $n_\downarrow U$ and $n_\uparrow U$ are not present. Single-electron Green functions which yield spectra (1) and (2) have two poles, and the damping of quasiparticle states is absent; therefore, it is desirable to go beyond the mentioned approximations. In this paper, we develop an approach which extends the region of validity of analytic procedures of papers [10–12] and removes the noted deficiencies.

2 Model Hamiltonian and Energy Spectrum

We begin with Hamiltonian of the generalized Hubbard model in configurational representation of X_i^{kl}-operators [10]:

$$H = H_0 + H_1 + H_1', \tag{3}$$

where

$$H_0 = -\mu \sum \left(X_i^{\uparrow} + X_i^{\downarrow} + 2X_i^2\right) + U \sum X_i^2, \tag{4}$$

$$H_1 = \sum_{ij\sigma}{}' t_{ij}(n) X_i^{\sigma 0} X_j^{0\sigma} + \sum_{ij\sigma}{}' \tilde{t}_{ij}(n) X_i^{2\sigma} X_j^{\sigma 2}, \tag{5}$$

$$H_1' = \sum_{ij\sigma} \left(t_{ij}'(n) \left(X_i^{\downarrow 0} X_j^{\uparrow 2} - X_i^{\uparrow 0} X_j^{\downarrow 2}\right) + h.c.\right). \tag{6}$$

Here σ denotes a spin projection ($\sigma = \uparrow, \downarrow$), $t_{ij}(n)$,$t_{ij}'(n)$,$\tilde{t}_{ij}(n)$ are concentration-dependent integrals of electron hopping between the nearest neighbors for the lower subband, the upper subband, and subband hybridization [12]. Using the relation between electron creation and annihilation operators and X_p^{kl}-operators [13],

$$\begin{aligned} a_{p\uparrow}^+ &= X_p^{\uparrow 0} - X_p^{2\downarrow}, \quad a_{p\uparrow} = X_p^{0\uparrow} - X_p^{\downarrow 2}, \\ a_{p\downarrow}^+ &= X_p^{\downarrow 0} + X_p^{2\uparrow}, \quad a_{p\downarrow} = X_p^{0\downarrow} + X_p^{\uparrow 2}, \end{aligned} \tag{7}$$

we present the single-electron Green function as

$$G_{ps}(E) = \langle\langle X_p^{0\uparrow} \mid X_s^{\uparrow 0}\rangle\rangle - \langle\langle X_p^{\downarrow 2} \mid X_s^{\uparrow 0}\rangle\rangle + \langle\langle X_p^{\downarrow 2} \mid X_s^{2\downarrow}\rangle\rangle - \langle\langle X_p^{0\uparrow} \mid X_s^{2\downarrow}\rangle\rangle. \quad (8)$$

Equations for Green function $\left\langle\left\langle X_p^{0\uparrow} \mid X_s^{\uparrow 0}\right\rangle\right\rangle$ are

$$\begin{aligned}(E+\mu)\langle\langle X_p^{0\uparrow} \mid X_s^{\uparrow 0}\rangle\rangle =& \frac{< X_p^{\uparrow} + X_p^{0} > \delta_{ps}}{2\pi} + \langle\langle [X_p^{0\uparrow}, H_1]_- \mid X_s^{\uparrow 0}\rangle\rangle + \\ &+ \langle\langle [X_p^{0\uparrow}, H_1'] \mid X_s^{\uparrow 0}\rangle\rangle,\end{aligned} \quad (9)$$

where $[A, B]_-$ denotes the commutation procedure. Solving (9) for the single-electron Green function, we follow papers [14, 15] and apply a variant of projection method

$$[X_p^{0\uparrow}, H_1]_- = \sum_i \varepsilon_1^{\uparrow}(pi) X_i^{0\uparrow}, \quad (10)$$

where $\varepsilon(pi)$ is a non-operator expression. After anticommutation of both sides of the above equation with $X_k^{\uparrow 0}$ operator, one can obtain an equation for $\varepsilon(pi)$. For $n = 1$ (here n is the electron concentration per site) at the absence of magnetic ordering

$$\varepsilon_1^{\uparrow}(pi) = \varepsilon_1^{\downarrow}(pi) = \varepsilon_1(pi) = (1 - 2d)t(pi), \quad (11)$$

where d is the doublon concentration.

Explicit expression for the last term in (9) is

$$-\sum_{i \neq p} t(pi) \left\langle\left\langle (X_p^{\uparrow} + X_p^{0}) X_i^{\downarrow 2} \mid X_s^{\uparrow 0}\right\rangle\right\rangle + \left\langle\left\langle X_p^{02} X_i^{\downarrow 0} \mid X_s^{\uparrow 0}\right\rangle\right\rangle - \left\langle\left\langle X_p^{\downarrow\uparrow} X_i^{\uparrow 2} \mid X_s^{\uparrow 0}\right\rangle\right\rangle. \quad (12)$$

Taking into account the peculiarities of hybridization hopping described by H_1', we use the following representation for the first Green function in (12):

$$\langle\langle X_p X_p^{\downarrow 2} \mid X_s^{\uparrow 0}\rangle\rangle + < X_p^{\uparrow} + X_p^{0} > \langle\langle X_p^{\downarrow 2} \mid X_s^{\uparrow 0}\rangle\rangle, \quad (13)$$

where $X_p = X_p^{\uparrow} + X_p^{0} - < X_p^{\uparrow} + X_p^{0} >$, and represent the latter two with

$$\sum \varepsilon^{\uparrow}(pi) \langle\langle X_p^{\downarrow 2} \mid X_s^{\uparrow 0}\rangle\rangle, \quad (14)$$

where $\varepsilon^{\uparrow}(pi)$ is calculated in the same way as $\varepsilon_1(pi)$.

Expression (12) is therefore rewritten as

$$-\sum_i \varepsilon_2^{\uparrow}(pi)\left\langle\left\langle X_i^{\downarrow 2} \mid X_s^{\uparrow 0}\right\rangle\right\rangle - \sum_i t(pi)\left\langle\left\langle X_p X_i^{\downarrow 2} \mid X_s^{\uparrow 0}\right\rangle\right\rangle, \tag{15}$$

where $\varepsilon_2^{\uparrow}(pi) = \varepsilon_2^{\downarrow}(pi) = \varepsilon_2(pi) = -2dt(pi)$ for paramagnetic case.

Now for the Green function $\left\langle\left\langle X_p^{0\uparrow} \mid X_s^{\uparrow 0}\right\rangle\right\rangle$, we have the following expression:

$$\begin{aligned}(E+\mu)\langle\langle X_p^{0\uparrow} \mid X_s^{\uparrow 0}\rangle\rangle = & \frac{< X_p^{\uparrow} + X_p^0 > \delta_{ps}}{2\pi} + \sum_i \varepsilon_1(pi)\left\langle\left\langle X_i^{0\uparrow} \mid X_s^{\uparrow 0}\right\rangle\right\rangle \\ & + \sum_i \varepsilon_2(pi)\left\langle\left\langle X_i^{\downarrow 2} \mid X_s^{\uparrow 0}\right\rangle\right\rangle - \sum_i t(pi)\left\langle\left\langle X_p X_i^{\downarrow 2} \mid X_s^{\uparrow 0}\right\rangle\right\rangle,\end{aligned} \tag{16}$$

where $X_p = X_p^{\uparrow} + X_p^0 - < X_p^{\uparrow} + X_p^0 >$. If in this equation approximation $X_p^{\uparrow} + X_p^{\downarrow} =< X_p^{\uparrow} + X_p^0 >$ is adopted (the last Green function is absent), then we obtain the equations obtained in papers [9, 10]. If Green function $\left\langle\left\langle X_p X_i^{\downarrow 2} \mid X_s^{\uparrow 0}\right\rangle\right\rangle$ is kept, hopping processes described by H_1' (pair creation of holes and doublons at neighboring sites) can be described consistently. Below, we show that this allows obtaining the correct kinetic exchange term in spectrum.

In a similar way for the Green function $\left\langle\left\langle X_i^{\downarrow 2} \mid X_s^{\uparrow 0}\right\rangle\right\rangle$, which entered (16), we write the following equation:

$$\begin{aligned}(E+\mu-U)\langle\langle X_r^{\downarrow 2} \mid X_s^{\uparrow 0}\rangle\rangle = & \sum_i \tilde{\varepsilon}_1(pi)\left\langle\left\langle X_i^{\downarrow 2} \mid X_s^{\uparrow 0}\right\rangle\right\rangle + \\ & + \sum_i \tilde{\varepsilon}_2(pi)\left\langle\left\langle X_i^{0\uparrow} \mid X_s^{\uparrow 0}\right\rangle\right\rangle - \sum_i t(pi)\left\langle\left\langle \tilde{X}_p X_i^{0\uparrow} \mid X_s^{\uparrow 0}\right\rangle\right\rangle,\end{aligned} \tag{17}$$

where $\tilde{\varepsilon}_1(pi) = (1-2d)t(pi)$, $\tilde{\varepsilon}_2(pi) = -2dt(pi)$ (paramagnetic state), $\tilde{X}_p = X_p^{\downarrow} + X_p^2 - < X_p^{\downarrow} + X_p^2 >$.

If in (16) and (17) we neglect the last terms assuming these are of the second order of magnitude, we reproduce the results of work [10, 14].

For Green function $\left\langle\left\langle X_p X_r^{\downarrow 2} \mid X_s^{\uparrow 0}\right\rangle\right\rangle = T_{prs}(E)$, we obtain

$$\begin{aligned}(E+\mu-U)\langle\langle X_p X_r^{\downarrow 2} \mid X_s^{\uparrow 0}\rangle\rangle = & - t(rp)\langle\langle (X_r^{\downarrow} + X_r^2) X_p^{0\uparrow} \mid X_s^{\uparrow 0}\rangle\rangle + \\ & + \sum_i t(ri)\left\langle\left\langle X_p X_i^{\downarrow 2} \mid X_s^{\uparrow 0}\right\rangle\right\rangle + \Delta,\end{aligned} \tag{18}$$

where Δ denotes irreducible part which occurs at transition to (18).

If now we apply the mean-field approximation

$$X_r^{\downarrow} + X_r^{2} =< X_r^{\downarrow} + X_r^{2} >= n_{\downarrow},$$

$$X_p^{\uparrow} + X_p^{0} =< X_p^{\uparrow} + X_p^{0} >= 1 - n_{\downarrow} = n_{\uparrow},$$

then the Green functions in Δ containing operator expressions X_p and $\tilde{X}_p$ turn to zero. We also neglect processes of simultaneous creation and annihilation of two electrons on a site. As a result, we obtain from (18)

$$(E + \mu - U)\langle\langle X_p X_r^{\downarrow 2} \mid X_s^{\uparrow 0}\rangle\rangle = - n_{\downarrow} t(rp)\langle\langle X_p^{0\uparrow} \mid X_s^{\uparrow 0}\rangle\rangle + \\ + \sum_i t(ri)\langle\langle X_p X_r^{\downarrow 2} \mid X_s^{\uparrow 0}\rangle\rangle, \quad (19)$$

and from (17)

$$(E + \mu)\left\langle\left\langle \tilde{X}_p X_r^{0\uparrow} \middle| X_s^{\uparrow 0}\right\rangle\right\rangle =- < X_p^{\uparrow} + X_p^{0} > t(rp)\langle\langle X_p^{\downarrow 2} \mid X_s^{\uparrow 0}\rangle\rangle + \\ + \sum_i t(ri)\left\langle\left\langle \tilde{X}_p X_r^{0\uparrow} \middle| X_s^{\uparrow 0}\right\rangle\right\rangle. \quad (20)$$

In $\vec{k}$-space, we obtain the closed system of equations

$$\left(E + \mu - \varepsilon_1\left(\vec{k}\right)\right)G_{\vec{k}} = \frac{1}{2\pi} + \varepsilon_2\left(\vec{k}\right)T_{\vec{k}} - \frac{1}{N}\sum_{p\vec{\kappa}} T_{p\vec{\kappa}\vec{k}} t(\vec{\kappa}), \quad (21)$$

$$\left(E + \mu - U - \tilde{\varepsilon}_1\left(\vec{k}\right)\right)T_{\vec{k}} = \tilde{\varepsilon}_2\left(\vec{k}\right)G_{\vec{k}} - \frac{1}{N}\sum_{\vec{k}} t(\vec{\kappa})\tilde{T}_{p\vec{\kappa}\vec{k}}, \quad (22)$$

$$\left(E + \mu - U - t\left(\vec{k}\right)\right)T_{p\vec{\kappa}\vec{k}} = n_{\downarrow}\sum_r t(rp)\exp\left(i\vec{\kappa}r - i\vec{k}p\right)G_{\vec{k}}, \quad (23)$$

$$\left(E + \mu - t\left(\vec{k}\right)\right)\tilde{T}_{p\vec{\kappa}\vec{k}} = n_{\uparrow}\sum_r t(rp)\exp\left(i\vec{\kappa}r - i\vec{k}p\right)T_{\vec{k}}. \quad (24)$$

In the above equations

$$\varepsilon_1 = \frac{n_{\downarrow}}{N}\sum_{\vec{k}} \frac{t^2\left(\vec{k}\right)}{E + \mu - U - t\left(\vec{k}\right)}, \quad (25)$$

$$\varepsilon_2 = \frac{n_{\uparrow}}{N}\sum_{\vec{k}} \frac{t^2\left(\vec{k}\right)}{E + \mu - t\left(\vec{k}\right)}. \quad (26)$$

For paramagnetic state $\varepsilon_1(\vec{k}) = \tilde{\varepsilon}_1(\vec{k}) = (1-2d)t(\vec{k})$, $\varepsilon_2(\vec{k}) = \tilde{\varepsilon}_2(\vec{k}) = -2dt(\vec{k})$; therefore, we obtain from (21)–(24) the Green functions $\langle\langle X_p^{0\uparrow} \mid X_s^{\uparrow 0}\rangle\rangle$ and $\langle\langle X_p X_r^{\downarrow 2} \mid X_s^{\uparrow 0}\rangle\rangle = T_{prs}(E)$ in wave-vector representation as

$$G_{\vec{k}} = \frac{1}{4\pi}\left(\frac{A_{\vec{k}}}{E - E_1(\vec{k})} + \frac{B_{\vec{k}}}{E - E_2(\vec{k})}\right), \tag{27}$$

$$T_k = \frac{1}{4\pi}\frac{\varepsilon_2(\vec{k})}{E_2(\vec{k}) - E_1(\vec{k})}\left(\frac{1}{E - E_2(\vec{k})} - \frac{1}{E - E_1(\vec{k})}\right), \tag{28}$$

where spectral weights are

$$A_{\vec{k}} = \frac{1}{2}\left(1 + \frac{U + \varepsilon_2 - \varepsilon_1}{E_2(\vec{k}) - E_1(\vec{k})}\right),\ B_{\vec{k}} = 1 - A_{\vec{k}},$$

and the quasiparticle spectrum is

$$E_{1,2}(\vec{k}) = -\mu + \frac{U + \varepsilon_1 + \varepsilon_2}{2} + (1-2d)t(\vec{k}) \mp \frac{1}{2}\sqrt{(U + \varepsilon_2 - \varepsilon_1)^2 + \left(4dt(\vec{k})\right)^2}. \tag{29}$$

In analogous way, other Green functions in (8) can be obtained. The resulting single-electron Green function is

$$G_{\vec{k}}(E) = \frac{1}{2\pi}\left(\frac{C_{\vec{k}}}{E - E_1(\vec{k})} + \frac{D_{\vec{k}}}{E - E_2(\vec{k})}\right), \tag{30}$$

$$C_{\vec{k}} = \frac{1}{2} - \frac{2dt(\vec{k})}{E_2(\vec{k}) - E_1(\vec{k})},\ D_{\vec{k}} = 1 - C_{\vec{k}}.$$

The above results allow calculating the single-electron density of states (DOS), as shown in Figs. 1 and 2.

From Fig. 1, we conclude that the improved approximation enhances the effectiveness of Coulomb repulsion which is evident from central quasiparticle peak reduction and doublon concentration decrease. From this figure, we also see that the used procedure allows us to obtain a more realistic critical value for the metal–insulator transition and, thus, eliminate shortcomings of the approximation I in comparison to dynamical mean-field approach.

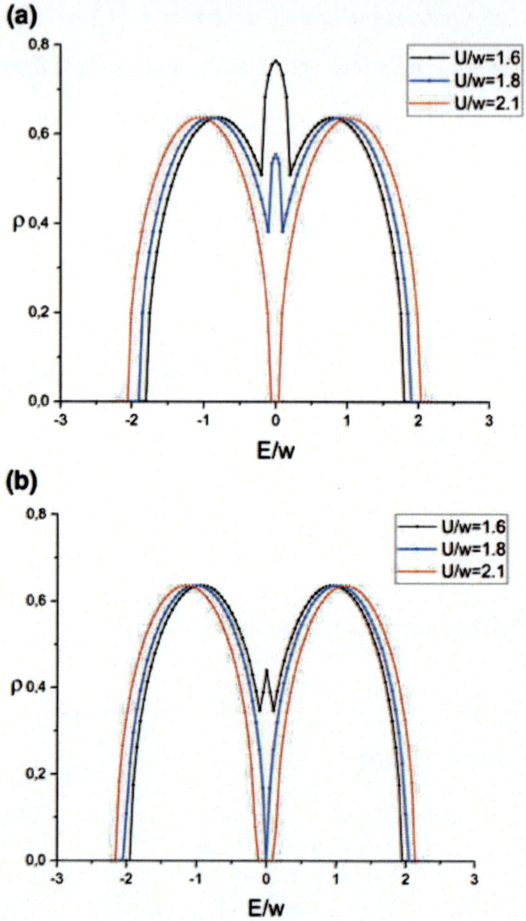

Fig. 1 The upper panel displays DOS obtained in approximation I and the lower one—obtained in the present approximation for Hubbard model (correlated hopping is absent). Semi-elliptic bare DOS is used. Black (the most inner) line with thin quasiparticle peak is for $U/w = 1.6$; blue curve is for $U/w = 1.8$ and red (the most outer) curve for $U/w = 2.1$

In a system with electron–hole asymmetry (Fig. 2), the proposed approximation gives lower critical value for metal-to-insulator transition, wider upper subband. By comparing plots for near-critical values of Coulomb correlations with correlated hopping taken into account, we note that the improved approximation estimates the effect of the correlated hopping more conservatively. Subband widths decrease by far slower and, correspondingly, maximum DOS values are lower than in the approximation I. Whatever approximation is used, the correlated hopping of electrons enhances localization, which is apparent from the spectacular reduction of the upper bandwidth.

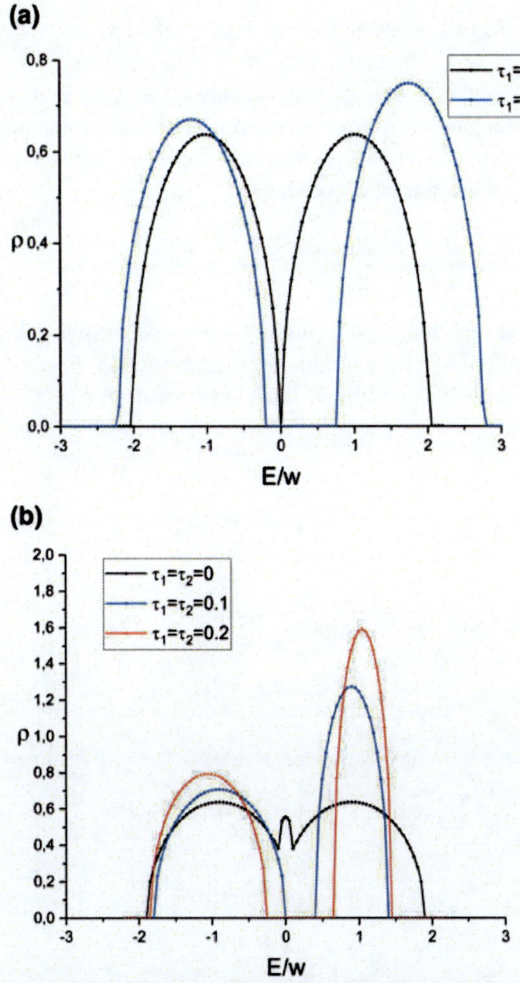

Fig. 2 DOS for the generalized Hubbard model with correlated hopping of electrons in approximation I at $U/w = 2.1$ (the upper panel) and the proposed approximation at $U/w = 1.8$ (the lower panel), semi-elliptic bare DOS. Black (the lowest) line is for $\tau_1 = \tau_2 = 0$; blue line is for $\tau_1 = \tau_2 = 0.05$ (the upper panel), for $\tau_1 = \tau_2 = 0.1$ (the lower panel); red line corresponds to $\tau_1 = \tau_2 = 0.2$

3 Results and Discussion

Let us consider the energy spectrum (29) at condition $(U + \varepsilon_2 - \varepsilon_1)^2 >> (4dt(k))$ which corresponds to the state of Mott–Hubbard insulator or semimetal. Then

$$E_1\left(\vec{k}\right) = -\mu + \varepsilon_1 + (1 - 2d)t\left(\vec{k}\right), \tag{31}$$

$$E_2\left(\vec{k}\right) = -\mu + U + \varepsilon_2 + (1 - 2d)t\left(\vec{k}\right), \tag{32}$$

for the lower and the upper subbands, respectively ($\mu = U/2$ at $n = 1$). One can see that taking into account transitions between $|i\sigma\rangle$-states and the upper Hubbard subband and transitions between $|i\uparrow\downarrow\rangle$-states and the lower Hubbard subbands lead to a renormalization of the atomic energy levels:

$$-\frac{U}{2} \to -\frac{U}{2} + \varepsilon_1, \quad \frac{U}{2} \to \frac{U}{2} + \varepsilon_2.$$

For the model rectangular DOS, one can solve for ε_1 and ε_2 using the exact solutions of the atomic limit. Substituting $E = -\mu = -U/2$ into (25) and $E = -\mu + U$ into (26), we have in paramagnetic case

$$\varepsilon_1 = -\frac{1}{2N}\sum_{\vec{k}} \frac{t^2\left(\vec{k}\right)}{U + t\left(\vec{k}\right)},$$

$$\varepsilon_2 = \frac{1}{2N}\sum_{\vec{k}} \frac{t^2\left(\vec{k}\right)}{U - t\left(\vec{k}\right)}.$$

To obtain an analytical expression, we use the model rectangular DOS with half bandwidth w

$$\frac{1}{N}\sum_{\vec{k}} \delta\left(E - t\left(\vec{k}\right)\right) = \frac{1}{2w}\theta\left(w^2 - E^2\right), \tag{33}$$

where $\theta(x) = 1$ for $x > 0$, $\theta(x) = 0$ for $x < 0$. Then

$$\varepsilon_1 = \frac{U}{2} - \frac{U^2}{4w}\ln\frac{U + w}{|U - w|}, \quad \varepsilon_2 = -\varepsilon_1. \tag{34}$$

At condition $U >> w$ (Mott–Hubbard insulator state),

$$\varepsilon_1 = -\varepsilon_2 = -\frac{w^2}{6U}. \tag{35}$$

The energy spectrum (29) with expression (35) reproduces the result of the effective Hamiltonian of Hubbard model obtained in the case of $U >> w$ [15, 16] if the kinetic exchange is taken into account in the mean-field approximation.

2. If the system becomes spin-polarized (ferromagnetically ordered [17, 18] or under the external magnetic field [19]), then in the spectrum for electrons of spin $\sigma =\uparrow$ atomic levels shifts will take values

$$\varepsilon_1^{\uparrow} = -n_{\downarrow}\frac{w^2}{6U},\ \varepsilon_2^{\uparrow} = n_{\uparrow}\frac{w^2}{6U}, \tag{36}$$

and for spin-down electrons

$$\varepsilon_1^{\downarrow} = -n_{\uparrow}\frac{w^2}{6U},\ \varepsilon_2^{\downarrow} = n_{\downarrow}\frac{w^2}{6U}. \tag{37}$$

3. The energy gap between the subbands, obtained from (29), is

$$\Delta E = -2w(1-2d) + \sqrt{(U+\varepsilon_2-\varepsilon_1)^2 + (4dw)^2}, \tag{38}$$

(quantities ε_1 and ε_2 are given by expression (34) and differ from that obtained in [10] by the presence of renormalized activation energy of the hole–doublon pair ($U \to U+\varepsilon_2-\varepsilon_1$). The term $\varepsilon_2-\varepsilon_1$ has clear physical meaning: this is the increase of the activation energy with taking into account transitions "site–Hubbard subband" (needed to overcome antiferromagnetic exchange energy of the site with its neighbors in a Mott–Hubbard insulator).

4. Self-consistent equation for doublon concentration can be obtained from function $\left\langle\left\langle X_p^{\downarrow 2} \mid X_s^{2\downarrow}\right\rangle\right\rangle_{\vec{k}}$ analytically as

$$d = \frac{1}{4} + \frac{U+\varepsilon_2-\varepsilon_1}{32dw}\ln(1-4d)\theta(2w-U+\varepsilon_1-\varepsilon_2), \tag{39}$$

if we use the model rectangular DOS at zero temperature. In distinction from paper [10], where the critical value of the Coulomb repulsion which corresponds to the metal–insulator transition $U_c = 2w$, here it is close to 1.8.

5. In consequence of "site–Hubbard subband" transitions, not only the atomic levels shifts exist, but also the levels are widened. This important peculiarity is lost with the transition from expressions (25)–(26) to expression (34). Indeed, we can rewrite formulae (25) and (26) as

$$\lim_{s\to 0}\frac{1}{N}\sum_{\vec{k}}\frac{t^2(\vec{k})}{E+\mu-U-t(\vec{k})+is} = P\frac{1}{N}\sum_{\vec{k}}\frac{t^2(\vec{k})}{E+\mu-U-t(\vec{k})} - \frac{i\pi}{N}\sum_{\vec{k}}t^2(\vec{k})\delta\left(E+\mu-U-t(\vec{k})\right), \tag{40}$$

$$\lim_{s\to 0}\frac{1}{N}\sum_{\vec{k}}\frac{t^2(\vec{k})}{E+\mu-t(\vec{k})+is} = P\frac{1}{N}\sum_{\vec{k}}\frac{t^2(\vec{k})}{E+\mu-t(\vec{k})} - \frac{i\pi}{N}\sum_{\vec{k}}t^2(\vec{k})\delta\left(E+\mu-U-t(\vec{k})\right), \tag{41}$$

where P denotes principal value. Each of the right-hand side terms can be expressed as $\delta_1 - i\Delta_1,\quad \delta_2 - i\Delta_2$, where δ_1 and δ_2 are virtual energy level shifts, and Δ_1 and Δ_2 are their widths. Here, an analogy with single-site Anderson model and auxiliary single-site problem in DMFT can be seen. Consequences of representing ε_1 and ε_2 in forms (40) and (41) will be discussed elsewhere.

4 Conclusions

In a new variant of Hartree–Fock approximation for the calculation of electrical and magnetic properties of strongly correlated electron systems in nanoscale material, the quasiparticle energy spectrum is obtained. The spectrum is temperature- and concentration-dependent, exact in atomic and band limits. The electron–hole asymmetry in the model reflects a peculiar behavior of specific nanomaterials, generalizing the Hubbard model and t–J model. Hopping integrals are renormalized by electron correlations and appear to be concentration-dependent. As a consequence, at a change of conductance type, the bandwidth, activation energy, and conductivity are found to undergo sharp changes, kinetic energy increases, and Fermi surface symmetry broke down. At the external pressure and decreasing temperature, the transition from insulator to metallic phase occurs due to the reduction of electron correlation. The proposed approach will be developed further for studies of conductance and effective masses of carriers in ferromagnetic phase within the ideology of works [18, 20].

The above-described mechanisms, revealed by the approach we used in the present work, will be even more pronounced in the degenerated case, for which similar studies have been done in papers [21–23].

References

1. Fesenko O, Yatsenko L (2019) Nanocomposites, nanostructures, and their applications. NANO 2018. Springer Proceedings in Physics, vol 221. Springer, Cham
2. Wei J, Wang Z, Chen W, Cobden DH (2009) New aspects of the metal–insulator transition in single-domain vanadium dioxide nanobeams. Nat Nanotechnol 4:420–424
3. Jager M, Ott C, Kraus P et al (2017) Tracking the insulator-to-metal phase transition in VO2 with few-femtosecond extreme UV transient absorption spectroscopy. PNAS 114:9558–9563
4. Lantz G, Mansart B, Grieger D et al (2017) Ultrafast evolution and transient phases of a prototype out-of-equilibrium Mott-Hubbard material. Nat Commun 8:13917–13924
5. Majid S, Gautam K, Ahad A et al (2019) Mott-Hubbard insulator-metal transition in the VO2 thin film: a combined XAS and resonant PES study. arXiv:1907.12746
6. Pang S, Kim Y, Shin YJ et al (2017) Electronic characteristics of ultrathin SrRuO3 films and their relationship with the metal—insulator transition. arXiv:1712.05137
7. Lee J, Kwon SG, Park J-G et al (2015) Size dependence of metal-insulator transition in stoichiometric Fe3O4 Nanocrystals. Nano Lett 7:4337–4342
8. Aleshin AN (2007) Quasi-one-dimensional transport in conducting polymer nanowires. Phys Solid State 49:2015–2033
9. Didukh L (1997) Model of narrow-band material with electron-hole asymmetry. J Phys Stud 1:241–250 [In Ukrainian]
10. Didukh L (1998) Energy spectrum of electrons in the Hubbard model: a new mean-field approximation. Physica Status Solidi (B) 206:R5–R6
11. Didukh L (1998) A modified form of the polar model of crystals. Condens Matter Phys 1:125–144
12. Didukh L (2000) A modified form of the polar model of crystals. Acta Phys Pol B 31:3097–3133
13. Didukh LD, Stasyuk IV (1968) Effective Hamiltonian in the Anderson model. Fiz Met Metalov 26:582–588 [In Russian]

14. Didukh L, Skorenkyy Yu, Kramar O (2008) Electron correlations in narrow energy bands: modified polar model approach. Condens Matter Phys 11:443–454
15. Skorenkyy Y, Didukh L, Kramar O, Dovhopyaty Y (2012) Phase diagram of metal-insulator transition in system with Anderson-Hubbard centers. Acta Phys Pol A 122:532–534
16. Skorenkyy Y, Didukh L, Kramar O, Dovhopyaty Y (2007) Mott transition, ferromagnetism and conductivity in the generalized Hubbard model. Acta Phys Pol A 111:635–644
17. Didukh L, Kramar O, Skorenkyy Y (2002) Ground state energy of metallic ferromagnet in a generalized Hubbard model. Physica Status Solidi (B) 229:1241–1254
18. Didukh L, Kramar O (2002) Metallic ferromagnetism in a generalized Hubbard model. Fizika Nizkikh Temperatur (Kharkov) 28:42–50
19. Didukh L, Skorenkyy Y, Kramar O, Dovhopyaty Y (2006) Effect of magnetic field, pressure and correlated hopping of electrons on conductivity of Mott-Hubbard material. Physica B 378–380:321–322
20. Didukh L, Kramar O (2005) Metallic ferromagnetism in the systems with strongly correlated electrons. Condens Matter Phys 8:547–564
21. Didukh L, Skorenkyy Y, Hankevych V, Kramar O (2001) Ground state ferromagnetism in a doubly orbitally degenerate model. Phys Rev B 64:144428
22. Didukh L, Hankevych V, Kramar O, Skorenkyy Y (2002) Itinerant ferromagnetism of systems with orbital degeneracy. J Phys: Condens Matter 14:827–835
23. Skorenkyy Y, Kramar O, Didukh L, Dovhopyaty Y (2018) Electron correlation effects in theoretical model of doped fullerides. In: Fesenko O, Yatsenko L (eds) Nanooptics, Nanophotonics, Nanostructures, and Their Applications. NANO 2017. Springer Proceedings in Physics, vol 210. Springer, Cham

Ising-Like Model of Nanosize Spin-Crossover Molecular Crystals

Iurii Gudyma and Artur Maksymov

1 Introduction

The transition coordination complexes with pseudo-octahedral (O_h) symmetry containing metal ion surrounded by nitrogen ligands under external stimuli show a so-called spin-crossover (SCO) transition between low-spin (LS) and high-spin (HS) states with diamagnetic and paramagnetic properties, respectively. Such a transition could be induced by weak external perturbations like temperature, pressure, light, magnetic, or electrical field. The properties of spin states in SCO compounds are determined by the redistribution of the electrons on the d-shell of the metal ion that can be controlled by external driving force. Therefore, the LS and HS states of an SCO compound demonstrate not only various magnetic properties but different optical, structural, and electrical properties as well. Due to increased energy gap of the SCO molecule in HS state, it is characterized by higher effective degeneracy and, therefore, by larger volume than the LS molecule. The different molecular volumes of HS and LS states could provoke the breaking of equilibrium of local forces and cause the localized elastic distortions. These effects could lead to the appearance of effective long-range elastic interactions. Physically, to such a size, differences may, at least partially, contribute the Jahn–Teller effect that also can be the reason for reduced degeneracy of the spin state and can change the symmetry of SCO molecule [1]. However, the Jahn–Teller effect may be very different for various SCO compounds, i.e., the complexes with $3d^4$, $3d^5$, and $3d^7$ electronic configurations manifest weak Jahn–Teller distortion for LS state, whereas the strong Jahn–Teller effect is observed for HS and LS states of the compounds with $3d^4$ and $3d^7$ electronic configurations,

I. Gudyma (✉)
Department of General Physics, Institute of Physical, Technical and Computer Sciences, Yuriy Fedkovych Chernivtsi National University, 2 Kotsjubynskyi Str., Chernivtsi 58012, Ukraine
e-mail: yugudyma@gmail.com

A. Maksymov
Marian Smoluchowski Institute of Physics, Jagiellonian University, Łojasiewicza 11, 30-348 Kraków, Poland
e-mail: maxyartur@gmail.com

O. Fesenko and L. Yatsenko (eds.), *Nanomaterials and Nanocomposites, Nanostructure Surfaces, and Their Applications*, Springer Proceedings in Physics 246, https://doi.org/10.1007/978-3-030-51905-6_11

respectively. If the SCO sample is exposed to controllable varying fields, the effective long-range elastic interaction may induce first-order phase transition which, in turn, can interconvert the initially LS system into metastable HS state and leads to hysteresis [2, 3]. These crucial properties have a big potential to be applied as a base for a new generation of SCO-based nanodevices. For scientific studies, the most interesting are the SCO complexes based on $3d^6$ iron(II) ions due to pronounced difference in magnetic response for LS and HS states. In these compounds, the SCO transition occurs between the diamagnetic LS and paramagnetic HS states that are characterized by the total spin number $S = 0$ and $S = 2$, respectively.

The general trend toward miniaturization of electronic and magnetic devices has shifted the focus of interest of scientific community from the macroscopic bulk compounds to the nanomaterials [4–9]. In this context, it becomes especially important to the relationship between size effects and cooperativity, which in most of the cases is studied on SCO nanoparticles incorporated into coordination networks. Other important and unique features that are observed by increasing their surface-to-volume ratio are the vanishing of hysteresis because of decreasing the cooperativity in the lattice, the incompleteness of the spin-state switching, and the downshift of the transition temperature. These effects are induced by peculiar features of nanoparticle's surface (strains, energies) and consequently by its interfaces with external environment. A particular property of surface-to-volume ratio effects is the vanishing of hysteresis loop below a critical value of a particle size.

The theoretical researches of transition metal complexes at microscopic approach widely involve the Ising-like model which was introduced for SCO compounds by Wajnflasz and Pick in 1971 [10] and is one of the most useful tools for studying the magnetic phenomena in these complexes. In the original formulation given by Wajnflasz and Pick, the SCO Ising-like model considers only the short-range interaction between the two-state sites. In the present paper, the SCO Ising-like model is completed by the different interaction energies in a bulk and on a surface of a nanocrystal that could play crucial role in explaining the experimental results. We believe that the main reason of size effects in SCO nanocrystals is related to the difference in intermolecular interaction. Our results demonstrate good agreement with the experimental data.

2 Ising-Like Model of Spin-Crossover Nanocrystal

The magnetic properties of molecular spin-crossover nanocrystals can be modeled phenomenologically by the Ising-like Hamiltonian [11]

$$H = -J \sum_{\langle ij \rangle} s_i s_j - h_0 \sum_i s_i . \tag{1}$$

Here, s_i is a pseudospin operator with two eigenvalues ± 1 that effectively corresponds to the HS and LS states of respective i-s molecule, and $< ij >$ denotes the summation over spin sites involved in interactions. The interaction described by parameter J from Ising-like Hamiltonian (1) has elastic nature and describes the coupling between the spins of nearest-neighbor two-level units. The second term in (1) corresponds to the interaction of degenerated pseudospin s_i with ligand field. The effective external field describing the result of surrounding action on single molecular magnet is

$$h_0 = -\frac{1}{2}(\Delta - k_B T \ln g), \tag{2}$$

where Δ indicates the energy gap between HS and LS states (the enthalpy change associated with the $LS \rightarrow HS$ conversion) for an individual spin-crossover molecule, k_B is the Boltzmann constant, and $g = g_{HS}/g_{LS}$ is the degeneracy ratio between HS and LS energy levels, respectively. At less rigorous (qualitative) approach, the strength of ligand field Δ is determined only by the ligand environment of the transition metal ion. In this case, the external effective field is homogeneous and is controlled by a single parameter which is temperature. Therefore, in the system without interactions, the equilibrium temperature T_{eq}^0 (the one at which the fractions of LS and HS molecules are the same) corresponds to a zero effective field and, consequently, $T_{eq}^0 = \Delta/(k_B T \ln g)$. We notice that the T_{eq}^0 coincides with the transition temperature in the bulk material in the present study. The behavior of SCO system with temperature T as a control parameter is effectively described by the Hamiltonian (1). The relation between critical temperature T_c and equilibrium temperatures T_{eq} in purely Ising model determines the type of spin transition: for $T_c < T_{eq}$ by increasing the temperature, the gradual spin transition from LS to HS is observed; in other case, the spin transition can be considered as a first-order phase transition due to its discontinuous type. The functional dependence of the ligand field on the molecules' positions in the lattice is crucial for studying the surface effects of SCO nanoparticles that lead to unusual behavior of thermal hysteresis with changing their sizes.

However, the model (1) can be improved by introducing more rigorous approach with "breathing" crystal field. This means that the external site-dependent field h_i varies in time according to the relation $-h_i = -h_0 + \xi_i(t)$, which leads to the Hamiltonian of the next form [12]

$$H = -J \sum_{\langle ij \rangle} s_i s_j - \sum_i [\Delta - kT \ln g + \xi_i(t)]\, s_i. \tag{3}$$

This situation is more natural and better corresponds to real physical system for which various kinds of randomness that perturb instantaneously the effective field are common. At the approach in which the Ising model is applicable, the stochastic process $\xi_i(t)$ corresponds to the statistically perturbed effective local random field and is interpreted in the manner presented in the works [13, 14]. Due to this fact, the system's behavior during heating and cooling processes significantly differs from the one for the systems without fluctuations [12]. From (3), one can see that the stochastic term becomes especially relevant at critical temperatures.

The simplest fluctuations are the uncorrelated ones and can be described as a white Gaussian stochastic process. We started our analysis with this type of fluctuations. In this approach, each site fluctuates separately and is insensitive to the influence of any other sites from the system. Such type of fluctuations considered for separate SCO compounds can be characterized statistically in the following manner [15–17]:

$$\langle \xi(t) \rangle = 0, \ \langle \xi(t)\xi(t') \rangle = 2\varepsilon^2 \delta(t - t'), \tag{4}$$

where the fluctuation strength ε characterizes the spectral density of stochastic variable $\xi_i(t)$, and t and t' are distinct times. The type of fluctuations introduced here is more common for classical macroscopic systems; however, they can be successfully used in microscopic models as well.

The completely uncorrelated outcomes of white fluctuations serve as a convenient mathematical model of the real noise, but they also bear some physical inconsistencies, such as the infinite white noise energy. In reality, the autocorrelation time of noise may be very small but different from zero. Thus, the system is under colored noise action, which may be described in terms of the Ornstein–Uhlenbeck (OU) process with an exponential correlation function

$$\langle \xi(t)\xi(t') \rangle = \frac{\varepsilon^2}{\tau} \exp\left(\frac{|t - t'|}{\tau}\right), \tag{5}$$

where τ is the autocorrelation time which is also related to the cutoff frequency characteristic to the Lorentzian power spectrum of OU noise.

With decreasing the system size of spin-crossover crystals, the surface effects become crucial for determining their properties. The main difference that drastically impacts the magnetic properties of spin-crossover nanocrystal is the different couplings for molecules inside the lattice and on the surface. For obtaining the model for such spin-crossover lattice, we split the coupling term in (1) into three separate couplings: the one describing the interaction of bulk molecules, the second one describing the coupling of molecules on the surface, and additional one that is responsible for linking the bulk of nanocrystal with its surface. The Hamiltonian of the model is following [18–20]:

$$\begin{aligned} \mathcal{H} = &- \textstyle\sum_i h_i s_i - \sum_{<ij>}^{b} J_{ij}^{b} s_i s_j \\ &- \textstyle\sum_{<ij>}^{s} J_{ij}^{s} s_i s_j - \sum_{<ij>}^{bs} J_{ij}^{b-s} s_i s_j. \end{aligned} \tag{6}$$

Here, J_{ij}^{α} is short-range coupling between nearest-neighbor sites, given in energy units, and is considered as a parameter of the theory. The superscript index $\alpha = s, bs, b$ describes three different interaction cases displayed on the simplified illustration of the model shown in Fig. 1. The three kinds of bonds shown in Fig. 1 arise from surrounding the bulk sites with a surface of one site thick and corresponds to (i) the coupling of molecules on the surface J^s (blue); (ii) the coupling of entire surface with bulk part of nanocrystal J^{bs} (green) which is realized through some intermedi-

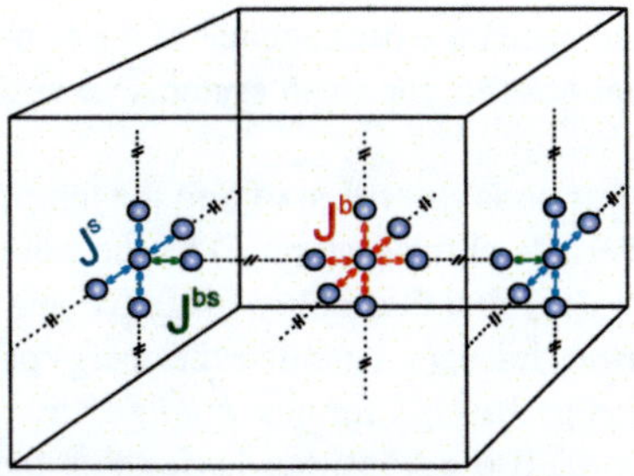

Fig. 1 The spatial structure (simplified) of a 3D SCO crystal where the difference between the interactions of magnetic molecules situated on the surface and inside the lattice is presented (adopted from [20]). Here is additionally shown the coupling between surface and bulk molecules

ate layer of one site thick (in text it is called interface layer and respective coupling is considered as interface one); and (iii) the coupling between the sites inside the lattice J^b (red) which is considered as the strongest one.

3 Numerical Simulation of Magnetic Behavior in Ising-Like Model of Finite-Size Spin-Crossover 3D Structures

We have studied two main effects that impact the behavior of spin-crossover nanocrystal: the properties of a spin-crossover system with fluctuating external filed (in text called as breathing crystal filed) given by Hamiltonian (3) with statistical characteristics of fluctuations described in (4) and (5), and the influence of coupling of molecules from the surface of nanocrystal on system magnetization following the model (6). In this study, we analyze only the impact of weaker interaction of surface's molecules as well as their magnetic order (ferromagnetic or antiferromagnetic). The small particles are modeled with "open boundary conditions" which means that the coupling of surface's sites with environment is zero. In the current research, we focus on three-dimensional spin-crossover models with open boundary conditions and cubic lattice. For three-dimensional system, the first term of Hamiltonian (3) is considered with the contribution of a large enough number of nearest neighbors in order to remain in the valid frame of the mean-field approximation. We investigate our three-dimensional Ising-like systems numerically, since this is almost the only options taking into account the difficulties to obtain the analytical solution for considered dimension and the presence of stochastic processes [21].

The provided numerical studies of spin-crossover system given by Ising-like Hamiltonian, whether of type (3) or (6), involve the Monte Carlo (MC) simulations, which we implemented by using standard Metropolis algorithm. The properties of the system are characterized by the behavior of transition temperatures during heating and cooling processes. Besides the influence of size effects on the transition between spin states, we have investigated the dependence of the temperature transition point on the strength of fluctuations that for small systems and open boundaries becomes very relevant.

We apply the next Metropolis algorithm to find the temperature transition curves that are the main characteristic of the system: (i) fix the value of temperature; (ii) fix initial spin configuration; (iii) find the energy for initial spin configuration; (iv) flip arbitrarily one spin from the lattice; (v) calculate the energy of new spin configuration; (vi) evaluate the transition probability for new configuration and decide if transition is possible or not; and (vii) if the transition is possible the system magnetization (and other necessary parameters) may be found; otherwise, the spin configuration remains unchanged and the next Monte Carlo step starts by repeating the algorithm from step (i). The spin transition probability on step (vi) for arbitrary chosen spin s_i is given by the following expression:

$$P(s_i \to -s_i) = \min\left[1, \exp\left(-\frac{\Delta H\{s_i\}}{kT}\right)\right], \tag{7}$$

where $\Delta H\{s_i\}$ is the energy difference when a spin changes between s_i and $-s_i$:

$$\Delta H\{s_i\} = 2Js_i \sum_j s_j - 2h_i s_i. \tag{8}$$

For provided simulations, the Boltzmann constant is fixed at $k = 1$.

We start the calculations of transition curves from high temperature that corresponds to HS state where all spins are "up." The thermal cycle is considered finished if a temperature sweep from the HS state to the LS one, where all spins are "down," and back to initial HS state takes place. The final system's magnetization is calculated at every temperature value by averaging over an ensemble of pseudospin values s_i taken from the stationary regime of MC trajectory. The magnetization $m = \langle s_i \rangle$ of Ising model and the order parameter of spin-crossover system, i.e., the HS fraction n_{HS}, are related by the following expression:

$$n_{HS} = \frac{m+1}{2}. \tag{9}$$

The temperature at which the fraction of HS and LS molecules is the same, i.e., $n_{HS} = 0.5$, during cooling and heating of spin-crossover system is defined as transition temperature.

4 Cooperative Behavior in 3D Spin-Crossover Ising-Like Model with Correlated Fluctuations

We have studied the thermal phase transition and the bistable behavior of 3D Ising-like spin-crossover system (3) for external random field with various statistical prop-

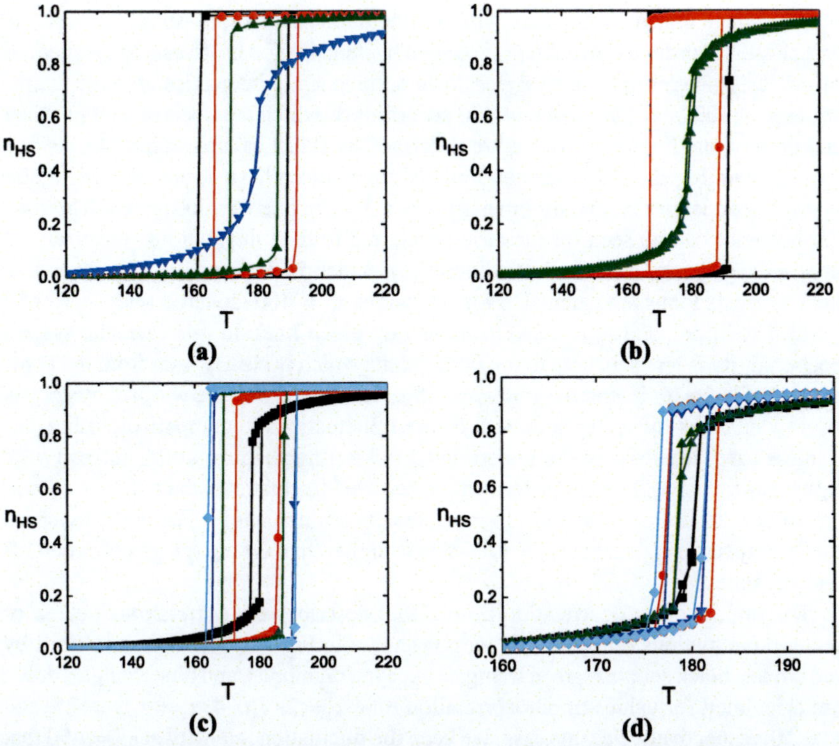

Fig. 2 The temperature transition curves for considered statistical characteristics of fluctuations for $J = 85$ in **a** providing the drastic transition and $J = 55$ in **d** for gradual transition. The hysteresis loops in (a) correspond to fluctuationless system and to the system with uncorrelated fluctuations; in **b**, the hysteresis is reproduced for the system with correlated in time (colored) fluctuations with autocorrelation time $\tau = 10$; in **c**, the impact of autocorrelation time of hysteretic behavior is analyzed for $\varepsilon_c = 450$. Here, the system size is $10 \times 10 \times 10$. The description and details for curves are given in the text

erties. The outcome of investigations is represented in Fig. 2. The computations in this subsection were performed for 1000 MC steps for the cubic model with $L = 10$ and open boundary condition. If the cooperativity magnitude in noiseless system is higher than a threshold one, the advantageous conditions for the occurrence of hysteresis are established. We chose the values of spin–spin interaction $J = 85$, energy gap $\Delta = 900$, and states degeneracy $g = 150$ giving the critical temperature about 180 K and are not related to specific compounds, though similar parameters are often used for numerical computations. The LS and HS steady states become ordered for the range of saturated values on the lower and upper parts of transition curves accordingly.

The thermal behavior of order parameter of spin-crossover system n_{HS} under the influence of white fluctuations is depicted in Fig. 2a. The transition curve marked by squares is calculated for the case without fluctuations. The spin–spin interaction J was chosen in order to have the pronounced hysteresis. The hysteresis loops marked by circles, triangles up, and triangles down correspond to intensities of white fluctuations $\varepsilon = 100$, $\varepsilon = 250$, and $\varepsilon = 450$ accordingly, which are uncorrelated in space and time (white fluctuations) and are described by statistical conditions (4). As one can see from Fig. 2a, the white uncorrelated fluctuations have destructive impact on system's cooperativity, and for intensity $\varepsilon = 450$ the hysteresis disappears. The similar behavior can be seen for the case of colored in time fluctuations (uncorrelated in space), given by statistical conditions (5), which is displayed in Fig. 2b where the hysteresis loops are calculated for intensities $\varepsilon_c = 100$ (squared line), $\varepsilon_c = 450$ (circled red line), and $\varepsilon_c = 1200$ (triangled up green line). In this case, the system cooperativity is less sensitive to the fluctuations which is clearly seen from the comparison of loops for intensities $\varepsilon = \varepsilon_c = 450$. At this intensity, the width of hysteresis slightly changes for the system with colored fluctuations in comparison to fluctuationless case, whereas for the uncorrelated (white) fluctuations at this intensity the hysteresis vanishes. For the system with correlated fluctuations, the collapse of hysteresis width takes place at much higher intensity of fluctuations. For studied system, the hysteresis disappears at $\varepsilon_c = 1200$ that is higher than the energy gap between LS and HS states.

The impact of autocorrelation time τ that describes the correlations degree of colored fluctuations on hysteresis loop is shown in Fig. 2c. The curves marked by red circles, black squares, green triangles up, cyan diamonds, and blue triangles down are calculated for values of autocorrelation time $\tau = 2$, $\tau = 4$, $\tau = 8$, $\tau = 25$, and $\tau = 50$, respectively. For this case, we keep the fluctuation intensity at $\varepsilon_c = 450$ that in the limit of $\tau \to 0$ corresponds to white fluctuations and therefore provides the collapse of hysteresis (see Fig. 2a). With increase of degree of autocorrelation τ, one can see the nonlinear enlargement of hysteresis loop (the hysteresis widths in Fig. 2c slightly change for $\tau = 8; 25; 50$ in comparison to ones for $\tau = 2; 4; 8$).

If a spin–spin interaction J is lower than threshold one between non-hysteresis and hysteresis behavior, a gradual phase transition is predicted. Nonetheless, the fluctuations of crystal field effectually change the behavior of the spin-crossover system. The characteristic transition curves for threshold interaction $J = 55$ are presented in Fig. 2d for deterministic system and for the system with the fluctuating fields that are assumed to be a white Gaussian and colored stochastic processes. In this case, the squared line corresponds to the deterministic system, the circled and triangled up lines are for the system with white fluctuations with intensities $\varepsilon = 20$ and $\varepsilon = 140$, respectively, and the lines indicated by triangles down and diamonds describe the system with time-correlated fluctuations with intensity $\varepsilon_c = 140$ and different degrees of correlations $\tau = 5$ and $\tau = 50$, respectively.

The peculiar feature of temperature curves for the system located on the boundary between non-hysteretic and hysteretic phase space regions is a hysteresis induced by fluctuations that appears at small fluctuation strength and vanishes with increase of their intensity. As one can observe from Fig. 2d, the system shows hysteretic properties in case of relatively small intensity of white fluctuations ($\varepsilon = 20$) which collapses with further increase in fluctuation strength. In studied case, the threshold intensity of white fluctuations is at which hysteresis collapse is $\varepsilon = 140$. We keep this intensity for the system with colored fluctuations in order to provide the qualitative comparison of the impact of degree of autocorrelation on hysteresis width. As shown in Fig. 2d, the hysteresis width for $\tau = 5$ and $\tau = 50$ is slightly changed. Therefore, in the system with fluctuations, there are such conditions in which the action of fluctuations stimulates the manifestation of first-order phase transition and the shifting of cooperativity threshold for the first-order phase transition toward lower values.

For the first-order phase transition, the fluctuations of external field provoke the narrowing of the hysteresis loop despite the correlation properties. The hysteresis loop for the system with time-correlated fluctuations envelops the one for system with white Gaussian stochastic process. With narrowing the spectrum of fluctuations, the hysteresis loop is narrowing but with much slower rate than for white fluctuations. The opposite behavior is noticed with the rising of the autocorrelation time of OU random process, i.e., the degree of correlation. For strongly correlated fluctuations, the width of hysteresis loop rises with an increase in degree of correlation and, at last, it may become wider than without fluctuations (deterministic system).

The relationship between microscopic Ising-like approach and macroscopic models was previously given in our paper [22], where analytical results obtained for classical macroscopic system and supported by numerical simulations show the same trends of vanishing system's bistability under the noise action.

The transition temperatures T_{up} and T_{down} are related to the heating and cooling branches which are initiated from the low-temperature stable LS state and the high-temperature metastable HS state, respectively. Altogether, the competition between energy of the molar entropy change $\Delta S = R \ln g$ and octahedral ligand field splitting energy Δ identifies the preferred (LS or HS) state but the breathing of crystal field changes this situation significantly. If Δ is larger than $k_B T \ln g$ (the strong ligand field), the ground is electronic configuration t_{2g}^6 associated to LS state; otherwise, the $t_{2g}^4 e_g^2$ electronic configuration which correspond to HS state of iron(II) spin-crossover system becomes ground. Our results demonstrate that the transition temperatures of the system strongly depend on the breathing crystal field.

5 The Manifestation of Hysteresis in 3D SCO Nanocrystals with Ferromagnetic and Antiferromagnetic Surface

The influence of surface molecule interaction on hysteretic behavior of whole 3D SCO nanocrystal is evaluated via temperature transition graphs following the same manner used in paper [23], additionally taking into account the possibility of fluctuations of external field. Therefore, the basic model of SCO with ferromagnetic and antiferromagnetic surface from [23] has been adapted for nonzero fluctuation case as is given in (6). In the present studies, the initial system has been prepared in first-order phase transition mode that was reached by picking up the positive coupling value of molecules inside the lattice at $J^b = 85$.

Since the manifestation of first-order phase transition is size-dependent for the studies in the current sections, we choose the reference system size $L = 8$, which is also optimal to observe evident functional dependence on fluctuation strength. Other parameters of the system, the energy gap Δ between LS and HS state, the degeneracy ratio g, and the number of MC steps used in simulations are kept as in Sect. 4. We focused our analysis on the influence of fluctuation strength ε on manifestation of hysteresis, as well as on the role of size effects in fluctuationless system and in the system with nonzero fluctuations. The results are shown in Fig. 3.

At the beginning, the nature of cooperativity of the molecules on the surface is considered ferromagnetic (Fig. 3a–c) with coupling of surface's molecules $J^s = 0.5J^b$. On the next step, the impact of antiferromagnetic nature of surface is investigated by changing the sign of coupling for surface molecules to opposite ($J^s = -0.5J^b$) (Fig. 3d–f). We notice that the interface coupling between the surface and the bulk part of nanocrystal J^{bs} is chosen to be the same by absolute value as J^s for both cases.

The influence of fluctuations on temperature transition is shown in Fig. 3a, d for the system with ferromagnetic and antiferromagnetic interactions on surfaces, respectively. For the model (6) considered here, only the influence of Gaussian white fluctuations described by statistical conditions (4) has been analyzed. The system size was chosen in such a way that it provides the clearly observed hysteresis in the absence of fluctuations; therefore, as one can see, from both cases, the increase of fluctuation strength narrows the width of hysteresis. However, the changing of hysteresis width is not the same for both cases which speaks about different sensitivities of nanocrystal with ferromagnetic and antiferromagnetic surfaces on action of fluctuations. Such a behavior matches well the classical consideration about destructive role of noise in the systems with hysteresis. The keystone difference between the model reported in paper [12], for which the enlargement of hysteresis with increase of fluctuation is observed, and the model studied here is the way of introducing the external randomness. For the transition curves shown in Fig. 3, the external fluctuations were updated at each temperature step, whereas for the results reported in paper [12] the external randomness is updated just after the spin-flip. Physically, such a setup determines the nature of fluctuations: the changing of fluctuations after spin-flip relates their nature rather to the structural disorder of the lattice, in comparison to the latter case in which the fluctuations have thermal origin.

Besides the narrowing of hysteresis loop due to the increase of fluctuation strength, from Fig. 3a, b, one can also conclude that the reaching of saturation values of magnetization takes place for stronger control field (here temperature) for fully ordered HS state and weaker one for fully ordered LS configuration in the system with nonzero fluctuation in comparison to the fluctuationless model. The effect of incomplete transition for the model with antiferromagnetic surface reported in work [23] that favor the remanent magnetization in the system expectedly is also observed in this case, but now one can analyze the role of fluctuations into incompleteness of temperature transition curves. We notice that, since the magnetization of the system is determined by the fractions of HS molecules n_{HS}, the remanent magnetization is inherent only to the incomplete transition from HS to LS phase, whereas the incompleteness of LS

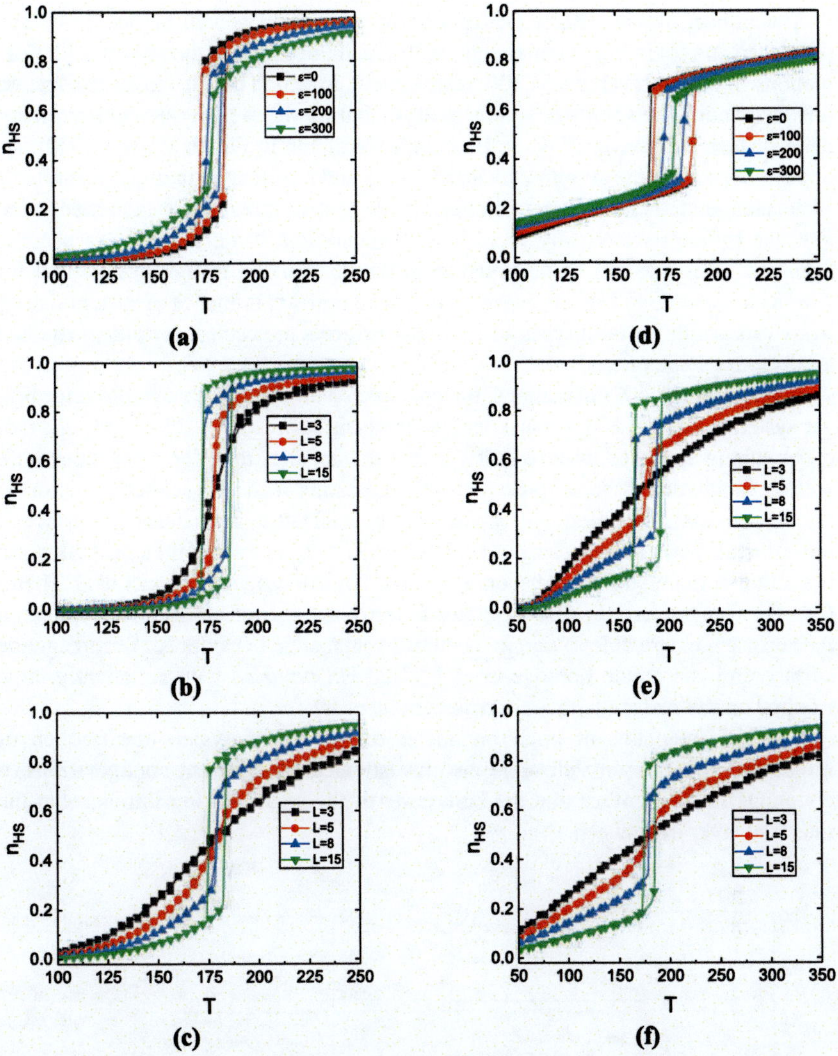

Fig. 3 The behavior of hysteresis loop for 3D spin-crossover system with various configurations of couplings for surface and for interface layer between bulk and surface parts of nanocrystal: **a**, **b**, **c** are for $J^s = J^{bs} = 0.5J^b$; **d**, **e**, **f** are for $J^s = -0.5J^b$; $J^{bs} = 0.5J^b$. The panels (b) and (e) are for fluctuationless system, whereas the plots from (c) and (f) are obtained for $\varepsilon = 300$

to HS transition demagnetize the system due to the presence of remanent fractions of LS molecules that are diamagnetic. As is displayed in Fig. 3d, with increase of fluctuation strength, the slope of the tails of transition curves decreases indicating about contributive role of fluctuations into remanent magnetization for HS to LS transition and remanent demagnetization for LS to HS interconversion.

The impact of size effects on the system behavior is shown in Fig. 3b–f. The panels (b) and (c) in Fig. 3 correspond to fluctuationless and to the system with fluctuations of strength value $\varepsilon = 300$, respectively, in which the surface is considered ferromagnetic. The results of similar analysis but for the system with antiferromagnetic surface are given in Fig. 3e (fluctuationless) and in Fig. 3f (for $\varepsilon = 300$). As one can expect for the system with sizes $L \leq 5$ and ferromagnetic order of surface's molecules, there is no hysteretic behavior. For system sizes bigger than mentioned one, the hysteresis is clearly seen; however, its width has a tendency to become saturated. This one can be observed from the comparison of hysteresis width for $L = 8$ and $L = 15$ in Fig. 2b, e. We notice about uniform influence of system size on hysteresis width for both types of order of surface's molecules: ferromagnetic and antiferromagnetic one.

In the system with fluctuations, the appearance of hysteresis takes place for bigger system's sizes $L > 8$. From a detailed analysis between Fig. 3c, f, we found that difference in shape of hysteresis loops is much smaller than for other cases with weaker fluctuations. We suppose that this interesting feature is related to action of fluctuations that could randomly flip the spin over all lattice inclusively the surface of nanocrystal. Since the coupling of surface's molecules is weaker than the bulk ones, the action of fluctuations can become dominant on the surface of nanocrystal. Therefore, they may reduce the effects related to ferro- and antiferromagnetic coupling on the surface, i.e., are able to mimic the antiferromagnetic behavior for ferromagnetic surface and can create homogeneous LS and HS domains with sufficiently large lifetime on the surface with antiferromagnetic order.

The characteristics of order parameter of SCO nanocrystal—the fraction of molecules in high-spin state—are also investigated for different configurations of couplings for the surface and the bulk parts of the crystal taking into account the interface layer in between.

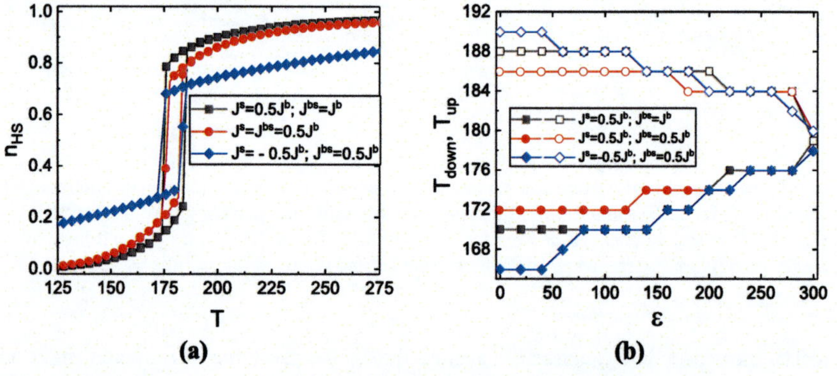

Fig. 4 The hysteresis loops for different configurations of interactions surface-to-link layer-to-bulk for fluctuation strength $\varepsilon = 200$ **a** and the behavior of transition temperatures with increase of fluctuation strength **b**. The data are obtained for 3D cubic lattice with side $L = 8$

We considered three main types of setups that are shown in Fig. 4a: (i) the coupling of surface's molecules is one half from coupling of bulk sites $J^s = 0.5J^b$, whereas the coupling of interface layer is the same as bulk one $J^{bs} = J^b$ (squared curve); (ii) the coupling of surface's molecules and the coupling of interface layer is one half from the bulk coupling $J^s = J^{bs} = 0.5J^b$ (red circled curve); (iii) the coupling of surface molecules is one half from bulk coupling, but is of antiferromagnetic nature, and the coupling of interface layer J^{bs} is the same as in (ii) (blue diamonded curve). As one can observe the role of coupling with interface layer is rather small, it slightly influences the reaching of saturated HS and LS states. The changing of the nature of surface coupling from ferromagnetic to antiferromagnetic one changes the nonlinear slopes of transition curves to linear. In the system with antiferromagnetic surface, the transition is incomplete in comparison to ferromagnetic coupling of surface's molecules; however, the degree of incompleteness for the same parameters of the system is bigger than in work [23] due to the presence of fluctuations.

The changing of transition temperatures for cooling and heating branches of hysteresis as a function of fluctuation strength is plotted in Fig. 4b for all three configurations of couplings considered in Fig. 4a. Here, the curves with filled markers are for temperatures of cooling branch, and the ones with empty markers are the transition temperatures during heating. The general tendency for all three configurations is the collapsing of hysteresis width with increase in fluctuation strength. However, the narrowing of hysteresis is non-uniform as a function of fluctuation strength which is shown by ranges of constant values on temperature curves. The impact of chosen configuration of couplings on hysteresis width is much more visible for small fluctuations. In this case, the coupling configurations (iii) described above give the larger hysteresis width, but it is much sensitive to increase of fluctuation strength. The hysteresis width is the smallest one among analyzed cases for coupling configuration (ii); however, this configuration as well as the configuration (i) shows almost stable hysteresis width in large interval of fluctuation strength ($0 < \varepsilon \lesssim 150$).

The presence of fluctuations in the model with ferromagnetic and antiferromagnetic coupling for surface's molecules emphasizes the role of size effects by shifting the collapse of hysteresis toward larger values of lattice side and reduce the role of nature of coupling on the surface in reaching the saturated values of HS fractions n_{HS} for temperature transition curves.

6 Summary and Conclusion

In the present study, the conception of breathing crystal field was applied to spin-crossover materials in the context of Ising-like Hamiltonian. The existence of an external (relative to magnetic ion) crystal field of random statistical nature is the crucial feature of the model. We have provided the study of spin-crossover magnetic nanoparticles that are depicted by Ising-like model with the accounting of correlated fluctuations.

The transition curves in three-dimensional spin-crossover crystal were computed by Monte Carlo simulations for various external fields: (i) without fluctuations; (ii) with white fluctuations; and (iii) with colored time-correlated fluctuations. It was shown that the strengthening of fluctuation of crystal field in 3D spin-crossover acts qualitatively different whether the first- or gradual phase transition occurs. In the case of first-order phase transition, the white and colored fluctuations tend to narrow the loop width; however, the rate of narrowing is very dependent on the statistical characteristics of fluctuations that we have probed in our study (the strength of fluctuations and the autocorrelation time). The presence of time correlations in fluctuations makes hysteresis to collapse at much higher fluctuation strength. For the system located on the boundary between non-hysteretic and hysteretic phase regions, characterized by threshold spin–spin coupling, the presence of fluctuations favors the appearance of hysteresis and, therefore, reduces the threshold spin–spin interaction.

The impact of the size and surface effects on the hysteresis loops of two-dimensional and three-dimensional spin-crossover system with temperature changes and perturbed by fluctuations has been examined as well. The conditions of mono- and bistable behavior of spin-crossover compounds were characterized, and respective regions of first-order and gradual phase transitions were found. Likewise, the detailed discussion of the relations of hysteresis width on the size effects and the strength of statistical fluctuations were given.

It has been found out that for spin-crossover system the increase of system size leads to the increase in hysteresis width, whereas the high enough fluctuations ($\varepsilon \gtrsim 15\%$ of ligand field strength) are destructive and move the system toward gradual transitions on phase space. It was shown the special role of fluctuations at the border between gradual and first-order phase transition which corresponds to the critical point of the system. In this regime, the presence of fluctuations of small strength ($\varepsilon \approx 1.5 - 5\%$ of ligand field strength) favors the manifestation of hysteresis, which therefore means the shifting of the system from the region of gradual transition to the one with first-order phase transition. The reduction of intersite interactions by increasing fluctuation strength emphasizes the role of fluctuations in hysteretic behavior of spin-crossover systems and makes possible the obtaining of spin-crossover solids with bistable properties with weaker coupling between their sites.

We have analyzed the properties of spin-crossover 3D nanocrystals in terms of cooperativity accounting for changing from the surface of nanocrystal to the bulk of the lattice in the presence of Gaussian random white fluctuations. The implied change of cooperativity represents better approach for the system with nanoscopic size that is exposed to the environment. It demonstrates the effect of cooperativity of surface molecules on the shifting of critical temperatures of spin transition under fluctuations. We have determined also that the critical temperature of establishing the targeted phase (whether LS or HS) depends on the effective cooperativity defined by all system sites.

It was shown that the nature of cooperativity for sites on the surface has the drastic impact on the completeness of phase transition. For a system with antiferromagnetic coupling of the sites on the surface, the temperature transition curves demonstrate a premature saturation with smaller magnetization value for HS state and greater one for LS configuration in contrast to the completely ferromagnetic system. Physically, it is explained by the antiferromagnetic ordering of molecules on the surface that can impact the ordering of the next layer. The last plays a role of interface between the surface and the bulk parts of the nanocrystal system.

The determination of critical particle size necessary to preserve the first-order transition in SCO materials is imperative because bistability depends on the collective behavior of molecules in the SCO lattice. The critical temperature of the spin transition shifts toward equilibrium temperature as the nanocrystal size approaches to the critical one, where the hysteresis width vanishes.

It was an established good agreement between the obtained numerical results and the experimental studies. Most of the experimentally obtained hysteresis loops for spin-crossover compounds, in which the shifting of critical temperatures, the incompleteness of transitions, and the vanishing of hysteresis with decreasing of system size are observed, might be explained within the system with the change in cooperativity from the surface to the bulk of the lattice, taking into account the anti- and ferromagnetic nature of surface's sites.

It is expected that obtained results will be useful for designing new electronic devices based on spin-crossover compounds, including sensors, data storage, and recording systems with much better characteristics. Understanding the role of size effects in spin-crossover materials with statistical fluctuations may improve the functioning of any system based on molecular materials with spin transition.

Acknowledgements The research of A.M. was partly supported by the project No.2015/19/B/ST2/01028 financed by National Science Centre (Poland).

References

1. Atkins P, Overton T, Rourke JP, Weller MT, Armstrong FA (2010) Shriver and Atkins inorganic chemistry. Oxford University Press, New York
2. Halcrow MA (ed) (2013) Spin-crossover materials: properties and applications. Wiley, Chichester
3. Gudyma Iu, Enachescu C, Maksymov A (2015) Kinetics of nonequilibrium transition in spin-crossover compounds. In: Nanocomposites, Nanophotonics, Nanobiotechnology, and Applications, Springer, pp 375–401
4. Kumar KS, Ruben M (2017) Emerging trends in spin crossover (SCO) based functional materials and devices. Coordin Chem Rev 346:176–205
5. Iasco O, Boillot M-L, Bellec A, Guillot R, Riviere E, Mazerat S, Nowak S, Morineau D, Brosseau A, Miserque F (2017) The disentangling of hysteretic spin transition, polymorphism and metastability in bistable thin films formed by sublimation of bis (scorpionate) Fe (II) molecules. J Mater Chem C 5(42):11067–11075
6. Molnár G, Rat S, Salmon L, Nicolazzi W, Bousseksou A (2018) Spin crossover nanomaterials: from fundamental concepts to devices. Adv Mater 30(5):1703862
7. Simon-Yarza T, Mielcarek A, Couvreur P, Serre C (2018) Nanoparticles of metal-organic frameworks: On the road to in vivo efficacy in biomedicine. Adv Mater 2018:1707365
8. Salmon L, Catala L (2018) Spin-crossover nanoparticles and nanocomposite materials. Comptes Rendus Chimie 21(12):1230–1269

9. Mikolasek M, Ridier K, Bessas D, Cerantola V, Félix G, Chaboussant G, Piedrahita-Bello M, Angulo-Cervera E, Godard L, Nicolazzi W, Salmon L, Molnár BA (2019) Phase stability of spin-crossover nanoparticles investigated by synchrotron Mössbauer spectroscopy and small-angle neutron scattering. J Phys Chem Lett 10(7):1511–1515
10. Wajnflasz J, Pick R (1971) Transitions Low spin—High spin dans les complexes de Fe^{2+}. J de Physique Colloques 32:C1–91
11. Gudyma Iu, Maksymov A, Bobák A (2017) Modeling problems of spin crossover nanocrystals. In: Nanophysics, nanomaterials, interface studies, and applications, Springer, pp 63–77
12. Gudyma I, Maksymov A, Enachescu C (2014) Phase transition in spincrossover compounds in the breathing crystal field model. Phys Rev B 89:224412
13. Gudyma YV (2004) Nonequilibrium first-order phase transition in semiconductor system driven by colored noise. Physica A 331(1–2):61–68
14. Gudyma I, Maksymov A, Enachescu C (2010) Decay of a metastable high-spin state in spin-crossover compounds: mean first passage time analysis. Eur Phys J B 78(2):167–172
15. Gudyma Y, Semenko O (2004) Nonequilibrium kinetics in spin-crossover compounds. Phys Status Solidi B 241(2):370–376
16. Gudyma I, Maksymov A (2011) High spin metastable state relaxation of spin-crossover solids driven by white noise. J Phys Chem Solids 72:73–77
17. Gudyma I, Maksymov A, Dimian M (2013) Stochastic kinetics of photoinduced phase transitions in spin-crossover solids. Phys Rev E 88(4):042111
18. Gudyma I, Maksymov A, Spinu L (2015) Size effects in spin-crossover nanoparticles in framework of 2D and 3D Ising-like breathing crystal field model. Appl Surface Sci 352:60–65
19. Gudyma I, Ivashko V, Bobak A (2017) Surface and size effects in spin-crossover nanocrystals. Nanoscale Res Lett 12(1):101
20. Gudyma I, Maksymov A (2017) Surface-environment effects in spin-crossover solids. Appl Surface Sci 407:93–98
21. Gudyma I, Maksymov A, Miyashita S (2011) Noise effects in a finite-size Ising-like model. Phys Rev E 84(3):031126
22. Gudyma I, Maksymov A (2012) Optically induced switching in spin-crossover compounds: microscopic and macroscopic models and their relationship. Applied Opt 51(10):C55–C61
23. Gudyma I, Maksymov A (2019) The cooperativity in 3D spin-crossover nanocrystals with ferromagnetic and antiferromagnetic surface. Appl Surface Sci 489:779–784

Ballistic Transmission of the Relativistic Quasielectrons Through the Potential Barrier in the Alfa-T_3 Model

A. M. Korol, N. V. Medvid', A. I. Sokolenko, and O. Yu. Shevchenko

1 Introduction

Some modern physical structures can be conveniently described using the so-called α-T_3 model [1–8]. This model can rightly be attributed to a new class of objects that have received the name of Dirac materials in recent years [9]. These include very different objects in their structure, in particular, the low- and high-temperature d-wave superconductors, superfluid phases 3He, graphene, two- and three-dimensional insulators, etc. [9]. The key concept that unites these different objects is a linear dispersion relation that describes the low-energy excitations of the quasiparticles. Due to the fact that the Dirac materials have a number of non-trivial, interesting properties, they are actively studied in the last time. Under low energies, the quasiparticle states of the Dirac materials are described by a massless Dirac equation in one or two dimensions, analogous to the equation for the quasielectrons in graphene. The dispersion relation for the Dirac particles relates to a cone in the three-dimensional case. Some properties of the quasiparticle states are expressed in terms of topologically invariant quantities and, importantly, are protected from the influence of moderate perturbations due to the symmetry of inversion of time in the corresponding Hamiltonian.

The α-T_3 model is an intermediate structure between a dice lattice and graphene. It is characterized by the parameter α, which determines the coupling strength between the central atom of the hexagonal lattice and the atoms in the hexagon vertices [1–8]. It is clear that different values of α correspond to different physical states of the α-T_3 model, and it was successfully applied to various physical structures [1–8].

A. M. Korol (✉)
Laboratory on Quantum Theory in Linkoping, ISIR, P.O. Box 8017, 580 Linkoping, Sweden
e-mail: korolam@ukr.net

A. M. Korol · N. V. Medvid' · A. I. Sokolenko · O. Yu. Shevchenko
National University for Food Technologies, Volodymyrska Str. 68, Kiev, Ukraine

O. Fesenko and L. Yatsenko (eds.), *Nanomaterials and Nanocomposites, Nanostructure Surfaces, and Their Applications*, Springer Proceedings in Physics 246, https://doi.org/10.1007/978-3-030-51905-6_12

At the same time, it is known that the characteristics of structures based on Dirac materials are significantly influenced by the difference in the values of the Fermi velocity in different parts of the structure [10–20]. A lot of various structures with non-equal Fermi velocities in different regions of the given structure were studied in last years. They comprise the graphene-based single- and double-barrier structures, various types of superlattices including the quasiperiodic ones, superconducting junctions, structures based on the topological insulators, etc. [10–20].

Motivated by the above considerations, in this paper, we study the ballistic transmission of quasielectrons through a rectangular potential barrier in the α-T_3 model and show that it depends strongly on the relation between the parameters α and β, where β is equal to the ratio of the Fermi velocities in the barrier and out-of-barrier areas. By changing the values of the parameters α and β, one can flexibly control the transmission properties of the structure under consideration within a wide range.

2 Model and Formulae

The Dirac-like equation for the considered model can be represented as follows [1–8]:

$$\begin{pmatrix} 0 & f\cos\varphi & 0 \\ f^*\cos\varphi & 0 & f\sin\varphi \\ 0 & f^*\sin\varphi & 0 \end{pmatrix}\psi + UI_0\psi = E\psi, \tag{1}$$

where U is the external potential which corresponds to the rectangular barrier and is equal to

$$U(x) = \begin{cases} 0, x \le 0 \\ U, 0 < x < D \\ 0, D << x \end{cases} \tag{2}$$

in different regions of the given structure; I_0 is the identity matrix.

The quantity f in (1) is equal to

$$f = v_F(k_x - ik_y). \tag{3}$$

For our purpose, it suffices to take into consideration only one K valley in the hexagonal Brillouin zone. The quantities v_F and k_x acquire different values in the barrier and out-of-barrier regions. The parameter φ is introduced for convenience: $\varphi = arctg\alpha$, α is a parameter showing the coupling strength of the central atom with the atom at the hexagon vertices; for the dice lattice $\alpha = 1$, for graphene $\alpha = 0$.

The eigenfunctions in the (1) can be represented as follows:

$$\psi_I = \frac{1}{\sqrt{2}}\begin{pmatrix} \cos\phi e^{i\varphi} \\ 1 \\ \sin\phi e^{-i\varphi} \end{pmatrix} e^{ik_x x} e^{ik_y y} + \frac{r}{\sqrt{2}}\begin{pmatrix} -\cos\phi e^{i\varphi} \\ 1 \\ -\sin\phi e^{i\varphi} \end{pmatrix} e^{-ik_x x} e^{ik_y y}$$

$$\psi_{II} = \frac{a}{\sqrt{2}}\begin{pmatrix} \cos\phi e^{i\theta} \\ -1 \\ \sin\phi e^{-i\theta} \end{pmatrix} e^{iq_x x} e^{ik_y y} - \frac{b}{\sqrt{2}}\begin{pmatrix} \cos\phi e^{-i\theta} \\ 1 \\ \sin\phi e^{i\theta} \end{pmatrix} e^{-iq_x x} e^{ik_y y}$$

$$\psi_{III} = \frac{t}{\sqrt{2}}\begin{pmatrix} \cos\varphi e^{-i\phi} \\ 1 \\ \sin\varphi e^{-i\phi} \end{pmatrix} e^{ik_x x} e^{ik_y y}, \quad (4)$$

where ϕ and θ are the angle of incidence and the refraction angle, respectively; the quasimomentums in the out-of-barrier region k_x and in the barrier region q_x are equal to

$$k_x = \sqrt{E^2 - k_y^2}; \quad q_x = \sqrt{\frac{(U-E)^2}{\beta} - k_y^2}; \quad (5)$$

$tg\theta = \frac{q_y}{q_x}$; $q_y = k_y$; the units with $v_{F1} = 1$; $h = 1$ are adopted.

Using the appropriate matching conditions [21, 22]

$$\sqrt{v_{F1}}[\cos\varphi\psi_I(-\varepsilon) + \sin\varphi\psi_{III}(-\varepsilon)] = \sqrt{v_{F2}}[\cos\varphi\psi_I(\varepsilon) + \sin\varphi\psi_{III}(\varepsilon)],\ \varepsilon \to 0, \quad (6)$$

we can deduce the expression for the transmission coefficient T

$$\begin{aligned} T &= 16\cos^2\theta\cos^2\phi / \left(f_p^2 + f_m^2 - 2f_p f_m \cos(2q_x D)\right) \\ f_p &= 2 - 2\cos(\theta+\phi) - \sin^2(2\varphi)(\sin\theta + \sin\phi)^2; \\ f_m &= 2 + 2\cos(\theta-\phi) - \sin^2(2\varphi)(\sin\theta + \sin\phi)^2. \end{aligned} \quad (7)$$

3 Results and Discussion

Figures 1, 2, and 3 show the dependence of the transmission coefficient T on the angle of incidence of the quasielectrons on the structure considered for such values of parameters:

for Fig. 1: $E = 2$; $U = 3$; $\beta = 0.5$;
for Fig. 2: $E = 3$; $U = 4$; $\beta = 1$;
for Fig. 3: $E = 3$; $U = 4$; $\beta = 5$; other parameters as shown in Fig. 1.

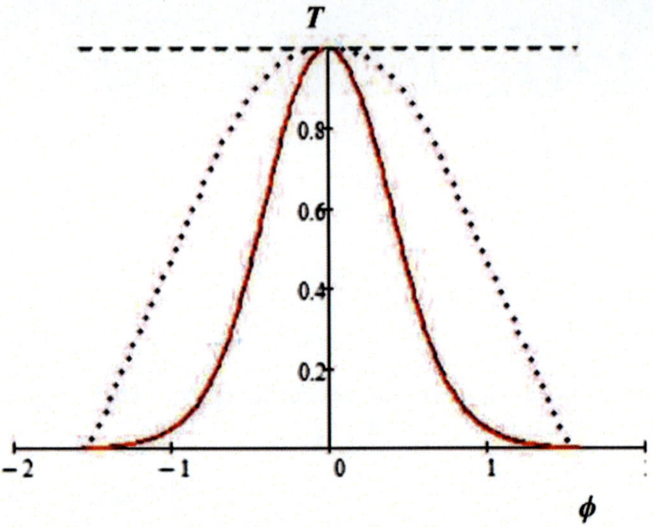

Fig. 1 Dependence of the transmission coefficient T on an incidence angle ϕ for the following set of the parameters involved: $E = 2$; $U = 3$; $\beta = 0.5$

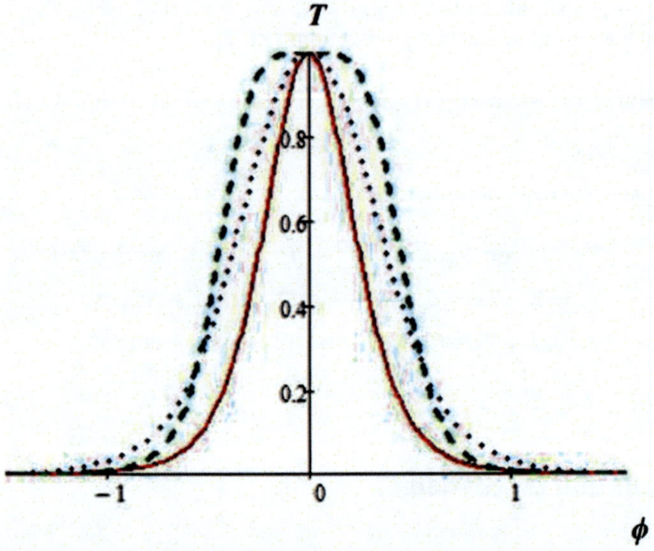

Fig. 2 Same as in Fig. 1 with the parameters: $E = 3$; $U = 4$; $\beta = 1$

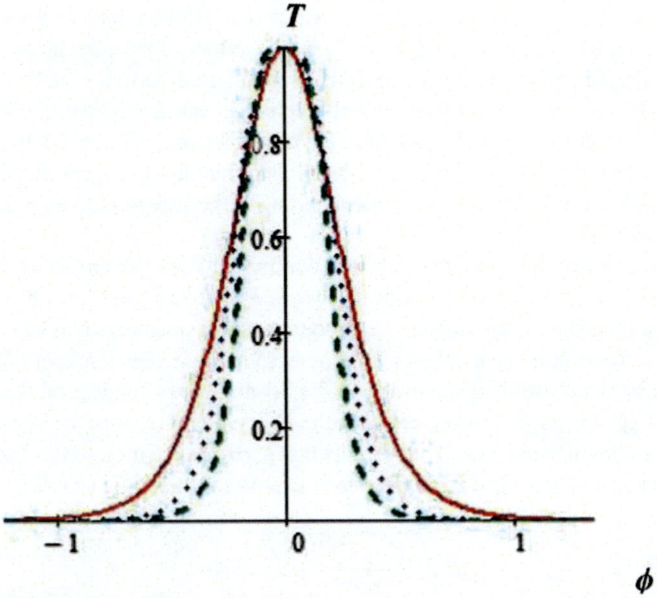

Fig. 3 Same as in Fig. 2 but the value of β is equal to 5

The barrier thickness for all these figures is equal to $D = 1$. The angle values are given in radians.

In all figures of this article, except Fig. 7, the solid line corresponds to the value $\alpha = 0.2$, the dotted lines to $\alpha = 0.7$, and the dashed lines to $\alpha = 1$.

Analyzing the spectra presented in Figs. 1, 2 and 3, first of all, note that the transmission of quasielectrons through a given structure depends essentially on the values of the parameter α, as well as on the values of the parameter β, that is, substantially depends on the interplay between α and β. For most values of the parameters of this problem, the region of the incidence angles with a high value of T is wider for larger α and smaller β, and the values of the coefficient T for a fixed angle increase with increasing of α and decreasing in β.

At the same time, as shown in Fig. 3, for sufficiently large values of parameter β, for example, for $\beta = 5$, the range of angles ϕ with large values of T is slightly dependent on parameter α (curves with different values of α almost overlap) and are smaller than for smaller values of β, for example, for $\beta = 0.5$. In addition, for some angles of incidence ϕ, the values of the coefficient T for smaller values of α are greater than its values for larger α—this indicates the non-monotonic nature of the dependence of the transmission coefficient T on the coupling parameter α.

All the above figures show that in the structure considered there is an effect similar to the Klein paradox: for the normal incidence of quasiparticles on the barrier, it becomes perfectly transparent, that is, the value of the coefficient T is equal to unity, regardless of the value of the height of the barrier U and its thickness D. We emphasize that in this case $T = 1$ also independently on the values of the parameters α and β.

A remarkable property of the transmission of quasiparticles in this structure is that it exhibits an effect called supertunneling in literature, namely, for certain values of the parameters of the problem, the transmission coefficient T is maximum ($T = 1$) in the whole range of values of the angle of incidence ϕ—see the dashed line in Fig. 1 (the formula for energies corresponding to the supertunneling phenomenon is given below).

Figure 4 shows the dependence of the coefficient T on the angles ϕ for thicker barriers; the parameters are as follows: $D = 8$, $E = 2.5$, $U = 4$, and $\beta = 0.5$. The peaks observed for certain angles correspond to the resonance states of the Fabry–Perot type. Their number increases with increase in the barrier thickness D, and this number, like the position of the peaks on the 0ϕ-axis, does not depend on α; but the value of T for the angles placed between the Fabry–Perot resonances depends on α, increasing with increase in α. The formula that gives the position of the Fabry–Perot resonances can be obtained from the above expressions and it is as follows:

$$D\sqrt{\frac{(E-U)^2}{\alpha^2} - k_y^2} = n\pi. \quad (8)$$

Figure 5 depicts the dependence $T\ (\phi)$ for the case of the overbarrier passage (parameters $D = 8$; $E = 5$; $U = 1$; $\beta = 0.5$) of the electron wave through the given structure. Note that (1) the position of the resonant peaks on the 0ϕ-axis is independent of α; (2) the value of T is large even for the angles of incidence ϕ close to $\pi/2$.

Figure 6 shows the dependence of the transmission coefficient T on the quasielectron energy E for the parameters $U = 4$, $\phi = 1$, $\beta = 0.5$, and $D = 1$.

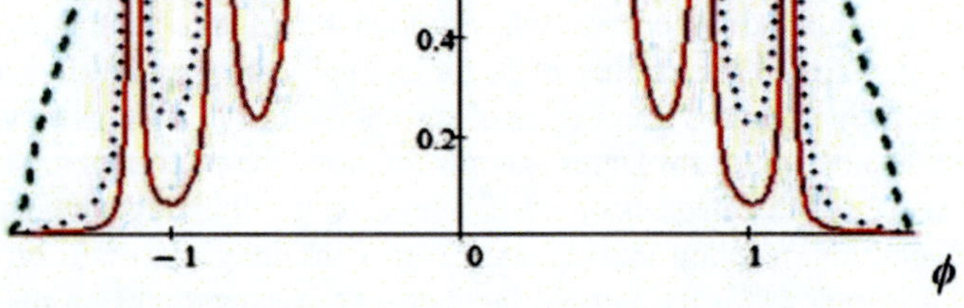

Fig. 4 T vs. ϕ plot for larger barrier width: $D = 8$, $E = 2.5$; $U = 4$; $\beta = 0.5$

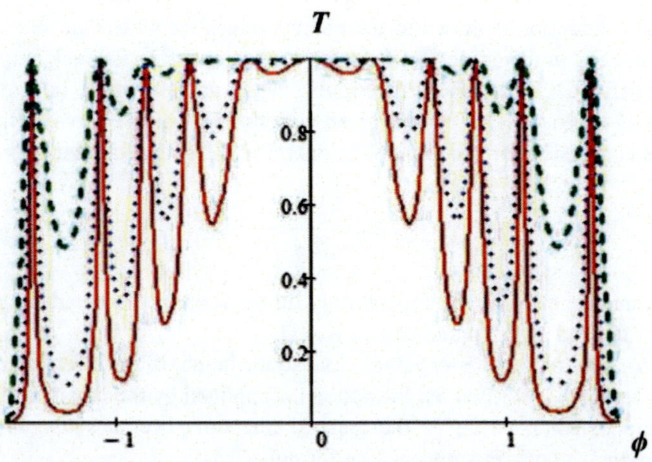

Fig. 5 T vs. ϕ plot for the overbarrier passage of the quasielectrons; $D = 8$; $E = 5$; $U = 1$; $\beta = 0.5$

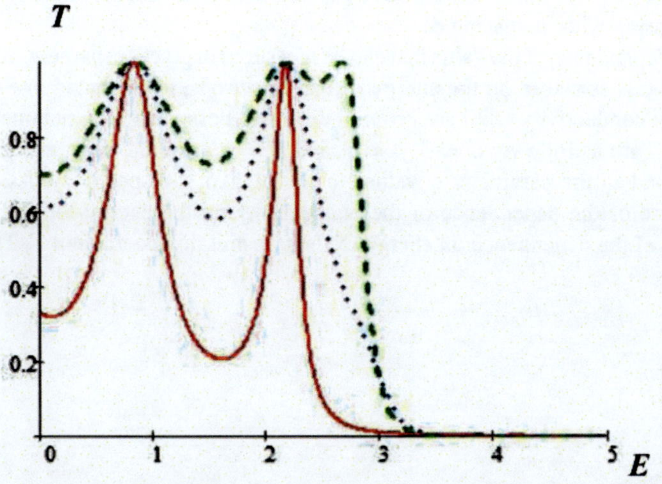

Fig. 6 The dependence of the transmission coefficient T on the quasielectron energy E for the parameters: $U = 4$; $\phi = 1$; $\beta = 0.5$; $D = 1$

In this dependence, it is particularly noteworthy that it has a pronounced tunneling resonance character. This indicates that the structure considered in this paper is a tunneling resonance structure. Most of the resonance peaks shown correspond to Fabry–Perot resonances. Their energy position can be found from the above expressions and is determined by formula (8).

The energy values for the Fabry–Perot resonances depend significantly on the relation between the parameters U, D, and β, but do not depend on the parameter α.

Particular attention is drawn to the energy values for which the supertunneling effect is realized in this structure: for them the transmission is ideal, i.e., the transmission coefficient T equals the maximum value, namely, unity, for any angle of incidence of quasiparticles on the given structure. From the above formulas, we obtain the expression for supertunneling energies; it has the following form:

$$E_s = \frac{U}{1+\beta}. \quad (9)$$

Therefore, the energies E_s depend only on the ratio of the parameters U and β and do not depend on the parameters α and D.

In the vicinity of the energy values close to the height of the electrostatic barrier: $E = U$, a bandgap is formed. Its formation is explained by the fact that for energies $E \sim U$ for most incidence angles, the quasi-momentum q becomes imaginary and the electron wave becomes evanescent (see formula (5)).

The bandgap of energies depends on the parameters α and β, namely, it increases with decreasing α and increasing β, and is much more sensitive to the change in parameter β. Also, the forbidden band depends essentially on the angle of incidence ϕ, increasing with increasing ϕ.

The dependence of the value that can be measured in practice, namely, the conductivity of this structure on the energy of quasielectrons, is presented in Fig. 7. The values of conductivity $G(E)$ are given in dimensionless units. The parameter values for Fig. 7 are as follows: $U = 3$, $\beta = 1$, and $D = 1$; here red, blue, and green lines correspond to the parameter α values: $\alpha = 0.2$; 0.6;1, respectively. As expected, the nature of the dependence of the conductivity on the energy reflects the main features of the dependence on energy of the transmission coefficient T. The energy

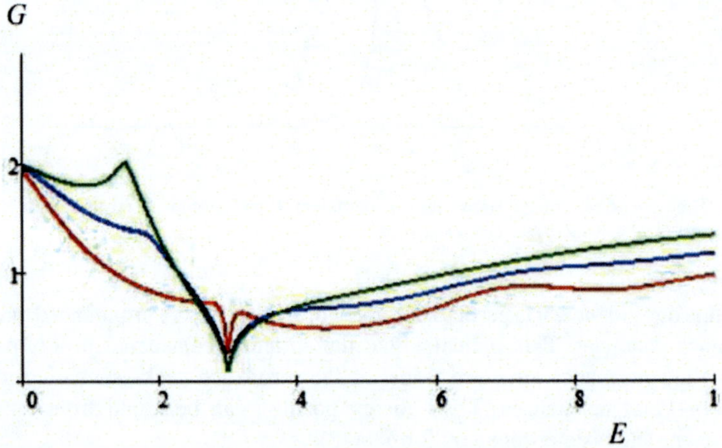

Fig. 7 Dependence of the normalized conductivity G on energy E for the parameters: $U = 3$; $\beta = 1$; $D = 1$

for the maximum values of the function *G(E)* corresponds to the supertunneling phenomenon. The maxima of the function *G(E)* also hold for the energies for which the Fabry–Perot resonances are observed. There is a region of energies in which the function *G(E)* has minimal values—it is related to the bandgap in the dependence of *T(E)*. The magnitude of *G(E)* is an oscillatory function of *E*, and the amplitude and period of the oscillations depend on the parameters α, β, and *D*. As α increases or β increases, the conductivity values of *G(E)* increase for most parameters. As the thickness of the barrier increases, the number of oscillations (peaks) depending on *G(E)* in the fixed energy interval increases.

4 Conclusions

We consider here one of the modern structures: the so-called alfa-T_3 model which interpolates between the dice lattice and the graphene one with the help of the parameter α that allows to change the coupling strength between the honeycomb lattice (HCL, graphene) and the HCL with the central cite (dice lattice) and varies from zero to unity.

It is shown that this object, in the general case, is a resonant-tunneling structure, that is, its transmission spectra are represented by a set of resonance peaks with the values of the transmission coefficient *T* close to unity. In particular, for the values $\beta < 1$ (β is equal to the ratio of the Fermi velocities in the barrier and out-of-barrier regions), there is a clear structure of the resonance peaks of the Fabry–Perot type for any α values; their position on the energy axis depends on the value of β.

In the case of a zero incidence angle, there is a Klein paradox phenomenon which occurs for all values of β and α and for any values of the height and width of the potential barrier. In the vicinity of the energy values close to the height of the potential barrier, there is a forbidden band (gap) whose width and position are regulated by the values of the quantities β and ϕ and do not depend on α.

As the value of α increases, a significant increase of the transmission coefficient for all values of *E* and β is observed. For the magnitude $\alpha = 1$, there is a phenomenon of supertunneling. The energy for which the supertunneling takes place is the function of β and is expressed by the formula $E = U/(1 + \beta)$, $\beta < 1$, provided that $\alpha = 1$.

The conductivity of the given structure *G(E)* is evaluated with the help of the Landauer–Buttiker formalism and, as shown, essentially depends both on the value of α and on the value of β. Specifically, the conductivity increases with increase in α and has a complex dependence on β. In particular, for values $\beta > 1$, the quantity *G(E)* decreases with increasing β for all α. For $\beta < 1$, the function *G(E)* is characterized by regions with the conduction oscillations, as well as with the regions with a wide minimum in the vicinity of the energy close to the potential barrier height.

References

1. Raoux A, Morigi M, Fuchs J-N, Piéchon F, Montambaux G (2014) From Dia- to paramagnetic orbital susceptibility of massless fermions. Phys Rev Lett 112:026402
2. Piéchon F, Fuchs J-N, Raoux A, and Montambaux G (2015) Tunable orbital susceptibility in α-T_3 tight-binding models. J Phys Conf Ser 603:012001
3. Malcolm JD, Nicol EJ (2015) Magneto-optics of massless kane fermions: role of the flat band and unusual Berry phase. Phys Rev B 92:035118
4. Illes E, Nicol EJ (2016) Magnetic properties of the α-T3 model: Magneto-optical conductivity and the Hofstadter butterfly. Phys Rev B 94:125435
5. Kovács ÁD, Dávid G, Dóra B, Cserti J (2017) Frequencydependent magneto-optical conductivity in the generalized α-T3 model. Phys Rev B 95:035414
6. Biswas T, Ghosh TK (2015) Magnetotransport properties of the α-T3 model. J Phys Condens Matter 28:95302
7. Illes E, Carbotte JP, Nicol EJ (2015) Hall quantization and optical conductivity evolution with variable Berry phase in the α-T3 model. Phys Rev B 92:245410
8. Illes EE, Nicol EJ (2017) Klein tunneling in the alfa-T_3 model. Phys Rev B 95:235432
9. Wehling TO, Black-Schaffer AM, Balatsky AV (2014) Dirac materials. Adv Phys 63:1
10. Liu L, Li Y-X, Liu J (2012) Transport properties of Dirac electrons in graphene based double velocity-barrier structures in electric and magnetic fields. Phys Lett A 376:3342–3350
11. Wang Y, Liu Y, Wang B (2013) Resonant tunneling and enhanced Goos-Hänchen shift in graphene double velocity barrier structure. Physica E 53:186–192
12. Sun L, Fang C, Liang T (2013) Novel transport properties in monolayer graphene with velocity modulation. Chin PhysLett 30(4):047201
13. Raoux A, Polini M, Asgari R, Hamilton AR, Fasio R, MacDonald AH (2009) Velocity–modulation control of electron-wave propagation in graphene. arXiv:0912.2608v1 [cond-mat.mes-holl]
14. Concha A, Tešanović Z (2010) Effect of a velocity barrier on the ballistic transport of Dirac fermions. Phys Rev B 82:033413
15. Yuan JH, Zhang JJ, Zeng QJ, Zhang JP, Cheng Z (2011) Tunneling of Dirac fermions in graphene through a velocity barrier with modulated by magnetic fields. Phys B 406:4214–4220
16. Krstajic PM, Vasilopoulos P (2011) Ballistic transport through graphene nanostructures of velocity and potential barriers. J Phys Condens Matter 23:000000(8 pp)
17. Korol AM, Medvid' NV, Litvynchuk SI (2015) Transport properties of the Dirac-Weyl electrons through the graphene-based superlattice modulated by the Fermi velocity barriers. In: Proceedings in Physics, vol 167. Springer, pp 215–221
18. Korol AM, Medvid' NV, Sokolenko AI (2018) Transmission of the relativistic fermions with the Pseudospin equal to one through the quasi-periodic barriers. Physica Status Solidi (B) Basic Res 255(9):1800046
19. Korol AM, Medvid' NV, Sokolenko AI, Sokolenko IV (2019) Ballistic transmission of the Dirac quasielectrons through the barrier in the 3D topological insulators. In: Proceedings in Physics, vol 221. Springer, pp 517–525
20. Korol AM (2019) Tunneling conductance of the s-wave and d-wave pairing superconductive graphene-normal graphene junction. Low Temp Phys 45(5):A48
21. Takahashi R, Murakami S (2011) Gapless interface states between topological insulators with opposite dirac velocities. Phys Rev 107:166805
22. Sen D, Deb O (2012) Junction between surfaces of two topological insulators. Phys Rev B 85:245402

Laser-Driven Behavior of the Nanosystem Used for the Photodynamic Therapy

Yuliia Kuziv, Valeriy Pavlov, Elena Mokrinskaya, Irina Davidenko, Nikolay Davidenko, Vitalii Smokal, Vasyl Chumachenko, and Nataliya Kutsevol

1 Introduction

Photothermal therapy (PTT) shows great promise for treating tumors [1]. In PTT, the injected photothermal agent converts photo-energy into thermal energy under light excitation, leading to the thermal ablation of cancer cells [2]. Combined therapy, like PTT with photodynamic therapy (PDT), could enhance therapeutic efficacy and reduce the recurrence rates of tumors [2, 3]. Both of these two methods involve light irradiation and the use of photosensitive drugs [2]. Therefore, there remains a strong need for a single agent that could be used as both a PTT and PDT agent and that could

Y. Kuziv (✉) · V. Pavlov · E. Mokrinskaya · I. Davidenko · N. Davidenko · V. Smokal · V. Chumachenko · N. Kutsevol
Taras Shevchenko National University of Kyiv, Volodymyrska St., 60, 010601 Kiev, Ukraine
e-mail: garaguts.yulia.fox@gmail.com

V. Pavlov
e-mail: valeriy.pavlov@gmail.com

E. Mokrinskaya
e-mail: mokrinskaya@ukr.net

I. Davidenko
e-mail: irynadavydenko@gmail.com

N. Davidenko
e-mail: ndav@univ.kiev.ua

V. Smokal
e-mail: vitaliismokal@gmail.com

V. Chumachenko
e-mail: chumachenko_va@ukr.net

N. Kutsevol
e-mail: kutsevol@ukr.net

O. Fesenko and L. Yatsenko (eds.), *Nanomaterials and Nanocomposites, Nanostructure Surfaces, and Their Applications*, Springer Proceedings in Physics 246, https://doi.org/10.1007/978-3-030-51905-6_13

particularly work under single-wavelength light irradiation [2]. For this purpose, it is perspective to combine inorganic nanostructures (PTT) and photosensitizers (PDT) [4, 5].

Recent advancements in nanomedicine have recognized the use of Au in the therapeutic delivery of drugs or as an agent for PTT [6]. Gold nanoparticles have many benefits that make them suitable for the photothermal treatment of cancer. AuNPs can be administered into the local tumor area, while minimizing non-specific distribution, they can be activated via near-infrared (NIR) laser light, creating the ability to penetrate deep into biological tissues, and they can be modulated to create multifaceted cancer PTT and drug delivery systems [7].

Chlorin e6 (Ce6) is a widely used photosensitizer for PDT [8, 9] and is activated by laser irradiation to generate highly cytotoxic reactive oxygen species, especially singlet oxygen.

Previously, it was shown that the combination of AuNPs and Ce6 in polymer matrix D-g-PAA was successfully used for the photodynamic anticancer therapy with laser illumination [10].

In this study, changes of the refractive index were analyzed for the systems of Au sol synthesized into the solutions of Dextran-graft-Polyacrylamide (D-g-PAA) star-like copolymers in nonionic and anionic forms with and without photosensitizer Chlorine e6 (Ce6), and for the system D-g-PAA/Ce6 without Au nanoparticles. The polymer molecular characteristics were described in [3]. Au nanoparticles (AuNPs) were 1.5–6 nm in size. The size of individual polymer molecule in solution was equal to 50–60 nm. The aim of the present study was to understand the role of photosensitizer and AuNPs in the nanocomposite which was efficiently used for photodynamic therapy [11].

In our investigation, the laser was used with $\lambda_{irr} = 650$ nm which is close to the long-wavelength maximum absorption of Au nanoparticles.

It was shown that over time (t_{irr}) after the illumination begins, the number of interference bands increases testifying an increase of Δn. After ceasing the light, the interference pattern relaxes and the bands disappear. The interference bands are localized only in the volume limited by the laser beam, and there is no observed diffusion of the heated solution into other parts of the cuvette. To explain the experimental results, a phenomenological model is proposed which takes into account the features of the molecular structure of the star-like D-g-PAA copolymer with incorporated Au nanoparticles and photosensitizer Chlorine e6.

2 Polymer Matrix

In our investigation, a branched copolymer was used obtained by grafting polyacrylamide (PAA) chains onto dextran ($M_w = 7 \times 10^4$, g mol^{-1}) backbone [12] using a ceric-ion-reduce initiation method. This redox process initiates free radical sites exclusively on the polysaccharide backbone, thus preventing the formation of homopolymer (PAA).

Anionic form of the copolymer was obtained via alkaline hydrolysis of Dextran-graft-Polyacrylamide for 30 min by using sodium hydroxide. The fraction of mers bearing carboxylate groups evaluated by potentiometric titration was equal to approximately 37% [13].

The synthesis and analysis of internal polymer structure in detail were given a description in [12, 14].

3 Gold Nanoparticles Synthesis

The AuNPs were synthesized by the chemical reduction of Au precursor (Tetrachloroauric acid). All synthesized polymers play a role of matrices capable to act as nucleating, capping, and stabilizing agents simultaneously.

0.012 ml of tetrachloroauric acid aqueous solution (C = 0.1 M) was added to 0.5 ml of aqueous polymer solution ($C = 1 \times 10^{-3}$ g/cm3) and stirred during 20 min. Then, 0.047 ml of 0.1 M aqueous solution of sodium borohydride was added. The final solution was stirred for 30 min. It turned ruby red in color, and thus the formation of Au NPs was indicated. The reduction process was performed at T = 25 °C. The size of AuNPs was 1.5–7 nm [11].

4 Synthesis Of Hybrid Nanocomposite Polymer/AuNPs/Ce6

Photosensitizer Chlorine e6 (Ce6) was obtained from Santa Cruz Biotechnology (USA).

All polymer and polymer/AuNPs composites with Ce6 were prepared ex tempore. Stock solution of Ce6 (1 mg/ml) was prepared in DMSO. Then, photosensitizer solution was 100-fold diluted in HBSS with constant stirring. The resulting Ce6 solution was gently mixed with D-g-PAA(PE)/AuNPs or D-g-PAA(PE) and incubated at room temperature for 5 min before use [10, 11].

5 Experiment

Holographic interferometry technique was used to explore the change of the refractive index Δn. The measurement procedure was described in detail in [15]. If a change of n occurs in the cuvette under the light illumination, then light and dark interference stripes are observed in the image of the cuvette (Figs. 1 and 2). The larger change of n provides the greater number of interference stripes. To measure the exact value of Δn, we used specially developed technique [15].

Standard cuvettes filled with aqueous solutions of nanocomposites were used as the samples for investigations. The changes of the refractive index of the solutions Δn depending on the time of illumination t_{irr} with a laser (radiation wavelength was 650 nm, power was 10 mW, beam diameter was ~2 mm) and on the time t_{rel}

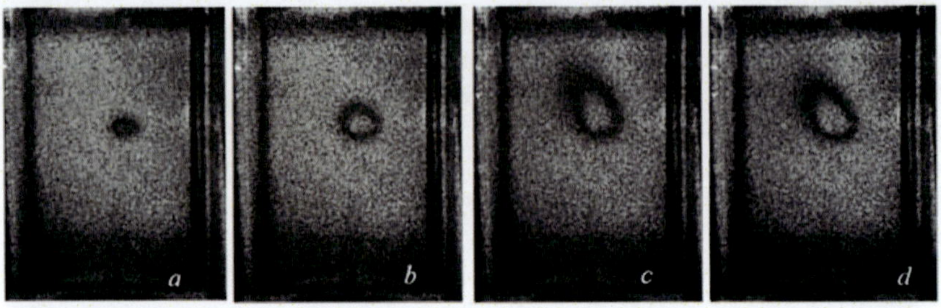

Fig. 1 Photographs of cuvettes with the solution D-g-PAA/AuNPs during illumination with laser light $\lambda_{irr} = 650$ nm with power 10 mW: $t_{irr} = 2$ s (**a**), $t_{irr} = 4$ s (**b**), $t_{irr} = 11$ s (**c**), $t_{irr} = 15$ s (**d**) [15]

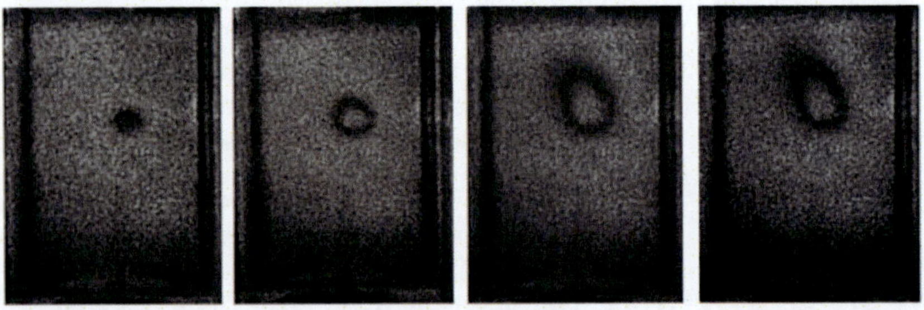

Fig. 2 Photographs of cuvettes with the solution D-g-PAA(PE)/AuNPs during illumination with laser light $\lambda_{irr} = 650$ nm with power 10 mW: $t_{irr} = 2$ s (**a**), $t_{irr} = 4$ s (**b**), $t_{irr} = 11$ s (**c**), $t_{irr} = 15$ s (**d**)

after turning off this laser were measured [15]. The measurement process in samples D-g-PAA/AuNPs and D-g-PAA(PE)/AuNPs is represented by selective photographs in Figs. 1 and 2.

It was investigated that over time (t_{irr}) after the illumination begins, the number of interference bands increases testifying an increase of Δn. After ceasing the light, the interference pattern relaxes and the bands disappear. The interference bands are localized only in the volume limited by the laser beam, and there is no observed diffusion of the heated solution into other parts of the cuvette. The localization of interference rings only in the region of light absorption is unusual. Therefore, there is no heat propagation after the absorption of laser radiation to other areas of the cuvette.

Photographs of cuvettes with solution D-g-PAA/Ce6 are presented in Fig. 3. The increase in the number of interference bands compared to samples Polymer/AuNPs indicates a significant change in the refractive index of the solution in the laser beam region. It was shown that there was heat propagation after the light absorption to the top area of the cuvette. This might indicate heating of the solvent.

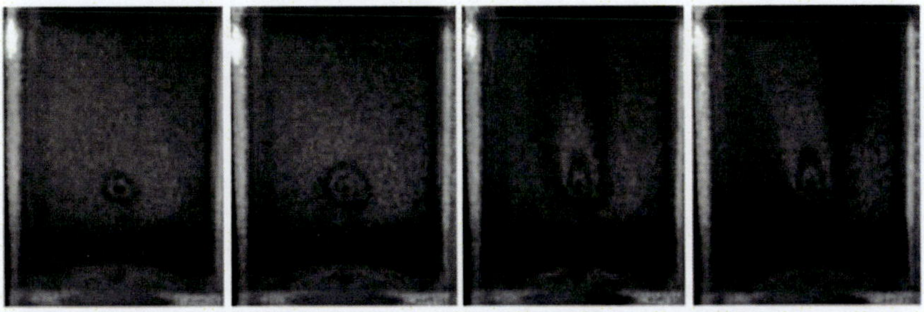

Fig. 3 Photographs of cuvettes with the solution D-g-PAA/Ce6 during illumination with laser light $\lambda_{irr} = 650$ nm with power 10 mW: $t_{irr} = 1$ s **(a)**, $t_{irr} = 2$ s **(b)**, $t_{irr} = 10$ s **(c)**, $t_{irr} = 30$ s **(d)**

Fig. 4 Photograph of cuvette with the solution D-PAA(PE)/AuNPs/Ce6 during illumination with laser light $\lambda_{irr} = 650$ nm with power 10 mW: $t_{irr} = 60$ s

It was investigated that the behavior of the ternary polymer/AuNPs/Ce6 systems was similar to binary polymer/AuNPs systems, and even after the longtime of irradiation ($t_{irr} = 60$ s) there was no heat propagation to the upper area of the cuvette (Fig. 4).

The results of measurements for all samples containing AuNPs or both AuNPs and Ce6 into nonionic and anionic polymer matrices are presented in Figs. 5, 6, 7, and 8.

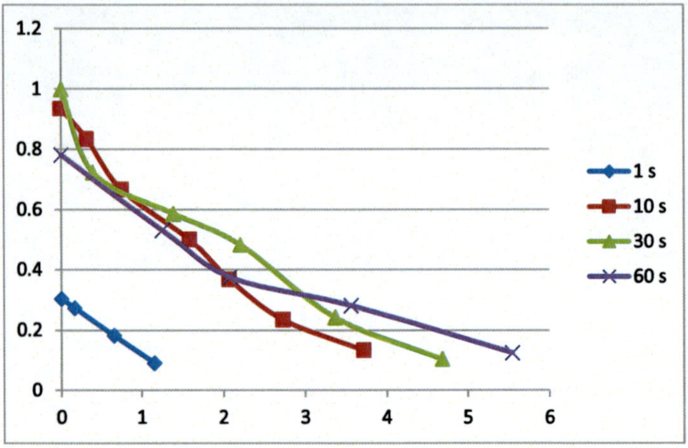

Fig. 5 Dependencies of changes of the refractive index Δn on the time t_{rel} after switching off the laser of nanosystem D-PAA/AuNPs

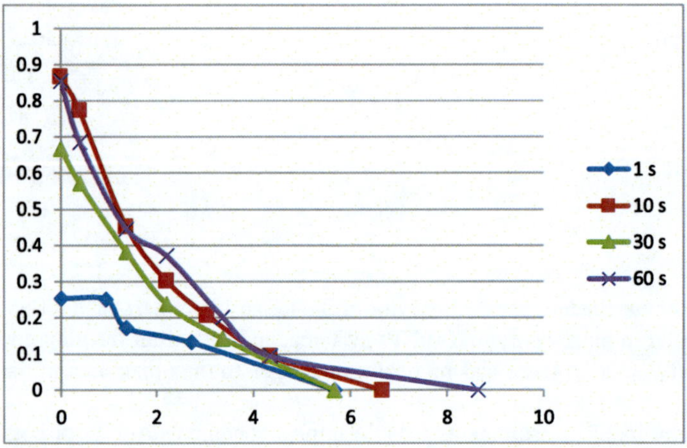

Fig. 6 Dependencies of changes of the refractive index Δn on the time $\mathbf{t_{rel}}$ after switching off the laser of nanosystem D-PAA(PE)/AuNPs

The characteristic relaxation time of the refractive index $\Delta n/\Delta n_{max}$ change within the area of laser irradiation after switching off the laser was the same for all examined samples. Besides, the characteristic relaxation time did not depend on the irradiation time t_{irr}, i.e., did not depend on the absorbed energy.

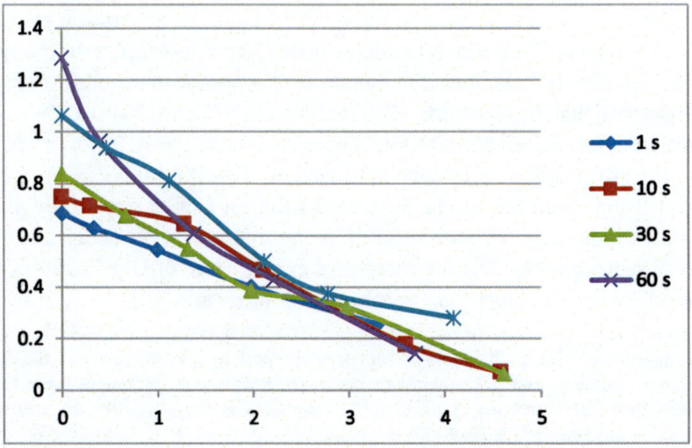

Fig. 7 Dependencies of changes of the refractive index Δn on the time t_{rel} after switching off the laser of nanosystem D-PAA/AuNPs/Ce6

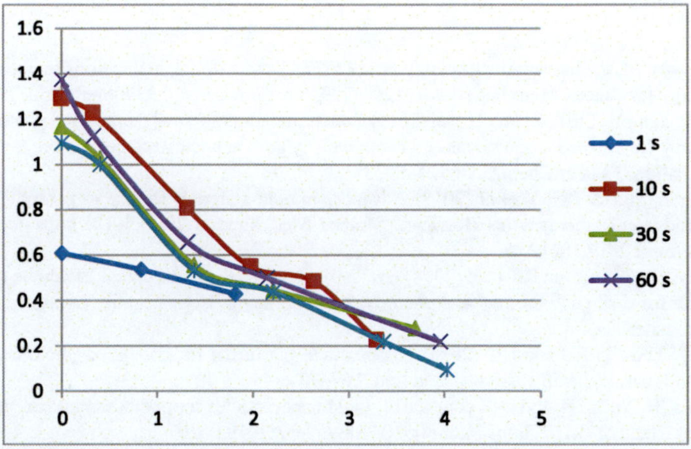

Fig. 8 Dependencies of changes of the refractive index Δn on the time t_{rel} after switching off the laser of nanosystem D-PAA(PE)/AuNPs/Ce6

6 Conclusions

To explain the results obtained, a phenomenological model is offered which takes into account the peculiarities of the molecular structure of star-like Dextran-graft-Polyacrylamide copolymer with incorporated Au nanoparticles. It has been shown that adding of the photosensitizer has almost no effect on the processes in ternary nanosystem Polymer/AuNPs/Ce6 during irradiation. Under illumination with light with $\lambda_{irr} = 650$ nm, the variation of n is observed mainly in the small volume of the cell near the laser beam transmission region [15]. Thus, the absorption of laser light $\lambda_{irr} = 650$ nm occurs in Au nanoparticles. Since the fluorescence in the analyzed samples is not observed and there are no reasons for photochemical transformations, it can be assumed that the scattering of all absorbed light energy happens by excitation of oscillations. Namely, all absorbed light energy is transformed into thermal energy.

Thus, the advantage of a ternary nanosystem D-g-PAA/AuNPs/Ce6 for photodynamic therapy based on two factors: production of singlet oxygen by photosensitizer and local heating (hypertermia) of solution near AuNPs under laser irradiation. This conclusion explains the successful examination of D-g-PAA/AuNPs/Ce6 nanocomposite for photodynamic treatment of tumor cells [10].

Acknowledgements This publication is supported in part by the Ministry of the Education and Science of Ukraine: joint Ukrainian-Belarusian research and development projects "Design and physico-chemical properties of novel multicomponent nanosystems for the treatment and diagnostics of solid tumors" (2019–2020).

References

1. Doughty ACV, Hoover AR, Layton E et al (2019) Nanomaterial applications in photothermal therapy for cancer. Materials (Basel) 12(5):779. https://doi.org/10.3390/ma12050779
2. Liu P, Yang W, Shi L, Zhang H et al (2019) Concurrent photothermal therapy and photodynamic therapy for cutaneous squamous cell carcinoma by gold nanoclusters under a single NIR laser irradiation. J Mater Chem B 7:6924–6933
3. Yang Z, Sun Z, Ren Y et al (2019) Advances in nanomaterials for use in photothermal and photodynamic therapeutics (Review). Molecu Med Reports 20(1):5–15. https://doi.org/10.3892/mmr.2019.10218
4. Riley RS, O'Sullivan RK et al (2018) Evaluating Nanoshells and a potent Biladiene photosensitizer for dual photothermal and photodynamic therapy of triple negative breast cancer cells. Nanomaterials
5. Oh J, Yoon H-J, Park J-H (2013) Nanoparticle platforms for combined photothermal and photodynamic therapy. Biomed Eng Lett 3:67–73
6. Vines JB, Yoon JH, Ryu NE et al (2019) Gold nanoparticles for photothermal cancer therapy. Front Chem 5(7):167. https://doi.org/10.3389/fchem.2019.00167
7. Kennedy LC, Bickford LR, Lewinski NA et al (2011) A new era for cancer treatment: gold-nanoparticle-mediated thermal therapies. Small 7:169–183. https://doi.org/10.1002/smll.201000134
8. Juzeniene A (2009) Chlorin e6-based photosensitizers for photodynamic therapy and photodiagnosis. Photodiagnosis Photodyn Therapy 6(2):94–96. https://doi.org/10.1016/j.pdpdt.2009.06.001
9. Liang H, Zhou Z, Luo R et al (2018) Tumor-specific activated photodynamic therapy with an oxidation-regulated strategy for enhancing anti-tumor efficacy. Theranostics 8(18):5059–5071. https://doi.org/10.7150/thno.28344
10. Kutsevol N, Naumenko A, Harahuts Yu et al (2018) New hybrid composites for photodynamic therapy: synthesis, characterization and biological study. Appl Nanosci 2018:1–8. https://doi.org/10.1007/s13204-018-0768-y

11. Chumachenko VA, Shton IO, Shishko ED et al (2016) Branched copolymers dextran-graft-polyacrylamide as Nanocarriers for delivery of gold nanoparticles and Photosensitizers to tumor Cells. In: Fesenko O, Yatsenko L (ed) Nanophysics, Nanophotonics, Surface Studies, and Applications, vol 183 Springer Proceedings in Physics, pp 379–390. https://doi.org/10.1007/978-3-319-30737-4_32
12. Kutsevol NV, Chumachenko VA, Rawiso M et al (2015) Star-like polymers dextran-polyacrylamide: the prospects of application for nanotechnology. J Struct Chem 56(5):1016–1023
13. Kutsevol N, Bezuglyi M, Rawiso M, Bezugla T (2014) Star-like Destran-graft-(polyacrylamide-co-polyacrylic acid) Copolymers. Macromol Symp 335:12–16
14. Kutsevol N, Bezugla T, Bezuglyi M, Rawiso M (2012) Branched dextran-graft-polyacrylamide copolymers as perspective materials for nanotechnology. Macromol Symp 317–318(1):82–90
15. Harahuts Y, Pavlov V, Mokrinskaya E et al (2018) Anomalous change of refractive index for au sols under laser illumination. In: Nanophotonics, Nanooptics, Nanobiotechnology, and Their Applications, Selected Proceedings of the 6th International Conference Nanotechnology and Nanomaterials (NANO2018), pp 53–71 https://doi.org/10.1007/978-3-030-17755-3_3

New Effective Filter in the Spatial Domain for Speckle Noise Reduction

Yu. Kotsiuba, V. M. Fitio, H. Petrovska, and Ya. V. Bobitski

1 Introduction

Digital holography (DH) is an effective and robust method that is used in metrology to study deformation fields, retrieve surface topography, and visualize phase images [1, 2]. In the DH method, the difference in the optical path length between the reference beam and the object beam is measured. The object beam can pass through transparent samples (transmission configuration) or being reflected by opaque surfaces (reflection configuration). After digital reconstruction, one can calculate the corresponding phase field of the object.

Digital holographic interferometry (DHI), in particular, the method of two exposures, is one of the submethods of digital holography (DH) which is commonly used in metrology. In this method, two reconstructed phase fields of the same object recorded at different moments of time are compared. Thus, a phase difference map containing 2π discontinuities is obtained as a result of the direct phase subtraction. After that, the phase unwrapping is carried out to obtain a quantitative representation of the measured value (displacement field, deformation field, vibration field, etc.).

A significant drawback of the DHI is speckle decorrelation. It occurs when comparing phase fields obtained from two different digital holograms and leads to the

Yu. Kotsiuba
Karpenko Physico-Mechanical Institute of the NAS of Ukraine, Naukova Str. 5, 79000 Lviv, Ukraine

V. M. Fitio · H. Petrovska (✉) · Ya. V. Bobitski
Department of Photonics, Lviv Polytechnic National University, S. Bandera Str. 12, 79013 Lviv, Ukraine
e-mail: halyna.a.petrovska@lpnu.ua

Ya. V. Bobitski
Faculty of Mathematics and Natural Sciences, University of Rzeszow, Pigonia Str.1, 35959 Rzeszow, Poland

O. Fesenko and L. Yatsenko (eds.), *Nanomaterials and Nanocomposites, Nanostructure Surfaces, and Their Applications*, Springer Proceedings in Physics 246, https://doi.org/10.1007/978-3-030-51905-6_14

noise formation on the obtained phase map [3–5]. The mechanism of the speckle noise formation and its nature are described in [1–3]. It should be noticed that phase decorrelation can have several sources in the DH [4]. It occurs when coherent light passes through an inhomogeneous medium or being reflected from a rough surface. Both cases take place during mechanical characteristics study of materials [1]. Speckle noise can significantly distort the quality of the obtained phase image. As a result, the spatial resolution, signal-to-noise ratio, and measurement accuracy decrease [1]. Therefore, it becomes impossible to carry out the unwrapping correctly.

Thus, it is necessary to carry out the noise filtering. The main task of filtering is to increase the signal-to-noise ratio without removing the 2π phase gaps [6]. Today, a lot of approaches to reduce the noise level are known [7–18]. All methods can be divided into optical and digital. Optical methods are based on increasing the amount of information received about the object by increasing the number of exposures. In turn, the digital filtering methods using the mathematical algorithms extract information about the object which is obtained in a single exposure [7]. The main idea of optical noise reduction techniques is the average of the several holograms which were recorded under various object illumination conditions: wavelength [8], phase [9], the angle of incidence [10], and the polarization [11]. Other optical techniques for reducing speckle noise in a DH are based on the use of partially coherent light sources [12].

In digital filtering methods, the image processing is usually done on the sines and cosines images calculated from the phase to avoid the 2π phase jumps distortion on the wrapped phase map [13, 14]. Today, several digital filtering methods for reducing the speckle noise in DHI are known [7, 15–18], for example, wavelet filtering [16] and filtering in the frequency domain using the Fourier transform [18].

In this work, we propose a new filter based on Chebyshev polynomials of the first kind. The evaluation of the filtration errors is done using phase maps obtained as a result of numerical simulation of the physical processes in the DHI.

2 Mathematical Description of a Filter Based on Chebyshev Polynomials

This section describes the mathematical representation of Chebyshev polynomials of the first kind, showing their application for filtering narrow peaks, as well as the optimal number of polynomials for representing a function as a sum of polynomials.

2.1 Mathematical Representation of Chebyshev Polynomials of the First Kind

Chebyshev polynomials of the first kind $T_n(x)$ can be determined using recurrence relations [19, 20]:

$$\begin{aligned} T_0(x) &= 1 \\ T_1(x) &= x \\ T_{n+1}(x) &= 2xT_n(x) - T_{n-1}(x) \end{aligned} \tag{1}$$

or from the relation [19, 20]

$$T_n(x) = \cos(n \arccos x). \tag{2}$$

The first kind Chebyshev polynomials are orthogonal with respect to scalar product with weight $1/\sqrt{1-x^2}$ in the interval $[-1, 1]$, i.e., the following relation is valid [19–21]:

$$\int_{-1}^{1} T_n(x)T_m(x)\frac{dx}{\sqrt{1-x^2}} = \begin{cases} 0, & n \neq m \\ \pi, & n = m = 0 \\ \frac{\pi}{2}, & n = m \neq 0 \end{cases} \tag{3}$$

Thus, any continuous function *F(x)* on the interval [− 1, 1] can be represented as a sum of infinite elements [20]:

$$F(x) = \sum_{n=0}^{\infty} f_n T_n(x), \tag{4}$$

where coefficients f_n are determined as follows [20]:

$$f_0 = \frac{1}{\pi}\int_{-1}^{1} F(x)T_0(x)\frac{dx}{\sqrt{1-x^2}} \tag{5}$$

$$f_{n\neq 0} = \frac{2}{\pi}\int_{-1}^{1} F(x)T_n(x)\frac{dx}{\sqrt{1-x^2}}. \tag{6}$$

Figure 1a shows the graphical representation of the first five Chebyshev polynomials on the interval [−1,1]. Figure 1b shows the function $F(x) = 2 + x + \sin x$(red line) and its representation as sum of 11 elements using (4) (blue rings).

We see the quite good overlay of both curves in Fig. 1b.

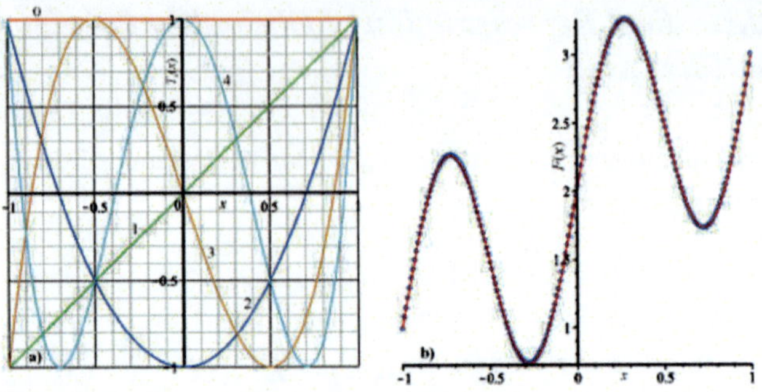

Fig. 1 **a** The first five Chebyshev polynomials on the interval $[-1, 1]$: $T_0(x)$—red line, $T_1(x)$—green line, $T_2(x)$—blue line, $T_3(x)$—brown line, $T_4(x)$—cyan line; **b** function $F(x) = 2 + x + \sin x$—continuous red line, representation of $F(x)$ as sum of 11 elements—blue rings

2.2 *Examples of Narrow Peak Filtration with Chebyshev Polynomials of the First Kind*

In digital holographic interferometry, during the study of objects diffusely scattering light, numerous narrow peaks of significant intensity are formed in the resulting phase fringe patterns. As was mentioned earlier, filtration of the wrapped phase map $\Delta\varphi$ is done by filtration of phase map terms *cos(*$\Delta\varphi$*)* and *sin(*$\Delta\varphi$*)*. For investigating the possibility of elimination of narrow peaks from phase map terms with Chebyshev polynomials, let us consider the following function:

$$F(x) = 2 + x + \sin(2\pi x) + 20\exp\left\{-[400(x - 0.3)]^2\right\}. \tag{7}$$

Here, the fourth term denotes a narrow peak. This function is shown in Fig. 2a (red line). If we expand this function in Chebyshev polynomials (using the first 11 polynomials), we get a function (blue rings) without a narrow peak. That is, a narrow peak was cut off. Obviously, with an increase in the number of terms in the sum (4),

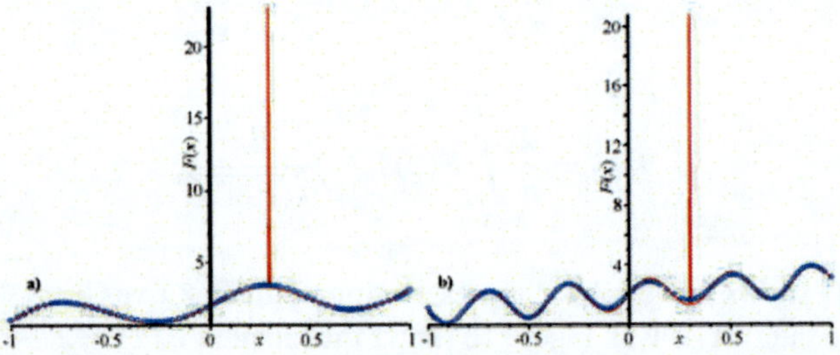

Fig. 2 Function $F(x)$, described by (7) and its representation in the form of (4) (**a**); Function $F(x)$, described by (8) and its representation in the form of (4) (**b**)

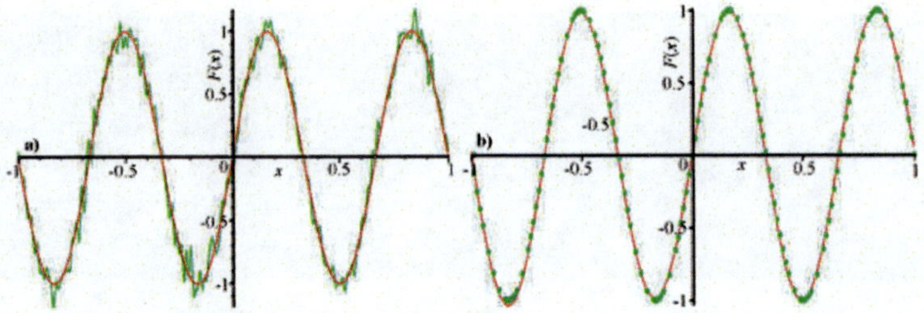

Fig. 3 Filtering of the function $F(x) = \sin(3\pi x) + \text{noise}$ by 16 Chebyshev polynomials: **a** continuous red line represents $F(x) = \sin(3\pi x)$, continuous green line—$F(x) = \sin(3\pi x) + \text{noise}$; **b** continuous red line represents the result of filtration, green circles—$F(x) = \sin(3\pi x)$

the intensity peak will increase in magnitude, and its width will decrease. It should also be noted that the blue rings fit well on the red curve in the entire interval [−1,1], which is described by the first three terms of (4).

Figure 2b shows the following function:

$$F(x) = 2 + x + \sin(5\pi x) + 20\exp\{-[400(x - 0.3)]^2\} \quad (8)$$

and its representation as a sum (4) with the first 21 Chebyshev polynomials (blue rings). We see the quite good overlay of blue rings on the continuous curve, except areas near a narrow peak. However, representation in the form of (4) eliminates a narrow peak and do not distort the curve, described by the function $F(x) = 2 + x + \sin(5\pi x)$.

It should be noted that the result depends on the number of terms in (4), as seen when comparing Fig. 2a, b. The required number of polynomials can be determined experimentally.

Now consider the case when noise is present on the entire interval of the sine pattern, but with an amplitude lower than the amplitude of the signal, as shown in Fig. 3a (continuous green curve). Figure 3b shows the result of filtering noisy curve by the Chebyshev polynomials (continuous red curve). We see that the red curve fits quite well on the green circles being the function $F(x) = \sin(3\pi x)$.

Looking at the Figs. 2 and 3, one can conclude that filtering the noisy sine pattern by Chebyshev polynomials is quite successful.

2.3 Determination of the Optimal Number of Polynomials for the Function Representation

Obviously, for the proper representation of the phase components, it is necessary to select the correct number of Chebyshev polynomials, since it has a strong influence on the quality of the result. The determination of the optimal number of polynomials was carried out experimentally. During the experiment, a series of sinusoids with an

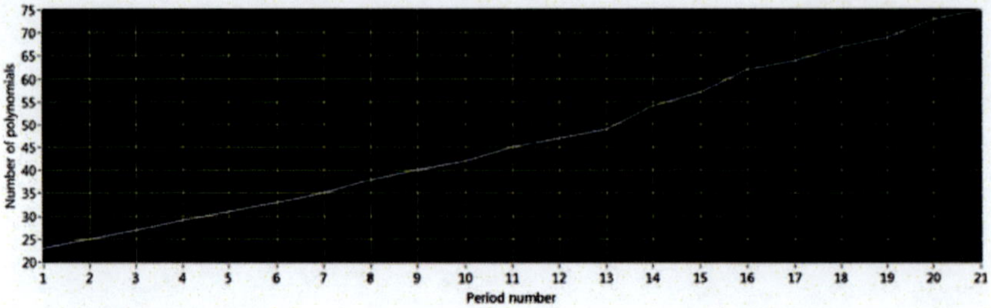

Fig. 4 The dependence of the optimal number of Chebyshev polynomials on the number of a sinusoids' period

integer number of periods was generated. By increasing the number of polynomials, while considering the errors of representation, we searched for the optimal number where the minimum error occurs. Figure 4 shows the dependence of the optimal number of Chebyshev polynomials on the number of a sinusoids' period.

From Fig. 4, one can see that the optimal number of polynomials linearly increases with an increase of sine period.

3 Calculation of the Phase Fringe Patterns of the Rough Object Deformation

A mathematical evaluation of the proposed filter was carried out by filtering test phase fringe maps. For this purpose, we calculated a series of difference deformation phase maps of an object with different fringe densities. The complex amplitude of an object was obtained using a computer implementation of the optical scheme shown in Fig. 5.

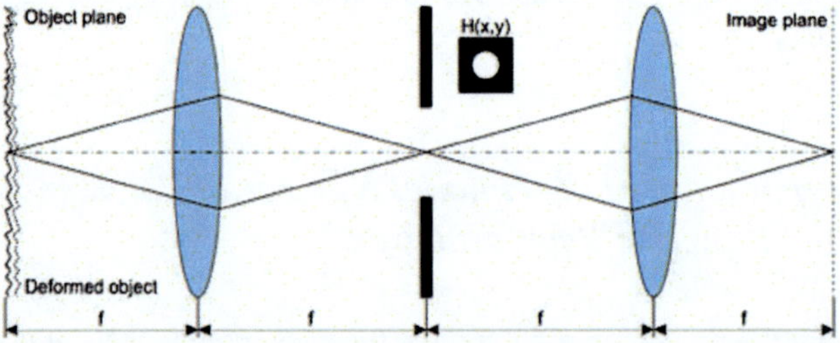

Fig. 5 The 4f optical system with a spatial filter to adjust the speckle size. The complex amplitude of the object beam is formed at the image plane. Some phase field related to the object deformation is added to the initial rough object to obtain phase fringe pattern

A computer model of a rough surface with the dimension 500 × 500 and pixel size $\Delta x = 10$ μm was chosen as an object. When illuminating this object with a coherent light beam ($\lambda = 0.5$ μm), a field with a unit intensity and an arbitrary phase distribution is obtained. Simulation of the propagation of an object wave through lenses is done by the Fourier transform. The aperture diaphragm located at the focal plane serves as a spatial filter with which one can adjust the speckle size. In the computer model, this is done by multiplying the spectrum of the object by the function circ(*x*,*y*). After that, some reference wave is added to obtain the digital hologram. Having calculated digital hologram in the initial state and with the addition of simple deformation, we received a series of 100 phase fringe patterns of the object deformation with deformation step $\delta = 0.1$ μm. In Fig. 6, some examples of obtained phase maps are shown.

By selecting the size of the diaphragm, we achieved the speckle size $p = 5$ μm. The dependence of the standard deviation of noise on the deformation value with this value of p is shown in Fig. 7.

One can see a nonlinear increase in the noise level with the increase in the fringe density.

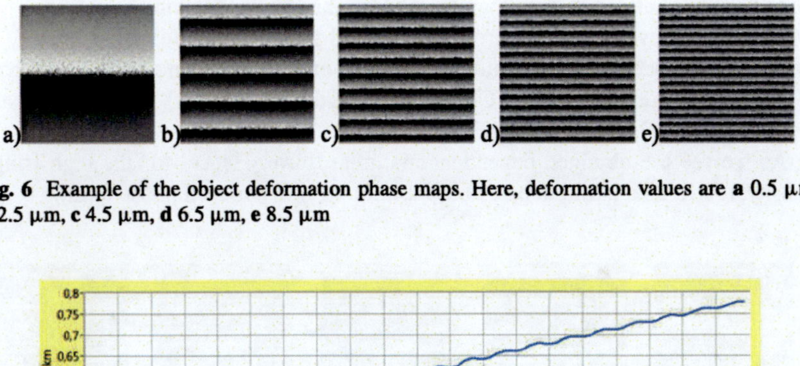

Fig. 6 Example of the object deformation phase maps. Here, deformation values are **a** 0.5 μm, **b** 2.5 μm, **c** 4.5 μm, **d** 6.5 μm, **e** 8.5 μm

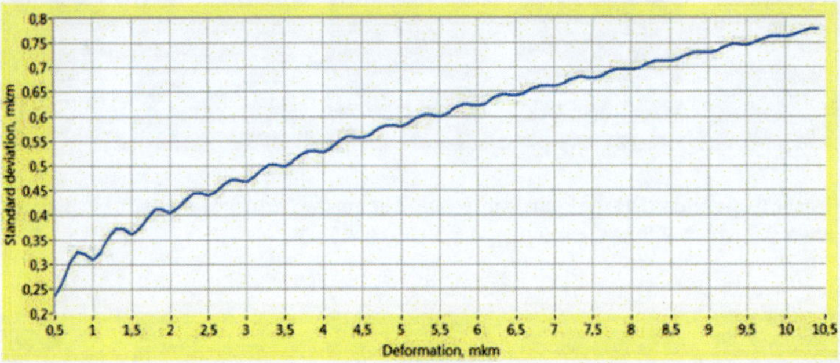

Fig. 7 The dependence of the standard deviation of noise on the deformation value with pixel size $p = 5$ μm

4 Results and Discussion

This section shows the results of filtering the obtained phase maps with two filters. In addition to the proposed filter, we consider here the filter described in [22]. To compare both filters, the standard deviation value obtained from the filtering error was calculated.

In work [22], the filtration is carried out in the frequency domain using the Fourier transform. A threshold filter is applied to the spectrum of the sine pattern. The threshold value is determined statistically, calculating a percentile value below which $n\%$ of the whole spectrum intensity values may be found. Figure 8 shows the result of filtration, taking $n = 99.9\%$.

From Fig. 8, quite good filtering quality is observed. However, phase discontinuities are fuzzy. Moreover, with an increase in the fringe density, they are more distorted. This problem does not occur when using the filter with Chebyshev polynomials (see Fig. 9).

Comparing Figs. 8 and 9, we can conclude that the proposed filter has an advantage in the quality of the output phase maps. The calculation of the standard deviation of filtration errors for two filters is shown in Fig. 10. Obviously, the proposed filter (blue line) has better accuracy than the filter in the frequency domain (red line).

The main drawback of the proposed method is the calculation time. Figure 11 shows the dependence of calculation time on the number of polynomials for an array of dimension 1000×1000. The CPU parameters are Intel Core i3-2350 M with base frequency of 2.30 GHz.

As we see the obtained dependence is quite linear. Hence, for the high fringe density, it will take 2 min to carry out filtration (1 min per each phase term).

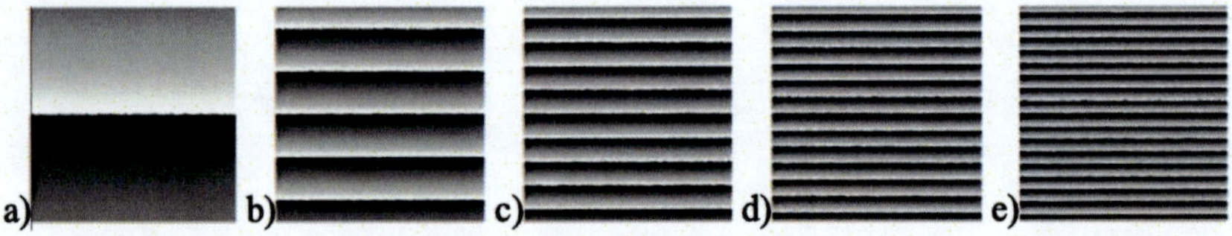

Fig. 8 Phase maps filtered by the Fourier threshold filter in the frequency domain. Here, deformation values are **a** 0.5 μm, **b** 2.5 μm, **c** 4.5 μm, **d** 6.5 μm, **e** 8.5 μm

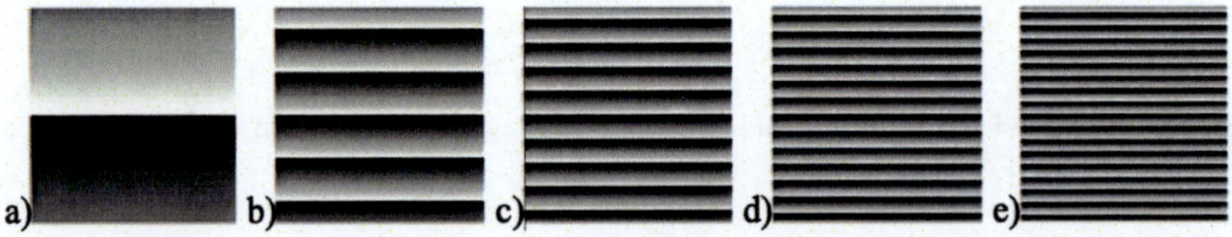

Fig. 9 Phase maps filtered by the Chebyshev polynomials. Here, deformation values are **a** 0.5 μm, **b** 2.5 μm, **c** 4.5 μm, **d** 6.5 μm, **e** 8.5 μm

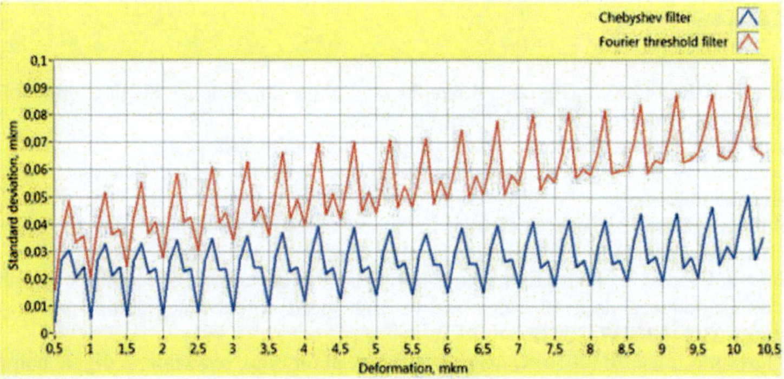

Fig. 10 The dependence of the standard deviation of filtration errors on the deformation value

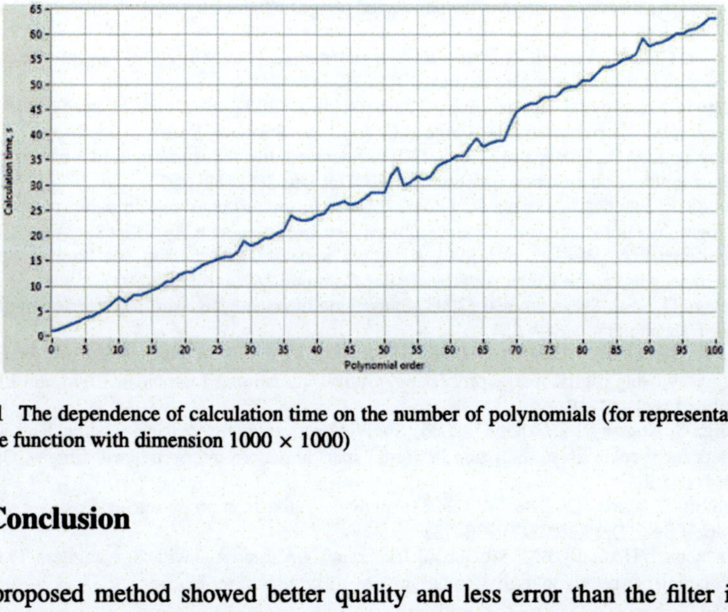

Fig. 11 The dependence of calculation time on the number of polynomials (for representation of 2D sine function with dimension 1000 × 1000)

5 Conclusion

The proposed method showed better quality and less error than the filter in the frequency domain, but it requires more processing time. However, this drawback is not so critical. The ability to cut off narrow intense peaks eliminates speckle noise even from periodic high-frequency structures with virtually no distortion. The next step of our work is to improve the capabilities of the new filter by introducing two-dimensional Chebyshev polynomials to filter complex phase maps. In addition, the possibility of applying the proposed method to filter digital holograms from the speckle structure to eliminate the diffraction halo is considered.

References

1. Poon TC (Ed) (2006) Digital holography and three-dimensional display: principles and applications. Springer Science & Business Media
2. Picart P (Ed) (2015) New techniques in digital holography. Wiley
3. Goodman JW (2007) Speckle phenomena in optics: theory and applications. Roberts and Company Publishers
4. Poittevin J, Picart P, Faure C, Gautier F, Pézerat C (2015) Multi-point vibrometer based on high-speed digital in-line holography. Appl Opt 54(11):3185–3196
5. Poittevin J, Picart P, Gautier F, Pezerat C (2015) Quality assessment of combined quantization-shot-noise-induced decorrelation noise in high-speed digital holographic metrology. Opt Express 23(24):30917–30932
6. Montresor S, Picart P (2016) Quantitative appraisal for noise reduction in digital holographic phase imaging. Opt Express 24(13):14322–14343
7. Uzan A, Rivenson Y, Stern A (2013) Speckle denoising in digital holography by nonlocal means filtering. Appl Opt 52(1):A195–A200
8. Nomura T, Okamura M, Nitanai E, Numata T (2008) Image quality improvement of digital holography by superposition of reconstructed images obtained by multiple wavelengths. Appl Opt 47(19):D38–D43
9. Park Y, Choi W, Yaqoob Z, Dasari R, Badizadegan K, Feld MS (2009) Speckle-field digital holographic microscopy. Opt Express 17(15):12285–12292
10. Hertwig S, Babovsky H, Kiessling A, Kowarschik R (2009). Reduction of speckles in digital holographic interferometry. In Fringe 2009 (pp 1–5). Springer, Berlin, Heidelberg
11. Rong L, Xiao W, Pan F, Liu S, Li R (2010) Speckle noise reduction in digital holography by use of multiple polarization holograms. Chin Opt Lett 8(7):653–655
12. Dubois F, Joannes L, Legros JC (1999) Improved three-dimensional imaging with a digital holography microscope with a source of partial spatial coherence. Appl Opt 38(34):7085–7094
13. Aebischer HA, Waldner S (1999) A simple and effective method for filtering speckle-interferometric phase fringe patterns. Optics Commun 162(4–6):205–210
14. Kemao Q, Soon SH, Asundi A (2003) Smoothing filters in phase-shifting interferometry. Opt Laser Technol 35(8):649–654
15. Garcia-Sucerquia J, Ramírez JAH, Prieto DV (2005) Reduction of speckle noise in digital holography by using digital image processing. Optik-International Journal for Light and Electron Optics 116(1):44–48
16. Mirza S, Kumar R, Shakher C (2005) Study of various preprocessing schemes and wavelet filters for speckle noise reduction in digital speckle pattern interferometric fringes. Opt Eng 44(4):045603
17. Almoro P, Pedrini G, Osten W (2007) Aperture synthesis in phase retrieval using a volume-speckle field. Opt Lett 32(7):733–735
18. Maycock J, Hennelly BM, McDonald JB, Frauel Y, Castro A, Javidi B, Naughton TJ (2007) Reduction of speckle in digital holography by discrete Fourier filtering. JOSA A 24(6):1617–1622
19. Abramowitz M, Stegun IA (1972) Handbook of Mathematical Functions with Formulas, Graphs, and Mathematical Tables. National Bureau of Standards Applied Mathematics Series 55. Tenth Printing
20. Elliott D (1964) The evaluation and estimation of the coefficients in the Chebyshev series expansion of a function. Math Comput 18(86):274–284
21. Hernández MA (2001) Chebyshev's approximation algorithms and applications. Comput Math Appl 41(3–4):433–445
22. Kotsiuba YM, Petrovska HA, Fitio VM, Bobitski YV (2016, September) Improving digital holographic interferogram quality by frequency filtering. In: 2016 IEEE 7th international conference on advanced optoelectronics and lasers (CAOL) (pp. 67-68). IEEE

Configuration of Charge Waves in Polymethine Linear Dye Systems

S. V. Vasylyuk, A. D. Suprun, L. V. Shmeleva, and O. D. Kachkovsky

In the middle of the essential particular properties of linear conjugate systems **1**, it is an ability to exhibit quasimetallic conductivity. The basis polymers **1** looks like [R—(CH) $_m$—R]^n, where n = 0, ±1, ±2. This discovery, awarded with the Nobel Prize in 2000, led to the revision of many theoretical models based on representations of the action of π-electrons in the conjugated molecules, primarily in anions and cations of polymethine dyes, in cation or anion radicals of polyenes, and thus inspired the improvement of original theoretical model of the electronic structure of similar special type of systems, taking into account the features of charge distribution and molecular geometry [1–4]. In addition to the use of conjugated systems as molecular conductors [1], they find numerous applications in nonlinear optics: as a result, polymethine dyes with uneven number of carbon group CH (system **1,** where m = 2 k + 1, n = ±1, and R—various heterocyclic residues) are capable to intensively absorb and emanate a quant of light in visible part of spectrum and near-infrared part of the spectrum [5, 6]. Such unique spectral properties are suitable to the definite charge distribution in a main and in the excited state, as well as an arrangement of the lengths of carbon–carbon bonds [1–8]. According to modern notions, the carbon atom in the conjugate system is in the sp^3-hybridized state and accordingly, forms three σ-bonds, resulting in the molecules **1** being flat, as can be seen in Fig. 1, which shows an example of polymethine ion at **5**, n = + 1, R = CH_2. The last $2p_z$-electron together with the same electrons of other carbon atoms forms a common collective π-electron system. The transition from a neutral conjugate molecule to ions that is the injection of an electron or a hole into a high polarizable common collective π-electronic system is accompanied by the manifestation of supposed soliton level (level of impurity) in the energy gap in addition to significant shift in both the conduction band base and the valence band top [2–4]. The injection charge is not

S. V. Vasylyuk (✉) · A. D. Suprun · L. V. Shmeleva · O. D. Kachkovsky
Physics Faculty of Kyiv National Taras Shevchenko University, Volodymirska 64, Kiev, Ukraine
e-mail: svetlanafvasylyuk@gmail.com; bladewalker@ukr.net

O. Fesenko and L. Yatsenko (eds.), *Nanomaterials and Nanocomposites, Nanostructure Surfaces, and Their Applications*, Springer Proceedings in Physics 246, https://doi.org/10.1007/978-3-030-51905-6_15

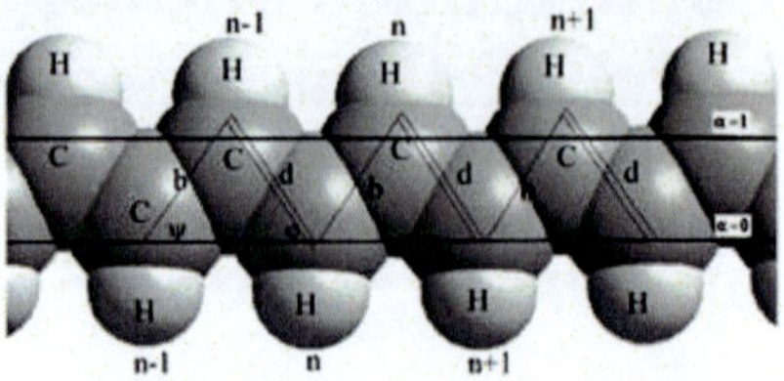

Fig. 1 The established model of polymethine dye

uniformly delocalized in the length of the complete conjugation structure; nevertheless, it is self-localized [4], that is, it generates a charge wave of finite sizes of the soliton type. Quantum-chemical calculations show that the length of a soliton-like wave is 15–17 carbon–carbon bonds [2–4]. Here, the case of **1** molecular systems with a relatively extended chain of carbon groups while the length of the π-system goes over the charge wavelength, then the soliton turns out to be transportable and might locate at any desired fragment of the chain without changing molecule whole energy. Similar transfer of charge wave in polymethine dyes cation can lead to the symmetry breakage of an electron structure, which is experimentally manifested in a severe change in the IR spectra polymethine dyes' spectral band [9, 10]. Simultaneously, the wave of bond lengths is produced with the charge wave in molecular π-electron system; otherwise, a topological (or else geometric) soliton beforehand a kink; it is obvious that localization of charge and topological waves centers coincide in the equilibrium state. Calculations have shown that in the unchanged polymethine systems the form and localization of both types of soliton-like waves are not determined by the molecule value [3–7, 11]. The influence of finite groups, R, can lead to the soliton transfer. Similar quantum-chemical studies of sufficiently long molecules have been performed within semi-empirical approximations.

To identify the polyene polymer chain with change of double- and single-carbon–carbon groups, recognized as conjugate π-electron system is used [7] (Fig. 1). Axis X is going to alongside the chain, and axis Y is perpendicular to it. In this case, the part of chain and (XY) part of coordinates are overlapped with each other; two carbon chains differ in the index $\alpha = \{0; 1\}$. In this case, X-axis is observed as coinciding with a sequence in the bottom and equivalent to the $\alpha = 0$ value, and the higher sequence is connected with the $\alpha = 1$ index (Fig. 1).

"Triangles, shown in Fig. 1, are formed by three atoms of carbon. Here d is the length of the double bond and b is the length of the single bond. It is known [3–7] that these values are approximately equal: $d = 1,36$ Å, $b = 1,44$ Å, $d - b = 0,08$ Å. Such insignificant difference enables one to use further equilibrium model of the carbon–carbon chain, in which the lengths of bonds are considered equal and are accepted equal to $1,4$ Å" [2–7]. The established method of the lattice cell division is shown in Fig. 1. There are three patterns the cell could be exposed in fact: $n - 1$, $n, n + 1$. "It is seen that each lattice cell contains a couple of carbon atoms or, to be more precise, a couple of CH- groups, which refer to as single bond between carbon atoms. At such a choice of the coordinate system radius vectors $\mathbf{R}_{\alpha,n}$ for positions of CH- groups in each of chains under conditions of equilibrium are determined by the formula" [7]:

$$\mathbf{R}_{\alpha n} = \mathbf{e}_x(R_0 n + \alpha b \cos\psi) + \mathbf{e}_y \alpha d \sin\varphi;\ \alpha = \{0; 1\}, \tag{14.1}$$

wherever $R_0 = d\cos\varphi + b\cos\psi$.

To explain the electron dynamics, when the electron is inserted inside polymethylene chain, we use the Hamiltonian [7] that has this form with considering peculiarity of this item:

$$E(\{a\}) = E_c + W^{(0)} + \frac{1}{2}\sum_{\alpha n}\sum_{\beta m} w_{\alpha n,\beta m} + \frac{1}{2}\sum_{\alpha n}(D_{\alpha n} + W^{(1)}_{\alpha n})|a_{\alpha n}|^2 + \sum_{\alpha n}\sum_{\beta m} M_{\alpha n,\beta m}\, a^*_{\alpha n} a_{\beta m}. \tag{14.2}$$

It is typical for condensates of the crystalline type with the semiconductor or dielectric zone structure [7]. The first part of it E_c of (14.2) shows the energy of N isolated molecular groups electrons (in the case of polymethines it could be the carbon CH groups), to which the energy of the injected electron that interacts only with the electrons of such isolated molecular group is added. "$W^{(0)}$ is the energy of interaction of the external field with the electrons of the valence zone. The energy $w_{\alpha n,\beta m}$ in (14.2) plays the main role in incorporation of the isolated molecular groups into the united bonded system and determines interaction between them under conditions of the absence of excitation. Energy $D_{\alpha n}$ in the Hamiltonian (14.2) by its physical sense reflects the change in the interaction of the exited molecular group with non-exited surrounding in comparison with the similar interaction, when all molecular groups are not exited. The summand $W^{(1)}_{\alpha n}$ represents the influence of the external field on the injected electron. Value $a_{\alpha n}$ is, in fact, the wave function of the variables α, n, which determines the distribution in the molecular chain of the electron, injected in the conduction zone, and $|a_{\alpha n}|^2$ is usually interpreted as the probability of the localization of the electron on the knot with the number αn. At last, the energy $M_{\alpha n,\beta m}$, which is included in the Hamiltonian (14.2), is called the energy of the resonance exchange interaction. By its physical sense it determines not so much additional interaction among molecular groups, as the dynamics of the considered charge excitation in the polyene chain" [7].

"Hereinafter, only varying part of the Hamiltonian (14.2) [7] is considered:

$$\Delta E(\{a\}) \equiv E(\{a\}) - E_c - W^{(0)}.$$

Below nearest-neighbor approximation, we resolute (14.1–14.7) [7]; it is made of

$$\begin{aligned}\Delta E(\{a\}) = \frac{1}{2}\sum_n \Big\{ & w(\mathbf{R}_{1n} - \mathbf{R}_{0n}) + w(\mathbf{R}_{0,n+1} - \mathbf{R}_{1n}) + \left[W^{(1)}_{0n} + D(\mathbf{R}_{1n} - \mathbf{R}_{0n})\right]|a_{0n}|^2 + \\ & + \left[W^{(1)}_{1n} + D(\mathbf{R}_{0,n+1} - \mathbf{R}_{1n})\right]|a_{1n}|^2 + M(\mathbf{R}_{1,n+1} - \mathbf{R}_{0n})\left(a^*_{0n} a_{1,n+1} + a^*_{1,n+1} a_{0n}\right) + \\ & + M(\mathbf{R}_{0,n+1} - \mathbf{R}_{0n})\left(a^*_{0n} a_{0,n+1} + a^*_{0,n+1} a_{0n}\right) + M(\mathbf{R}_{1,n+1} - \mathbf{R}_{1n})\left(a^*_{1n} a_{1,n+1} + a^*_{1,n+1} a_{1n}\right) + \\ & + M(\mathbf{R}_{1n} - \mathbf{R}_{0n})(a^*_{0n} a_{1n} + a^*_{1n} a_{0n}) + M(\mathbf{R}_{0,n+1} - \mathbf{R}_{1n})\left(a^*_{1n} a_{0,n+1} + a^*_{0,n+1} a_{1n}\right)\Big\}.\end{aligned}$$

In this case in the zero approximation by the external field (or at the field switched off), both these equations for each branch "±" coincide and reduce themselves to the form:

$$\frac{1}{2\mu}[1+\tau_{\mathrm{p}}v_{\perp}]\frac{\partial^2\Phi_{\pm}(\xi_{\pm})}{\partial\xi_{\pm}^2}+(2g_{11}(\xi_{\pm}+x_c^{\pm})+g_{10}(\xi_{\pm}+x_c^{\pm}))\Phi_{\pm}^3(\xi_{\pm})$$
$$+\left[\Omega+\left(\prod_x+\prod_y\right)+\frac{1}{\mu}+\tau_{\mathrm{p}}v_{\perp}\left(\frac{1}{2}+\frac{1}{\mu}\right)\right]\Phi_{\pm}(\xi_{\pm})=0. \quad (14.49)$$

In this case, we designate $g_{00}\equiv g_{11}\equiv g_{\mathrm{p}}$; (14.50) $g_{00}\equiv g_{11}\equiv g_{\mathrm{p}}$" (14.51).

That is, under conditions of the absence of the external field or in zero approximation by it, parameters g_{ij}, as follows from their definitions (14.20) ÷ (14.22), (14.25), do not depend on the variable $x=\xi_{\pm}+x_c^{\pm}$ (id est, they are constants) and coincide with pairs according to definitions (14.50), (14.51).

In the approximation considered ($|\beta(p)|\to 0$, $|p|\to 0$ $\{\Pi\}\to 0$) the dependence of μ on p is rather weak and can be neglected. In this case, the functions $\Phi_{\pm}^{(0)}(\xi_{\pm})$ do not obtain additional, besides $x_c^{\pm}(\tau)$, dependence on τ, which is forbidden by the system (14.39) [7].

For simplifications of computations, make $2g_{\parallel}+g_{\perp}\equiv g$, and also

$$\Omega+\frac{1}{\mu}\pm\tau_{\mathrm{p}}v_{\perp}\left[\frac{1}{2}+\frac{1}{\mu}\right]\equiv\varepsilon_{\pm};\ (14.52)\ 1\pm\tau_{\parallel}v_{\perp}\equiv s_{\pm}. \quad (14.53)$$

Make (14.49) to the more compact form [26–29]:

$$\frac{s_{\pm}}{2\mu}\overset{(0)}{\Phi''_{\pm}}(\xi_{\pm})+g\overset{(0)}{\Phi^3_{\pm}}(\xi_{\pm})+\varepsilon_{\pm}\underset{\pm}{\overset{(0)}{\Phi}}(\xi_{\pm})=0. \quad (14.54)$$

This equation is well known [7] as the soliton equation.

Using the results of the first part of this work, let us consider now in the approximation [8], zero by field (this approximation in the physical sense corresponds to the external field switched off), and the conductivity, caused by the transfer of the injected charge. From the following formulae of the first part [7]

$$\left(\Omega+\frac{1}{\mu}\pm\tau_{\parallel}v_{\perp}\left[\frac{1}{2}+\frac{1}{\mu}\right]\equiv\varepsilon_{\pm}\right) \quad (1.52)$$

$$(1\pm\tau_{\parallel}v_{\perp}\equiv s_{\pm}). \quad (1.53)$$

"From (2.8) for each mode of the $I_{\pm}$, we can, at last, write [8] current, caused by the transfer of the injected charge

$$I_{+}=\frac{e|M_{\parallel}|}{\hbar N}\left[(1+\tau_{\parallel}v_{\perp})\sin(p)-\frac{g^2\sin^4(\phi)\sin(p)}{8(1+\tau_{\parallel}v_{\perp})\cos^2(p)}\right] \quad (2.9)$$

$$I_{-} = \frac{e|M_{\parallel}|}{\hbar N}\left[(1-\tau_{\parallel}\nu_{\perp})\sin(p) - \frac{g^2\cos^4(\phi)\sin(p)}{8(1-\tau_{\parallel}\nu_{\perp})\cos^2(p)}\right]. \quad (2.10)$$

Substituting these current modes into (2.11), we obtain

$$I = \frac{e|M_{\perp}|\tilde{g}^2}{8\hbar N}\frac{\sin(p)}{\cos^2(p)}\cos(2\phi), \quad (2.13)$$

where parameter $\tilde{g}$ differs from the parameter g by the fact that it is reduced to the dimensionless form using the constant $|M_{\perp}|$, not $|M_{\parallel}|$" [7].

"In this case the current is determined by the final expression:

$$I = \frac{2e|M_{\parallel}|}{\hbar N}\sin(p_0)\left[1 - \frac{g^2}{32\cos^2(p_0)}\cdot\frac{1}{[1-(\tau_{\parallel}\nu_{\perp})^2]}\right] \quad (2.20)$$

This result means that if the dissipative losses, which lead to the decrease of p_0 to zero during the time of the charge flight, are absent or are not high enough, the effect, similar to the superconductivity arises, when the current different from zero at the absence of the external field takes place" [7]. Thus, this type of organization of the "crystalline lattice" can be referred to the metallic type as the most physically substantiated value of the current there is expression (2.20), which allows the existence of the current also in the absence of the external field.

1 Quantum-Chemical Calculation Experiment

The article idea was to revise the charge wave configurations and localizations by non-empirical quantum mechanics methods, to evaluate the effect on soliton wave's form of the finite group's fundamentality, and to confirm the pattern of previous theoretical charge distribution by experimental methods, in particular, by measuring NMR spectra of ^{13}C.

The quantum-chemical calculations were made for mono- and dis-replaced cationic polymethines, as well as the corresponding unsubstituted ions **2–4**:

$H_2C^+–(CH=CH)_n–CH=CH_2$ **2**

$H_2C^+–(CH=CH)_n–CH=CH–R$ **3**

$CH–(CH=CH)_n–CH=CH–R$ **4**

The fragments of OH (OCH_3) and NH_2 ($N(CH_3)$ were used as model groups: the former developed end groups of low basicity, and the second ones are high-base groups, in accordance with the orbital electronegativity of non-separated pair, respectively, of oxygen atoms or nitrogen atoms.

Earlier [11] on the example of molecules **2**, it was shown that charges on carbon atoms and carbon–carbon bond lengths practically coincide with calculated by means of using the HF/ 6-31G ** and HF/ 3-21G ** biasses. Therefore, in carrying out this work, we confine ourselves to a smaller basis in calculations of relatively long molecules. In our calculations, we had used the HYPERCHEM 7.0 program package.

Calculations of ions **2–4** had shown a significant electron density alternation of carbon atoms in polymethine π-electronic collective system, and our calculations had shown an alternation phase in cations, as established in previous papers [3–11], which is the opposite of the alternation of electron density in anions. The alternation amplitude was decreased gradually from center of the polymethine to its boundaries; therefore, the charges form a wave package described by the hyperbolic soliton function [12]. To analyze the dependence of charges and the localization of soliton waves from the topology of polymethine molecules, one could use electronic density q_μ but much more functional difference of their values of charge Δq_μ on neighboring atoms (1) that are calculated by the following formula:

$$Dq_m = (-1)^\mu (q_\mu - q_{\mu-1}). \tag{1}$$

Similarly, instead of the length of the bands, it is more convenient to analyze the amplitude of their alternation:

$$DL_v = (-1)^v (L_v - L_{v-1}), \tag{2}$$

where the value L_ν is the distance between two carbon atoms of ν-bond number.

Scalar parameters are often used: $|\Delta q_\mu|$ and $|\Delta L_\nu|$.

Consequently, this special type of the "crystalline lattice" configuration of polymer π- electron system of polymethine dye that is defined by (2.18) could be reflected as "metallic lattice" in the polymer as confirmed current value in (2.20) [7] that permits the current presence in deficiency of the external electric field.

Figure 2 shows the graph $|\Delta q_\mu| = f(\mu)$ for ions **2–4** with a relatively long polymethine chain (n = 15), which significantly exceeds the size of soliton waves. As can be seen, the charge wave $|\Delta q_\mu|$ is autolocalized in the middle of the conjugate system, separating from its edges. True, as you can see from Fig. 2a, the calculations do not give a zero value of the amplitude of the alternation, and a certain constant value (≈0,17), which is connected with the nonequivalence of the polarization of the links of C and H onto even in addition to odd location in the polymethine chain with opposite sign charges. Figure 2b shows a graph of alternating lengths **C–C** bonds obtained in optimization of molecular geometry (with precision up to 0.001 Å). As can be seen, the graph $|\Delta l_\nu| = f(\mu)$ describes the typical shock wave or the tip. At the edges of the polymethine chain, the value of the alternating amplitude reaches a constant value, which is close to the value of the alternation in the unaligned neutral polyenes [11].

We will think about a soliton as a charge or bond length distribution like single wave, where its size should not connect with the dimension of collective the π-system and not rely on it. In the case of conjugate systems with a chain length shorter than a soliton wave, only a part of the charge or topological wave is projected onto a molecule [11]. Now let us consider the effect of the nature of the terminal group in the mono-substituted polymethine ion **2**. As the model finite residues, groups were used, the main element of which is an atom with a non-allocated electron pair (NEP),

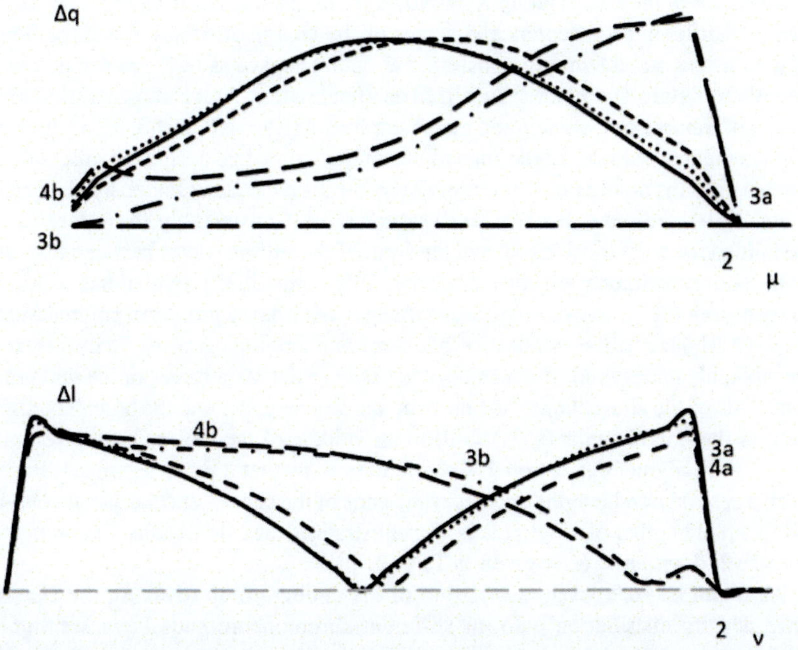

Fig. 2 The charge distribution (**a**) and lengths of CH bonds (**b**) in polymethine systems: **2**—H_2C^+–$(CH = CH)_{15}$–$CH = CH_2$; **3a**—H_2C^+–$(CH = CH)_{15}$–$CH = CH$–OH; **3b**—H_2C^+–$(CH = CH)_{15}$–$CH = CH$–NH_2; **4a**—HO^+=CH–$(CH = CH)_{15}$–$CH = CH$–OH; **4b**—H_2N^+=CH–$(CH = CH)_{15}$–$CH = CH$–NH_2

which is conjugate to the main π-electron system. Previously, it was shown that the level of the NEP of the oxygen atom is located below the top of the valence band of polymethine cation 2, while the level of NAP nitrogen falls into the bandgap; consequently, the effects of these two substituents, R = $-OCH_3$ and R = $-N(CH_3)_2$ on the localization of charge and topological waves, are significantly different.

The graph plots of the values $|\Delta q_\mu| = f(\mu)$ and $|\Delta L_\nu| = f(\mu)$, calculated for ions **3** and **4** with the equivalent vinyl groups number (n = 15), and for unchanged systems **2**, and for conjugated systems have been shown in Fig. 2, respectively. Here, the insertion of one group OCH_3 with low donor properties (low location of the corresponding level of the substituent) can cause the displacement of charge wave peak (Fig. 2a), as well as the minimum wavelength of the bonds (Fig. 2b) to near 11–12 atoms toward the substituent—an atom or group of atoms taking the place of another group or occupying a specified position in a molecule, whereas the extremum value of charge distribution amplitude stored as well as maintaining the alignment of two neighboring bond lengths that are under the charge soliton peak. Due to the displacement as well as the limited size of the soliton, the wing of the soliton waves that is closer to the substituent turns out to be truncated.

Influence of the amino group R = -$N(CH_3)_2$, as can be seen from Fig. 1, is fundamentally different: the centers and waves of the charge distribution and the bonds lengths waves are not simply transferred at definite space as of the molecule center nevertheless almost completely localized on a heteroatom. As it is shown by the calculations of monoamino-substituted molecular ions **3** (R = -$N(CH_3)_2$), such a limiting displacement of solitons on the edge of the conjugate system occurs regardless of the vinylene groups number **n** in polymethine π-electronic chain. Obviously, the graphs in Fig. 2a, b represent only half of each soliton wave. Considerably, the differences in the influence on the localization and the form of the soliton waves between low- and high-donor substituents are given more carefully for distilled polymethines **4**. As can be seen from Fig. 2, the simultaneous influence of two oxy groups in polymethine **4** (R = -OCH_3) leaves the localization of charge and topological waves in the center of the molecule unchanged, and also does not change the magnitudes of the maximum amplitude of the alternation of the electron density (Fig. 2a) and the boundary alignment of the bond lengths ($|\Delta L_\nu| = 0$) in the middle of the molecule relative to the unchanged polymethine cation **2**. Calculations show that a more significant impact on charges and bond lengths appears at the ends of the chain, especially in molecules with a small number of vinyl groups. For illustration, the calculations for conjugated systems **2–4** with n = 6 are given in Table 1.

As it can be seen, even at insertion of the ending group (dyes **3**), the electron wave density distribution onto the different chromophore ends increases significantly, although there is no significant difference between the hydroxy (R = -OH) and the amino-substituted (R = -NH_2) polymethine dye results. At the end of the chromophore near the finite group R, the atom charge alteration is more substantial, so that the charge sign is even changed to a positive carbon atom, which is bound to a terminal heteroatom, oxygen, or nitrogen. Accordingly, both carbon atoms are positively charged at both ends of the polymethine chain in the case of dispersed dyes **4**. Unlike charges, the length of carbon–carbon bonds at the ends of chromophore is less sensitive to the introduction of substituents. As it can be seen from the table, the calculations give a significant decrease in alternating length amplitude of neighboring CH bonds at ending boundaries of the chromophore, bringing it closer to ΔL_ν in the interior the polymethine chains.

In the case of molecules with short polymethine chains that are found to be physical, the effects of finite groups can become dominant. This can lead to significant deformation of the soliton-shaped charge waves.

Assuming that the finite groups influence the distribution of the electron density in the main chromophore primarily due to both σ-inductive and π-mesomeric effects, then in the theory of perturbations of molecular orbits [13] quantitatively such an effect can be estimated by changing the Coulomb integral of the terminal atom. At the same time, the electron density changes for a random μ -s of the polymethine chain atom: $q_\mu = q_\mu^o + \pi_{1,\mu}h$, where h is the correction for the Coulomb integral, and $\pi_{1,\mu}$-the atom–atomic polarizability. For linear conjugate systems **1**, when R is a heteroatom, the parameter $\pi_{1,\mu}$ can be written analytically [13] as

$$\pi_{1,\mu} \approx (-1)^\mu/[\pi(2^\mu - 1)]. \tag{3}$$

Table 1 Charge of the atoms and bond lengths in the polymethines 2–4 (n = 6)

№ Atom/bonds	2	3	4		
		R = OH	R = NH_2	R = OH	R = NH_2
Charge of atoms, q_μ (a.u.)					
1	−0.350	−0.385	−0.390	0.226	0.273
2	−0.181	−0.261	−0.260	−0.254	−0.360
3	−0.098	−0.208	−0.214	−0.025	−0.145
4	−0.251	−0.226	−0.226	−0.227	−0.288
5	−0.067	−0.210	−0.219	−0.020	−0.127
6	−0.269	−0.218	−0.220	−0.224	−0.297
7	−0.055	−0.202	−0.217	−0.018	−0.120
8	−0.269	−0.214	−0.217	−0.224	−0.297
9	−0.067	−0.176	−0.204	−0.020	−0.127
10	−0.251	−0.226	−0.225	−0.227	−0.288
11	−0.098	−0.151	−0.163	−0.025	−0.145
12	−0.181	−0.338	−0.319	−0.254	−0.360
13	−0.350	0.307	0.298	0.226	0.273
Bond lengths, L_υ Å					
1	1.3299	1.3644	1.3662	1.3576	1.3348
2	1.4525	1.4207	1.4140	1.4153	1.4350
3	1.3516	1.3865	1.3868	1.3648	1.3552
4	1.4257	1.4065	1.3993	1.4055	1.4196
5	1.3729	1.3918	1.3904	1.3770	1.3762
6	1.3987	1.4022	1.3977	1.3917	1.3975
7	1.3987	1.3886	1.3887	1.3917	1.3975
8	1.3729	1.3987	1.3997	1.3770	1.3762
9	1.4257	1.3853	1.3812	1.4055	1.4196
10	1.3516	1.3891	1.4028	1.3648	1.3552
11	1.4525	1.3967	1.3717	1.4153	1.4350
12	1.3299	1.3617	1.3994	1.3576	1.3348

As a result, disturbance at the ends of the chain caused by the insertion of the final groups R1 and R2 generates its own wave of alternating (due to the factor $(-1)^\mu$) positive and negative charges at both ends of the collective π-electron system near the heteroatoms. However, the amplitude of this wave rapidly decays, thanks to the factor: $1/[\pi(2\mu-1)]$. Thus, the resulting curve of the graph $|\Delta q_\mu| = f(\mu)$ for the polymethine chain of short molecules **1** should be interpreted as a superposition of three waves: soliton charges in a circuit generated by the self-charged π-electron system, as well as two waves that arise under the influence of finite groups.

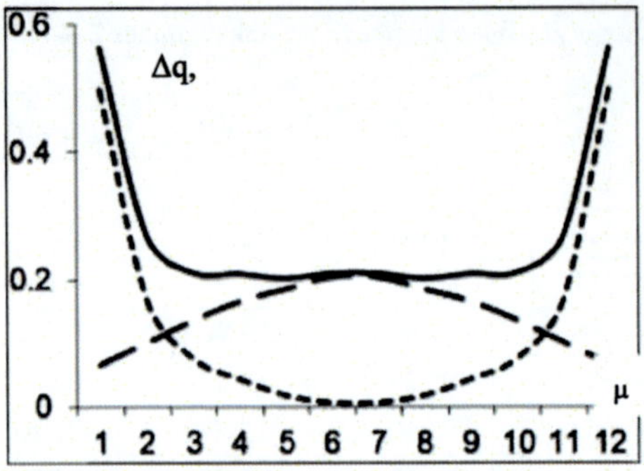

Fig. 3 Scheme of superposition (–) of a charge wave (Δq_μ), generated by a polyene chain (——) and ending groups (– –) in the symmetric cyanine dye **1** (C1 = C2, n = 6)

As an example, Fig. 3 shows a similar superposition of three charge waves in streptocyanine **5** with n = 6.

Apparently, due to the imposition of waves of different origins, the total charge wave has no minimum. In the case of dyes **5** or **6** with even shorter chains, calculations give even reducing the amplitude of the charge alternation, as can be seen from Fig. 4a, b.

This dependence of the charge distribution is confirmed experimentally, for example, NMR spectra of ^{13}C. As you know, there is dependence between chemical shifts, δ_μ, and electron density: $\delta_\mu = aq_\mu + b$ [13]. Then, similar to formula (1), we can write the formula for the amplitude of the alternation of chemical shifts.

$$\Delta\delta_\mu = (-1)^\mu(\delta_\mu - \delta_{\mu+1}). \quad (4)$$

From Fig. 4b, one could see that amplitude of chemical shift alternation $\Delta\delta_\mu$ inside the polymethine chain of cyanine dyes **5** and **6** is smaller than at the ends of the chain. This clearly indicates that the effects of finite groups R on the borders of the chain are principal as in [14].

The main idea of article about the electron charge transport in the case of electron injected in the polymethine chain conduction zone is achieved. Whereas the "crystal lattice" response of crystal chain is explained, the explanation equation systems of the charge transport are gained, and we found its solutions in the zero approximation virtual case of weak fields so as to approximately correlate to the charge soliton-transferring model. It was found that the structure of the "crystal lattice" initiates two excitation modes, so task with the polymer chain soliton nanoconductivity of the polymethine type was examined.

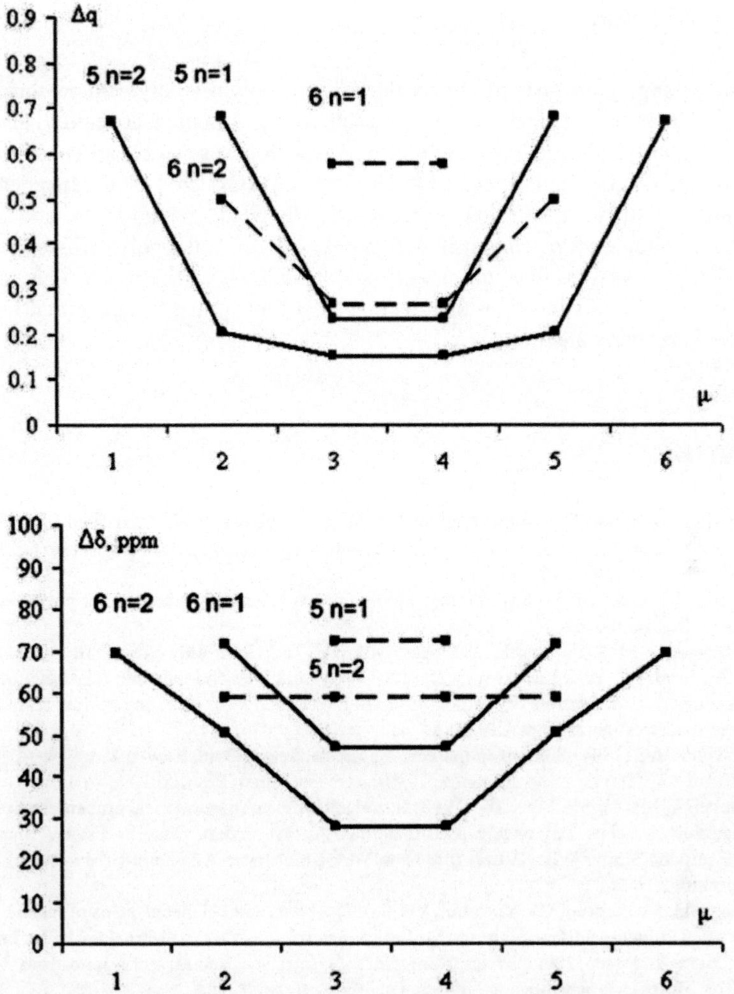

Fig. 4 Alteration of charges **(a)** and ^{13}C NMR experimental shift **(b)** of carbon atoms in the conjugated chain of cyanines **5** and **6**

2 Conclusions

The wide-ranging analysis of the conductivity of polymer polymethine chain was made in this work, and the common expression for the current that caused by electron injected in polymethine dye with the soliton excitation was made. Particular variants that are equivalent to special conductivity types are considered. This dependence of charge distribution is confirmed experimentally by NMR spectra of ^{13}C and shown that the amplitude of the chemical shift alternation inside the polymethine chain of long cyanine dyes was less important than at the ends of the chain. This, in our opinion, undoubtedly indicates that the effects of finite groups on the borders of the chain are preponderated.

References

1. Breda JL, Belionne D, Cornil J, Calhert JPh, Shuai Z, Silbey R (2002) The electronic structure of π-conjugated oligomers and polymers: a quantum-chemical approach to transport properties. Synth Met 125:107–116
2. Breda JL, Street GB (1985) Polaron, bipolaron and solitons in conductiong polymers. Acc Chem Res 18:309–315
3. Sevryukova MM, Piryatinski YuP, Vasylyuk SV, Yashchuk VM, Viniychuk OO, Gerasov AO, Slominskii YuL, Kachkovsky OD (2012) Cyanine-like and polyenic relaxation paths of merocyanines derivatives of malodinytryle in the excited state detecting by low temperature time-resolved fluorescence\UJP 57(8)
4. Davydov AC (1984) Solitons in molecular systems. Scient Opin, Kiev, p 288
5. Mishra A (2000) Cyanine during the 1990s: a review. Chem Rev 100:1973–2011
6. Meyers F, Marder SR, Perry JW (1998) Introducing the nonlinear optical properties of organic materials. In: chemistry of advanced materials. An overreview. Chapt 6. Ed. LV Interrante, J Hampden-Smith. Wiley-VCH. Inc. New York-Chicherster-Weinheim-Brisbane-Singapore-Toronto, 207–268
7. Vasylyuk SV, Suprun AD, Yashchuk VN (2017) About possible mechanisms of Nanoconductivity in Polyenes polymers: the charge Solitons at extremely weak external fields. In: Fesenko O, Yatsenko L (eds) Nanophysics, Nanomaterials, Interface Studies, and Applications. NANO 2016. Springer Proceedings in Physics, vol 195. Springer, Cham
8. Suprun AD, Vasylyuk SV, Yashchuk VN (2018) The possible mechanisms of conductivity in polyene-like polymers and types of conductivity in maximally feeble external fields. In: Fesenko O, Yatsenko L (eds) Nanooptics, Nanophotonics, Nanostructures, and Their Applications. NANO 2017. Springer Proceedings in Physics, vol 210. Springer, Cham
9. Lepkowich RS, Przhonska OV, Hales JM, Hagan DJ, Van Sryland EW, Bondar MV, Slominski YL, Kachkovski AD (2004) Nature of electron trasitions in thiacyanines with a long polymethine chain. Chem Phys 305:259–270
10. Kachkovski AD, Tolmachev AI, Slominski YL, Kudinova MA, Derevyanko NA, Zhukova OO (2005) Electronic properties of polymethine systems. Soliton symmetry breaking and spectral features of dyes with a long chain. Dyes and Pigments 64:207–216
11. Kachkovskiy AD, Przhonska OV, Ryabitzki AB (2007) Symmetry breaking in cationic and anionic polymethine dyes. J Molec Struct (THEOCHEM) 802:75–83
12. Reimers JR, Hush NS (1993) Hole, Electron and energy transfer through bridged systems. VIII Soliton molecular switching in symmetry-broken Brooker (polymethinecyanine) cations. Chem Phys 176:407–420

13. Dewar M (1972) Theory of molecular orbitals in organic chemistry. Mir, 590
14. Vasylyuk SV, Yashchuk VM, Viniychuk OO, Piryatinski YP, Sevryukova MM, Gerasov AO, Zyabrev KV, Kovtun YP, Shandura MP, Kachkovsky OD (2011) The investigation of relaxation paths in oxyborine anionic polymethine dyes detected by low-temperature time-resolved fluorescence. Molecul Cryst Liquid Cryst 535(1):123–131

Thermodynamic Calculation of Vortex Granulator Operation for Producing of Ammonium Nitrate with Nanoporous Structure

A. E. Artyukhov, A. V. Lytvynenko, J. Krmela, and V. Krmelova

1 Introduction

The granulated ammonium nitrate has been widely used in agriculture as an effective nitrogen fertilizer. The native industry produces an ordinary ammonium nitrate by the proven technologies, which have been improved for more than 50 years. At the same time, the ammonium nitrate is used as a component of the industrial explosive substance ANFO (ammonium nitrate/fuel oil) in the industry [1–3]. Such ammonium nitrate, called porous ammonium nitrate (PAN), differs significantly by its properties from the ordinary ammonium nitrate. The ordinary ammonium nitrate can also be a component of ANFO; however, it has several disadvantages:

1. The ordinary ammonium nitrate is covered with a special shell, which prevents the premature dissolution of the granules after its addition to the soil. The protective shell prevents access to the inner pores of the granules.
2. The ordinary ammonium nitrate can be chalked, causing a significant reduction (and, in fact, loss) of the ability to explode in a mixture with diesel fuel.

A. E. Artyukhov (✉) · A. V. Lytvynenko
Processes and Equipment of Chemical and Petroleum-Refineries Department, Sumy State University, Rymskogo-Korsakova Str 2, 40007 Sumy, Ukraine
e-mail: a.artyukhov@pohnp.sumdu.edu.ua

J. Krmela
Department of Numerical Methods and Computational Modeling, Alexander Dubcek University of Trencin, 491/30 I. Krasku, 02001 Puchov, Slovak Republic
e-mail: jan.krmela@fpt.tnuni.sk

V. Krmelova
Department of Materials Technologies and Environment, Alexander Dubcek University of Trencin, 491/30 I. Krasku, 02001 Puchov, Slovak Republic
e-mail: vladimira.krmelova@fpt.tnuni.sk

O. Fesenko and L. Yatsenko (eds.), *Nanomaterials and Nanocomposites, Nanostructure Surfaces, and Their Applications*, Springer Proceedings in Physics 246, https://doi.org/10.1007/978-3-030-51905-6_16

3. The ordinary ammonium nitrate does not have sufficient porosity for successful retention of the diesel fuel distillate inside the granule.
4. The ordinary ammonium nitrate has a certain number of pores, but their nature is mechanical, not modifying. These are the pores that are obtained as a result of the destruction of the granules (cracks, chips, caverns). Such pores are large enough (10^{-6}–10^{-4} m), and it is difficult to hold the diesel fuel distillate (when transporting ANFO to the blasting places, the diesel fuel distillate flows out of the granule and ideally stays on the surface, at worst—leaves the surface of the granule).

The production of PAN from the ordinary ammonium nitrate by the method of humidification and heat treatment (the method is one of the most energy-efficient and environmentally safe, which is proved in a number of works [4–6]) requires special hydrodynamic and thermodynamic conditions, as well as rational selection of the type and construction of the main technological equipment in granulation. The main task of the technology to produce PAN from the ordinary ammonium nitrate is to form a network of nanopores on the granule surface and in the surface layers, keeping the strength of the core [7–9]. Vortex granulators are proposed as granulation equipment in the works [10–13]. They have the following advantages:

- the possibility significantly to reduce the overall dimensions (in particular, the height) of the workspace due to the variable height of the cross-sectional area and the possibility of internal circulation of the seeding agent;
- control of the granules' residence time in the workspace;
- the ability to control the motion of granules in the workspace;
- the possibility to create intensive turbulence in the workspace;
- universality (the ability to carry out granulation and drying processes in one device);
- processability and ease of production;
- possibility of fast adjustment and change of constructive and technological parameters if necessary.

The aim of this work is to study the thermodynamic conditions to form PAN granules with nanoporous structure and to evaluate the influence of thermodynamic parameters on the features of the nanopores structure.

2 Theoretical Principles of Thermodynamics Regarding PAN Granules Drying

According to the previous studies in the field of PAN granule production [7, 9, 11], thermodynamic and hydrodynamic conditions (temperature, gas velocity, time of contact of the granules with hot gas (air) environment) are the main factors which form the porous structure. In addition, having removed the bulk of the moisture and having formed the pores, the granules need to undergo some drying process to ensure the required strength. Otherwise, the granules can significantly lose their strength.

Thus, in the calculations, it is necessary to predict the kinetics description of granule heating and to determine the minimum drying time, the value of which exceeds the required time to warm the granules to the desired temperature.

The granule is a spherical body with pores, the peculiar size of which is the radius R and in terms of the physicochemical properties in the calculations—the structure of the granule's substance (including moisture content as the main index).

In order to describe the influence of the thermodynamic parameters of the gas flow on the humidified granule, a physical model was investigated. It is based on the following principles:

- The selected diameter of the granules is determinant for the entire set of granules in the workspace, i.e., the formation of the porous structure in the granules with the selected diameter preferably defines the porous structure of the entire granules flow;
- The granule is washed by a stream of gas (air) which has a constant temperature.

Thus, one may accurately assume that the surface temperature of the granules during the whole residence time in the workspace will be constant;

- Due to the fact that the granules are small in size from 1 mm to 4 mm, and the vast majority of granules obtained at nitrogen plants are granules in size from 2 mm to 3 mm, we can suppose that due to the hygroscopicity of the granules, the humidification process should be minimized. So, one may presuppose that moisture saturation occurs along the radius according to a linear law;
- The pore formation starts consecutively from the top layer. The pore formation front moves along the radius of the granule to its center, representing a spherical surface;
- Evaporating moisture is freely removed from the granule through the formed pores in the previous layers of the substance with a larger diameter, "external" layers (these layers are located closer to the surface of the granule) which form the granule;
- The proportionality of the formed pores in the granules with the steam extracted from moisture is the reason for the fact that the retention capacity of the granules is greater than the moisture content.

Based on the above assumptions, it is possible to create a mathematical model to determine the basic hydro- and thermodynamic parameters, the effect of which enables it to create a porous structure in the granules.

The temperature distribution in the PAN granule is described by the differential equation [14]:

$$\frac{d}{d\tau}(rT(r,\tau)) = a\left(\frac{d^2}{dr^2}(rT(r,\tau))\right), \tag{1}$$

where r—current radius, and a—thermal diffusivity coefficient.

The solution of this equation will be the following:

$$T(r,\tau) = T_C - (T_C - T_0)\times$$
$$\times\left(\sum_{n=1}^{\infty}\left(\frac{2(\sin(n\pi) - n\pi\cos(n\pi))R\sin\left(\frac{n\pi r}{R}\right)e^{\left(\frac{n^2\pi^2 a\tau}{R^2}\right)}}{(n\pi - \sin(n\pi)\cos(n\pi))rn\pi}\right)\right). \quad (2)$$

As a result of the drying process, the diameter of the granules increases due to the modification transitions [7] and the porous structure formation.

Thus, the granule together with the heat transfer agent will be heated to a predetermined temperature, and only then the moisture will be removed. This factor increases the total required residence time of the granules in the device. The calculations of the above mathematical model showed that 8 s are required to complete the heating process of the granules with d = 2 mm at a temperature of 20 °C–120 °C in a stream of heat transfer agent with a temperature up to 120°. If we perform a gradual calculation taking into account that with the introduction of granules or moistening the heat transfer agent is cooled, then gradual heating of the granules to a temperature of 120° will take a time interval of 3–3.5 times longer than in the previous case. The results of experimental studies of the workspace heating kinetics in the vortex granulator under different conditions, which are given below, make the basis to determine the heat transfer agent's temperature with the gradual heating of the granules simultaneously with the heat transfer agent's flow.

Based on the value *dm* of dry substance, the weight of the elemental volume *dV* of granules, taking into account the presence of moisture, is

$$dm = \rho_{gr}(1 + U(r,\tau))r^2\sin\theta d\theta d\phi dr, \quad (3)$$

where $dV = r^2\sin\theta d\theta d\phi dr$—elementary volume value.

On the other hand, the mass of "dry" granule with radius *R* in general is

$$Msr = 2\int_0^R\int_0^\pi\int_0^\pi \rho_{gr}r^2\sin(\Theta)d\Theta d\phi dr, \quad (4)$$

or after integrating

$$Ms = \frac{4}{3}\cdot\rho_{gr}\cdot\pi\cdot R^3. \quad (5)$$

Given the fact that *U* is a function from *r* and drying time τ

$$Mg_{n=\infty}(\tau) = \frac{4}{3}\rho\times Mg_{n=\infty}(\tau) = \frac{4}{3}\rho_z\times$$

$$\times\left(1+\frac{\int_0^R U_p+(U_o-U_p)\left(\sum_{n=1}^{\infty}\left(\frac{2(\sin(n\pi)-n\pi\cos(n\pi))R\sin\left(\frac{n\pi r}{R}\right)e^{\left(-\frac{n\pi m\tau}{R}\right)}}{(n\pi-\sin(n\pi)\cos(n\pi))rn\pi}\right)\right)dr}{R}\right) \quad (6)$$

where *m*—diffusion coefficient.

Besides, the above equation takes into account the fact that the moisture content of the granule takes some average value, which is calculated based on the integral properties.

The calculations show that the sum of the first two components is significant in solving the differential equation describing the mass transfer during the drying process. Further calculation of the sum leads to a change in the result obtained in the seventh digit of the fractional part. Based on this conclusion, it is possible to calculate the first two coefficients of the sum. Thus, the equation to determine the change in humidity over time takes the following form:

$$Us(\tau)=\frac{\int_0^R U_p+(U_0-U_p\left(\frac{2R\sin\left(\frac{\pi r}{R}\right)e^{\left(-\frac{\pi^2 m\tau}{R^2}\right)}}{\pi r}-\frac{R\sin\left(\frac{2\pi r}{R}\right)e^{\left(-\frac{4\pi^2 m\tau}{R^2}\right)}}{\pi r}\right)dr}{R} \quad (7)$$

Integrating the obtained formula, it is possible to receive a dependency to find the mass of the granules at any time during the drying process.

$$Mg(\tau)=\frac{4}{3}\rho_z\pi R^3\times\sqrt{b^2-4ac}$$

$$\times\left(1+\frac{\int_0^R U_p+(U_0-U_p)\left(\frac{2R(\sin\left(\frac{\pi r}{R}\right)e^{\left(-\frac{\pi^2 m\tau}{R^2}\right)}}{\pi r}-\frac{R(\sin\left(\frac{2\pi r}{R}\right)e^{\left(-\frac{4\pi^2 m\tau}{R^2}\right)}}{\pi r}\right)dr}{R}\right)$$

$$+2R\left(\sum_{n=1}^{\infty}\left(\frac{\sin\left(\frac{n\pi r}{R}\right)(-\sin(n\pi)+n\pi\cos(n\pi)e^{\left(-\frac{n^2\pi^2 m\tau}{R^2}\right)}}{(n\pi-\sin(n\pi)\cos(n\pi))n}\right)\right)U_p)dr \quad (8)$$

3 Experimental Studies of Thermodynamic Indices of the Workspace in a Vortex Granulator

Within the framework of the scientific and research works "Investigation of hydrodynamic and heat-mass transfer features of devices with vortex and highly turbulent single and two-phase flows," "Hydrodynamic parameters of two-phase flows of heat-mass transfer, granulation and separation equipment," at the "Processes and Equipment of Chemical and Petroleum Refinery Department" vortex granulator stand to obtain PAN granules (Fig. 1) is constructed.

Devices and equipment:

- determine the hydrodynamic features of the flow motion—TES 1340 Hot-Wire Anemometer;
- temperature measurement in air heater—self-recording-register potentiometer;
- measurement of granulator workspace temperature—thermal imager Fluke Ti25;
- measurement of moisture granules—Multimeter DT-838;

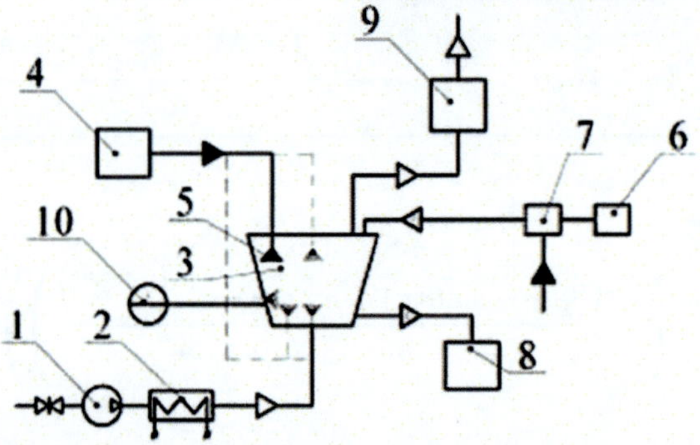

Fig. 1 The unit of the study on hydrodynamics of the vortex granulator workspace, thermodynamic conditions to obtain granules (with different configurations of the workspace): 1—gas blower; 2—heater; 3—vortex granulator; 4—container for preparation of the humidifier; 5—nozzle; 6—container of granules; 7—container of pre-humidification (used when moistening); 8—container for finished granules; 9—the degree of purification of outgoing gases (bubbler); 10—recording potentiometer. The dashed line shows the variants of the nozzle installation

- study of the microstructure of granules—microscope KONUSPIX-450X KONUS, scanning electron microscope VEGA3 XM and X-ray spectrometer with an energy dispersion.

The results of the experimental study of the drying agent's temperature field in the vortex granulator workspace are presented in Figs. 2 and 3. It should be noted that with the increase of the flow twisting degree one can observe the alignment of the temperature field in the granulator workspace. Intensification of the flow motion (turbulization due to the directed vortex movement) lets to distinguish an "active" zone, where the heat treatment process occurs with maximum intensity in the device.

Based on [15, 16], the heat-mass transfer processes occur with the highest intensity in the so-called "active" zone (heat-mass transfer zone [17]), directly above the gas distribution device of the granulator or dryer with a fluidized bed. The height of this zone with the same features of the heat transfer agent depends on the type of gas-distributing device. Varying the design of the gas-distributing device (degree of

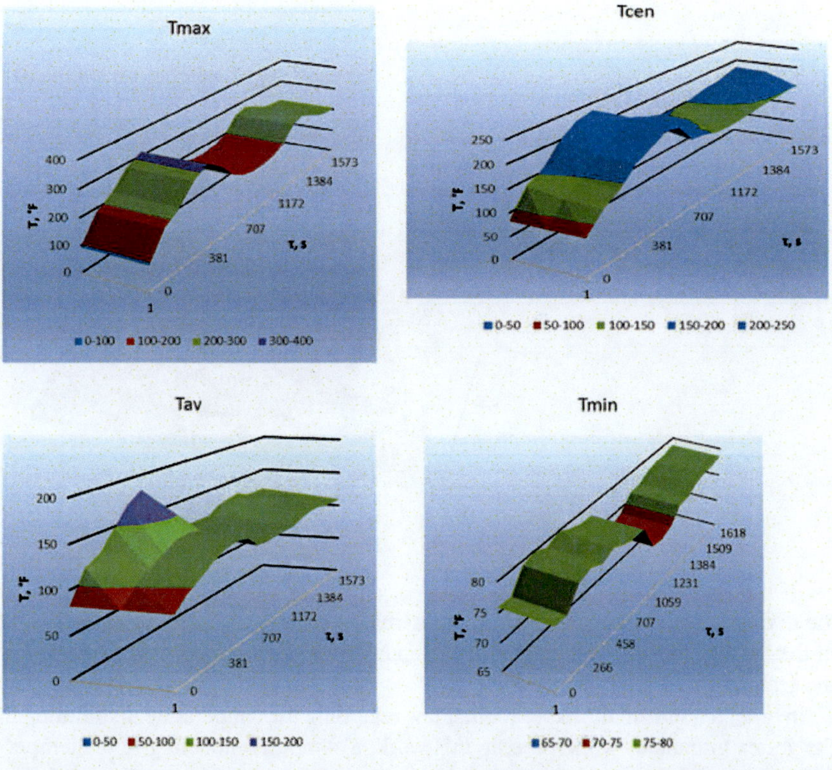

Fig. 2 Dynamics of the temperature change in the vortex granulator workspace depending on the intensity of solubilizer (drying agent) twisting (circular velocity of the drying agent 5.2–8.1 m/s) Temperature of the liquid agent: Tmin—minimum temperature; Tmax—maximum temperature; Tav—average temperature; Tcen—temperature in the center. Granule injection time is 650 s

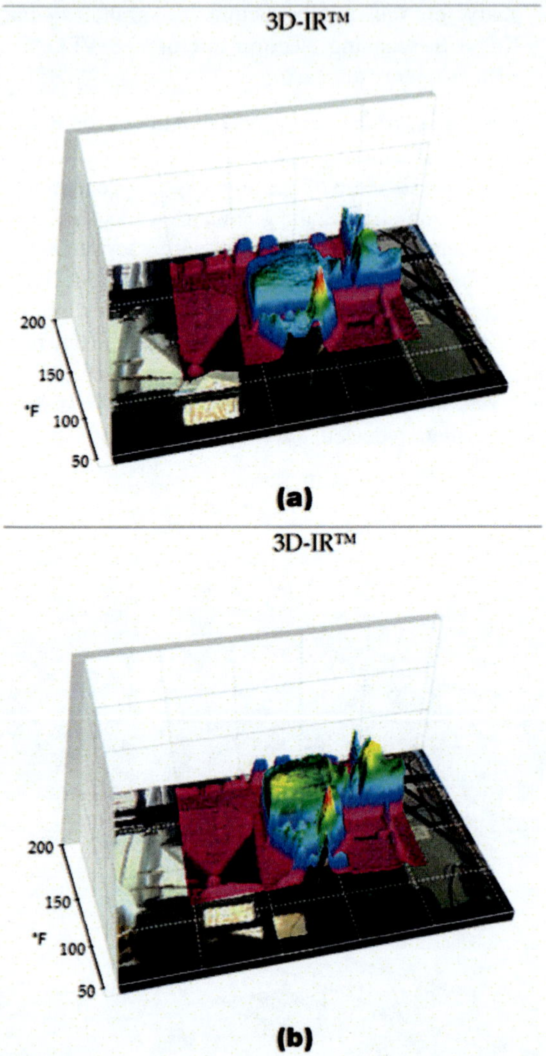

Fig. 3 Data of thermal imaging studies of the vortex granulator workspace: **a** access to the operating mode; **b** operating mode

the drying agent's flow twisting), it is possible to achieve an increase in the height of the "active" zone to that height of the zone of the predominant vortex motion of the granules.

In order to confirm the experimental law regarding the temperature distribution in the vortex granulator workspace, in this work, a computer modeling of the temperature field of the twisted flow of a drying agent was carried out. The modeling results are presented in Fig. 4. Satisfactory convergence between the experiment and modeling results in the operating (steady) mode of the vortex granulator operation is pointed out.

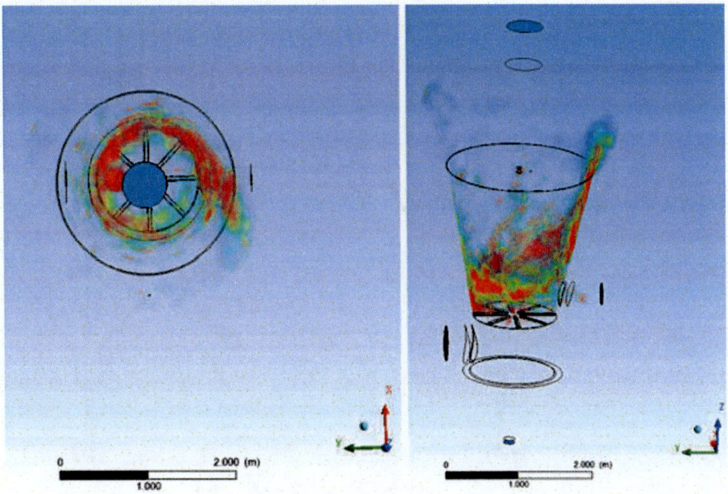

Fig. 4 The computer modeling results regarding the drying agent temperature field in the vortex granulator workspace

The model for calculation of the granule heating and drying kinetics proposed in the first section of the chapter, and the results to determine the drying agent's temperature field in a vortex granulator, let to carry out the thermodynamic calculation of the nanoporous structure obtaining process in the PAN granule. The received data of the temperature field are introduced to the calculation formulas in order to define the required time of the heat treatment process of PAN granules. Further, this paper presents constructive solutions regarding the methods to stabilize the drying agent's vortex flow, as well as the research findings of the nanoporous granule structure (nanoporous structure obtained at different twisting velocities of the drying agent).

4 Methods to Stabilize the Drying Agent's Vortex Flow: The Microstructure of PAN Granules

Figure 5 demonstrates the construction of the gas-distributing unit in the vortex granulator without stabilization and with the stabilization of the drying agent's flow. The introduction of additional knots in the vortex granulator's construction practically does not complicate the production of the device and does not increase its price. However, as you can see below, such constructive solutions have a significant effect on the quality of the nanoporous surface in the PAN granules.

The nanoporous surface microscopy is performed in two steps. At the first stage, the effect of the temperature field (at different degrees of the drying agent's twisted flow in the same range as in Fig. 2) on the nanoporous structure is investigated. The

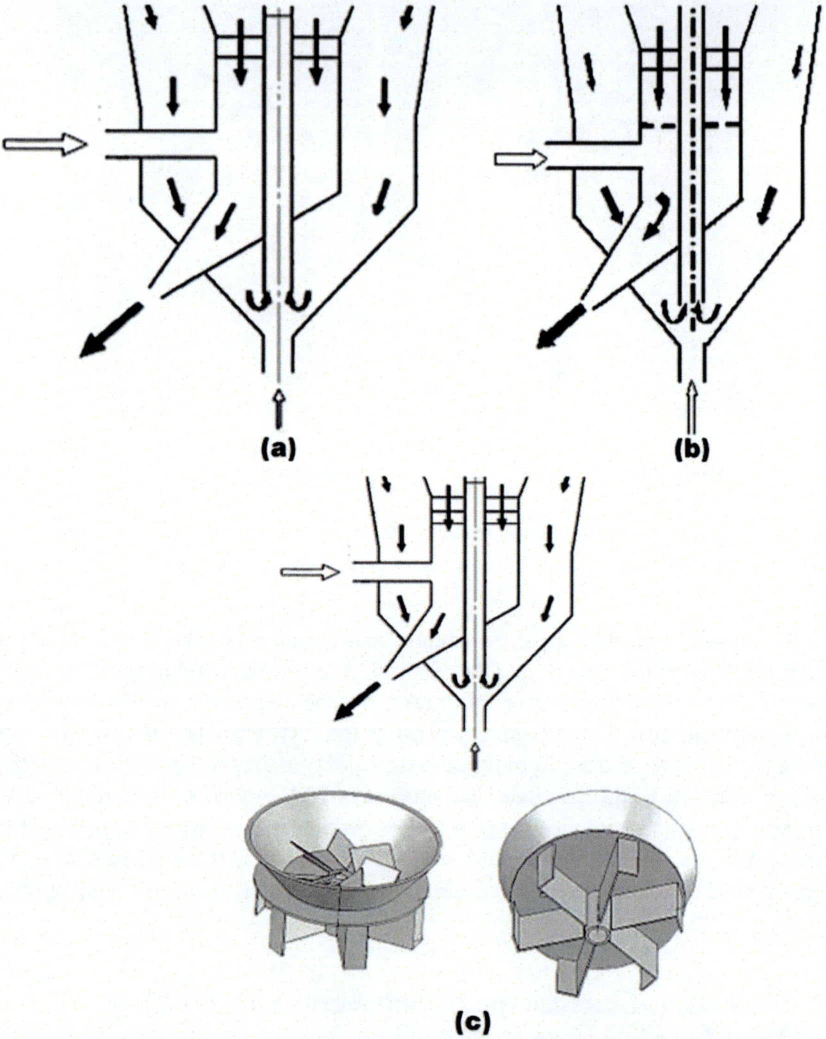

Fig. 5 The construction of the gas-distributing units in the vortex granulator: **a** without stabilization of the drying agent's flow motion [18]; **b** with stabilization grid under the vortex gas-distributing unit [19]; **c** with a two-stage swirler [20]

results of this microscopy are shown in Fig. 6. The granules are obtained in a vortex granulator without stabilizing the drying agent's vortex flow.

Data analysis in Fig. 6 confirms the assumption that the best quality of the nanoporous layer is obtained in the vortex granulator's "active" zone. You can also clearly see an increase in the relative area of the nanoporous surface of the PAN granule with an increase in the degree of the drying agent flow twisting. A further

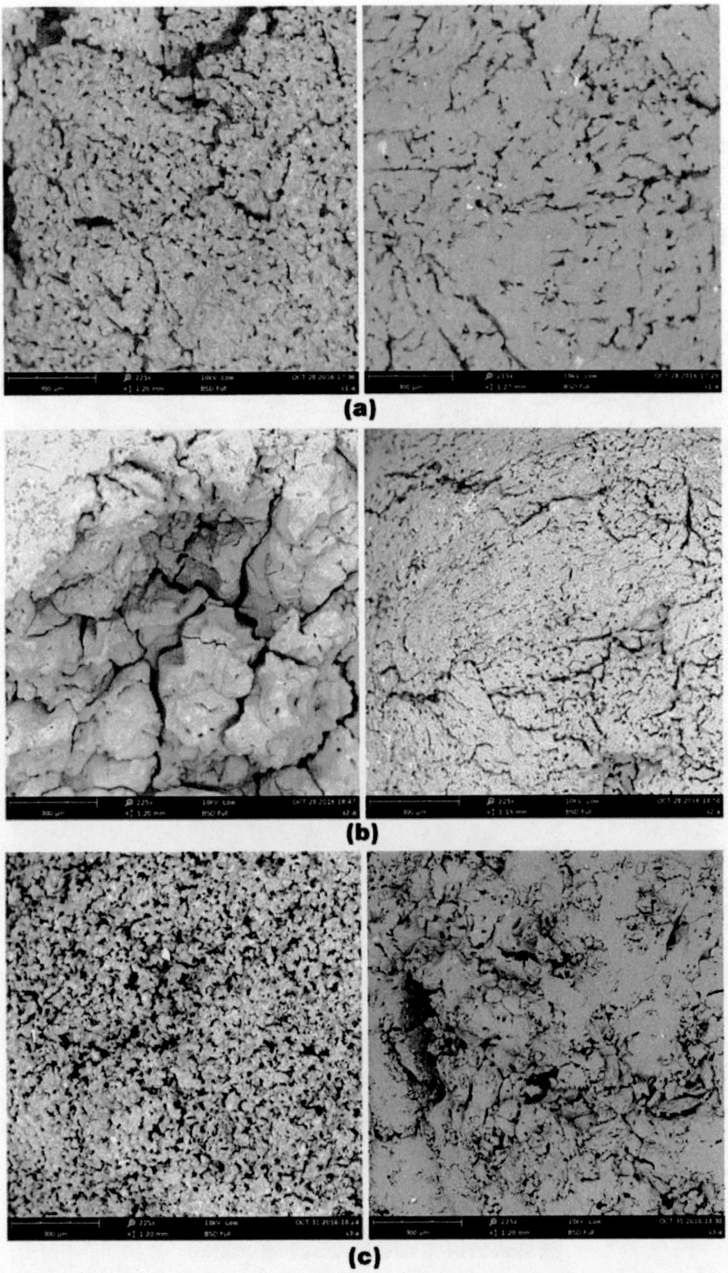

Fig. 6 Microscopy of the nanoporous surface in the granules obtained at different temperature distributions in the vortex granulator (according to Fig. 2): **a** circumferential velocity of 5.2 m/s; **b** circumferential velocity of 6.5 m/s; **c** the circumferential velocity 8.1 m/s. Left photo—PAN granule obtained in the "active" zone of the granulator, right photo—PAN granule obtained over the "active" granulator zone

increase in the drying agent's twisting causes damages in the granule caused by the intensive collision of the granules with each other and with the walls in the granulator. Besides, an increase in the drying agent twisting degree over 8.1 m/s leads to an increase in heat loss in the granulator's "active" zone through the wall of the device. Data from Fig. 6 shows that the relative area of the nanoporous surface can be increased. Therefore, the second stage of the research is devoted to the microscopy of granule samples obtained in granulators with stabilization of the drying agent's vortex flow. The results of these studies are presented in Fig. 7.

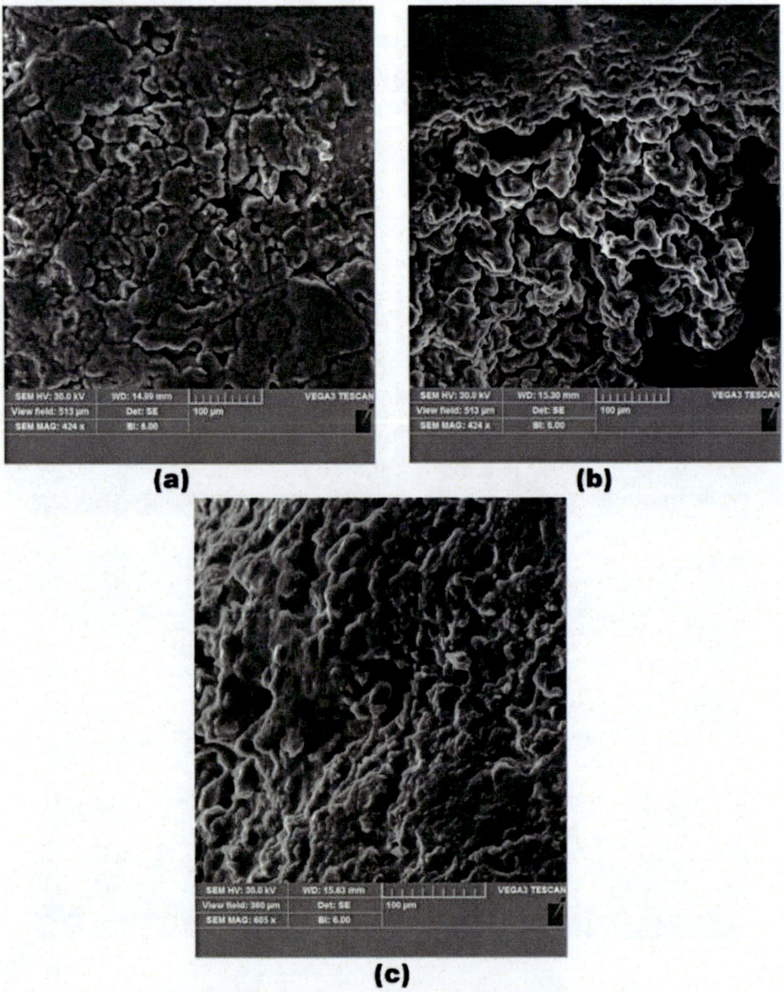

Fig. 7 Microscopy of the nanoporous surface of the granules obtained by using various methods of stabilizing the drying agent's vortex flow: **a** granulator with a stabilization grid under the vortex gas-distributing unit; **b** granulator with a two-stage swirler

Table 1 Identification of the relative area in the nanoporous surface of the PAN granule

Stabilization type	The relative area of the porous surface, m^2/m^2, at					
	V = 5,2 m/s		V = 6,5 m/s		V = 8,1 m/s	
	In the "active" zone of the granulator	Above the "active" zone of the granulator	In the "active" zone of the granulator	Above the "active" zone of the granulator	In the "active" zone of the granulator	Above the "active" zone of the granulator
Without stabilization	0.36	0.25	0.39	0.28	0.41	0.3
Granulator with stabilization grid under the vortex gas-distributing unit	–	–	–	–	0.45	–
Granulator with two-stage swirler	–	–	–	–	0.48	–

Note the "—" sign indicates the fields for which the studies were not carried out in view of determining the optimal conditions for the process to obtain the nanoporous layer on the PAN granule at the previous stages in the research

Figure 7 confirms the assumption about improving the nanoporous surface quality in the PAN granule due to the rational selection of a method to stabilize the vortex flow of a drying agent.

The data of experimental studies to define the relative area of the nanoporous surface in the PAN granule are summarized in Table 1.

5 Conclusions

The nanoporous structure quality of the PAN granules largely depends on the thermodynamic operation mode of the vortex granulator. Rational selection of the drying agent's flow twisting degree and the twisting stabilization method lets us achieve an extended network of nanopores on the surface in the granules, which provide access to the internal pores (where the diesel distillate will be retained). During the creation of an optimization calculation algorithm, the change in the nanoporous structure of the ammonium nitrate granule in the specified temperature ranges and the intensity of the fluidizing agent motion were taken into account. An optimal thermodynamic range (main modes) of the vortex granulator work which enables to obtain NH_4NO_3 granules with branchy net of pores with required size on the surface of the sample and inside it is proposed.

Acknowledgements This research work had been supported by the Ministry of Science and Education of Ukraine under the project « Small-scale energy-saving modules with the use of multifunctional devices with intensive hydrodynamics for the production, modification and encapsulation of granules » , project No. 0119U100834 and by the Cultural and Educational Grant Agency of the Slovak Republic (KEGA), project No. KEGA 002TnUAD-4/2019.

References

1. Kubota N (2015) Propellants and explosives: thermochemical aspects of combustion, 3rd edn. Wiley-VCH Verlag & Co., Weinheim
2. Janssen TJ (2011) Explosive materials: classification, composition and properties. Nova Science Publishers, Inc.
3. Martin G, Barbour W (2003) Industrial nitrogen compounds and explosives, chemical manufacture and analysis. Watchmaker Publishing
4. Artyukhov AE, Sklabinskyi VI (2013) Experimental and industrial implementation of porous ammonium nitrate producing process in vortex granulators. Nauk Visnyk Nats Hirnychoho Univ 6:42–48
5. Artyukhov AE, Sklabinskyi VI (2017) Investigation of the temperature field of coolant in the installations for obtaining 3D nanostructured porous surface layer on the granules of ammonium nitrate. J Nano Electron Phys 9(1)01015–1–01015–4
6. Ivaniia AV, Artyukhov AY, Olkhovyk AI (2019) Hydrodynamic and thermodynamic conditions for obtaining a nanoporous structure of ammonium nitrate granules in vortex granulators. Springer Proc Phys 221:257–268
7. Artyukhov AE, Krmela J, Gavrylenko OM (2019) Evaluation of the impact made by the hydrodynamic regime of the granulation equipment operation on the nanoporous structure of N_4HNO_3 granules. J Nano Electron Phys 11(4):03033
8. Artyukhov A, Gabrusenoks J (2018) Phase composition and nanoporous structure of core and surface in the modified granules of NH_4NO_3. Springer Proc Phys 210:301–309
9. Artyukhov AE, Sklabinskyi VI (2016) Thermodynamic conditions for obtaining 3D nanostructured porous surface layer on the granules of ammonium nitrate. J Nano Electron Phys 8(4):04083–1–04083–5
10. Artyukhov AE, Artyukhova NO (2019) Technology and the main technological equipment of the process to obtain N_4HNO_3 with Nanoporous Structure. Springer Proc Phys 221:585–594
11. Artyukhov AE, Ivaniia AV (2017) Obtaining of porous ammonium nitrate in multistage and multifunctional vortex granulators. Naukovyi Visnyk Natsionalnoho Hirnychoho Universytetu 6:68–75
12. Artyukhov AE, Sklabinskyi VI (2016) 3D nanostructured porous layer of ammonium nitrate: influence of the moisturizing method on the layer's structure. J Nano Electron Phys 8(4):04051–1–04051–5
13. Artyukhova NO, Krmela J (2019) Nanoporous structure of the ammonium nitrate granules at the final drying: the effect of the dryer operation mode. J Nano Electron Phys 11(4):04006
14. Kudra T, Mujumdar AS (2002) Advanced drying technologies. Marcel Dekker, New York
15. Vengateson U, Mohan R (2016) Experimental and modeling study of fluidized bed granulation: Effect of binder flow rate and fluidizing air velocity. Resour Eff Technol 2(1):S1124–S1135
16. Mujumdar AS (2006) Handbook of industrial drying. Taylor & Francis Group, Boca Raton
17. Maronga (1998) On the Optimization of the Fluidized Bed Particulate Coating Process. PhD Thesis, Department of Chemical Engineering and Technology Royal Institute of Technology
18. Artyukhov AE et al (2017) Vykhrovyy hranulyator zvazhenoho sharu (Suspended layer vortex granulator) UA Patent 114521, 3 Mar 2017

19. Artyukhov AE et al (2017) Vykhrovyy hranulyator zvazhenoho sharu (Suspended layer vortex granulator) UA Patent 114517, 3 Mar 2017
20. Artyukhov AE et al (2016) Prystriy dlya hranulyuvannya u vykhrovomu zvazhenomu shari (Device for granulation in the vortex suspended layer) UA Patent 111592, 10 Nov 2016

Nanostructural Changes in a Ni/NiO Cermet During High-Temperature Reduction and Reoxidation

Yevhen Kharchenko, Zinoviy Blikharskyy, Volodymyr Vira, B. D. Vasyliv, V. Ya. Podhurska, Andriy Kalynovskyy, and Vitaliy Korendiy

1 Introduction

A solid oxide fuel cell (SOFC) is used for the direct conversion of the chemical energy of hydrogen or another fuel directly into electrical energy. It is considered a promising clean technology for power generation. Nickel-containing cermets are the most used materials for anode substrates of SOFCs. During the operation of an SOFC, the metallic nickel phase of the YSZ–Ni anode undergoes transformations due to high operating temperature and other auxiliary factors [1–3]. Oxygen available in an operating gas atmosphere can substantially affect the cermet structure. At high temperature (600 °C and more), nickel oxidizes rapidly if the oxygen partial pressure is high enough. The electrolyte and anode substrate can be destroyed as a result of the volume increase during oxidation.

The reasons for reoxidation of nickel at the operating temperature can be as follows: (1) in the case of intense fuel inlet, the oxygen partial pressure critically increases in local microvolumes [4]; (2) small air leakage toward the anode can occur while increasing the oxygen partial pressure in air channels [5]; (3) there can

Y. Kharchenko · Z. Blikharskyy · V. Vira
Institute of Building and Environmental Engineering, Lviv Polytechnic National University, 12 S. Bandera str., Lviv 79000, Ukraine

B. D. Vasyliv (✉) · V. Ya. Podhurska
Karpenko Physico-Mechanical Institute, NAS of Ukraine, 5 Naukova str., Lviv 79060, Ukraine
e-mail: mechengin1111@gmail.com

A. Kalynovskyy
Department of Foreign and Custom Activities, Lviv Polytechnic National University, 12 S. Bandera str., Lviv 79000, Ukraine

B. D. Vasyliv · V. Korendiy
Department of Mechanics and Automation Engineering, Lviv Polytechnic National University, 12 S. Bandera str., Lviv 79000, Ukraine

O. Fesenko and L. Yatsenko (eds.), *Nanomaterials and Nanocomposites, Nanostructure Surfaces, and Their Applications*, Springer Proceedings in Physics 246, https://doi.org/10.1007/978-3-030-51905-6_17

be accidental interruptions in fuel supply; (4) performing shut down of the fuel cell and its start-up without inlet of a neutral gas. The abovementioned reasons cause deterioration of the structure and mechanical and physical properties of the YSZ–Ni cermet anode [6]. This requires a significant effort of researchers to be applied in order to improve SOFC technology.

Before considering the YSZ–Ni cermet material, one should study the reduction and oxidation of pure nickel.

During the reduction of NiO in hydrogen, Ni clusters are nucleated. The cluster growth rate is almost constant for the larger part of the reduction process. It is finally lowered because of the formation of porous Ni layer which complicates the hydrogen diffusion.

In the case of metallic nickel behavior in air, the oxidation forms on its top NiO layer according to the reaction $Ni + ½O_2 = NiO$. This layer separates the oxygen-containing gas and the metallic Ni [7]. For the case when NiO layer thickness is less than 0.1 μm, the oxidation process is performed according to the anodic reaction ($Ni = Ni^{2+} + 2e^-$) and the cathodic one ($½O_2 + 2e^- = O^{2-}$). For thicker layers, this process is due to ion diffusion through the NiO scale.

At high temperatures, the Ni^{2+} cation diffusion in NiO is much faster than the O^{2-} anion diffusion. At these conditions, voids are formed inside the NiO layer during its growth. Microcracking occurred in the bulk of the material promotes the oxygen inward diffusion. The microstructure of the NiO layer may vary depending on the oxidation temperature and the layer thickness.

When Ni particles are small, the difference in diffusion coefficients of Ni^{2+} and O^{2-} causes the growth of multiple internal nanopores in the NiO layer. At the temperatures in the range of 800–1000 °C, it is observed that the outward Ni^{2+} cation diffusion is faster than the inward O^{2-} anion diffusion. This is a cause of NiO internal porosity. It is known that smaller nanoparticles show single pores. In the case when Ni particles are larger, the Ni self-diffusion is too slow and the multiple nanopores cannot be condensed into a single pore [8]. So, the larger particles present multiple pores.

It was found by some researchers that the oxidation rate decreases with decreasing the oxygen partial pressure [9–12]. This may be used as the main principle of microstructural modification during oxidation.

At low temperature (about 600 °C), diffusion of the Ni^{2+} ions through boundaries of the NiO grains occurs [13]. In this case, the oxidation rate depends on the grain size in the NiO layer.

This work is aimed to perform a study on micro- and nanostructural changes in nickel-containing cermets for SOFC anodes during high-temperature (600 °C) reduction and reoxidation, highlighting both the positive and negative effects on their physical and mechanical properties.

2 Materials and Methods

The disk-shaped NiO ceramic specimens were sintered from pure NiO powders [14]. The specimens manufacturing procedure comprised of the following stages: the NiO powder was milled in a rotating mill for 24 h, then it was dried, compacted by applying the pressure 20 MPa, and sintered for 2 h at 1400 °C in air. Finally, we obtained disks of a thickness of 1.8 mm, and their diameter was 27 mm. To measure their porosity, we used a hydrostatic method. The average value of porosity based on measurements of all sintered specimens was in a range of 41–42%.

Series of disks were subjected to one-time reduction in hydrogen of 99.99 vol% H_2 purity and Ar–5 vol% H_2 mixture (modes 2 and 3 in Table 1, respectively). They were heated in vacuum from room temperature to 600 °C, held in a hydrogenous atmosphere at 600 °C under a pressure 0.15 MPa for 4 h, and cooled down in pure Ar to 20 °C (Fig. 1a) [14]. The redox treatment (mode 4) was carried out in Ar–5 vol% H_2 mixture and in air for five cycles using to the following technique (Fig. 1b): the specimens were heated in vacuum from room temperature to 600 °C and reduced in Ar–5 vol% H_2 mixture at 600 °C for 1 h under a pressure 0.15 MPa, then the test chamber was degassed; after that, the specimens were oxidized for 1 h at 600 °C in air and cooled down to 20 °C [15]. The heating/cooling rate was 20 °C/min. The treatment was followed by heating of the material in vacuum to 600 °C, reduction in Ar–5 vol% H_2 mixture at 600 °C for 1 h under a pressure 0.15 MPa and cooling down in pure Ar to 20 °C.

Biaxial bend tests of the specimens were carried out at 20 °C in air using the "ring-on-ring" scheme [16]. We determined the fracture stresses in the as-sintered material (σ_{f0}) and treated one (σ_f) based on the "stress–flexure" diagrams [17] using [16, 18]

Table 1 Effect of the treatment modes on the mechanical and physical properties of the material

Mode marking	The material treatment mode	Mean values			
		σ_f (MPa)	σ_f/σ_{f0}	E/E_0	σ (S/m)
			(%)		
1	As-sintered	13.3	100	100	([a])
2	Reduction in hydrogen (99.99 vol% H_2)	18.8	141	53	2.85×10^6
3	Reduction in Ar–5 vol% H_2 mixture	15.8	121	91	3.25×10^5
4	Redox treatment (5 cycles in Ar–5 vol% H_2 mixture/air) followed by reduction in Ar–5 vol% H_2 mixture	25.7	192	107	1.85×10^5

[a]Electrical conductivity is very low

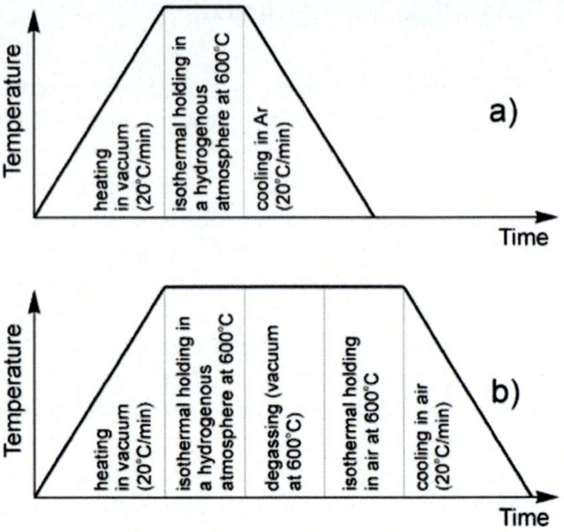

Fig. 1 Diagrams of **a** a one-time reduction mode and **b** a redox cycle mode

$$\sigma_{\mathrm{f}} = \frac{3P\mathrm{max}}{2\pi \cdot t^2}\left[(1-\nu)\cdot\frac{D_S^2 - D_L^2}{2\cdot D^2} + (1+\nu)\cdot\ln\frac{D_{\mathrm{S}}}{D_{\mathrm{L}}}\right], \tag{1}$$

where $P_{\max}$ is the maximum load [*N*], *D* and *t* are the specimen diameter and thickness, respectively [mm], ν is Poisson's ratio, D_{L} is the diameter of the loading ring [mm], and D_{S} is the diameter of the supporting ring [mm].

For the quantitative evaluation of changes in the mechanical properties of materials, the relative stiffness characteristic E/E_0 was proposed. Here, E_0 and E are values of Young's modulus for the material in as-sintered state and after corresponding treatment. Using the slopes of linear sections of "stress–flexure" diagrams, we determined the characteristic E/E_0 as the tangent ratio for the material in the treated and as-sintered state.

We used the four-point method [19] for determining the specific electrical conductivity σ of specimens at 20 °C in air.

The morphology of the fracture surface and the material microstructure were studied using a scanning electron microscope Carl Zeiss EVO-40XVP. It was equipped with an INCA Energy 350 system. This system allows determining the chemical homogeneity of materials by X-ray energy-dispersive spectroscopy (EDS) microanalysis. The detailed microstructural analysis was performed using a scanning transmission electron microscope (STEM) Hitachi-HD2700 with the Cs corrector, which was used to study nanostructural changes in materials. Imaging was performed using an annular dark field (ADF). The electron energy loss spectroscopy (EELS) technique was used for chemical analysis of nanoparticles.

Based on the analysis of nano- and microstructure, the changes in the nanostructure of the material during the high-temperature (600 °C) reduction and reoxidation have been modeled and explained.

3 Results and Discussion

As-sintered NiO ceramic specimens (of mode 1) exhibit quite low strength (Table 1) corresponding to weak bonds between nickel oxide particles. Truncated hexagonal pyramid-shaped particles are seen on the specimen fracture surface (Fig. 2a). They are mostly unbroken exhibiting a distinct intergranular fracture micromechanism (Fig. 3a).

The intense reduction course of as-sintered ceramic material in hydrogen at 600 °C under the pressure 0.15 MPa (mode 2) is displayed in the metallic nickel structure of the former hexagonal NiO particles (Fig. 2b). Each spongy-like particle has smoothed edges, as compared to the initial material. The particle shrinkage occurs due to the appearance of numerous nanopores in them decreasing thus their contact surface area. Also, the interparticle porosity increased, which, however, did not violate the material integrity. As compared to the as-sintered ceramics, the material stiffness decreased by 47% (see Table 1). This, however, did not cause the lowering of its strength. It was concluded that the material strength is affected by the quality of the reduced nickel as well as of interparticle contacts and grows to 141% (see Table 1). Completely reduced grains of metallic nickel with small pores inside can be observed in the structure of the material (Fig. 2c). By analyzing the specimen fracture surface, it was noted the intergranular fracture micromechanism with signs of plastic deformation (Fig. 3b).

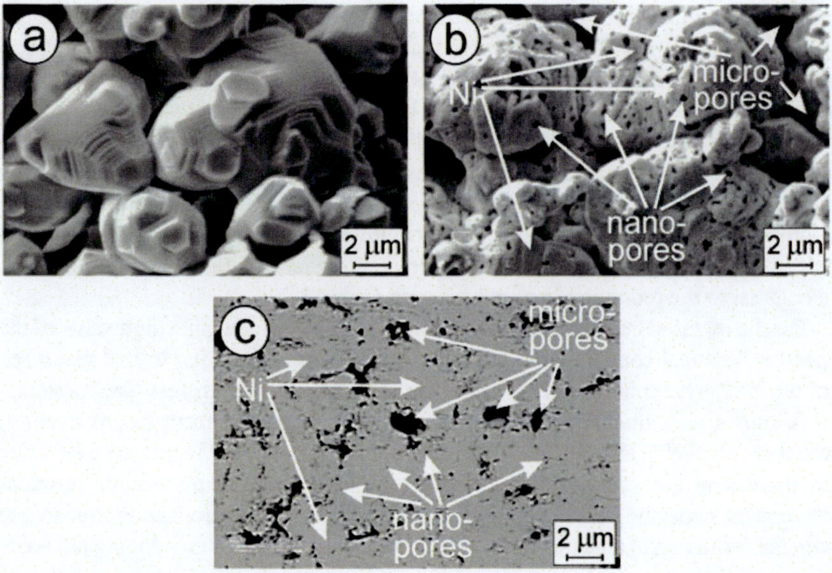

Fig. 2 SEM **a**, **b** fractography and **c** structure of specimens in (**a**) as-sintered state and (**b**, **c**) reduced in pure hydrogen according to mode 2 (Table 1)

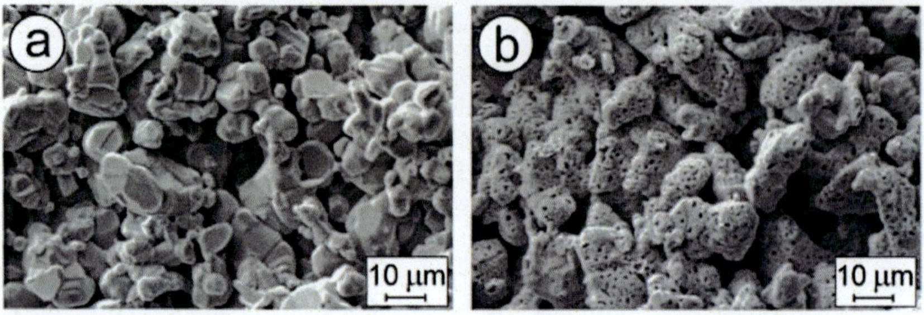

Fig. 3 SEM fractography of specimens in **a** as-sintered state and **b** reduced in pure hydrogen according to mode 2 (Table 1)

It was revealed that the average value of the specific electrical conductivity for the material reduced according to mode 2 is 2.85 × 10^6 S/m. This value is high enough for materials to be used for manufacturing SOFC anodes because the electrical conductivity of traditional materials, e.g., YSZ–Ni, is in a range of 1 × 10^5 S/m to 9 × 10^5 S/m [20, 21]. However, electrical conductivity of pure nickel is five times higher [20]. Such a difference is caused by imperfect contacts between nickel phase particles in the cermet and the heterogeneity of its structure, including porosity.

The cermet reduced in Ar–5 vol% H_2 mixture (mode 3, see Table 1) mainly comprises particles which have non-reduced NiO core (dark-gray areas in Fig. 4a) with fringes of reduced Ni (light-gray areas). The structure of the former hexagonal NiO particles with clear unsmoothed, like in the as-sintered material, edges (Fig. 4c) demonstrates the quite slow reduction course in the gas mixture at a temperature of 600 °C under a pressure 0.15 MPa. No shrinkage occurs because of the absence of nanopores in the particles. Therefore, their contact surface area does not decrease substantially. The elongated crests are formed on nickel phase particles due to ductile fracture of nickel fringes demonstrating energy dissipation during the particles debonding (Fig. 4c). This causes the partly ductile intergranular fracture micromechanism noted on the specimen fracture surface (Fig. 5a), which is followed by corresponding characteristics of the relative strength (120%) and stiffness (90%), in contrast to the reduction in pure hydrogen (Table 1).

For the material treated according to mode 3, we measured quite high value of the specific electrical conductivity (3.25 × 10^5 S/m) which is similar to that measured for the YSZ–Ni cermet [14, 20], where the network of nickel fringes was formed.

A known technique of high-temperature cyclic redox treatment (redox cycling) tested on YSZ–NiO and ScCeSZ–NiO anode materials [21, 22] seems to be effective for increasing the strength of pure NiO ceramics. The treatment mode provides consequent steps: the partial reduction of NiO particles in a hydrogenous atmosphere with the formation of a metallic nickel layer of certain thickness enveloping NiO core of each nickel phase particle; the oxidation of this layer in air with the formation of nano-grained structure. Therefore, the redox procedure of a certain mode provides the reduction of the nickel phase particle size and an increase in the electrical conductivity of cermets [15, 23–27]. Reduction of the nickel oxide ceramic material is much

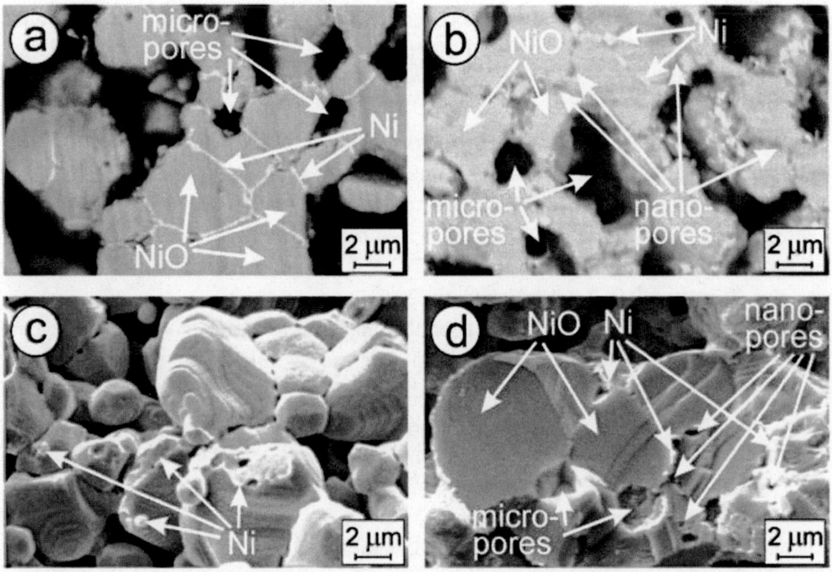

Fig. 4 SEM **a**, **b** structures and **c**, **d** fractography of specimens (**a**, **c**) reduced in the gas mixture according to mode 3 and (**b**, **d**) after redox treatment in the gas mixture/air according to mode 4 (Table 1)

Fig. 5 SEM fractography of specimens **a** reduced in the gas mixture according to mode 3 and **b**, **c** after redox treatment in the gas mixture/air according to mode 4 (Table 1)

faster than the YSZ–NiO ceramics, and its reoxidation is intensified by intermediate degassing [15, 20, 28]. We estimated that in contrast to optimal reduction/oxidation stage duration of 4 h for YSZ–NiO composite [28], its value for pure NiO ceramics is 1 h.

After redox cycling of NiO ceramics according to mode 4 (Table 1), we revealed that its microstructure has some resemblance to that formed by mode 3. Most of the coarse particles have reduced fringes and unreduced cores. Tiny particles of metallic nickel of sizes in a range of 0.2–0.5 μm and nanopores in a range of 0.1–0.8 μm can be found on the grain boundaries (Fig. 4b). Besides, completely reduced fine particles (less than 2.0–2.3 μm) can be seen.

The coarse particles have the hexagonal shape similar to the as-sintered, material but their edges are smoothed and the contours are distorted. The particle surfaces are covered with the nanoparticles of reduced nickel (Fig. 5c). For this treatment mode, the transgranular cleavage is predominant (Fig. 5b). It is accompanied by the formation of elongated nickel fragments surrounding each cleavage facet. Most of the facets are oriented perpendicularly to the direction of the maximum tensile stresses.

The changes in the nanostructure of the material during the high-temperature (600 °C) reduction and reoxidation have been modeled and explained based on the analysis of nano- and microstructure. For this purpose, a concept of stepwise outward loss and inward gain of metallic Ni during a redox cycle has been proposed. A cermet reduced in the first half-cycle mainly comprises particles which have non-reduced NiO core (dark-gray area in Fig. 6a) with fringes of reduced nickel (light-gray areas).

We analyzed the ADF STEM image of the Ni/NiO nanostructure of the cermet reduced in Ar–5 vol% H_2 mixture according to mode 3 that should be the same as in the case of treatment according to mode 4 for 1 cycle (Fig. 6a). No discernible shrinkage of the nickel phase was found that can be explained by its partial reduction. Based on EELS and EDS analyses, it was confirmed that each particle of the size of 1 μm or larger consists of non-reduced NiO core (dark-gray area, Fig. 6a) with an envelope of reduced nickel (light-gray area). We can see randomly oriented subgrains as constituents of the Ni envelope, whereas NiO core looks like a single grain.

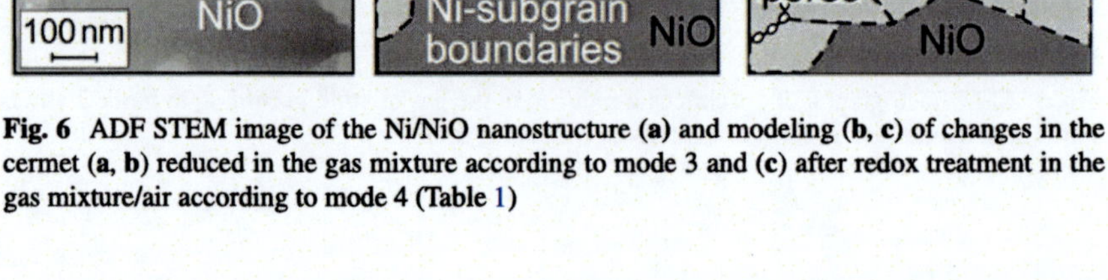

Fig. 6 ADF STEM image of the Ni/NiO nanostructure (**a**) and modeling (**b, c**) of changes in the cermet (**a, b**) reduced in the gas mixture according to mode 3 and (**c**) after redox treatment in the gas mixture/air according to mode 4 (Table 1)

Subgrain sizes are in a range of 50–250 nm. We did not found any pore inside the Ni grain or shrinkage (Fig. 6a). Such a microstructural peculiarity can be used to explain quite high strength of the cermet.

A cermet oxidized in the next half-cycle mainly comprises particles that are characterized by completely oxidized nickel subgrain envelope formed in the previous half-cycle. Due to the degassing, partial pressure inside the cermet is lowered what is a cause of intense inward diffusion of oxygen [29–31].

The closer the subgrains are to the surface of a grain, the more their boundaries are distinguished with the increasing number of redox cycles, because a network of nanopores is formed in the external layer on the boundaries [20]. Therefore, tiny pieces consisting of the subgrains of reduced Ni are being separated stepwise. The gradual outward Ni losses occur after each reduction half-cycle. On the other hand, a stepwise inward Ni gain occurs due to the decrease in Ni particle size, intensifying the particle reduction (Fig. 6c).

Tiny pieces of reduced nickel are separated stepwise and coagulate in the places of former boundaries forming "bridges" between nickel phase particles (Fig. 6c). This is also a cause of some decrease in material porosity. The "bridges," along with the nickel fringes, form the cermet structure of an electrically conducted nickel network. The measured value of specific electrical conductivity (1.85×10^5 S/m) is satisfactory for SOFC applications and agreed well with the values presented in [20, 32]. Besides, it may be improved because there is a possibility of further reduction of the anode in a hydrogen-containing atmosphere during SOFC operation.

The abovementioned peculiarities of fracture surface morphology (Fig. 5b) show that the adhesion strength between nickel phase particles is higher than the ultimate cleavage stresses. The characteristics of relative stiffness and strength (see Table 1), which reflect the rearrangement of structural components on the boundaries of the particles, also confirmed this hypothesis. We can characterize this phenomenon as the clustering of particles with the formation of nanopores and "bridges" at the sites of former boundaries (Fig. 5c). Such structural rearrangement followed by the "bridging" effect is due to the intermediate degassing before starting the oxidation stage. The "bridges" play an important role in providing high bond strength between the particles. If cleavage planes of a nickel phase particle are perpendicular to the direction of the maximum tensile stresses, a clear transgranular cleavage is observed, but if they are not then a stepwise alternation of the transgranular cleavage and transverse crack occurs till the particle boundary is reached (Fig. 4d). In any case, the bond strength between the particles is high enough to provide such fracture micromechanism.

Therefore, the following structural changes in the course of redox cycling may be marked for mode 4: (1) a network of nanopores was formed in the external layer of particles of the nickel phase; (2) thin pieces of reduced Ni subgrains were separated causing a reduction in the size of particles of Ni-phase; (3) tiny Ni pieces were coagulated causing partial reduction of the porosity and formation of "bridges"; (4) the initial particles of the nickel phase were united in clusters with the formation of nanopores at the sites of former boundaries; and (5) a nickel network was formed that consisted of Ni fringes and "bridges." Finally, the specific electrical conductivity of the material increased along with its strength.

The trend of changes in the characteristics of relative stiffness (107%) and strength (192%) during redox treatment differs absolutely from those obtained for one-time reduced material (see Table 1). This peculiarity shows the efficiency of the proposed technique in providing the established levels of the physical and mechanical characteristics of the YSZ–NiO and ScCeSZ–NiO materials for manufacturing SOFC anodes.

4 Conclusions

In this work, after applying of the developed redox treatment mode, the following micro- and nanostructural changes in the as-sintered NiO ceramics were found: a network of nanopores was formed in the external layer of particles of the nickel phase; thin pieces of reduced Ni subgrains were separated causing a reduction in the size of particles of Ni-phase; tiny Ni pieces were coagulated causing partial reduction of the porosity and formation of "bridges"; the initial particles of the nickel phase were united in clusters with the formation of nanopores at the sites of former boundaries; and a nickel network was formed that consisted of Ni fringes and "bridges." All these changes allow providing the required mechanical and physical properties of the Ni/NiO cermet. The authors suggest such treatment technique be applicable for the YSZ–NiO and ScCeSZ–NiO anode materials while manufacturing SOFCs.

References

1. Thydén K (2008) Microstructural degradation of Ni-YSZ anodes for solid oxide fuel cells. Ph.D. thesis, Technical University of Denmark, Roskilde, Denmark
2. Simwonis D, Tietz F, Stoever D (2000) Nickel coarsening in annealed Ni/8YSZ anode substrates for solid oxide fuel cells. Solid State Ionics 132:241–251
3. Vassen R, Simwonis D, Stoever D (2001) Modelling of the agglomeration of Ni-particles in anodes of solid oxide fuel cells. J Mater Sci 36:147–151
4. Van Herle J, Larrain D, Autissier N et al (2005) Modeling and experimental validation of solid oxide fuel cell materials and stacks. J Eur Ceram Soc 25:2627–2632
5. Wuillemin Z, Autissier N, Van Herle J et al (2005) Modeling and study of the influence of sealing on a solid oxide fuel cell. In: Proceedings of the 1st European fuel cell technology and applications conference, Rome, Italy, 14–16 Dec 2005
6. Steele BCH, Heinzel A (2001) Materials for fuel-cell technologies. Nature 414:345–352
7. Faes A, Hessler-Wyser A, Zryd A et al (2012) A review of RedOx cycling of solid oxide fuel cells anode. Membranes 2(3):585–664. https://doi.org/10.3390/membranes2030585
8. Railsback JG, Johnston-Peck AC, Wang J et al (2010) Size-dependent nanoscale Kirkendall effect during the oxidation of nickel nanoparticles. ACS Nano 4:1913–1920
9. Gmelin L (1968) Gmelin handbook of inorganic chemistry, 8th edn. Springer, Berlin
10. Karmhag R, Niklasson GA, Nygren M (1999) Oxidation kinetics of large nickel particles. J Mater Res 14:3051–3058

11. Karmhag R, Niklasson GA, Nygren M (1999) Oxidation kinetics of small nickel particles. J Appl Phys 85:1186–1191
12. Karmhag R, Niklasson GA, Nygren M (2001) Oxidation kinetics of nickel nanoparticles. J Appl Phys 89:3012–3017
13. Atkinson A (1985) Transport processes during the growth of oxide films at elevated temperature. Rev Mod Phys 57:437–470
14. Podhurs'ka VY, Vasyliv BD, Ostash OP et al (2014) Structural transformations in the NiO-containing anode of ceramic fuel cells in the course of its reduction and oxidation. Mater Sci 49(6):805–811
15. Vasyliv BD (2010) Improvement of the electric conductivity of the material of anode in a fuel cell by the cyclic redox thermal treatment. Mater Sci 46(2):260–264
16. Radovic M, Lara-Curzio E (2004) Mechanical properties of tape cast nickel-based anode materials for solid oxide fuel cells before and after reduction in hydrogen. Acta Mater 52:5747–5756
17. Vasyliv BD (2009) A procedure for the investigation of mechanical and physical properties of ceramics under the conditions of biaxial bending of a disk specimen according to the ring–ring scheme. Mater Sci 45(4):571–575
18. Wang Y, Walter ME, Sabolsky K et al (2006) Effects of powder sizes and reduction parameters on the strength of Ni–YSZ anodes. Solid State Ionics 177:1517–1527
19. Van der Pauw LJ (1958) A method of measuring specific resistivity and hall effect of discs of arbitrary shape. Philips Res Rep 13:1–9
20. Clemmer RMC, Corbin SF (2009) The influence of pore and Ni morphology on the electrical conductivity of porous Ni/YSZ composite anodes for use in solid oxide fuel cell applications. Solid State Ionics 180:721–730
21. Vasyliv BD, Podhurs'ka VY, Ostash OP et al (2013) Influence of reducing and oxidizing media on the physicomechanical properties of ScCeSZ–NiO and YSZ–NiO ceramics. Mater Sci 49(2):135–144
22. Vasyliv BD, Ostash OP, Podhurska VY et al (2013) Method of treatment of NiO-containing anodes for a solid oxide fuel cell. Patent of Ukraine No. 78992. Published on 10.04.13, Bulletin No. 7 (in Ukrainian)
23. Podhurska V, Vasyliv B (2012) Influence of NiO reduction on microstructure and properties of porous Ni–ZrO_2 substrates. In: Proceedings of the 3rd international conference on oxide materials for electronic engineering (OMEE-2012), Lviv, Ukraine, 3–7 Sept 2012
24. Vasyliv B, Podhurska V, Ostash O (2017) Preconditioning of the YSZ-NiO fuel cell anode in hydrogenous atmospheres containing water vapor. Nanoscale Res Lett 12:265. https://doi.org/10.1186/s11671-017-2038-4
25. Ettler M, Blaβ G, Menzler NH (2007) Characterization of Ni–YSZ-cermets with respect to redox stability. Fuel Cells 5:349–355
26. Zhang Y, Liu B, Tu B et al (2009) Understanding of redox behavior of Ni–YSZ cermets. Solid State Ionics 180:1580–1586
27. Faes A, Nakajo A, Hessler-Wyser A et al (2009) Redox study of anode-supported solid oxide fuel cell. J Power Sources 193:55–64
28. Waldbillig D, Wood A, Ivey DG (2005) Electrochemical and microstructural characterization of the redox tolerance of solid oxide fuel cell anodes. J Power Sources 145:206–215
29. Wood A, Waldbillig D (2011) Preconditioning treatment to enhance redox tolerance of solid oxide fuel cells. US Patent 8,029,946 B2, 4 Oct 2011
30. Peraldi R, Monceau D, Pieraggi B (2002) Correlations between growth kinetics and microstructure for scales formed by high-temperature oxidation of pure nickel. I. Morphologies and microstructures. Oxid Met 58:249–273
31. Podhurska V, Vasyliv B, Ostash O et al (2016) Influence of treatment temperature on microstructure and properties of YSZ–NiO anode materials. Nanoscale Res Lett 11:93. https://doi.org/10.1186/s11671-016-1306-z
32. Yu JH, Park GW, Lee S et al (2007) Microstructural effects on the electrical and mechanical properties of Ni–YSZ cermet for SOFC anode. J Power Sources 163:926–932

Electrodeposition of Nanostructured Silicon Coatings onto Different Materials from Halide and Halide–Oxide Melts

Oleksandr Yasko, D. B. Shakhnin, Angelina Gab, Viktor Malyshev, and Marcelle Gaune-Escard

1 Introduction

Crystalline silicon is one of the most promising materials for terrestrial solar cells, and the results of further studies on its obtaining should be connected to the lowering of its price. Production of solar cells using LiF-KF-K_2SiF_6 solution in fluoride melts is a very attractive and promising method in terms of obtaining silicon of desired purity level. Well-adherent and relatively pure silicon can be electrodeposited onto silver surface from similar fluoride melts at 1023 K [2]. The use of silver as the cathode material is unprofitable in this case.

Deposition of silicon coatings onto conductive substrates of lower prices such as steel, brass, nickel, tungsten, or graphite can be very promising in terms of reducing

O. Yasko
Student, Open International University of Human Development "Ukraine", 23 Lvivska St., 03115 Kyiv, Ukraine
e-mail: o.yasko@meta.ua

D. B. Shakhnin (✉) · A. Gab
Assoc. Prof., Open International University of Human Development "Ukraine", 23 Lvivska St., 03115 Kyiv, Ukraine
e-mail: shakhnin@ukr.net

A. Gab
e-mail: lina_gab@ukr.net

V. Malyshev
Director of Institute of Engineering & Technology, Open International University of Human Development "Ukraine", 23 Lvivska St., 03115 Kyiv, Ukraine
e-mail: viktor.malyshev.igic@gmail.com

M. Gaune-Escard
Directeur de Recherche, Ecole Polytechnique, CNRS, UMR 6595, Technopôle deChateau Gombert, 5 rue Enrico Fermi, 13453 Marseille Cedex 13, France
e-mail: mgescard@gmail.com

O. Fesenko and L. Yatsenko (eds.), *Nanomaterials and Nanocomposites, Nanostructure Surfaces, and Their Applications*, Springer Proceedings in Physics 246, https://doi.org/10.1007/978-3-030-51905-6_18

the cost of solar cells. Unfortunately, most of these substrates are characterized by silicide formation at 1023 K. Therefore, graphite is more promising material because of its lower reactivity.

The further development of methods for silicon electrodeposition onto different materials was made in fluoride [1, 3] and chloride–fluoride [4, 6] melts.

The purpose of the present study is the electrodeposition of high-purity silicon onto graphite substrate as adherent coating and the investigation of the deposits morphology dependence on the conditions of electrodeposition.

2 Materials and Methods

Instruments, chemicals, and main experimental conditions used during this work were similar to those used [5]. Rod cathode was made of PGM-7 graphite. Previous studies have shown that the optimal concentration of K2SiF6 is 8–14 mol% in binary eutectic KF-LiF mixture (765 K). Electrodeposition was carried out at 1023 K in a graphite crucible, which was also used as the anode.

Morphology of deposits was studied by methods of optical and scanning electron microscopy. Analysis of impurities was made using electron microprobes and emission spectroscopy methods.

List of abbreviations

CNRS the National Center for Scientific Research of the French Ministry of Education and Research

3 Results and Discussion

Electrodeposition

For dense silicon coating deposition, concentration of K_2SiF_6 was maintained within limits of 8–14 mol% at a constant potential. Electrodeposition was carried out at potential -0.75 ± 0.05 V versus Pt or Ag reference electrode. Current strength was varied within limits of 10–100 mA/cm^2. During the deposition, uniform dense deposits of silicon were obtained.

Morphology

Silicon coating with grains of wedge shape up to 1 mm thick was obtained by electrolysis during 5–6 h. Electrodeposited silicon was easily separable from the graphite substrate. The cross section of deposits (Fig. 1) indicates nodular or dendritic growth up to a few millimeters at the surface of main adherent layer of silicon coating. The grain size was about 250 nm (Fig. 2). Larger grains (up to 750 nm) were formed by the electrolysis for more than 5–6 h.

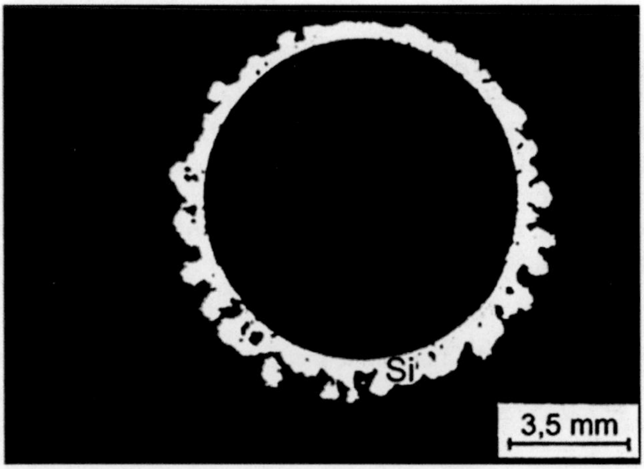

Fig. 1 Cross section of silicon deposit obtained from the LiF-KF—10 mol% K_2SiF_6 melt. $T = 1023$ K; $i_k = 50$ mA/cm^2; $\tau = 6$ h

Fig. 2 Electron micrograph of electrodeposited silicon

Purity

X-ray phase analysis showed the presence of only polycrystalline silicon in the deposit. In addition, the electronic microprobe studies and X-ray scattering energy analysis revealed no impurities. However, levels of Li, Cr, Ni, Fe, Cu, Ag, Mn, Pb, and Al impurities up to 0.02% were detected by emission spectroscopy method. The most common impurities are Cu, Fe, Ni, and Ag, but the overall level of their concentrations would not impact significantly on the performance of solar cells. Deposits' purity level was generally over 99.99%.

Specific resistivity

Results of measurements by four-probe method showed disagreement of specific resistivity values at different places of electrodeposited silicon. At room temperature, measured resistance of silicon electrodeposited from this system was always greater than 1 Ω cm.

4 Conclusions

1. From molten LiF-KF-K_2SiF_6 (8–14 mol%) fluoride electrolytes at 1023 K, silicon coatings were obtained at graphite substrates.
2. Silicon layer deposited by prolonged electrolysis consists of two parts, namely, coating with wedge-shaped grains and dendrites.

Competing Interests
Authors of the present paper have no competing interests.

Funding
Authors of the present paper have not used any external sources of funding in addition to regular financing for scientific investigations provided by CNRS and the University "Ukraine."

Authors' Contributions
Authors of the present paper have made equal contributions in conducting scientific researches and writing this paper.

Acknowledgements Authors of the present paper acknowledge support of all the Ecole Polytechnique and the University "Ukraine" staff in conducting scientific researches and writing this paper.

References

1. Bieber AL, Massot L, Gibilaro M, Cassayre L, Taxil P, Chamelot P (2011) Silicon electrodeposition in molten fluorides. Electrochim Acta 62:282–289
2. Elwell D, Rao GM (1988) Electrolytic production of silicon. J Appl Electrochem 18:15–22
3. Haarberg GM, Famiyeh L, Martinez AM, Osen KS (2013) Electrodeposition of silicon from fluoride melts. Electrochim Acta 100:226–228
4. Maeda K, Yasuda K, Nohira T, Hagiwara R, Homma T (2015) Silicon electrodeposition in water-soluble KF-KCl molten salt: investigations on the reduction of Si(IV) ions. J Electrochem Soc 162:D444–D448
5. Olson JM, Carleton KL (1981) A semipermeable anode for silicon electrorefining. J Electrochem Soc 128:2698–2699
6. Zhuk SI, Isaev VA, Grishenkova OV, Isakov AV, Apisarov AP, Zaykov YD (2017) Silicon electrodeposition from chloride-fluoride melts containing K_2SiF_6 and SiO_2. J Serbian Chem Soc 82:51–62

Nanostructured Interfaces and Surfaces

Femtosecond Laser Surface Micro- and Nanotexturing of Metals, Alloys, and Ceramics Perspective for Biomedical Applications

I. M. Dmitruk, N. I. Berezovska, R. S. Kolodka, A. M. Dmytruk, I. V. Blonskiy, O. M. Mishchenko, and M. V. Pogorielov

1 An Application of Femtosecond Laser Treatment for Functionalization of Surfaces for Biomedical Purposes

During the last decades, the surfaces of different materials modified with power laser pulses find their applications in various fields of technology and medicine. In particular, the character and quality of the surface structuring are the important factors

I. M. Dmitruk · N. I. Berezovska (✉) · R. S. Kolodka
Physics Faculty, Taras Shevchenko National University of Kyiv, Volodymyrska Street, 64/13, Kiev 01601, Ukraine
e-mail: n_berezovska@univ.kiev.ua

I. M. Dmitruk
e-mail: igor_dmitruk@univ.kiev.ua

R. S. Kolodka
e-mail: rkolodka@gmail.com

A. M. Dmytruk · I. V. Blonskiy
Photon Processes Department, Institute of Physics of the NAS of Ukraine, Prospect Nauky, 46, Kiev 03039, Ukraine
e-mail: admytruk@iop.kiev.ua

I. V. Blonskiy
e-mail: blon@iop.kiev.ua

O. M. Mishchenko
Medical Faculty no. 3, Zaporizhzhia State Medical University. Maiakovskyi Prospect, 26, Zaporizhzhia 69035, Ukraine
e-mail: dr.m@i.ua

M. V. Pogorielov
Department of Public Health, Medical Institute of Sumy State University, Pratsi Street, 3, Sumy 40000, Ukraine
e-mail: m.pogorielov@gmail.com

O. Fesenko and L. Yatsenko (eds.), *Nanomaterials and Nanocomposites, Nanostructure Surfaces, and Their Applications*, Springer Proceedings in Physics 246, https://doi.org/10.1007/978-3-030-51905-6_19

when material is used in implant, especially for an enhancement of its biocompatibility. In recent years, significant advances in the manufacturing and quality of dental and orthopedic implants have been achieved in the modern surgery. However, there are many problems to be solved to improve the functioning of different implants. Nevertheless, it should be noted that the majority of implants still have a limited lifetime (up to 15 years). Different complications often occur due to an inflammation, infections, allergic reactions, corrosion, wear, fractures, and lack of osteointegration. It is recognized that the improvement of biological response of implants depends on their surface functionality. The surface topography and chemical composition of the implant determine such properties as surface roughness and wettability, which are essential for the formation of a solid implant–bone interface and for low risk of infection. All these acute problems make the implant surface engineering an extremely important area of research. Thus, the usage of nano- and microstructuring of the implant surfaces is to be one of the promising methods for an enhancement of implant functionality [1, 2]. Such methods as plasma spraying, sandblasting, acid and anodic etching have been commercially successful for the implant treatment. However, some of these methods have been distinguished by such disadvantages as the surface contamination, the change of the structure and properties of the basic material, etc. [2]. More advanced methods, such as photolithography, electron beam lithography, and nanoprinted lithography, allow one to create surfaces with ordered or random nanoscale and microscale structures without altering the chemical composition of the surface [3]. Recent achievements show that the laser treatment of the implant surfaces provides both specific surface topography and less surface contamination compared to other methods [4]. Comparative studies of material processing using different types of lasers have shown that femtosecond lasers have the advantages over nanosecond lasers in terms of higher accuracy of treatment, the reduction of the area affected by heat, and the reduction of contamination of the modified area [5].

Many scientific groups study laser-induced periodic surface structures (LIPSS) formed on the surface of metals and alloys perspective for the dental [6, 7]. Thus, authors of [8] studied the grooves with different periodicities and "lotus-like" structures on a titanium surface induced by femtosecond laser pulses from the point of view of their influence on the behavior of human fibroblast and MG-63 osteoblast cells. "Lotus-like" structures have been found to be superhydrophobic. These structures promote osteoblast proliferation but inhibit fibroblast proliferation that is useful for orthopedic titanium implants. Therefore, there is possible correlation between the laser-induced hydrophobicity and cell growth, namely, the higher hydrophobicity reduces the fibroblast proliferation. Thus, simple wetting tests on the structured surfaces can predict cell response on laser-modified surfaces.

Authors of [9] reveal specific morphological changes of titanium surface under the femtosecond laser treatment with different laser fluencies in vacuum. Chemical analysis of the treated Ti surface indicates the local surface cleaning under the influence of high-intensity femtosecond radiation.

An extensive proliferation and spreading of MG 63 osteoblast-like cells cultured on the treated Ti–35Nb–xZr alloy has been reported in [10]. In this research, Ti–35Nb–xZr alloy was initially processed with femtosecond laser radiation with subsequent surface treatment using a potentiostat and a 1M H_3PO_4 solution containing 0.8 wt% NaF with an applied cell voltage of 10 V for 2 h to produce nanotubes on the alloy surface.

In the study [11], authors evaluate the biofilm formation by supragingival flora on a laser micro-textured titanium surface using an in vitro oral microcosm model grown using a bioreactor, and an in situ model based on intraoral trays. In vitro experiments show the lowest biofilm formation, in particular, microorganisms cover the edges of the laser-created pits, and minimal effect of the biofilm formation has been observed inside the pits. The in situ results demonstrate that the laser treatment also reduces the biofilm formation with a maximal effect when the surface is blasted orthogonally by the laser beam. Thus, an accurate optimization of surface laser texturing could promote effective prevention of the biofilm formation.

Authors of [12] successfully use the method of generation of highly regular LIPSS (HR-LIPSS) [13] that combines micron-scale low-spatial-frequency LIPSS (LSFL) with nanoscale roughness on the surfaces of the titanium alloy (Ti6Al4V) and zirconium (Zr) dental implants to improve the osteointegration processes. They emphasize the strong dependence of cell proliferation on the topography and oxidation degree of the surface in the absence of dependence of cell proliferation on the type of metal.

Recently, the zirconia implants have been suggested as an alternative to titanium ones. Dental ceramic, especially zirconia (zirconium dioxide, ZrO_2), is an excellent material for dental restoration, as it is characterized by high biocompatibility, wear resistance, high fracture toughness and high compression resistance, excellent esthetic properties, etc. All mentioned advantages stimulate active study of the processing of materials based on zirconia ceramics [14–16]. However, an ideal treatment of the zirconia surface providing sufficient bond strength to minimize the detachment of orthodontic holders from the zirconia surface is still being developed. The ceramic surface treatment by femtosecond laser irradiation has been proposed as an alternative in an attempt to improve adhesion of dental cements and orthodontic holders [17–21].

The results reported in [22] concerning the study of zirconia implants of the same geometry but with differences in the surface treatment have shown that the surface treatment of zirconia dental implants by femtosecond laser increases implant stability and bone-to-implant contact.

In [23], researchers studied the influence of four different techniques, including femtosecond laser irradiation, on shear bond strength of metallic and ceramic brackets to zirconia. The highest bond strength values of ceramic and metallic brackets to zirconia were obtained for the combined process of sandblasting + femtosecond laser treatment confirming the high effectiveness of ultra-short laser irradiation of zirconia surface.

In the present research, we analyze the modification of the surfaces of zirconium metal (Zr), Ti–Zr alloys, and zirconia ceramics under the influence of laser pulses of femtosecond duration. Additional surface pre-treatment procedures have been fulfilled. The topography features of the obtained laser-induced periodic/disordered surfaces are examined from the point of view of efficiency of cell culture viability on such surfaces.

2 Experimental Details

For the laser treatment of specimen surfaces, the technological capabilities of the Femtosecond Laser Centre for Collective Use of NAS of Ukraine have been used, namely, a femtosecond laser composed of a Ti:sapphire laser set "Mira Optima 900-F" and regenerative amplifier "Legend HE" by "Coherent" (USA). The main parameters of a femtosecond laser are a pulse duration 150 fs–3 ps (with chirp), a repetition rate of 1 kHz, an average power up to 1 W, a pulse energy up to 1 mJ, and the main wavelength of the radiation 800 nm. The generation of the second (400 nm) and third (266 nm) harmonics of the femtosecond laser was also available due to the usage of corresponding optical borate crystals. For the generation of the third harmonic, the plate compensator of group velocity (calcite) and plate $(\lambda/2)_{800}$ which rotated the polarization plane of the radiation of the fundamental harmonic (800 nm) were put after the crystal generator of the second harmonic in the experimental setup. The laser beam focused on the surface of the irradiated sample with a lens. The power density of the laser radiation could be varied by adjusting the distance of the sample from the lens focus. To measure the average power (up to 10 W), the laser power meter "Field Master GS" with the detector head "LM-10" ("Coherent," USA) was used. A computer-controlled three-coordinate platform (Standa, Lithuania) has been used for precision moving-scanning of the samples. All laser treatment procedures have been performed in air.

The morphology of the studied specimen surfaces has been analyzed by means of scanning electron microscopy (SEM) using Tescan VEGA 3 microscope.

The samples in our study are of three types related to dental implants, namely, metal, two-component alloy, and ceramics. The samples of Zr and Ti–Zr alloys are in the form of ground or polished washers. Some of the specimen surfaces were modified by the method of sandblasting with large grit with subsequent electroetching—SLE surface (Sand-blasted, Large grit, and Electro-etched). Sintered yttrium-stabilized tetragonal zirconia polycrystalline (Y-TZP) samples have been also examined.

For the biocompatibility study, the samples were sterilized in autoclave (121 °C during 1 h) and placed in 24-well plates. Each well was filled up with 20% of Fetal Bovine Serum (FBS; Invitrogen) in Dulbecco's Modified Eagle Medium (DMEM; Invitrogen, cat. no. 11960) overnight. This procedure was performed to mimic protein adhesion to the implant surface after implantation. Next day media removed from well and 2 ml of DMEM supplemented with 10% Fetal Bovine Serum (FBS; Invitrogen), 2 mM L-glutamine (Invitrogen, cat. no. 25030), 0.1 mM 2-mercaptoethanol (Sigma, cat. no. M7522), 50 units/mL penicillin, and 50 g/mL streptomycin (Invitrogen, cat. no. 15070) was added to each well. 10_5 Rats Osteoblast and Dermal Fibroblast (in separate well) were seeded on the top surface of each sample and incubated at 37 °C in a humidified environment with 5% CO_2. Cultural media was changed every 3 days during a 7-day culture period. All experiments were triplicate.

3 Morphology Study of Laser-Treated Surfaces of Specimens of Dental Type (Zr, Ti–Zr Alloys)

The scanning electron microscopy has been used for the characterization of the topography of the samples of Zr, Ti–Zr alloys, and zirconia ceramics. After femtosecond laser treatment, almost all specimen surfaces demonstrate the surface periodic structures.

The commonly accepted mechanism of the laser-induced formation of periodic surface structures on metals considers the interference of the incident wave and surface electromagnetic waves (SEW) or surface plasmon polaritons excited at the metal surface by laser radiation. Thus, the quasiperiodic structures with a period close to the wavelength of the laser and oriented along the perpendicular to the polarization plane of the incident light begin to form.

Figure 1 shows the SEM images of sample No. 1-0. The front surface of the specimen is completely structured ("hatched") by the laser. However, the adjacent lines do not overlap but only easily touch each other at some points due to the small size of the laser beam cross section caused by the sample position near the focus of the converging lens. The sub-micron structures formed are regular with a period of approximately 570 nm.

We used the 2D Fourier transform of SEM images of the sample surface to quantify the period of the formed structures (see Figs. 2 and 3). Thus, for the sample No. 2

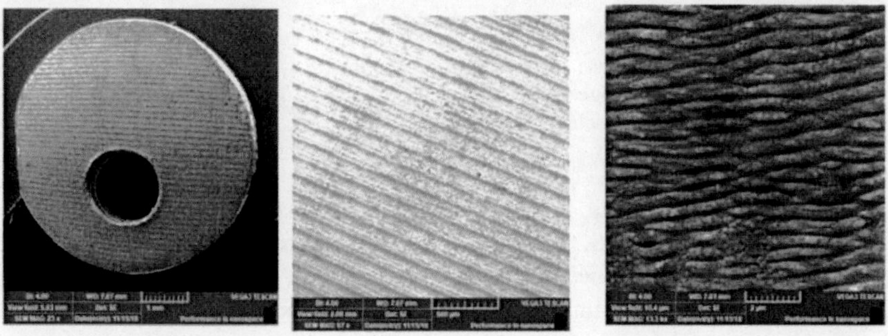

a) b) c)

Fig. 1 SEM images (**a–c**) with different magnifications of the surface of Zr sample (No. 1-0) treated with femtosecond laser with the wavelength of 800 nm and the pulse irradiation energy density of 1.4 J/cm^2

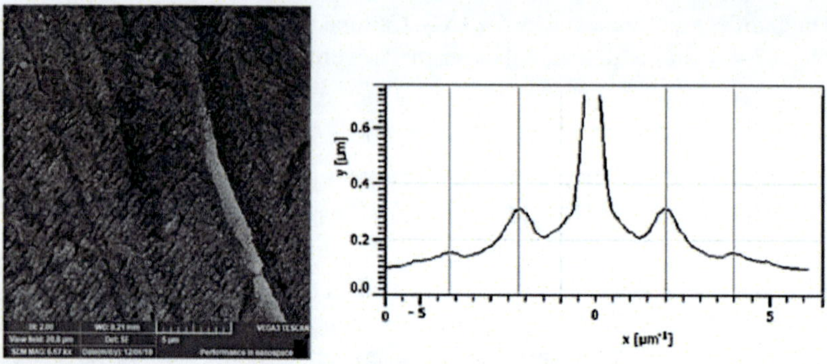

a) b)

Fig. 2 SEM image of the surface of Zr sample (No. 2) treated with femtosecond laser with the wavelength of 800 nm and the pulse irradiation energy density of 0.46 J/cm^2 (**a**), the cross section of 2D Fourier transform of presented SEM image (**b**)

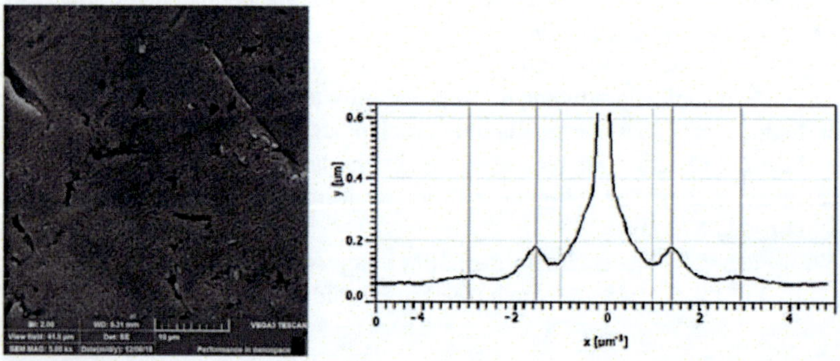

a) b)

Fig. 3 SEM image of the surface of Zr sample (No. 3) treated with femtosecond laser with the wavelength of 800 nm and the pulse irradiation energy density of 0.35 J/cm^2 (**a**), the cross section of 2D Fourier transform of presented SEM image (**b**)

LIPSS, two different periods, 475 and 245 nm, have been discovered. For the sample No. 3, two periods of LIPSS, namely, 673 and 336 nm, have been revealed.

It should be also mentioned so-called "coarse ripples," which have characteristic dimensions of the order of microns and exhibit quasiperiodicity in a direction parallel to the polarization plane of the laser beam.

It has been revealed the structures known as "fine ripples" which have a period in the range of several tens to 200 nm and exhibit a well-pronounced periodicity in the direction perpendicular to the polarization plane of the laser beam. The appearance of "fine ripples" is irregular, and as usual, they cover a small percentage of the total area of the treated surface (see Fig. 4).

a) b)

Fig. 4 SEM images with different magnifications of the surface of Ti–Zr alloy (No. 32) treated with femtosecond laser with the wavelength of 800 nm

Thus, the homogeneity of the structure on the surface at the macroscopic scale of dimensions (micrometer size) should be also taken into consideration while analyzing the specimen surface morphology. This is important for verifying the reliability of the biocompatibility of the treated surfaces, as well as for the elaboration of further processing improvement to ensure optimum quality over the entire surface area. The examples of macroscopic homogeneous and inhomogeneous surfaces are presented (Fig. 5). There are several reasons for the appearance of macroscopic nonuniformity: the manifestation of the scanning lines of laser beam on a surface, quality of the surface pre-treatment, its residual roughness, the presence of scratches, possible instability of power density of a laser beam due to the fluctuations of power with time, or changes of a distance of a surface from the lens focus in the experimental setup. At high scanning velocities, a "spotty" structure of the treated surface is also observed (see Fig. 5b). This type of structure can be caused by the small overlapping of spots irradiated by the sequential laser pulses and the discreteness of the translational movement of the sample.

Minimizing all these factors will certainly increase the macroscopic uniformity of the surface morphology but may negatively affect the speed of treatment or the cost and complexity of preliminary preparation of implants. Therefore, the actual objective is to select laser treatment modes that minimize the negative influence of the above factors under other identical conditions.

We have also examined the surface modification of dental implants with the third harmonic (266 nm) of the femtosecond laser. Figure 6 shows the SEM image of the surface of Ti–Zr sample (No. 33) treated with the third harmonic and the cross section of 2D Fourier transform of this SEM image. The obtained LIPSS uniformly fills the sample surface. LIPSS period is 177 nm.

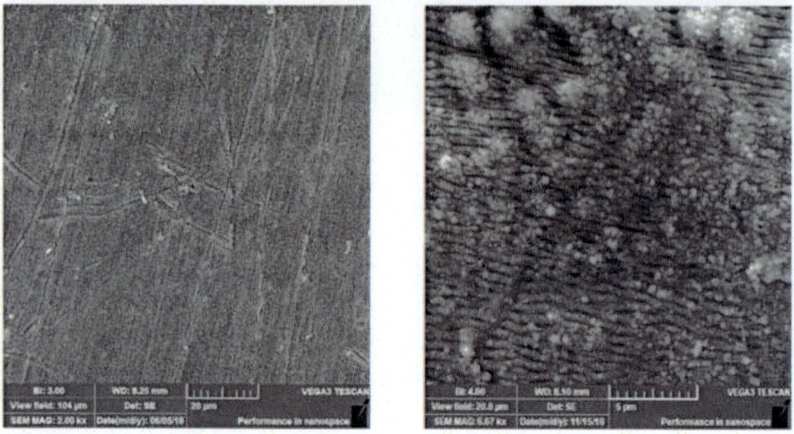

a) b)

Fig. 5 SEM images of the surfaces of Zr samples treated with femtosecond laser with the wavelength of 800 nm representing examples of macroscopic homogeneous (**a**, Sample No. 26) and inhomogeneous (**b**, Sample No. 2-0) surfaces

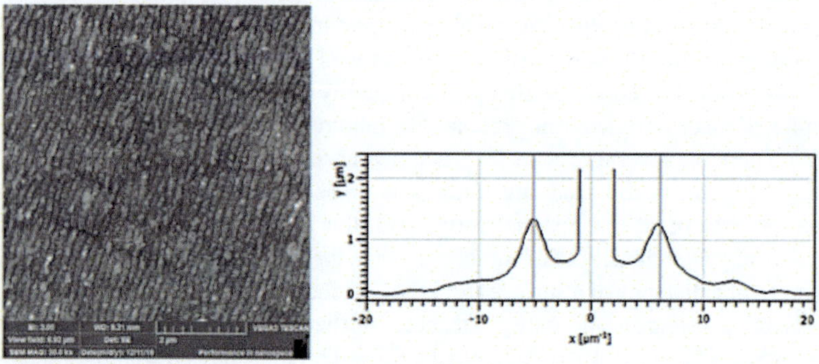

a) b)

Fig. 6 SEM image of the surface of Ti–Zr alloy (No. 33) treated with femtosecond laser with the wavelength of 266 nm (**a**), the cross section of 2D Fourier transform of presented SEM image (**b**)

4 Morphology Study of Laser-Treated Surfaces of Zirconia Ceramics

The specimens of zirconia ceramics with or without preliminary mechanical processing have been treated with femtosecond laser with a wavelength of 800 nm at different laser fluencies and scanning velocities. SEM images of the surfaces of some zirconia samples are presented in Figs. 7 and 8. A distinctive feature of this

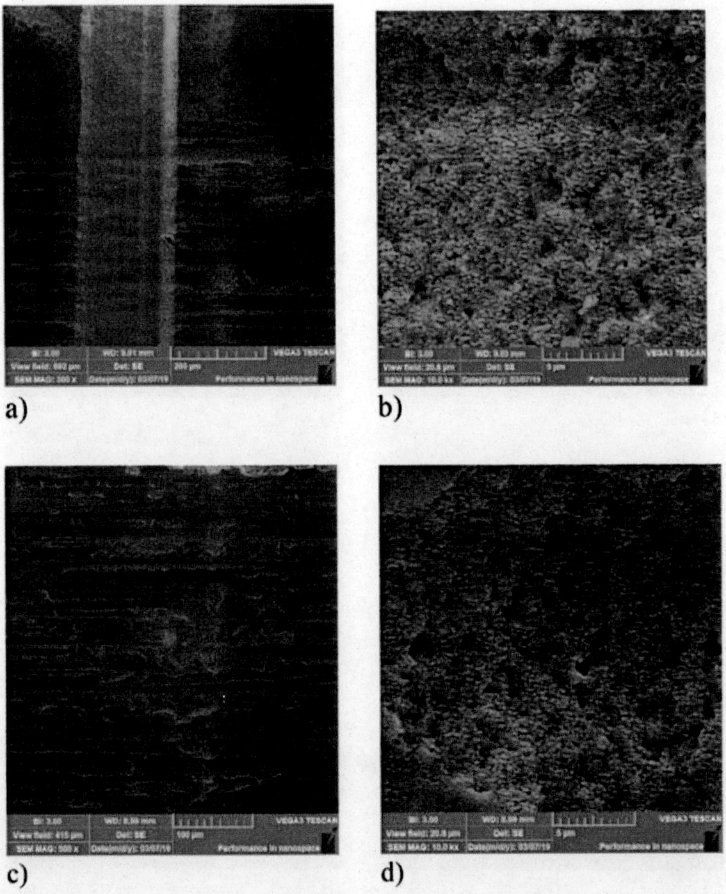

Fig. 7 SEM images with different magnifications of unpolished surface of zirconium ceramics treated with femtosecond laser with the wavelength of 800 nm at scanning velocity of 1 mm/s **a**, **b** at pulse irradiation energy density of 1.13 J/cm^2; **c**, **d** at pulse irradiation energy density of 0.65 J/cm^2

processing mode is rather gentle but precise treatment without producing mechanical degradation of the material.

An analysis of the morphology of obtained structures on polished and unpolished zirconia ceramics reveals in the laser-modified parts of the samples the features that resemble pores, cavities, and tracks of initial processing. Diameters of cavities are in the range of 240–600 nm. At nanoscale range, the fine ripples of 120–140 nm in diameter are observed. Femtosecond laser processing of the polished zirconia surface can also lead to the smoothing of the surface, as well as to the formation of cavities with a complex nanostructure inside (see Fig. 8c, d).

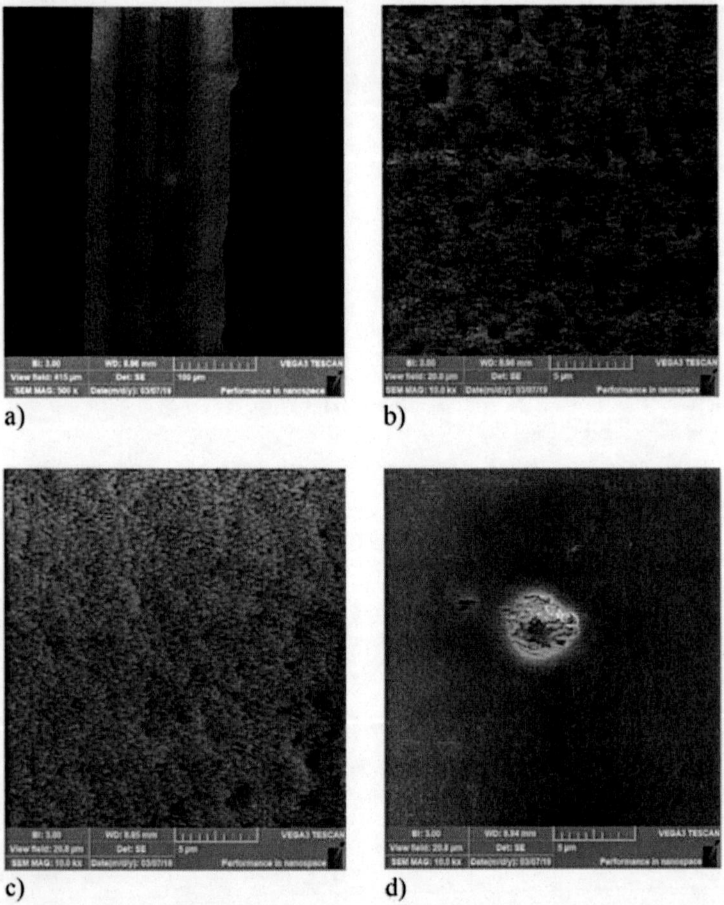

Fig. 8 SEM images of polished surface of zirconium ceramics treated with femtosecond laser with the wavelength of 800 nm at scanning velocity of 0.5 mm/s **a**, **b** at pulse irradiation energy density of 1.13 J/cm^2; **c** at pulse irradiation energy density of 0.94 J/cm^2; at pulse irradiation energy density of 0.64 J/cm^2

5 Wettability and Biocompatibility of Laser-Treated Surfaces of Zr and Ti–Zr Alloys

One of the important characteristics which determines the biological response of the implants is to be the surface energy of implants and consecutively wettability of its surfaces. The wettability of the implant surfaces can influence the protein adhesion, the bacterial adhesion, and subsequent biofilm formation, as well as the rate of osteointegration processes. It is characterized by liquid–solid contact angle (CA) which is defined as an angle between the tangent line to a liquid drop surface at the three-phase boundary (solid, liquid, and fluid) and the horizontal solid surface. CA is measured on the side of the liquid. We used the sessile drop method for CA evaluation, when the distilled water droplet is deposited with a syringe onto the sample surface. CA is measured from the drop shape surface using the image analysis software.

Figure 9 presents the examples of droplets on surfaces of Zr and Ti–Zr alloys mechanically polished or SLE processed before femtosecond laser treatment.

It is found that the surfaces with greater roughness exhibit greater hydrophobicity (contact angles in the range of 123°–128°). For defect-free surfaces and quasiperiodic structures, the contact angles are to be in the range of 56°–61° that indicates a greater hydrophilicity of such surfaces. The increase in the surface roughness, surface energy, and chemical composition (formation of surface oxides) increases the wettability of metal surfaces.

Resazurin sodium salt assay has been used to access cell viability and proliferation rate on days 1, 3, and 7 after cell seeding. Media was removed from each well and washed with PBS twice. 1 ml of resazurin sodium salt solution was added to each scaffold and incubated for 4 h. Three aliquots of 200 μl of resazurin sodium salt solution were collected from each scaffold and read at a wavelength of 620 nm in a colorimetric plate reader (Multiskan FC Microplate Photometer, Thermo Fisher Scientific) to obtain baseline values of colorimetric absorbance. Cell viability was calculated as a % of resazurin reduction compared to pure resazurin sodium salt solution. We selected the most promising Zr samples and Ti–Zr alloy samples for the biocompatibility (see Figs. 10 and 11). We can suppose that the biological response of both metal and alloy samples is similar to the studied parameters of treatment. Meanwhile, the obtained data are promising for the usage of laser-treated surfaces for dental applications.

6 Conclusions

Periodic and ununiform rough structures have been obtained on the surfaces of metal (Zr), alloy (Ti–Zr), and zirconia ceramics under a powerful femtosecond laser irradiation of fundamental (800 nm) and third harmonics (266 nm). The highest quality processing of studied implant specimens has been achieved at higher radiation power densities at higher scanning velocities. At an increase of the distance between the specimen and the lens focus, the height of the grooves profile and the grating period decrease. Laser texturing can change the wettability of the samples. The preliminary experiments on cell viability confirm the effect of biofunctionalization of treated surfaces.

Hence, among commercially available techniques used for the implant surface modification, the femtosecond laser treatment is promising due to its flexibility, simplicity, high reproducibility, and application for a wide range of materials such as metals, metal alloys, and ceramics.

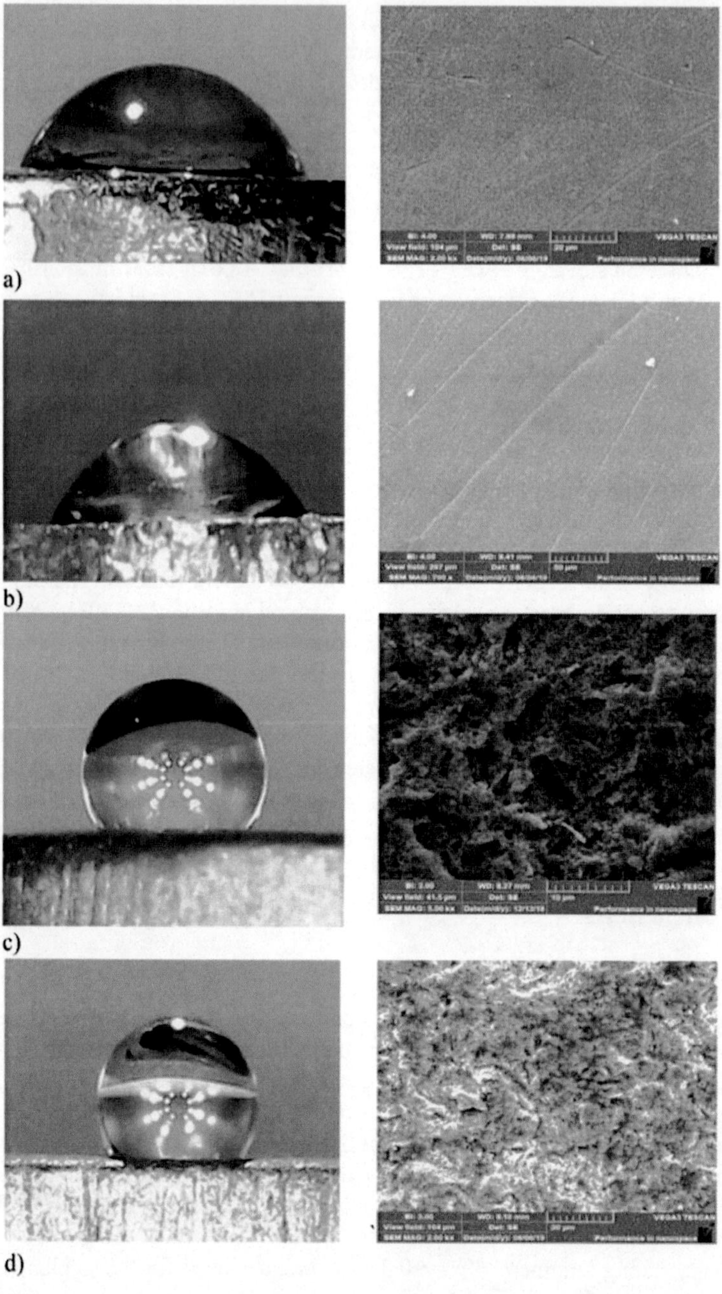

Fig. 9 Photos of water droplet on sample surface and corresponding SEM images of this surface: **a** Zr sample (No. 6, polished), CA = 61°; **b** Ti–Zr alloy sample (No. 34, polished), CA = 59°; **c** Ti–Zr alloy sample (No. 43, SLE), CA = 123°; **d** Zr sample (No. 44, SLE), CA = 125°

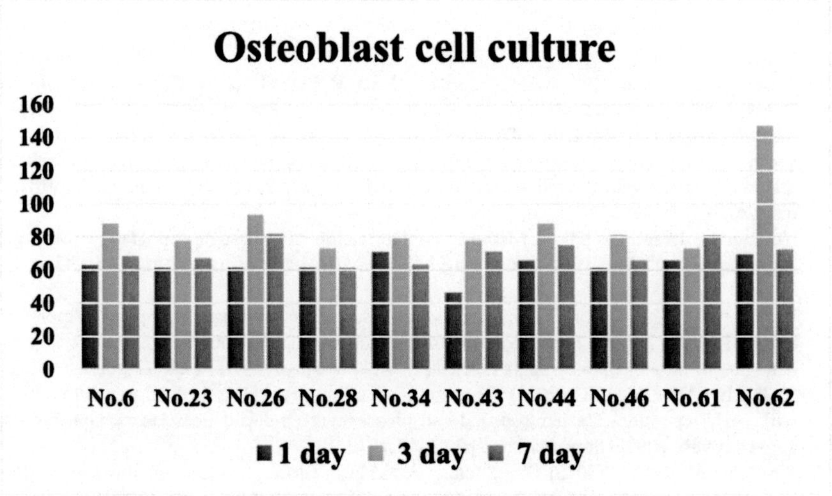

Fig. 10 Cell viability for osteoblast cultures for polished Zr samples—No. 6, 23, 26, 28, SLE Zr samples—No. 44, 46, 61, 62; polished Ti–Zr alloy samples—No. 34, 43

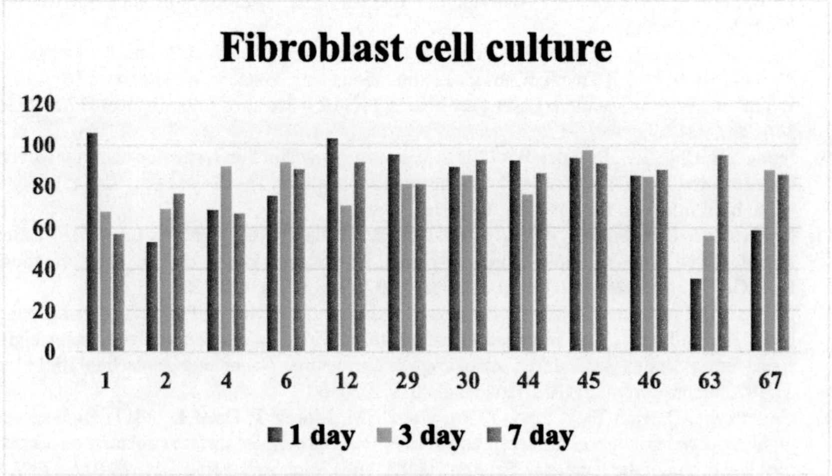

Fig. 11 Cell viability for dermal fibroblast cultures for polished Zr samples—No. 1, 2, 4, 6, 12, 30, SLE Zr samples—No. 44, 45, 46, 63, 67; polished Ti–Zr alloy sample No. 29

Acknowledgements Authors acknowledge the support from Ministry of Education and Science of Ukraine, project No. 19BF051-04, and technical support of Femtosecond Laser Center for Collective Use of NAS of Ukraine. Authors appreciate Dr. Teselko P. for the SEM measurements.

References

1. Anselme K, Davidson P, Popa AM, Giazzon M, Liley M, Ploux L (2010) The interaction of cells and bacteria with surfaces structured at the nanometre scale. Acta Biomater 6(10):3824–3846. https://doi.org/10.1016/j.actbio.2010.04.001
2. Liu X, Chu PK, Ding C (2004) Surface modification of titanium, titanium alloys, and related materials for biomedical applications. Mater Sci Eng R 47:49–121. https://doi.org/10.1016/j.mser.2004.11.001
3. Norman JJ, Desai TA (2006) Methods for fabrication of nanoscale topography for tissue engineering scaffolds. Ann Biomed Eng 34(1):89–101. https://doi.org/10.1007/s10439-005-9005-4
4. Vorobyev AY, Guo C (2007) Femtosecond laser structuring of titanium implants. Appl Surf Sci 253(17):7272–7280. https://doi.org/10.1016/j.apsusc.2007.03.006
5. Erdoğan M, Öktem B, Kalaycıoğlu H, Yavaş S, Mukhopadhyay PK, Eken K, Özgören K, Aykaç Y, Tazebay UH, Ilday FÖ (2011) Texturing of titanium (Ti6Al4V) medical implant surfaces with MHz-repetition-rate femtosecond and picosecond Yb-doped fiber lasers. Opt Express 19(11):10986–10996. https://doi.org/10.1364/OE.19.010986
6. Vorobyev AY, Guo C (2013) Direct femtosecond laser surface nano/microstructuring and its applications. Laser Photon Rev 7:385–407. https://doi.org/10.1002/lpor.201200017
7. Öktem B, Pavlov I, Ilday S, Kalaycioglu H, Rybak A, Yavas S, Erdogan M, Ilday Ilday FÖ (2013) Nonlinear laser lithography for indefinitely large-area nanostructuring with femtosecond pulses. Nat Photon 7:897–901. https://doi.org/10.1038/nphoton.2013.272
8. Fadeeva E, Schlie S, Koch J, Chichkov BN (2010) Selective cell control by surface structuring for orthopedic applications. J Adhesion Sci Technol 24:2257–2270. https://doi.org/10.1163/016942410X508000
9. Trtica M, Batani D, Redaelli R, Limpouch J, Kmetik V, Ciganovic J, Stasic J, Gakovic B, Momcilovic M (2013) Titanium surface modification using femtosecond laser with 10^{13}–10^{15} W/cm^2 intensity in vacuum. Laser Part Beams 31(1):29–36. https://doi.org/10.1017/S0263034612000924
10. Jeong YH, Choe HC, Brantley WA (2011) Nanostructured thin film formation on femtosecond laser-textured Ti–35Nb–xZr alloy for biomedical applications. Thin Solid Films 519(15):4668–4675. https://doi.org/10.1016/j.tsf.2011.01.014
11. Ionescu AC, Brambilla E, Azzola F, Ottobelli M, Pellegrini G, Francetti LA (2018) Laser microtextured titanium implant surfaces reduce in vitro and in situ oral biofilm formation. PLoSONE 13(9):e0202262. https://doi.org/10.1371/journal.pone.0202262
12. Gnilitskyi I, Pogorielov M, Viter R, Ferraria AM, Carapeto AP, Oleshko O, Orazi L, Mishchenko O (2019) Cell and tissue response to nanotextured Ti6Al4V and Zr implants using high-speed femtosecond laser-induced periodic surface structures. Nanomed Nanotechnol Biol Med 21:102036. https://doi.org/10.1016/j.nano.2019.102036
13. Gnilitskyi I, Derrien TJ-Y, Levy Y, Bugakova NM, Mocek T, Orazi L (2017) High-speed, highly regular femtosecond laser printing of laser-induced periodic surface structures on metals: physical origin of the regularity. Sci Rep 7:8485. https://doi.org/10.1038/s41598-017-08788-z
14. Papadopoulou EL, Samara A, Barberoglou M, Manousaki A, Pagakis SN, Anastasiadou E, Fotakis C, Stratakis E (2010) Silicon scaffolds promoting three-dimensional neuronal web of cytoplasmic processes. Tissue Eng C Methods 16:497–502. https://doi.org/10.1089/ten.TEC.2009.0216
15. Poosti M, Jahanbin A, Mahdavi P, Mehrnoush S (2012) Porcelain conditioning with Nd:YAG and Er:YAG laser for bracket bonding in orthodontics. Lasers Med Sci 27(2):321–324. https://doi.org/10.1007/s10103-010-0878-6
16. Mosharraf R, Mansour R, Savabi O, Ashtiani AH (2011) Influence of surface modification techniques on shear bond strength between different zirconia cores and veneering ceramics. J Adv Prosthodont 3(4):221–228. https://doi.org/10.4047/jap.2011.3.4.221

17. Usumez A, Hamdemirci N, Koroglu BY, Simsek I, Parlar O, Sari T (2013) Bond strength of resin cement to zirconia ceramic with different surface treatments. Lasers Med Sci 28(1):259–266. https://doi.org/10.1007/s10103-012-1136-x
18. Erdur EA, Basciftci FA (2015) Effect of Ti:sapphire laser on shear bond strength of orthodontic brackets to ceramic surfaces. Lasers Surg Med 47(6):512–519. https://doi.org/10.1002/lsm.22371
19. Akpinar YZ, Irgin C, Yavuz T, Aslan MA, Kilic HS, Usumez A (2015) Effect of femtosecond laser treatment on the shear bond strength of a metal bracket to prepared porcelain surface. Photomed Laser Surg 33(4):206–212. https://doi.org/10.1089/pho.2014.3791
20. Vicente M, Gomes AL, Montero J, Rosel E, Seoane V, Albaladejo A (2016) Influence of cyclic loading on the adhesive effectiveness of resin-zirconia interface after femtosecond laser irradiation and conventional surface treatments. Lasers Surg Med 48(1):36–44. https://doi.org/10.1002/lsm.22442
21. Vicente Prieto M, Gomes ALC, Montero Martín J, Alvarado Lorenzo A, Seoane Mato V, Albaladejo Martínez A (2016) The effect of femtosecond laser treatment on the effectiveness of resin-zirconia adhesive: an in vitro study. J Lasers Med Sci 7(4):214–219. https://doi.org/10.15171/jlms.2016.38
22. Calvo-Guirado Jl, Ramos-Oltra ML, Negri B, Delgado-Ruíz RA, Ramirez-Fernández P, Mate-Sánchez JE, Abooud M, Gargallo Albiol J, Satorres Nieto M, Romanos G (2013) Osseointegration of zirconia dental implants modified by femtosecond laser vs. zirconia implants in healed bone: a histomorphometric study in dogs with three-month follow-up. J. Osseointegr 5(3):39–44. https://doi.org/10.23805/jo.2013.05.03.01
23. García-Sanz V, Paredes-Gallardo V, Bellot-Arcís C, Mendoza-Yero O, Doñate-Buendía C, Montero J, Albaladejo A (2017) Effects of femtosecond laser and other surface treatments on the bond strength of metallic and ceramic orthodontic brackets to zirconia. PLoS ONE 12(10):e0186796. https://doi.org/10.1371/journal.pone.0186796

Thin Layer of Cyclodextrins on Graphene—MD Simulations

D. Makieła and Z. Gburski

1 Introduction

Graphene is a monolayer of tightly packed carbon hexagons forming honeycomb lattice. It has attracted significant attention both in fundamental studies and advanced applied research [1–11]. Here, we present the results of our research on graphene sheet covered by β-cyclodextrin (βCD) molecules. The motivation was perspective of future applications in nanotechnology, suggested in several experimental reports [12–21]. The cyclic β-cyclodextrin molecule consists of seven (α-1,4)-linked α-D-glucopyranose units. Due to the chair conformation of the glucopyranose, the βCD is shaped like a truncated cone. In aqueous solutions, β-cyclodextrin can form inclusion complexes with various drugs, by taking up a drug molecule into βCD central cavity. This important and desirable feature of βCD underlies the widespread use of this molecule in the pharmaceutical industry [22–27]. In our previous report [28], we investigated the graphene sheet surrounded by β-cyclodextrin molecules, placed in water. Now we present the molecular dynamics (MD) computer simulations of β-cyclodextrin molecules locate on a graphene surface, with no water added. Our preliminary studies reported here may contribute to future use of βCD decorated graphene in nanotechnology.

D. Makieła · Z. Gburski (✉)
Silesian Centre for Education and Interdisciplinary Research, Institute of Physics, University of Silesia in Katowice, 75 Pułku Piechoty 1a, 41-500 Chorzów, Poland
e-mail: zygmunt.gburski@us.edu.pl

D. Makieła
e-mail: damian.makiela@smcebi.edu.pl

Z. Gburski
Katowice Institute of Information Technologies, Mickiewicza 29, 40-085 Katowice, Poland

O. Fesenko and L. Yatsenko (eds.), *Nanomaterials and Nanocomposites, Nanostructure Surfaces, and Their Applications*, Springer Proceedings in Physics 246, https://doi.org/10.1007/978-3-030-51905-6_20

2 Calculation Details

The NAMD 2.11 software [29] was used to perform the molecular dynamics (MD) computer simulations with the periodic boundary conditions. The common CHARMM interaction potential [30] was used with VMD 1.9.2 [31] software for visualization of the time evolution of the studied ensemble. The system was placed in 66 × 68 × 140 Å cuboid simulation box. The β-cyclodextrins were located on the graphene in the following manners, (a) OH group (first order) of βCD (site 1) opposite the graphene layer, (b) OH group (second order) of βCD (site 2) opposite the graphene surface, and (c) both first and second order–OH groups of βCD settled alternately down (site 3) opposite the graphene sheet. The two densities of βCD were investigated, low (nine β-cyclodextrins) and high with sixteen β-cyclodextrin molecules. The ensembles were studied at six temperatures $T = 280, 290, 300, 310,$ 320, and 330 K. The temperature was controlled via Langevin thermostat, a damping coefficient of 1 ps^{-1} was applied. The non-bonding interactions were cut off at 10 Å. The Particle Mesh Ewald method was used to calculate the contribution of long-range electrostatic interactions. The time step of the integration of equations of motion was $\Delta t = 1.0$ fs for all calculations. The system was equilibrate in NVT [29] ensemble. The equilibration phase was conducted up to 5×10^6 time steps, then the simulation data were collected over the 30 ns, i.e., 30×10^6 integration time steps. The NAMD software integration procedure based on BBK algorithm [32] was applied. The data were collected in the "production" phase of simulation, every 2000 integration time steps (2 ps).

3 Results

Figure 1 shows an instantaneous settlement of β-cyclodextrins on graphene honeycomb lattice.

The inspection of the simulated trajectories of β-cyclodextrin molecules indicates that they are pretty mobile, can migrate all over a graphene layer. Nonetheless, βCDs move only in near proximity of graphene, and they do not escape up or out of the graphene surface. The motion of β-cyclodextrin molecules, perpendicular to graphene sheet, is restricted by the attractive influence of graphene. In addition, the characteristics of the movement of molecules depend on whether these molecules point toward the plain of graphene with: site (1) OH group of the first order, or site (2) OH group of the second order (Fig. 2). The above statements find justification while inspecting the forthcoming figure.

The next figure (Fig. 3) shows the calculated mean square displacement $\langle|\Delta\vec{r}_\perp(t)|^2\rangle$ of the center of mass of βCD molecule, where $\Delta\vec{r}_\perp(t) = \vec{r}_\perp(t) - \vec{r}_\perp(0)$ and $\vec{r}_\perp$ is the component of the total displacement vector, perpendicular to graphene plane.

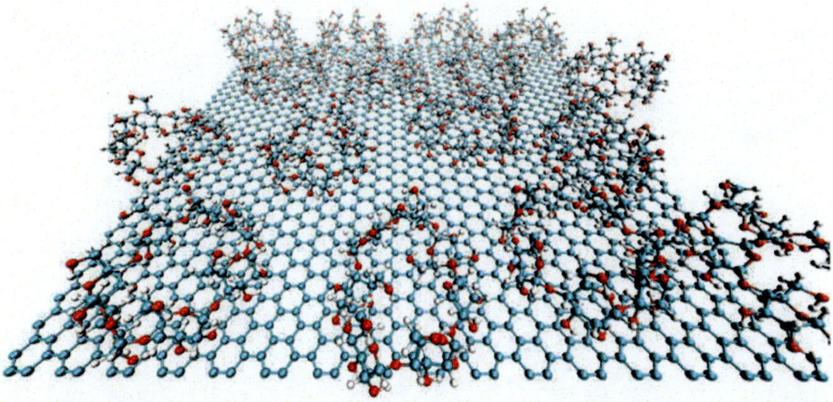

Fig. 1 Example of the snapshot of sixteen β-cyclodextrin molecules on the surface of graphene (equilibrium phase)

Fig. 2 The position of β-cyclodextrin molecule relative to the graphene surface **a** site 1, **b** site 2

The nonzero slope of $\langle|\Delta\vec{r}_{\perp}(t)|^2\rangle$ shows that βCD molecules are mobile, while placed on the surface of graphene. Their movement in the direction perpendicular to graphene sheet is substantially weaker for case 1 comparing to site 2. The perpendicular to graphene plain displacement of βCD molecule only slightly depends on the variation of density. Figure 4 presents the time evolution of the function $\langle\left|\Delta\vec{r}_{\parallel}(t)\right|^2\rangle$, where $\Delta\vec{r}_{\parallel}(t) = \vec{r}_{\parallel}(t) - \vec{r}_{\parallel}(0)$ and $\vec{r}_{\parallel}$ is the parallel to graphene surface ingredient of the total displacement.

One can observe that the in-plane movement of β-cyclodextrin slows down for the high density. Figure 5 clearly indicates the drastic contrast (three orders of magnitude) between the great ease of βCD mobility over the graphene surface (in-plane motion), contrary to restricted motion in the direction perpendicular to graphene sheet. In other words, the β-cyclodextrin molecules develop a thin 2D liquid phase layer on graphene.

Figure 6 shows the obtained simulated trajectories of the diffusion coefficient D of β-cyclodextrin, associated with its translational motion perpendicular to graphene sheet. The Einstein formula for the determination of D reads $\langle|\Delta\vec{r}_{\perp}(t)|^2\rangle = 2Dt$.

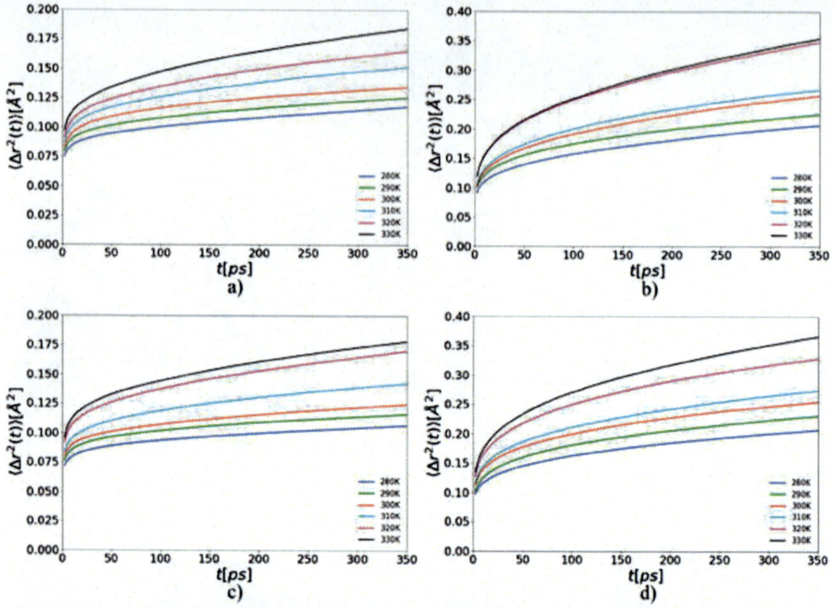

Fig. 3 The mean square displacement of βCD, perpendicular to graphene surface: **a** low density (site 1), **b** low density (site 2), **c** high density (site 1), **d** high density (site 2)

We calculated also the diffusion coefficient of βCD molecules, connected with its translation in the direction perpendicular to graphene plane. The translational diffusion coefficient D for one-dimensional movement can be obtained via Einstein relation $\langle|\Delta\vec{r}_{\perp}(t)|^2\rangle = 2Dt$. The sign of very low mobility of βCDs in the direction perpendicular to graphene surface is the nonzero slope determined from a linear part (from 100 to 250 ps) of $\langle|\Delta\vec{r}_{\perp}(t)|^2\rangle$, as shown in Fig. 6.

Note, that the out of graphene plane motion of βCD is more energetic when it is facing graphene with the second order –OH group (site 2). Thus, βCD molecules are stronger attracted by graphene when they initially face graphene layer via the first order –OH group (site 1). In case of translation perpendicular to graphene layer, only a slight sensitivity of the diffusion coefficient to the initial position of βCD molecules is visible.

Figure 7 shows the obtained diffusion coefficient of βCD for the motion over graphene surface. The diffusion coefficient D for translation over graphene layer can be determined from Einstein formula $\langle\left|\Delta\vec{r}_{\parallel}(t)\right|^2\rangle = 4Dt$.

The diffusion coefficient is larger for low density of βCDs on graphene. That is the result of more free space accessible for the translational displacement of graphene layer. The translational diffusion of βCD parallel to graphene layer only slightly depends on temperature, and it does not follow Arrhenius law.

Figure 8 shows the Lindemann index δL of βCDs located on graphene surface, for several temperatures.

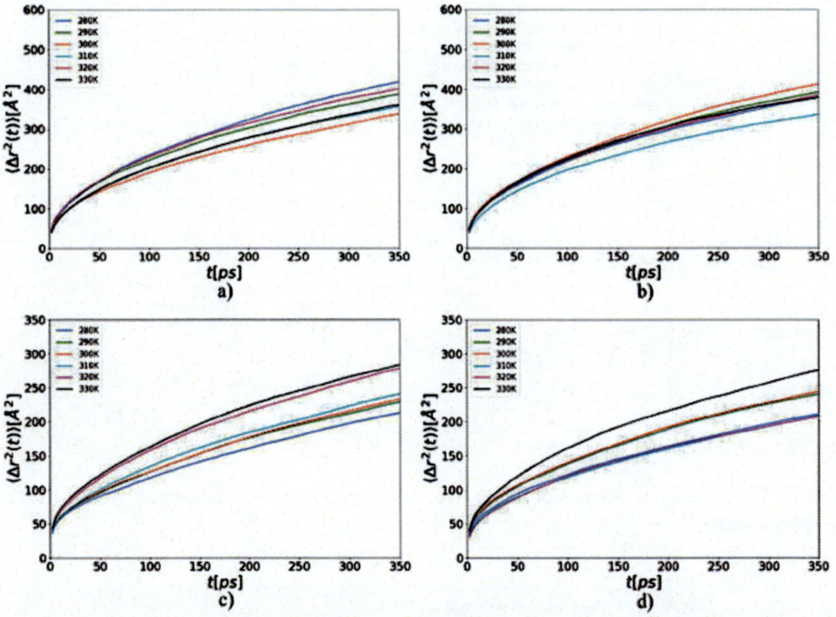

Fig. 4 The parallel to graphene layer component of the mean square displacement of βCD: **a** low density (site 1), **b** low density (site 2), **c** high density (site 1), **d** high density (site 2)

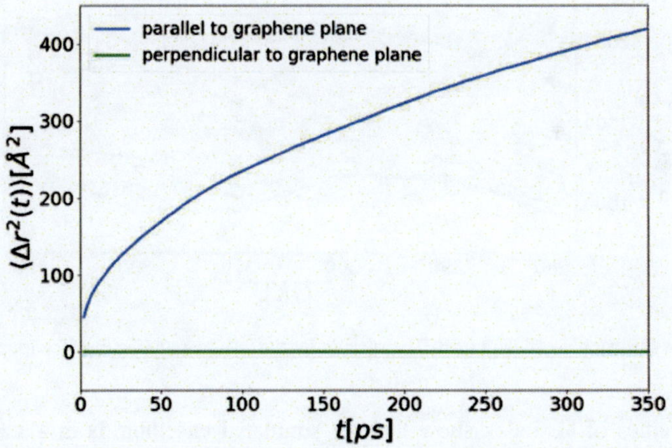

Fig. 5 The mean square displacement of β-cyclodextrin, both perpendicular and parallel to graphene surface

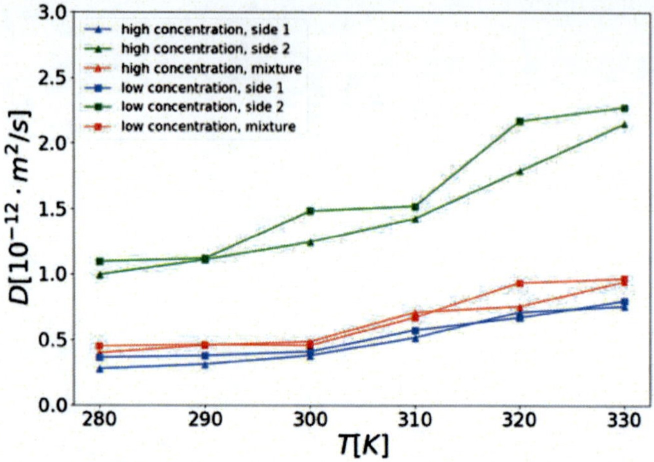

Fig. 6 The diffusion coefficient of βCD molecule associates with its motion perpendicular to the graphene sheet

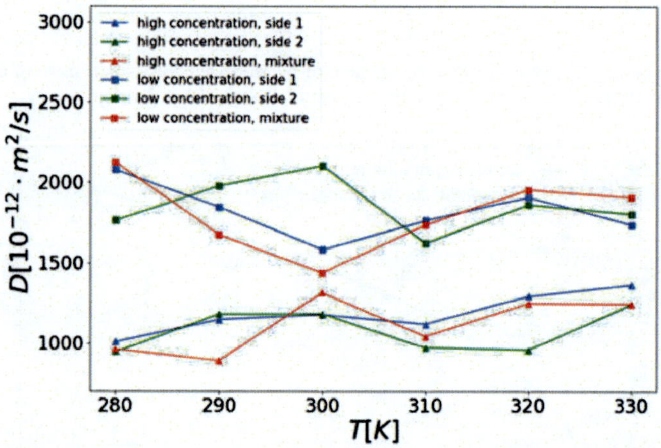

Fig. 7 The diffusion coefficient of βCD molecule for the in-plane motion over the graphene sheet

The values of δL index show that the simulated ensemble is in a stable, two-dimensional liquid state on graphene; no phase transition appears for the studied temperatures 280 K < T < 330 K. The major contribution to δL comes from the motion of βCD molecules in the ultra-thin molecular layer naturally formed on the surface of graphene.

In conclusion, the molecular dynamic (MD) computer simulations of β-cyclodextrins—graphene systems show that βCD molecules spread all over the

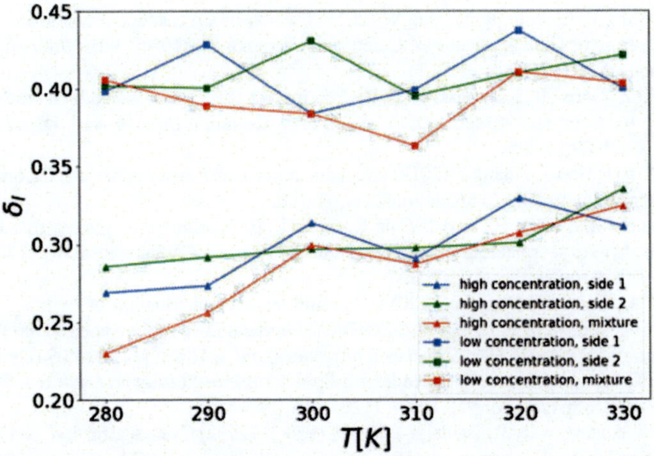

Fig. 8 The Lindemann index for β-cyclodextrins on graphene surface

graphene layer. The physisorption of β-cyclodextrin molecules to any carbon absorption sites on graphene was not observed. The dynamics of βCD molecules in the studied system is essentially their two-dimensional (2D) translation all over graphene layer. The displacements of β-cyclodextrins directed perpendicular to graphene sheet is strongly limited. The stable 2D liquid phase of β-cyclodextrins exists on graphene sheet. These observations documented above set a ground for future use of βCD in the context of searching for new applications of decorated graphene in nanotechnology.

Acknowledgements We would like to thank the PL Grid supercomputers network for sharing computational resources.

References

1. Lawal AT (2019) Graphene-based nano composites and their applications. A review. Biosens Bioelectron 141(1–17): 111384. https://doi.org/10.1016/j.bios.2019.111384
2. Nandanapalli KR, Mudusu D, Lee S (2019) Functionalization of graphene layers and advancements in device applications. Carbon 152:954–985. https://doi.org/10.1016/j.carbon.2019.06.081
3. Bahiraei M, Heshmatian S (2019) Graphene family nanofluids: a critical review and future research directions. Energy Convers Manag 196:1222–1256. https://doi.org/10.1016/j.enconman.2019.06.076
4. Wang J, Mu X, Sun M, Mu T (2019) Optoelectronic properties and applications of graphene-based hybrid nanomaterials and van der Waals heterostructures. Appl Mater Today 16:1–20. https://doi.org/10.1016/j.apmt.2019.03.006
5. Iqbal MZ, Rehman AU, Siddique S (2019) Prospects and challenges of graphene based fuel cells. J Energy Chem 39:217–234. https://doi.org/10.1016/j.jechem.2019.02.009
6. Sun C, Liu M, Bai B (2019) Molecular simulations on graphene-based membranes. Carbon 153:481–494. https://doi.org/10.1016/j.carbon.2019.07.052

7. Li Z, Wang L, Li Y, Feng Y, Feng W (2019) Carbon-based functional nanomaterials: preparation, properties and applications, Compos Sci Technol 179:10–40. https://doi.org/10.1016/j.compscitech.2019.04.028
8. Gburski Z, Górny K, Raczyński P (2010) The impact of a carbon nanotube on the cholesterol domain localized on a protein surface. Solid State Commun 150:415–418. https://doi.org/10.1016/j.ssc.2009.12.005
9. Li G, Xiao P, Hou S, Huang Y (2019) Graphene based self-healing materials. Carbon 146:371–387. https://doi.org/10.1016/j.carbon.2019.02.011
10. Fernandez MA, Silva F, Vico RV, de Rossi RH (2019) Complex systems that incorporate cyclodextrins to get materials for some specific applications. Carbohydr Res 480:12–34. https://doi.org/10.1016/j.carres.2019.05.006
11. Dawid A, Górny K, Gburski Z (2015) The influence of distribution of hydroxyl groups on vibrational spectra of fullerenol C-60(OH)(24) isomers: DFT study. Spectrochim Acta Part A Mol Biomol Spectrosc 136:1993–1997. https://doi.org/10.1016/j.saa.2014.08.023
12. Szente L, Fenyvesi E (2017) Cyclodextrin-lipid complexes: cavity size matters. Struct Chem 28:479–492. https://doi.org/10.1007/s11224-016-0884-9
13. Roux M, Auzely-Velty R, Djedaini-Pilard F, Perly B (2002) Cyclodextrin-induced lipid lateral separation in DMPC membranes: ^{2}H nuclear magnetic resonance study. Biophys J 82:813–822. https://doi.org/10.1016/s0006-3495(02)75443-x
14. Grachev MK, Malenkovskaya MA, Vasyanina LK (2015) NMR study of inclusion complexes formation between amphiphilic dimeric β-cyclodextrin derivative and some pharmacologically important compounds. J Incl Phenom Macrocycl Chem 83:209–214. https://doi.org/10.1007/s10847-015-0548-1
15. Dawid A, Gburski Z (2003) Rayleigh light scattering in fullerene covered by a spherical argon film—a molecular dynamics study. J Phys Condens Matter 15:2399–2405. https://doi.org/10.1088/0953-8984/15/14/315
16. Litz JP, Thakkar N, Portet T, Keller SL (2016) Depletion with cyclodextrin reveals two populations of cholesterol in model lipid membranes. Biophys J 110:635–645. https://doi.org/10.1016/j.bpj.2015.11.021
17. Joset A, Grammenos A, Hoebeke M, Leyh B (2015) Investigation of the interaction between a β-cyclodextrin and DMPC liposomes: a small angle neutron scattering study. J Incl Phenom Macrocycl Chem 83:227–238. https://doi.org/10.1007/s10847-015-0558-z
18. Gburski Z, Gray CG, Sullivan DE (1983) Information theory of line shape in collision-induced absorption. Chem Phys Lett 100:383–386. https://doi.org/10.1016/0009-2614(83)80292-9
19. Mascetti J, Castano S, Cavagnat D, Desbat B (2008) Organization of β-cyclodextrin under pure cholesterol, DMPC, or DMPG and mixed cholesterol/phospholipid monolayers. Langmuir 24:9616–9622. https://doi.org/10.1021/la8004294
20. Tsamaloukas A, Szadkowska H, Slotte PJ, Heerklotz H (2005) Interactions of cholesterol with lipid membranes and cyclodextrin characterized by calorimetry. Biophys J 89:1109–1119. https://doi.org/10.1529/biophysj.105.061846
21. Gwizdała W, Górny K, Gburski Z (2008) Molecular dynamics and dielectric loss in 4-cyano-4-n-pentylbiphenyl (5CB) mesogene film surrounding carbon nanotube—computer simulation. J Mol Struct 887:148–151. https://doi.org/10.1016/j.molstruc.2007.12.045
22. Loftsson T, Vogensen SB, Brewster ME, Konráðsdóttir F (2007) Effects of cyclodextrins on drug delivery through biological membranes. J Pharm Sci 96:2532–2546. https://doi.org/10.1002/jps.20992
23. Puglisi G, Fresta M, Ventura CA (1996) Interaction of natural and modified β-cyclodextrins with a biological membrane model of dipalmitoylphosphatidylcholine. J Colloid Interface Sci 180:542–547. https://doi.org/10.1006/jcis.1996.0335
24. Piatek A, Dawid A, Gburski Z (2011) The properties of small fullerenol cluster (C-60(OH)(24))(7): computer simulation. Spectrochim Acta Part A Mol Biomol Spectrosc 79:819–823. https://doi.org/10.1016/j.saa.2010.08.059
25. Menezes P, Andrade T, Frank LA, Souza EPBS, Trindade GG, Trindade IAS, Serafinni MS, Guterres SS, Araujo AA (2019) Advances of nanosystems containing cyclodextrins and their

applications in pharmaceuticals. Int J Pharm 559:312–328. https://doi.org/10.1016/j.ijpharm.2019.01.041

26. Dendzik Z, Górny K, Gburski Z (2009) Cooperative dipolar relaxation of a glycerol molecular cluster in nanoscale confinement—a computer simulation study. J Phys Condensed Matter 21:425101 (7 pp.). https://doi.org/10.1088/0953-8984/21/42/425101./l
27. Hammoud Z, Khreich, N, Auezova L, Fourmentin S, Elaissari A, Greige-Gerges H (2019) Cyclodextrin-membrane interaction in drug delivery and membrane structure maintenance. Int J Pharm 564:59–76. https://doi.org/10.1016/j.ijpharm.2019.03.063
28. Makieła D, Janus-Zygmunt I, Górny K, Gburski Z (2019) The dynamics of beta-cyclodextrin molecules on graphene sheet. A molecular dynamics simulation study. J Mol Liq 288(1–7):110974. https://doi.org/10.1016/j.molliq.2019.110974
29. Phillips JC, Braun R, Wang W, Gumbart J, Tajkhorshid E, Villa E, Chipot C, Skeel RD, Kalé L, Schulten K (2005) Scalable molecular dynamics with NAMD. J Comput Chem 26:1781–1802. https://doi.org/10.1002/jcc.20289
30. Brooks BR, Brooks CL, Mackerell AD, Nilsson L, Petrella RJ, Roux B, Won Y, Archontis G, Bartels C, Boresch S et al (2009) CHARMM: the biomolecular simulation program. J Comput Chem 30:1545–1614. https://doi.org/10.1002/jcc.21287
31. Humphrey W, Dalke A, Schulten K (1996) VMD—visual molecular dynamics. J Mol Graph 14:33–38. https://doi.org/10.1016/0263-7855(96)00018-5
32. Brünger A, Brooks CL, Karplus M (1984) Stochastic boundary conditions for molecular dynamics simulations of ST2 water. Chem Phys Lett 105:495–500. https://doi.org/10.1016/0009-2614(84)80098-6

The Physical–Chemical Model of Nanoscaled Metal Component Formation on the Surface of Graphite Supporter

Luidmila Yu. Matzui, Iryna V. Ovsiienko, Luidmila L. Vovchenko, Tatiana L. Tsaregradskaya, Galina V. Saenko, Oleg D. Marinin, and Nataliia B. Bielousova

1 Introduction

Research on nanocomposite materials is the aim of new branches of science that are developing rapidly. The use of a chemical modification of the surface of a graphite support for the transfer of metals or their compounds to the nanocrystalline state seems very promising [1]. For the further development of the synthesis of nanocomposite materials with a given set of structural, physical, and chemical properties, it is necessary to create a theoretical justification for describing the basic mechanisms of the formation of metal components on the surface of structurally different graphite. Investigations showed the formation of nanoscaled component on the surface of graphite supporter to be the multistage process, each stage of it being determined by a variety of factors [2].

The use of the thermodynamic method allows you to create a physical-chemical model of the formation of nanoscale metal components on the surface of a graphite carrier [3]. That is why the aim of the current research was the creation of a physical-chemical model model of nanoscaled metal component formation on the surface of graphite supporter. Finding the correlation between structural peculiarities of the

L. Yu. Matzui · I. V. Ovsiienko · L. L. Vovchenko · T. L. Tsaregradskaya · G. V. Saenko (✉) · O. D. Marinin
Departments of Physics, Taras Shevchenko National University of Kyiv, Volodymyrska 64/13, Kiev 01601, Ukraine
e-mail: g_saenko@ukr.net

L. Yu. Matzui
e-mail: tsar_grd@ukr.net

N. B. Bielousova
Departments of International Information, Taras Shevchenko National University of Kyiv, Volodymyrska 64/13, Kiev 01601, Ukraine

O. Fesenko and L. Yatsenko (eds.), *Nanomaterials and Nanocomposites, Nanostructure Surfaces, and Their Applications*, Springer Proceedings in Physics 246, https://doi.org/10.1007/978-3-030-51905-6_21

source graphite components, synthesis conditions, structure and phase composition of obtained materials is one of the tasks of our investigations.

2 Creation of the Physical Model of Metal Oxide Formation on the Surface of Graphite Materials

The reaction of thermolysis of the metal salt (which is preliminarily deposited from solution on the graphite surface) is the basis of metal oxide formation on the graphite surface. Iron, nickel, and cobalt nitrates and acetates were used in our investigation in order to impregnate graphite material. The solid-phase reactions of thermolysis in a generalized form for these salts may be presented as

$$A_{\mathrm{SOLID}} \xrightarrow{T} B_{\mathrm{SOLID}} + C_{\mathrm{GASEOUS}}, \tag{1}$$

where A is the metal salt (crystalline hydrate), B is the crystalline metal oxide, and C is the gaseous-like products of the reaction of thermolysis.

The physical–chemical model of the metal oxide formation on the graphite surface must combine the chemical aspect of thermolysis (decomposition of individual salt molecules with formation of metal oxide molecules), solid-phase specific character of this process (formation of the oxide crystalline phase, interphase interaction, mechanism of the removal of the gaseous-like products from the reaction zone), and the geometry peculiarities of the "graphite-salt" system, where oxide phase formation is taking place. The layered "graphite-metal oxide" system is illustrated schematically by Fig. 1.

The salt layer is formed on the surface of graphite particles (thermoexfoliated or disperse graphite particles) after impregnation by salt solution and further drying. According to the performed studies, the typical thickness of graphite and salt layers are close to each other. The typical thickness of graphite layer is equal to $h_{\mathrm{Gr}} = z_1 =$

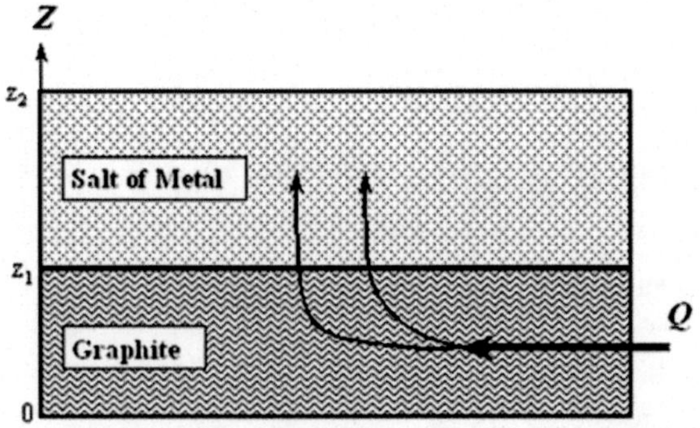

Fig. 1 Schematic image of the "graphite-salt" system

15 nm and the thickness of salt layer varies within $h_{\text{Salt}} = z_2 - z_1 = (14\text{–}30)$ nm depending on solution concentration.

To start the passing of the reaction of thermolysis, it is necessary to convert metal salt into nonequilibrium state through temperature increasing. Let us analyze the duct of heat delivering to the salt layer in thermoexfoliated graphite.

It is necessary to point out three important conditions determining the character of heat delivering to the salt layer. The salt and graphite layers are in direct contact with an ambient medium through their outer ends, which outcrop on the surface of graphite particle.

The thickness of graphite and salt layers are close to each other.

Hear conductivity coefficient within graphite layer is essentially higher than that of salt crystalline hydrate.

This permits to state that on heating of "graphite-salt" layered system, the heat predominantly spreads along the graphite layers and enters the salt layer through the graphite-salt boundary. This means the arising of temperature gradient along graphite layer is directed toward the outer end of plane; the temperature gradient in the bulk of salt layer $\partial T/\partial z$ is directed to the boundary of graphite and salt layers. The magnitudes of these gradients are determined by the heating regime, i.e., by the maximal heating temperature and by the heating rate. In any case, the maximal temperature in the salt layer is believed to reach the boundary of graphite and salt layers.

The condition of salt thermal stability is the following in equation

$$G_{\text{Salt}}(T) \le G_{\text{Oxide}}(T), \tag{2}$$

where $G_{\text{Salt}}(T)$ and $G_{\text{Oxide}}(T)$ are the Gibbs thermodynamic potentials for metal salt and metal oxide, respectively. The typical behavior of $G_{\text{Salt}}(T)$ and $G_{\text{Oxide}}(T)$ functions is presented in Fig. 2.

The temperature T_0 that corresponds to the equality of $G_{\text{Salt}}(T)$ and $G_{\text{Oxide}}(T)$ determines the limit of salt thermal stability: salt is thermodynamically stable at $T < T_0$, and at $T > T_0$ it decomposes with the formation of oxide.

Reasoning from the presented qualitative analysis of temperature distribution over the salt layer thickness, one can conclude that during heating, first of all, the salt molecules directly adjoining to the boundary of salt and graphite layer turn to thermodynamically nonequilibrium state since the temperature in this zone is the highest. The nuclei of the new phase, i.e., metal oxide nuclei, will form just there. Then each nucleus grows due to thermolysis at "salt-oxide" interphase boundary.

The dominant formation of the oxide phase nuclei at the boundary of graphite and salt layers is the typical feature of heterogeneous character of oxide phase formation in the system under consideration.

Selection of the model for the heterogeneous formation of the metal oxide phase permits to turn to the description of the kinetics of phase formation and, finally, to determine the influence of these process parameters (graphite surface characteristics, type of salt, parameters of the reaction, i.e., heating regime, etc.) on the composition and structure of the oxide phase.

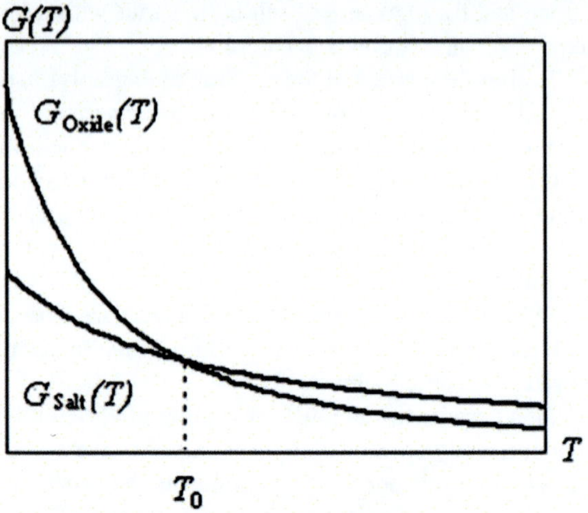

Fig. 2 Temperature dependences of the Gibbs potentials for metal salt $G_{Salt}(T)$ and metal oxide $G_{Oxide}(T)$

Figure 3 presents the part of graphite-salt boundary, where the oxide nucleus has been formed. The decomposition of salt molecules with the formation of the molecules, which "build up" the oxide phase, takes place in the monomolecular salt layer at the boundary of salt and oxide phases.

The kinetics of heterogeneous formation of the oxide phase is determined by a set of parameters such as the frequency of nucleus formation (per unit square of the graphite and salt boundary), the linear rate of oxide phase growth, and (in our case) the geometry of salt layer.

According to the statements of the general theory of heterogeneous phase formation, the frequency of nucleus formation J is determined by the following expression [4]:

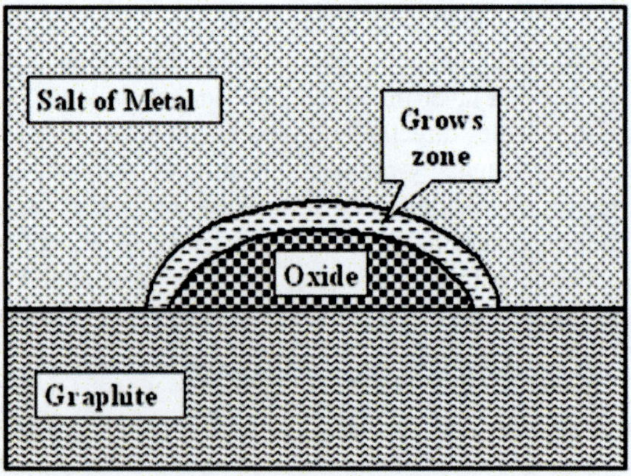

Fig. 3 Schematic image of the part of graphite-salt boundary

$$J = K_\nu \exp\left[-\frac{A_C}{kT}\right], \tag{3}$$

where K_ν is the kinetic coefficient of heterogeneous nucleation, A_C is the work of the formation of critical oxide nucleus, which is determined as

$$A_c = \frac{16}{3}\pi \frac{\sigma^3 \psi(\theta)}{(\Delta g_{\text{salt-Oxide}})^2} \quad g_{\text{salt}} = \frac{G_{\text{Salt}}}{V^M_{\text{Salt}}};\ g_{\text{Oxide}} = \frac{G_{\text{Oxide}}}{V^M_{\text{Oxide}}}. \tag{4}$$

V^M_{Salt}; V^M_{Oxide} in (4) are the molar volumes of salt and oxide, respectively, θ is the limiting wetting angle, σ is the coefficient of interphase tension at the "salt-oxide" boundary, $\psi(\theta_i) = \frac{1}{4}(1 - \cos\theta_i)^2(2 + \cos\theta_i)$ is the function of limiting wetting angle.

It is convenient to write the kinetic coefficient of heterogeneous nucleation as

$$K_\nu = a\, n\, Z\, \nu, \tag{5}$$

where a is the thickness of monomolecular layer of salt crystalline hydrate, $n = N/V$ is the numerical density, ν is the characteristic frequency of decomposition of salt molecule to metal oxide, and Z is the nonequilibrium Zeldovitch's factor.

$$Z = \frac{1}{N_C}\left[\frac{A_C}{3\pi\, k\, T}\right]^{\frac{1}{2}}. \tag{6}$$

In (6), N_C denotes a number of atoms in a critical oxide phase nucleus, i.e., $N_C = n V_C$ and the kinetic coefficient could be expressed as

$$K_\nu = \frac{a\,\nu}{V_C} \cdot \left[\frac{A_C}{3\pi\, k\, T}\right]^{\frac{1}{2}}. \tag{7}$$

Let us designate

$$\nu = \frac{1}{\tau}, \tag{8}$$

where τ is the characteristic time of oxide molecule formation from the metal salt molecule.

The reaction of thermolysis is of threshold character, hence

$$\tau = \tau_o \exp\left[\frac{E_a}{kT}\right], \tag{9}$$

where E_a is the activation energy of thermolysis. In a case of multistage character of the reaction of thermolysis, (9) is approximate. Here, the characteristic time is determined by the slowest stage of decomposition.

From (8) and (9) one can deduce that

$$\nu = \nu_o \exp\left[-\frac{E_a}{kT}\right]. \tag{10}$$

Now one could write the final expressions for the kinetic coefficient and for the frequency of heterogeneous nucleation

$$K_\nu = \frac{a\,\nu_o}{V_C}\cdot\left[\frac{A_C}{3\pi\,k\,T}\right]^{\frac{1}{2}}\cdot\exp\left[-\frac{E_a}{kT}\right]; \tag{11}$$

$$J = \frac{a\,\nu_o}{V_C}\cdot\left[\frac{A_C}{3\pi\,k\,T}\right]^{\frac{1}{2}}\cdot\exp\left[-\frac{E_a}{kT}\right]\cdot\exp\left[-\frac{A_C}{kT}\right]. \tag{12}$$

As it was stated above, the growth of oxide phase nucleus passes due to the thermolysis of salt molecules in a monomolecular salt layer neighboring to the boundary of oxide and salt phases.

The linear rate u of oxide phase growth is described by the expression:

$$u = a\nu_o \exp\left[-\frac{E_a}{kT}\right]\cdot\left\{1-\exp\left[-\frac{\Delta G_{\text{Salt-Oxide}}}{RT}\right]\right\}. \tag{13}$$

The process of phase formation in "graphite-salt" system is of "quasi-2D" character. So, the linear velocity of the front of oxide phase formation u_P in the plane of "graphite-metal salt" boundary becomes more important parameter for the description of this process than u. The elementary geometric considerations results in

$$u_P = \frac{u}{\sin\theta} = \frac{a\nu_o}{\sin\theta}\exp\left[-\frac{E_a}{kT}\right]\cdot\left\{1-\exp\left[-\frac{\Delta\,G_{\text{Salt-Oxide}}}{RT}\right]\right\}. \tag{14}$$

Let us denote the portion of graphite surface, where the metal oxide is formed in a point of time t, as $X(t)$:

$$X(t) = \frac{S_{\text{Oxide}}(t)}{S}, \tag{15}$$

where $S_{\text{Oxide}}(t)$ is the square of graphite-oxide boundary in a point of time t, and S is the square of graphite surface. Temporal dependence $X(t)$ in a case of 2D phase formation is described by the equation:

$$X(t) = 1-\exp\left[-\alpha J u_P^2 t^3\right]. \tag{16}$$

The number of oxide phase nuclei per unit square of graphite surface $N(t)$ is determined by the expression:

$$N(t) = J \int_0^t \exp\left[-\alpha J u_{\mathrm{P}}^2 \tau^3\right] \mathrm{d}\tau. \quad (17)$$

The mean square of nucleus $S_o(t)$ is equal to $N^{-1}(t)$ or

$$S_o(t) = J^{-1} \left\{ \int_0^t \exp\left[-\alpha J u_{\mathrm{P}}^2 \tau^3\right] \mathrm{d}\tau \right\}^{-1}. \quad (18)$$

It is necessary to determine the mean square of nuclei at the final stages $X(t) \propto 1$. In this case (17) and (18) could be reduced to

$$N = J^{\frac{2}{3}} \frac{0.893}{\left(\alpha\, u_{\mathrm{P}}^2\right)^{\frac{1}{3}}}; \quad S_o = J^{-\frac{2}{3}} \frac{\left(\alpha\, u_{\mathrm{P}}^2\right)^{\frac{1}{3}}}{0.893}. \quad (19)$$

Using the expression (19) for the mean square of nuclei, the mean linear size of crystallite $\langle r \rangle_{XY}$ in the plane of graphite surface may be expressed as

$$\langle r \rangle_{XY} \approx \sqrt{S_o} \approx \left[\frac{u_{\mathrm{P}}}{J}\right]^{\frac{1}{3}}. \quad (20)$$

By substitution of (12) and (14) to (20) one can obtain

$$\begin{aligned} \langle r \rangle_{XY} \approx & \left[\frac{V_{\mathrm{C}}\, \Delta\, g_{\text{Salt-Oxide}}}{\sin\theta}\right]^{\frac{1}{3}} \cdot \left\{1 - \exp\left[-\frac{\Delta G_{\text{Salt-Oxide}}}{RT}\right]\right\}^{\frac{1}{3}} \\ & \times \exp\left[\frac{16\pi}{9kT} \frac{\sigma^3 \psi(\theta)}{(\Delta\, g_{\text{Salt-Oxide}})^2}\right] \left[\frac{9kT}{16\sigma^3 \psi(\theta)}\right]^{\frac{1}{6}} \end{aligned}. \quad (21)$$

It should be emphasized that (21) corresponds to the point of time, when metal salt totally transformed into metal oxide at graphite surface.

Let us define the parameters of (21) influencing on the mean size of metal oxide crystallite, which is formed at graphite surface.

The limiting wetting angle θ characterizes the state of graphite supporter surface. The limiting wetting angle is determined, firstly, by the type of graphite material and, secondly, by the type of oxidant that activates the graphite supporter surface before its impregnation by the metal salt. Parameters $\Delta\, g_{\text{Salt-Oxide}}$; $\Delta G_{\text{Salt-Oxide}}$ and their temperature dependences, coefficient of the interphase tension σ is the parameters that are determined by the type of salt being used. Temperature T is the external parameter of the thermolysis.

To analyze the type of salt influence on the oxide phase properties, it is necessary to use not only (21) characterizing the final state of the oxide phase but also (12) and (14), which define the kinetics of phase formation.

The crystallite size of oxides on thermoexfoliated graphite surface depends on parameters such as difference and specific difference of Gibbs potential of salt and oxides $\Delta\, g_{\text{Salt-Oxide}};\ \ \Delta G_{\text{Salt-Oxide}}$, i.e., on the used salt type. At the analysis of salt type influence on oxide phase properties, it is necessary to use not only (21), characterizing final state of oxide phase but also (12) and (14) that determine the kinetics of phase formation.

Equation (1) takes into account in an explicit form such external parameters as the temperature of the reaction as well as temperature dependences of $\Delta\, g_{\text{Salt-Oxide}};\ \ \Delta G_{\text{Salt-Oxide}}$.

The analysis of (21) shows that the main factor defining the temperature dependence $\langle r \rangle_{XY} = f(T, \theta)$ is

$$\exp\left[\frac{16\pi}{9kT}\frac{\sigma^3\psi(\theta)}{(\Delta\, g_{\text{Salt-Oxide}})^2}\right]. \tag{22}$$

Indeed, the value of $T\,(\Delta\, g_{\text{Salt-Oxide}})^2$ increases with temperature quite fast. That is why factor $\exp\left[\frac{16\pi}{9kT}\frac{\sigma^3\psi(\theta)}{(\Delta\, g_{\text{Salt-Oxide}})^2}\right]$ together with $\langle r \rangle_{XY}$ decreases monotonously with temperature increase.

It is necessary to note that such dependence of the mean crystallite size on temperature does not consider that the probability of the formation of big agglomerates consisting of small oxide particles due to surface diffusion is increased with the temperature.

As it is as follows from (21), the mean crystallite size depends in a complicated manner on the temperature of thermolysis T and limiting wetting angle θ. For analysis of $\langle r \rangle_{XY}$ dependence on T and θ, we have performed the calculations of $\langle r \rangle_{XY} = f(T, \theta)$ and $\langle r \rangle_{XY} = f(T, \theta)$ for Co oxide crystals, which are formed on the graphite surface during the salt thermolysis. Time exposure at T was chosen as the time of complete salt thermolysis. Calculation results are presented graphically in Figs. 4 and 5.

At fixed thermolysis temperature, the mean size of oxides increases under increase of limiting wetting angle θ. The limiting wetting angle θ influences on mean crystallite size decreases with the increase of thermolysis temperature. The described above general regularities of $\langle r \rangle_{XY} = f(T, \theta)$ behavior are enough universally and correct for metals acetates.

The analysis of obtained analytical and numerical dependence $\langle r \rangle_{XY} = f(T, \theta)$ allows to confirm the following.

The previous activation of graphite surface for decrease of limiting wetting angle θ is very important at the formation of metal oxide layer under salt thermolysis.

The thermolysis temperature must be high; however, it is limited by the process of secondary growth of oxides crystallites that enhanced sharply under temperature increase.

One of the important parameters of the reaction of thermolysis, which is not explicitly considered in (21) is the mechanism of gaseous-like products removal from the reaction zone. It is known that slow removal of gaseous-like products of

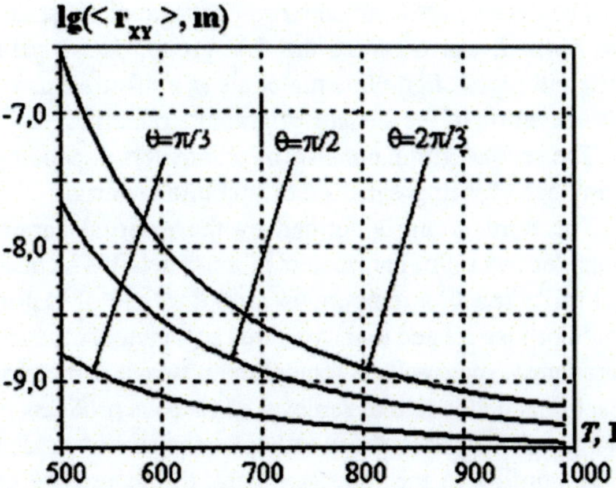

Fig. 4 Temperature dependence of the mean crystallite size of CoO on the graphite surface at different values of limiting wetting angle θ

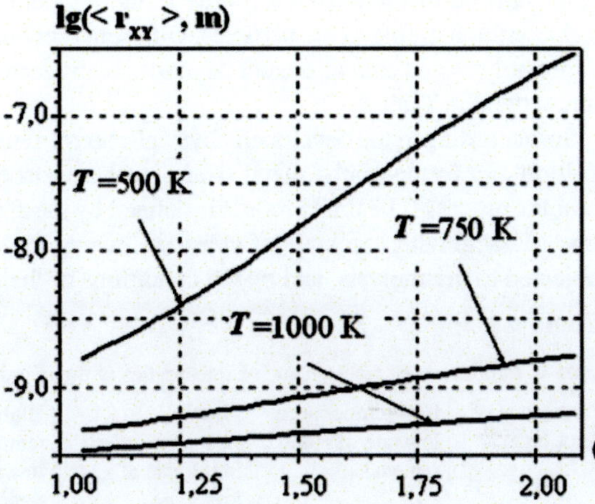

Fig. 5 Dependence of the mean crystallite size of CoO on graphite surface versus limiting wetting angle θ at different temperatures of thermolysis T

reaction from the reaction zone leads to thermolysis inhibition, and thus, the system is exposed to an enhanced temperature for more extended time and, finally, results in oxide crystals growing. The removal of gaseous-like products of reaction from the reaction zone could be passed through two mechanisms: gas diffusion through metal salt layer; gas outlet through the ducks made by defects, creeps, etc.

According to the obtained data, the removal of gaseous-like products from reaction zone passes predominantly by the second mechanism. So, the pressure of gaseous-like products in vicinity of reaction zone is determined by external conditions of thermolysis.

Thus, according to (21), all factors defining sizes and dispersion of metal oxide particles formed on graphite surface due to thermolysis may be divided into three groups.

The parameters characterizing the state of substrate, where metal oxide particles are formed, can organize the first group. These parameters will change at using different types of graphite materials as a substrate as well as using different oxidants for activation of the surface of graphite material.

The second group consists of parameters depending on the type of salt, which is being used for impregnation of graphite material.

The third group is formed by the external parameters of thermolysis. Let us consider this set of parameters in more detail. The heating rate of the reacting material up to reaction temperature and the time of exposition at reaction temperature could be also related to the external parameters of reaction as well as the listed above parameters (thermolysis temperature, which is the most important parameter of the reaction, and the rate of removal of gaseous products from the reaction zone). These two parameters essentially influence the size of oxide particles being formed.

According to the reference data, the increasing of the heating rate causes the decreasing of size and enhancement of imperfections of these crystallites whereas the increasing of the exposition time at the temperature of reaction promotes the agglomeration of fine oxide particles. Different combinations of external parameters of thermolysis and characteristics of oxides being formed under these conditions are summarized in Table 1.

So, according to the developed physical-chemical model, thermolytic salt decomposition and formation of metal oxide on the surface of graphite supporter is the complex process. The final product is defined by the action of three groups of parameters characterizing the state of substrate, where the oxide is formed, type of salt subjected to thermolysis, and by the conditions of the reaction itself.

Table 1 Parameters of reaction and characteristics of the appropriate oxides

Parameters of reaction	Low temperature, low rate of gas removal	Low temperature, high rate of gas removal	High temperature, low rate of gas removal	High temperature, high rate of gas removal
Performance conditions	Slow heating, flow of gaseous product of reaction	Rapid heating over decomposition temperature, blown with external gas, "boiling layer"	Flow of gaseous products of reaction	Small portions, rapid heating, blown with external gas
Formed products	Fine particles of regular shape, high level of microstresses and defects	Fine particles, microstresses, deformations, defects, twinning, associations of defects	Relatively big particles, no twins and microstresses	Relatively big particles, twins are the main type of defects

3 Conclusion

The generalized physical–chemical model of the process of the nanoscaled metal phase formation on the surface of graphite materials has been developed. This model allows to describing each stage of these processes under the conditions of different types of graphite materials, different salt, and different regimes of each stage.

The mean size of oxide particles forming on graphite surface due to salt thermolysis has been estimated within the framework of this model. This size was shown to be determined by three factors: the topology of graphite material surface, the properties of salt, and the thermodynamic parameters of thermolysis reaction. The most important factor determining salt thermolysis and characteristics of the final products is the temperature of the reaction.

The comparative analysis of the structure and phase composition of graphite-metal nanocomposites prepared under different temperature-temporal conditions showed the thermolysis decomposition of salt on the surface of graphite supporter being complex multistage process. The duration of each stage and characteristics of the final product are determined by the simultaneous action of series of factors: the type of graphite material, the state of its surface, the type of salt, and the thermodynamic kinetic conditions of reactions. The desirable size, homogeneity, and uniformity of distribution of the final product could be obtained by the variation of these parameters.

References

1. Borovij MO, Kunic'kij YUA, Kalenyk OO, Ovsiienko IV, Tsaregradskaya TL (2015) Nanomateriali, nanotekhnologiï, nanopristroï. Interservis, Kiïv, p 350
2. Matsui LY, Vovchenko LL, Ovsiienko IV, Tsaregradskaya TL, Saenko GV (2019) Springer Proc Phys 221:333–348
3. Shpak AP, Lysov VI, Kunitskij YuA, Tzaregradska TL (2002) Crystallization and amorphization of metallic systems. Akademperiodika, Kyiv, p 207
4. Matsuy LYu, Ovsiienko IV, Fedorov VE (2006) Modeling of the formation processes of nanocomposite material "graphite-metal" oxide. Metallofiz Noveishie Tekhnol 28(4):521–533

Theoretical Modeling of Laser-Stimulated Nanostructures

L. V. Shmeleva, A. D. Suprun, S. M. Yezhov, and V. V. Datsyuk

1 Introduction

The structure of the surface and its physics and chemical properties are important factors influencing the processes occurring at the interface. By changing the structure and these properties of the surface, one can control these processes. The lasers with femtosecond pulses have become widely used for the modification of the surface. The efficiency of practical use of femtosecond laser pulses is due to the very short pulse length (5–500 fs), high peak power in the pulse (up to 10^{15} W), and when focused, high laser intensity (10^{11}–10^{22} W/cm^2). Powerful parameters of femtosecond laser pulses are actively used to solve various fundamental and applied problems.

A special place for the use of femtosecond laser pulses is assigned to the technological field, which is associated with micro-processing, as well as nano- and microstructuring of solid-state materials. Surface treatment of solid materials by femtosecond laser is an efficient and highly co-productive technology (up to 500 m/s). It allows to create a huge variety of modified surfaces. Due to the effect of nonlinear absorption of powerful femtosecond laser pulses, not only surface but also 3D processing of transparent materials for laser radiation of materials is possible. With the help of femtosecond laser pulse effects, one can process almost any material including metals, semiconductors, dielectrics, and various alloys, including both massive and multilayer thin film structures (layer thicknesses 5–100 nm).

However, it is clear that the result of treatment depends on the characteristics of the irradiated material (Fig. 1). An important role is played by the environment in which the irradiation process takes place (Fig. 2). Also significant is the duration and, accordingly, the impulse power (Fig. 3).

L. V. Shmeleva (✉) · A. D. Suprun · S. M. Yezhov · V. V. Datsyuk
Kiev 01601, Ukraine
e-mail: lshmel@univ.kiev.ua

O. Fesenko and L. Yatsenko (eds.), *Nanomaterials and Nanocomposites, Nanostructure Surfaces, and Their Applications*, Springer Proceedings in Physics 246, https://doi.org/10.1007/978-3-030-51905-6_22

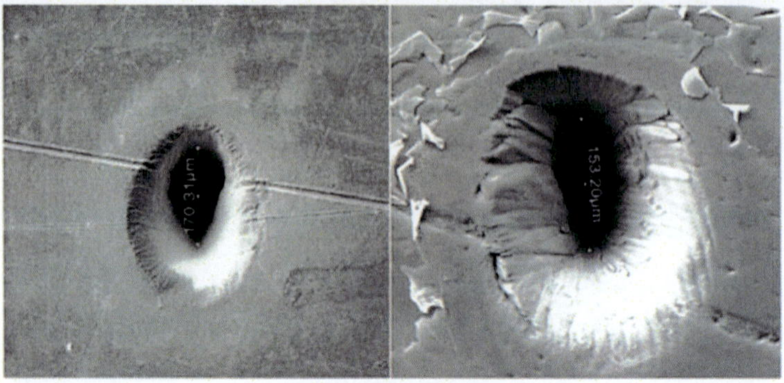

Fig. 1 (left) A laser cavity on surface of the sample of nickel (characteristic size 170 μm); (right) a laser cavity on surface of the sample glass carbon (characteristic size 153 μm) [1]

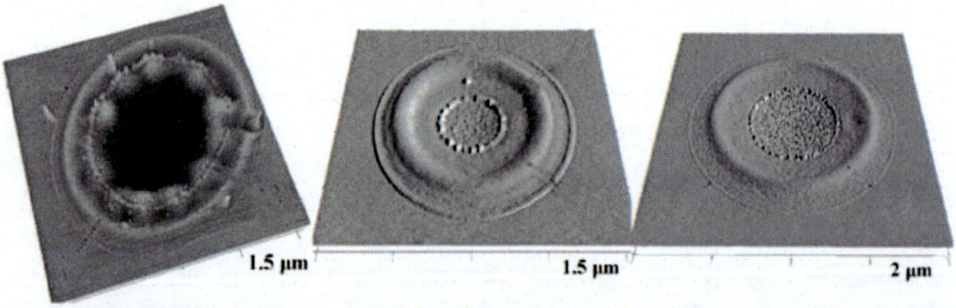

Fig. 2 Images of modified silicon surfaces produced by single femtosecond laser pulses in air (left), water (center) and oil (right) [2]

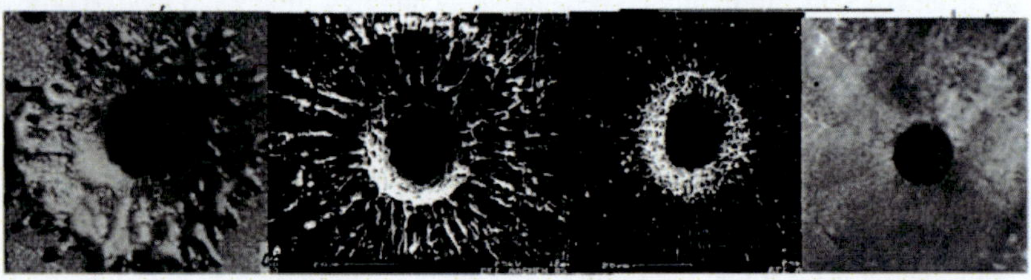

Fig. 3 General view of laser craters at ms, μs, ns, fs pulses duration (left to right) [3]

Therefore, consideration of radiation, either irradiated material or environmental characteristic is important for specific applications.

As it is shown in Fig. 3, reducing the length of the irradiated pulse reduces the melting of the crater edges. That is, the influence of the femtosecond pulse is the most interesting from a technological point of view [4]. That is why a large number of scientific studies [5–14] are devoted to the study of femtosecond pulses.

2 Features of the Boundary Conditions for the Problem of Surface Destruction by a Laser Pulse

In this research, the interaction of powerful pulses of femtosecond duration with the solid surface is theoretically investigated. The condition that provides a crater formation without melting edges, as we received earlier, has the form: $q > \frac{\rho_0 R_g T_c c}{\mu} \cdot \frac{1}{1-R^2}$. Here q—power of electromagnetic wave falling on the matter surface, R_g—the universal gas constant, T_c—the critical temperature which the surface region reached under irradiation as the phase state changes, μ—the molecular weight, ρ—the density of the irradiated material, R—light reflectivity.

On the basis of an inhomogeneous system of the continuous medium equations, boundary conditions at the interface between two phases, which coexist during a crater formation on this surface were formulated [13]:

- condition of the mass flow balance

$$\rho_s\,(\boldsymbol{v}_s \cdot \boldsymbol{n}) - \rho_{0s}(\boldsymbol{v}_{0s} \cdot \boldsymbol{n}) = 0, \tag{1}$$

- condition of the momentum flow balance

$$\rho_s v_s^i(\boldsymbol{n} \cdot \boldsymbol{v}_s) - P_{ij}^s n_j - \rho_{0s} v_{0s}^i(\boldsymbol{n} \cdot \boldsymbol{v}_{0s}) + P_{ij}^{(0)s} n_j = 0, \tag{2}$$

- the condition of the energy flow balance:

$$\rho_s(\boldsymbol{n} \cdot \boldsymbol{v}_s)\left(H_s + \frac{\left(v_s^{(\zeta)}\right)^2}{2} + \varphi_0\right) + L_d\left(\boldsymbol{n} \cdot \boldsymbol{q}_s\right) - L\lambda_s(\boldsymbol{n} \cdot \mathrm{grad}T)_s = 0. \tag{3}$$

Under the boundary conditions (1)–(3) ρ, $\boldsymbol{v}$ are a density and a convective velocity vector of a continuous medium (with index "s" of the gas phase, with index "0 s" of the solid phase), $v_s^{(\zeta)}$ is the normal component of the velocity vector, $\boldsymbol{q}_s$ is a superficial part of the flux that has reached the surface, $-\lambda_s \mathrm{grad}T$ defines a superficial part of the heat flow in the gas phase, L_d and L are generalized coefficients of losses in the transition of light and heat fluxes, respectively, through the boundary matter—gas, H_s is Gibbs heat function (enthalpy), φ_0 is the specific heat of the phase transition of condensate-gas, and P_{ij}^s, $P_{ij}^{(0)s}$ are the surface values of a stress tensor in the gas and solid phases.

All boundary conditions are formulated in the local reference system, located on the boundary between the condensed media and the gas phase in the area of the momentum action. It is clear that when the solid surface does not collapse, the reference system will remain stationarily relative to the observer. But when destruction occurs, then the chosen system in the area of this destruction is rigidly associated with a certain point of the surface of the solid medium, and moves along with it. That is, such systems describe the events directly at each individual point of the surface. So far as process observation passes in a laboratory system of coordinates, it is necessary to establish a relation between these systems. The laboratory system of coordinates is immobile relative to the surface before and after its destruction, as well as, for example, relative to the point of center of a cross section of the stimulating radiation flux.

3 The Relationship Between the Coordinates of the Local and Laboratory Reference Systems

To establish a connection between the two reference systems, we consider Fig. 4, which geometrically explains the problem.

We arrange the laboratory reference system in such a way that its origin is at point O. Then the surface of the crater, which is formed under the influence of the flow, will be determined by some surface that changes in time. In Fig. 4, this surface is determined by the radius vector $\boldsymbol{r}$. Obviously, $\boldsymbol{r}(t) = x(t)\boldsymbol{e}_x + y(t)\boldsymbol{e}_y + z(t)\boldsymbol{e}_z$, moreover, by the general definition of a surface: $z(t) = S(t, x(t), y(t))$, where $S(t, x(t), y(t))$ is the form function of this surface (for each moment of time, it determines a certain dependence of z on x and y). In the future, this function will be called the "crater form function." In Fig. 4, the radius vector $\boldsymbol{r}$ ends at the starting point of the vector $\boldsymbol{n}$. With this point, we will associate some local reference frame.

In order to establish a connection between a laboratory system with a center at point O and a local system with a center at the point, where vector $\boldsymbol{n}$ begins, it is first necessary to determine the relationship between the unit vectors of these systems. If the normal $\boldsymbol{n}$ to the surface is identified with the unit vector $\boldsymbol{e}_\zeta$ of the local reference frame, that is, if put $\boldsymbol{n} \equiv \boldsymbol{e}_\zeta$, then by definition [15] this normal in the laboratory reference frame will be determined by the relation $\boldsymbol{n} \equiv \boldsymbol{e}_\zeta = \boldsymbol{N}/|\boldsymbol{N}|$, where

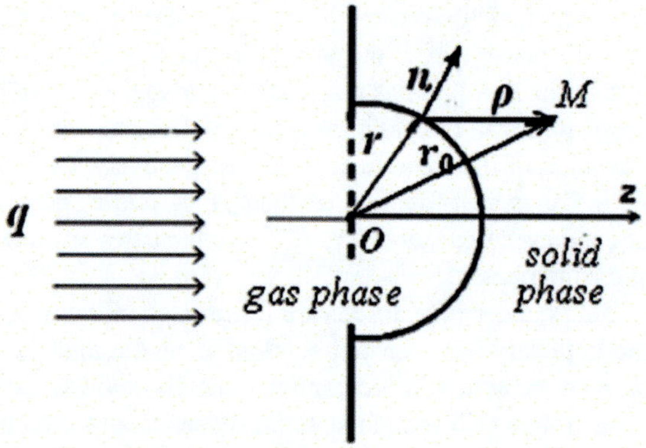

Fig. 4 Illustration of the relationship between local and laboratory coordinate systems

$$N = \begin{vmatrix} e_x & e_y & e_z \\ x_x & y_x & z_x \\ x_y & y_y & z_y \end{vmatrix} = \begin{vmatrix} e_x & e_y & e_z \\ 1 & 0 & S_x \\ 0 & 1 & S_y \end{vmatrix} = -e_x S_x - e_y S_y + e_z.$$

The notation x_x, x_y, y_x, y_y means the partial derivatives $\partial x/\partial x, \partial x/\partial y, \partial y/\partial x, \partial y/\partial y$. From here for the normal vector $\boldsymbol{n}$ in the coordinates of the laboratory system we get

$$\boldsymbol{n} \equiv e_\zeta = \frac{-e_x S_x - e_y S_y + e_z}{\sqrt{1 + S_x^2 + S_y^2}}. \tag{4}$$

The other two (tangential) unit vectors: e_ξ and e_η, should be placed in the plane perpendicular to the normal vector. Such can be the vectors [15]: $r_x = e_x + S_x e_z$ or $r_y = e_y + S_y e_z$. We will assume:

$$e_\xi = \frac{e_x + S_x e_z}{\sqrt{1 + S_x^2}}. \tag{5}$$

Then the property holds: $(e_\xi \cdot e_\zeta) = 0$, that is, the unit vectors e_ξ and e_ζ are perpendicular to each other. Using (4) and (5), based on the conditions that the unit vectors e_ξ, e_η, e_ζ must satisfy the relations: $e_\eta = [e_\zeta, e_\xi], e_\xi = [e_\eta, e_\zeta], , e_\zeta = [e_\xi, e_\eta]$, and the scalar product of two more pairs of vectors must be zero: $(e_\xi \cdot e_\eta) = (e_\zeta \cdot e_\eta) = 0$, one can find:

$$e_\eta = \frac{-S_x S_y e_x - (1 + S_x^2) e_y + S_y e_z}{\sqrt{1 + S_x^2 + S_y^2} \cdot \sqrt{1 + S_x^2}} \tag{6}$$

Definitions (4)–(6) provide the full connection between two coordinate systems: a local one, moving along with the border, and a laboratory one, which has a beginning at point O, and makes it possible to observe the dynamics of processes.

If direct relations are known, then it is easy to obtain inverse ones:

$$\begin{aligned} e_x &= -\frac{S_x}{N} e_\zeta - \frac{S_x S_y}{NT} e_\eta + \frac{1}{T} e_\xi, \\ e_y &= -\frac{S_y}{N} e_\zeta + \frac{T}{N} e_\eta, \\ e_z &= \frac{1}{N} e_\zeta - \frac{S_y}{NT} e_\eta + \frac{S_x}{T} e_\xi, \end{aligned} \tag{7}$$

де $N \equiv \sqrt{1 + S_x^2 + S_y^2}$, and $T \equiv \sqrt{1 + S_x^2}$.

Any vector can be decomposed into components of both the local reference system and the laboratory system. Now, we will consider only the vector of convective velocity. Obviously, some vector $\boldsymbol{v}$ can be represented in the form: $\boldsymbol{v} = v^{(\alpha)}\boldsymbol{e}_\alpha$, where $\alpha = \xi, \eta, \zeta$—for the local system or in the form: $\boldsymbol{v} = v^{(i)}\boldsymbol{e}_i$, where $i = x, y, z$—for the laboratory system. If, for example, we know the components $v^{(\alpha)}$ of this vector in the local system, then we can replace in the representation $\boldsymbol{v} = v^{(\xi)}\boldsymbol{e}_\xi + v^{(\eta)}\boldsymbol{e}_\eta + v^{(\zeta)}\boldsymbol{e}_\zeta$ unit vectors $\boldsymbol{e}_\xi, \boldsymbol{e}_\eta, \boldsymbol{e}_\zeta$ using relations (4)–(6). Then, grouping this to a form $\boldsymbol{v} = v^{(x)}\boldsymbol{e}_x + v^{(y)}\boldsymbol{e}_y + v^{(z)}\boldsymbol{e}_z$, we can find the components of this vector in the laboratory frame of reference:

$$\begin{aligned} v^{(x)} &= \frac{1}{T}v^{(\xi)} - \frac{S_x S_y}{NT}v^{(\eta)} - \frac{S_x}{N}v^{(\zeta)}, \\ v^{(y)} &= \frac{T}{N}v^{(\eta)} - \frac{S_y}{N}v^{(\zeta)}, \\ v^{(z)} &= \frac{S_x}{T}v^{(\xi)} - \frac{S_y}{NT}v^{(\eta)} + \frac{1}{N}v^{(\zeta)}. \end{aligned} \tag{8}$$

Inverse relations can also be obtained by solving this system with respect to the components $v^{(\alpha)}$ ($\alpha = \xi, \eta, \zeta$):

$$\begin{aligned} v^{(\xi)} &= \frac{v^{(x)} + S_x v^{(z)}}{T}; \\ v^{(\eta)} &= \frac{-S_x S_y v^{(x)} + T^2 v^{(y)} - S_y v^{(z)}}{NT}; \\ v^{(\zeta)} &= \frac{-S_x v^{(x)} - S_y v^{(y)} + v^{(z)}}{N}. \end{aligned} \tag{9}$$

Obviously, the components of the vector $v^{(\alpha)}$ in the local frame of reference depend on the variables ξ, η, ζ, while in the laboratory frame they depend on x, y, z. Therefore, in order to compare these two systems, it is necessary to supplement the transformations for velocities with the transformations for coordinates. To do this, let us turn back to Fig. 4.

Let us choose some arbitrary observation point M, which in Fig. 4 is located in the middle of the solid medium for convenience of the image (it is clear that it can also be in the gas part) and does not depend on the surface processes that are considered. Its position relative to the laboratory reference system will be described by the vector $\boldsymbol{r}_0 = x_0\boldsymbol{e}_x + y_0\boldsymbol{e}_y + z_0\boldsymbol{e}_z$. On the other hand, a vector describing the position of the same point relative to the local reference system located at point N will have the form:

$$\boldsymbol{\rho} = \xi_0\boldsymbol{e}_\xi + \eta_0\boldsymbol{e}_\eta + \zeta_0\boldsymbol{e}_\zeta. \tag{10}$$

Since, as can be seen from Fig. 4, $\boldsymbol{\rho} = \boldsymbol{r}_0 - \boldsymbol{r}$, we can obtain the values of the vector $\boldsymbol{\rho}$ in the coordinates of the laboratory reference frame:

$$\boldsymbol{\rho}(t) = \boldsymbol{e}_x|x_0 - x| + \boldsymbol{e}_y|y_0 - y| + \boldsymbol{e}_z|z_0 - z|. \tag{11}$$

Let us introduce the following notation: $x_0 - x \equiv X$; $y_0 - y \equiv Y$; $z_0 - z \equiv Z$. To obtain the relationship between the coordinates of the two systems, we use both of the above definitions of the vector $\boldsymbol{\rho}$. Substituting into the expression (11), the values $\boldsymbol{e}_x, \boldsymbol{e}_y, \boldsymbol{e}_z$ from (7) and comparing that is obtained with (10), one can get

$$\begin{cases} \xi_0 = \frac{1}{T}X + \frac{S_x}{T}Z, \\ \eta_0 = -\frac{S_x S_y}{NT}X + \frac{T}{N}Y + \frac{S_y}{NT}Z, \\ \zeta_0 = -\frac{S_x}{N}X - \frac{S_y}{N}Y + \frac{1}{NT}Z. \end{cases} \tag{12}$$

Now we return to the relations (8). In it, the components $v^{(\xi)}$, $v^{(\eta)}$, $v^{(\zeta)}$ depend on ξ_0, η_0, ζ_0. We can replace in these components the variables ξ_0, η_0, ζ_0 in accordance with (12). Then the components of the laboratory system $\boldsymbol{v} = v^{(i)}\boldsymbol{e}_i$ will depend both on the coordinates of the observation point $\boldsymbol{r}_0$ and on the coordinates of some local system $\boldsymbol{r}$. Since there are many such local systems on the surface, then in order to find the resulting value of the velocity vector at the observation point $\boldsymbol{r}_0$ at a certain point in time in the laboratory reference frame, it is necessary to carry out continuous summation over all surface points (i.e., over the variable $\boldsymbol{r}$), or rather integration over the entire surface with a weight factor equal to the area of this surface at the same moment of time.

The obtained relations allow us to go to the laboratory reference system, since in it all the processes associated with destructive processing can be observed. Theoretical calculations are more convenient to do in the local reference system.

We also note that assuming $\boldsymbol{r} = \boldsymbol{r}_0$, and accordingly $\boldsymbol{\rho} = 0$, we can obtain all the above relations for the velocities at the condensate-gas interface. Relation (12) in this case turns into identities of the form $0 = 0$.

4 Crater Dynamics Equations

To formulate the equation for the crater form function in the laboratory coordinate system, we use the mass flow balance condition (1). This condition in the local reference system can be written as

$$\rho_s v_s^{(\zeta)} = \rho_{0s} v_{0s}^{(\zeta)}. \tag{13}$$

The surface values of the velocity $v_s^{(\zeta)}$ and the density of the gas phase are determined [13] by the relations: $v_s^{(\zeta)} = -\frac{P_s^\sigma}{\sigma\sqrt{\kappa a}}$ and $\rho_s = aP^\eta$, based on the conditions of the balance of the flow of energy (3) and the conditions of the balance of the flow of pulses (2). In these definitions, P_s is the surface pressure at the phase boundary, κ is the polytropic exponent, which varies from 1 to 5/3, $\eta \equiv 1/\kappa$, $\sigma \equiv (\kappa - 1)/2\kappa$. In addition, P_s depends on time t and x, y coordinates, if we consider them in a laboratory frame of reference. That is, the left side of (13) is already formulated in the variables of the laboratory system. As for the right-hand side of (13), it should be taken into account that from the point of view of the local reference system, the velocity v_{0s} looks like the speed of "flowing through" of the solid phase via the interface. Thus, its component $v_{0s}^{(\zeta)}$, as well as the gas component $v_{s}^{(\zeta)}$, are negative. In this case, the transformations (9) for $v_{0s}^{(\zeta)}$ can be used immediately in the form:

$$v_{0s}^{(\zeta)} = \frac{S_x}{N} v_{0s}^{(x)} + \frac{S_y}{N} v_{0s}^{(y)} - \frac{1}{N} v_{0s}^{(z)}, \tag{14}$$

where $v_{0s}^{(i)}$ are the components of the velocity vector of surface points.

Since $z(t) = S(t, x(t), y(t))$, then the velocity component $v_{0s}^{(z)}$ is equal to

$$\frac{dz}{dt} = v_{0s}^{(z)} = S_t + S_x \frac{dx}{dt} + S_y \frac{dy}{dt} = S_t + S_x v_{0s}^{(x)} + S_y v_{0s}^{(y)}.$$

Substituting the result obtained in (14) we have

$$v_{0s}^{(\zeta)} = -S_t / N. \tag{15}$$

Substituting this into (13) and taking into account $v_s^{(\zeta)} = -\frac{P_s^{\sigma}}{\sigma\sqrt{\kappa a}}$ and $\rho_s = a P^{\eta}$, as well as the explicit form of the coefficient N defined after formula (7), we obtain

$$S_t = \frac{2\sqrt{a\kappa}}{\gamma \rho_{0s}} P_s^{\beta} \sqrt{1 + S_x^2 + S_y^2} \tag{16}$$

or in dimensionless form:

$$\frac{\partial \Sigma}{\partial \theta} = \Pi^{\beta} \sqrt{1 + \left(\frac{\partial \Sigma}{\partial X}\right)^2 + \left(\frac{\partial \Sigma}{\partial Y}\right)^2}. \tag{17}$$

Here $\gamma \equiv \kappa - 1$, $\beta \equiv \frac{\kappa+1}{2\kappa}$, $\Pi = P/P_0$ is the dimensionless pressure, $\theta = t/t_0$ is the dimensionless time, $\Sigma = S/S_0$ is the dimensionless function of the crater form and dimensionless coordinates $X = x/S_0$, $Y = y/S_0$. During the dimensionless procedure, it was found that

$$S_0 = \frac{b\gamma^4 L}{\kappa^{5/2} \rho_{0s} \alpha^2 L_d}, \quad P_0 = \frac{\gamma^2 L_d q_{in}}{2\alpha^{1/2} \kappa \sqrt{\varphi_0}}$$

$$t_0 = \frac{b\gamma^4 \varphi_0 L}{q_{in} \alpha^2 \kappa^{5/2} L_d}, \quad a = \frac{\kappa \alpha}{\varphi_0 \gamma^2} P_0^{2\sigma}$$

where $\alpha \equiv \kappa + 1$, and parameter b, which includes important characteristics of the vaporized substance, such as the R_g gas constant, μ_a is the molecular mass of the gas, d_a is the effective diameter of the molecule, m_a is the mass of the atom of the vaporized substance, k_B is the constant Boltzmann, and has the form:

$$b \equiv \frac{2R_g m_a^2}{3\pi\sqrt{\pi}\gamma\mu_a d_a^2 k_B}.$$

Equation (16) and (17) are differential equations of the shape of the crater, in dimensional and dimensionless form. The equation includes the parameter P, or Π, and requires its determination, since it changes during the irradiation of a solid surface, and a consequence of its change is the formation of a crater on this surface. The change of pressure and its dynamics is determined by a nonlinear differential equation [13]. That is, the solution to the problem of surface destruction by a laser pulse is a system of partial differential equations. Asymptotic estimates of the pressure behavior and the crater shape function [13] confirmed the consistency of the presented model.

The next step in checking the model was a partial numerical calculation. The equation of crater dynamics was presented in a cylindrical frame of reference:

$$\frac{\partial\Sigma}{\partial\theta} = \Pi\sqrt{1 + \left(\frac{\partial\Sigma}{\partial R}\right)^2}. \tag{18}$$

The boundary conditions for this equation are obvious $\Sigma(0, R) = 0$, $\Sigma(\theta, \infty) = 0$. A typical model pressure form caused by a laser pulse was specified:

$$\Pi = \frac{1 - \exp(-\theta^n)}{1 + \exp(10(\theta - \tau))}\exp\left(\frac{R^2}{R_0^2}\right),$$

where τ is the laser impulse length, R_0 is beam width. The impulse lengths had femtosecond values at the values of the radiation flux $q \sim 10\,\mathrm{TW/cm^2}$. The function of the crater shape $\Sigma(\theta, R)$) was calculated in nanometers.

Figure 5 shows the crater profiles for different time points. The origin coincides with the center of the laser impulse section.

As can be seen from the figure, the formation of a crater on the surface is clearly followed.

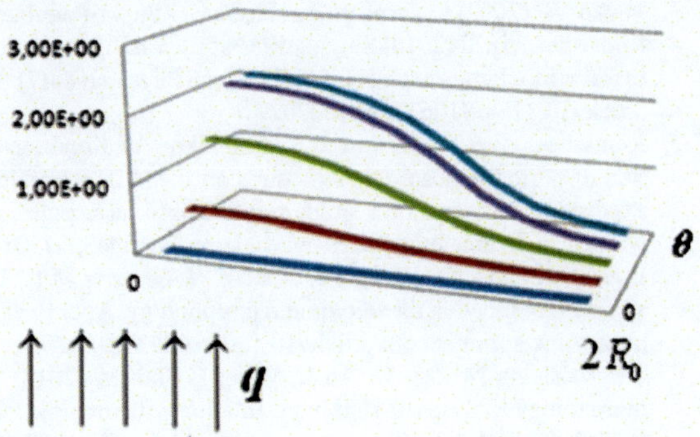

Fig. 5 Dynamics of crater development over time

5 Conclusions

The work describes in detail the procedure that allows us to establish a relationship between a local coordinate system associated with an arbitrary point on the moving surface of a laser-stimulated crater; and a laboratory coordinate system associated, generally speaking, with an arbitrary point on an immovable part of the surface (which does not destruction). This connection allows one to determine the velocities of the movement of surface points and points of material, and also the coordinates of any point of material both in the laboratory coordinate system and in local systems associated with a specific surface point. The relationship between the local and laboratory systems makes it possible to relate the results of theoretical calculations that are convenient to perform in the local system, with experimental observations that are performed in the laboratory system. Based on these transformations, a partial differential equation is formulated in the work, which allows one to observe the development of a nanocrater on an irradiated surface. A numerical simulation of the development of a nanocrater was performed for a set typical model form of pressure dynamics stimulated by an active laser pulse. It was confirmed that the formation of a crater on the surface of a solid substance begins during the action of a pulse and stops almost immediately after its completion. The test calculations carried out in the study correspond to experimental observation and theoretical ideas about the development of a corrosion crater under pulsed laser irradiation.

References

1. Gerke MN, Khorkov KS, Telushko OB, Bolshakova ON, Prokoshev VG, Arakelian SM (2010) Formation of nanostructures at laser ablation under the action of ultrashort laser impulses on a surface of solid states. Phys Proc 5(A):213–219. https://doi.org/10.1016/j.phpro.2010.08.139. https://www.sciencedirect.com/science/article/pii/S1875389210005651
2. Romashevskiy SA, Ashitkov SI, Dmitriev AS (2016) Formation of ordered nano- and mesostructures in silicon irradiated with a single femtosecond laser pulse in different environments. Tech Phys Lett 42(8):810–813. https://link.springer.com/article/10.1134/S1063785016080150
3. Veyko VP (2007) Lazerna mikroobrobka - - Sankt-Peterburh: SPbGU ITMO, 111 pp.
4. Romashevskiy SI (2018) Singularities of silicon surface nanostructuring due to ultrafast heating in water by a femtosecond laser pulse. Tech Phys Lett 44(7):630–633. https://link.springer.com/article/10.1134/S1063785018070271
5. Romashevskiy SA, Ashitkov SI, Ovchinnikov AV, Kondratenko PS, Agranat MB (2016) Formation of periodic mesoscale structures arranged in a circular symmetry at the silicon surface exposed to radiation of a single femtosecond laser pulse, Appl Surf Sci 374:12–18. https://www.sciencedirect.com/science/article/pii/S0169433215016797
6. Bonse J, Baudach S, Kruger J, Kautek W, Lenzner M (2002) Femtosecond laser ablation of silicon-modification thresholds and morphology. Appl Phys A Mater Sci Process 74(1):19–25. https://link.springer.com/article/10.1007/s003390100893
7. Wu Z, Zhang N, Zhu X, An L, Wang G, Tan M (2018) Time-resolved shadowgraphs and morphology analyses of aluminum ablation with multiple femtosecond laser pulses. Chin Phys B 27(7):077901. http://iopscience.iop.org/article/10.1088/1674-1056/27/7/077901/meta

8. Mann T, Mathieson R, Murray M, Richards B, Jose G (2018) Femtosecond laser ablation properties of Er^{3+} ion doped zinc-sodium tellurite glass. J Appl Phys 124(4):044903. https://aip.scitation.org/doi/abs/10.1063/1.5040947
9. Zhang N, Yang J, Zhu X (2012) Investigation of the ultrafast process of femtosecond laser ablation of highly oriented pyrolytic graphite. Chin J Lasers 39(5):0503002 http://en.cnki.com.cn/Article_en/CJFDTOTAL-JJZZ201205018.htm
10. Berezovska N, Dmitruk I, Kalyuzhnyy A, Dmytruk A, Blonskyi I (2018) Self-organized structuring of surface of metal-semiconductor composite by femtosecond laser processing. Ukrainian J Phys 63(5):406–412
11. Berezovska N, Dmitruk I, Vovdenko S, Yeshchenko O, Teselko P, Dmytruk A, Blonskyi I (2019) Sub-micron and nano sized features in laser-induced periodic surface structures. Ind J Phys 93(4):495–502. https://link.springer.com/article/10.1007/s12648-018-1323-0
12. Suprun AD, Shmeleva LV, Razumova MA (2011) The influence of bulk absorption of substance on the threshold of destruction by the intensive pulse of electromagnetic radiation. Funct Mater 18(2):237–243. http://functmaterials.org.ua/contents/18-2/
13. Shmeleva LV, Suprun AD, Yezhov SM (2019) Simulation of the formation of a surface nanocrater under the action of high-power pulsed radiation—nanocomposites, nanostructures, and their applications part of the Springer proceedings in physics book series, vol 221, no 34, pp 505–515 https://link.springer.com/chapter/10.1007%2F978-3-030-17759-1_34
14. Pavlyniuk OR, Datsyuk VV (2016) Electrostrictive mechanism of nanostructure formation at solid surfaces irradiated by femtosecond laser pulses. Nanoscale Res Lett 11(1):1–6. http://nanoscaleresletts.springeropen.com/articles/10.1186/s11671%2D015%2D1224%2D5, https://doi.org/10.1186/s11671-015-1224-5
15. Korn GA, Korn TM (2013) Mathematical handbook for scientists and engineers: definitions, theorems, and formulas for reference and review. Courier Corporation, pp 1132

Specificity of Boundary Conditions for Laser-Stimulated Destructive Surface Treatment Without Melting

L. V. Shmeleva, A. D. Suprun, S. M. Yezhov, and V. V. Datsyuk

1 Introduction

The femtosecond laser technique is an important instrument of fundamental researches of nonequilibrium processes in the condensed environments, actively used in modern laser technologies for high-precision materials processing, as well as for the creation of new functional nanomaterials. Unlike the typical evaporation process for influence on metals and semiconductors of laser pulses of microsecond and nanosecond duration, in the case of femtosecond pulses, the main mechanism is thermomechanical ablation, which leads to the removal of the surface nanolayer as a result of the appearance of a powerful substance tensile stresses [1]. For laser pulses, the duration of which is less than the relaxation time of thermally elastic loads or the speed of sound in the substance being heated, and the density energy above the ablation threshold, stress concentration can lead to a pressure higher than gigapascal level [2]. Such a powerful action on the surface of the irradiated substances leads to damages in the form of microcraters that is demonstrated in Fig. 1. The study of nanostructures arising on the surface of a solid under the influence of destructive radiation is given considerable attention [4–10].

According to experimental estimations characteristic level of energy, from which active destruction of surface begins practically without formation of liquid phase exceeds 0.5–20 PW/cm^2 [2] depending on the material. If the flow is substantially higher than this value, a considerable part of energy of laser radiation is outlaid on a direct phase solid–gas transition. A liquid phase in the area of treatment is practically absent in this case. Such streams represent the main interest of this research.

L. V. Shmeleva (✉) · A. D. Suprun · S. M. Yezhov · V. V. Datsyuk
Kiev 01601, Ukraine
e-mail: lshmel@univ.kiev.ua

O. Fesenko and L. Yatsenko (eds.), *Nanomaterials and Nanocomposites, Nanostructure Surfaces, and Their Applications*, Springer Proceedings in Physics 246, https://doi.org/10.1007/978-3-030-51905-6_23

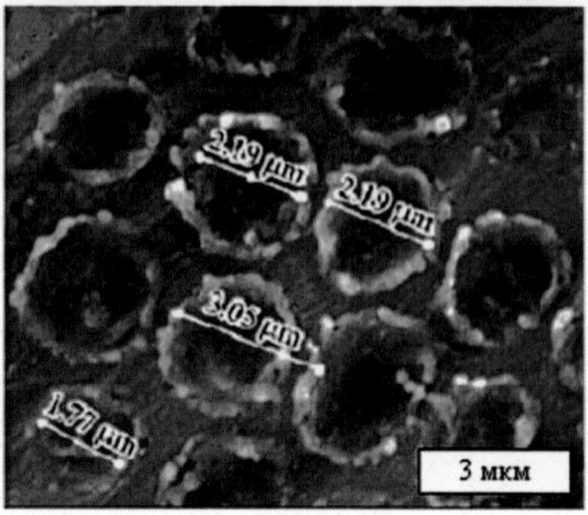

Fig. 1 Microcraters obtained by femtosecond laser titanium irradiation [3]

2 Analysis of the Considered Processes in Terms of the Equation of State of Matter

To define limit, which separates treatment with formation of the liquid state and treatment which is not accompanied with surface melting, we will consider the superficial layer of matter on an area with the size equal to the cross section of laser impulse. To prevent liquid phase appearing on the solid surface, its temperature during impulse action has to amount, at least, critical value T_c, without going out the borders of the matter state phase surface which belongs exceptionally to the solid state. Thus pressure in this near-surface area of solid state must also exceed some critical value P_c. As a laser impulse is considered, a necessary condition at which the system gets in the noted area is $P_c < q_s(1 + R)/c$, where q_s—power of electromagnetic wave falling on the matter surface, R—light reflectivity, and c—light speed.

Taking into account that on the critical isotherm of gas phase for arbitrary pressure P and volume V_Γ, the state equation $PV_\Gamma = \nu R_\Gamma T_c$ is holding, where ν is a molar gas amount, and it is possible to estimate the value of pressure P_c, which provides implementation of the above condition. This value can be estimated from the next considering. In order that a liquid phase did not appear, a hard phase after a phase transition must get on an isotherm which is with a temperature not below than critical isotherm with a temperature T_c. Examining a critical isotherm, we will get the sought threshold value P_c for which gas state equation will acquire a kind $PV_\Gamma^{(c)} = \nu R_\Gamma T_c$. In this equation remains indefinite volume $V_\Gamma^{(c)}$ so far. For its estimation, it is possible to take advantage of the fact that in the moment of phase state transition from solid to gas, the change of volume $V_T \rightarrow V_\Gamma$ is not so substantial comparing to the similar change for other points of isotherm T_c (where this change is large). That is why a volume $V_\Gamma^{(c)}$ can be estimated counting it practically equal to the proper volume of hard phase $V_T^{(c)}$. As one mol volume of hard matter V_μ is determined by ratio: $V_\mu = \mu/\rho_0$, where μ is molecular mass and ρ_0 is density, the volume of ν molls is obviously equal to $V_T^{(c)} = \nu\mu/\rho_0$. Using this value of volume in state equation

$P_c V_\Gamma^{(c)} = \nu R_\Gamma T_c$ in place of $V_\Gamma^{(c)}$ we will get an estimation: $P_c = \rho_0 R_\Gamma T_c/\mu$, and a sufficient condition for a flow q_s will look in this case like

$$q_s > \frac{\rho_0 R_\Gamma T_c c}{\mu} \cdot \frac{1}{1+R}.$$

In practice, such estimation is interesting for an initial flow q, which is related to the flow q_s which passed in a solid by correlation $q_s = (1-R)q$. Then for the initial value of flow, we get

$$q > \frac{\rho_0 R_\Gamma T_c c}{\mu} \cdot \frac{1}{1-R^2}.$$

After this equation, estimations were made for concrete materials. They show that, for example, for aluminum the superficial value of stream must meet condition $q \geq \frac{6\times10^{13}}{1-R^2}$ W/cm^2, for copper and chrome $q \geq \frac{9\times10^{13}}{1-R^2}$ W/cm^2 for tungsten $q \geq \frac{1.5\times10^{14}}{1-R^2}$ W/cm^2, for titan $q \geq \frac{7.6\times10^{13}}{1-R^2}$ W/cm^2, and for potassium $q \geq \frac{4\times10^{12}}{1-R^2}$ W/cm^2. As evidently, such estimations do not contradict experimental information [11]. Consequently, here we will name flows intensive, if they meet condition got before.

3 Dynamics of Formation of Processes Caused by Intense Laser Irradiation on the Surface of the Substance

Traditionally, the problems of the destructive and non-destructive interaction of a powerful impulse in the flow of energy from the surface are regarded as different. However, assuming that each of the coexisting phases is already a continuous medium, the system of equations, describing them, will be common to both, namely [12]

$$\frac{\partial \rho}{\partial t} + \mathrm{div}(\rho v) = R_m, \tag{1}$$

$$\frac{dv^i}{dt} + (v \cdot \nabla)v^i - \frac{1}{\rho}\frac{\partial P_{ij}}{\partial x_j} = f^i, \tag{2}$$

$$\rho\frac{dU}{dt} + \rho(v \cdot \nabla)U + \mathrm{div}Q = P_{ij}v_{ij}. \tag{3}$$

In this system, U, ρ, v are, respectively, intrinsic energy of the continuous medium mass unit, density of this medium, and a convection velocity vector, Q is general energy flow in the medium, P_{ij} is a stress mechanical tensor pressure, $v_{ij} = \frac{\partial v_i}{\partial x_j}$ is strain velocity tensor, $\{i, j\} = \{1, 2, 3\}$.

System (1)–(3) is heterogeneous due to the presence of bulk sources and drains in the gas phase at the phase transition of the solid into gas. To account for them in (1) the function of sources—drains of mass R_m is introduced, which has dimension of mass of unit of volume per unit time [12], and in (2) the mass density of force f acting on the flow gaseous substance from the radiation side [13] has an acceleration dimension. The function R_m and the force f are given by a specific problem statement.

Since system (1)–(3) is a system of differential equations that simultaneously describe both phases—solid and gas, then to ensure the uniqueness of its solutions' initial and boundary conditions should be formulated at the interface of these phases. We state general boundary conditions of destruction and non-destruction of the surface.

The first boundary condition—the condition of the balance of mass flow across the boundary is obtained by (1).

To do this, we will integrate the continuity equation over the volume that covers the surface phase section. Then, we tighten the volume of integration to the surface so that, perpendicular to the surface, this volume became infinitely narrow. Local is infinite the narrow volume of matter that crosses the surface can be considered constant over time. The factor R_m has a physical meaning only in volume. The term $\iiint_V \operatorname{div}(\rho v)\mathrm{d}V$ remains important. Using the Gauss–Ostrogradsky theorem, we can obtain the first boundary condition in integral form:

$$\iiint\limits_V \operatorname{div}(\rho \boldsymbol{v})\mathrm{d}V = \oint\!\!\!\oint\limits_S (\rho_s \boldsymbol{v}_s \cdot \boldsymbol{n}) = 0..$$

By integrating over a closed surface such that normal $\boldsymbol{n}$ is external in relation to the volume covered by this surface, and considering that in each case each point of the interface will have opposite directions in the middle of the gas and solid phases where we will have [14]:

$$\rho_s(\boldsymbol{v}_s \cdot \boldsymbol{n}) - \rho_{0s}(\boldsymbol{v}_{0s} \cdot \boldsymbol{n}) = 0. \tag{4}$$

Here ρ_s v_s are the surface values of density and convective velocity in the gas phase, ρ_{0s}, v_{0s} are the density and rate of leakage inflow to the interface phases. The relation (4) is formulated in a local reference system moving with the surface phase section. If the surface of the solid does not collapse under the action of radiation, then the gas characteristics in (4) are absent. Then (4) is simplified to the form: $(\boldsymbol{v}_{0s} \cdot \boldsymbol{n}) = 0$.

The following is the impulse flow balance condition. It is obtained from the second equation of the system (1)–(3). Motion (2) multiplied by density ρ, and continuity (1), multiplied by the velocity component of the substance v^i, add up. Such a combination of equations leads to

$$\frac{\partial(\rho v^i)}{\partial t} - \frac{\partial P_{ij}}{\partial x_j} + \operatorname{div}(\rho v \cdot v^i) = \rho f^i + R_m \cdot v^i.$$

By integrating over a closed surface around the phase boundary, and taking into account the opposite of the normals $\boldsymbol{n}$ to the distribution surface of both media, we obtain

$$\rho_s v_s^i (\boldsymbol{n} \cdot \boldsymbol{v}_s) - P_{ij}^s n_j - \rho_{0s} v_{0s}^i (\boldsymbol{n} \cdot \boldsymbol{v}_{0s}) + P_{ij}^{(0)s} n_j = 0. \tag{5}$$

Here P_{ij}^s, $P_{ij}^{(0)s}$ are the surface values of the stress tensor in the gas and solid phases, v_s^i, v_{0s}^i are the velocity vector components of the gas and condensed media at the boundary of their distribution.

The stress tensor for the gas phase, which is infinitely close to the surface, submits as follows:

$$P_{ij}^s = -P_s \delta_{ij} - \frac{1}{c} (\boldsymbol{n} \cdot \boldsymbol{q}_s) \delta_{ij}, \tag{6}$$

where the first appendix describes the elastic portion of the stress tensor for gases [15] (gas pressure), and the second additive corresponds to the additional light pressure on the surface of the substance. In the above expression c is the speed of light, q_s is the external flux of radiation, and δ_{ij} is the symbol of Kronecker. We obtain from (5) with (6):

$$P_s n_i + \frac{1}{c} (\boldsymbol{n} \cdot \boldsymbol{q}_s) n_i + \rho_s (\boldsymbol{n} \cdot \boldsymbol{v}_s) v_s^i - \rho_{0s} (\boldsymbol{n} \cdot \boldsymbol{v}_{0s}) v_{0s}^i + P_{ij}^{(0)s} n_j = 0. \tag{7}$$

If the external flow of q_s is not strong enough to cause surface fracture, then condition (7) is simplified by the fact that $\boldsymbol{v}_{0s} \equiv 0$, $P_s \equiv 0$, $\rho_s \equiv 0$, $\boldsymbol{v}_s \equiv 0$, that is, all additions are not related to destruction. In this case, the impulse flow balance condition is transformed to one that characterizes the elastic response of the condensed medium to the actual mechanical shock flow action q

$$\frac{1}{c} (\boldsymbol{n} \cdot \boldsymbol{q}_s) n_i + P_{ij}^{(0)s} n_j = 0. \tag{8}$$

Condition (8), which is a partial case of condition (7), is a boundary condition for processes, which are related to the propagation of sound, including intense, in a continuous environment.

In the general case, when the processes of destruction of a solid surface are present, on the contrary: (7), which have gas characteristics and were absent without destruction, are beginning to dominate. Therefore, in the presence of fracture, all components of the obtained equation should remain. In view of (4), the boundary condition (7) will be

$$P_s n_i + \frac{1}{c} (\boldsymbol{n} \cdot \boldsymbol{q}_s) n_i + \rho_s (\boldsymbol{n} \cdot \boldsymbol{v}_s)(v_s^i - v_{0s}^i) + P_{ij}^{(0)s} n_j = 0. \tag{9}$$

Supplements $P_{ij}^{(0)s} n_j$ are small relative to other additions in (9) in the presence of fracture, and, in some approximation, can be neglected in the presence of fracture.

Equation (9) is a boundary condition for the balance of the impulse flow for the problem of the influence of the powerful pulsed radiation to the surface of the medium in the presence of surface destruction. It is formulated in a local coordinate system associated with the interface. For a non-destructive process, this surface coincides with the standard laboratory system, since it is motionless. In the case when there is destruction, the system (9) is "attached" to the surface at each local point and moves with it. Let the local frame of reference be characterized by the unit vectors $\boldsymbol{e}_\xi, \boldsymbol{e}_\eta, \boldsymbol{e}_\zeta$,, where the unit vector $\boldsymbol{e}_\zeta$ is the unit vector normal to the surface, and the unit vectors $\boldsymbol{e}_\xi$ and $\boldsymbol{e}_\eta$ lie in the plane tangent to the surface at the point from which the normal vector $\boldsymbol{n} \equiv \boldsymbol{e}_\zeta$ emerges. Due to this, in each local reference frame of the projections $\boldsymbol{e}_\xi$ and $\boldsymbol{e}_\eta$ are equal to zero, and the projection $\boldsymbol{e}_\zeta = 1$. Then, in explicit form, (9) are reduced to the system:

$$P_s + \frac{1}{c} q_s^{(\zeta)} + \rho_s v_s^{(\zeta)} \left(v_s^{(\zeta)} - v_{0s}^{(\zeta)} \right) + P_{\zeta\zeta}^{(0)s} = 0,$$
$$\rho_s v_s^{(\zeta)} \left(v_s^{(\xi)} - v_{0s}^{(\xi)} \right) + P_{\xi\zeta}^{(0)s} = 0,$$
$$\rho_s v_s^{(\zeta)} \left(v_s^{(\eta)} - v_{0s}^{(\eta)} \right) + P_{\eta\zeta}^{(0)s} = 0.$$

If the solid is amorphous, or liquid, then the off-diagonal components of the tensor $P_{ij}^{(0)s}$ are equal to zero. In this case, the last two equations of the system are simplified to the condition of continuity of the tangential velocity components:

$$v_s^{(\xi)} - v_{0s}^{(\xi)} = 0, \quad v_s^{(\eta)} - v_{0s}^{(\eta)} = 0.$$

The energy flow balance condition is formulated on the basis of the equation for energy, the third equation of the system (1)–(3). To formulate it, we use the sum of equations: (3), (1) multiplied by U, and (2) multiplied by ρv^i, taking into account summation over i. Substituting into the obtained value of derivative ρ_t from the first equation of the original system, we obtain

$$\left(\rho \left[U + \frac{v^2}{2} \right] \right)_t + \operatorname{div} \left(\rho U \boldsymbol{v} + \boldsymbol{Q} + \rho \frac{v^2}{2} \boldsymbol{v} \right) - \frac{\partial \left(v^i P_{ij} \right)}{\partial x_j}$$
$$= \left(U + \frac{v^2}{2} \right) R_m + \rho v^i \boldsymbol{v} f^i.$$

Here, the subscript t means the partial derivative with respect to time.

This is the final expression, the structure of which allows the use of the same integrating procedure as in the first two cases. Using the condition (4), we obtain

$$\rho_s (\boldsymbol{n} \cdot \boldsymbol{v}_s) \left(U_s + \frac{v_s^2 - v_{0s}^2}{2} - U_{0s} \right) + (\boldsymbol{n} \cdot \boldsymbol{Q}_s) - (\boldsymbol{n} \cdot \boldsymbol{Q}_{0s})$$

$$-n_j v_s^i P_{ij}^s + n_j v_{0s}^i P_{ij}^{(0)s} = 0. \tag{10}$$

This shows that the kinetic energy that contributes to the equation from the side of the solid-state medium is determined by the difference in the velocities of the gas and solid phases. It is obvious that the gas velocity is significantly higher than the surface velocity, but this difference will remain until the final expression is obtained.

To exclude summands from (10) that contain the tensors P_{ij}^s and $P_{ij}^{(0)s}$, we use conditions (9), (6), whence

$$n_j v_s^i P_{ij}^s = -P_s(\boldsymbol{n} \cdot \boldsymbol{v}_s) - \frac{(\boldsymbol{n} \cdot \boldsymbol{v}_s)}{c}(\boldsymbol{n} \cdot \boldsymbol{q}_s),$$

$$\begin{aligned} n_j v_{0s}^i P_{ij}^{(0)s} =& \frac{\rho_s}{\rho_{0s}} P_s(\boldsymbol{n} \cdot \boldsymbol{v}_s) - \frac{\rho_s}{\rho_{0s}} \frac{(\boldsymbol{n} \cdot \boldsymbol{v}_s)}{c}(\boldsymbol{n} \cdot \boldsymbol{q}_s) \\ & - \rho_s(\boldsymbol{n} \cdot \boldsymbol{v}_s) v_{0s}^i (v_s^i - v_{0s}^i). \end{aligned}$$

By using the definitions of tensors obtained above, we can bring (10) to the form:

$$\begin{aligned} &(\boldsymbol{n} \cdot \boldsymbol{v}_s)\left(\rho_s U_s + \rho_s \sum_i \frac{\left(v_s^i - v_{0s}^i\right)^2}{2} - \rho_s U_{0s} + \left(1 - \frac{\rho_s}{\rho_{0s}}\right) P_s\right) \\ &+ \left(1 - \frac{\rho_s}{\rho_{0s}}\right) \frac{(\boldsymbol{n} \cdot \boldsymbol{v}_s)}{c}(\boldsymbol{n} \cdot \boldsymbol{q}_s) + (\boldsymbol{n} \cdot \boldsymbol{Q}_s) - (\boldsymbol{n} \cdot \boldsymbol{Q}_{0s}) = 0. \end{aligned} \tag{11}$$

Parameters $\boldsymbol{Q}_s$, $\boldsymbol{Q}_{0s}$ are generalized energy flows, include to itself both external (light) flux and heat fluxes. That is, $\boldsymbol{Q}_{0s} = \boldsymbol{q}_{0s} - \lambda_{0s}(\text{grad}T_0)_s$, where $\boldsymbol{q}_{0s}$ is the surface part of the flow that passed into the condensed matter, $-\lambda_{0s}(\text{grad}T_0)_s \equiv \boldsymbol{q}_{0s}^T$ defines the surface part of the heat flux in a solid state, $\boldsymbol{Q}_s = \boldsymbol{q}_s - \lambda_s(\text{grad}T)_s$, where $\boldsymbol{q}_s$ is the part of the flow which reached to the surface, and $-\lambda_s(\text{grad}T)_s \equiv \boldsymbol{q}_s^T$ defines the surface part of the heat flow in the gas phase.

Like the previous conditions, the energy flow balance condition can be used for two different problems: a problem related to surface destruction and a problem that is not destructive. For the case when the destruction of the surface does not occur, that is, all gas characteristics are absent, only the last two summands of (11) will be nonzero. The interaction of the environment with the surface of the substance occurs regardless of the presence of destruction of surface. In the absence of destruction, the heat fluxes $\boldsymbol{q}_{0s}^T$ are the fluxes from the environment. They can be defined as $\boldsymbol{q}_{\nu s}^T \equiv -\lambda_{\nu s}(\text{grad}T_\nu)_s$, where $\lambda_{\nu s}$ is the thermal conductivity of air or gas, and in vacuum they must be neglected.

That is, a non-destructive process can be described by the following equation:

$$(\boldsymbol{n} \cdot \boldsymbol{q}_s) - (\boldsymbol{n} \cdot \boldsymbol{q}_{0s}) + \lambda_{0s}(\boldsymbol{n} \cdot \text{grad}T_0)_s = 0.$$

The flow $\boldsymbol{q}_s$, which hit the surface of a substance, during its passage thru surface, will change due to the excitation of various kinds of surface effects: excitation of surface waves and states, scattering, etc. [16, 17]. All these losses occur during the transition $\boldsymbol{q}_s \to \boldsymbol{q}_{0s}$, and therefore they can be characterized by the coefficient of not heat dissipative losses $0 \le L_{\mathrm{d}} \le 1$. Then, the flow $\boldsymbol{q}_{0s}$ penetrating into the solid medium can be associated with the flow $\boldsymbol{q}_s$, which reaches the surface, by the relation: $\boldsymbol{q}_{0s} = (1 - L_{\mathrm{d}})\boldsymbol{q}_s$. Thus, the general equation of energy flow balance for a non-destructive problem will have the form:

$$L_{\mathrm{d}}\left(\boldsymbol{n} \cdot \boldsymbol{q}_s\right) + \lambda_{0\mathrm{s}}(\boldsymbol{n} \cdot \mathrm{grad}T_0)_{\mathrm{s}} = 0. \tag{12}$$

Two limiting cases are considered. The first, when the losses on the surface of the substance affected by the energy flux are large, that is, $L_{\mathrm{d}} \sim 1$ (the case of metals and opaque materials). In this case, (12) takes the form:

$$\left(\boldsymbol{n} \cdot \boldsymbol{q}_{\mathrm{s}}\right) + \lambda_{0\mathrm{s}}(\boldsymbol{n} \cdot \mathrm{grad}T_0)_{\mathrm{s}} = 0.$$

The second case, when the light flux nearly has no loss when passing through the surface of the material, that is, $L_{\mathrm{d}} \to 0$, leads (12) to the form:

$$\lambda_{0\mathrm{s}}(\boldsymbol{n} \cdot \mathrm{grad}T_0)_{\mathrm{s}} = 0.$$

When the surface is destroyed, due to local phase transitions, a strongly heated plasma-gas medium is formed in the irradiated substance. In this case, the flux $\boldsymbol{Q}_s$ when passing through the "solid state–gas" interface will change, both due to dissipative processes such as scattering, reflection, and transmission of the electromagnetic flux, and due to the properties of the substance associated with the thermal characteristics of each of the two bordering environments. The heat flux will undergo changes on the surface of the substance, for example, due to surface atomic vibrations and radiation from a heated surface. It is convenient to take these changes into account in the form of some loss coefficient $0 \le L \le 1$. That is, the transition $\lambda_{0\mathrm{s}}(\boldsymbol{n} \cdot \mathrm{grad}T_0)_{\mathrm{s}} \to \lambda_{\mathrm{s}}(\boldsymbol{n} \cdot \mathrm{grad}T)_{\mathrm{s}}$ can be defined as follows: $\lambda_{\mathrm{s}}(\boldsymbol{n} \cdot \mathrm{grad}T)_{\mathrm{s}} = (1 - L)\lambda_{\mathrm{s}}(\boldsymbol{n} \cdot \mathrm{grad}T)_{\mathrm{s}}$. Thus, the energy flow balance (11) for processes accompanied by destruction will have the form:

$$(\boldsymbol{n} \cdot \boldsymbol{v}_{\mathrm{s}})\left(\rho_{\mathrm{s}}U_{\mathrm{s}} + \rho_{\mathrm{s}}\sum_i \frac{\left(v_{\mathrm{s}}^i - v_{0\mathrm{s}}^i\right)^2}{2} - \rho_{\mathrm{s}}U_{0\mathrm{s}} + \left(1 - \frac{\rho_{\mathrm{s}}}{\rho_{0\mathrm{s}}}\right)P_{\mathrm{s}}\right)$$
$$+\left(1 - \frac{\rho_{\mathrm{s}}}{\rho_{0\mathrm{s}}}\right)\frac{(\boldsymbol{n} \cdot \boldsymbol{v}_{\mathrm{s}})}{c}\left(\boldsymbol{n} \cdot \boldsymbol{q}_{\mathrm{s}}\right) + L_{\mathrm{d}}\left(\boldsymbol{n} \cdot \boldsymbol{q}_{\mathrm{s}}\right) - L\lambda_{\mathrm{s}}(\boldsymbol{n} \cdot \mathrm{grad}T)_{\mathrm{s}} = 0.$$

To estimate the value of the summand that determines the kinetic energy of the process, we use the system of boundary conditions for the balance of the momentum flux, written in explicit form. As a result, we obtain $\sum_i \left(v_{\mathrm{s}}^i - v_{0\mathrm{s}}^i\right)^2 = \left(1 + \frac{\left(P_{\xi\zeta}^{(0)s}\right)^2 + \left(P_{\eta\zeta}^{(0)s}\right)^2}{\left(P_{\mathrm{s}} + q_{\mathrm{s}}^{(\zeta)}/c + P_{\zeta\zeta}^{(0)s}\right)^2}\right)\left(v_{\mathrm{s}}^{(\zeta)} - v_{0\mathrm{s}}^{(\zeta)}\right)^2$. Considering that the radiation flux is directed along the z-axis, it can be assumed that the shear components of the stress tensor are much smaller than the diagonal component of this tensor. It is as follows: $\frac{\left(P_{\xi\zeta}^{(0)s}\right)^2+\left(P_{\eta\zeta}^{(0)s}\right)^2}{\left(P_s+q_s^{(\zeta)}/c+P_{\zeta\zeta}^{(0)s}\right)^2} \ll 1$. But even if we assume that all three components of the stress tensor are quantities of the same order, then we can estimate the order of magnitude of the term $\frac{\left(P_{\xi\zeta}^{(0)s}\right)^2+\left(P_{\eta\zeta}^{(0)s}\right)^2}{\left(P_s+q_s^{(\zeta)}/c+P_{\zeta\zeta}^{(0)s}\right)^2}$. It will be ~0.22. This allows us to neglect this summand. Further, using the condition of balance of mass flux, we obtain $v_s^{(\zeta)} - v_{0s}^{(\zeta)} = \left(1 - \frac{\rho_s}{\rho_{0s}}\right)v_s^{(\zeta)}$. The density of the solid phase is much higher than the density of the gas $\left(\frac{\rho_s}{\rho_{0s}} \sim 10^{-3} \ll 1\right)$, therefore realize the equality $\sum_i \left(v_s^i - v_{0s}^i\right)^2 = \left(v_s^{(\zeta)}\right)^2$.

Consider the factor $\left(\left(1 - \frac{\rho_s}{\rho_{0s}}\right)\frac{(\boldsymbol{n}\cdot\boldsymbol{v}_s)}{c} + L_d\right)(\boldsymbol{n}\cdot\boldsymbol{q}_s)$, which is part of the boundary condition. Since the speed with which the gas can fly off the surface is of the order of the thermal velocity, then quantity $\left(1 - \frac{\rho_s}{\rho_{0s}}\right)\frac{(\boldsymbol{n}\cdot\boldsymbol{v}_s)}{c} \sim 10^{-5}$. Taking it into account makes sense when the coefficient of non-thermal dissipative losses L_d does not exceed this value, that is, in fact, when the radiation flux does not have losses on the surface and completely passes into the solid medium. Such a situation is unlikely when the surface is destroyed, so we will assume

$$\left(\left(1 - \frac{\rho_s}{\rho_{0s}}\right)\frac{(\boldsymbol{n}\cdot\boldsymbol{v}_s)}{c} + L_d\right)(\boldsymbol{n}\cdot\boldsymbol{q}_s) \approx L_d(\boldsymbol{n}\cdot\boldsymbol{q}_s).$$

Discarding all small summands in (11), we finally obtain

$$(\boldsymbol{n}\cdot\boldsymbol{v}_s)(\rho_s U_s + P_s) + (\boldsymbol{n}\cdot\boldsymbol{v}_s)\rho_s\frac{\left(v_s^{(\zeta)}\right)^2}{2} + L_d(\boldsymbol{n}\cdot\boldsymbol{q}_s)$$
$$- L\lambda_s(\boldsymbol{n}\cdot \mathrm{grad}T)_s - (\boldsymbol{n}\cdot\boldsymbol{v}_s)\rho_s U_{0s} = 0,$$

where $v_s^{(\zeta)}$ is the normal component of the velocity vector.

In this equation, the first term can be interpreted as the energy flux of a substance into the gas phase, the second as its convective outflow into the plasma-gas medium from the interface, the third and fourth terms are the flows associated with the transfer of stimulating (electromagnetic) energy (3-d summand) and thermal energy (summand 4) across the interface. The last summand determines the energy flux from the condensed phase to the interface between the phases, and U_{0s} is the internal energy per unit mass of the condensed medium, which is related to the specific heat of the condensate-gas phase transition φ_0 by the relation $U_{0s} = -\varphi_0$ [17]. This ratio is due to the fact that internal energy is determined by the binding energy between atoms in a solid, and by definition it is a negative value. Therefore, the condition for the balance of the energy flux will be considered in the form:

$$\rho_s(\boldsymbol{n} \cdot \boldsymbol{v}_s)\left(H_s + \frac{\left(v_s^{(\zeta)}\right)^2}{2} + \varphi_0\right) + L_d(\boldsymbol{n} \cdot \boldsymbol{q}_s) - L\lambda_s(\boldsymbol{n} \cdot \text{grad}T)_s = 0. \tag{13}$$

Here H_s is the Gibbs thermal function (enthalpy), which characterizes the state of the macroscopic system in thermodynamic equilibrium when entropy and pressure are chosen as the main independent variables. For a gaseous medium, $H \equiv U + \frac{kT}{m}$, where $\frac{kT}{m} = \frac{P}{\rho}$. This shows that $\rho_s U_s + P_s = \rho_s H_s$.

The energy flow balance condition (13) is final. This condition includes the dynamic parameters characteristic of the gas. The only parameter φ_0, which is a characteristic of the solid phase, remains dynamically inactive here.

4 Detailed Statement of the Problem of Destructive and Non-destructive Surface Treatment

Based on the basic system of volume (1)–(3) and the formulated boundary conditions, let define the problem statements separately for destructive and non-destructive surface treatment. When the destruction takes place, then there are both gas and solid-state phases. Then, from this system, it is easy to obtain a system of equations of gas dynamics. To do this, it suffices to use the expression for the stress tensor in the volume $P_{ij} = -P\delta_{ij}$ [13, 15]:

$$\begin{cases} \frac{\partial \rho}{\partial t} + \text{div}(\rho v) = R, \\ \frac{dv}{dt} + \frac{1}{\rho}\text{grad}P = \boldsymbol{f}, \\ \rho\frac{dU}{dt} + P\text{div}v + \text{div}\boldsymbol{Q} = 0. \end{cases} \tag{14}$$

All parameters that are included in the above system are gas characteristics. U, ρ, $\boldsymbol{v}$ and P are the internal energy per unit mass of the gaseous medium, its density, convective velocity vector, and pressure, respectively, $\boldsymbol{Q}$ is the generalized flow of external energy in a plasma-gas medium.

System (14), together with the boundary conditions obtained above, for the case when a solid surface is destroyed, completely describes the dynamics of the plasma-gas phase. But the boundary conditions also contain solid-state characteristics. In particular, the boundary value (9) includes the surface value of the stress tensor $P_{ij}^{(0)s}$, which is connected with the strain tensor u_{ij} or the displacement vector $\boldsymbol{u}$ by the corresponding Hooke's laws. It is known from [13, 15] that $P_{ij}^{(0)s} = \lambda_{ijkl}u_{lk} + \eta_{ijkl}v_{lk}$, where u_{lk}, v_{lk} are the strain tensor and strain rate tensor, and λ_{ijkl}, η_{ijkl} are tensors of elastic strains and internal viscosity. In solids, the main role is played by the elastic forces $P_{ij}^{(0)s} = \lambda_{ijkl}u_{lk}$, and the friction forces $\sigma_{lk} = \eta_{ijkl}v_{lk}$, manifest themselves in the flows of liquids and gases. For an isotropic solid $P_{ij}^{(0)s} = 2\mu u_{ij} + \delta_{ij}\lambda\text{div}\boldsymbol{u}$, where λ and μ are elastic constants (Lame coefficients).

Let us formulate the system (1)–(3) in terms of the solid phase using Hooke's laws. The continuity equation for the solid state of a substance, given the fact that the density here is a constant, and the function of the source-sinks is zero ($R_m = 0$), may be written in the form:

$$v_0 = 0 \tag{15}$$

The equations of motion (2) for a solid medium, taking into account the laws of Hooke [13], will have the form:

$$\rho_0 \ddot{\boldsymbol{u}} = \rho_0 \boldsymbol{f} + (\lambda + \mu)\nabla \mathrm{div}\boldsymbol{u} + \mu \cdot \boldsymbol{u}, \tag{16}$$

where $\boldsymbol{u}$ is the displacement vector, $\boldsymbol{f}$ is the mass density of the bulk force, $v_0 = \mathrm{d}u/\mathrm{d}t$. The above equation characterizes the response of the elastic properties of a solid to the action of a pulse, and condition (6) (taking Hooke's laws into account) is a boundary condition for it.

The third equation for energy in a condensed medium will have the form:

$$\rho_0 \frac{\mathrm{d}U_0}{\mathrm{d}t} + \mathrm{div}\,\boldsymbol{Q}_0 - P_{ij}^{(0)} v_{ij} = 0. \tag{17}$$

In (17), ρ_0 is the density of a condensed medium, U_0 is the internal energy of a unit mass of a solid, $P_{ij}^{(0)}$ is the elastic tensor, $\boldsymbol{Q}_0$ is the energy flux that has passed into the volume of a solid medium.

Equation (17) in fact is a heat equation. This can be easily verified, taking into account the definition of the generalized flux $\boldsymbol{Q}_0 = \boldsymbol{q}_0 - \lambda_0 \nabla T_0$ and the determination of the internal energy U_0 for temperatures lower than the phase transition temperature: $U_0 = -\varphi_0 + C_v^{(0)} T_0$.

In the latter formulas $\boldsymbol{q}_0$ is the external energy flux in the bulk of the solid phase; λ_0 is the coefficient of its thermal conductivity; T_0 is the temperature of the solid phase, which is a function of time and coordinates; $C_v^{(0)}$ is the mass specific heat (specific heat of a unit mass) φ_0 is the interatomic bond energy of a unit mass of a solid substance and this is in fact the specific energy of the phase transition.

Given all this, and also neglecting the terms $P_{ij}^{(0)} v_{ij}$, (17) can be reduced to the form:

$$\rho_0 C_v^{(0)} \frac{\partial T_0}{\partial t} = \mathrm{div}(\lambda_0 \nabla T_0) - \mathrm{div}\boldsymbol{q}_0. \tag{18}$$

It is taken into account that $\varphi_0 = \mathrm{const}$ and $\frac{\mathrm{d}T_0}{\mathrm{d}t} = \frac{\partial T_0}{\partial t}$, since there is no convective velocity in the solid phase. This equation must be considered together with the boundary condition (12).

Introducing the coefficients L_d and L allows us to separate the gas part of the problem, considering it with the boundary condition (12), from the heat conduction problem, which has the boundary condition:

$$\lambda_{0s}(\boldsymbol{n} \cdot \mathrm{grad}T_{0s}) = (1 - L)\lambda_s(\boldsymbol{n} \cdot \mathrm{grad}T_s).$$

This boundary condition is formulated by introducing the coefficient of total heat loss L [18]. In it, the right-hand side of the equation is completely determined by the gas problem with condition (13).

If surface destruction is absent, then the gas part of the problem is completely absent. In this case the solid-state part of the problem is reduced to (15), (16), and (18) with the corresponding boundary conditions. The boundary condition (4) to (15) reduces to the form $(\boldsymbol{v}_{0s} \cdot \boldsymbol{n}) = 0$, that is, to the condition of surface immobility. The boundary condition to (16) will have the form (6) taking into account Hooke's laws. Equation (18) is used with the boundary condition reduced to the form (12), where $\boldsymbol{q}_s$ is the meaning of the flow that has reached the surface and, in the absence of destruction, coincides with the output flow $\boldsymbol{q}_{\mathrm{in}}$ (in vacuum), since has no losses in a plasma-gas medium.

5 Conclusion

The paper formulates the problem of the interaction of high-power laser pulsed radiation with a solid surface on the basis of a complete system of equations of continuum mechanics. Particular attention is paid to the process of the sublimation type, when the gas phase arises immediately from the solid, bypassing the liquid phase. The criterion for the absence of surface melting is determined. Also, all the boundary conditions are formulated: the balance of the mass flow, the balance of the flow of momentum, and the balance of energy flow. Each of them, in the presence of destruction, takes into account the characteristics of both coexisting phases.

A methodically general approach to the formulation of problems arising in the modeling of destructive and non-destructive surface treatments by powerful energy flows has been developed. Application of this approach allows us to separate the system of equations and boundary conditions so that each of the parts of the system can be considered independently of the other.

References

1. Anisimov SI, Luk'yanchuk BS (2002) Selected problems of laser ablation theory. Phys Uspekhi 45(3):293. https://doi.org/10.1070/pu2002v045n03abeh000966
2. Ionin AA, Kudryashov SI, Seleznev LV, Sinitsyn DV (2009) Tunneling ionization of air in the strong field of femtosecond laser pulses. JETP Lett 90:181. https://link.springer.com/article/10.1134%2FS0021364009150053
3. Abramov DV, Arakelyan SM, Makov SA, Prokoshev VG, Khor'kov KS (2013) Formirovaniye sistemy mikrokraterov na poverkhnosti titana pri vozdeystvii femtosekundnym lazernym izlucheniyem v usloviyakh bystrogo okhlazhdeniya. Pis'ma v ZHTF 39(16):14–22 (in Russian)

4. Wu Z, Zhang N, Zhu X, An L, Wang G, Tan M (2018) Time-resolved shadowgraphs and morphology analyses of aluminum ablation with multiple femtosecond laser pulses. Chin Phys B 27(7):077901. http://iopscience.iop.org/article/10.1088/1674-1056/27/7/077901/meta
5. Mann T, Mathieson R, Murray M, Richards B, Jose G (2018) Femtosecond laser ablation properties of Er^{3+} ion doped zinc-sodium tellurite glass. J Appl Phys 124(4):044903. https://aip.scitation.org/doi/abs/10.1063/1.5040947
6. Shugaev MV, Gnilitskyi I, Bulgakova NM, Zhigilei LV (2017) Mechanism of single-pulse ablative generation of laser-induced periodic surface structures. Phys Rev B 96(20):205429. https://journals.aps.org/prb/abstract/10.1103/PhysRevB.96.205429
7. Zhang N, Yang J, Zhu X (2012) Investigation of the ultrafast process of femtosecond laser ablation of highly oriented pyrolytic graphite. Chin J Lasers 39(5):0503002. http://en.cnki.com.cn/Article_en/CJFDTOTAL-JJZZ201205018.htm
8. Berezovska N, Dmitruk I, Kalyuzhnyy A, Dmytruk A, Blonskyi I (2018) Self-organized structuring of surface of metal-semiconductor composite by femtosecond laser processing. Ukrainian J Phys 63(5):406–412
9. Shmeleva LV, Suprun AD, Yezhov SM (2019) Simulation of the formation of a surface nanocrater under the action of high-power pulsed radiation—nanocomposites, nanostructures, and their applications part of the Springer Proceedings in Physics book series, vol 221, no 34, pp 505–515. https://link.springer.com/chapter/10.1007%2F978-3-030-17759-1_34
10. Pavlyniuk OR, Datsyuk VV (2016) Electrostrictive mechanism of nanostructure formation at solid surfaces irradiated by femtosecond laser pulses. Nanoscale Res Lett 11(1):1–6. http://nanoscalereslett.springeropen.com/articles/10.1186/s11671%2D015%2D1224%2D5. https://doi.org/10.1186/s11671-015-1224-5
11. Zhigalov VS (1998) Laser technology. Krasnoyarsk, 1998, 114 pp. http://sibsauktf.ru/courses/hitech/html/index_htm (in Russian)
12. Alekseev BV (1982) Mathematical kinetics of reacting gases. Nauka, Moscow. 419 pp. https://scholar.google.com.ua/scholar?hl=ru&as_sdt=0%2C5&q=Alekseev+B.+V.+Mathematical+Kinetics+of+Reacting+Gases.+Nauka%2C+Moscow%2C+1982&btnG (in Russian)
13. Fedorchenko AM (1992) Teoretychna fizyka [Theoretical physics], Vol. 1. Kyiv, Vyshcha shkola. 535 pp. https://scholar.google.com.ua/scholar?hl=ru&as_sdt=0%2C5&q=Fedorchenko+A.+M.+%22Teoretychna+fizyka%22+Theoretical+physics%5D.+Kyiv%2C+Vyshcha+shkola+1+%281992%29%3A+535&btnG (in Ukrainian)
14. Landau LD, Lifshits EM (1986) Theoretical Physics, Vol. VI, Hydrodynamics. Science. 736 pp. https://scholar.google.com.ua/scholar?hl=ru&as_sdt=0%2C5&q=Landau+L.+D.%2C+Lifshits+E.+M.+Theoretical+Physics%2C+V.+VI%2C+Hydrodynamics.+Science.+1986&btnG (in Russian)
15. Landau L.D., Lifshits Ye.M. Teoriya uprugosti. Tom VII. – M. Nauka, 1987. – 248p. (in Russian)
16. Datsyuk VV, Tovkach OM (2011) Optical properties of a metal nanosphere with spatially dispersive permittivity. J Opt Soc Am B: Opt Phys 28(5):1224–1230. https://doi.org/10.1364/JOSAB.28.001224
17. Ashkroft N., Mermin N. Fizika tverdogo tela. Tom 2. – M.: Mir, 1979. 424 pp. http://mat.net.ua/mat/biblioteka-fizika/Ashkroft-tverdoe-telo-t1.pdf (in Russian)
18. Suprun AD, Shmeleva LV, Razumova MA (2011) The influence of bulk absorption of substance on the threshold of destruction by the intensive pulse of electromagnetic radiation. Funct Mater 18(2):237–243. http://functmaterials.org.ua/contents/18-2/

Formation and Transient Photovoltaic Properties of ZnO/Si Isotype Heterojunctions by Magnetron Sputtering

V. Melnik, B. Romanyuk, V. Kladko, V. Popov, O. Gudymenko, O. Liubchenko, T. Sabov, O. Oberemok, O. Dubikovskyi, JU. Gomeniuk, O. Kosulya, V. Shmid, A. Podolian, A. Nadtochiy, and O. Korotchenkov

1 Introduction

The integration of thin films with Si technology implies serious limitations of the minimum thickness of the film, allowing to create an isotype heterojunction with high quality. This is a particularly competitive task as numerous thermal treatment steps are usually involved. One of the most important problems in this respect is the need to avoid the formation of large defect concentrations at the film/substrate boundary. It is known that zinc oxide exhibits unique properties, which make it interesting for study and numerous applications [1].

V. Melnik · B. Romanyuk (✉) · V. Kladko · V. Popov · O. Gudymenko · O. Liubchenko · T. Sabov · O. Oberemok · O. Dubikovskyi · JU. Gomeniuk · O. Kosulya
Lashkaryov Institute of Semiconductor Physics, Prospect Nauki, 41, Kiev 03028, Ukraine
e-mail: romb@isp.kiev.ua

V. Melnik
e-mail: romb@isp.kiev.ua

V. Kladko
e-mail: kladko@isp.kiev.ua

V. Popov
e-mail: romb@isp.kiev.ua

O. Gudymenko
e-mail: gudymen@ukr.net

O. Liubchenko
e-mail: lubchenco.a@gmail.com

T. Sabov
e-mail: tsabov92@gmail.com

O. Oberemok
e-mail: ober@isp.kiev.ua

O. Fesenko and L. Yatsenko (eds.), *Nanomaterials and Nanocomposites, Nanostructure Surfaces, and Their Applications*, Springer Proceedings in Physics 246, https://doi.org/10.1007/978-3-030-51905-6_24

One promising type of ZnO-based structures suitable for photovoltaic applications employs undoped ZnO layers grown on Si wafers [2]. Zinc oxide is also a piezoelectric and optical waveguide material, which was used in sensor and ultraviolet (UV) detector, light emitting diode (LED), surface acoustic wave (SAW), and solar cell technologies, as well as transparent electrodes, particularly, carrier-selective contacts with wide band gaps for Si heterojunction solar cells [3–9]. Zinc oxide was also extensively explored for using in transparent thin film transistors (TFTs) [10–12]. High optical transparency in the visible range and low resistivity make ZnO an important material for its using as heat mirrors in stoves, conducting coatings in aircraft glasses to avoid surface icing [13].

The native defects in ZnO, oxygen vacancies V_O and zinc interstitials Zn_i, form donor levels in the forbidden gap, so that zinc oxide is naturally an *n*-type semiconductor [14]. The Si/ZnO heterojunction is also widely employed. The presence of SiO_x native oxide and defects at the interface affects the charge carrier generation and recombination and drastically changes the electrical and photoelectric properties of the heterojunction.

Physical properties of ZnO depend on growth method, impurities, temperature, and other factors, which in turn are important for the effective performance of practical devices. One important prerequisite for the LED applications is the ability to achieve *n* and *p* types of electrical conduction in ZnO [15]. This allowed to fabricate and realize *p-n* junction on ZnO and observe electroluminescence in the junction [16, 17]. However, as the acceptor impurities are hardly dissolved, it is still quite difficult to produce *p*-type ZnO [18–20]. Moreover, because of a strong polarization field, the quantum efficiency of ZnO-based LEDs is reduced [21]. It was shown that growing nonpolar films, such as GaN-based layers, can improve the efficiency [22].

Nickel oxide offers an interesting alternative for fabricating ZnO-based heterojunctions, since it is also a wide direct bandgap semiconductor ($E_g = 3.7$ eV) and

O. Dubikovskyi
e-mail: dubikovsky_o@ukr.net

JU. Gomeniuk
e-mail: yurigom@lab15.kiev.ua

O. Kosulya
e-mail: alexandr250990@gmail.com

V. Shmid (✉) · A. Podolian · A. Nadtochiy · O. Korotchenkov
Faculty of Physics, Taras Shevchenko Kyiv National University, Volodymyrska 64/13, Kiev 01601, Ukraine
e-mail: shmdvi@gmail.com

A. Podolian
e-mail: summer.podolian@gmail.com

A. Nadtochiy
e-mail: namobem@gmail.com

O. Korotchenkov
e-mail: olegk@univ.kiev.ua

can much simpler be obtained with a *p*-type conductivity [23, 24]. Consequently, *p-n* junctions utilizing NiO/ZnO and MgZnO/NiO were successfully fabricated [25–27]. In this respect, metal oxide semiconductor nanocomposite heterojunctions, which combined *n*-type zinc oxide and *p*-type nickel oxide, offered an interesting approach for enlarging the free charge separation lengths increasing their lifetimes [28–34].

Different low-dimensional structures with ZnO, e.g., thin films, nanotubes, nanowires, nanofibers, were fabricated using various deposition techniques, such as MBE, CVD, sol–gel method [35, 36]. Microwaves and sonochemical synthesis were also used [37–40].

Using magnetron sputtering technique, ZnO-on-*p*-Si LEDs were obtained [41]. It was reported in this work that the electroluminescence behavior of the ZnO/Si depends sensitively on the growth conditions. Morphology aspects in thin films of ZnO synthesized with magnetron sputtering method were recently addressed by Nagabharana et al. [42], and a crucial role of the deposition rate in the substrate-ZnO interface bonding was demonstrated.

As reported by Wang et al., changing the deposition temperature of wurtzite ZnO FAZO films, co-doped with fluorine and aluminum, modified the film surface morphology such that "pyramid" and "crater" shapes were observed [43]. These transformations were detected by combining different techniques, e.g., XRD, SEM, AFM, XPS, and optical spectroscopy.

When doping ZnO with Mg, Al, and Ga (AGMZO), the properties of the films, such as structure, morphology, resistivity, and optical transmittance, were found to depend upon the Mg content, as reported by Liu and Zhu [44].

Kim et al. studied ZnO films integrated in the ITO cathodes as transparent buffer layers in inverted polymer solar cells (IPSCs) [45]. Comparing magnetron sputtering and solution-based growth techniques, the authors demonstrated that the power conversion efficiency was much lower in the former growth method. Meanwhile, it exhibited improved performance of ZnO buffers due to the formation of a suitable heterojunction structure

Photoelectrical properties of Cr-doped magnetron-sputtered ZnO layers were studied by Fareed et al. [46]. It was shown that the barrier height in ZnO/Pt Schottky diodes was lowered due to such a doping of ZnO. Moreover, the ideality factor of the diode was made better. Furthermore, the doping induced the diode responsivity in the visible region, although an undoped ZnO film in the photodiode did not show appropriate response. This effect was attributed to the reduction in E_g due to Cr doping.

Frequently, ZnO is doped by the IIIA group elements to obtain a layer with modified properties. Aluminum is the most suitable for this purpose because Al dopant atoms offer an enhanced concentration of free electrons, which increases the electrical conductivity [47]. Besides, adding aluminum into ZnO (AZO films) changes E_g. ZnO:Al/Si heterojunctions have a higher switching speed and lower temperature budget in comparison with typical silicon *p-n* or *p-i-n* junctions. The heterojunction quality depends on the number and type of defects on the interface and in the thin film. Ion implantation with the next temperature treatments allows to change the point defect concentration in ZnO film and obtains the ZnO/Si structures with desired properties.

Here, the charge transport properties related to the different magnetron-sputtered ZnO and NiO thin film and interface morphologies, current-voltage characteristics, and surface photovoltage will be discussed and favorable deposition conditions will be concluded.

Different AZO films with thicknesses varying from 10 to 140 nm were prepared on Si substrates by DC reactive magnetron sputtering. Some of them were modified by using Ar^+ ion implantation (50 and 100 keV). The samples were annealed by rapid thermal and furnace annealing at temperatures ranging from 450 to 750 °C.

The resulting films were studied using XRD, SIMS, SEM, as well as surface photovoltage (SPV) decays. The diode structures were manufactured for investigation of current-voltage (*I*–*V*) and capacitance-voltage (*C*–*V*) characteristics.

2 Formation of ZnO/Si Isotype Heterojunction at Room Temperature and Its Structural and Morphological Properties

Typical ZnO fingerprints with the diffraction lines from (100), (002), and (101) planes are observed in the XRD spectra of AZO grown on Si (see Fig. 1). It is seen that, at film thicknesses smaller than 67 nm, only one peak (002) is seen in the XRD spectra. The (002) peak gains in intensity upon enhancement of the AZO film thickness from 11 to 120 nm. This is indicative of the fact that ZnO columns appear having crystallographic c-axis normal to the substrate surface. Moreover, the (002) peak position shifts to higher diffraction angles for the film thicknesses of 11 and 22 nm, which can be attributed to decreasing the film density with increasing its thickness.

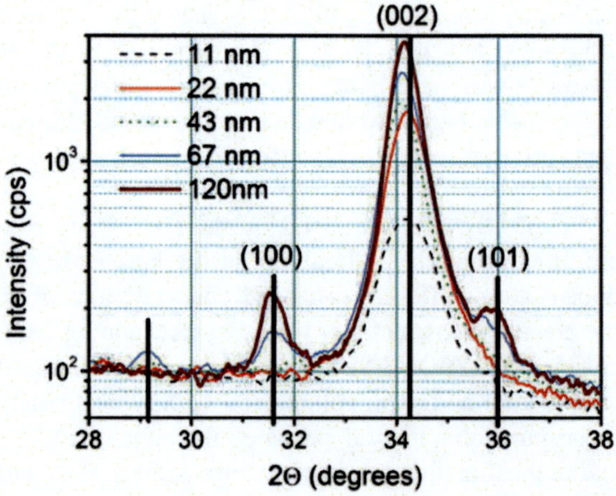

Fig. 1 XRD patterns of different ZnO films grown on Si

Upon increasing the thickness to about 67 nm, the three ZnO fingerprints could be clearly resolved on a semi-log scale of the XRD spectrum. This may be due to the weakening of the influence of the substrate, which thus changes the internal structure of the films. At the film thicknesses of 120 nm, the (101) and (100) peaks gain in intensity.

The above effect of the film loosening with increasing the film thickness is furthermore illustrated by the AFM data shown in Fig. 2. A less tight stacking of the columns in the film, accompanied by an enhanced nanograin density and size, is observed. Indeed, at the film thickness of 11 nm, the surface is smooth and matches with the Si substrate surface relief, as observed in image a in Fig. 2. This trend toward nanograin formation with the size from 10 to 15 nm is clearly visible as bright white spots distributed across the image a. The size and the density of these features gradually increase with increasing the film thickness (b and c in Fig. 2) until the tight consolidated layer is formed when it reaches a critical size greater than ≈65 nm (d and e in Fig. 2). In the latter case, the grain diameter reaches the value from 10 to 60 nm. It has been found that a bimodal grain size distribution peaked at 20 and 50 nm is observed in the 120-nm film.

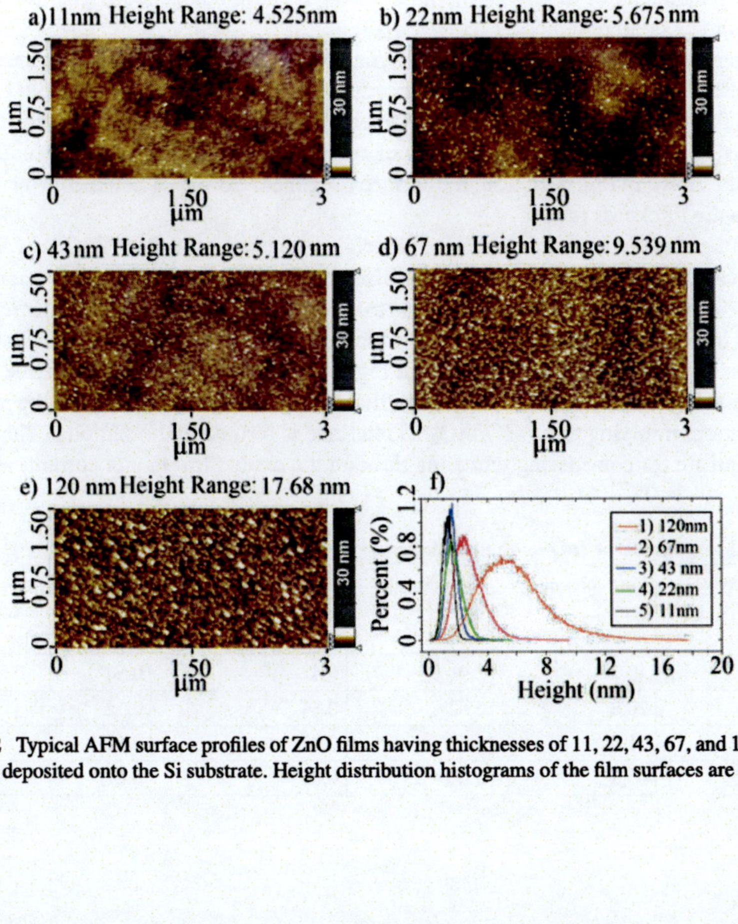

Fig. 2 Typical AFM surface profiles of ZnO films having thicknesses of 11, 22, 43, 67, and 120 nm (a–e) deposited onto the Si substrate. Height distribution histograms of the film surfaces are shown in (f)

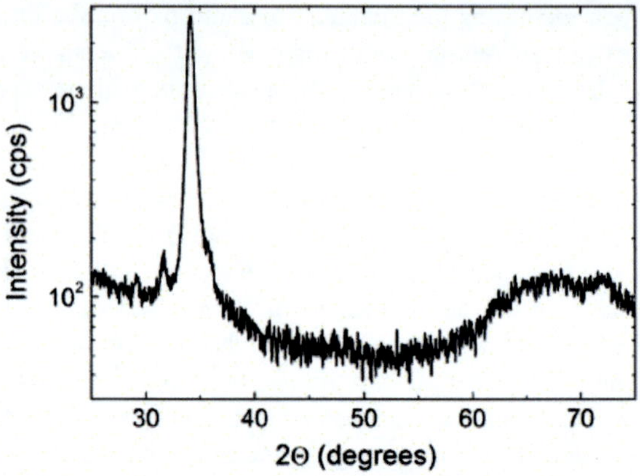

Fig. 3 XRD pattern of a 65.8-nm thick ZnO film

Of further significance is the fact that the ZnO films with thicknesses less than 50 nm are textured. At greater thicknesses, a polycrystalline structure of hexagonal modification is formed (Fig. 3). This may originate from the large difference in the lattice parameters of Si and ZnO, which leads to an inhomogeneous growth of the ZnO film at small thicknesses.

From XRR studies, thickness, density, and root mean squared (rms) roughness of deposited films are determined. This procedure yields the values given in Table 1. It is found that the roughness of the Si/ZnO interface does not significantly change upon increasing the film thickness, for the thicknesses less than ≈70 nm, as evidenced by the data shown in Fig. 4. In turn, the surface roughness exhibits a noticeable increase within the thickness range.

Further insight into the processes associated with the growth of ZnO:Al (AZO) films can be obtained by analyzing the SEM images shown in Fig. 5. It is seen that the surface relief has an irregular grainy structure character, whereas the inner region of the film forms a consolidated layer, which can be due to Al inclusions in ZnO.

Mass spectrometric studies given in Fig. 6 shows an ultra-thin SiO_2 oxide film layer having the thickness of about 2 nm that is formed on the boundary of the phase separation, implying that AZO/SiO_2/Si interface is formed in the samples. This is a prerequisite for considering tunneling through the oxide film (tunnel current) as the

Table 1 Parameters of ZnO-on-Si films discussed in this work

Sample No.	Film thickness (nm)	Film density (g/cm^3)	ZnO/Si interface rms roughness measured by XRR (nm)	ZnO surface rms roughness measured by XRR (nm)
1	10.2	5.9	0.4	0.54
2	20.8	5.62	0.34	0.52
3	41.6	5.59	0.02	0.51
4	65.8	5.60	0.05	0.6

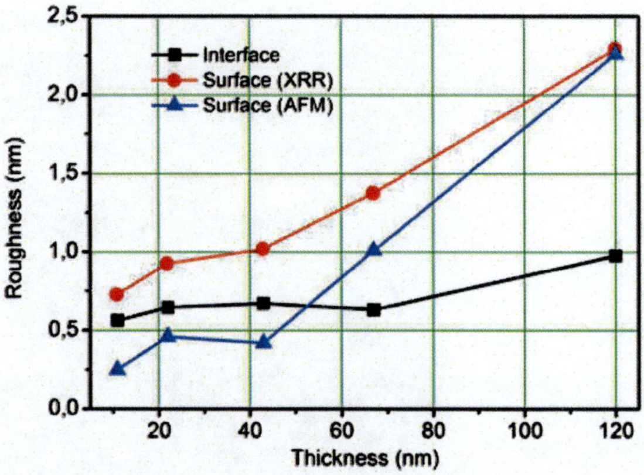

Fig. 4 Surface and interface roughness of different ZnO-on-Si films inferred from the XRR and AFM data

dominant free electron transport mechanism in these structures. It is also shown in Fig. 6 that the Al impurity is homogeneously distributed over the thickness of the film.

It should be noted that some features of the distribution of SiO_2^- ions is shown in Fig. 6, namely the presence of two maxima near the boundary of phase separation. This is due to the peculiarities of the origin and growth of the ZnO film at the initial stages of magnetron sputtering. In order to clarify this issue and find out the role played by the above nanograins formed at small film thicknesses, processes of film recrystallization after their amorphization with argon ions and annealing have been studied. Some results are exemplified in Fig. 7. Comparing these data for an as-grown (ZnO) and ion implanted (ZnO + I) sample shows that the Ar^+ ion implantation causes the compressive strain in a crystal orientation perpendicular to the ZnO/Si interface (along the c-axis). Increasing the implantation dose and ion energy increases the strain in the film. It has also been verified that the thickness of the films decreases somewhat upon implantation due to decreased inter-grain spacing while their porosity concomitantly increases.

Adding Al during the growth tightens the film, enhances the electrical conductivity and varies the bandgap of ZnO. The thickness of an intermediate SiO_x layer developed at the ZnO/Si interface has been found to be nearly independent of the film thickness.

3 Current-Voltage Characteristics

Dark current-voltage characteristics are shown in Fig. 8a. It is seen that diode-like current-voltage $I(V)$ characteristics appear at film thicknesses greater than ≈60 nm. Reducing the film thickness leads to a significant increase in reverse currents and shortened diode. It has been found that ZnO/Si heterostructures are photosensitive under illumination with wavelengths from 400 nm to 2.5 μm, and a typical result is exemplified in Fig. 8b.

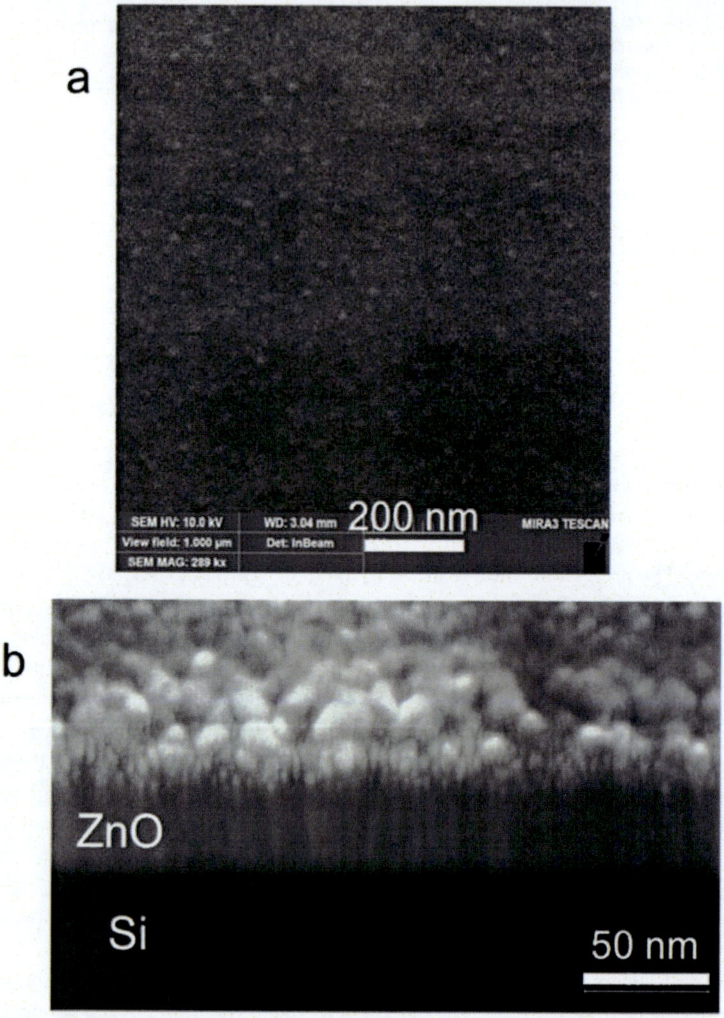

Fig. 5 Surface (**a**) and cross section (**b**) SEM images of a 67-nm thick ZnO film deposited onto the Si substrate

Voltage-capacitance $C(V)$ characteristics taken at frequencies of 1, 10, 100, and 1000 kHz are shown in Fig. 9a. One can see that the barrier capacitance depends upon the signal frequency. The capacitance decreases with increasing the frequency, illustrating the occurrence of interface states that cannot trace the quicker signal change with increasing the frequency.

In this respect, the effect of aluminum segregation at the AZO/SiO_2/Si interface was discussed, e.g., by Bikowski et al. [48] and Jaramillo et al. [11]. This may be due to the domination of either aluminum condensation or zinc re-evaporation at the

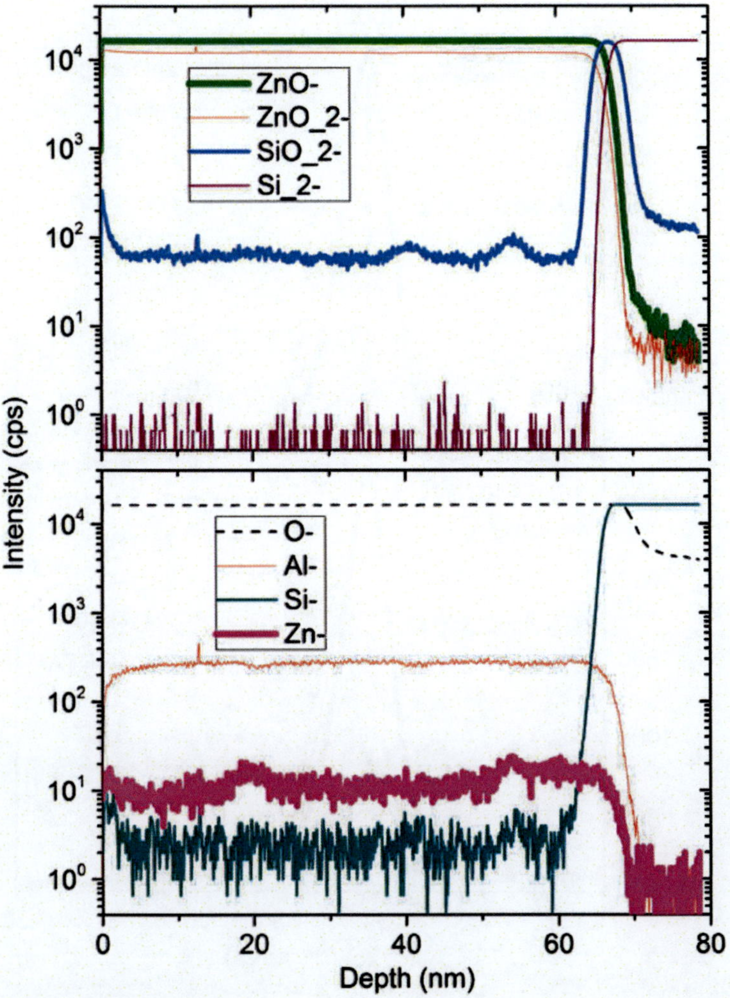

Fig. 6 In-depth elemental distributions in SIMS profiling analysis of a 67-nm thick ZnO film deposited onto the Si substrate

interface at the beginning of the deposition of AZO. As reported by Bikowski et al. [48], the condensation and re-evaporation effects can be related to the vapor pressure during the growth. Moreover, according to Alcock et al. [49], the evaporation rate of Zn is much greater than that of Al. In contrast, metal oxides tend to reside on the surface much stronger than pure metals since the pressure of vapor is considerably smaller in the former case. Furthermore, the formation of aluminum oxide or aluminum-doped zinc oxide is energetically favorable compared to formation of zinc oxide. Therefore, this can explain the domination of the segregated aluminum mentioned above during the deposition at high substrate temperatures.

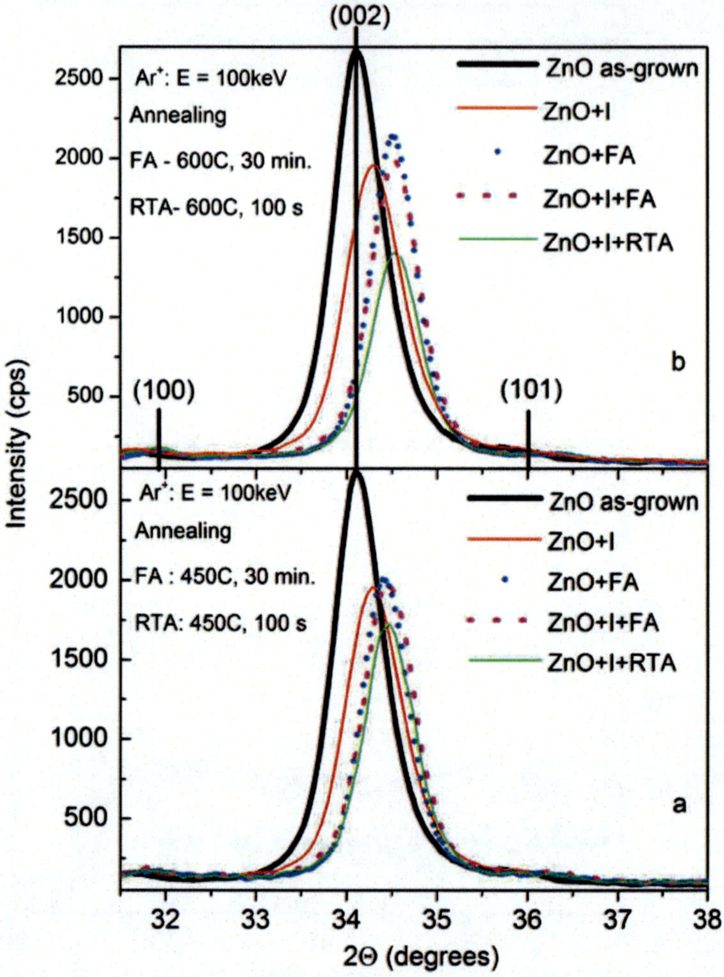

Fig. 7 XRD spectra of ZnO films, as-grown (ZnO as-grown) and processed either upon 450 °C (**a**) or 600 °C (**b**) annealing for 30 min (ZnO + FA) or implanting with 100 keV Ar^+ ions (ZnO + I) followed by annealing at 450 °C (**a**) or 600 °C (**b**) for 100 s (ZnO + I+RTA) or 30 min (ZnO + I + FA)

This in turn suggests that the AZO film is highly doped at the interface. Therefore, the AZO/Si junction is asymmetric with a thicker space charge region in the Si substrate compared with that in the ZnO film. Then the capacitance is simply $C \approx \varepsilon_{Si}/W$, where ε_{Si} is the dielectric constant of Si, and W is the total thickness of the space charge region. At reverse bias, the capacitance depends only weakly on the frequency. In turn, at forward bias, the free electrons in Si are captured at the interface states thus decreasing the capacitance with increasing the frequency.

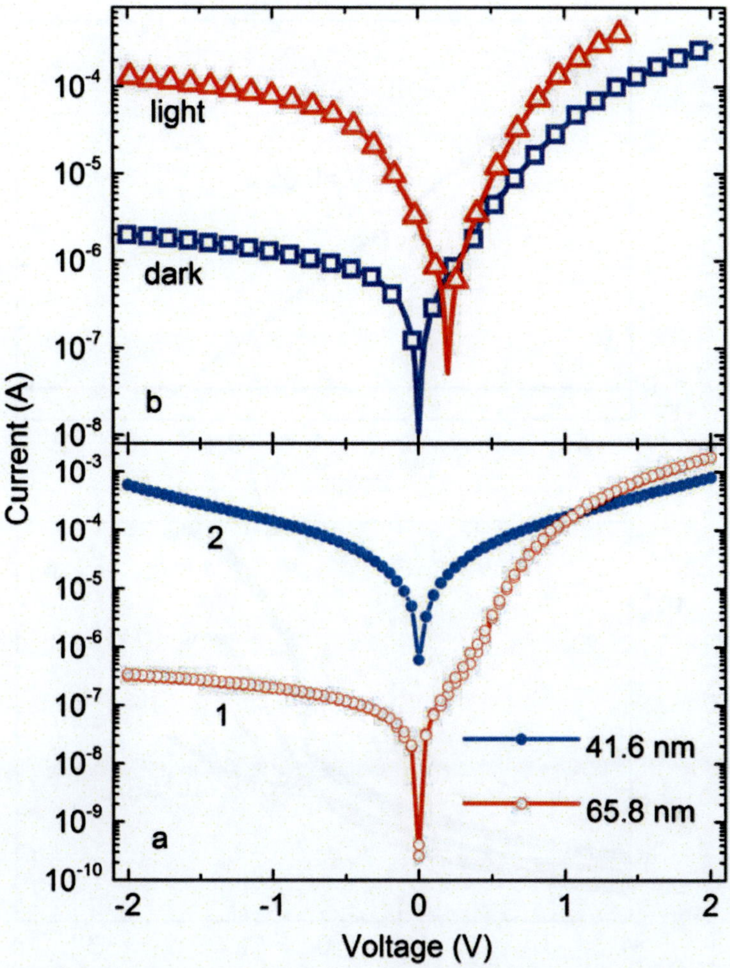

Fig. 8 **a** Dark $I(V)$ curves of ZnO-on-Si films having a thickness of 65.8 nm (1) and 41.6 nm (2). **b** $I(V)$ curves of ITO/ZnO/Si structure in the dark and at an illumination with white light

The appearance of peaks in the $C(V)$ curves shows evidence of tunneling through the interface states and oxide film discussed above. Replotting the data of Fig. 9a in the form of $1/C^2$ versus V in Fig. 9b yields the barrier height V_b of 0.66 eV. It has been observed that V_b depends on the annealing temperature of the implanted structures.

One thus concludes that the distribution of aluminum-rich phases has yet been studied insufficiently in magnetron-sputtered AZO films. In particular, Sieber et al. [50] addressed the microstructural properties of ZnO:Al, illustrating the occurrence of zinc–aluminum–oxygen phases, 10–20 nm beneath the Si substrate. However,

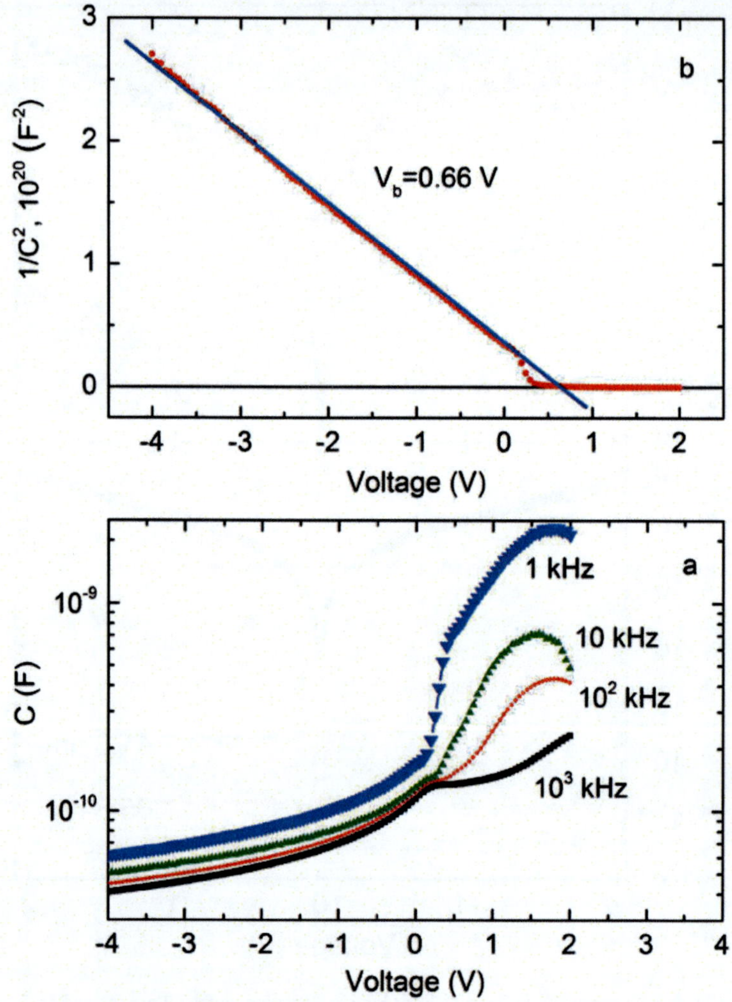

Fig. 9 **a** Voltage-capacitance characteristics of ZnO film with a thickness of 65.8 nm grown on Si (**a**) and determination of the potential barrier value in this ZnO/Si structure (**b**)

the film electrical conductivity remained uncovered in this study. Kinemuchi et al. reported aluminum-rich boundaries in ceramic AZO grains [51].

In the remaining part of this work, we, therefore, present photoelectrical properties of the films, and, particularly, spatially distributed SPV signals, which can shed light on some of the above segregation problems unsolved so far.

4 Transient Surface Photovoltage

Surface photovoltage (SPV) is formed by the separation of nonequilibrium carriers in the near-surface sample region [52]. In particular, the separation exists in the space charge regions that occur on the sample surfaces or at grain boundaries and interfaces. The magnitude and direction of the appropriate band bending in these regions determine the magnitude and sign of the SPV signal formed under light illumination. The band bending itself depends on the concentration and type of defects in surface and interface areas. These concentrations and type also determine the generation and recombination velocity of nonequilibrium free charges.

In our case of $ZnO/SiO_2/Si$ heterostructures, the SPV amplitude and decay times taken after the light is turned off can, in general, be related to varying defect content in ZnO films and in a narrow subsurface region of the Si substrate. The SPV characteristics would therefore depend on the deposition method and thickness of the ZnO layer. Measuring SPV can furthermore be interesting for improving the photovoltaic properties of $ZnO/SiO_2/Si$ and NiO/ZnO/Si heterostructures.

Varying the light wavelength, one can vary the light penetration depth into the semiconductor structure, thus appropriately changing the carrier generation depth spanning the junction regions of the heterostructures. This is particularly true in case of small pulse widths of the illuminating light, when the width remains much smaller than the carrier lifetime. Otherwise, if the light pulse is of the order of or greater than the lifetime of nonequilibrium carriers, then the carriers from a thick subsurface layer comparative with their diffusion length contribute to the SPV signal, even for the strongly absorbed light [53].

Taking the bandgap $E_g = 3.2$–3.3 eV in ZnO [54] and $E_g = 1.12$ eV in the Si substrate [55] one gets that exciting light with $\lambda > 375$ nm correspond to E_g of ZnO [54]. Therefore, the SPV signal generated by a visible light is formed only by free carriers excited in the Si substrate. For the light with a wavelength of less than $\approx$375 nm, the signal will come from nonequilibrium carriers generated both in the ZnO film and Si substrate. Strictly speaking, for both excitation conditions, one should take into account the carrier diffusion and drift from the Si substrate into the ZnO film and in the opposite direction.

The SPV data given below were measured in the capacitance arrangement and the measurement technique was described in detail in [56]. SPV decay curves taken in different ZnO-on-Si samples are shown in Fig. 10. The samples are excited by light pulses of 1 μs duration with three different wavelengths $\lambda = 405$ nm, 470 nm, and 860 nm. At these conditions, nonequilibrium charge carriers are generated only in the Si substrate at depths of $1/\alpha \approx 98$ nm, 312 nm, and 21 μm, respectively, where α is the light absorption coefficient in Si for the above three excitation wavelengths [55]. Negative SPV signals were experimentally detected in the circuit schematics used and, for convenience, they are shown in Fig. 10 as well as in the text that follows, the SPV decay curves are inverted.

It is shown in Fig. 10 that regardless of the wavelength of the exciting light, the SPV decay rate decreases monotonically with increasing the ZnO thickness from 11

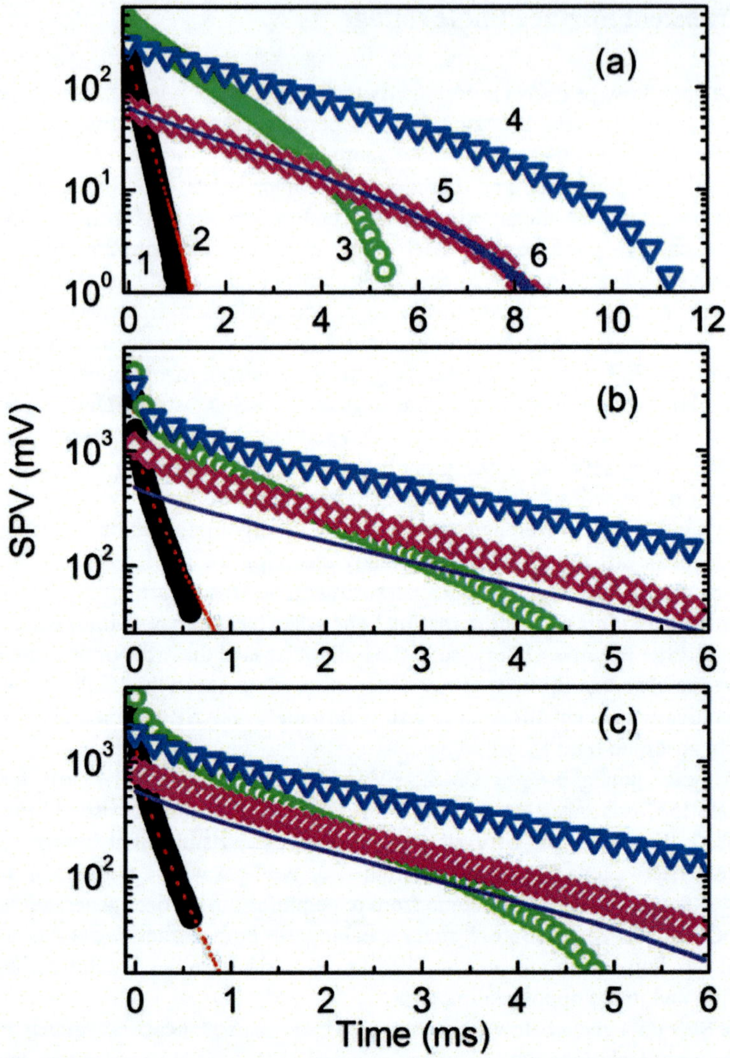

Fig. 10 SPV decay curves in ZnO/Si excited with LED light of wavelengths λ = 405 nm (**a**), 470 nm (**b**) and 860 nm (**c**). The light pulse width is 1 μs. The thickness of ZnO films is 11 (curve 1), 22 (2), 43 (3), 67 (4), 70 (5), and 120 nm (6)

to 67 nm. When further increasing the thickness to 120 nm, the decay rate of the SPV signal remains nearly unchanged. At the same time, when the thickness increases from 11 to 43 nm, the SPV amplitude value first increases monotonically and then begins to decrease for greater thicknesses (up to 120 nm).

If the same samples are excited by light pulses of 10 ns duration from a nitrogen laser (λ = 337.1 nm), the decay curves shown in Fig. 11 are observed. Under these

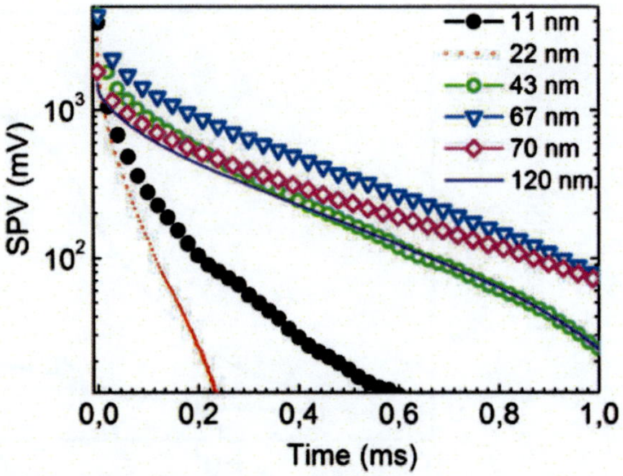

Fig. 11 SPV decay curves in ZnO films with different thicknesses grown on Si excited with N_2 laser light (λ = 337.1 nm) having pulse width of 10 ns

excitation conditions, nonequilibrium charge carriers are generated both in the ZnO film and Si substrate with a depth of $1/\lambda \approx 9$ nm [55]. It can be found that, in this case, the decay curves are described by a biexponential law. The appropriate time constants of both the first and second components increase with increasing the thickness of ZnO from 11 to 120 nm (see Fig. 12a). In contrast, the magnitude of the SPV signal first increases slightly at film thicknesses varying from 11 to 67 nm and then sharply decreases at the thicknesses greater ≈ 70 nm; see Fig. 12b.

Next, Fig. 13 shows the distributions of the SPV decay time constant τ across the sample surface taken for different thicknesses of the ZnO film. Appropriate distributions of the SPV amplitude are given in Fig. 14. The data of Figs. 13 and 14 are taken when scanning the surfaces of the samples with a narrow light beam of a laser diode with $\lambda = 630$ nm with a spatial resolution of 100 μm. This scanning SPV apparatus has been discussed elsewhere [57].

It is shown in Fig. 13 that the width of the distribution function for τ increases for the thicknesses from 11 to 120 nm. The maximum of the distribution shifts toward greater values of τ. At the same time, the width of the distribution function for the SPV amplitude shown in Fig. 14 first increases in the thickness range from 11 to 43 nm and then decreases very sharply for 67–120 nm thicknesses. In this case, the distribution maximum shifts toward greater values of the SPV amplitude for the thickness range from 11 to 67 nm and then, at greater thicknesses from 70 to 120 nm, shifts to almost 100% of its initial value.

These SPV results can be explained as follows by taking the change of the film morphology with its thickness into consideration. The film microstructure changes significantly with increasing the thickness (see Fig. 2). Thus, the size of nanograins and nanocolumns can vary from about 10 to 60 nm for the film thickness ≤67 nm, until the tight consolidated layer is formed in the range from about 65 to 120 nm.

Consistent with these observations, increased SPV amplitude values are detected primarily due to a larger spatial separation of photogenerated carriers in larger grains. Furthermore, the observed broadening of the SPV amplitude distribution can be

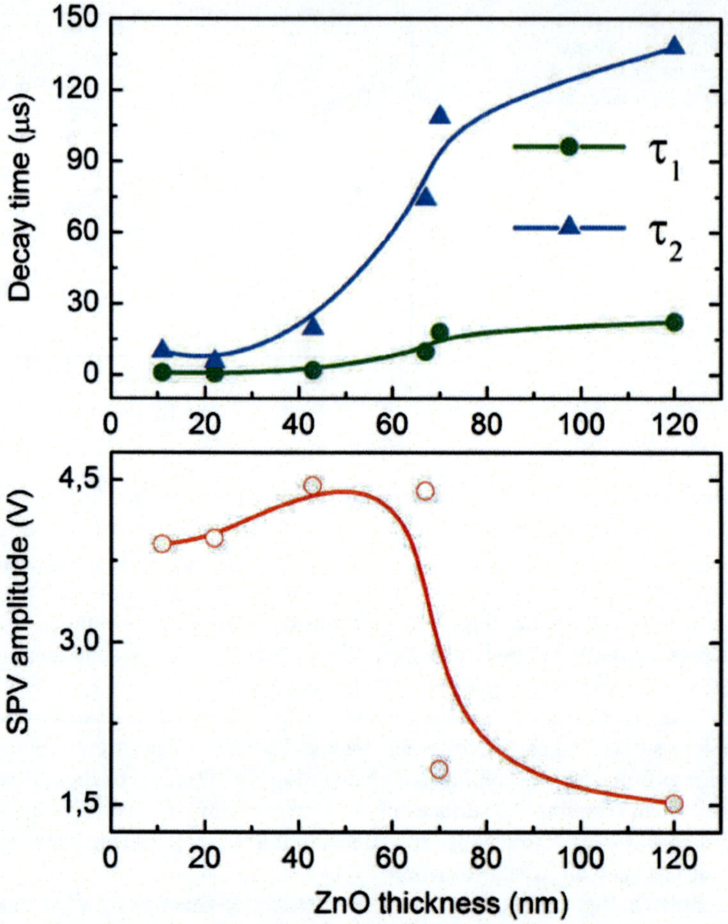

Fig. 12 SPV decay time (**a**) and amplitude (**b**) in ZnO/Si versus the thickness of ZnO films excited with N_2 laser light ($\lambda = 337.1$ nm) having a pulse width of 10 ns

explained by a concomitantly broader dispersion in grain sizes. It appears that in the thickness range from about 65 to 120 nm, when a dense consolidated layer occurs, the SPV amplitude decreases with a narrowing of its distribution function across the sample surface. Moreover, the observed increase in τ, broadening of the distribution of τ, and a shift of the maximum value of this distribution to greater τ are most likely due to the fact that increasing the grain size and grain density affects the composition and concentration of recombination-active defects at grain boundaries. Finally, carrier trapping at deep levels and grain surface or interface states may become dominating over fast carrier recombination processes, which also increases τ.

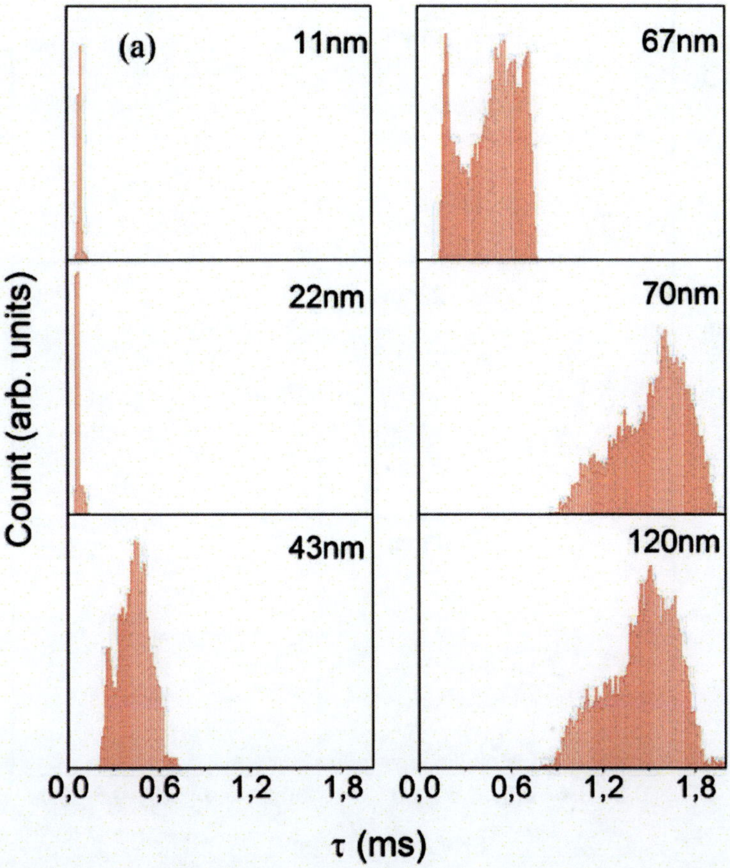

Fig. 13 Probability functions of the surface distribution of τ in ZnO/Si with different thickness of ZnO layer. See measurement details in [57]

For comparison, the SPV attenuation curves were also measured on a sample in which a *p*-NiO film, ≤70 nm thick, was additionally deposited on top of a 120 nm thick ZnO film. In this structure, the NiO/ZnO interface forms a *p*-*n* junction, which in turn should enlarge the SPV magnitude due to increased number of photoexcited carriers separated by the *p*-*n* junction field.

In Fig. 15, we show SPV decay curves for samples with a 120 nm ZnO film without (curve 1) and with an additional *p*-NiO layer (curve 2). These samples are excited by light pulses of 1 μs duration with wavelength of 405 nm.

Most significantly, the deposition of an additional *p*-NiO layer increases the SPV magnitude by about 3 times, while the SPV decay time also increases significantly. Again, this increase in the SPV amplitude comes from the *p*-NiO/*n*-ZnO junction. The shortened SPV decay is apparently due to enhanced recombination rate

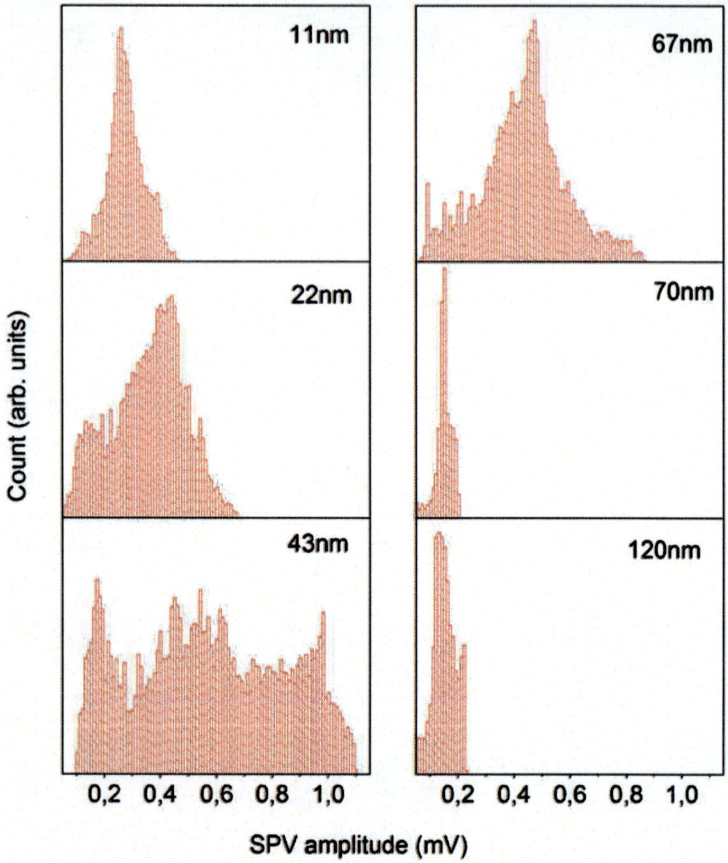

Fig. 14 Probability functions of the surface distribution of the SPV amplitude in ZnO/Si with different thickness of ZnO layer

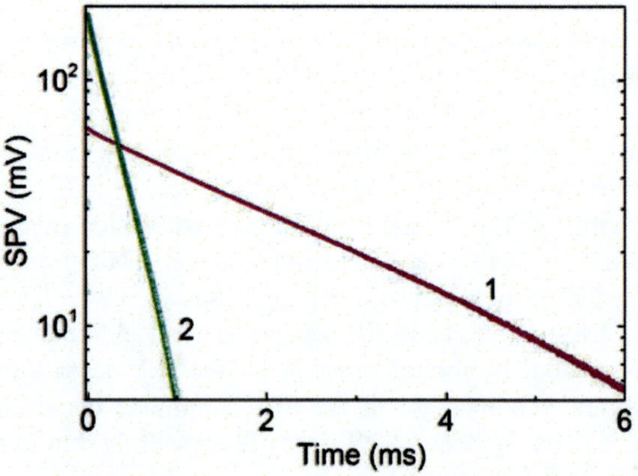

Fig. 15 SPV decay curves in ZnO/Si (1) and *p*-NiO/*n*-ZnO (2) excited with $\lambda = 405$ nm light having pulse width of 1 μs. The thickness of ZnO films is 120 nm in both cases, the thickness of NiO is about 70 nm

of photogenerated carriers at the additional recombination centers formed at this interface.

5 Conclusions

In summary, the structural and electro-physical characteristics of ZnO/Si heterostructures formed by the magnetron film deposition from aluminum-doped ZnO target are addressed. The film recrystallization and microcrystal structure restoration are observed. It is found that the ZnO/Si barrier is about 0.66 eV and depends on the annealing temperature of the implanted structures. Diode structures with a transparent electrode (ITO/ZnO/Si/Al) have photosensitivity in the wide spectral region of 0.4–2.5 μm. The photosensitivity mechanisms in different spectral bands require further refinement.

The deposited ZnO films with thicknesses of less than ≈70 nm have roughly three times greater photovoltage magnitude compared with that having greater thicknesses, and this coincide with the microstructural evolution of the film. The SPV decays slow down accordingly exhibiting decay times of the order 10 μs at the small film thicknesses and 100 μs at the ones exceeding ≈70 nm. These results can be interpreted in terms of the recombination and carrier trapping centers in the film and at the ZnO/Si boundary, whose concentration varies with the film thickness.

It is furthermore shown that forming *p*-NiO/*n*-ZnO/Si heterostructure increases the photovoltage magnitude, which is up to several times, and shortens the SPV decays. This can be effectively used in manufacturing rapid photosensitive elements.

References

1. Pearton SJ, Norton DP, Ip K, Heo YW, Steiner T (2005) Recent progress in processing and properties of ZnO. Prog Mater Sci 50(3):293–340
2. Bedia Z, Bedia A, Kherbouche D, Benyoucef B (2013) Electrical properties of ZnO/p-Si heterojunction for solar cell application. Int J Mater Eng 3:59–65
3. Purica M, Budianu E, Rusu E (2000) Heterojunction with ZnO polycrystalline thin films for optoelectronic devices applications. Microelectr Eng 51–52:425–431
4. Zheng ZQ, Yao JD, Wang B, Yang GW (2015) Light-controlling, flexible and transparent ethanol gas sensor based on ZnO nanoparticles for wearable devices. Sci Rep 5:11070
5. Shih CC, Lee WY, Chiu YC, Hsu HW, Chang HC, Liu CL, Chen WC (2016) Transistor memory devices using nano-floating gate of polymer/ZnO nanocomposites. Sci Rep 6:20129
6. Umit O, Daniel H, Hadis M (2010) ZnO devices and applications: a review of current status and future prospects. Proc IEEE 98(7):1255–1268
7. Battaglia C, de Nicolás SM, Wolf SD, Yin X, Zheng M, Ballif C, Javey A (2014) Silicon heterojunction solar cell with passivated hole selective MoO_x contact. Appl Phys Lett 104:113902
8. Islam R, Shine G, Saraswat KC (2014) Schottky barrier height reduction for holes by Fermi level depinning using metal/nickel oxide/silicon contacts. Appl Phys Lett 105:182103

9. Cole JJ, Wang X, Knuesel HO, Jacobs RJ (2008) Integration of ZnO microcrystals with tailored dimensions forming light emitting diodes and UV photovoltaic cells. Nano Lett 8:1477–1481
10. Faber H, Klaumünzer M, Voigt M, Galli D, Vieweg BF, Peukert W, Spiecker E, Halik M (2011) Morphological impact of zinc oxide layers on the device performance in thin-film transistors. Nanoscale 3:897–899
11. Jaramillo R, Youssef A, Akey A, Schoofs F, Ramanathan S, Buonassisi T (2016) Using atom-probe tomography to understand ZnO:Al/SiO_2/Si Schottky diodes. Phys Rev Appl 6:034016
12. Hubya N, Ferrari S, Guziewicz E, Godlewski M, Osinniy V (2008) Electrical behavior of zinc oxide layers grown by low temperature atomic layer deposition. Appl Phys Lett 92:023502
13. Wisz G, Virt I, Sagan P, Potera P, Yavorskyi R (2017) Structural, optical and electrical properties of zinc oxide layers produced by pulsed laser deposition method. Nanoscale Res Lett 12:253
14. Janotti A, Van de Walle CG (2009) Fundamentals of zinc oxide as a semiconductor. Rep Prog Phys 72:126501
15. Ozgür U, Alivov YI, Liu C, Teke A, Reshchikov MA, Doğan S, Avrutin V, Cho SJ, Morkoç HA (2005) Comprehensive review of ZnO materials and devices. J Appl Phys 98:041301
16. Chu S, Lim JH, Mandalapu LJ, Yang Z, Li L, Liu JL (2008) Sb-doped p-ZnO/Ga-doped n-ZnO homojunction ultraviolet light emitting diodes. Appl Phys Lett 92:152103
17. Xu WZ, Ye ZZ, Zeng YJ, Zhu LP, Zhao BH, Jiang L, Lu JG, He HP (2006) ZnO light-emitting diode grown by plasma-assisted metal organic chemical vapor deposition. Appl Phys Lett 88:173506
18. Liu W, Gu SL, Ye JD, Zhu SM, Liu SM, Zhou X, Zhang R, Shi Y (2006) Blue-yellow ZnO homostructural light-emitting diode realized by metalorganic chemical vapor deposition technique. Appl Phys Lett 88:092101
19. Xiao ZY, Liu YC, Mu R, Zhao DX, Zhang JY (2008) Stability of p-type conductivity in nitrogen-doped ZnO thin film. Appl Phys Lett 92:052106
20. Ma MJ, Lu B, Zhou TT, Ye ZZ, Lu JG, Pan XH (2013) Orientation dependent band alignment for p-NiO/n-ZnO heterojunctions. J Appl Phys 113:163704
21. Waltereit P, Brandt O, Trampert A, Grahn HT, Menniger J, Ramsteiner M, Reiche M, Ploog KH (2000) Nitride semiconductors free of electrostatic fields for efficient white light-emitting diodes. Nature 406:865
22. Pant P, Budai JD, Aggarwal R, Narayan R, Narayan J (2009) Thin film epitaxy and structure property correlations for non-polar ZnO films. Acta Mater 57:4426
23. Zhang Z, Zhao Y, Zhu M (2006) NiO films consisting of vertically aligned cone-shaped NiO rods. Appl Phys Lett 88:033101
24. Ohta H, Hirano M, Nakahara K, Maruta H, Tanabe T, Kamiya M, Kamiya T, Hosono H (2003) Fabrication and photoresponse of a pn-heterojunction diode composed of transparent oxide semiconductors, p-NiO and n-ZnO. Appl Phys Lett 83:1029
25. Long H, Fang GJ, Huang HH, Mo XM, Xia W, Dong BZ (2009) Ultraviolet electroluminescence from ZnO/NiO-based heterojunction light-emitting diodes. Appl Phys Lett 95:013509
26. Wang JY, Lee CY, Chen YT, Chen CT, Chen YL, Lin CF, Chen YF (2009) Double side electroluminescence from p-NiO/n-ZnO nanowire heterojunctions. Appl Phys Lett 95:131117
27. Deng R, Yao B, Li YF, Zhao YM, Li BH (2009) X-ray photoelectron spectroscopy measurement of n-ZnO/p-NiO heterostructure valence-band offset. Appl Phys Lett 94:022108
28. Xi YY, Hsu YF, Djurisic AB, Ng AMC, Chan WK, Tam HL, Cheah KW (2008) NiO/ZnO light emitting diodes by solution-based growth. Appl Phys Lett 92:113505
29. Wang H, Fang G, Huang H, Mo X, Xia W, Dong B, Meng X, Zhao X (2011) Ultraviolet electroluminescence properties of the p-NiO/n-GaN-based heterojunction diodes. Semicond Sci Technol 26:125015
30. Abbasi MA, Ibupoto ZH, Khan A, Nur O, Willander M (2013) Fabrication of UV photo-detector based on coral reef like p-NiO/n-ZnO nanocomposite structures. Mater Lett 108:149–152
31. Hasan MR, Xie T, Barron SC, Liu G, Nguyen NV, Motayed A, Rao MV, Debnath R (2015) Self-powered p-NiO/n-ZnO heterojunction ultraviolet photodetectors fabricated on plastic substrates. APL Mater 3:106101

32. Baraik K, Singh SD, Kumar Y, Ajimsha RS, Misra P, Jha SN, Ganguli T (2017) Epitaxial growth and band alignment properties of NiO/GaN heterojunction for light emitting diode applications. Appl Phys Lett 110:191603
33. Jumaand AO, Matibini A (2017) Synthesis and structural analysis of ZnO-NiO mixed oxide nanocomposite prepared by homogeneous precipitation. Ceram Int 43:15424–15430
34. Majeed AMA, Nusseif AD, Hameed NS (2019) Investigations of ZnO-NiO/PSi heterojunction for solar cell application. AIP Conf Proc 2144:030015
35. Li X, Shen R, Zhang B, Dong X, Chen B, Zhong H, Cheng L, Sun J, Du G (2011) Nitrogen doped ZnO thin films prepared by photo-assisted metalorganic chemical vapor deposition. J Nanosci Nanotechnol 11:9741–9744
36. Roshchina NM, Smertenko PS, Stepanov VG, Zavyalova LV, Lytvyn OS (2013) Some properties of thin film structures on the base of ZnO obtained by MOCVD method. Solid State Phenom 256:228–236
37. Chatel G, Macfarlane DR (2014) Ionic liquids and ultrasound in combination: synergies and challenges. Chem Soc Rev 43:8132–8149
38. Hosni M, Farhat S, Schoenstein F, Karmous F, Jouini N, Viana B, Mgaidi A (2014) Ultrasound assisted synthesis of nanocrystalline zinc oxide: experiments and modeling. J Alloy Compd 615:S472–S475
39. Zakirov MI, Korotchenkov OA (2017) Carrier recombination in sonochemically synthesized ZnO powders. Mater Sci Pol 35:211–216
40. Zakirov MI, Kuryliuk VV, Korotchenkov OA (2016) Optical properties of ZnO fabricated by hydrothermal and sonochemical synthesis. J Phys Conf Ser 741:012028
41. Ji PF, Li Y, Zhou FQ, Song YL, Huang HC (2019) Fabrication and electroluminescence of sheet-like ZnO/Si light-emitting diodes by radio frequency magnetron sputtering method. Mater Lett. http://www.doi.org/10.1016/j.matlet.2019.127028 (in Press)
42. Nagabharana RM, Kiran N, Guha P, Sundaravel B, Bhatta UM (2019) Structural characterization of magnetron sputtered ZnO thin films on Si(100) using RBS, scanning and high resolution transmission electron microscopy methods. Surf Interfaces 15:239–243
43. Wang Y, Song J, Zhang J, Zheng G, Duan X, Xie X, Han B, Meng X, Yang F, Wang G, Zhao Y, Li J (2019) Effect of substrate temperature on F and Al co-doped ZnO films deposited by radio frequency magnetron sputtering. Sol Energy 194:471–477
44. Liu Y, Zhu S (2019) Preparation and characterization of Mg, Al and Ga co-doped ZnO transparent conductive films deposited by magnetron sputtering. Results Phys 14:102514
45. Kim H-K, Chung K-B, Kal J (2019) Comparison of ZnO buffer layers prepared by spin coating or RF magnetron sputtering for application in inverted organic solar cells. J Alloys Compd 778:487–495
46. Fareed S, Jamil A, Tiwari N, Rafiq MA (2019) Influence of Cr doping on Schottky barrier height and visible light detection of ZnO thin films deposited by magnetron sputtering. Micro Nano Eng 2:48–52
47. Ellmer K, Bikowski A (2016) Intrinsic and extrinsic doping of ZnO and ZnO alloys. J Phys D Appl Phys 49:413002
48. Bikowski A, Rengachari M, Nie M, Wanderka N, Stender P, Schmitz G, Ellmer K (2015) Research update: Inhomogeneous aluminium dopant distribution in magnetron sputtered ZnO: Al thin films and its influence on their electrical properties. APL Mater 3:060701
49. Alcock CB, Itkin VP, Horrigan MK (1984) Vapour pressure equations for the metallic elements: 298–2500K. Can Metall Q 23:309–313
50. Sieber I, Wanderka N, Urban I, Dörfel I, Schierhorn E, Fenske F, Fuhs W (1998) Electron microscopic characterization of reactively sputtered ZnO films with different Al-doping levels. Thin Solid Films 330:108–113
51. Kinemuchi Y, Nakano H, Kaga H, Tanaka S, Uematsu K, Watari K (2011) Microstructural evidence of Hall mobility anisotropy in c-axis textured Al-doped ZnO. J Am Ceram Soc 94:2339–2343
52. Kronik L, Shapira Y (1999) Surface photovoltage phenomena: theory, experiment, and applications. Surf Sci Rep 37:1–206

53. Luke KL, Cheng LJ (1987) Analysis of the interaction of a laser pulse with a silicon wafer: determination of bulk lifetime and surface recombination velocity. J Appl Phys 61:2282–2293
54. Wang XC, Chen XM, Yang BH (2009) Microstructure and optical properties of polycrystalline ZnO films sputtered under different oxygen flow rates. J Alloys Compd 488:232–237
55. Edwards DF (1998) Silicon. In: Palik ED (ed) Handbook of optical constants of solids. Academic, San Diego, pp 547–569
56. Podolian A, Kozachenko V, Nadtochiy A, Borovoy N, Korotchenkov O (2010) Photovoltage transients at fullerene-metal interfaces. J Appl Phys 107:093706
57. Nadtochiy A, Podolian A, Korotchenkov O, Schmid J, Kancsar E, Schlosser V (2011) Water-based sonochemical cleaning in the manufacturing of high-efficiency photovoltaic silicon wafers. Phys Status Solidi C 8:2927–2930

The Sensitivity to Moisture Peculiarities of Nanoscale Tin Dioxide Films Obtained by Means of Polymers

A. P. Chebanenko, L. M. Filevska, V. S. Grinevych, and V. A. Smyntyna

1 Introduction

Tin dioxide, having a high chemical resistance and stability of the adsorption characteristics [1], is widely used as a material for gas sensors sensitive elements [2–4]. As a degenerate semiconductor, transparent in film form, it is also in demand as the electrode material of solar cells, optical converters, and sensors [5–7]. The atmosphere, in which devices with sensitive or electrode elements based on SnO_2 operate, as a rule, contains active chemical compounds. The most widespread of them is conventional water vapor. Its influence on the operation of devices (sensors, solar cells, optoelectronic devices, etc.) is possible when taking into account its impact on their active elements [8]. Therefore, researchers deal with the interaction of water vapor with tin dioxide, as well as with other materials of sensitive elements. The Japanese company "Figaro" produces water vapor sensors on thick porous sintered SnO_2 layers to monitor food vapor [9].

The electrical resistance and gas sensitivity of tin dioxide strongly depend on the size of its grains [10], since it determines the number of surface states that are important both for the adsorption interaction and the conductivity. From this point of view, exactly the nanoscale forms of SnO_2 show the greatest gas sensitivity, as a consequence of their high conductivity variability. On the other hand, the large relative surface of such forms leads to the active adsorption of water, which was shown in [11]. The presence of bound water in tin dioxide nanoscale forms up to almost 800 °C was also shown there. As a rule, active adsorption of water vapor from the atmosphere is already observed at room temperature. Moreover, as theoretically shown in [12, 13], it can be both associative and dissociative in nature, and both of these mechanisms affect the change in the electrical conductivity of tin dioxide.

A. P. Chebanenko · L. M. Filevska (✉) · V. S. Grinevych · V. A. Smyntyna
Odessa Mechnikov National University, Odessa, Ukraine
e-mail: lfilevska@gmail.com

O. Fesenko and L. Yatsenko (eds.), *Nanomaterials and Nanocomposites, Nanostructure Surfaces, and Their Applications*, Springer Proceedings in Physics 246, https://doi.org/10.1007/978-3-030-51905-6_25

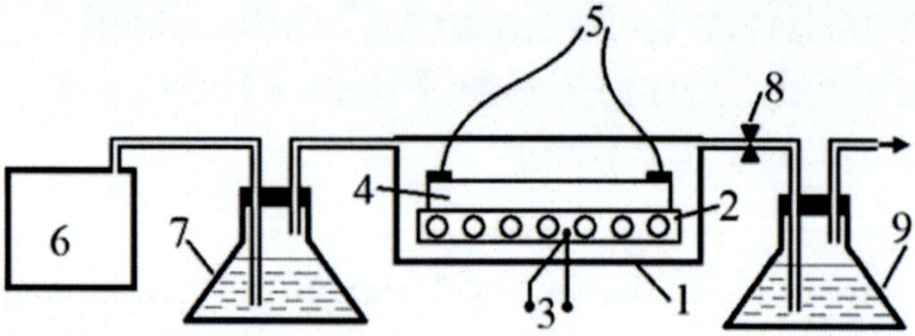

Fig. 1 Schematic illustration of the camera for measurements in water vapor. Here: 1 is the measuring chamber; 2—heating element; 3—thermocouple; 4—SnO_2 film; 5—tap to measuring installation; 6—compressor; 7—a source of vapors (water, ethyl alcohol, etc.); 8—tap; 9—water seal

Considering the necessity to create gas sensors operating without additional heating, it is advisable to study the moisture influence on the electrical properties of tin dioxide at room temperature.

In the proposed work, the moisture sensitivity kinetics features at room temperature of nanosized tin dioxide films, obtained using polymers are investigated.

2 Research Methods

Thin nanosized films of tin dioxide were obtained by the sol–gel method using polyvinyl acetate for structuring [14]. The bis(acetylacetonato) dichlorotin(IV) was used as a precursor for the preparation of SnO_2 layers, the properties of which were studied in [15, 16].

For measurements in water vapor, the studied SnO_2 films were placed in an installation schematically shown in Fig. 1. Water vapor and the dry air were periodically pumped into the chamber. During the inlet time (90 s), the sample managed to achieve maximum or minimum sensitivity depending on the atmosphere in the chamber. The measurements were carried out at a temperature of 20 °C.

3 Results and Discussion

The relaxation of the current in the SnO_2 film at the periodic inlet of water vapor or dry air into the chamber ($V = 250$ V) is shown in Fig. 2.

It is seen that, after the water vapor inlet into the chamber, the current in a relatively short period of time (about 10 s) almost doubles, and for the whole period of water vapor's presence in the chamber (90 s)—it increases by almost an order of magnitude. Moreover, after the water vapor removal, the current strength decreases

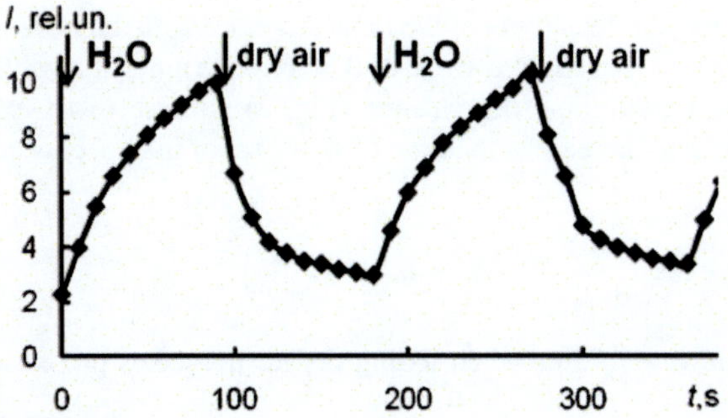

Fig. 2 Relaxation of the current in the SnO_2 film with periodic inlet of water vapor or dry air into the chamber ($V = 250$ V)

approximately to the initial value for the same period of time, that is, the change in conductivity is reversible.

Measurements of current relaxation were carried out in the range of applied voltages from 150 to 320 V. The magnitude of the sensitivity S at various voltages was calculated by the formula:

$$S = \frac{I_1 - I_0}{I_0} \quad (1)$$

where I_1 is the current in the presence of water vapor, I_0 is the current in the atmosphere of dry air. The results of the calculations are shown in Table 1 and Fig. 3.

Table 1 The sensitivity values of the studied SnO_2 films at various values of the applied voltage

V, V.	150	200	250	300	350
S, rel.un.	0,62	0,64	0,71	0,76	0,73

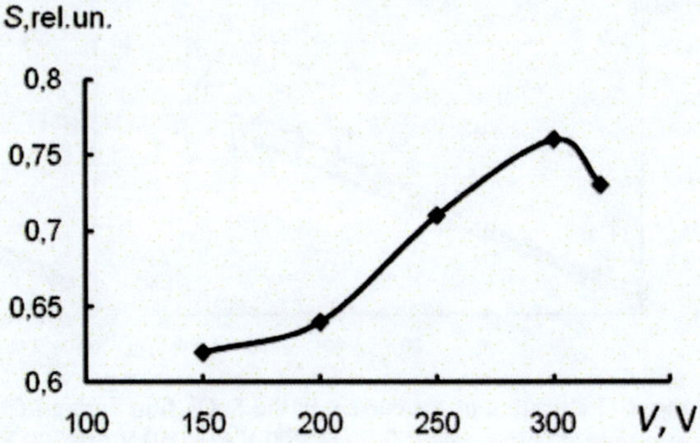

Fig. 3 The applied voltage dependence of the SnO_2 film sensitivity in the presence of water vapor

As can be seen, the sensitivity of the films increases with the increasing of applied voltage and reaches a maximum value of 0.76 relative units at a voltage of 300 V.

The type of current/time dependence (Fig. 2) is close to an exponential law. Namely, the current increase at the time of water vapor inlet occurs according to the law:

$$I = I_0\left(1 - e^{-\frac{t}{\tau_a}}\right) \tag{2}$$

A current decline in time when letting dry air in—takes place according to the law:

$$I = I_0 e^{-\frac{t}{\tau_d}} \tag{3}$$

Here τ_a and τ_d are some time constants which characterize the rate of adsorption (τ_a) and desorption (τ_d). In this case, the time dependences of the current growth, replotted in the coordinates $\ln \frac{I_0}{I_0 - I} = f(t)$, and the current decay, rebuilt in the coordinates $\ln \frac{I_0}{I} = f(t)$ should be straight lines. This is exactly what is observed in Figs. 4 and 5, showing the relaxation of the current in the SnO_2 film during the inlet of water vapor into the chamber, measured at voltages from 150 to 320 V, as well as during the inlet of dry air.

A satisfactory rectification of the experimental results in the indicated coordinates confirms the validity of the accepted assumption about the exponential nature of the change in current, and therefore also the adsorption and desorption processes. From the angular coefficients, the values of adsorption (τ_a) and desorption (τ_d) time constants were calculated (Table 2 and Fig. 5).

As can be seen, the value of the adsorption time constant τ_a is within 32–39 s, the desorption time constant τ_d is 22 s. This indicates the rates discrepancy of adsorption and desorption processes and a higher rate of the desorption process. Such character of these values indicates the predominance of associative adsorption of water on the surface of the studied films at room temperature and the indicated voltage values.

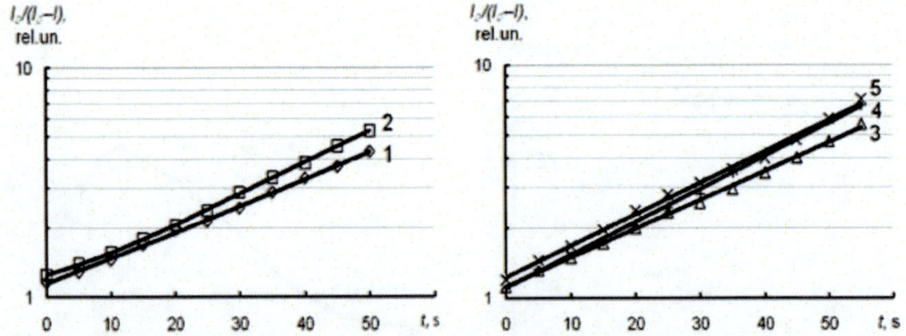

Fig. 4 Relaxation of the current in the SnO_2 film during inlet of water vapor into the chamber, measured at voltages of 150 V (1), 200 V (2), 250 V (3), 300 V (4), 320 V (5)

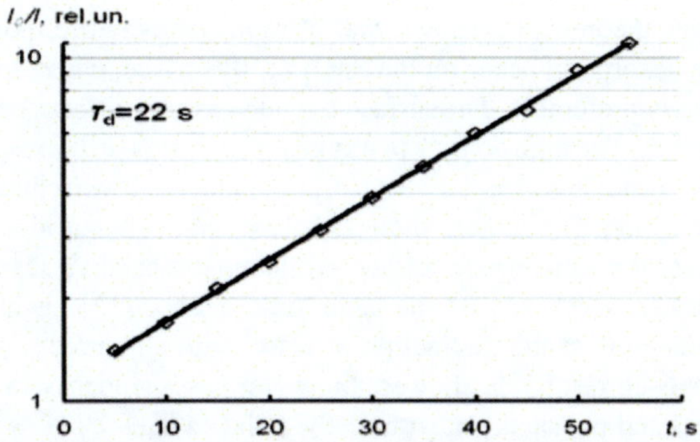

Fig. 5 Relaxation of the current in the SnO_2 film when letting dry air into the chamber ($V = 200$ V)

Table 2 The time constants of water vapor adsorption at different values of the applied voltage

V, V	150	200	250	300	320
τ_a, s	39	35	33	32	32

The time constant dependence of water vapor adsorption on the applied voltage is shown in Fig. 6.

It can be noticed that with the applied voltage growth, the adsorption time constant τ_a decreases, which indicates an increase in the rate of adsorption processes. As it was expected, the films' sensitivity magnitude to water vapor *S* in this region of voltages increases (Fig. 3). It is characteristic that at a voltage $V \geq 300$ V (when the sensitivity value reaches a maximum and then decreases) the adsorption time constant reaches $\tau_a = 32$ s and then does not depend on the voltage. Such dependence may indicate some saturation of the adsorption process associated with the water vapor centers' saturation on the film surface at the specific voltage. But the decrease in sensitivity may indicate the presence of a competitive process, the nature of which can be determined by reactions on the surface with the adsorbed water vapor participation.

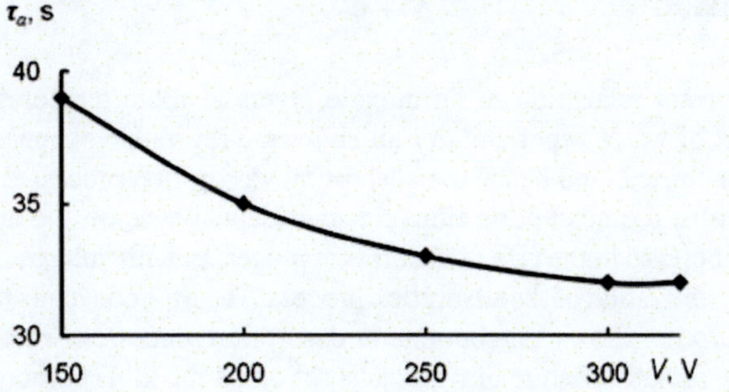

Fig. 6 The time/voltage dependence for the water vapor adsorption constant

In [12], it was theoretically shown that at room temperature, both associative and dissociative adsorption occur on the SnO_2 (110) surface, where the latter being thermodynamically preferable. The surface oxygen has a significant influence on the nature of adsorption. The high density of oxygen on the surface, corresponding to the oxidized surface, promotes associative adsorption, and the low (reduced surface)—dissociative adsorption. The authors showed that associative adsorption leads to electron enrichment of the metal oxide surface, while the dissociative adsorption causes surface depletion by electrons. At the same time, author [17] having carried out theoretical calculations of the interaction of water with all surfaces of tin dioxide, obtained the opposite result. He showed that water has a tendency to dissociate on oxidized surfaces and physically adsorb on the reduced ones.

The calculations using the density functional theory and the formalism of Gaussian and plane waves in [13], theoretically showed that dissociative adsorption of H_2O is usually observed on SnO_2 (110) surfaces, accompanied by stable associative configurations of H_2O. Both associative and dissociative adsorption of water was stabilized due to intermolecular hydrogen bonds with an increase of water amount on tin dioxide surface (110).

In [18, 19], the authors experimentally showed the presence of a noticeable amount of adsorbed oxygen and its effect on the electrical conductivity of the studied films at room temperature, as well as the dissociative nature of water adsorption on their surface as the temperature rises to 150 °C. Taking into account these data and theoretical calculations of authors [12, 13, 17], it is concluded about preferably associative adsorption of water at room temperature in the films, at least up to the voltage of 300 V. It is predictable that a further increase in voltage leads to a change in the adsorption character from associative to dissociative, due to strong measuring electric field contributing to the dissociation of water molecules. In [18], this is facilitated by increased temperature. This is reflected in the fall in sensitivity of the studied films. It can be assumed that the strong electric field applied to the samples promotes adsorption of both oxygen and water at room temperature, whereas an increased temperature contributes to a better sensitivity of tin dioxide to various adsorbates.

4 Conclusion

Studies of current relaxation in tin dioxide layers at room temperature under the alternate inlet of water vapor and dry air showed a reversible exponential nature of the change in current, and hence the adsorption–desorption processes.

The nature of the adsorption time constant dependence on the applied voltage indicates an increase in the rate of adsorption processes with voltage growth and the subsequent stabilization of the adsorption process. The time constants ratio of adsorption and desorption shows a higher rate of desorption processes, which witnesses in favor of the associative nature of water adsorption on the films at room temperature.

The results indicate that SnO_2 films can be used as sensitive elements for moisture sensors operating at room temperature.

References

1. Maksimovich NP, Matushko IP, Ripko OP, Derkachenko NM (2011) Doslidgennya stabilnosti dovgotryvaloji roboty adsorbcijno-napivprovidnykovykh gasovykh sesoriv na osnovi SnO_2 z dobavkamy cobaltu. Metody i ob'ecty khimicheskogo analiza 6(4): 198–201 (Maksimovich NP, Matushko IP, Ripko OP, Derkachenko NM (2011) Investigation of the stability of long-term operation of adsorption-semiconductor gas sensors based on SnO_2 with cobalt additives. Methods Objects Chem Analy 6(4):198–201)
2. Sun Y-F, Liu Sh-B, Meng F-L, Liu J-Y, Jin Zh, Kong L-T and Liu J-H (2012) Metal oxide nanostructures and their gas sensing properties: a review. Sensors 12:2610–2631
3. Chiu S-W, Tang K-T (2013) Towards a chemiresistive sensor-integrated electronic nose: a review. Sensors (Basel) 13(10):14214–47
4. Woo HS, Na CW, Lee JH (2016) Design of highly selective gas sensors via physicochemical modification of oxide nanowires: overview. Sensors 16(9):1531
5. Granqvist CG (2007) Transparent conductors as solar energy materials: a panoramic review. Sol Energy Mat Sol Cells 91(17):1529–1598
6. Ginley D, Hosono H, Paine DC (2011) Handbook of transparent conductors. Springer, London
7. Smyntyna V, Borshchak V, Brytavskyi I (2018) Nonideal heterojunctions for image sensors. Nova Publishers, New York
8. Gordienko YuO, Dzadevich SV, Druzhinin AA, Yevtukh AA, Lenkov SV, Lepikh YI, Melnik VG, Romanov VA (2010) Stvorenya mikroelectronnykh datchykiv novogo pokolinnya dlya intellektualnych system. Za redaccijeyu YI Lepikha. Odessa Astroprint 289 p. (Gordienko YuO, Dzadevich SV, Druzhinin AA, Yevtukh AA, Lenkov SV, Lepikh YI, Melnik VG, Romanov VA Creating next generation microelectronic sensors for intelligent systems. In: Lepikh II (ed). Odessa Astroprint 2010)
9. Ignatieva N (2005) Datchiki gasa firmy FIGARO, Elektronika: Nauka. Tekhnologii. Bisnes, 2: 34–37. (Ignatieva N (2005) FIGARO gas sensors, electronics: science technology. Business 2:34–37)
10. Xu C, Tamaki J, Miura N, Yamazoe N (1991) Grain size effects on gas sensitivity of porous SnO2—based elements. Sens Actuat B Chem 3(2):147–155
11. Rumyantseva MN, Bulova MN, Chareev DA, Ryabova LI, Akimov BA, Arkhangelsky IV, Gaskov AM (2001) Sintes i issledovanije nanokompositov na osnove poluprovodnicovykh oksidov SnO_2 i WO_3, Vestn Mosk Univ Ser 2 Khimiya 42(5): 348–355. (Rumyantseva MN, Bulova MN, Chareev DA, Ryabova LI, Akimov BA, Arkhangelsky IV, Gaskov AM (2001) Syntesys and research of nanocomposites based on semiconductor oxides SnO_2 and WO_3. Bull of Mosc Univ Ser 2 Chem 42(5):348–355)
12. Santarossa G, Hahn KR, Baiker A (2013) Free energy and electronic properties of water adsorption on the SnO_2 (110) surface. Langmuir 29(18):5487–5499
13. Hahn KR, Tricoli A, Santarossa G, Vargas A, Baiker A (2012) First principles analysis of H_2O adsorption on the (110) surfaces of SnO_2, TiO_2 and their solid solutions. Langmuir 28(2):1646–1656
14. Filevskaya LN, Smyntyna VA, Grinevich VS (2006) Morphology of nanostructured SnO_2 films prepared with polymers employment. Photoelectronics 15:11–14
15. Ulug B, Türkdemir HM, Ulug A, Büyükgüngör O, Yücel MB, Smyntyna VA, Grinevich VS, Filevskaya LN (2010) Structure, spectroscopic and thermal characterization of bis(acetylacetonato)dichlorotin (IV) synthesized in aqueous solution. Ukr Chem J 76(7):12–17
16. Grinevych V, Smyntyna V, Filevska L, Savin S, Ulug B (2017) Thermogravimetric study of nano SnO_2 precursors. In: Fesenko O, Yatsenko L (eds) Nanophysics, nanomaterials, interface studies, and applications, selected proceedings of the 4th international conference on nanotechnology and nanomaterials (NANO 2016), August 24–27, 2016, Lviv, Ukraine. Springer proceedings in physics 195. Springer, pp 53–61
17. Zakaryan H (2016) Adsorption of the H and H_2O on SnO_2 surfaces in an O_2 environment: density functional theory study. Armen J Phys 9(4):283–293

18. Chebanenko AP, Filevska LM, Grinevych VS, Simanovich NS, Smyntyna VA (2017) The humidity and structuring additives influence on electrophysical characteristics of tin dioxide films. Photoelectronics 26:5–10
19. Filevska LM, Chebanenko AP, Grinevych VS, Simanovych NS (2016) The electrical characteristics of nanoscale SnO_2 films, structured by polymers. Photoelectronics 25:62–67

Formation of Manganese-Containing PEO Coatings on Aluminum Alloys

Hanna Karakurkchi, Maryna Ved', and N. D. Sakhnenko

Increasing the complexity, accuracy, reliability, and versatility of technical devices and their elements requires the widespread use of functional coatings for various purposes. The development of eco-friendly novel and original technologies for the producing of coatings with a wide range of functional properties on metal carriers is today one of the urgent tasks [1, 2]. This is due to the toughening of operating conditions and aggressiveness of the applied technological environments, as well as a general increase in the requirements for the used structural materials. Plasma electrolytic oxidizing (PEO) provides coatings similar to ceramic [3]. Such systems have a unique complex of properties, in particular catalytic, wear-, heat- and corrosion-resistant, and electrical insulating [4, 5]. A PEO feature is the possibilities of implementing in one process both material surface treatment procedure and the synthesis of oxide coatings. Obtained oxide layer may include metals from the processing substrate as well as the electrolyte components derivatives, and the resulting coatings are significantly greater in thickness than traditional oxide films [6].

Aluminum and alloys based on it are widely applied as construction materials and carriers of various active systems. This is facilitated by the features of their physic-mechanical and chemical properties. The presence of intermetallic compounds with magnesium, manganese, iron, and copper in the alloys provides a combination of mechanical strength and lightness of products. At the same time, surface heterogeneity significantly reduces corrosion resistance in aggressive environments and complicates the PEO process. It is especially difficult to synthesize thick coatings on copper-containing aluminum alloys [7].

Formation of coatings with mixed oxides on multi-component aluminum alloys, in most cases, is carried out by impregnation pre-oxidized porous layer of substrate

H. Karakurkchi (✉) · M. Ved' · N. D. Sakhnenko
National Technical University, Kharkiv Polytechnic Institute, 2, Kyrpychova str., 61002 Kharkiv, Ukraine
e-mail: anyutikukr@gmail.com

O. Fesenko and L. Yatsenko (eds.), *Nanomaterials and Nanocomposites, Nanostructure Surfaces, and Their Applications*, Springer Proceedings in Physics 246, https://doi.org/10.1007/978-3-030-51905-6_26

with solutions containing doping components and their subsequent treatment at high temperatures. But this does not allow achieving high adhesion of mixed oxide layer. In this regard, technical modes are of great interest, which make it possible to combine the surface homogenization of Al alloys with the formation of mixed conversion oxide coatings containing transition metals [8, 9]. Suchwise, strong adhesion and even distribution of the oxide layer over the surface are ensured.

Metal oxides are used as catalysts in many technological processes in gas and liquid media, as electrode materials in electrochemical synthesis and chemical current sources [5, 10]. Of particular interest are non-stoichiometric oxides, because for both the electrochemical and chemical reactions rate increases with deviation from stoichiometry of surface oxides.

Recently, works of foreign and domestic researchers are devoted to the processes of oxidation of passive metals in high-energy fields (anode-spark or micro-arc modes).

Varying the composition of the working solutions and processing conditions allows flexible control over the process of producing oxide coatings and the composition of film materials. Also, PEO method allows to obtain a uniform catalytically active layer on parts of complex profile and shape, and significant sizes. This greatly expands conversion of mixed oxide systems usage and application. Eco-friendly processing solutions and fairly simple equipment for PEO allow considering such treatment as safe resource-saving and ecological technology for obtaining oxide coatings with a broad area of application [10, 11].

Conversion coatings on different valve metals, predominantly aluminum alloys, possess catalytic properties which were widely used in the heterogeneous catalysis [12].

Mixed metal-oxide systems are known as the most promising for environmental improvements. This is due to the catalytic properties of such compounds facilitated by the complex composition, in particular, different chemical substances formation, incorporation of dopants, deviation oxides from stoichiometric ratio. Above oxide systems are characterized by a long service life at high pressures and temperatures, which is extremely important for cleaning and neutralizing toxic emissions. High mechanical strength, corrosion resistance, reliability and ability to restore provide a long operational lifecycle for coated parts and carriers.

Manganese as well as its compounds occupies a special place among the catalysts. Particularly high interest in manganese (IV) oxide is due to the large number of its varieties differing by chemical and electrochemical properties and widespread use in many technological processes [7, 13].

The purpose of this study is to improve the technology of Aluminum alloys oxidation in alkaline solutions to obtain manganese-containing coatings with high corrosion resistance and activity in gaseous wastes purification.

1 Experimental

Oxide coatings were applied to the surface of commercial Al alloys. The chemical composition of Al alloys is given in Table 1.

For PEO alkaline (KOH) solutions with addition of diphosphate $P_2O_7^{4-}$ or manganate (VII) MnO_4^- oxygen-containing anions (oxo-anions) of varied concentration were used [14]. Oxidizing was performed at anodic polarization with direct current density of 5…20 A/dm^2 and a voltage 120…240 V. Oxidizing was forced cooling for keeping up the temperature of electrolytes in the interval 25…30 °C. Voltage time dependences as well as operating parameters were fixed by MTech PGP-550S connected with computer. Stages of PEO were recorded visually and refined after graphical processing of voltage chronograms and differential dependencies.

Polarization measurements were conducted close to the corrosion potential using potentiostat Model IPC Pro-M. Potential scan rate was varied within the interval $s = 10^{-3}\ldots10^{-1}$ V/s.

The coated with oxides samples were analyzed by scanning electron microscope ZEISS EVO 40XVP. Images of a surface were gained by means of recording secondary electrons (BSE) by a scan of an electron stream on sample's surface that allowed with high resolving ability and good contrast range to explore topography of a surface. Software environment SmartSEM was used for processing images/Chemical composition of a surface explored using the analysis of a characteristic roentgen spectrum filled by means of energy-dispersive spectrometer INCA Energy 350. Stimulation of X-rays was carried out bombarding radiation of samples by the bundle of electrons with energy 15 кeV.

Topography of the surface oxide coating was studied by the AFM method. Research was carried out using NT-206 microscope with the CSC-37 contact probe. The cantilever has a lateral resolution of the order of 3 nm [15].

A diffractometer DRON-2.0 was used for the investigation of the coatings' structure. The studies were carried out using monochromatic Co-Kα radiation with a wavelength of $\lambda = 1.7902$A. The phases in the coatings were identified using the PCPDFWIN electronic file system.

Corrosion resistance of samples (area of 1 cm^2) coated with mixed oxides was investigated by methods of electrode impedance spectroscopy (EIS) and polarization resistance [16]. EIS of the electrodes with the coatings were registered in the frequency range of 10^{-2}–10^6 Hz at the free corrosion potential using the electrochemical module Autolab-30 was equipped with the frequency response analyzer module (FRA-2). The module was controlled using the Autolab 4.9 program according to the standard procedure with the subsequent processing by the Zview 2.0 package. The corrosion tests were performed in 0.1 M Na_2SO_4 and 0.01 M NaOH aqueous solutions. Graphic-analytical method was applied to simulate the electric equivalent circuits of interfacial boundary metal–metal oxide.

The catalytic properties of samples coated with mixed oxides were established in the model reaction of CO conversion to CO_2. The equipment specification and

Table 1 The chemical composition of Al alloys* (in wt%)

Alloy grade	Al	Mg	Cu	Mn	Si	Fe	Ti	Zn	Others
A99, GOST 11069	99.99	to 0.001	to 0.002	to 0.002	to 0.003	to 0.003	to 0.002	to 0.003	Ga to 0.003
AMn (1400), GOST 4784	96.35–99.0	≤0.5	≤0.2	1.0–1.6	≤0.6	≤0.7	–	≤0.1	–
D16 (1160), GOST 4784	90.9–94.7	1.2–1.8	3.8–4.9	0.3–0.9	≤0.5	≤0.5	≤0.1	≤0.3	Ni ≤0.1
AL25, GOST 1583	79.5–85.55	0.8–1.3	1.5–3.0	0.3–0.6	11.0–13.0	≤0.8	≤0.1	≤0.5	Cr ≤0.2 Ni 0.8–1.3; Pb ≤0.1; Sn ≤0.02;

*The content of aluminum indicated approximately

experimental details are similar to that described in [17]. The indicators of catalytic activity were the ignition temperature (T_i) and the conversion degree (X, %).

2 Peculiar Properties of PEO of Aluminum Alloys in Alkaline Electrolytes

Properties of Al alloys essentially depend on the chemical composition and alloy components/impurity elements ratio. Intermetallic compounds (IMC) of enough various composition and amount (Table 2) presenting in aluminum alloys as hardening phases in many respects cause a difference in their properties.

The change of alloy electrochemical properties is caused by pointed intermetallic compounds localization in the surface layer. It is allowed for in parameter q_i—the specific surface area corresponding to the appropriate structure element localization. For example, it is known [19], that AA 2024-T alloy contain intermetallic compound Al-Cu-Mg with the considerable proportion of surface area, therefore as a first approximation it is possible to introduce such surface as

$$(\mathrm{AA\,2024-T3})_s \approx \theta_1(\mathrm{Al}) + \theta_2(\mathrm{Al-Cu-Mg}) + \ldots + \theta_i(\mathrm{IMC}),$$

or in limiting state:

$$\theta(\mathrm{IMC}) \gg 0, \theta(\mathrm{Al}) > 0, \theta(\mathrm{Cu}) \to 0, \theta(\mathrm{Mg}) \to 0,$$

at invariable fulfillment of normality condition $\Sigma\theta_i = 1$.

Manganese presence in the AMn alloy favorably influences its corrosion-electrochemical behavior due to the formation of intermetallic compounds with iron Al_6(Fe, Mn), AlFeSiMn, and others with enough negative electrode potential [19]. It eliminates the contribution of the cathode sections of a corrosion process localized, as a rule, on iron. However, at the common decrease of the surfaces hetero-resistivity, there is a propagation of its reactive capacity. Therefore, the variation of aluminum alloys and intermetallic compounds electrode potentials over a wide range (Table 3)

Table 2 The Intermetallic compounds in aluminum alloys [18, 19]

Alloy (ISO)	Intermetallic compounds
D16 (2024)	Hardening phases θ (Al_2Cu) i S (Al_2CuMg), intermetallic compounds (Fe, Mn)
AMn (AlMn1Cu)	Al_2Cu, $Al_{20}Cu_2(MnFe)_3$, Al-Cu-Mg, Al-Cu-Mn–Fe
AL25 (AlSi12CuNiMg)	$MnAl_6$, in Si and Fe presence—Al_6(Fe, Mn), AlFeSiMn, α (AlFeSi)

Table 3 Corrosion potentials in aqueous solutions

System	– E_{cor}, V, in solutions	
	0,5 M NaCl	NaCl—53 g/dm^3, H_2O_{2-3} g/dm^3
D16T, D1T	0.37…0.39	0.47…0.48
AA 2024 –T3	–	0.35…0.37
AMn	–	0.83
Al_2Cu	0.37…0.42	
Al_2CuMg	0.67…0.69	0.69
Al_6Mn	0.58…0.61	–
Al_3Mg_2	0.93…1.00	1.02
Al_2MgSi	1.15	–
Al_3Fe	0.14…0.33	0.34
Al_3Ni	0.21…0.49	0.30
Mn (Fe, Si)	0.52	–

is necessary to allow the substantiation of electrolyte composition and development of alloys anodic oxidation technological regimes.

In particular, at D16 alloy oxidizing the cuprum content increase negatively affects thickness and a micro-hardness of oxides, therefore the oxide films by thickness no more than 30…50 μm may be formed on such alloys. If cuprum content is more than 5%, thick layer oxidizing in standard electrolytes is not recommended due to the formed complication and low oxides electro-physical properties [20, 21]. It is known that there are a number of electrolytes for thick (up to 100 μm) films on D16 alloys formation, but the time of oxidizing in them increases till 4 h.

It is possible to suppose, that efficiency of Al–Cu–Mn–Fe and Al–Cu–Mg alloys anodic oxidation will be essentially above, if partial processes (aluminum oxidation and alloy components dissolution) become parallel during anodic processing. Such action may be presented by the scheme (Fig. 1), allowing the parallel proceeding of two reactions—anodic oxidation of aluminum both from the alloy matrix $(Al)_S$ and its IMC $(Al–Cu–Mg)_S$, and also alloy elements ionization with formation of complexes, capable to dissolve in electrolyte bulk.

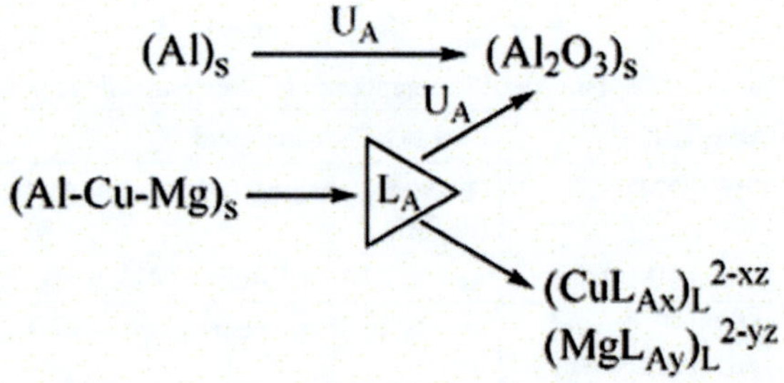

Fig. 1 The scheme of surface layers transformations of aluminum $(Al)_S$ and an intermetallic compound $(Al\text{-}Cu\text{-}Mg)_S$ at anodic polarization by voltage U_A in a solution with ligand L_A

For this purpose, it is necessary to add into an electrolyte ligand L_A, fulfilling to a series of demands among which the most relevant are ability to create enough strong $(M_xL_{Ay})_L$ complexes with IMC elements, such as Cu, Mn, and Mg; high solubility of the ligand and formed complexes; intoxicity of the ligand particles and products of probable transformations with its participant; compatibility with components of an oxidizing electrolyte; thermal and chemical stability, in particular, resistance to air oxidation; the low price.

Among the restricted number of the probable ligands, fulfilling to this list of contradictory demands, we have chosen [14, 22] diphosphate-ion of a various degree of protonation

$$L_A \in \left\{P_2O_7^{4-}, HP_2O_7^{3-}, \ldots\right\}.$$

In these requirements, the most informative and objective estimation of dynamics and efficiency of an oxide coating with high protective properties formation process will be a variation of a ratio of specific surface areas on which aluminum alloys structure parts and forming aluminum oxides are localized, in particular:

$$\theta(Al_2O_3) \rightarrow 1, \theta(Al_2Cu) \rightarrow 0, \theta(Al_2CuMg) \rightarrow 0, \theta(Cu) \rightarrow 0.$$

In favor of this hypothesis, the fact testifies also that instability constants of alloy elements diphosphate complexes (Table 4) are practically in one interval of values reflecting their enough high strength [23]. The high-level pH of diphosphate solutions (pH > 9) promotes dissolution of such elements as silicon and manganese as first it is unstable in alkaline conditions, and second shows the ability to dissolve with oxo-anion formation.

Diphosphate complexes stability, as a rule, is slashed in process of the ligand protoning, therefore it is necessary to processing alloys in solutions at enough high pH when the pointed anion is present in a completely deprotonated form (pH ≥ 8.5) for the achievement of this work aims. Thus, the anodic processing of AA differed by a composition, viewed it as possible to control the formed complexes strength by a solution pH variation.

The character of aluminum and its alloys anodic polarization dependences (APD) completely confirm the validity of pushed hypotheses. Appearance of additional, in relation to pure aluminum waves and peaks on AMn and D16 alloys APD in a

Table 4 The instability constants of D16 alloy formed elements diphosphate complexes

Central complexing ion	pK_1	$pK_{1,2}$
Cu^{2+}	7.6	12.45
Mg^{2+}	7.2	–
Ni^{2+}	5.82	7.19
Zn^{2+}	8.7	11.0
Fe^{3+}	–	5.55

background electrolyte (1 M Na_2SO_4) at presence of potassium diphosphate (Figs. 2 and 3), as well as essential magnification of passivation current, is caused by alloy elements and intermetallic compounds involvement in anodic reactions.

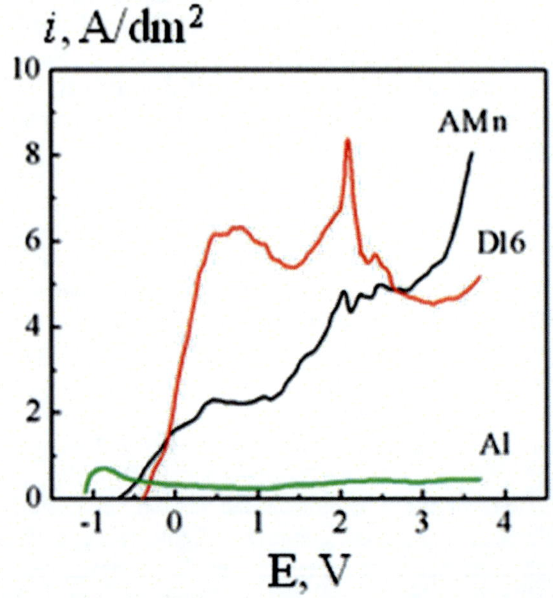

Fig. 2 Anodic polarization dependences of aluminum, alloys D16 and AMn in a solution 1 M Na_2SO_4, 0.05 M $K_4P_2O_7$. The potential scanning rate 0.002 V/s

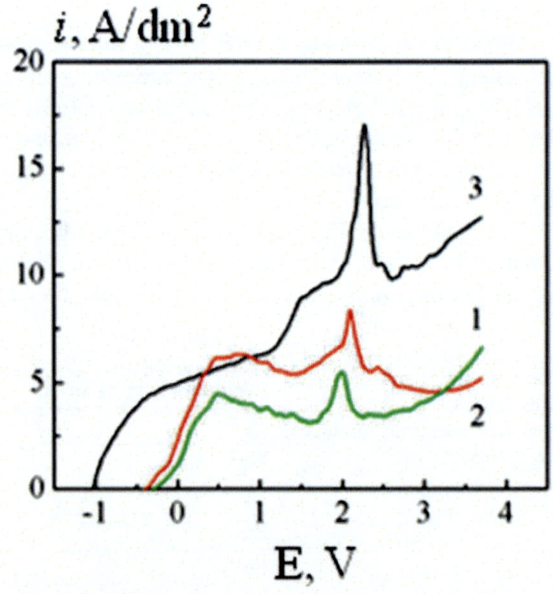

Fig. 3 Anodic polarization dependences of D16 alloy in 1 M Na_2SO_4 (1) and under addition 0.05 M (2) and 1 M (3) $K_4P_2O_7$. The potential scanning rate 0.002 V/s

Fig. 4 The content of alloying components (wt%) in the surface layer of D16 alloy (**a**) and after PEO within 30 min in electrolyte 1.0 mol/dm^3 $K_4P_2O_7$ (**b**)

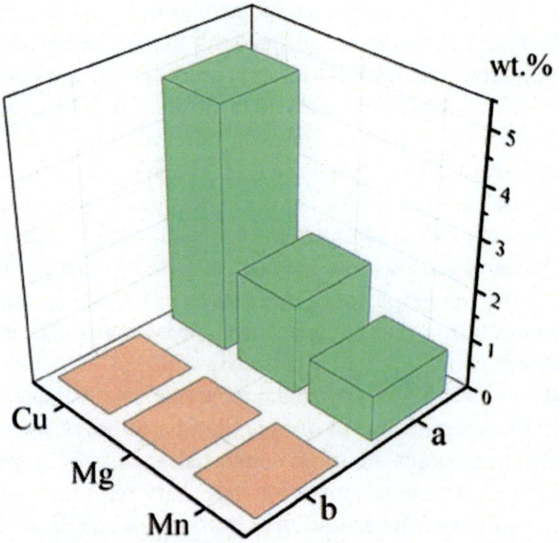

Simultaneously with aluminum oxide forming (Fig. 1) the alloy components pass in solution, and at the potential interval 2.0…2.1 V peak on both alloys anodic polarization dependences can be assigned to reactions of the deep manganese oxidizing (up to manganate VI or VII). In favor of the pointed deduction peaks also presence of peak on APD in background electrolyte, and also the fact that on the sample's surface oxidized in potassium diphosphate tracks of manganese it is not revealed from the roentgen spectra (Fig. 4).

Additional wave appearance at potentials 1.36…1.4 V on D16 alloy anodic polarization dependences in a potassium diphosphate solution testifies, in our opinion, the IMC dissolution and cuprum ions pass in the solution with its subsequent complexing with diphosphate under the scheme

$$Cu + P_2O_7^{4-} \rightarrow CuP_2O_7^{2-} + 2e. \quad (1)$$

In favor of this assumption, a wave current increase with ligand concentration raise testifies (Fig. 3), as well as an absence of characteristic sections on anodic polarization dependences of aluminum and AMn alloy.

Plurality of the obtained data creates backgrounds for development of scientific basis of aluminum alloys (such as D16) anodic oxidation process control within the limits of the tendered scheme (Fig. 1), and allows to choose a working solutions composition and electrolysis technological parameters.

Application of high-energy regimes of the plasma electrolytic oxidizing (PEO) [4, 24] is well known for obtaining the oxide coatings of the significant thickness with the educed surface, strong adherent to a substrate. D16 alloy PEO processing allows forming multifunction coatings with the wide complex of properties: strong and electrical insulating, catalytic active, and electric isolating, thermo- and corrosion-resistant. Application of the alkali metals diphosphate as the basic component of the electrolyte for PEO [22, 23] produced thick oxide coatings on aluminum alloys.

The comparative analysis of forming dependences (Fig. 5) testifies that barrier layer most promptly forms on pure aluminum, whereas the micro-arc oxidizing regime for alloys (AMn and D16) is attained during long time. It is quite explainable by the presence of IMC of the various natures in alloys surface layer (Table 2), thus the part of an anodic current expended for partial processes (Fig. 1) which produces the homogeneous equipotential surface structures of processed materials.

The topography of samples surface (Fig. 6) reflects a transition from almost juvenile surface (a) to the appearance of islands of oxide barrier Al oxide film forming and growing (b, c) simultaneously with dissolution of alloying components, then the thick coating forms in PEO regime (d). After complete oxidizing a branched surface is formed on which uniformly distributed pores and spheroidal structures alternate; and micropores are often connected among themselves (Fig. 6d). It is possible to confirm, that on a conversion layer surface there is a three-dimensional porous grid structure which becomes less expressed in the layer depth. Despite it, according to authors [23], the common porosity of such conversion layer is rather insignificant (about 5%). High pressure and temperature contribute to the presence of dissolved oxygen high concentrations in the remelted oxides and cavities. On the other hand, the porosity promotes obtaining of conversion coating of greater thickness owing to simplification of electric discharge between substrate and electrolyte.

Evidently, it is possible to describe the dynamics of basic alloying components content $\omega(t)$ in D16 alloy surface layer during PEO with usage of the chronograms of dimensionless coordinate $\chi = \omega(t)/\omega_s$, that is, the attitudes of alloy component value at a moment t to initial (maximal) one. The identical geometry of experimental

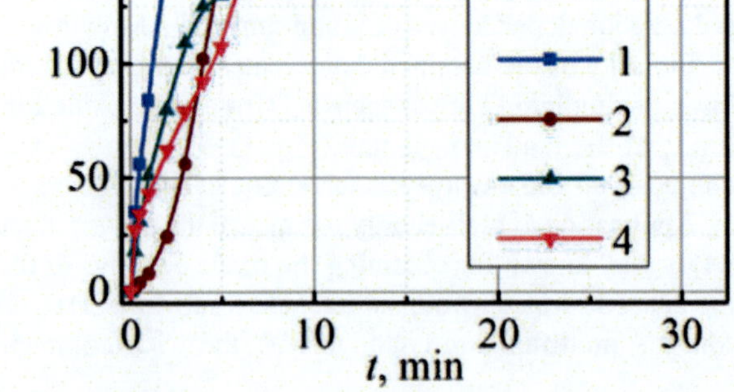

Fig. 5 Voltage chronograms for PEO of aluminum and its alloys at direct current density 5 A/dm^2 in 1 M $K_4P_2O_7$

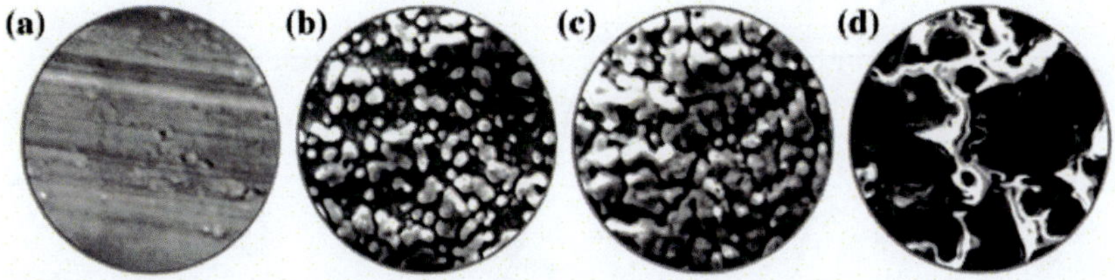

Fig. 6 Morphology of D16 alloy surface after 5 min (**a**); 7 min (**b**); 10 min (**c**); 30 min (**d**) of oxidizing in 1.0 mol/dm^3 $K_4P_2O_7$. Magnification × 1000

dependences of alloy components content from an electrolysis duration (Fig. 7a) adequately reflects a change of surface state, and some statistical dispersing, for example, for manganese (Fig. 7b), is caused by usage of ω(*t*) data measured on local sections of a samples' surface which differ a state and chemical composition.

We can see three independent sections on chronograms of the surface element composition: linear—at the exposure beginning, the subsequent decay (close to exponential), and the linear section at the significant duration of electrolysis. It is possible to introduce such chronograms ω(*t*) geometry by formalized asymptotic (Fig. 8)

$$\chi(t) \in \begin{cases} 1, 0 \le t \le t_0 \\ \lambda(t), t_0 \le t < \infty \end{cases} \quad (2)$$

where λ(*t*)—the failure function describing the alloy components losses by alloy surface layer during PEO; t_0—the parameter of localization, i.e., a time of electrode immune state before the beginning of surface layer composition varying at PEO.

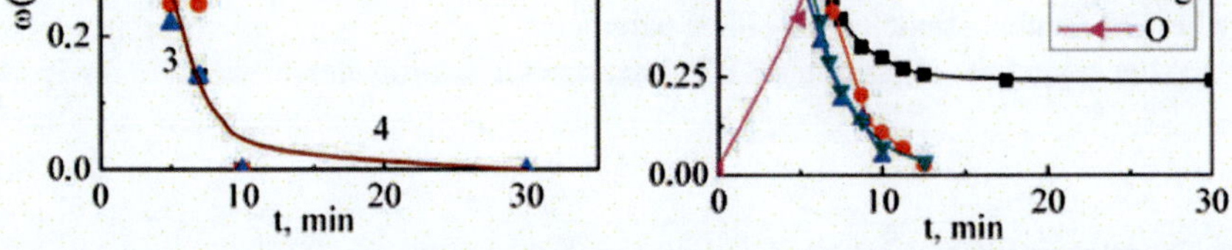

Fig. 7 Experimental (**a**) and formalized (**b**) chronograms of the alloy components contents in D16 alloy surface layer at PEO in a diphosphate solution. Numerals match to values measured on ridges (1); in valleys (2) and integral values (3). The continuous curve (4) is a result of calculation by the Equation (3)

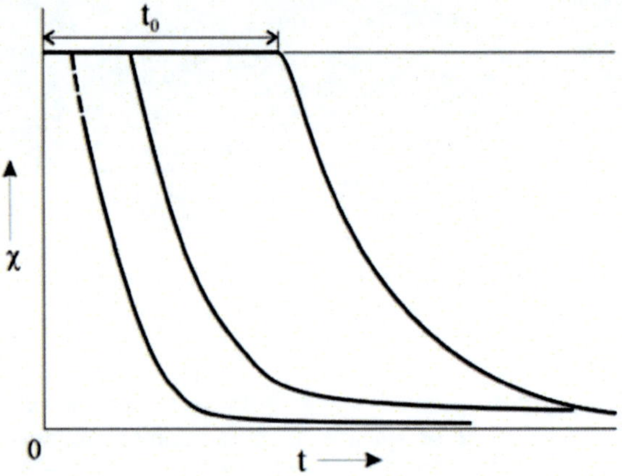

Fig. 8 Formalized chronograms of alloy components content in D16 alloy surface layer at PEO in diphosphate solutions

Naturally, both parameters $\lambda(t)$ and t_0 depend on the considerable number of factors and should be preliminarily determined for the objective specification of system state. Formalized description of the surface state (2) reflects the fact that the PEO termination time coincides with almost complete dissolution of alloy components from D16 surface layer (Fig. 4), i.e., limiting level of these quantities at $t \to \infty$ compounds $\omega(t) = 0$.

At the same time, it is known that at D16 alloy oxidizing in electrolytes which is distinct from diphosphate solutions, even in PEO regime the alloy elements content in a coating surface layer decrease slightly [25]. For example, Mn content relieves up to 0.25%, Fe—up to 0.2%, and Cu—only up to 3.0% weight.

The description of chronograms $\omega(t)$ dynamic section consists in determination of the $\lambda(t)$ function analytical view. Such problems we shall solve by leveling method on the assumption that in the elementary case $\lambda(t)$ dependence can be featured by the equation

$$\lambda(t) = a \cdot t^b. \tag{3}$$

where *a*—scale parameter, *b*—shape parameter.

After taking the logarithm (2) it is linearized in coordinates

$$Y = A + b\,X \tag{4}$$

with replacement $Y = \ln\lambda(t)$, $A = \ln a$, $X = \ln t$.

Experimental dependences $\omega(t)$ are linearized (Fig. 9) with parameters given in Table 5, and calculated from (3) $\omega(t)$ are identical to the experimental ones (Fig. 9a, a curve 4).

The shape and scale parameters (Table 5) depend on alloy components nature, electrolysis regime, and of an oxidizing electrolyte composition that provides the basis for their optimization.

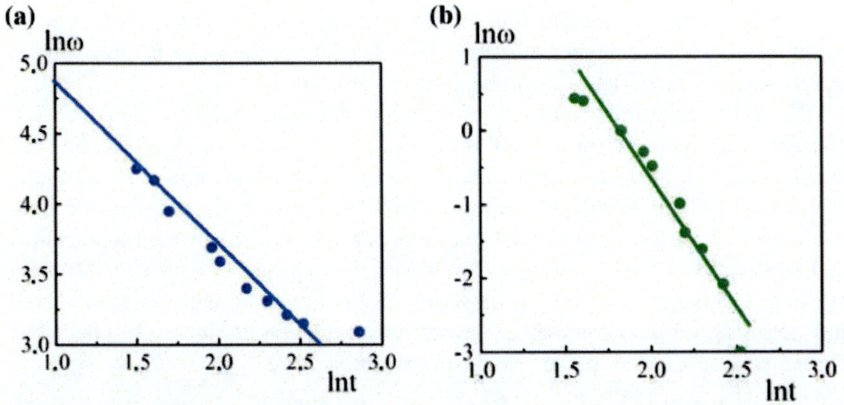

Fig. 9 Determination of failure function parameters of D16 alloy at PEO processing in diphosphate solutions: Al (**a**) and Cu (**b**)

Table 5 The parameters of failure function of D16 alloy at PEO in diphosphate solutions

Element in surface layer	a	b	t_0, min
Al	400	−1.15	5–10
Mg	121	−3.0	3–5
Cu	400	−3.3	3–5
O	25.8	0.36	≤5
Mn	5.0	−2.0	≤5

So, from the analysis of data (Figs. 5 and 8) follows, that the localization parameter t_0 corresponds to the kinetic section of forming dependences $U(t)$ duration, i.e., a time at which the voltage attains a level reasonable to start micro-electric arcs. Really, this characteristic value is predominantly stipulated by electrolysis technological parameters and an oxidizing electrolyte composition. It is possible to determine t_0 from the $\omega(t)$ chronograms: t_0 decreases with increment of oxidation current density, but in these requirements, the results are significantly dispersed concerning an average, therefore the linear segment $\omega(t) = \omega_s$ at $0 \leq t \leq t_0$ is used as the trend.

The scale parameters values for IMC elements form some bunches: approximately 4×10^2 for aluminum and cuprum, 1.2×10^2—for magnesium and 5.0—for manganese. The physical sense of parameter a accounts for an electrode reaction rate. So, the rate of Al and Cu oxidation is higher than the others that completely agree with kinetic examinations. However, to the most interesting conclusions it is possible to come on the basis of the shape parameter b analysis which, in our opinion, characterizes the reaction mechanism peculiarities. Really, similar values of the shape parameter for cuprum and magnesium match to their dissolution up to the two-charging cations M^{2+} with the subsequent formation of anionic complexes. Whereas a differing value b for manganese reflects the formation of oxo-anions MnO_4^- or MnO_4^{2-}. Deeper oxidizing of manganese, probably, causes a low rate of process, i.e., *a* value. An oxide film formation at aluminum oxidizing defines much lower *b* value, and the oxygen immobilization in a film composition is characterized by the positive fractional value of the shape parameter.

The other important peculiarity of studying aluminum alloys is the possibility of electrochemical co-deposition of multi-component oxide systems MOS on their surface [13, 26]. Such representations are formed within the limits of a hypothesis that during PEO of Al alloys in electrolytes with additives of cations or/and oxo-anions of the various natures, the requirements are created for thermo-chemical and electrochemical reactions and conversions as the result of which one strongly adherent coating is formed on a substrate. Proceeding from existing views about the mechanism of aluminum alloys electrolytic oxidation in high-energy fields we recognize four stages of oxide coatings formation at PEO:

(a) phase dielectric oxide forming obeying the faradeic mechanism before the electrical breakdown;
(b) a film breakaway and occurrence of a plasmoid in the discharge canal (*a cold light and a sparking*);
(c) thermo-chemical transformations and reactions involving gases, the electrolyte components and metal (*start of micro-electric arcs*);
(d) condensation and polymorphic transformations of oxide phases (*transition of a micro-electric arc in the arc*).

On the basis of the thermodynamic analysis of possible high-temperature chemical reactions in the discharge canal, at an electrode layer and on the oxide/electrolyte boundary the plurality of the processes proceeding in alkali solution at the oxo-anion (MnO_4^-) presence, it is possible to introduce by the scheme (Fig. 10).

Aluminum and oxygen electrochemical oxidation is tracked by thermolysis of water in a high-temperature area of the discharge canal with radicals OH· and molecules O_2 formation. Last, in view of resonance capture of an electron possibility, dissociate on corpuscles O and O^-, diffusing in an oxide phase and oxidizing a metal.

Simultaneously reactions of thermal manganate(VII)-ions intra-molecular oxidation-reduction proceed

$$2MnO_4^- \rightarrow MnO_4^{2-} + MnO_2 + O_2,\ \Delta G^0_{298} = -75.28\,\mathrm{k}J,$$

as well as disproportionation of unstable anions MnO_4^{2-} (Fig. 10).

The manganese dioxide formed in both events can incorporate in Al_2O_3 lattice as Al^{+3} and Mn^{+4} ionic radiuses are closely related or according to [8] can go through a series of high-temperature transformations:

$$MnO_2 \xrightarrow{900K} Mn_2O_3 \xrightarrow{1300K} Mn_3O_4$$

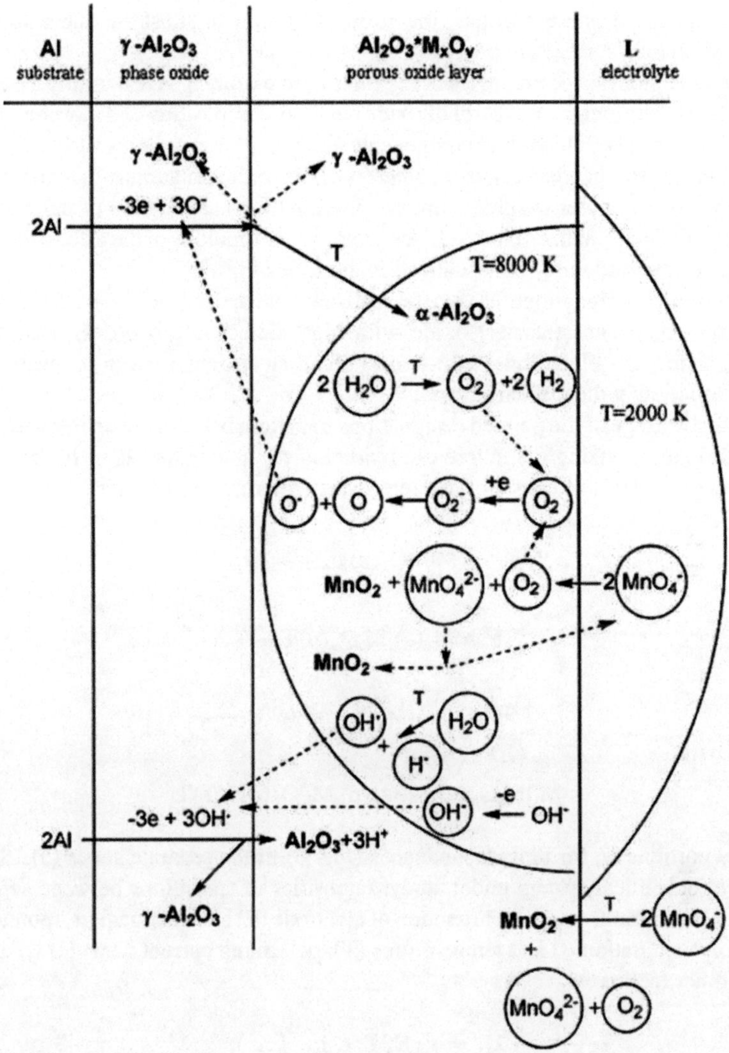

Fig. 10 The scheme of transformations at aluminum PEO treatment in a potassium permanganate (VII) solution

Thus, manganese dioxide, in our opinion, is formed predominantly in the low-temperature area and on the oxide/electrolyte boundary, and double manganese oxide of the mixed valence—in the high-temperature area of discharge canal.

Since specific electrical conductivity of MnO_2 and Mn_3O_4 on some orders exceed Al_2O_3 conductivity the voltage of sparking and forming in permanganate (VII) solution are less than in aluminum oxide formation in an alkaline electrolyte. Due to the crystallization of both oxides simultaneously with the Al_2O_3 which undergoing high-temperature phase changes, the strong adhesion to substrate and a uniform distribution on a surface are provided.

The held analysis of processes at the micro-arc oxidizing AA plurality allows to formulate backgrounds of control of oxide coatings composition and morphology by a justified choice of electrolyte components and a synthesis regimes variation. Wide opportunities for an electrolysis parameters optimization, materials diagnostic, and their functional properties prediction are provided with application of the systemic approach to the formalized electrode description. The topology of the electrochemical systems under study may be visualized in the following way.

Let us inlet designations: M—metal (substrate or carrier); MO_x—oxide of metal; MO_x (ε)—the phase (barrier) oxide with high dielectric properties; MO_x (p)—porous oxide; L—electrolyte; L(s)—solid phase electrolyte melt or products of its transformation in the discharge canal.

With the usage of the pointed designations transition between reversible states (1), pre-sparking oxidizing (2), micro-arc oxidizing (3) and synthesis of bi-functional systems (4) reflects sequence of discrete states with transitions intensity λ_{ij}

$$\mathrm{M|MO_x}(\varepsilon)|L \xrightarrow{\lambda_{12}} \tag{5}$$

$$\mathrm{M|MO_x}(\varepsilon)|MO_x\mathrm{(p)|L} \xrightarrow{\lambda_{23}} \tag{6}$$

$$\mathrm{M|M_{I}O_x}(\varepsilon)|M_{I}\mathrm{O_x(p),L(s)|L} \xrightarrow{\lambda_{34}} \tag{7}$$

$$\mathrm{M_{I}|M_{I}O_x}(\varepsilon)|M_{I}\mathrm{O_x(p),M_{II}O_y(p),L(s)|L} \tag{8}$$

The portions on forming dependences (Fig. 6) fit the separate states (5)…(8) of an electrochemical system under study. Intensities of transitions between separate states λ_{ij} are a multi-parameter function of electrode (E) and chemical (μ_i) potentials, electrolysis duration (t) and temperatures (T), polarizing current densities (j) and of some other factors

$$\lambda_{ij} = F(E, T, t, \mu_i, j \ldots). \tag{9}$$

Determination of the nature and intervals of control actions λ_{ij} variation is a starting prerequisite not only for optimization of bi-functional systems forming regimes, but also the (5)…(8) transitions realization as a whole. Regarding difference of MOS components physical properties and degrees of localization on a surface, it is possible to expect differentiating of their response on EIS alternative current signal in a wide frequency range. Therefore, each of the states (5)…(8) can be described by the equivalent circuit. The variation of its elements gives the information about the level and dynamics of the functional properties, and also the stability of a material.

3 Mixed Oxide Coatings on Aluminum Alloys

As previously shown, it is possible to irreversibly oxidize aluminum alloys in an alkaline environment. The products of oxidation are mainly aluminates and hydroxy-complexes of different compositions $Al(OH)_x^{3-x}$. Upon reaching the passivation potential, a film of phase aluminum oxides is formed on the surface, the perfection of which depends on the alloy composition. There is some contradiction associated with the fact that micro-alloying additives of copper and silicon contribute to the hardening of aluminum alloys. And at the same time, they complicate the surface treatment and violate the homogeneity of the oxide phase film. The formation of a thick protective oxide film on the surface of such alloys becomes possible only at high voltages [13].

Voltage chronogram for the PEO treatment of aluminum in the KOH solution (pH > 12.0) at low current densities *i* (5.0 A/dm^2) are sloped (Fig. 11, dependence 1). An increase in the PEO current density leads to a change in the shape of *U*—*t* curvers. An increasing in duration of PEO treatment causes the acceleration of voltage growth (Fig. 11), the dependences 2, 3).

Also there is no peak on the dependence d*U*/d*t*—*U* (Fig. 12, dependence 1), but with the increasing current density the peaks appear (Fig. 12, dependences 2, 3).

The differential curve d*U*/d*t*—*U* reaches its maximum at a voltage U_{max} = 200–215 V. Increase in voltage growth rate takes place in the range of voltages of 160–200 V. In this case, rising the oxidizing current, we observe more explicit peak on the curve. It is evident that the shaping rate is increased in comparison with that of the dissolution of phase oxide at a higher current density in the above voltage range. So, the conditions are created for the origination of film breakdown and change the oxidation route to the micro-arc region. The differential curves (oxidizing voltage

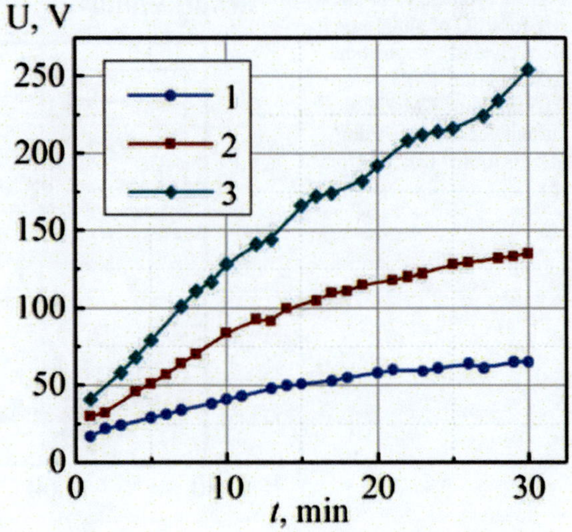

Fig. 11 Voltage chronograms for aluminum PEO in electrolyte 0.01 mol/dm^3 KOH at *i*: 5 A/dm^2 (1); 10 A/dm^2 (2); 20 A/dm^2 (3)

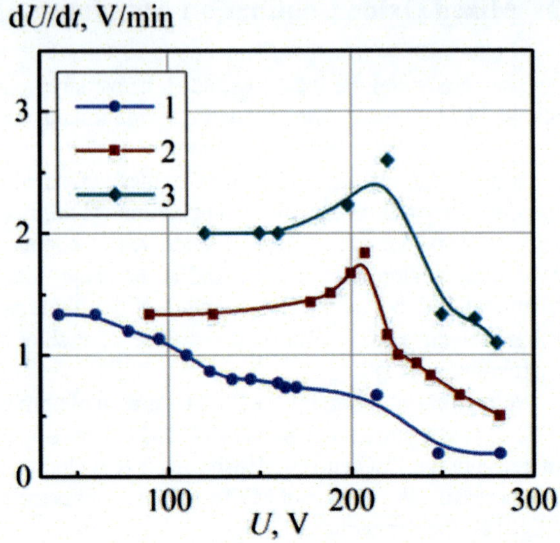

Fig. 12 Rate of voltage change for aluminum PEO in electrolyte 0.01 mol/dm^3 KOH at *i*: 5 A/dm^2 (1); 10 A/dm^2 (2); 20 A/dm^2 (3)

effect on voltage variation rate d*U*/d*t*) for the aluminum and alloys treatment when adding MnO_4^- into the alkaline electrolyte are considerably different from that in pure KOH solution (Fig. 13).

The voltage variation rate increases and such changes are more expressed with the oxidizing current rising; a maximum becomes more explicit already at the voltages of 80–95 V and as soon as a maximum is reached an abrupt drop in the voltage variation rate is observed. So, the PEO process is stabilized.

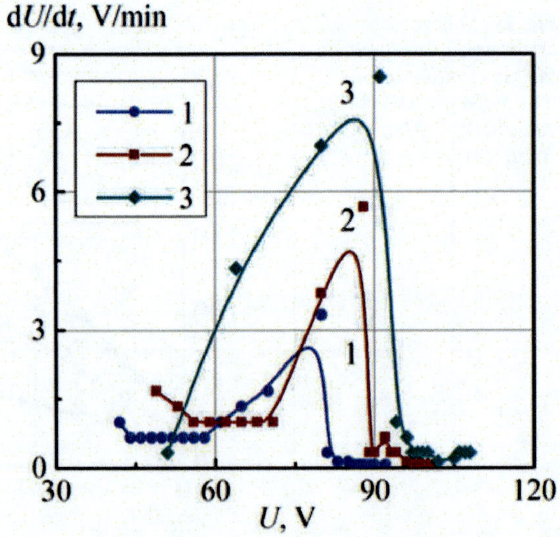

Fig. 13 Voltage variation rate for PEO of aluminum in electrolyte of composition 0.01 mol/dm^3 KOH + 0.05 mol/dm^3 $KMnO_4$ at current densities: 5 A/dm^2 (1); 10 A/dm^2 (2); 20 A/dm^2 (3)

The sparking voltage and operating one which corresponds to the formation of oxide films in manganese-containing electrolytes are lower than in solutions based on KOH. Such results can be explained by the inclusion of more conductive manganese oxides in the composition of the growing oxide layer, while in alkali almost pure dielectric alumina is formed [14].

It agrees with the previously proposed scheme and sequence of stages of mixed conversion oxide coatings formation in alkaline electrolytes. Aluminum alloys contain alloying components of various nature and intermetalides. This leads to certain features of the PEO process of such materials. Some micro-additives (for example, manganese, zinc, and silicon) can form soluble salts when the alloy is oxidized in an alkaline electrolyte. In this case, the surface layer is depleted in these components and it becomes possible to form a complete and perfect alumina film. The oxidation of copper alloyed alloys (for example, D16, AK12M2MgN) in an alkali solution is difficult due to the inability to remove copper from the surface layers. So, PEO for such alloys is not feasible to conduct in solutions of pure alkali. The oxidizing occurs with the formation of anodic oxide film containing copper, and the breakdown and sparking potentials in the PEO process are not achieved.

The addition of $KMnO_4$ to electrolyte leads to oxidizing of impurity alloy components. This contributes to the fact that the anodic oxidation process goes into modes of sparks and micro-arc discharges. As a result, mixed oxide films doped with Mn grows on aluminum alloys (Fig. 14).

It should be noted that at low current densities ($i \leq 10$ A/dm^2), sparking is almost absent, and the thickness of the forming film changes slowly, while the content of manganese in the coating changes slightly. To produce coatings that are enriched with Mn, it is necessary to maintain the current density of oxidizing in the range 15–20 A/dm^2.

The PEO of aluminum alloy AK12M2MgN with high silicon content is characterized by certain features [27, 28]. The high current densities (about 20 A/dm^2) are necessary to remove the alloying components from the outer layer and for an efficient conduction of the oxidizing.

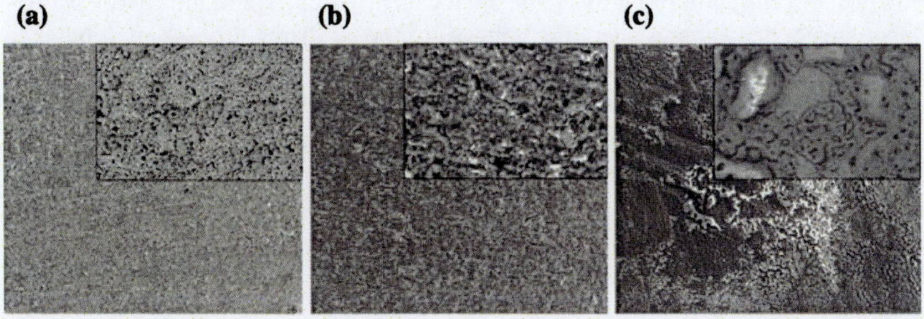

Fig. 14 Morphology of surface of Al|Al_2O_3, MnO_x coatings on alloys of aluminum: A99 (**a**), AMn (**b**), D16 (**c**). Magnification × 500 and 1000

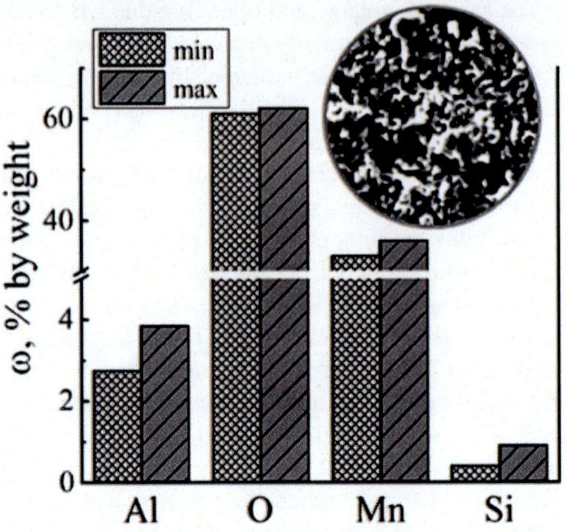

Fig. 15 Surface morphology and chemical composition of the oxide coating Al|Al_2O_3·MnO_x on AK12M2MgN. Magnification × 500

Two-step oxidizing was applied when we treated AK12M2MgN alloy by the following mode: first step at higher current density and when reaching the steady-state sparking, we decrease *i*. The research results show that the current density of 10 to15 A/dm^2 provides at the second stage, the maintenance of micro-arc mode during a long time, and the conversion mixed oxide containing manganese of 36.0 at.% is formed (Fig. 15).

The chemical analysis data are indicative of the fact that the surface consists of mixed oxides, majority of which are non-stoichiometric MnO_x, a few Al_2O_3 and silica traces (Fig. 15).

As a result of experiments, it was found that varying PEO current density and duration allows us to control the incorporation of manganese oxides into the matrix of Al_2O_3. A surface of mixed oxide layer becomes more branched and enriched with manganese when the sample is oxidized at higher current densities [10, 29]. So, not only the concentration of permanganate but especially change in the energy parameters (current density and operating voltage) of the oxidizing process contributes to the producing of catalytic-active surface.

Thereby according to research results, we consider the effect of both the PEO mode and electrolyte nature and concentration on the composition and surface state of mixed manganese-containing conversion coatings formed on the substrate made of AK12M2MgN.

Based on this, the following technological parameters can be recommended for the producing of manganese-containing oxide coatings on high alloyed alloys of aluminum (Table 6).

The above material on the chemical composition and surface character of aluminum alloy samples oxidized in alkaline electrolytes containing permanganate indicates the formation of a fairly uniformly developed surface structure. This

Table 6 Recommended parameters of PEO of aluminum alloys for producing mixed oxide coatings doped with manganese

Parameter	Coating $Al_2O_3{\cdot}MnO_x$
Electrolyte composition, mol/dm^3	KOH of 0.01–0.05; $KMnO_4$ of 0.005–0.05
Temperature, °C	20…25
Treatment mode	direct current, incident power
Current density, A/dm^2	15…20
Voltage, V	160…240
Treatment time, min	20…40
Content, at. % manganese	up to 40.0

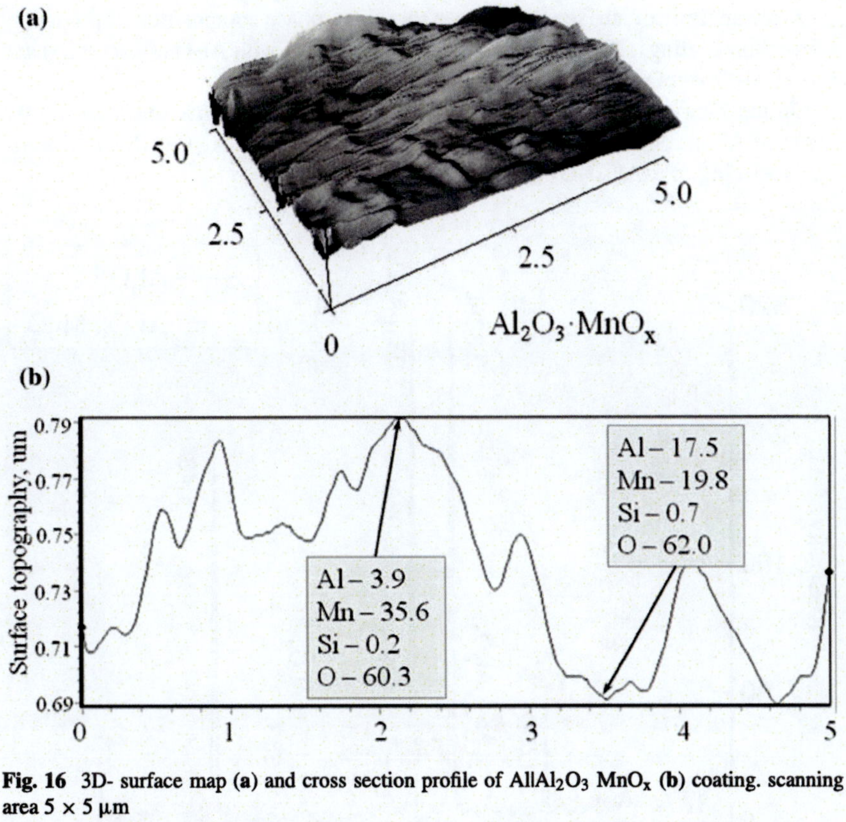

Fig. 16 3D- surface map (a) and cross section profile of Al|Al_2O_3 MnO_x (b) coating. scanning area 5 × 5 μm

assumption is confirmed by the results of surface layers topography analyzing with atomic force microscopy (Fig. 16).

We compared the surface topography of alloy samples with a monoxide coating Al_2O_3 and a mixed oxide layer Al_2O_3 MnO_x. In addition, the ratio of the hills and valleys of the relief of the coating and the distribution of the components of the oxide layer on them were established. As can be seen from Fig. 16, conical hills ranging in size from 200 nm to 1 μm alternate on the surface with valleys, and the height difference is in a wide range of 50–500 nm. In addition, it should be noted that the hills are enriched in manganese, which is consistent with the proposed mechanism for the incorporation of MnO_x into an alumina matrix. Thus, it was established that the manganese incorporation contributes to a significant increase in oxide coatings roughness in comparison to the Al_2O_3.

The surface of oxides $Al_2O_3{\cdot}MnO_x$ on the AK12M2MgN alloy has a microglobular structure [26]. So, oxides MnO_x are incorporated in the matrix of aluminum oxide and form the external coating layer.

The obtained results allow us to assume that oxides MnO_x are incorporated in the matrix of aluminum oxide and form the external coating layer.

X-ray analysis of surface oxide layer shows the phase composition of producing conversion coatings. X-ray patterns for samples coated with Al_2O_3 (black line) and mixed $Al_2O_3{\cdot}MnO_x$ (red line) are shown in Fig. 17.

On the X-ray diffraction pattern for Al|Al_2O_3 (Fig. 14, **black line**), we see the peaks of Al and Si, which are part of the base material and Al_2O_3 which indicates the producing of α-Al|Al_2O_3 during PEO [30].

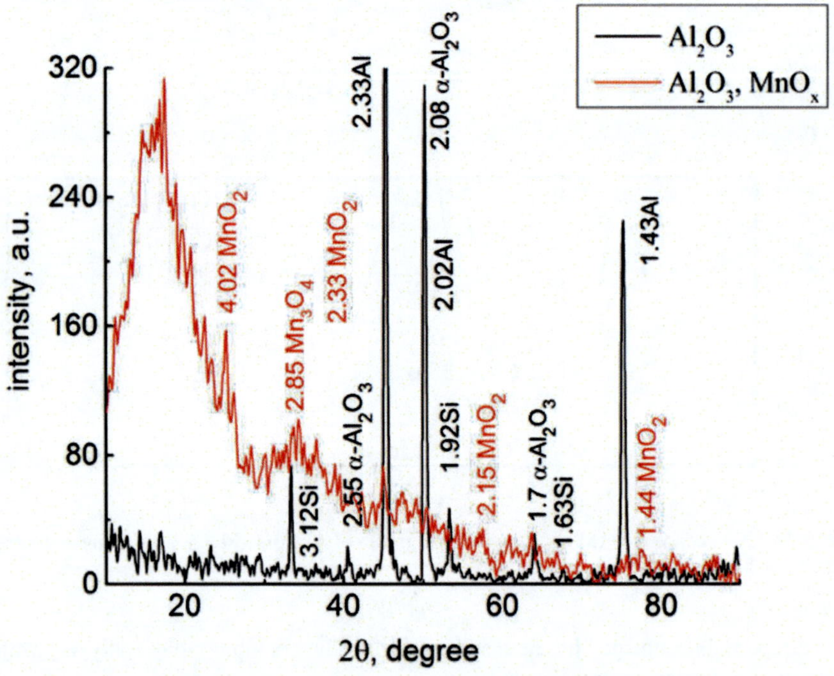

Fig. 17 X-ray patterns for samples coated with Al_2O_3 and Al_2O_3, MnO_x

The PEO of Al alloys in electrolytes with $KMnO_4$ addition leads to the producing of mixed oxide systems Al|Al_2O_3·MnO_x. The incorporation of the manganese oxides into the alumina matrix reflects on X-ray pattern (Fig. 17, red line). After that PEO Al and Si lines on X-ray pattern disappear. In the same time, the lines of the MnO_x are manifested. We note that along with manganese dioxide MnO_2, non-stoichiometric oxides Mn_3O_4 are present in the surface layers. Moreover, the higher intensity of these oxides' replicas is in the region of a fairly wide halo (at angles 2θ~20°). This is due to both the amorphous structure of oxides and PEO nonequilibrium conditions.

We can conclude that the obtained data for both X-ray and AFM analysis coincides with results for other oxide coatings on valve metals including Al alloys [14, 26].

The chemical composition, surface topography and morphology of the mixed conversion oxide coatings on alloys of aluminum are a prerequisite for different functional properties [5, 10].

4 Corrosion Properties and Catalytic Activity for Mixed Manganese and Aluminum Oxides

Quantitative parameters for corrosion of samples coated with mixed oxides Al_2O_3, MnO_x namely corrosion current density at open circuit potential j_c and appropriative depth index k_h (Table 7) were calculated using parameter $R3$ which reflects the polarization resistance in the electric equivalent circuit of studied systems. The corrosion resistance of mixed oxides is about 2–3 orders of magnitude higher as compared with conversion coatings Al_2O_3 obtained during traditional oxidizing in sulfate acid. Open circuit (corrosion) potentials of samples coated with mixed oxides become more stable and positive than that for samples with Al_2O_3 coatings. Braking of the corrosion process is achieved at long time exposition in model aerated solutions (Table 3) due to low porosity of mixed oxides which does not exceed 5%.

The phase composition of mixed oxide coatings Al_2O_3·MnO_x, as well as branched micro-porous globular surface with a high content of non-stoichiometric manganese oxides will predetermine catalytic properties in heterogeneous reactions of oxygen reduction/oxidation or destruction of the bond "oxygen—element."

The model reaction of CO flameless oxidation on the surface of mixed oxide Al_2O_3·MnO_x in the reactor at air flow starts at the temperature 345 K which is positioned as ignition temperature. This parameter is lower as compared not only than pure alumina but also than platinum catalyst (Table 8). The complete conversion temperature T_c which corresponds to the 100% transformation of CO to CO_2 at the surface of samples with mixed oxides is also indicative of their higher catalytic activity.

So the mixed oxide system, Al_2O_3·MnO_x exhibits outstanding catalytic activity not inferior even to the contacts that contain precious metals. Such behavior is due to some properties of non-stoichiometric manganese oxides. Firstly, among them are MnO_x acceptor affinity to electron reach intermediates such as double bond $O = O$

Table 7 Corrosion indicators of PEO coatings

Corrosion medium	Coating	Current index, $j_c \cdot 10^6$, A/cm^2		Deep index, $k_h \cdot 10^3$, mm/year	
		1 day	30 days	1 day	30 days
0,1 M Na_2SO_4	Al|Al_2O_3*	3.40	0.39	37.03	4.25
	Al|Al_2O_3, MnO_x	0.03	0.02	0.32	0.22
0,01 M NaOH	Al|Al_2O_3*	3.54	10.36	38.55	112.8
	Al|Al_2O_3, MnO_x	2.12	0.52	23.09	5.66

*Oxide formed in sulfate acid

Table 8 Characteristics of the CO conversion to CO_2

Material	Active metal content, at.%	Ignition temperature T_i, K	Complete conversion temperature T_c, K
Pt	100	490	570
Al|Al_2O_3	30.8	595	670
Al|$Al_2O_3 \cdot MnO_x$	Mn—33	345	505

which is favorable for redistribution of negative charge density and bond breakdown. Secondly, MnO_x also exhibits donor properties, being a source of mobile oxygen for the oxidation of intermediates. Ultimately the branched micro-porous globular surface of mixed oxides contains a large number of high-energy catalytic centers [31] that provides an increase in the number of elementary collisions and reactions [32].

Mixed conversion coatings $Al_2O_3 \cdot MnO_x$ were applied on the surface of the piston of the combustion chamber which was the part of a single-cylinder diesel engine. It was established that fuel economy when operating a diesel engine with a piston, the surface of which is coated with $Al_2O_3 \cdot MnO_x$, reaches 3%. The toxic emissions such as nitrogen oxides from engine operating under forced modes decrease significantly as was shown in [33]. So, mixed oxide systems applied on the aluminum alloys by PEO are promising in the environmental technologies for both air and water media purification [10, 11]. Also, such catalytic systems may be used for the intra-cylinder catalysis in order to reduce toxicity of engine emissions [33].

5 Conclusions

1. The plasma electrolytic oxidizing technique for producing mixed oxide coatings on aluminum alloys was developed. The composition of alkaline electrolytes containing potassium permanganate as well as oxidizing current density and regime was proposed to enrich the surface layer with manganese. It was shown that the designed method to be the most technologically advanced for producing

one process conversion coatings with developed surface and a high content of manganese. The incorporation of manganese oxides into the composition of Al_2O_3 matrix contributes to the enhancement of catalytic activity of mixed oxide systems.

2. The phase composition and morphology of mixed oxide covers surface have been analyzed. The highly branched micro-porous and globular surface of $Al_2O_3 \cdot MnO_x$ coatings applied onto aluminum alloys is characterized by uniform distribution of manganese which content reaches up to 36 wt%. The control of composition and surface state of conversion mixed coatings has been achieved by varying the PEO regimes.
3. The mixed manganese-aluminum oxide conversion coatings on the aluminum alloys are characterized by a high corrosion resistance and catalytic activity. Therefore, mixed oxide systems applied on the aluminum alloys are promising in the environmental technologies for both air and water media purification.

Acknowledgements This research was conducted within the confines of the project (Registration Number 0119U002568) with the support of the Ministry of Education and Science of Ukraine.

References

1. Heck RM, Farrauto RJ, Gulati ST (2009) Catalytic air pollution control: commercial technology. Wiley, p 544
2. Stiles AB (1987) Catalyst supports and supported catalysts: theoretical and applied concepts. Butterworth, Stoneham, MA
3. Yerokhin AL, Nie X, Leyland A, Matthews A, Dowey SJ (1999) Plasma electrolysis for surface engineering. Surface Coat Technol 122(2–3):73–93
4. https://doi.org/10.1016/s0257-8972(99)00441-7
5. Gupta P, Tenhundfeld G, Daigle EO, Ryabkov D (2007) Electrolytic plasma technology: science and engineering—an overview. Surf Coat Technol 201(21):8746–8760
6. Rudnev VS, Lukiyanchuk IV, Vasilyeva MS, Medkov MA, Adigamova MV, Sergienko VI (2016) Aluminum- and titanium-supported plasma electrolytic multicomponent coatings with magnetic, catalytic, biocide or biocompatible properties. Surf Coat Technol 307(Part C):1219–1235
7. Rakoch AG, Khokhlov VV, Bautin VA, Lebedeva NA, Magurova YuV, Bardin IV (2006) Model concepts on the mechanism of microarc oxidation of metal materials and the control over this process. Prot Met 42(2):158–169
8. Sakhnenko ND, Ved MV, Vestfrid YuV, Stepanova II (1996) Predicting the catalytic activity of metal oxide systems in treatment of exhaust gases to remove nitrogen oxides. Russ J Appl Chem 69(9):1346–1350
9. Sakhnenko N, Ved M, Karakurkchi A, Galak A. (2016) A study of synthesis and properties of manganese-containing oxide coatings on alloy VT1–0. East Eur J Enterpr Technol 3/5(81):37–43
10. Sakhnenko M, Karakurkchi A, Galak A, Menshov S, Matykin O (2017) Examining the formation and properties of TiO_2 oxide coatings with metals of iron triad. East Eur J Enterpr Technol 2(11/86):4–10

11. Karakurkchi A, Sakhnenko M, Ved M, Galak A, Petrukhin S (2017) Application of oxide-metallic catalysts on valve metals for ecological catalysis. East Eur J Enterp Technol 5/10(89):12–18
12. Lukiyanchuk IV, Rudnev VS, Tyrina LM (2016) Plasma electrolytic oxide layers as promising systems for catalysis. Surf Coat Technol 307(Part C): 1183–1193
13. Md Jani AM, Losic D, Voelcker NH (2013) Nanoporous anodic aluminium oxide: Advances in surface engineering and emerging applications. Prog Mater Sci 58(5):636–704
14. Sakhnenko ND, Ved' MV, Androshchuk DS, Korniy SA. (2016) Formation of coatings of mixed aluminum and manganese oxides on the AL25 alloy. Surf Eng Appl Electrochem 52(2):145–151
15. Karakurkchi AV, Sakhnenko ND, Ved MV, Luhovskyi IS, Drobakha HA, Mayba MV (2019) Features of plasma electrolytic formation of manganese- and cobalt-containing composites on aluminum alloys. Adv Mater Sci Eng 19(Article ID 6381291)
16. Labardi M, Allegrini M, Salerno M, Fredriani C, Ascoli C (1994) Dynamical friction coefficient map using a scanning force and friction force microscope. Appl Phys 59
17. Arbizzani C, Borghini M, Mastragostino M, Meneghello L, Zanelli A (1994) Impedance spectroscopy in electrode/electrolyte interface investigations. Solid State Ionics 72(2):115–121
18. Ved' MV, Sakhnenko ND, Karakurkchi AV, Myrna TYu (2017) Functional mixed cobalt and aluminum oxide coatings for environmental safety. Funct Mater 24(2):303–310
19. Glazoff MV, Zolotorevsky VS, Belov NA (2007) Casting aluminum alloys. Elsiever, Oxford, p 544
20. Davis JR (1993) ASM specialty handbook: aluminum and aluminum alloys. ASM International, Cleveland, OH
21. Dong H. (2010) Surface engineering of light alloys: aluminium, magnesium and titanium alloys. Elsevier, p 680
22. Egorkin VS, Vyaliy IE, Sinebryukhov SL, Gnedenkov SV (2017) Composition, morphology and tribological properties of PEO-coatings formed on an aluminum alloy D16 at different duty cycles of the polarizing signal. Non-ferrous Met 42(1):12–16
23. Sakhnenko ND, Ved MV, Karakurkchi AV (2017) Nanoscale oxide PEO coatings forming from diphosphate electrolytes. In: Fesenko O, Yatsenko L (eds) Nanophysics, nanomaterials, interface studies, and applications. In: NANO 2016, Proceedings in physics, vol 195. Springer, pp 159–184
24. Boguta DL, Rudnev VS, Yarovaya TP, Kaidalova TA, Gordienko PS (2002) On composition of anodic-spark coatings formed on aluminum alloys in electrolytes with polyphosphate complexes of metals. Russ J Appl Chem 75(10):1605–1608
25. Ayday A, Durman M (2015) Growth characteristics of plasma electrolytic oxidation coatings on aluminum alloys. Acta Physic Polon A 127(4):886–887
26. Ponomarev IS, Krivonosova EA, Gorchakov AI (2015) Investigation of discharge dynamics in microarc oxidation of D16 aluminium alloy. Weld Intern 30(3):244–246
27. Rudnev VS, Vasilyeva MS, Kondrikov NB, Tyrina LM (2005) Plasma-electrolytic formation, composition and catalytic activity of manganese oxide containing structures on titanium. Appl Surf Sci 252(5):1211–1220
28. Sakhnenko ND, Ved MV, Karakurkchi AV (2019) Nanostructured mixed oxide coatings on silumin incorporated by cobalt. Nanophys Nanomater Interface Studies Appl 221:269–291
29. Dudareva NYu, Abramova MM (2016) The structure of plasma-electrolytic coating formed on Al–Si alloys by the micro-arc oxidation method. Prot Met Phys Chem Surf 52(1):128–132
30. Vasilyeva MS, Rudnev VS (2014) Composition, surface structure and catalytic properties of manganese- and cobalt-containing oxide layers on titanium. Adv Mater Res 875–877:351–355
31. Yar-Mukhamedova GSh, Ved MV, Karakurkchi AV, Sakhnenko ND (2017) Mixed alumina and cobalt containing plasma electrolytic oxide coatings. IOP Conf Ser Mater Sci Eng 213
32. Bykanova VV, Sakhnenko ND, Ved MV (2015) Synthesis and photocatalytic activity of coatings based on the TixZnyOz system. Surf Eng Appl Elechtrochem 51(3):276–282

33. Ved M, Glushkova M, Sakhnenko N (2013) Catalytic properties of binary and ternary alloys based on silver. Func Mater 20(1):87–91
34. Parsadanov IV, Sakhnenko ND, Ved' MV, Rykova I V, Khyzhniak VA, Karakurkchi AV, Gorokhivskiy AS (2017) Increasing the efficiency of intra-cylinder catalysis in diesel engines. Vopr Khim Khim Tekhnol 6:75–81

The Influence of La Doping on Structural, Optical, and Photocatalytic Properties of TiO_2 in Dyes Destruction and Hydrogen Evolution

T. A. Khalyavka, V. V. Shymanovska, E. V. Manuilov, N. D. Shcherban, O. Y. Khyzhun, G. V. Korzhak, and V. V. Permyakov

1 Introduction

Currently, the photocatalytic properties of titanium dioxide (TiO_2) and TiO_2-based nanocomposites in the destruction of organic substances are actively being studied. Titanium dioxide is a wide-gap semiconductor and is used as photocatalyst due to its high photocatalytic activity and chemical stability. TiO_2 exhibits high photoactivity in various reactions under UV irradiation [1–7].

To increase the photocatalytic activity of titanium dioxide and to reduce the bandgap, its doping is applied. Under the action of irradiation, electrons and holes are generated and participate in redox reactions with adsorbed molecules at TiO_2 surface. Dopants can trap photoexcited electrons before they recombine with holes. The efficiency of a photocatalyst is determined by the fact, how efficiently electrons and holes are involved in the oxidation and reduction reactions prior to their

T. A. Khalyavka (✉)
Institute for Sorption and Problems of Endoecology, NAS of Ukraine, General Naumov str., 13, Kyiv 03164, Ukraine
e-mail: takhalyavka@ukr.net

V. V. Shymanovska · E. V. Manuilov
Institute of Physics, NAS of Ukraine, 46 Nauki Ave, Kyiv 03028, Ukraine

N. D. Shcherban · G. V. Korzhak
L.V. Pisarzhevskii Institute of Physical Chemistry, NAS of Ukraine, Prospekt Nauky, 31, Kyiv 03028, Ukraine

O. Y. Khyzhun
Institute for Problems of Materials Science, NAS of Ukraine, 3, Krzhizhanovsky str., Kyiv 03142, Ukraine

V. V. Permyakov
Institute of Geological Sciences, NAS of Ukraine, O. Gonchar str., 55-B, Kyiv 01054, Ukraine

O. Fesenko and L. Yatsenko (eds.), *Nanomaterials and Nanocomposites, Nanostructure Surfaces, and Their Applications*, Springer Proceedings in Physics 246, https://doi.org/10.1007/978-3-030-51905-6_27

recombination. It was shown that the lifetime of trapped electrons correlates with the photocatalytic activity of pure and doped TiO_2; the higher photoactivity of the samples, the lower recombination rate [8, 9]. The mutual recombination of electrons and holes reduces the activity of the photocatalyst. A dopant can trap electrons and transfer them to an electron acceptor which reduces the rate of recombination of an electron-hole pair [10]. Oxygen adsorbed on the surface of the photocatalyst captures electrons and forms a superoxide anion as well as the holes can oxidize adsorbed hydroxide ions to hydroxyl radicals [11, 12]. Lanthanum and its compounds have numerous applications as catalysts, various additives in glasses, in ignition elements, electronic cathodes, in medicine, etc. [5, 13–15].

Lanthanum has an electronic configuration [Xe] $5d^1$ $6s^2$, very effective to increase electron-hole carrier lifetime [16]. The size of the La atom radius is 1.87 Å and that of La^{3+} ion is 1.17 Å, which is almost two times bigger than the radius of Ti^{4+} (0.64 Å) and Ti^{3+} (0.67 Å). In the ground state, the La atom is monovalent. Transferring it to the trivalent state requires an expense of 8 kcal/g-atom. The successive ionization potentials for La are 5.61, 11.43, and 19.17 eV. In chemical reactions, lanthanum usually gives away three valence electrons from *5d* and *6s* subshells in order to form an oxidation state of +3, reaching a stable configuration. In an oxygen atmosphere at 450 °C, lanthanum is oxidized to A-La_2O_3.

The aim of this study was to investigate the effect of La on the structural and optical properties of titanium dioxide, to evaluate Safranine T (ST) destruction (decolorization) and H_2 evolution over pure and La-doped TiO_2 under UV irradiation. For this purpose, we have synthesized TiO_2 samples with different lanthanum content by the sol–gel method. The sol–gel process is a widespread method used to obtain titanium dioxide-based nanocomposite materials. This process is a low-temperature synthesis which allows one to obtain polycrystalline powders of different chemical composition and various physicochemical properties [2, 3, 9].

2 Experimental Section

2.1 *Synthesis of TiO_2 and La/TiO_2 Samples*

All the samples were synthesized by sol–gel method. All the chemical reagents were of analytical grade and used directly without further purification. The experimental procedure was as follows.

Preparation of pure TiO_2. Titanium (IV) tetrabutoxide ($C_{16}H_{36}O_4Ti \geq 99.9\%$, Aldrich, 3 g), citric acid ($C_6H_8O_7$, 0.06 g), and glycerol ($C_3H_8O_3$, 2 g) were mixed at room temperature and calcinated at 500 °C for 2 h in the presence of air oxygen.

Preparation of La/TiO_2 samples. The mixture of titanium (IV) tetrabutoxide, $La(NO_3)_3 \cdot 6H_2O$, citric acid, and glycerol was carefully stirred in order to obtain uniform mass and further was calcined at 500 °C for 2 h in the presence of air. Lanthanum was doped at various atomic percentages (1.44 and 6.03% at., Table 1).

Table 1 The atomic and weight percentage of the elements of La/TiO_2 powders obtained by EDS technique

Sample	Ti	O	La
	Atomic, %		
1La/TiO_2	32.44	66.48	1.44
2La/TiO_2	28.31	65.66	6.03

After cooling, the resulting powders were triturated until smooth. The samples were designated as 1La/TiO_2 and 2La/TiO_2, respectively.

2.2 Methods and Instrumentation

For analysis of the sample composition (elemental analysis) and their morphology, a scanning electron microscope (SEM JSM 6490 LV, JEOL, Japan) with an integrated system for electron microprobe analysis INCA Energy based on energy-dispersive and wavelength-dispersive spectrometers (EDS + WDS, OXFORD, United Kingdom) with HKL Channel system was used.

Transmission electron microscopy (TEM) JEM-1200 EX (JEOL, Japan) for the prepared materials was applied.

Presence of chemical elements and chemical bonds features in the samples were analyzed using X-ray photoelectron spectroscopy (XPS) with the UHV-Analysis-System equipment produced by SPECS Surface Nano Analysis Company (Berlin, Germany). The instrument was equipped with semi-spherical analyzer PHOIBOS 150.

XPS spectra of core-level and valence electrons were analyzed in an UHV-Analysis-System chamber under residual pressure not higher than 7×10^{-8} Pa. XPS spectra were activated by X-ray Mg Kα-irradiation ($E = 1253.6$ eV) and recorded at a constant pass energy of 30 eV. The energy scale of the device was graded by the method [17] with using reference metals Au and Cu. Surface charge of the samples was taken into account in reference to the binding energy of the C 1s-line from hydrocarbon adsorbates which was set to 284.6 eV as recommended for transition metal oxides [18, 19].

Phase composition of the samples was determined by X-ray diffraction analysis (XRD). A computerized Bruker D8 Advance diffractometer was equipped with Cu Kα ($\lambda = 0.15406$ nm) radiation. All XRD peaks were checked and assigned to known crystalline phases. The average crystallite size was determined using broadening the most intensive reflex following the Debye–Scherrer equation: $D = 0.9\lambda/B\cos\theta$, where 0.9 is a constant, λ is a wavelength, nm. Interplanar distance (d, nm) was calculated using Wulff–Bragg's equation: $n\lambda = 2d\sin\theta$, where $n = 1$ is the order of reflection, $\lambda = 0.154$ nm is the wavelength, θ is the scattering angle, degrees. Thereby, $d = n\lambda/2\sin\theta$.

Measurements of UV-vis diffuse reflection spectra (DRUV) of the powders were carried out at room temperature using a Perkin Elmer Lambda Bio 35 spectrophotometer in the range from 1000 to 200 nm, which allowed converting data of the corresponding spectra using the Kubelka–Munch equation.

Recording the diffuse reflection spectra in the coordinates F(R) = f (λ, nm), where F(R) is the Kubelka–Munch function, was carried out with a special rate in cuvettes with a layer thickness of the investigated powder material of 3 mm and a reference sample (MgO). All the materials were ground directly in an agate mortar before recording the DRUV spectra and obtaining constant optical characteristics. The absolute and relative errors were ±0.01 eV and ±0.3%, respectively.

The values of the specific surface area (S_{sp}) of the samples as well as pore size distribution were determined using a Quantachrom Nova Win 2 device. The specific surface area of the samples was determined based on nitrogen adsorption–desorption isotherms using the Brunauer–Emmet–Teller (BET) approach. The pore radius (R_{max}) and the pore volume (V_{tot}) were calculated from the desorption branches of the isotherms using the Barret–Joiner–Halenda method [20].

Room temperature FT-IR spectra were recorded with a Perkin Elmer Spectrum One spectrometer in the spectral region of 4000–400 cm^{-1} with spectral resolution of 4 cm^{-1}.

2.3 *Photocatalytic Experiments*

2.3.1 Photocatalytic Degradation Experiments

Photocatalytic activity of the samples was evaluated for dye Safranine T ($C_{init.}$ = 0.03 g/l) destruction. Before irradiation, catalyst suspension (2 g/l) in aqueous solution was kept in dark up to achieving the adsorption equilibrium. The time of the sorption equilibrium establishment in the system powder—ST does not exceed 2 h for all samples. Irradiation of the dye aqueous solutions was performed at room temperature in a quartz reactor in the presence of air. A standard BUV-30 UV lamp (30 W, λ = 254 nm, exposure time = 1.5 h) was used for UV irradiation of the dye aqueous solutions.

Concentrations of the substrate were measured using a Shimadzu UV-2450 spectrophotometer at λ = 520 nm for ST. Discoloration of ST was calculated from the differences between the initial (100%) and final concentrations of the dye.

For comparative analysis, we conducted an experiment with the well-known and well-proven P25 photocatalyst (Evonic, S_{sp}~50 m^2/g).

2.3.2 Photocatalytic Hydrogen Evolution

Photocatalytic hydrogen evolution was studied in cylindrical glass thermostatted reactors. In a typical experiment, a sample of TiO_2 and La/TiO_2 photocatalyst (0.03 g) was dispersed in 10.0 ml of aqueous solution containing 2 mol/l ethanol (electron donor). Co-catalyst Pd/SiO_2 (0.01 g,) was obtained employing $PdCl_2$ reduction by hydrogen on the surface of commercial silica in alcohol solution (1 wt% palladium on SiO_2) [21], and was added into the mixture. The reaction mixture was degassed before irradiation using a forvacuum pump. The suspension was stirred using a magnetic stirrer and irradiated with a high-pressure mercury lamp (DRSH 1000; 1000 W; λ max = 365 nm; 50 mW/cm^2). The composition of the gas phase above the solution was determined every 30 min chromatographically.

3 Results and Discussion

3.1 Surface Morphology and Elemental Analysis

Investigation of the obtained powders by means of energy-dispersive spectroscopy based on energy-dispersive technique proves that these materials include the elements Ti, O, and La (Figs. 1, 2c) and do not contain other impurities. The EDS elemental mappings from a selected area show that the La, O. and Ti atoms are distributed uniformly in the materials (Fig. 1).

Analysis of SEM images of the samples shows that they consist of fragmented agglomerates (Fig. 2a, b, c).

TEM analysis results are presented in Fig. 3 and prove that agglomerates of pure TiO_2 consist of ca. 10 nm size crystallites (Fig. 3a) and La/TiO_2 composites—ca. 8–10 nm size crystallites (Fig. 3b).

3.2 XRD Analysis

Crystal structure of the powders was identified using XRD. XRD-pattern of pure titanium dioxide shows intensive peaks at $2\theta = 25.4$ (101); 37.8 (004); 48.0 (200); 53.9 (105); 55.1 (211); 62.75 (204); 75.1 (215), which belong to tetragonal modification of TiO_2 anatase and at $2\theta = 27.4$ (110) being characteristic of rutile (Fig. 4, curve 1). The content of anatase in pure TiO_2 significantly exceeds the content of rutile. With the introduction of La in TiO_2, the anatase content decreases, while that of rutile increases. The content of rutile in the 2La/TiO_2 composite increases by two times as compared to pure TiO_2 (Table 2).

Different authors have opposing conclusions about the role of lanthanum in the anatase—rutile phase transformations. Some authors [15, 22, 23] reported that La

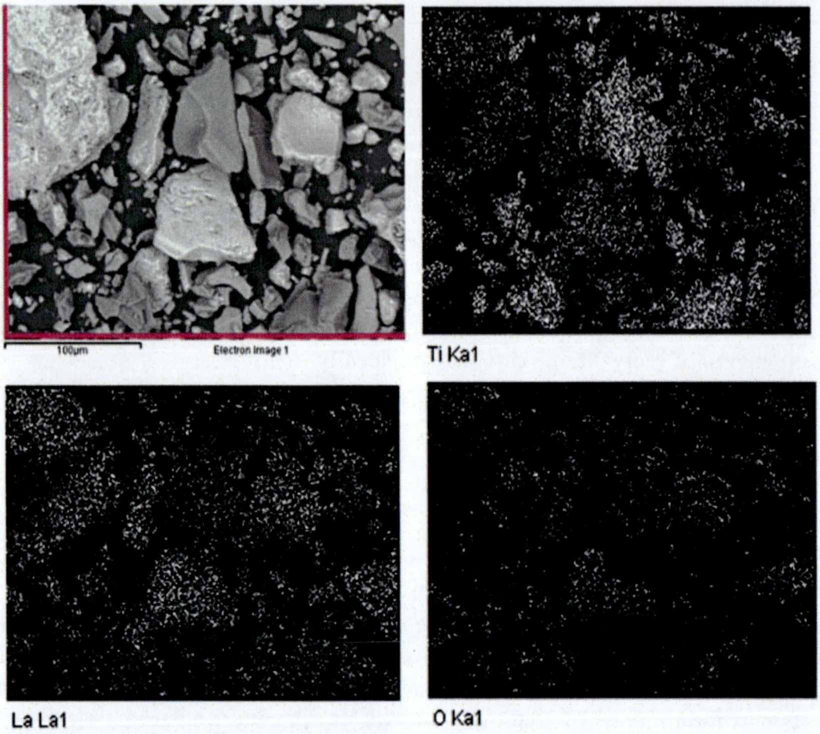

Fig. 1 SEM image and EDS elemental mapping of $2La/TiO_2$

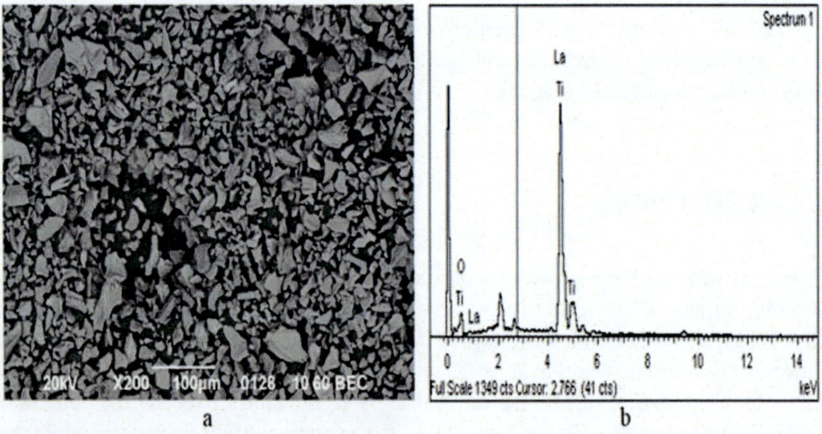

Fig. 2 SEM image (**a**) and the energy-dispersive spectrum (**b**) of $2La/TiO_2$

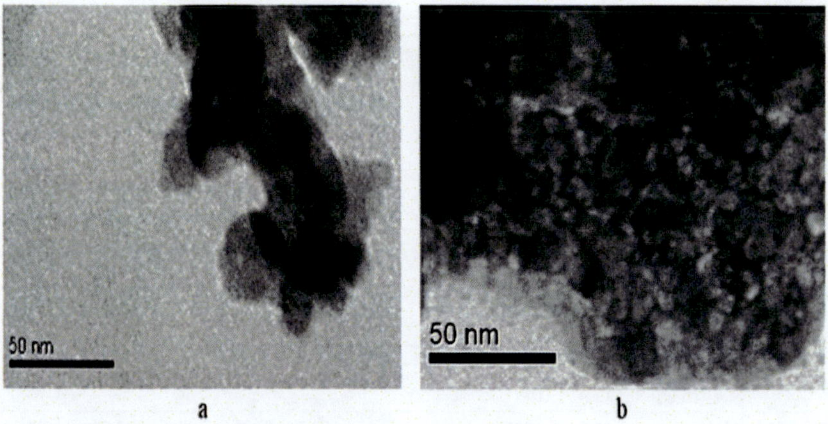

Fig. 3 TEM images of the samples: **a** TiO_2 and **b** $2La/TiO_2$

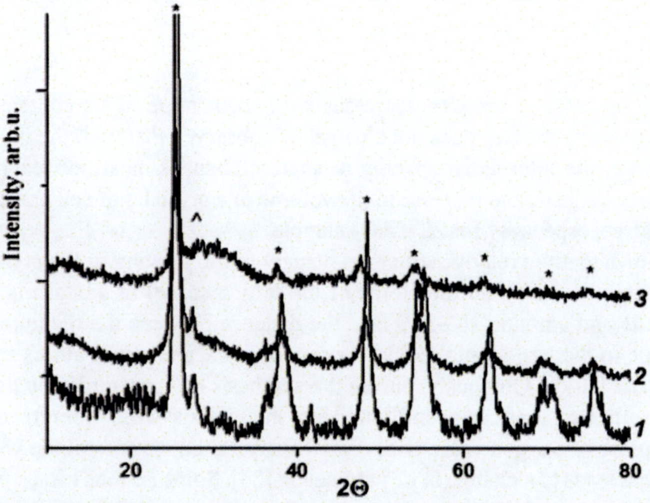

Fig. 4 XRD patterns of (1) TiO_2, (2) $1La/TiO_2$ and (3) $2La/TiO_2$ (*—anatase, ^—rutile)

ions inhibit the A–R phase transformation, others argued that La promotes the rutile formation. In our case, doping with La ions promotes the rutile formation even at lower temperatures (500 °C).

The crystallites size of the samples as calculated through Debye–Scherrer equation in the pure TiO_2 is equal to 10 nm for anatase (101) and 7 nm for rutile (110) (Table 2). Doping with La leads to a decrease in the anatase crystallite sizes from 10 to 8 nm and in the rutile crystallite sizes from 7 to 5 nm (Table 2). This statement is supported by our studies using TEM (Fig. 3). It is interesting to note that the size of rutile

Table 2 Phase composition and structural characteristics of the samples

Sample	Phase composition, hkl	Phase content, %	Interplanar spacing, d, nm	Lattice parameters, Å	Crystallite size, D_{hkl}, nm	Volume molecular unit TiO_2 $V = a^2c$, $Å^3$	Axial ratio, c/a
TiO_2	A (101)	87.2	0.348	$a = 3.7774$ $c = 9.2034$	10	131.32	2.44
	R (110)	12.8	0.322	$a = 4.5740$ $c = 2.9577$	7	61.88	0.65
$1La/TiO_2$	A (101)	66.7	0.349	$a = 3.7616$ $c = 9.4853$	8	134.21	2.52
	R (110)	20.9	0.324	–	6	–	–
	$La_xTi_yO_z$	12.4	–	–	–	–	–
$2La/TiO_2$	A (101)	46.8	0.351	$a = 3.8135$ $c = 12.5968$	8	183.19	3.30
	R (110)	26.4	0.317	–	5	–	–
	$La_xTi_yO_z$	26.8	–	–	–	–	–

crystallites in La/TiO_2 samples decreases with an increase of La content, and the size of anatase crystallites does not change. We observed that with the introduction of lanthanum, the interplanar spacing in anatase changes most noticeably. In this case, there is a substantial increase in the volume of a crystal unit cell and axial ratio c/a for anatase, especially for $2La/TiO_2$ sample.

It is known that the relative structural density of the packing of the crystal lattice (or the degree of filling by the atoms of the lattice space) of anatase is less than that of rutile and amounts to ~34% and the distance between the titanium atoms is bigger. Due to the large ionic radius, La cannot replace titanium atoms at the sites of the crystal lattice. Obviously, during the synthesis of the samples by the sol–gel method, lanthanum ions can be infiltrated into interstitials or into intergranular space or on the place of oxygen vacancy of the TiO_2 crystal lattice and fill the voids in the package that leads to a change in its parameters [24]. Some La ions can be present on the surface of the particles in the form of La_2O_3 oxide or complex compounds of the $La_xTi_yO_z$ type [15, 25, 26]. In the XRD spectra in the region of $2\theta = 30$–$31°$, a broad weak peak is observed, which can be attributed to the compounds $La_2(TiO_3)_3$ or La_2TiO_5, in accordance with the published data [25]. Such weak peaks are explained by poor crystallization of lanthanum compounds due to the low temperature of the sample processing and low La concentration.

3.3 Textural Studies

Analysis of nitrogen adsorption–desorption isotherms obtained at 77 K for the synthesized samples shows the presence of a hysteresis loop [27] (Fig. 5a). It was established that for all the samples nitrogen sorption isotherms belong to type IV in accordance with IUPAC classification. The H3 type of the hysteresis loop for TiO_2 and H2 type of the hysteresis loop for La/TiO_2 samples indicate that mesopores are present in all the synthesized samples [28].

The pore size distribution for the samples is shown in Fig. 5b. As can be seen from this figure, there is a bimodal pore size distribution in the regions of 1–2 nm and 2–6 nm with a predominance of a larger pore radius. Pore radius and pore volume are increased for La/TiO_2 samples. The specific surface area of La/TiO_2 composites is two times bigger than that one of pure TiO_2 (Table 3). Such a porous structure can provide a larger number of active sites for dyes adsorption.

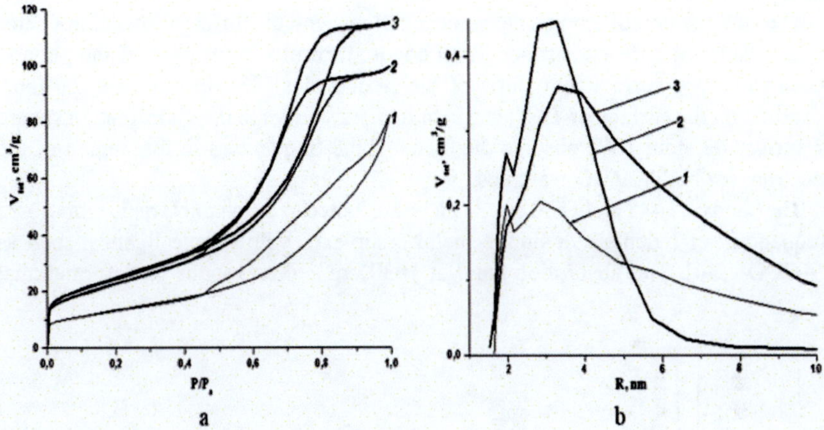

Fig. 5 **a** Nitrogen adsorption–desorption isotherms (77 K) and **b** pore size distribution for the samples: (1) TiO_2, (2) 1La/TiO_2, and (3) 2La/TiO_2

Table 3 Texture characteristics of the samples

Sample	Specific surface area, S_{sp}, m^2/g	Pore volume, V_{tot}, cm^3/g	Pore radius, R_{max}, nm
TiO_2	43	0.12	1.9; 2.8
1La/TiO_2	73	0.14	1.9; 3.2
2La/TiO_2	82	0.17	3.3

3.4 FTIR Studies

The functional groups of the synthesized samples were characterized by FTIR transmittance spectroscopy and the spectra are shown in Fig. 6. The bands near 400–500 cm^{-1} correspond to the Ti-O-Ti stretching vibrations, those ones at 550–700 cm^{-1} to the Ti-O bonds [29, 30] and they are detected in all the samples.

The presence of stretching vibration bands in the region of 500–1300 cm^{-1} is explained by the strengthening of –Ti–O– border or breaking bonds and the formation of double –Ti = O bonds. In accordance with the literature data [31] for polycrystalline metal oxides, these bands correspond to valence vibrations of surface oxygen atoms –T = O_s. Appearance of such bands is probably caused by breaking –Ti–O–OH bonds [32]. Compared to the spectrum of the undoped TiO_2, the peaks in the range of 700–1000 cm^{-1} in the spectrum of La/TiO_2 samples are shifted to lower wavenumbers. Incorporation of different molecules or impurity atoms to the coordination sphere of surface Ti atom leads to the deformation of the surface TiO_6 octahedrons.

The stabilization of oxygen atoms near cations with the different coordinate saturation results in different surface Ti–O bonds. It results in changes of the phonon spectra of TiO_2 surface and shifts of the frequencies [33]. In our case, La ions, adsorbed on the surface of TiO_2, can lead to disturbances in the coordination sphere of surface Ti atoms. We observe the shifts of the frequencies in this region of the spectrum for both La/TiO_2 samples.

The bands at 1447 and 1540 cm^{-1} can be attributed to the characteristic vibrational frequencies of bidentate or bridge metal complexes with simple ligands, such as Ti–OCO_2 [34]. The absorption band at 1630 cm^{-1} corresponds to deformational

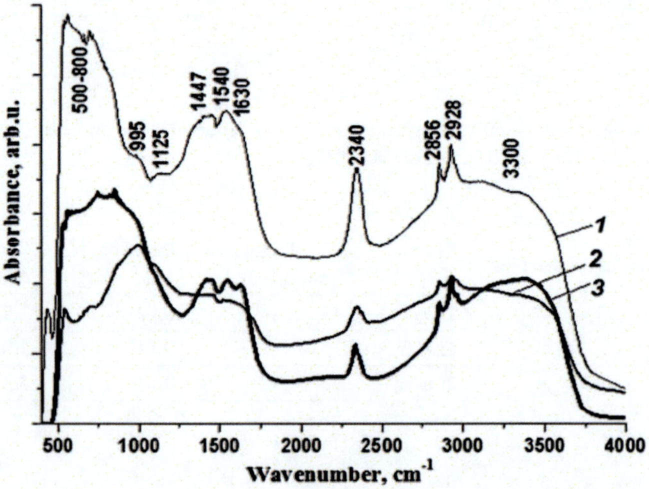

Fig. 6 FTIR spectra of the nanocomposites: (1) TiO_2, (2) 1La/TiO_2, and (3) 2La/TiO_2

vibrations in adsorbed water. Peak around 1600–1700 cm^{-1} represents the bending vibrations of the O–H bonds. The adsorbed hydroxyl ions on the surface of catalyst play an important role in increasing photocatalytic activity [33]. The absorption band at 2340 cm^{-1} corresponds to the physically adsorbed carbon dioxide [35–37]. The peaks at 2856 and 2928 cm^{-1} refer to the valence fluctuations of C–H and $-CH_2$ bonds [37]. The small absorption peaks around 2938 cm^{-1} are probably due to the stretching mode of unreacted organic groups such as $Ti–OC_xH_y$ [38]. In the region of about 3200–3400 cm^{-1} intensive broad absorption bands of ν(OH) valence vibrations and adsorbed water molecules coordinated on the TiO_2 surface are observed for all the samples. These OH groups after irradiation with light form •OH radicals are directly involved in the process of photodegradation of organic compounds.

3.5 XPS Studies

The method of X-ray photoelectron spectroscopy (XPS) was used to verify the elemental content and charge state of atoms composing the TiO_2, 1La/TiO_2, and 2La/TiO_2 samples. The survey XPS spectra are shown in Fig. 7.

The binding energy spectra of the Ti 2*p* core-level electrons appear as spin-orbit doublet with the typical interval 5.7 eV between its two peaks, which corresponds to Ti^{4+} in a tetragonal structure (Ti $2p_{3/2}$ and Ti $2p_{1/2}$). The binding energy were found to be 458.6 eV ($2p_{3/2}$) and 464.3 eV ($2p_{1/2}$) (Fig. 8a).

As can be seen from XPS data (Fig. 8a, Table 4), the binding energy value of Ti in the 1La/TiO_2 sample is detected to be corresponding to the charge state +4 (Ti) [39, 40].

However, when going from 1La/TiO_2 to 2La/TiO_2, the binding energy of the Ti 2*p* spectra decreases by about 0.8 eV (Fig. 8a; Table 4), indicating that in the 2La/TiO_2

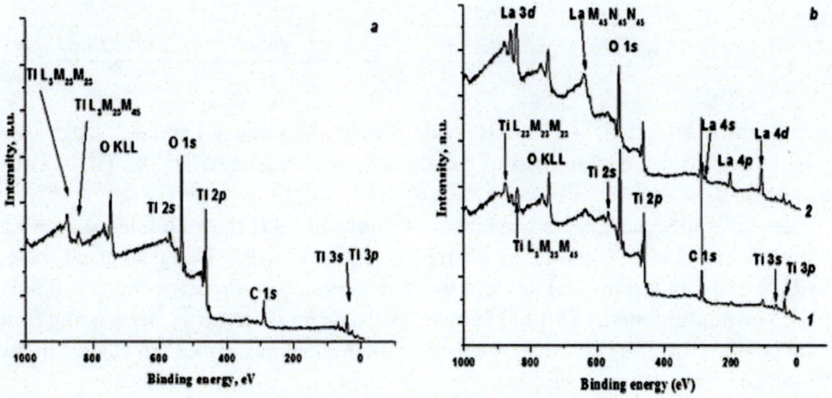

Fig. 7 XPS spectra of (a) TiO_2 and (b) 1La/TiO_2 (1) and 2La/TiO_2 (2)

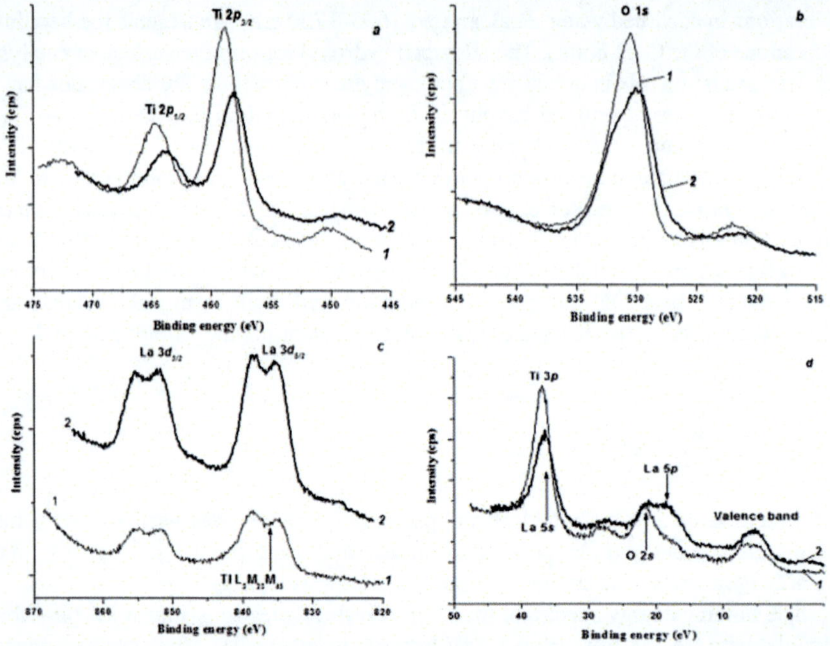

Fig. 8 XPS **a** Ti 2*p*, **b** O 1*s* and **c** La 3*d* core-level spectra and **d** XPS valence band spectra (including some higher core-level lines) of (1) 1La/TiO_2 and (2) 2La/TiO_2

Table 4 Binding energies (eV) of TiO_2, 1La/TiO_2, and 2La/TiO_2 (precision of determination is ±0.1 eV)

Core-level	TiO_2	1La/TiO_2	2La/TiO_2
Ti $2p_{3/2}$	458.6	458.9	458.1
Ti $2p_{1/2}$	464.3	464.6	463.8
O 1*s*	530.0	530.5	529.9
La $3d_{5/2}$	–	835.1	835.3
La $3d_{3/2}$	–	851.9	852.1

sample the charge state of Ti atoms is intermediate between +4 and +2. Obviously, titanium atoms can capture part of the electrons and reduce to Ti^{3+} or Ti^{2+} ($Ti^{4+} + e^- \rightarrow Ti^{3+}$).

The XPS spectra of O 1 *s* relevant to O^{2-} ions in TiO_2 show an intense signal at 530.5 eV for 1La/TiO_2 and 529.9 eV for 2La/TiO_2 (Fig. 8b). This peak is attributed to Ti-O and OH groups in TiO_2 and the shift of this core-level spectrum is 0.6 eV [41]. The binding energy of the O 1s core-level spectrum decreases when going from 1La/TiO_2 to 2La/TiO_2 (Table 1), while the charge state of La does not change in this sequence of compounds (Fig. 8c).

The charge state of La in the samples La/TiO_2 corresponds to +3 (La_2O_3) (Table 4) [39].

In the case of $2La/TiO_2$, the intensity of the XPS spectrum of La 3*d* increases in comparison with $1La/TiO_2$. In the sequence $1La/TiO_2 \rightarrow 2La/TiO_2$, the half-width of the main XPS valence band spectrum decreases by about 0.3 eV (Fig. 8d).

Some changes of the binding energy of the La $3d_{5/2}$ core-level spectrum (Table 4) can be explained by superposition of this spectrum with the Ti $L_3M_{23}M_{45}$ Auger line in such a case (Fig. 8c). In the investigated samples, the Ti 3*p* core-level spectrum superimposes the La 5*s* spectrum, while superposition of the O 2*s* and La 5*p* spectra is also observed (Fig. 8d).

Two peaks of C1s spectra with binding energy values located at ~290.0 and 284.85 eV, correspond to C–O and C–C bonds, respectively [5]. These data testify that no additional admixtures except for adsorbed hydrocarbons are detected.

3.6 DRUV Spectra Determination of the Catalyst Band Gap

The diffuse reflection spectra in the coordinates F(R) = *f* (λ, nm), where F(R) is the Kubelka–Munch function, were recorded. All the materials were grounded directly in an agate mortar before recording the DRUV spectra and obtaining constant optical characteristics.

The fundamental band gap E_g is an important characteristic of semiconductors. It is attributed to the inter-band transitions of electrons between the highest occupied 2*p* states of O in the valence band and the lowest unoccupied 3*d* states of Ti in the conduction band. The electron transitions in the bandgap of TiO_2 can be attributed to the direct or indirect transitions depending on the crystal structure, material dispersion, etc. [42]. These two different types of transitions can be distinguished by the energy dependence of the corresponding absorption edge. The value of E_g was estimated using the method proposed by Wood and Tauc by extrapolation of the linear part of the plot $(h\nu * \alpha(h\nu))^{1/n}$ versus hν toward energy axis at $\alpha(h\nu) = 0$ (n = ½ for direct allowed transitions, $n = 2$ for indirect allowed transitions). We determined E_g for direct and indirect electronic transitions for all the samples (Table 5).

Experimental DRUV spectra of pure TiO_2 powder and La/TiO_2 composites measured in the photon energy range from 700 nm (1.77 eV) to 300 nm (4.13 eV) are shown in Fig. 9.

As shown in Fig. 9, the absorption edge of La/TiO_2 samples is shifted to a shorter wavelength (blue shift) compared to pure TiO_2 that indicates an increase in the optical bandgap E_g of the composites. With an increase of the lanthanum content in TiO_2, the bandgap E_g for the direct and indirect transitions increases. The lower bandgap value

Table 5 The band gap values for the samples

Sample	E_g direct, eV	E_g indirect, eV
TiO_2	3.10	2.90
$1La/TiO_2$	3.12	2.95
$2La/TiO_2$	3.13	3.01

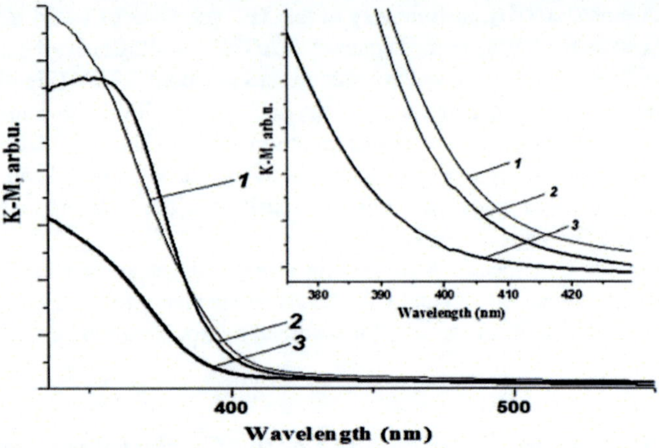

Fig. 9 DRUV spectra of (1) TiO_2, (2) $1La/TiO_2$, and (3) $2La/TiO_2$

for pure TiO_2 compared with those ones reported in the literature for the bulk anatase (3.2–3.3 eV) can be explained by high content of rutile (12.8%) in our samples.

Unlike our data, in [43] the authors observed a red shift of the absorption edge and narrowing of the bandgap of La–TiO_2 compared with pure TiO_2. This phenomenon was explained by a charge-transfer transition between 4*f*-orbital electrons of rare-earth ions and the conduction or valence band of TiO_2. In [15], the authors observed a slight blue shift, which they explained by a decrease in particle sizes caused by La doping, which confirms the quantum restriction. Such discrepancies in the experimental results can be explained by some differences in the method of synthesis of the samples and their structural characteristics.

It is known that the doping of semiconductors can lead to a change in their band structure [44]. One of these effects (red shift) is a decrease in the bandgap due to the formation of the so-called "tails" of the density of electronic states. This effect is a result of the non-uniform distribution of the impurity in a semiconductor.

Another phenomenon (blue shift) is associated with an increase in the bandgap of a semiconductor during its strong doping. Such an increase in the energy of inter-band transitions is explained by the filling of the conduction band by unoccupied electrons and is called the Burshtein–Moss effect [45]. Usually, these two effects compete with each other. The data of our XPS studies are consistent with these results.

3.7 Photocatalytic Studies

Photocatalysis testing was carried out to establish the effect of La on TiO_2 photoactivity. Safranine T photodegradation has been tested as a model reaction. The adsorption–desorption equilibrium was achieved within 120 min. The values of ST adsorption for TiO_2, 1La/TiO_2, 2La/TiO_2 were 5.2, 8.1, 15.3%, respectively. Each photocatalytic experiment was performed for 2 h by sampling every 30 min. The degradation percentages of ST at the corresponding concentration were calculated using the formula presented in (1).

$$\text{Degradation\%} = ((C_0 - C - t)/C_0)x100 \quad (1)$$

In our experiment, La/TiO_2 nanocomposites show higher photocatalytic activity than pure TiO_2 and P25 (Fig. 10).

As known, the dye molecules are able to absorb UV light with their following chemical transformations. The blank experiment (irradiation of the dye aqueous solution without a photocatalyst) showed that less than 10 (after 1 h) and 20% (after 2 h) of SF dye molecules undergo photolysis processes under UV light. The direct photocatalysis along with indirect photocatalytic processes can take place in the systems containing a photocatalyst and dye molecules with appropriate excited state reduction potential. However, direct photocatalysis occurs mainly in the presence of a semiconductor catalyst, as can be seen, doping of TiO_2 with lanthanum affects the percent decolorization of the dye.

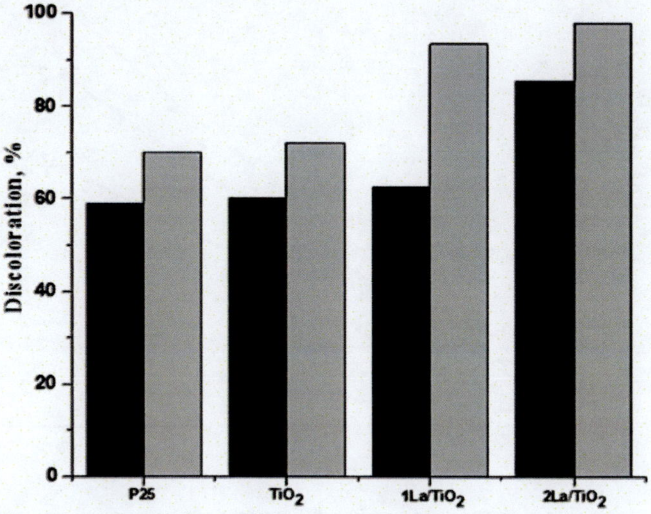

Fig. 10 Discoloration of ST after 1 (dark bars) and 2 (gray bars) hours

Investigation of the structural and textural properties of the obtained samples showed that with the introduction of La in TiO_2, the specific surface area, pore volume, and pore radius of the samples increase that leads to an increase in the number of active sites of the catalyst surface and its adsorption and photocatalytic capacity. That is, the number of reaction sites for the absorption and decomposition of dye molecules increases.

The photocatalytic activity of TiO_2 depends on the competition between the rate of transfer of surface charge carriers from volume to surface and the rate of recombination of photogenerated electrons and holes. It was shown in [46–49] that doping with TiO_2 lanthanide ions suppresses the rate of electron–hole recombination during the photocatalytic reaction.

With an increase in the amount of La in TiO_2, changes in the crystal lattice parameters are observed (an increase in the unit cell volume and the c/a ratio), as well as an increase in the bandgap for direct and indirect electronic transitions. The distorted titanium dioxide crystal lattice can take in more photoexcited holes and form stronger surface free radicals to oxidize adsorbed molecules [25]. All these factors explain the higher photocatalytic activity of La/TiO_2 composites compared to pure titanium dioxide in the decomposition of safranin dye.

Figure 11 presents kinetic curves of the photocatalytic hydrogen evolution from water–ethanol mixture under UV irradiation in the presence of TiO_2 and La/TiO_2 samples.

Hydrogen is not formed in dark (without irradiation), in the absence of photocatalyst, co-catalyst, and electron donor. The obtained results indicate that hydrogen

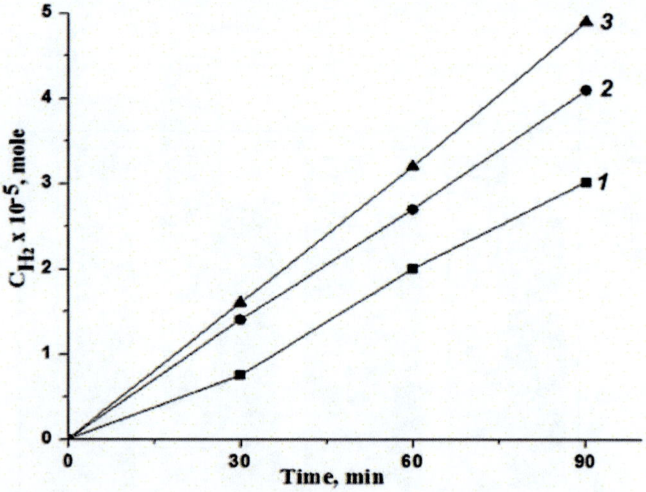

Fig. 11 Kinetic curves of photocatalytic hydrogen evolution from aqueous/ethanol solution in the presence of (1) TiO_2, (2) 1La/TiO_2, and (3) 2La/TiO_2

Table 6 The rates (υ_{H2}) of photocatalytic hydrogen evolution in the presence of photocatalysts

Sample	TiO_2	$1La/TiO_2$	$2La/TiO_2$
$\upsilon_{H2} \times 10^{-7}$, mol/min	3.3	4.7	5.6

evolution is caused by photocatalytic water reduction involving TiO_2 and La/TiO_2 in the presence of co-catalyst Pd/SiO_2 and an electron donor (ethanol).

As can be seen from Fig. 11 and Table 6, the rate of photocatalytic H_2 production in the presence of La/TiO_2 composites is 1.4–1.7 times higher compared to pure titanium dioxide.

Our results are in good agreement with the fact that La decreases the recombination rate of photogenerated charges by collecting electrons from TiO_2 while holes remained on TiO_2 [50]. In combination with the other structural and textural factors noted above, La/TiO_2 composites demonstrated higher photoactivity in the reaction of Safranin T photodestruction and H_2 photoproduction in comparison with pure TiO_2.

4 Conclusions

Mesoporous nanocomposite materials La/TiO_2 with different lanthanum contents were synthesized by sol–gel method. It was established that the content of rutile in the La/TiO_2 composites is twice higher compared to pure TiO_2. It was found that doping with La leads to a decrease in the anatase crystallite sizes from 10 to 8 nm and rutile crystallite sizes from 7 to 5 nm. The specific surface area of La/TiO_2 composites increases by two times compared with pure TiO_2. The absorption edge of La/TiO_2 samples shifts to shorter wavelengths (blue shift) compared with pure TiO_2. The banggap energy values of the samples increase compared with pure titanium dioxide. It was found that La/TiO_2 nanocomposites exhibit higher photocatalytic activity than pure TiO_2 in Safranine T photodestruction (decolorization) and H_2 photoproduction. The enhanced photocatalytic activity of the La/TiO_2 samples is mainly determined by a decrease in the recombination rate of photogenerated charges, small particle size, and larger surface area due to the doping with La.

Acknowledgements Support of the National Academy of Sciences of Ukraine under the Research Programs "Development of alternative methods for the preparation of nanosized oxide catalysts and composites based on Mo, Ti, Zr, Nb and Sn" 2018–2022, "Fundamental problems of nanostructured systems, nanomaterials and nanotechnologies" 2016–2020 and VC-180 "Optical, electrophysical and structural properties of disordered molecular systems and nanocomposites" 2016–2020 is gratefully acknowledged.

Availability of data and materials

The present research results have not been published before. Data and Materials are all in the main text, figures, and tables. Compliance with ethical standards.

Conflict of interests The authors declare that they have no competing interests.

References

1. Stroyuk OL, Ermokhina NI, Korzhak GV, Andryushina NS, Shvalagin VV, Kozytskiy AV, Manoryk PA, Barakov RYu, Kuchmiy SYa, Shcherbatyuk MS, Sapsay VI, Puziy AM (2017) Photocatalytic and photoelectrochemical properties of hierarchical mesoporous TiO_2 microspheres produced using a crown template J Photochem Photobiol A Chem 334:26–35
2. Sahoo C, Gupta AK (2012) Optimization of photocatalytic degradation of methylene blue using silver ion doped titanium dioxide by combination of experimental design and response surface approach. J Hazard Mater 215–216:302–310
3. Sun J, Liu N, Zhai S, Xiao Z, An Q, Huang D (2014) Gold-Titania/protonated zeolite nanocomposite photocatalysts for methyl orange degradation under ultraviolet and visible irradiation. Mater Sci Semicond Proces 25:286–293
4. Bondarenko MV, Khalyavka TA, Shcherban ND, Tsyba NN (2017) Mesoporous nanocomposites based on titanium dioxide and carbon as perspective photocatalysts for water purification. Nanosist Nanomater Nanotehn 15(1):99–112
5. Azam MU, Tahir M, Umer M, Jaffar MM, Nawawi MGM (2019) Engineering approach to enhance photocatalytic water splitting for dynamic H_2 production using La_2O_3/TiO_2 nanocatalyst in a monolith photoreactor. Appl Surf Sci 484:1089–1101
6. Linnik O, Chorna N, Smirnova N (2017) Nonporous iron titanate thin films doped with nitrogen: optical, structural and photocatalytic properties. Nanosci Res Lett 12:249–258
7. Gaponenko NV, Kortov VS, Smirnova NP, Orekhovskaya TI, Nikolaenko IA, Pustovarov VA, Zvonarev SV, Slesarev AI, Linnik OP, Zhukovskii MA, Borisenko VE (2012) Sol-gel derived structures for optical design and photocatalytic application. Microelectr Eng 90:131–137
8. Ohtani B, Kominami H, Bowman RM, Colombo DP, Jr, Noguchi H, Uosaki K (1998) Femtosecond diffuse reflectance spectroscopy of aqueous titanium(IV) oxide suspension: correlation of electron-hole recombination kinetics with photocatalytic activity. Chem Lett 27:579–580
9. Mogal SI, Mishra M, Gandhi VG, Tayade RJ (2013) Metal doped titanium dioxide: synthesis and effect of metal ions on physico-chemical and photocatalytic properties. Mater Sci Forum 734:364–378
10. Niu P (2013) Photocatalytic degradation of methyl orange in aqueous TiO_2 suspensions. Asian J Chem 25:1103–1106
11. Milenova K, Zaharieva K, Stambolova I, Blaskov V, Eliyas A, Dimitrov L (2017) Photocatalytic performance of TiO_2, CeO_2, ZnO and TiO_2-CeO_2-ZnO in the course of methyl orange dye degradation. J Chem Tech Met. 52:13–19
12. Baeissa ES (2016) Environmental remediation of aqueous methyl orange dye solution via photocatalytic oxidation using $AgGdFeO_3$ nanoparticles. J Alloy Com 678:267–272
13. Jing LQ, Sun XJ, Xin BF, Wang BQ, Cai WM, Fu HG (2004) The preparation and characterization of La doped TiO_2 nanoparticles and their photocatalytic activity. J Solid State Chem 177:3375–3382
14. Nguyen-Phan T-D, Song MB, Kim EJ, Shin EW (2009) The role of rare earth metals in lanthanide-incorporated mesoporous titania. Micropor Mesopor Mater 119:290–298
15. Priyanka KP, Revathy VR, Rosmin P, Thrivedu B, Elsa KM, Nimmymol J, Balakrishna KM, Varghese T (2016) Influence of La doping on structural and optical properties of TiO_2 nanocrystals. Mater Charact 113:144–151
16. Meksi M, Berhault G, Guillard C, Kochkar H (2015) Design of TiO_2 nanorods and nanotubes doped with lanthanum and comparative kinetic study in the photodegradation of formic acid. Catal Commun 61:107–111
17. Rajagopal S, Nataraj D, Khyzhun OYu, Djaoued Y, Robichaud J, Senthil K, Mangalaraj D (2011) Systematic synthesis and analysis of change in morphology, electronic structure and photoluminescence properties of pyrazine intercalated MoO_3 hybrid nanostructures. Cryst Eng Comm 13:2358–2368
18. Henrich VE, Cox PA (1994) The surface science of metal oxides. Cambridge University Press, Cambridge

19. Khyzhun OY, Solonin YM, Dobrovolsky VD (2001) Electronic structure of hexagonal tungsten trioxide: XPS, XES, and XAS studies. J Alloys Comp 320:1–6
20. Barret EP, Joyner LG, Halenda PP (1951) The determination of pore volume and area distributions in porous substances. I. Computations from nitrogen isotherms. J Am Chem Soc 73(3):373–380
21. Korzhak AV, Ermokhina NI, Stroyuk OL, Bukhtiyarov VK, Raevskaya AE, Litvin VI, Kuchmiy SYa, Ilyin VG, Manorik PA (2008) Photocatalytic hydrogen evolution over mesoporous TiO_2/Metal nanocomposites. J Photochem Photobiol A 198(2–3):126–134
22. Wu H-H, Deng L-X, Wang S-R, Zhu B-L, Huang W-P, Wu S-H, Zhang S-M (2010) The preparation and characterization of La doped TiO_2 nanotubes and their photocatalytic activity. J Dispers Sci Technol 31:1311–1316
23. Zhu X, Pei L, Zhu R, Jiao Y, Tang R, Feng W (2018) Preparation and characterization of Sn/La co-doped TiO_2 nanomaterials and their phase transformation and photocatalytic activity. Sci Rep 8(1):12387
24. Anandan S, Ikuma Y, Murugesan V (2012) Highly active rare-earth-metal La-doped photocatalysts: fabrication, characterization, and their photocatalytic activity. Int J Photoen 2012(Article ID 921412):10
25. Gerasumenko YuV, Logacheva VA, Babushkina EV, Khoviv AM (2010) Structure and optical properties of titanium dioxide films, modified with lanthanum. Condensed Matter Interphases 12(4):348–354. (in Russian)
26. Iwasaki M, Masaki H, Ito S, Park W (2007) Fabrication of La_2O_3-TiO_2-SiO_2 system glass derived from a sol-gel process. J Kor Ceram Soc 44(3):137–141
27. Lowell S, Shields JE (1998) Powder surface area and porosity. Chapman & Hall, London
28. Sing KSW, Everett DH, Haul RAW, Moscou L, Pierotti RA, Rouquerol J, Siemieniewska T (1985) Reporting physisorption data for gas/solid systems with special reference to the determination of surface area and porosity. Pure Appl Chem 57(4):603–619
29. Ansón-Casaos A, Tacchini I, Unzue A, Martínez MT (2013) Combined modification of a TiO_2 photocatalyst with two different carbon forms. Appl Surf Sci 270:675–684
30. Qu ZW, Kroes GJ (2006) Theoretical study of adsorption of $O(^3P)$ and H_2O on the rutile TiO_2(110) surface. J Phys Chem B 110(46):23306–23314
31. Fateley WG, Dollish FR, McDevitt NT, Bentley FF (1972) Infrared and raman selection rules for molecular and lattice vibrations. Wiley-Interscience, New York
32. Davydov A (1984) IK-Spektroskopiya v Khimii Poverkhnosti Okislov. Nauka, Novosibirsk. (in Russian)
33. Bezrodna T, Puchkovska G, Shymanovska V, Baran J, Ratajczak H (2004) IR-analysis of H-bonded H_2O on the pure TiO_2 surface. J Mol Struct 700:175–181
34. Nakamoto K (1963) Infrared spectra of inorganic and coordination compounds. Wiley, New York, London, 328 p
35. Bender ET, Katta P, Lotus A, Park SJ, Chase GG, Ramsier RD (2006) Identification of CO_2 sequestered in electrospun metal oxide nanofibers. Chem Phys Lett 423(4):302–305
36. Davies LE, Bonini NA, Locatelli S, Gonzo EE (2005) Characterization and catalytic activity of zirconium dioxide prepared by sol-gel. Latin Am Appl Res 35:23–28
37. Cheng CH, Lehmann J, Thies JE, Burton SD, Engelhard MH (2006) Oxidation of black carbon by biotic and abiotic processes. Org Geochem 37:1477–1488
38. Wang JA, Limas-Ballesteros R, Lopez T (2001) Quantitative determination of titanium lattice defects and solid-state reaction mechanism in iron-doped TiO_2 photocatalysts. J Phys Chem 105(40):9692–9698
39. Handbook of surface and interface analysis: methods for problems solving. In: Riviere JC, Myhra S, 2nd edn. CRC Press, Boca Raton/London/New York
40. (1990) Practical surface analysis. In: Briggs D, Seach PM (eds) Vol. 1: auger and X-Ray photoelectron spectroscopy, 2nd edn. Wiley, Chichester
41. Yao S, Jia X, Jiao L, Zhu C, Shi Z (2012) La-doped TiO_2 hollow fibers and their photocatalytic activity under UV and visible light. Indian J Chem 51A:1049–1056

42. Mo SD, Ching WY (1995) Electronic and optical properties of three phases of titanium dioxide: rutile, anatase, and brookite. Phys Rev 51:13023–13032
43. Huang Y, Cao J-J, Kang F, You S-J, Chang C-W, Wang Y-F (2017) High selectivity of visible-light-driven La-doped TiO_2 photocatalysts for NO removal aerosol and air quality research. Aeros Air Qual Res 17:2555–2565
44. Casey HC, Stern F (1976) Concentration-dependent absorption and spontaneous emission of heavily doped GaAs. J Appl Phys 47(2):631–643
45. Burstein E (1954) Anomalous optical absorption limit in InSb. Phys Rev 93(3):632–633
46. Kim HR, Lee TG, Shul YG (2007) Photoluminescence of La/Ti mixed oxides prepared using sol-gel process and their pCBA photodecomposition. J Photochem Photobiol A Chem 185:156–160
47. Li FB, Li XZ, Hou MF (2004) Photocatalytic degradation of 2-mercaptobenzothiazole in aqueous La^{3+}–TiO_2 suspension for odor control. Appl Catal B Environ 48:185–194
48. Li FB, Li XZ, Ao CH, Lee SC, Hou MF (2005) Enhanced photocatalytic degradation of VOCs using Ln^{3+}-TiO_2 catalysts for indoor air purification. Chemosphere 59(6):787–800
49. Liqiang J, Xiaojun S, Baifu X, Baiqi W, Weimin C, Hongganget F (2004) The preparation and characterization of la doped TiO_2 nanoparticles and their photocatalytic activity. J Solid Chem 177(10):3375—3382
50. Low J, Cheng B, Yu J (2017) Surface modification and enhanced photocatalytic CO_2 reduction performance of TiO_2: a review. Appl Surf Sci 392:658–686

Catastrophic Phenomena on Marine Slopes and in Artificial Dams in a Presence of Nanostructured Iron-Aluminosilicates

A. V. Panko, I. G. Kovzun, V. A. Prokopenko, O. M. Nikipelova, O. A. Tsyganovich, and V. O. Oliinyk

1 Introduction

It is known that the advances in science and technics are the basis of scientific-technical development. They always allowed the application of different universal knowledge, theories, and discoveries in all spheres of man's activity. They allow to develop new economically sounding, eco-safe, and effective methods, technologies, and ways of solving the arising problems. Thus, scientific–technical progress, for example, in spheres of physicochemical geomechanics, geophysics, geochemistry, biocolloidal chemistry, nanochemistry, eco-safety, medicine and balneology, created a wide range of possibilities for high efficient application of modern fine-dispersed iron-aluminosilicate materials of the Earth crust and compositions on their basis (IASMs) in eco-biotechnologies, geotechnologies, and in medical and cosmetic practice [1–6]. It was established recently, that the most effective application of IASMs is in the form of nanostructured systems (NIASSs) and compositions (NIASMs). They are the basis of complex technological structures like dams on water reservoirs, antifiltration bottom shields, and antistress (anticatastrophic) additions in composition of above- and underwater natural and anthropogenic slopes [5]. The most interesting in the modern stage of development is the not yet studied formation of

A. V. Panko (✉) · I. G. Kovzun · V. A. Prokopenko · O. A. Tsyganovich · V. O. Oliinyk
F.D, Ovcharenko Institute of Biocolloid Chemistry of NAS of Ukraine, Ak.Vernadskogo Blvd. 42, Kiev 03680, Ukraine
e-mail: gr.k.ibcc@ukr.net; phd.wiz@gmail.com

V. A. Prokopenko · O. A. Tsyganovich
National Technical University of Ukraine «KPI», Peremohy Ave., 37, Building 4, Kiev 03056, Ukraine

O. M. Nikipelova
SA «Ukrainian Research Institute of Medical Rehabilitation and Balneology, Ministry of Health of Ukraine», Lermontovskiy Lane 6, Odessa 65014, Ukraine

O. Fesenko and L. Yatsenko (eds.), *Nanomaterials and Nanocomposites, Nanostructure Surfaces, and Their Applications*, Springer Proceedings in Physics 246, https://doi.org/10.1007/978-3-030-51905-6_28

initial conditions for stress shears of fine-dispersed compacted sediments and soils on underwater marine slopes and artificial dams with following formation of fast mud currents with huge catastrophic consequences. The scientific basis of mechanisms of influences of viscous ultraanomaly and hyperanomaly effects [6] provoking initial laminar flow of NIASSs' and NIASMs' dispersions are not considered. The processes, which follow such flow and occur in phase nanosized zones of particle contacts in NIASSs' dispersions aided by siloxane, iron hydroxide, and carbonate structures are not considered too. The important role of wind waves in the formation of marine phaseolinic coastal oozes and further action of physicochemical, biocolloidal, and geomechanical factors resulting in formation of turbidite-ooze compositions on basis of IASMs and NIASMs on marine slopes was not shown. There are lack of investigations concerning the mechanisms of contact nano- and microstructured dispersion transformation of NIASSs and NIASMs in conditions of their transition from elastic to plastic state and vice versa. The respective models for the mentioned processes are not given. Thus, the importance and actuality of stated investigations are evident.

2 Physicochemical Characteristics of Investigated Materials

Preliminary initial investigations of transformation processes of polymineral microdisperse and nanostructured iron-aluminosilicate systems and materials (IASSs and IASMs) of natural and technogenic origin (iron ores, soils, pelitic sediments, peloids, natural, and thermodestructed clays, sands, etc.) under the influence of geomechanical and nanochemical dispersion were done. They showed that as a result of change of factors of influence on microdisperse and nanodisperse IASSs and NIASMs, there can occur various interactions between their components [1–6]. Such interactions can include reduction or oxidation, further dispersion or contact interparticle compaction, Ostwald ripening, change of surface interparticle or interaggregate nanostructured phase composition, chemical leaching, adsorption, adhesion, cohesion, ion exchange, etc. Such interactions give new properties and change characteristics of IASSs and NIASMs. Such changes mostly appear in the Earth crust and on its surface as a result of natural or technogenic processes. However, the processes, which occur in such systems and compositions of materials are still estimated only by final parameters. But that does not explain the source and the mechanisms of initial transformations, which cannot be observed easily anytime. There are still no unique, scientifically proved colloid–chemical ideas concerning the transformation of polydisperse IASSs and NIASMs in conditions of their geomechanical and nanochemical dispersion in natural geological conditions. And there are also no principal models of biocolloid–chemical interactions in such systems that can give a possibility to explain the processes of controlling IASSs and IASMs' behavior and characteristics. Concerning the said above, the choice of study materials was based on samples with shared colloid–chemical properties typical for most of IASSs and NIASMs. As an example of the iron-aluminosilicate materials, they were chosen earlier [4–6]: different soils and clays; polymineral iron-aluminosilicate compositions extracted from iron ores; pelagic shallow-water and deep-water sediments and peloids of the Black Sea and the Azov Sea; bentonite and montmorillonite used in other publications [1–3] too. The chemical composition of typical averaged material is given in Table 1.

The averaged by composition clay and other materials, and compositions used in the paper were concentrated by standard methods [7] and powdered up to particle sizes of 63 μm. XRD and X-ray fluorescence investigations of polymineral disperse systems and compositions (Table 1; Figs. 1 and 2) showed that their structure includes minerals of kaolinite, hydromica, montmorillonite, glauconite, saponite, goethite, and others. The composition of fine fractions has lower content of mixed layer formations of montmorillonite-hydromica and glauconite type with the advantage of hydromica-type formations.

Infrared spectrums were characterized by absorption bands with maximums at 480, 546, 700, 805, 925, 950, 1020, 1047, and 1126 cm^{-1}. According to the data [7, 8], they belong to the area of valence oscillations of Si–O(Si) and Si–O (1020, 1047, 1126 cm^{-1}). Maximums at 925 and 950 cm^{-1} belong to deformational oscillations of structural hydroxide groups bonded with octahedral Al^{3+} cations. The band at 546 cm^{-1} belongs to deformational oscillations of hydroxide groups in kaolinite structure [7]. A common view of triplet in valence oscillation area with maximums at 1020, 1047, and 1126 cm^{-1} is more characteristic for kaolinites than for hydromicas.

Further processing of IASSs and NIASMs was done according to the recommendations from [9, 10]. Nanodisperse calcium carbonate was made according to the methods described in [4].

Table 1 Chemical composition of averaged typical iron-aluminosilicate materials (IASMs and NIASMs)

Oxide content wt%	Polymineral iron-aluminosilicate composition	Saponite-goethite composition	The Black Sea ooze	Montmorillonite
SiO_2	19.5	46.4	54.6	49.5
Al_2O_3	2.9	5.2	11.4	21.1
FeO	2.6	2.8	–	2.7
Fe_2O_3	50.1	22.2	5.5	–
CaO	2.87	4.1	3.7	5.7
MgO	0.6	0.5	1.5	1.6
MnO	7.9	0.4	0.1	–
Na_2O	0.3	0.9	0.2	0.4
K_2O	0.3	0.4	0.9	0.3
LOI	13.9	17.1	12.1	18.2

LOI—loss of ignition

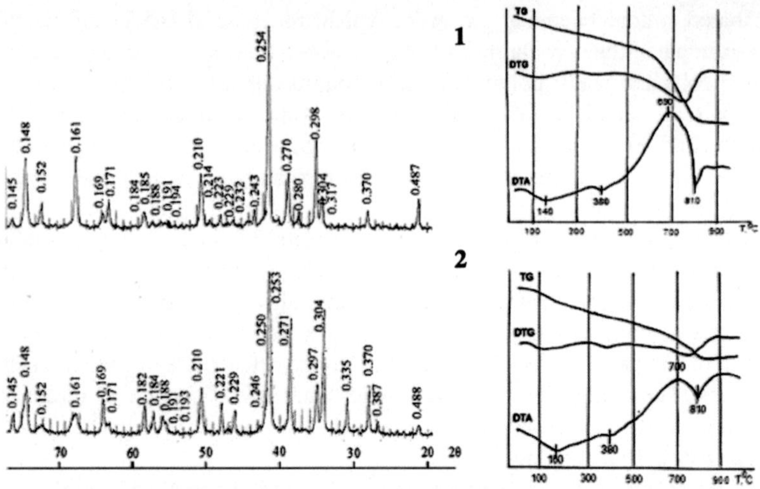

Fig. 1 XRD images and thermograms of natural (1) and washed (2) clay minerals and their nanoparticles iron-aluminosilicate material

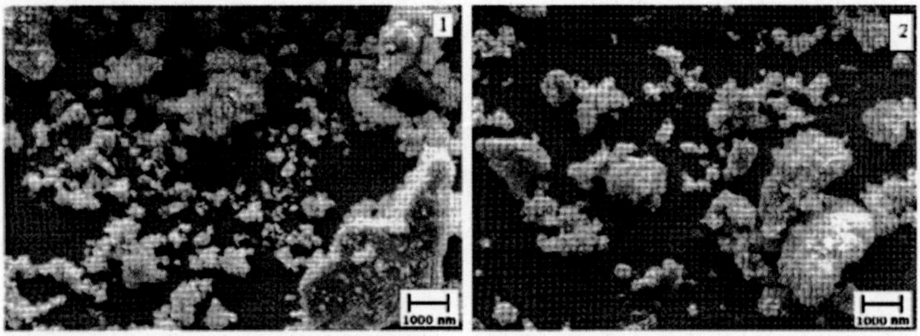

Fig. 2 SEM images of natural (1) and washed (2) clay minerals and their nanoparticles iron-aluminosilicate material

Chemically pure sodium carbonate and hydroxide were also used in investigations. Sodium silicate was made from sodium hydroxide and amorphous silica by well-known methods. Suspensions of tested materials were milled in porcelain ball mill with chemical additions containing nanoclusters and nanoparticles.

3 Methods of Investigation

Investigations of IASSs and NIASMs included using theoretical modeling methods and experimental methods: chemical, rheological, XRD, thermogravimetric, SEM and IR-spectroscopic methods of analysis, and when needed—gravity and magnetic separation, and biomedical methods [4–10]. Chemical and X-ray fluorescence analysis used in sample tests were done by the known methodology of qualitative and quantitative methods of analysis of iron-aluminosilicate rocks. X-ray diffraction (XRD) sample analysis was done using diffractometer Dron-UM-1 with CoK_{α} emission and nickel filter at room temperature. Diffractometer was connected to recorder KSP-4 and to a computer for diffraction pattern recording. Recording was conducted at 1 degree per minute speed. Registration of X-Ray emission was done by scintillation counter BDS-6. Electron microscopy of tested samples: electron microscopic images were received on electron microscope Selmi PEMU in light field mode. Scanning electron microscopes TESLA BN, JEOL NeoScope JCM-5000, and JEOL JSM6490 LV with INCA ENERGY-450 (Oxford) energy-dispersion device were also used. Thermogravimetric analysis (TGA): it used thermogravimetric analyzer MOM Q-1500 D (Hungary). Rheological characteristics of investigated systems and materials were determined by rotational viscometer with coaxial cylinders Rheotest-2 connected to a computer. Structure and mechanical properties of dispersions were investigated on Weiler–Rehbinder device with automatic recording of deformation-time curves.

4 Results and Discussion

4.1 Experimental and Model Conceptions About Nanoformation and Transformation of Iron Oxide-Hydroxide Aluminosilicate Phases

Problem alike for silicates and aluminosilicates with impurities of iron oxide-hydroxides has been already partly considered [4–6, 11–17]. It lies close to the problem of biocolloidal interaction of microorganisms with metallic and non-metallic surfaces, and the problem of biological transformation of iron oxide-aluminosilicate materials in the process of formation of NIASSs [11–17]. It is known, in general, that biocolloidal interaction can be conducted by both microorganisms and their aggressive metabolites—acids, alkalis, amphiphilic surfactants, and other substances. At that, the colonies of bacteria can form excrescences of mycelium or slime on iron surface reduced from IASMs or in the structure of oxide-hydroxides and iron-aluminosilicate polymineral formations. Under them, there occur appropriate microbiological and chemical biocolloidal processes of geomechanical and physico-chemical dispersion and nanochemical transformation of iron-aluminosilicates. Such phenomena are characteristics for dispersion and nanochemical transformation on the surface of iron oxide-hydroxides and iron-aluminosilicate structures, sulfides, and other compounds. Thus, for example, it is known that American gold-mining corporation in Denver (state of Colorado) processes sulfide ores in mixture with iron oxide-aluminosilicate rocks using sulfo-bacteria of Thiobacillus genus [11]. And jaspilite (iron quartzites) was formed 2 billion years ago with an aid of cyanobacteria [12]. It is also known that silicate bacteria help leaching of metals from ores [13]. That testifies about the potential of biocolloidal microbiological hydrometallurgy [14]. In works [15, 16], it is shown that while the process of using of microbiological association of T. Ferrooxidans M1 bacteria (mixture of bacteria of T. Thiooxidans, Leptospirillum sp. with dominating content of T. Ferrooxidans), while oxidation, the dispersion of iron-aluminosilicate ore occurred. But the mechanisms of such processes from the point of view of colloidal, biocolloidal, and nanolevel were not considered.

A proposition was made that in deposit formation of metals, the microorganisms could have a certain part. There were grown specific cenosises of microorganisms and algae inhabiting the Sea of Okhotsk for two months in lab conditions on nutrient medium with gold in solution and in suspended (colloid) state. Spectral analysis showed that cenosises transfer the gold from the solution and suspension into sediment. Its particles are of 3–9 μm sizes and its content is 37–70% of the sediment. Thus, it was shown that communities (cenosises) of microorganisms can have a part in the process reverse to dispersion, i.e., in gold particle growth and its concretion in deposits [17]. And it is possible for other materials.

Green rust (GR) plays a significant role as intermediates in the processes of dispersion of IASMs and their transformation to NIASMs [18–20]. It appears with the help of microorganisms in technogenic conditions, in lab practice, and in environment. Modeling of GR phase-formation process in abiotic conditions [18] or in conditions of sea-water corrosion was done with the help of microorganisms. The interaction of microorganisms with minerals and organic substances in natural ecosystems has been investigated in the work [20]. Another not yet studied aspect of such investigation could be the determination of colloid–chemical mechanisms of formation of various minerals including disperse iron oxide-aluminosilicate minerals by biogenic way with the use of microorganisms. This confirms by studying iron oxide-aluminosilicate pelagic sediments (Fig. 3).

As we can see from the data (Fig. 3), the concentration of bivalent iron increases in pelagic suspension in the conditions of microbiological–biocolloidal reduction process. So is the dispersion of averaged suspension particles at reduction of iron oxide-aluminosilicate structures by previously described mechanism [4–6] for not averaged specific samples. The data presented in [4–6] and in Fig. 3 differ not much and that indicates the practical independence of the process from changing of the content in specified deviations. According to the data of XRD, the oxide-hydroxide material extracted from the suspension had the structure of α-FeOOH (goethite), which transformed to magnetite (Fe_3O_4) in certain conditions.

Thus, it was shown that microbiological leaching of iron oxide-aluminosilicate structures leads to the formation of α-FeOOH or Fe_3O_4. And that agrees with the classical ideas of mineralogy [21]. The biocolloidal process of goethite or magnetite formation occurs due to inorganic chemical reactions with the help of ferment surfactant products of microorganism metabolism. The role of ferments, which count in this process reaches 3000 [22], is reduced to their action as surfactants according to the laws of physicochemical geomechanics [4, 6, 17]. It is noticeable, that the increase

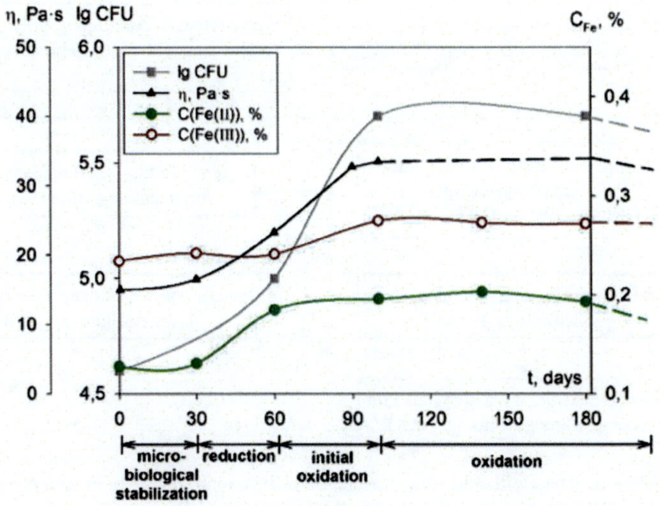

Fig. 3 Change dynamic of viscosity (η) and microbiological index (lgCFU) of partly deactivated averaged suspensions of The Black Sea ooze in the process of its balneological activity increasing, and change of Fe(II) and Fe(III) concentration in disperse medium while the Red-Ox processes

of viscosity (Fig. 3) of suspension is the highest in the period of Fe(II) concentration increasing in disperse medium in an interval of 30–60 days at concurrent decrease of Fe(III) content. It testifies that Fe(III) transits from disperse medium into contact zones of aluminosilicate microparticles in the form of phase nanostructures. And it is responsible there for the strength of bonds and rheological characteristics of suspensions. This phenomenon needs further investigations and their analysis.

It was also experimentally determined, that the intensive process of NIASSs formation lasts for 50–100 days, then the system goes to a stationary state (Fig. 3) in interval of 100–180 days, and a start of not enough studied interactions (dashed line). It can be concluded from this data, that the most probable here is a model of biocolloidal nanochemical nanostructural process, which occurs by the next scheme:

$$Fe_2O_3 \rightarrow GR \rightarrow Fe_3O_4\left(\begin{array}{c}\text{nano}\\ \text{magnetite}\end{array}\right) \rightarrow \gamma FeOOH\left(\begin{array}{c}\text{nano}\\ \text{lepidocrocite}\end{array}\right) \rightarrow \alpha FeOOH\left(\begin{array}{c}\text{micro- and macro-}\\ \text{crystalline}\\ \text{goethite}\end{array}\right) \quad (1)$$

biotransformation into nano- and micro-formations — geological epochs

It goes out from this scheme, that green rust (GR), which lifetime does not exceed 50–100 h [23], is an intermediate substance, and later on it transforms to nano-magnetite (Fe_3O_4), nanolepidocrocite (γ-FeOOH), and microgoethite (α-FeOOH). But the specific mechanisms of these processes are almost not developed and not explained, because according to known chemical models [23–25], the X-Ray amorphous $Fe(OH)_2$, plus FeO on initial stage in a form of nanoclusters, nanofilms or

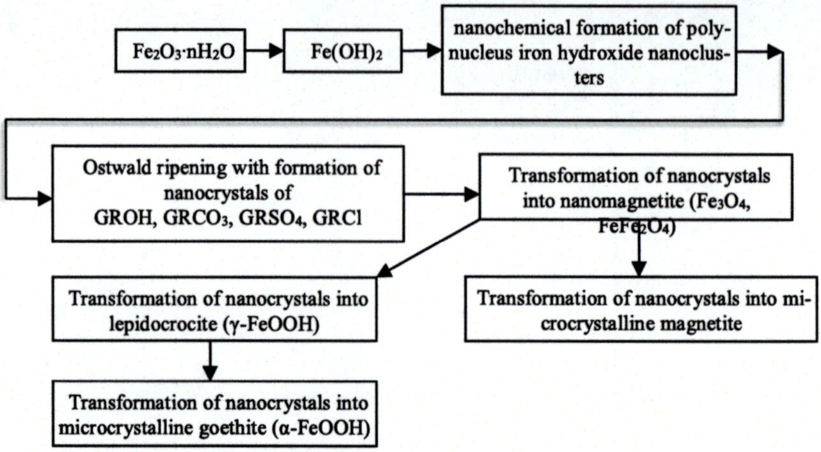

Fig. 4 Scheme of biocolloidal nanotransformation of iron oxide-hydroxides in a composition of IASMs and NIASMs in contact zones of colloid microparticles

separate nanoparticles with sizes below 10 nm (the area of nanochemical interaction) [26] are participating. And that is why the presented model (1) cannot be named full and it needs further complex and continuous detailed investigations.

In general view, these features of the processes can be explained with the application of general nanochemical [26] and colloid–chemical and biocolloidal models [4–7, 27], and by the ideas of physicochemical mechanics and geomechanics in dispersion of technogenic and ore materials [28].

Theoretically, the fuller scheme (1) has to be like next [4–7, 23–28]:

The scheme (Fig. 4) is confirmed also by the analysis of data [29] presented in Fig. 5.

As it can be seen, the solubility of $Fe(OH)_2$, which is a source of Fe^{2+} for GR formation, is the highest in alkaline medium (pH 8–11) while the solubility of $Fe(OH)_3$ in this interval is the lowest. That is why the process in total is limited by the formation of hydroxide ions that compensate parallel formation of H^+ ions [23]. This results in shifting of the equilibrium to the formation of lepidocrocite (α-FeOOH). But this conclusion given in [23] is not proved by the conclusions from [21] and [29].

By the known conceptions ([29]; Fig. 5), the most probable continuous existence including in geological lapse is of goethite—not lepidocrocite, which is an intermediate chemical compound [21]. Then the most stable from mineralogical point of view [21] magnetite (Fe_3O_4, $FeFe_2O_4$) is forming with the help of green rust, for example, by the scheme:

$$\begin{gathered}
Fe_4^{II}Fe_2^{III}(OH)_{12}SO_4 \cdot 8H_2O + 0,5O_2 \rightarrow Fe_2O_3 \cdot 3FeO \cdot 13H_2O + FeSO_4 + 2OH^- \\
Fe_2O_3 \cdot 3FeO \cdot 13H_2O \xrightarrow{OH^-} Fe_3O_{4\,\frac{nano}{micro}} + 3\underset{cluster}{[FeO \cdot H_2O]} + 10H_2O
\end{gathered} \tag{2}$$

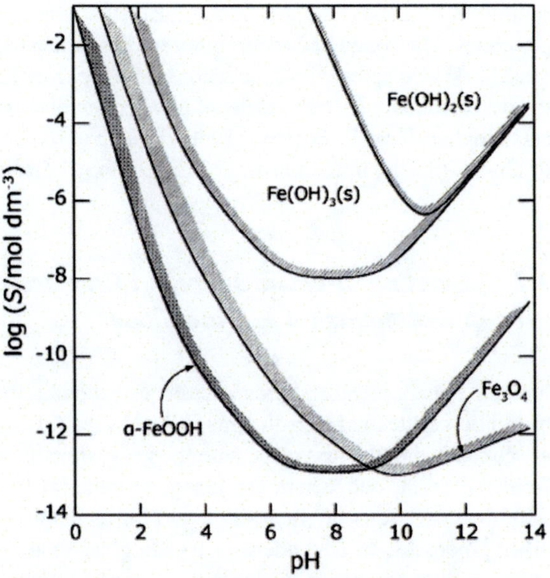

Fig. 5 Existence of iron compounds in dependence of pH [29]

According to Fig. 5, the magnetite is the most stable in an interval of pH 8–12, at the same time the goethite is the most stable at pH 6–10. The explanation of this is that the studied [23] process occurs in more acidic medium in a specific model apparatus with rotating iron disk. This leads primarily to lepidocrocite formation, which in the conditions of experiments in which the process is limited by the rate of diffusion, turns into goethite. At the same time, the magnetite forms noticeably later, after the formation of necessary quantity of OH^- ions and their diffusional delivery into reaction zone.

The most characteristic chemical reactions at that are also

$$FeO_{(nano)}+H_2O \rightarrow Fe(OH)_2. \tag{3}$$

$$6Fe(OH)_2+H_2CO_3+0.5O_2+2H_2O \rightarrow Fe_4^{II}Fe_2^{III}(OH)_{12} \cdot CO_3 \cdot 3H_2O. \tag{4}$$

$$3Fe(OH)_2+2O_2 \longrightarrow OH^-Fe_3O_{4(nano)}+6OH^-. \tag{5}$$

$$4Fe(OH)_2+O_2 \longrightarrow H_+4(\alpha - FeOOH) + 2H_2O. \tag{6}$$

Thus, the conducted model study and analysis of the data and also literature records showed that the process of contact interactions in NIASSs and NIASMs depends on their nanochemical, biocolloidal, and colloid–chemical transformations associated to iron oxides. These oxides are the basis of mechanisms of techno-genic and geological transformations of contents of artificial dams and underwater sediments on marine slopes. Such transformations can lead to stress (catastrophic) processes. The conducted preliminary studies of mineralogical processes in the Earth crust [4–7] also agree with conclusions from the work [30], and they are important for the development of similar ideas of physicochemical geomechanics in the theory of catastrophes. It can be developed with the help of rheological method and phenomena of elastic–plastic, ultraanomal, and hyperanomal dispersion flow [4–6, 21–28].

4.2 *Modeling of Geomechanical Catastrophic Processes in Sediments on Marine Slopes*

Indicated phenomena and processes occur in conditions of accumulating of disperse turbiditic-pelitic marine sediments [4–6, 21–28, 30]. The conditions of accumulation are connected in the first place with the geography of sedimentary basin, its distance from the coast, and where the terrigene material is mixing. It is also connected with the conditions of water circulation in dependence of wind, thermophilic and other processes. In dependence of geological factors, the sediments are accumulated on different areas of slopes and marine bottom: shelf (depth up to 200 m), continental slopes, on the bottom of deep-water channels and basins, abyssal (deep-water) plains with abyssal hills with height from 50 to 500 m, and on the tops of more high underwater rises (mountains). Overall, with the change of natural and geological conditions, especially with the change of the character of surface and underwater thermophilic waves under the influence of wind current forces that can reach the level of typhoon and tsunami, the sediments are accumulating on the bottom of deep-water channels, basins, and on abyssal plains. In other cases, the turbiditic-pelitic sediments sooner or later will also move to indicated areas of marine bottom. As a rule, such processes occur for tens of thousands and millions of years. In the geological epochs according to the laws of colloid chemistry and physicochemical geomechanics, all of them can compact in conditions of initial coagulation–condensation processes within thousands of years, and in the course of time, they can compact up to solid rocky state. Thus, the most interesting from the point of view of formation of catastrophic processes of fast shifts are sediments, which accumulate in quite huge quantities within thousands of years on continental slopes and on slopes of underwater mountains of volcanic origin on abyssal plains [31, 32]. The model scheme of such processes is given in Fig. 6.

The most widespread between sediments, which supply the slopes and marine bottom, are the pelagic oozes with additives of biogenic (biocolloidal) sediments, and the turbidity deposits that are forming at intensive turbulent wind-wave circulations, for example, typhoons. At this, the dispersive action of formed sea waves can reach on the depth of a few kilometers instead of the usual 500–700 m 500–700 м [31].

Pelagic sediments along with turbidities in most cases were accumulated for the last 18 millions of years in shelf sea zones during the denudation processes, and on the slopes of underwater hills and mountains (pelitic deposits) [31, 32]. The character

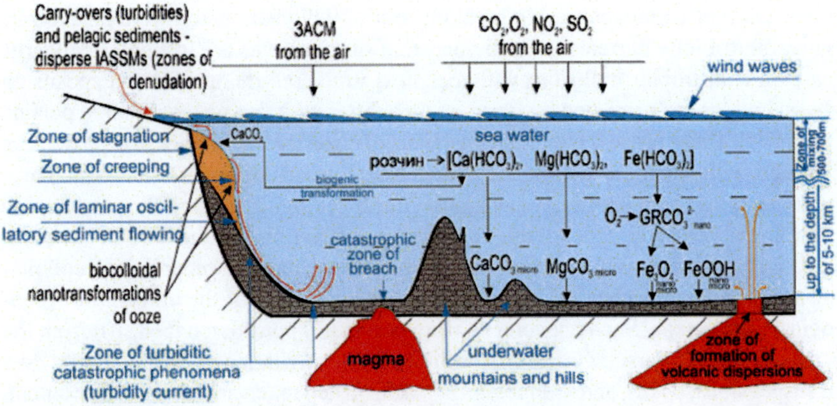

Fig. 6 The model of modern nanostructural marine transformations of IASMs and deposits of pelitic sediments and turbidities on the slopes and abyssal plains that can be accompanied by catastrophic phenomena

of sediments accumulated in shelf sea zones is connected with the characteristics of terrigene material that comes from the land and with the river ooze, and it depends on wind circulation in the marine basin. There, where the terrigenous material comes fast, the polydisperse aleuric and clay rocks including particle size up to 40–60 nm are depositing. During the transportation under the wind action, they are depositing in the form of aleuric-clay material on the surface of proturbiditic sand. The latter is a product of wind waves' treatment of barrier islands, which accordingly recede to the land. If there are a low quantity of debris come from the land, then the biocolloidal (biogenic) and chemical carbonate sedimentation on a surface of sand particles in shelf is possible. The addition of cold water from the depth rich in $Ca(Mg)(HCO_3)_2$ heats and the bicarbonates transforms into carbonates, generally into $CaCO_3$ [4, 9]. The latter precipitates from the water in a form of unstable small crystals of aragonite, including, due to an action of biocolloid processes while algae decaying [37]. Under the action of weak waves and currents, the precipitated $CaCO_3$ microconcretions on a surface of iron-aluminosilicate debris form huge above-water or underwater dunes from calcareous polymineral sand of IASM type. During the storms, such turbidity sands are moving on to continental slopes, then they are creeping further and sometimes they are forming powerful catastrophic turbidity currents with energy strength of huge values.

In most cases, the sediments accumulated on marine slopes and deep-water ocean bed are presented mainly by superfine pelagic oozes with addition of turbidities. The turbidities are formed during a long time including the dust carried with a wind. That is why their accumulation and movement rate are low [31, 32]. Thus, a priori it was considered that turbidity currents have no significant part in movement of marine sediments. This wrong idea is based on the fact that catastrophic phenomena with movement of huge amounts of sediments are registered rarely [33]. Their energetic potential can reach the level of nuclear explosion [34–36]. Thus, there were registered a few cases of underwater cable breaking since 1929 year, which were induced by powerful turbidity currents. It is obvious, that such currents are forming frequently, but their catastrophic collisions with technical structures are rare. Thus, deposits on abyssal plains are presented by layers of turbidities with fine nanostructured pelagic sediments. These processes explain big layers of oozes and turbidities in deep oceanic channels with thickness up to 30 km, which were formed during 1–3 mln years. Thus, it turns out that both in our days and in the Earth geological history, the catastrophic phenomena have a significant role, but their investigation had the lack of attention. Especially, it concerns the influence on them of nanochemical, biocolloidal, colloid–chemical and geomechanical processes, and exceptionally of the influence of rheological phenomena. Such phenomena were previously considered [5, 6, 10] from the point of view of character transformation under the gravitation of elastic–plastic flow of suspensions, oozes, and sediments, including in dependence of their water content, conditions of mixing, and movement speed. Averaged results of these investigations are presented in Fig. 7. It turns out from them that ultraanomaly of viscosity (Fig. 7f) reveals at minimal humidity and in conditions of elastic–plastic deformation. That is, these sediment movement conditions (Fig. 6) need the largest gravitational forces. Thus, the dispersion, suspension, ooze, or sediment would be resistant to stress influences if the current will not transform from laminar into turbulent in conditions of initial ultraanomal viscous behavior under the action of gravitational forces. And in addition to humidity, such viscosity is concerned with features of disperse phase, its

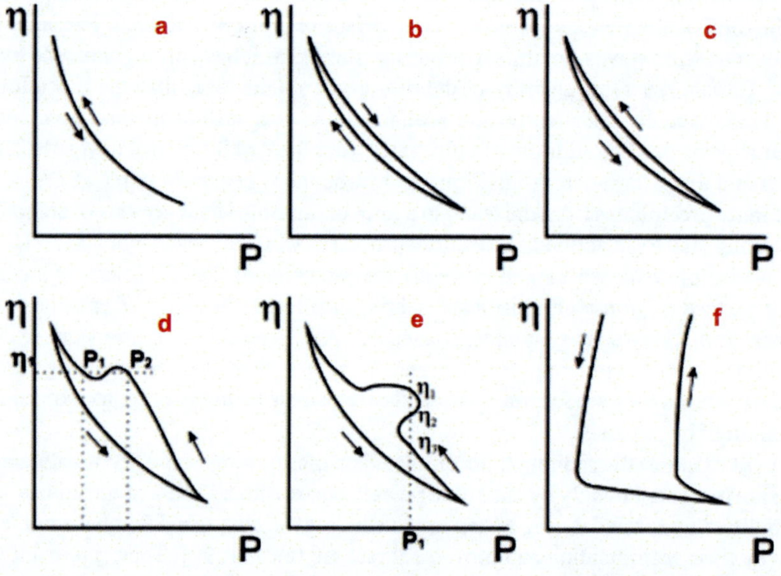

Fig. 7 The curves of increasing of the flow of dispersions, oozes, or sediments in direction from f to a, and its decreasing in direction from a to f: **a** non-Newtonian; **b** thixotropic; **c** rheopexic; **d** dilatant; **e** hyperanomal; **f** ultraanomal

granulometric composition, proportions of nano-, colloid, and macroparticles, mineralization of disperse medium, biocolloidal, geomechanical, chemical, wind, and other factors. These dependences were already partly showed in other investigations [4–6, 9, 10] and they need further studying for its generalization and classification.

The analysis of literature and presented above data (Fig. 1, 2, 3, 4, 5, 6 and 7; Table 1) showed that catastrophic processes on marine slopes in the presence of nanostructured iron-aluminosilicates depend on many factors. And first of all, it considers nanostructural, biocolloidal, colloid–chemical, mechanochemical, mechanical, and other factors that are summarily fixed in rheological phenomena studied by physicochemical mechanics and geomechanics. Correspondingly, the analysis of specific data received for marine slopes' sediment samples can allow to predict the behavior of specific sediments on marine slopes, and on underwater dams, slopes, and to prevent their catastrophic shifts due to stress influences (earthquakes, typhoons, tsunami waves, big technogenic and gravitational strains, etc.) using presented ideas from this study.

5 Conclusions

The influence of physical, colloid–chemical, nanochemical, geomechanical, and microbiological–biocolloidal factors on nanostructural contact formation and rheological behavior of real and model iron-aluminosilicate mineral formations was generalized and tested on theoretical model and experimental levels. It was shown that the formation of different NIASSs and NIASMs from IASSs and IASMs has similar characters, depends on transformation of iron compounds besides silicate and carbonate transformations, and it results in the final formation of structures like goethite (α-FeOOH) or magnetite Fe_3O_4. It agrees with the conclusions of geological–mineralogical investigations. It was considered that the role of rheological phenomena and physicochemical geomechanics in nanostructural transformations of iron-aluminosilicate formations were at conditions close to stress ones. The role of mechanical and wind factors, and also the dispersion of IASSs and IASMs in stress conditions at their initial nanostructural and mineralogical contact transformations was indicated on model level. It was also noted the fact that the existence of compounds of green rust type exceeds not 75–100 h, it needs the special formation conditions, so that their role in the transformation processes is not determinative. It was shown that microbiological processes in iron-aluminosilicate structures lead to the final formation of goethite or magnetite in dependence of pH. And it was shown that biocolloidal transformations occur through inorganic chemical reactions and in the presence of microorganism metabolic products.

For the first time, it showed that nanochemical, mechanochemical, and physicomechanical contact interactions in nanostructured IASMs in the presence of biogeocenosis are controlled by biocolloidal processes conditioned by bacterial and chemical reactions. And that the flow of NIASMs' dispersions reduces in the next direction: thixotropy → dilatancy ⇄ rheopexy → hyperanomaly → ultraanomaly of viscosity.

But in the conditions of catastrophic creeping of sediments from the slopes, the flow increases in the opposite way with increasing velocity, changing from laminar flowing into avalanche-like (catastrophic) currents of oozes and turbidities mix.

References

1. Gomes C, Carretero M-I, Pozo M, Maraver F et al (2013) Peloids and pelotherapy: historical evolution, classification and glossary. Appl Clay Sci 75–76:28–38
2. Rosa M et al (2011) Development of an anti-cellulite peloid containing bentonite of Porto Santo island, Madeira Archipelago. In Abstracts of geoMed 2011—4th international conference on medical geology, Italy
3. Manuel P et al (2013) Comparison of physical and physical-chemical properties of compositionally different peloids used in Spanish spas. In: Abstracts of XV international clay conference, Rio de Janeiro
4. Panko AV, Kovzun IG, Ulberg ZR, Oleinik VA, Nikipelova EM, Babov KD (2016) Colloid-chemical modification of peloids with nano- and microparticles of natural minerals and their practical use. In: Fesenko O, Yatsenko L (eds) Nanophysics, nanophotonics, surface studies, and applications, vol 183. Springer proceedings in physics, Cham, pp 163–177. https://doi.org/10.1007/978-3-319-30737-4_14
5. Panko AV, Kovzun IG, Nikipelova OM et al (2019) Nanostructural Effects in iron oxide silicate materials of the Earth's crust. In: Fesenko O, Yatsenko L (eds) Nanophysics, nanophotonics, surface studies and applications. Springer proceedings in physics, vol 221. Springer, Cham, pp 367–386. https://doi.org/10.1007/978-3-030-17759-1_25
6. Panko AV, Kovzun IG, Nikipelova OM et al (2018) Nanostructural and nanochemical processes in peloid sediments aided with biogeocenosis. In: Fesenko O, Yatsenko L (eds) Nanophysics, nanophotonics, surface studies and applications. NANO 2017. Springer proceedings in physics, vol 214. Springer, Cham, pp 215–230. https://doi.org/10.1007/978-3-319-92567-7_13
7. Tarasevich YuI, Ovcharenko FD (1975) Adsorbtsiya na glinistyih mineralah (Adsorption on clay minerals). Naukova dumka, Kyiv
8. Plusnina NI (1967) Infrakrasnyie spektryi silikatov (Infrared spectrums of silicates). Izdatelstvo MGU, Moscow
9. Kovzun IG, Ulberg ZR, Panko AV, Prokopenko VA, Oleinik VA, Nikipelova EM (2015) Colloid-chemical and nanochemical processes in peloids on basis of ferrous clay minerals. In: Fesenko O, Yatsenko L (eds) Nanoplasmonics, nano-optics, nanocomposites, and surface studies. Springer proceedings in physics, vol 167. Springer, Cham, pp 233–243. https://doi.org/10.1007/978-3-319-18543-9_15
10. Panko AV, Kovzun IG, Prokopenko VA, Tsyganovich OA, Oliinyk VO, Nikipelova OM (2017) Nano- and microdisperse structures in processes of metamorphism, reduction sintering, and component separation of iron-oxide-silicate materials. In: Fesenko O, Yatsenko L (eds) Nanophysics, nanomaterials, interface studies, and applications. NANO 2016. Springer proceedings in physics, vol 195. Springer, Cham, pp 743–755. https://doi.org/10.1007/978-3-319-56422-7_57
11. Pehtasheva EL (2013) Biopovrezhdeniya neprodovolstvennyih tovarov (Biodeterioration of nonfoods). Dashkov and K, Moscow
12. Rozanov AYu, Zavarzin GA (1997) Bacterial paleontology. Herald Russ Acad Sci 67(6):241–245
13. Ulberg ZR, Pertsov NV, Garbara SV, Nechaev SV, Stepanenko VG (1990) Silikatnyie bakterii v protsessah mikrobiologicheskogo vyischelachivaniya metallov iz rud (Silicate bacteria in the processes of microbiological leaching of metals from ores). Dokl AN USSR Ser B 5:80–83
14. Brierly IA (2007) Biohidrometallyrgy—this microbiologist's perspective. Adv Mater Res 20–21:3–10

15. Podolska VI, Ermolenko AI, Ulberg ZR, Pertsov NV (2003) Vplyv mahnetytu na shvydkist mikrobnoho vyluzhuvannia halenitu iz zolotovmisnoi rudy (Effect of magnetite on the rate of microbial leaching of galena from gold-containing ore). Ekotehnologii i resursosberezhenie 6:22–27
16. Ulberh Z, Podolska V (2011) Biotechnologies in gold-extracting industry. Visnyk Nat Acad Sci Ukraine 3:18–29
17. Prokopenko VA, Kovzun IG, Ulberg ZR (2014) The creative potential of scientific discovery. Visnyk Nat Acad Sci Ukraine (10):52–61. https://doi.org/10.15407/visn2014.10.052
18. Grassian VH (2005) Environmental catalysis. Taylor & Francis Group, New York
19. Waseda Y, Suzuki Sh (2006) Characterization of corrosion products on steel surfaces. Springer
20. Huang PM, Bollag J-M, Senesi N (2002) Interactions between soil particles and microorganisms: impact on the terrestrial ecosystem. Wiley. https://doi.org/10.1515/ci.2002.24.4.26a
21. Sherbak NP, Pavlishin VI, Litvin AL et al (1990) Mineralyi Ukrainyi: Kratkiy spravochnik (Minerals of Ukraine: a quick reference). Naukova dumka, Kyiv
22. Kessler YuM, Zaitsev AL (1989) Solvofobnyie effekty (Solvophobic effects). Himiya, Leningrad
23. Lavrynenko OM, Kovalchuk VI, Netreba SV, Ulberg ZR (2013) New rotation-corrosion dispergation method for obtaining of iron-oxygen nanoparticles. Nanostudies 7:295–322
24. Sugimoto T, Matijevic E (1979) Formation of uniform spherical magnetite particles by crystallization from ferrous hydroxide gels. J Colloid Interface Sci 74(1):227–243
25. Karapetjants MH (1970) Vvedenie v teoriyu himicheskih protsessov (Introduction to the theory of chemical processes). Vishaja shkola, Moskow
26. Sergeev GB (2003) Nanohimiya (Nanochemistry). Izdatelstvo MGU, Moscow
27. Pozin ME (1970) Tehnologiya mineralnyih soley, 3-e izdanie (The technology of mineral salts, 3rd edn). Himiya, Leningrad
28. Pertsov NV (1998) The Rehbinder effect in the earth's crust (Physicochemical geomechanics). Colloid J 60(5):578–588
29. Tamura H (2008) The role of rusts in corrosion and corrosion protection of iron and steel/H. Tamura. Corros Sci 50(7):1872–1883
30. Usman M, Abdelmoula M, Hanna K, Gregoire B, Faure P, Raby C (2012) FeII induced mineralogical transformations of ferric oxyhydroxides into magnetite of variable stoichiometry and morphology. J Solid State Chem 194:328–335. https://doi.org/10.1016/j.jssc.2012.05.022
31. Verhoogen I, Turner FI, Weiss LE, Wahrhaftig C, Fyfe WS (1970) The earth. An introduction to physical geology. Holt Rinehart and Winston, New York-Montreal-London-Sydney
32. Weyl PK (1970) An introduction to the marine environment. Wiley, New York
33. Scheidegger AE (1975) Physical aspects of natural catastrophes. Elsevier, Amsterdam, Oxford, New York
34. Asada T, Isibasi K et al (1984) Metodyi prognoza zemletryaseniy. Ih primenenie v Yaponii (Earthquake prediction methods. Their use in Japan). Nedra, Moscow
35. Riazanov IL (1980) Velikie katastrofyi v istorii Zemli (Great disasters in the history of the earth). Nauka, Moscow
36. Li Devis (2008) Natural disasters. Infobase Publishing, New York
37. Horne RA (1969) Marine chemistry. Wiley Interscience, New York

The Effect of the Intensity of Ammonium Nitrate Granules Humidification on the Quantitative and Qualitative Composition of the Final Granules Nanoporous Structure

A. V. Ivaniia, A. E. Artyukhov, A. I. Olkhovyk, and D. R. Potapov

1 Introduction

Porous ammonium nitrate is a valuable agrotechnical fertilizer. It is one of the components of the common explosive ANFO, widely used in the mining industry [1]. The quantitative and qualitative indices of the porous structure of ammonium nitrate granules (namely, porous ammonium nitrate PAN, according to [2–5] is the most effective component of ANFO) have a decisive impact on the properties of common explosive ANFO. The larger the number of pores, the greater the absorptivity of the PAN granules toward diesel fuel. The balance between macropores and mesopores affects the retentivity of PAN granules.

The multistage process of the ordinary ammonium nitrate granules humidification with various humidifiers and the subsequent heat treatment stages in vortex granulator is one of the most promising ways to obtain porous ammonium nitrate (PAN) as well as multilayer granules [6, 7].

The pore structure of the finished PAN granules is effected by such indices as the type of humidifier, the degree of granules humidification and the heat treatment mode. In [8–10], the dependence of the qualitative and quantitative features of the PAN granules porous structure on the type of humidifier is investigated. For laboratory tests, a laboratory a vortex granulator unit is used (Fig. 1). The batch of the initial ammonium nitrate granules is heated to a predetermined temperature, then the granules are moistened (humidifier—water) and subsequent heat treatment is carried out (8 min at a temperature of a layer—108° C, thermodynamic parameters are taken according to the results of studies [11, 12]). Samples of finished PAN granules are studied using microscope. (scanning electron microscope eTescan Vega 3, formation

A. V. Ivaniia (✉) · A. E. Artyukhov · A. I. Olkhovyk · D. R. Potapov
Processes and Equipment of Chemical and Petroleum-Refineries Department, Sumy State University, 2, Rymskogo-Korsakova str., 40007 Sumy, Ukraine
e-mail: andrey.ivaniy@gmail.com

O. Fesenko and L. Yatsenko (eds.), *Nanomaterials and Nanocomposites, Nanostructure Surfaces, and Their Applications*, Springer Proceedings in Physics 246, https://doi.org/10.1007/978-3-030-51905-6_29

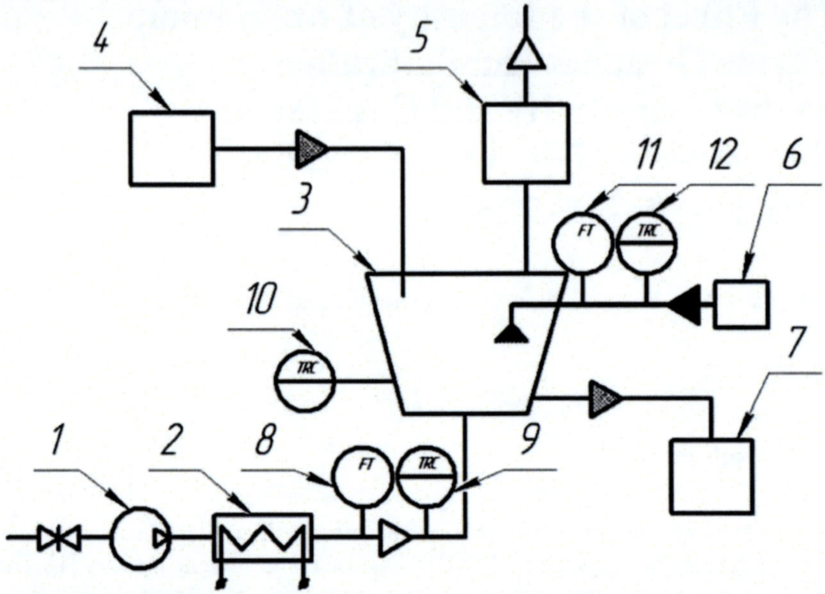

Fig. 1 The scheme of the laboratory vortex granulator unit: 1—gas blower; 2—heater; 3—vortex granulator; 4—container for granules; 5—waste gas cleaning stage; 6—container for preparation of humidifier; 7—container of finished granules; 8—automation device for measuring the air flow; 9—automation device for measuring the temperature of air; 10—automation device for measuring the temperature of layer; 11—automation device for measuring the humidifier flow; 12—automation device for measuring the temperature of humidifier

of the carbon film on the granule surface—Carbon evaporation head CA7625 and SC7620 high resolution, manual Sputter Coater).

2 The Effect of the Humidification Intensity on the Nanoporous Structure of PAN Granules

It is important to determine the first and second critical velocities (the rate of fluidization beginning and the rate of the granules ablation) for the fluidized bed [13, 14]. The first critical velocity is calculated by the formula [15]:

$$\omega_1 = \frac{\mathrm{Re} \cdot \mu_{fa}}{\rho_{fa} \cdot d}, \tag{1}$$

where *Re*—the Reynolds criterion, μ_{fa}—the viscosity of the fluidizing agent, ρ_{fa}—the density of the fluidizing agent, and *d*—the diameter of the fluidized bed particles (PAN granules).

The Reynolds criterion is calculated by the formula [15, 16]:

$$\mathrm{Re} = \frac{Ar}{1400 + 5,22 \cdot \sqrt{Ar}}, \tag{2}$$

where Ar—Archimedes criterion.

Archimedes criterion is calculated by the formula [17]:

$$Ar = \frac{d^3 \cdot \rho_{fa} \cdot g \cdot \rho_{fbp}}{\mu_{fa}^2}, \tag{3}$$

where ρ_{fbp}—density of the fluidized bed particles (PAN granules).

The second critical velocity is calculated by the formula [15]:

$$\omega_2 = \frac{\mu_{fa}}{d \cdot \rho_{fa}} \cdot \left(\frac{Ar}{18 + 0,575 \cdot \sqrt{Ar}} \right). \tag{4}$$

At the stage of the granules humidification, their diameter remains the same, but the mass increases. It leads to an increase in the density of the granules ρ_{fbp}, to an increase in the value of the Archimedes criterion Ar, an increase in the values of the first and second critical velocities. Thus, humidifying the granules, it is necessary gradually to increase the solubilizer's motion velocity.

The range of the granulator stable operation (critical velocities of fluidization) for a vortex fluidized bed of dry granules and granules humidified to a predetermined humidity, is calculated using the author's software product (Fig. 2) [18]. The works [19, 20] also show the main hydrodynamic indices of a vortex granulator operation, used in the experimental and industrial production line of PAN.

Depending on the granules humidification mode during the modification of the PAN granules in the vortex granulator, the qualitative and quantitative features of the porous structure in the finished granules are changed.

During the tests and analysis of their results, the following humidification modes are distinguished (Fig. 3):

- The uneven humidification mode (less than 50 g of water per 1 kg of ammonium nitrate granules);
- Optimal humidification mode (50–150 g of water per 1 kg of ammonium nitrate granules);
- High humidification mode (150–300 g of water per 1 kg of ammonium nitrate granules);
- Excessive humidification mode (more than 300 g of water per 1 kg of ammonium nitrate granules).

When using the uneven humidification mode, the finished PAN granules have an undeveloped porous structure (Fig. 4). The number and size of pores of the finished granules do not almost differ from the initial granules. Under conditions of low

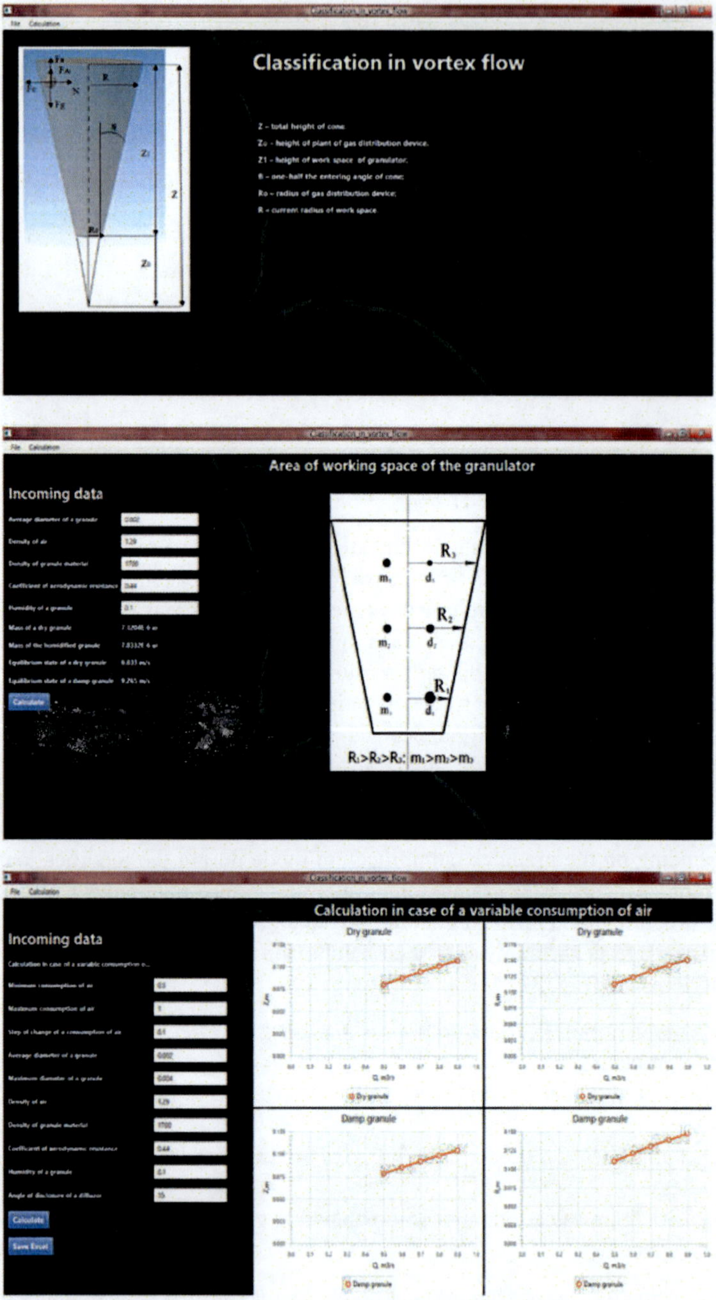

Fig. 2 The software product classification in vortex flow© [18]: calculation block for the vortex granulator stable operation range

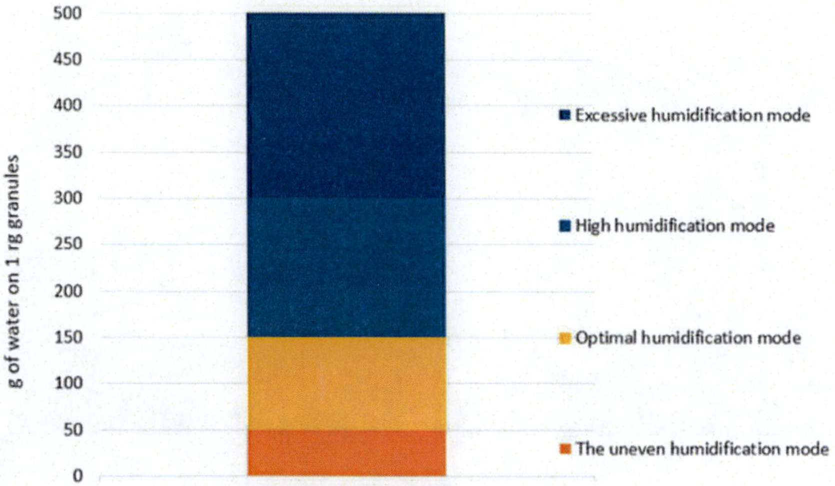

Fig. 3 Diagram of humidification modes (depending on the amount of humidifier)

humidity of the granules and their subsequent heat treatment, there are local areas with the melted surface.

Under conditions of the optimal humidification mode, the finished PAN granules have a developed porous structure with a large number of macropores (more than 50 nm) and mesopores (2–50 nm) (Fig. 5). There are no areas with melted surface or local sections with undeveloped porous structure.

Further increase in the intensity of humidification (the high humidification mode) leads to the over-humidification of PAN granules. During subsequent heat treatment, significant destructive phenomena (cracks and chips) occur due to considerable pore-formation (Fig. 6). Besides, increasing the moisture intensity, the tendency of granules to agglomeration increases, which impairs the qualitative features of the finished product.

When the intensity of the granules humidification increases, their destruction and partial dissolution occur. The fluidized bed of granules in the vortex granulator loses its stability and "falls," causing the necessity to stop the unit and clean it.

The graphical dependence of the qualitative and quantitative porosity features of the PAN granules on the moisture intensity is shown in Fig. 7 (surface layers) and Fig. 8 (inner layers).

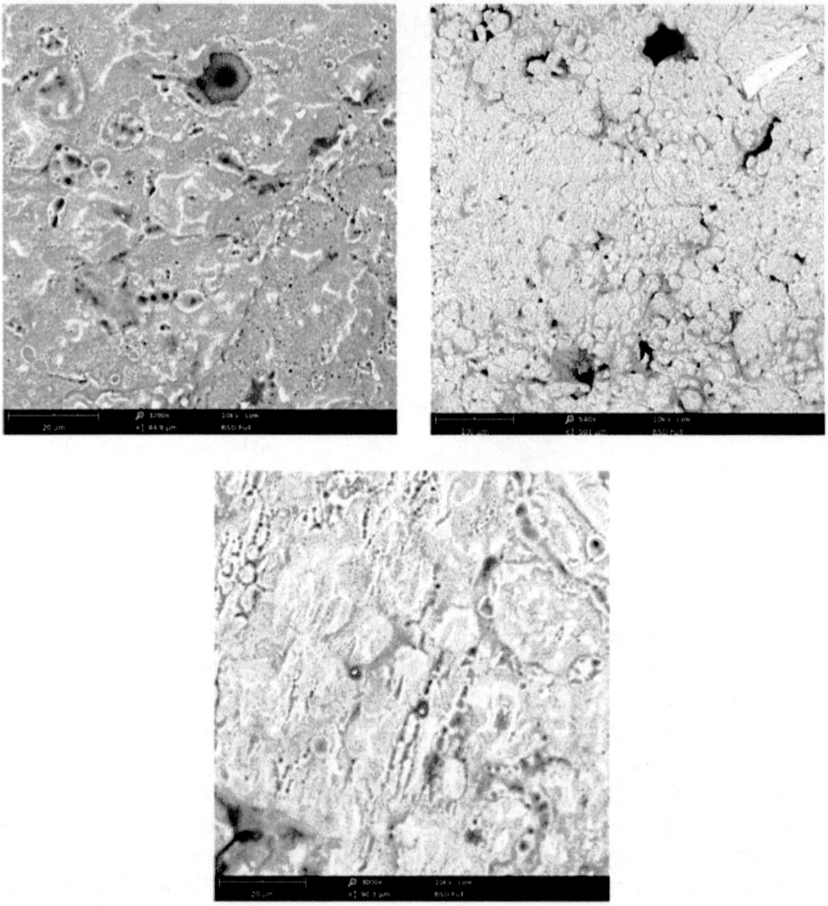

Fig. 4 The surface of the PAN granules (uneven humidification mode) with undeveloped porosity and local areas with a melted surface

3 Conclusions

In order to obtain PAN granules with a developed porous structure, it is necessary to use the optimal humidification mode, technological parameters of which are demonstrated in this research. If this mode is not kept, the humidification and heat treatment stages will be ineffective (the uneven humidification mode), or the commodity qualities of the finished product will be deteriorated (high humidification mode).

The task for further research is to work out the humidification stage with other humidifiers to define technological parameters for this process implementation.

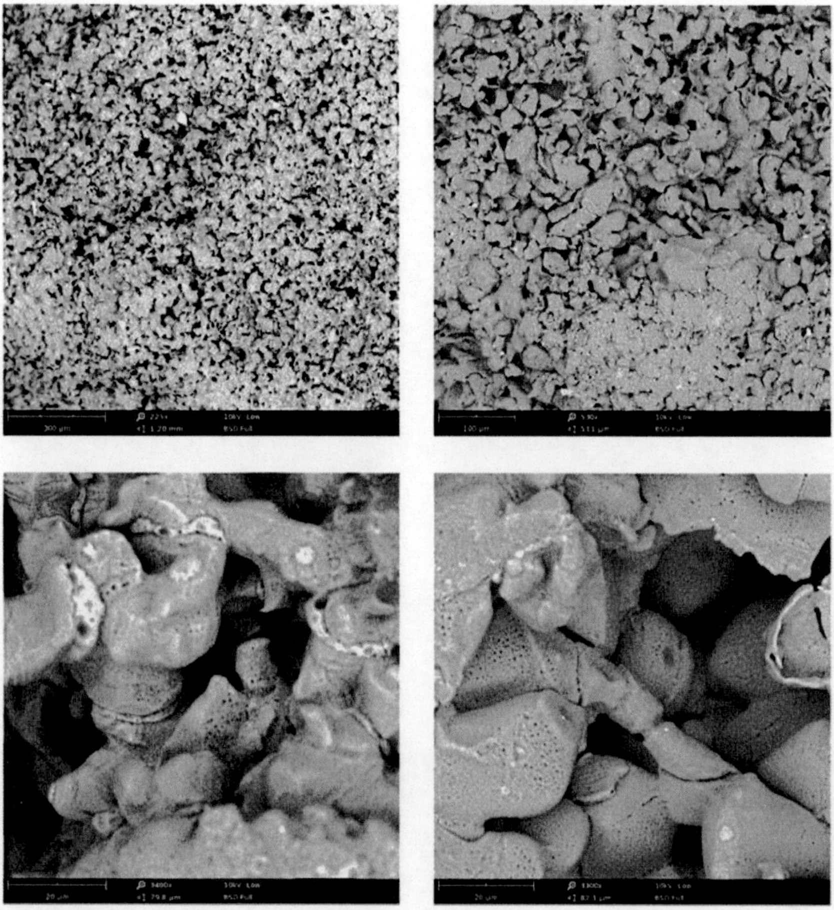

Fig. 5 The surface of the PAN granules (optimal humidification mode) with developed porous structure with macropores and mesopores

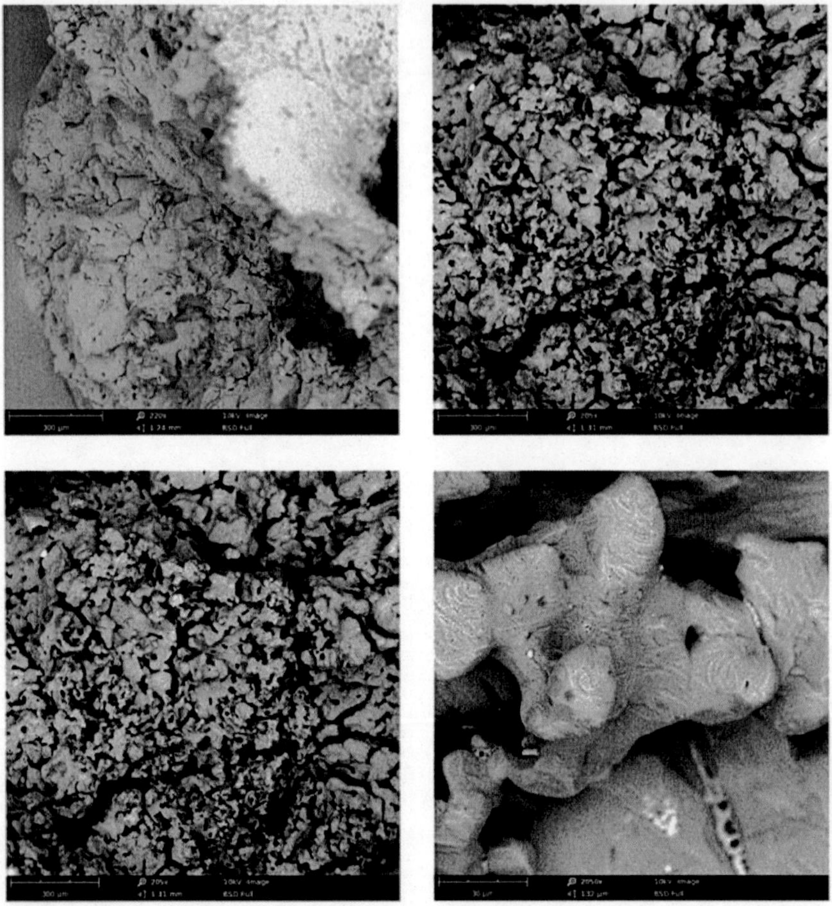

Fig. 6 The surface of the PAN granules (high humidification mode) with a lot of chips and cracks, granule agglomeration tendency

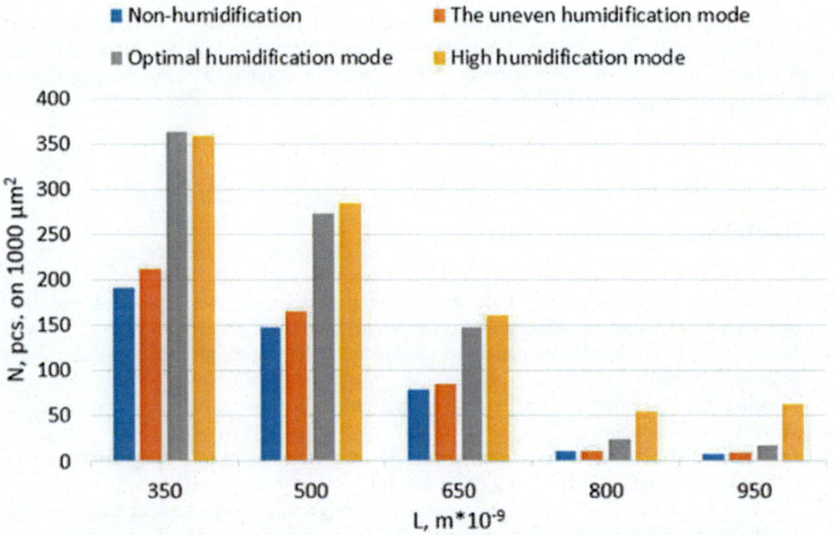

Fig. 7 Dependence of qualitative and quantitative features of PAN granules porosity on moisture intensity (surface layers)

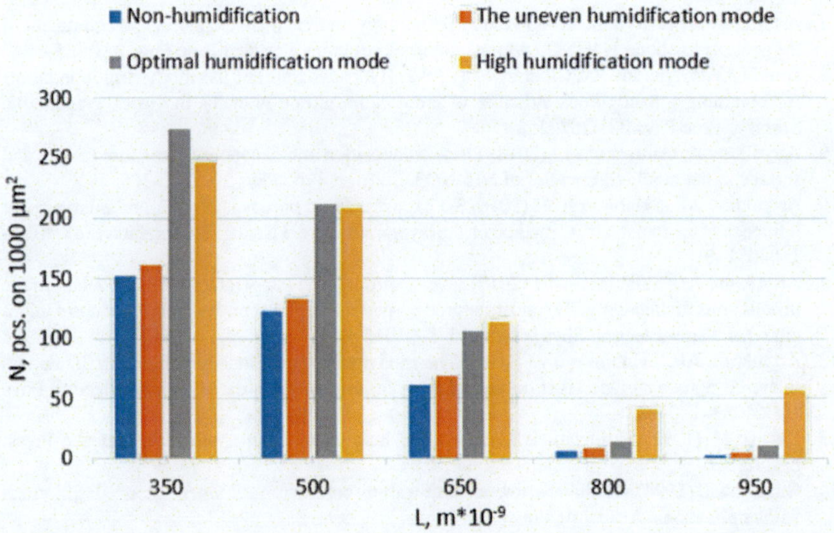

Fig. 8 Dependence of qualitative and quantitative features of PAN granules porosity on moisture intensity (inner layers)

Acknowledgements This research work has been supported by the Ministry of Science and Education of Ukraine under the project «Small-scale energy-saving modules with the use of multifunctional devices with intensive hydrodynamics for the production, modification and encapsulation of granules», project No. 0119U100834.

References

1. Martin G, Barbour W (2003) Industrial nitrogen compounds and explosives, chemical manufacture and analysis. Watchmaker Publishing
2. Janssen TJ (2011) Explosive materials: classification, composition and properties. Nova Science Publishers
3. Kubota N (2015) Propellants and explosives: thermochemical aspects of combustion, 3rd edn. Wiley-VCH, Weinheim
4. Jackson S (2016) The dependence of ammonium-nitrate fuel-oil (ANFO) detonation on confinement. Proc Combust Inst 36(2):2791–2798
5. Lipinska K et al (2005) Demilitarized propellants as ingredients in commercial explosives. In: European federation of explosives engineers: Brighton conference proceedings, Brighton, pp 493–498
6. Artyukhov AE, Sklabinskyi VI (2013) Experimental and industrial implementation of porous ammonium nitrate producing process in vortex granulators. Nauk Visnyk Nats Hirnychoho Univ 6:42–48
7. Artyukhov AE, Artyukhova NO (2019) Technology and the main technological equipment of the process to obtain N4HNO3 with nanoporous structure. Springer Proc Phys 221:585–594
8. Ivaniia AV, Artyukhov AY, Olkhovyk AI (2019) Hydrodynamic and thermodynamic conditions for obtaining a nanoporous structure of ammonium nitrate granules in vortex granulators. Springer Proc Phys 221:257–268
9. Artyukhov A, Gabrusenoks J (2018) Phase composition and nanoporous structure of core and surface in the modified granules of NH4NO3. Springer Proc Phys 210:301–309
10. Artyukhov AE, Sklabinskyi VI (2016) 3D nanostructured porous layer of ammonium nitrate: influence of the moisturizing method on the layer's structure. J Nano- Electron Phys 8(4)04051-1–04051-5
11. Artyukhov AE, Sklabinskyi VI (2017) Investigation of the temperature field of coolant in the installations for obtaining 3D nanostructured porous surface layer on the granules of ammonium nitrate. J Nano- Electron Phys 9(1)01015-1–01015-4
12. Artyukhov AE, Sklabinskyi VI (2016) Thermodynamic conditions for obtaining 3D nanostructured porous surface layer on the granules of ammonium nitrate. J Nano- Electron Phys 8(4)04083-1–04083-5
13. Kwauk M (1992) Fluidization: idealized and bubbleless, with application. Science Press, Beijing
14. Gidaspow D (1994) Multiphase flow and fluidization: continuum and kinetic theory descriptions with applications. Academic Press, San Diego
15. Yang W-C (2003) Handbook of fluidizfition and fluid-particle systems. Marcel Dekke, New York
16. Hiltunen K, Jäsberg A, Kallio S (2009) Multiphase flow dynamics. Theory and numerics. Edita Prima Oy, Helsinki
17. Crowe CT (2006) Multiphase flow handbook. Taylor & Francis Group, Boca Raton

18. Artyukhov AE et al (2018) Computer program «classification in vortex flow» UA Certificate for registration of copyright 67472, 26 August 2016
19. Artyukhov AE, Sklabinskyi VI (2015) Theoretical analysis of granules movement hydrodynamics in the vortex granulators of ammonium nitrate and carbamide production. Chemistry and Chemical Technology 9(2):175–180
20. Artyukhov AE, Sklabinskyi VI (2015) Hydrodynamics of gas flow in small-sized vortex granulators in the production of nitrogen fertilizers. Chem Chem Technol 9(3):337–342

Features of Phase Formation Processes in Amorphous Alloys of Fe–Zr System

Tatiana L. Tsaregradskaya, Inna V. Plyushchay, Olexandr O. Kalenyk, Olexandr I. Plyushchay, and Galina V. Saenko

1 Introduction

Alloys of the Fe–Zr system are widely used as functional materials, especially in the nuclear industry. Stainless steel zirconium was developed for the manufacture of containers for the storage of nuclear energy wastes [1, 2] and materials for nuclear reactors and fuel cells [3]. The search for ways to improve the functional properties of materials of the Fe–Zr system continues. One possible way of modifying the properties of materials is to change their atomic structure, in particular amorphization. It is known that homogeneity of amorphous materials improves their functional properties, for example, corrosion resistance, mechanical, and magnetic properties. Therefore, the amorphous metal alloys of the Fe–Zr system are characterized by a wide range of unique physical properties, widely used as the latest functional materials; so the study of the peculiarities of their crystallization process has an important fundamental and applied value. Since amorphous alloys are thermodynamically unstable systems, it is necessary to investigate the regularities of the crystallization process under the influence of various external factors.

One of these factors is impulse laser treatment. The use of pulsed laser treatment can significantly increase the rate of thermal impact on metals and alloys: up to (10^4–10^6) K/s using the most common pulsed and continuous technological lasers, and in some cases up to (10^9–10^{10}) K/s. Under such influence, crystallization must occur instantaneously throughout the volume of the amorphous tape (explosive crystallization) [4]. The phenomenon of explosive crystallization in amorphous metallic alloys is characterized by the intensive isolation of the latent heat of crystallization

T. L. Tsaregradskaya (✉) · I. V. Plyushchay · O. O. Kalenyk · O. I. Plyushchay · G. V. Saenko
Departments of Physics, Taras Shevchenko National University of Kyiv, Volodymyrska 64/13, 01601 Kiev, Ukraine
e-mail: tsar_grd@ukr.net

O. Fesenko and L. Yatsenko (eds.), *Nanomaterials and Nanocomposites, Nanostructure Surfaces, and Their Applications*, Springer Proceedings in Physics 246, https://doi.org/10.1007/978-3-030-51905-6_30

at the boundary of the phase separation, which is accompanied by the self-heating of the crystallization front.

The phenomenon of explosive crystallization is intensively studied both theoretically and experimentally. In addition to the apparent physical interest in this phenomenon, the study of the nature of explosive crystallization is also important in connection with the increasing use of amorphous materials in modern technology. Indeed, explosive crystallization, initiated by a rather strong local energy momentum, can extend to the entire volume of amorphous material at high velocity [5]. It is very important that the appearance of explosive crystallization can occur at temperatures substantially lower than the temperature of crystallization during isothermal annealing. This fact must be taken into account when choosing amorphous materials for practical application. The problem of research is related to the theoretical analysis of the possibilities of the process of explosive crystallization during laser treatment of binary amorphous alloys of the Fe–Zr system.

2 Equation of the Description of the Kinetics of the Crystallization Process in the Fe–Zr System

A series of theoretical calculations were carried out to analyze the possibility of the process of explosive crystallization during laser treatment of binary amorphous alloys of the Fe–Zr system. The calculations were carried out within the framework of the modified theory of homogeneous crystallization for binary alloys, which takes into account the work associated with the concentration fluctuation. An important characteristic of the crystallization process of amorphous alloys is the volume part of the crystalline phase $X(t)$ at a given time, which for the binary systems is described by the differential equation [6]:

$$\frac{dX(t)}{dt} = (1 - \sum_{i=1}^{k} X_i(t))\frac{4}{3}\pi\langle J_{0i} U_i^3\rangle(t - t_{0i})^3,$$

$$\langle J_{0i}, U_i^3\rangle = \int_0^1 J'_{0i}(C_2)U_i^3(C_2)dC_2, \tag{1}$$

where k—the number of crystalline phases; U_i—linear growth rate of crystals of the i phase; J_{0i}—initial nucleation frequency, which corresponds $\sum_{i=1}^{k} X_i = 0$; t_{0i}—the start time of the crystallization of the i phase.

The nucleation rate for binary systems J_i' is described by the distribution functions of the crystalline nuclei that arise, at certain concentrations, that is,

$$J_i'(c_2) = N\frac{D}{a^2}\exp\left[-\frac{16\pi\sigma^3 V_0^2}{3kT\left(\Delta G^{\alpha-i}(c_2) - 2|\Delta G^{\alpha-\alpha}(c_2)|\right)^2}\right],$$

$$\Delta G^{\alpha-i}(c_2) = \Delta G^{\alpha}(c_2) - \Delta G^{i}(c_2),$$
$$\Delta G^{\alpha-\alpha}(c_2) = \Delta G^{\alpha}(c_2) - \Delta G^{\alpha}(c_{20}), \tag{2}$$

where $\Delta G^{\alpha}(c_2)$ and $\Delta G^{i}(c_2)$—concentration dependences of the relative integral Gibbs free energy in the amorphous (α) and crystalline (i) phases; c_{20}—initial concentration of the second component in the α-phase; c_2—arbitrary concentration of the second component in the α- and i-phases, σ—surface tension at the boundary of two phases, V—molar volume, N—number of atoms per unit volume; a—atomic diameter, D—diffusion coefficient.

Taking into account that during crystallization in the binary system several stable formations are formed, as well as the possible occurrence of metastable phases, (2) is generalized to the

$$J_i = \int_0^1 J_i'(c_2)dc_2, , \tag{3}$$

where $i = \beta, \gamma, \delta, \varepsilon$-phases.

The total frequency of nucleation J of different types will be written in the form

$$J = \sum_{i=1}^{k} J_i, \tag{4}$$

where k—the number of stable and metastable phases occurring during crystallization.

The linear growth rate of crystals in the case of binary alloys is described by the equation

$$U_i = \frac{D}{a}\left(1 - \exp\left(-\frac{\Delta G^{\alpha-i} - \left|\Delta G^{\alpha-\alpha}\right|}{RT}\right)\right). \tag{5}$$

Equation (1) is solved by methods of numerical integration, from which all values are determined $X_i(t)$ for the system; $X(t) = \sum_{i=1}^{k} X_i$.

Let us consider the possibilities and thermodynamic preconditions of the phase stratification process in amorphous binary alloys. As in the case of liquid systems, the possibility of stratification in amorphous alloys is associated with the behavior of the concentration dependence of the relative integral free Gibbs energy. For most binary amorphous alloys there is a concentration region in which the free energy of a mixture of two amorphous phases is less than that of one amorphous phase. In this concentration region, for some amorphous metal alloys, there is a phase stratification, which occurs by the type of spinodal decay without the origin of crystals. This process is a special case of the initial stage of phase transformation when the system passes into a labile, that is, unstable with respect to the fluctuations in density and composition of the state, with the relaxation of the system accompanied by an increase in the fluctuation of the concentration of components. In binary systems there is an area of unstable homogeneous states; disturbance of stability occurs in relation to local deviations from equilibrium concentration. The thermodynamic prerequisite for bundle binary systems is the S-shaped form of the concentration dependence of the relative free Gibbs energy $\Delta G(c, T)$ [7].

In most amorphous alloys, the relative integral enthalpy $\Delta H(c, T)$ has a large negative value and weakly varies depending on temperature, but the value of the relative integral entropy $\Delta S(c, T)$ can also have a large negative value due to the dependence $\Delta S(c, T)$ not only from concentration but also from volume change $\frac{\Delta V}{V}(c, T)$ at the formation of the alloy.

Consequently, the competing behavior of the quantities $\Delta H(c, T)$ and $\Delta S(c, T)$ leads to the fact that the concentration dependence $\Delta G(c, T)$ takes the S-shape form, which reflects the tendency of the alloy to stratification process.

Experiments on low-temperature annealing of alloys of the Fe–Zr system showed that when heating, there is a stratification within the amorphous state: initially, 2 amorphous phases are formed with a concentration of zirconium of 5% and 20%, respectively (we denote their β and δ), then on the background of these phases are formed 2 crystalline phases: Zr solution in α-Fe and chemical compound Fe_3Zr (denote them γ and ε) [8].

The main parameter that determines the kinetics of phase formation of binary alloys is the concentration dependences of the free energies of the Gibbs of the amorphous and the crystalline phases. The relative integral Gibbs free energy of the amorphous phase is described by the following expression [9].

$$\Delta G^{\alpha}(c_{Zr}, T) = f_{\Delta H} + RT(c_{Zr}\ln c_{Zr} + (1 - c_{Zr})\ln(1 - c_{Zr}) - \\ -10,7(0,5 + \frac{3,5 \times 10^5}{T^2})f_{\Delta V/V_0}\Big) + \frac{\Delta H_0^{Fe}(T_0^{Fe} - T)}{T_0^{Fe}}(1 - c_{Zr}) + \\ + \frac{\Delta H_0^{Zr}(T_0^{Zr} - T)}{T_0^{Zr}}c_{Zr}. \quad (6)$$

In (1) the function ΔH_0^{Fe}, ΔH_0^{Zr}, T_0^{Fe}, T_0^{Zr} are the enthalpy and melting temperature of iron and zirconium. Functions $f_{\Delta H}$ and $f_{\Delta V/V_0}$ give an opportunity to take into account experimental dependencies $\Delta H(c_{Zr}, T)$ and $\frac{\Delta V}{V_0}(c_{Zr}, T)$ for different temperatures and concentrations, and are analytically described by the following relationships:

$$f_{\Delta H,\, \Delta V/V_0} = \frac{a_{\Delta H,\, \Delta V/V_0}(1 - c_{Zr})c_{Zr}}{1 + b_{\Delta H,\, \Delta V/V_0}c_{Zr}},$$

$$a_{\Delta H,\, \Delta V/V_0} = \frac{\beta_{\Delta H,\, \Delta V/V_0}}{\alpha^2_{\Delta H,\, \Delta V/V_0}},\ b_{\Delta H,\, \Delta V/V_0} = \frac{1 - 2\alpha_{\Delta H,\, \Delta V/V_0}}{\alpha^2_{\Delta H,\, \Delta V/V_0}}, \quad (7)$$

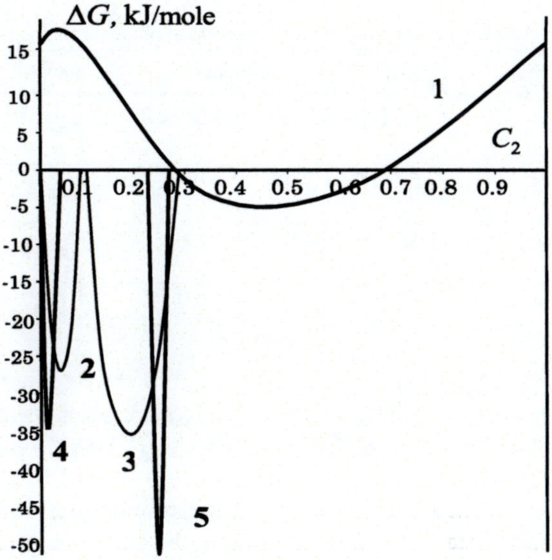

Fig. 1 Concentration dependence of the relative integral Gibbs free energy of the initial amorphous α-phase (1), amorphous phases β (2) and δ (3), which are formed when stratification of the initial amorphous phase and crystalline phases γ (4) and ε (5), which are formed at subsequent phase formation

where $\alpha_{\Delta H}$—position of the concentration dependence of enthalpy extremum $\Delta H(c_{Zr}, T)$, $\beta_{\Delta H}$—the absolute value of this function at the extremum point, $\alpha_{\Delta V/V_0}$—position of the concentration dependence of extremum $\frac{\Delta V}{V_0}(c_{Zr}, T)$, $\beta_{\Delta V/V_0}$—the absolute value of this function at the extremum point. These parameters are determined from experimental and calculation data [8–10]. Parameters dependence of the concentration of the relative integral Gibbs free energy of the amorphous phase in the Fe–Zr system are obtained on the basis of the analysis of the state diagram and experimental data, and were chosen as $\alpha_{\Delta H} = 0{,}35$, $\beta_{\Delta H} = -28, 8 \times 10^3$, $a_{\Delta H} = -235{,}1 \cdot 10^3$, $b_{\Delta H} = 2{,}448$, $\alpha_{\Delta V/V_0} = 0{,}15$, $\beta_{\Delta V/V_0} = -0{,}3$, $a_{\Delta V/V_0} = -13{,}3$, $b_{\Delta V/V_0} = 31{,}11$.

Figure 1 shows the concentration dependences of the relative integral Gibbs free energies $\Delta G(c_{Zr}, T)$ for the α-phase and, respectively, for the two amorphous phases (β and δ), which are formed during stratification of the initial amorphous phase and two crystalline phases (γ and ε), which are formed at subsequent phase formation. The concentration dependence of the relative integral free energy of the Gibbs of the initial amorphous phase has a specific S-shaped form due to the large value $\frac{\Delta V}{V}(c, T)$ and negative value $\Delta S(c, T)$, which indicates the propensity of the alloy to stratification.

The relative integral Gibbs free energy of the β, γ, δ, ε-phases, which are formed in the Fe–Zr system, are described by the following expression

$$\Delta G^{\beta,\gamma,\delta,\varepsilon}(c_{Zr}, T) = A^{\beta,\gamma,\delta,\varepsilon}(T)(c_{zr} - c_m^{\beta,\gamma,\delta,\varepsilon})^2 + \Delta G_0^{\beta,\gamma,\delta,\varepsilon}(T), \quad (8)$$

Table 1 Parameters for calculating the relative integral Gibbs free energy for amorphous and crystalline phases observed in the process of phase formation in alloys of the Fe–Zr system

Phase	$\Delta G_0^{\beta,\gamma,\delta,\varepsilon}(C_B, T)$, J/mole	$A^{\beta,\gamma,\delta,\varepsilon}$	$C_m^{\beta,\gamma,\delta,\varepsilon}$
β	$\Delta G_0^{\beta}(T) = -10 \cdot 10^3 - 1,3RT$	$A^{\beta} = -\frac{\Delta G_0^{\beta}(T)}{(C_m^{\beta})^2}$	$C_m^{\beta} = 0,05$
γ	$\Delta G_0^{\gamma}(T) = -30 \cdot 10^3 + 0,1RT$	$A^{\gamma} = -\frac{\Delta G_0^{\gamma}(T)}{(C_m^{\gamma})^2}$	$C_m^{\gamma} = 0,02$
δ	$\Delta G_0^{\delta}(T) = -21 \cdot 10^3 - 2RT$	$A^{\delta} = -\frac{\Delta G_0^{\delta}(T)}{(C_m^{\delta})^2}$	$C_m^{\delta} = 0,2$
ε	$\Delta G_0^{\varepsilon}(T) = -44 \cdot 10^3 + 0,1RT$	$A^{\varepsilon} = -\frac{\Delta G_0^{\varepsilon}(T)}{(C_m^{\varepsilon})^2}$	$C_m^{\varepsilon} = 0,25$

where $c_m^{\beta,\gamma,\delta,\varepsilon}$—concentration of zirconium, which corresponds to the minimum value of the function $\Delta G^{\beta,\gamma,\delta,\varepsilon}(C_{Zr}, T)$, $\Delta G_0^{\beta,\gamma,\delta,\varepsilon}(C_{Zr}, T)$—the minimum value of this function.

Parameters for calculating the relative integral Gibbs free energy for amorphous and crystalline phases observed in the process of phase formation in alloys of the Fe–Zr system are presented in Table 1.

According to (1–8), the concentration dependences of the relative integral Gibbs free energies of Fe–Zr system α-phase (amorphous phase) and, respectively, the amorphous (β, δ) and crystalline (γ, ε) phases were calculated and constructed. The obtained equations allow us to calculate the characteristics of the process of the phase formation process of alloys of the Fe–Zr system.

3 Results of Calculations of Parameters Phase Formation Process in Amorphous Alloys of Fe–Zr System

Figure 2 shows the results of calculating the temperature dependences of the total volume part of the crystalline phase $X(T)$ and the volumetric parts of the amorphous and crystalline phases $X_i(T)$ formed during the heating of the amorphous $Fe_{90}Zr_{10}$ alloy carried out according to (1–8).

The calculation showed that at a temperature 550 K, there is a stratification within the amorphous state: two amorphous phases with different concentrations of zirconium are released. With further heating, each of the amorphous phases is formed as a result of stratification that begins to crystallize, with the part of the amorphous phase in the sample decreasing.

Calculations have shown that the crystallization process of the $Fe_{90}Zr_{10}$ alloy takes place in two stages in accordance with the state diagram of the binary Fe–Zr system [10]. At a temperature of 600 K, solution of zirconium in the α-Fe is crystallized, while the concentration of zirconium in the amorphous matrix increases. With further

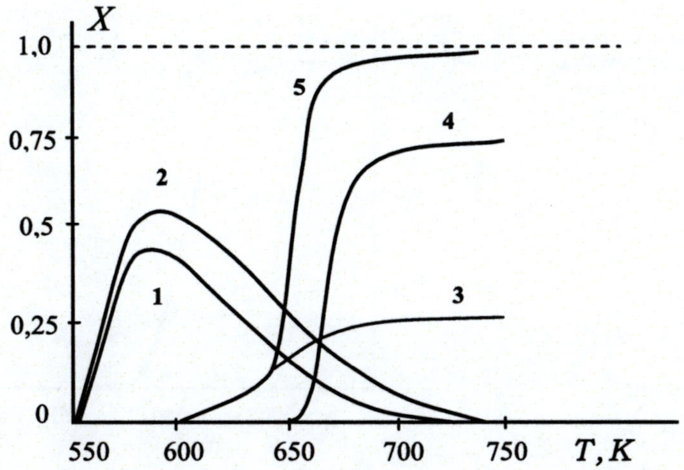

Fig. 2 Temperature dependence of volume part of phases that appear in the process of continuous heating of amorphous alloy $Fe_{90}Zr_{10}$: 1—part of amorphous β-phase X_β, 2—part of amorphous δ-phase X_δ, 3—part of crystalline γ-phase X_γ, 4—part of crystalline ε-phase X_ε, 5—total volume part of the crystalline phase $X(T)$

heating at 650 K, crystallization of a solid solution of zirconium and iron begins in the chemical compound Fe_3Zr.

The integral curve of the temperature dependence of the volumetric part of the crystalline phase is characterized by the presence of a «shelf», which indicates the two-stage crystallization process. The temperature range of crystallization at slow heating (0.16 K/s) is 90 K.

Let us consider the laws of the kinetics of crystallization in the high-temperature region, which determine the possibility of obtaining alloys of the Fe–Zr system in an amorphous state. Figure 3 shows the results of theoretical calculations carried out within the framework of the modified theory of homogeneous nucleation for the temperature dependence of the volume part of the crystalline phase $X(T)$, which reflects the kinetics of phase formation of the alloy with an initial concentration 10% (at) zirconium at quenching rate $dT/dt = 10^6$ K/s.

Figure 3 shows that the region of intensive crystallization is (600–700) K that lies in the narrow temperature interval $\Delta T \approx 100$ K, the total volume part of the crystalline phase at the same time is $X = 10^{-12,}$ that is, the quenching makes it possible to fix the amorphous state of the alloy, since the total criterion of an amorphous state: $X \leq 10^{-6}$. The results of calculations for determining the concentration area C_{Zr0} in which the quenching forms an amorphous state of the alloy is shown in Fig. 4.

The analysis of the calculation results presented in Fig. 4 allows us to conclude that the region of amorphization of alloys of the Fe–Zr system is within (3–15)% at. zirconium, which is well in agreement with experimental data.

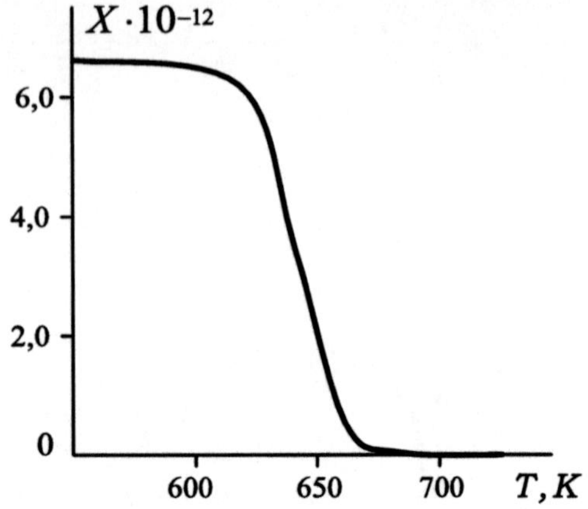

Fig. 3 The kinetics of crystallization of alloys of the Fe–Zr system with an initial concentration of zirconium $c_{zr} = 10\%$ (at.)

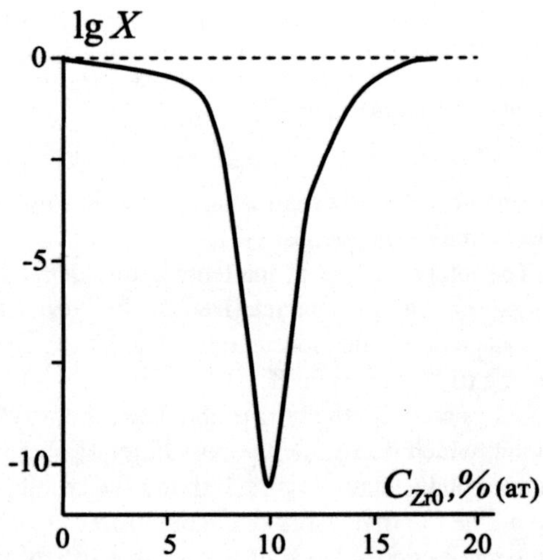

Fig. 4 Concentration dependence of the volumetric part of the crystalline phase in the Fe–Zr system at a quenching rate 10^6 K/s

We analyze the peculiarities of the crystallization process of amorphous alloys under the influence of laser treatment.

Calculation of the characteristics of the crystallization process was calculated for instant laser pulse, while it was considered that the amorphous tape was heated by a laser beam to a certain temperature within 10^{-6} s.

Figure 5 shows the calculated temperature dependence of the volume fraction of the crystalline phase under pulsed laser heating.

The high heating rates, achieved by the laser treatment of amorphous alloys, create a number of differences in the course of crystallization processes compared with crystallization during slow isothermal annealing. According to the calculations, impulse heating to temperatures below 550 K does not cause crystallization.

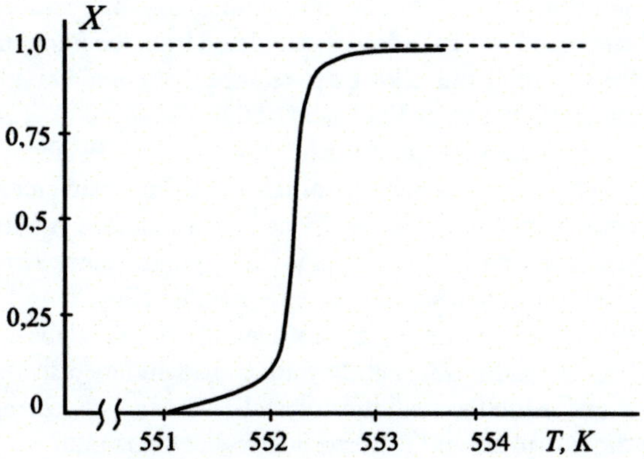

Fig. 5 Temperature dependence of the volumetric part of the crystalline phase under pulsed laser heating of an amorphous alloy $Fe_{90}Zr_{10}$

After 551 K, there are significant changes in the kinetics of crystallization. At 552 K, the crystalline phase part has reached the value 8%.

In the interval from 552 K to 553 K, there is a sharp jump in the value of the crystal phase proportion from 8 to 99%, that is, a complete crystallization of the amorphous alloy occurs.

The calculations carried out have shown that explosive crystallization may occur in the pulsed laser annealing in binary amorphous alloys of the Fe–Zr system. The temperature at which explosive crystallization is possible due to the laser pulse was less than 60 K for the temperature of the beginning of intensive crystallization at slow heating.

4 Molecular Dynamics Simulations of Phase Formation Processes in Amorphous Alloys of the Fe–Zr System

By the ab initio method of molecular dynamics, we carried out a simulation of the structure and electron spectra of the supercell $Fe_{29}Zr_3$ ($C_{Zr} = 9.375$ at.%) under a liquid-amorphous-crystalline phase transition. The supercell is selected so that the zirconium content is close to its content in the eutectic alloy of the Fe–Zr system in the iron-enriched Fe—9.8 atom% Zr. Studies of the Fe–Zr systems [8] show that alloys with a composition close to eutectic show good amorphization.

The calculation was carried out by the density functional method in a generalized gradient approximation [11] with the help of the program ABINIT [12].

To numerically simulate the process of obtaining the amorphous phase from the melt, a supercell was used containing 32 atoms of the high-temperature γ-Fe modification (FCC). Then, 3 Fe atoms were replaced by Zr atoms, so that the latter were located at the maximum possible distance from each other. The latter is necessary in order to avoid clustering. For relaxation of internal stresses in the supercell, numerical annealing of the atomic positions was performed according to the algorithm described in [13]. At numerical annealing, the positions of the atoms were relaxed in accordance with the forces calculated from the first principles. At that, the volume of the supercell was allowed to change while maintaining its shape (cubic). It is quite expected that the replacement of three Fe atoms ($z = 26$) with Zr atoms ($z = 40$) will increase the volume of the original supercell by about 0.3%.

Thus, we obtained the initial "crystalline" supercell $Fe_{29}Zr_3$ close to the eutectic composition (C_{Zr} = 9.375 at.%), which has an fcc structure, as in the high-temperature modification of γ-Fe. Of course, since Zr is insoluble in Fe, it is not a true crystalline solid solution or intermetallic alloy. The closest stable intermetallic alloy is Fe_2Zr [3]. Some researchers have also reported the existence of intermetallic alloys $Fe_{23}Zr_6$ and Fe_3Zr, but they are considered metastable [1]. The supercell described above, which we call the "crystalline" one, is necessary for further modeling of the "liquid" phase of the corresponding composition.

The electronic spectrum of the original "crystalline" supercell $Fe_{29}Zr_3$ is presented in Fig. 6a. The Fermi level is hereinafter indicated by a vertical line. It can be

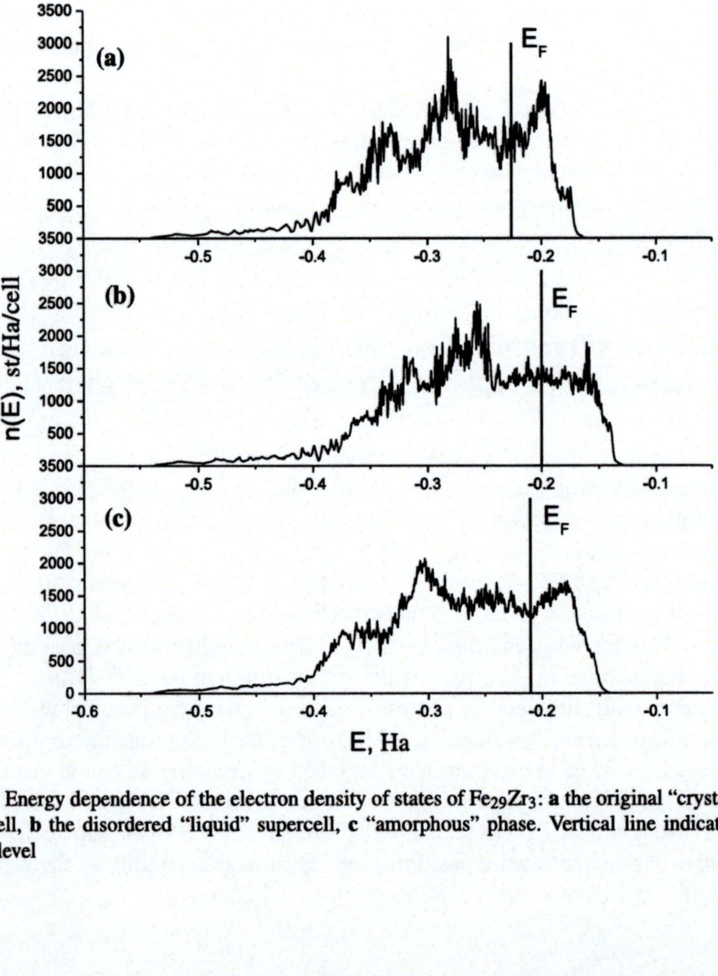

Fig. 6 Energy dependence of the electron density of states of $Fe_{29}Zr_3$: **a** the original "crystalline" supercell, **b** the disordered "liquid" supercell, **c** "amorphous" phase. Vertical line indicated the Fermi level

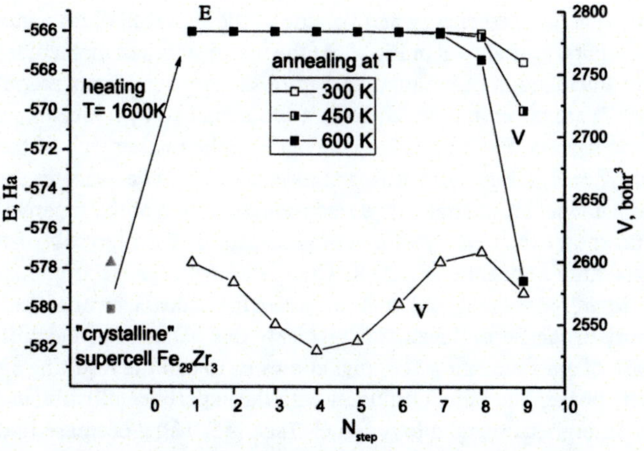

Fig. 7 The total energy of electron subsystem E (left scale) and volume V (right scale) of $Fe_{29}Zr_3$ supercell. The original "crystalline" supercell represented in gray

seen that we have a standard electronic spectrum typical of crystals with an fcc structure, with some "fine structure" (local maxima and minima) near the Fermi level. When comparing the obtained spectrum with the spectrum of the Fe_2Zr intermetallic compound, it can be seen that the aforementioned "fine structure" of the electronic spectrum inherits the main features of the intermetallic spectrum in the vicinity of the Fermi level. The total energy and volume of the original "crystalline" supercell are shown in Fig. 7 in gray.

To model the "liquid" (disordered) phase, we performed numerical heating of the original supercell by ab initio molecular dynamics in canonical (NVT) ensemble at 1600 K close to the eutectic. The energy dependence of the density of electronic states of the "liquid" phase is shown in Fig. 6b. It is seen that the result of isothermal annealing at 1600 K is the smearing of the aforementioned "fine structure" in the vicinity of the Fermi level. The quite expected expansion of the electron band by ~9% is also observed. Such a modification of the electronic spectrum is very characteristic with increasing temperature.

To simulate the process of amorphization and crystallization using the ab initio molecular dynamics method, we chose an isothermal/isoenthalpic ensemble, since structural phase transformations are often accompanied by a change in intrinsic volume. In the process of numerical annealing of the positions of atoms in contact with a thermostat, we completely optimized the geometry of the supercell: both a change in volume and a change in the shape of the supercell were allowed. As the initial one, we used the supercell of the "liquid" phase obtained after isothermal annealing at a temperature of 1600 K as the most disordered ($N = 1$ in Fig. 7).

The change in the total energy and volume of the supercell during annealing at a temperature of 300 K is shown in Fig. 7. At the initial stages of molecular dynamics, insignificant fluctuations of the total energy of the supercell (imperceptible on the scale of Fig. 7) are observed, which are not accompanied by noticeable changes in the electronic spectra. Further, a sharp decrease in the total energy of the supercell is observed ($N = 9$ in Fig. 7), which is accompanied by a decrease in the supercell volume and a noticeable change in the electronic spectrum of the supercell (Fig. 6c).

Figure 6c shows the density of electronic states of the supercell of the "amorphous" phase after annealing at 300 K. Compared with the electronic spectrum of the initial "liquid" phase (Fig. 6b), one can notice the appearance of a local minimum in the vicinity of the Fermi level. It is precisely this behavior of electronic spectra characteristic of amorphization [14] that allows us to call the resulting supercell an "amorphous" phase. The latter correlates with the experimentally obtained stability criterion of amorphous metal alloys Nagel–Tauc [15, 16]; a decrease in the density of electronic states at the Fermi level stabilizes the amorphous state.

It should be noted that the obtained electronic spectrum of the "amorphous" phase does not coincide with the electronic spectra of the stable crystalline phases of the Fe–Zr system at a temperature of 300 K (α-Fe and the Fe_2Zr intermetallic alloy). In addition, if we calculate the total energy of the mixture of stable crystalline phases ($23(\alpha\text{-Fe}) + 3(Fe_2Zr) = Fe_{29}Zr_3$), which corresponds to the supercell chosen by us, then we obtain a value of ~−680 Hartree. The latter, of course, is much less than the total energy of the obtained supercell of the "amorphous" phase, which is metastable.

It should be noted that during numerical annealing at a temperature of 300 K we did not observe signs of crystallization of our system: for example, the order of the mutual arrangement of atoms in the supercell did not change (atoms did not transfer from one coordination sphere to another). The latter correlates with the results of an experimental study of the amorphization and crystallization process in the Fe–Zr system: in [8], it was shown that the temporary annealing of the amorphous $Fe_{90}Zr_{10}$ alloy at temperatures below 500 K does not lead to crystallization.

It was possible to simulate processes similar to crystallization of amorphous alloys with increasing temperature of numerical annealing of our disordered supercell. For example, upon annealing at a temperature of 600 K, the formation of a local minimum is first observed in the vicinity of the Fermi level, similar to that discussed above for the "amorphous" phase. Then a more large-scale restructuring of the electronic spectra is observed (Fig. 8), which is accompanied by a change in the order of the mutual arrangement of atoms in the supercell—coordination spheres are mixed. Such a change in the arrangement of atoms allows us to talk about the beginning of crystallization processes in the structure under consideration.

Figure 8 shows the energy dependences of the electronic states density of the $Fe_{29}Zr_3$ supercell annealed at temperatures of 300 K (a), 450 K (b), and 600 K (c). It is seen that with increasing temperature of numerical annealing, the electronic spectrum of the system undergoes significant changes. With a further increase in the annealing temperature, the electronic spectra do not qualitatively differ from those

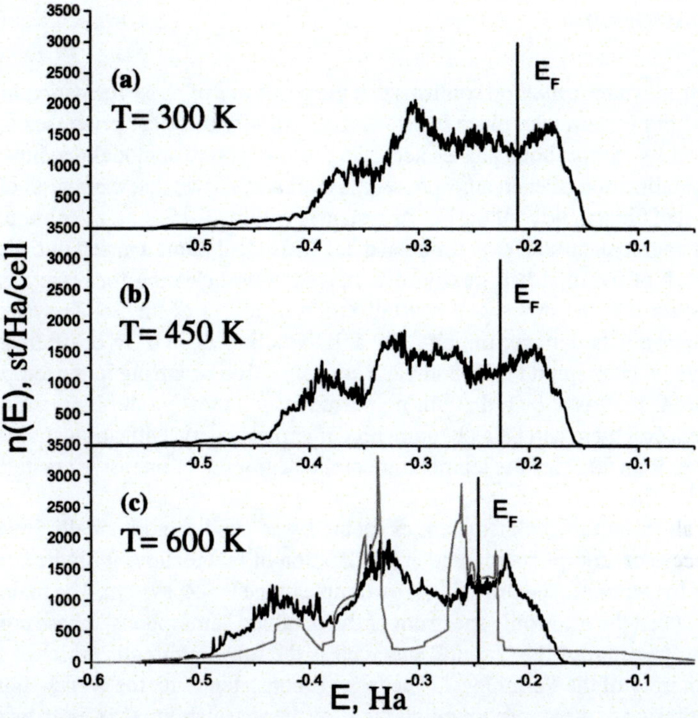

Fig. 8 Energy dependence of the electron density of states of $Fe_{29}Zr_3$ supercell after annealing at temperatures: **a** 300 K, **b** 450 K, **c** 600 K. The electronic spectrum of α-Fe presented in gray

obtained at a temperature of 600 K (Fig. 8c). If we compare this spectrum with the spectrum of the α-Fe phase stable at the indicated temperature (shown in Fig. 8c in gray), we can see that the three main peaks of the electronic spectrum of the supercell after annealing at 600 K generally correspond to the main three peaks α-Fe, but the width of these peaks is much larger (see Fig. 8c).

The latter is understandable if we take into account the results of experimental studies of the crystallization of amorphous alloys of the Fe–Zr system [8]—at the first stage of crystallization, a supersaturated solid solution of Zr in α-Fe is formed. So we can conclude that the obtained electronic spectra correspond to an unstable phase—a supersaturated solid solution of Zr in α-Fe. It should be noted that the energy of the obtained "unstable crystalline phase" (Fig. 7) is still much higher than the total energy of the corresponding mixture of stable phases, calculated above.

5 Conclusion

The thermodynamic method confirms that the processes of phase formation in binary alloys of the system take place in two stages: initially there are processes of phase bundle on two amorphous phases according to the type of spinodal decomposition, then crystallization of each amorphous phase. It was shown that the region of amorphization of binary alloys of the Fe–Zr system is within (3–15)% at. zirconium, which is well in agreement with experimental data. The calculations, carried out within the framework of the modified theory of homogeneous nucleation for binary systems, have shown that the process of crystallization of alloys of the Fe–Zr system with slow heating is two-stage: initially, the solid solution Zr in α-Fe crystallizes, then the chemical compound Fe_3Zr complete crystallization occurring in the temperature range of 90 K. It is shown that during pulsed laser annealing in binary amorphous alloy $Fe_{90}Zr_{10}$ there can be a phenomenon of explosive crystallization at a temperature lower than 60 K for the temperature of the beginning of intensive crystallization at slow heating.

The ab initio molecular dynamics method was used to numerically imitate the real process of amorphization and crystallization of metal alloys with the control of changes in electronic spectra using the example of the Fe–Zr system. The main difference between the electronic spectrum of the obtained "amorphous" phase compared with the spectrum of the "liquid" supercell is the local minimum, which is formed in the vicinity of the Fermi level. The latter is consistent with the Nagel–Tauc electronic criterion of the amorphous metallic alloy thermal stability [15]. With increasing temperature of numerical annealing, crystallization of the system is observed, accompanied by a rearrangement of the atomic order and a noticeable change in the electronic spectrum. The spectrum obtained corresponds to the electronic spectrum of a supersaturated solid solution of Zr in α-Fe, which is actually formed at the first stage of crystallization of amorphous alloys of the Fe–Zr system.

References

1. Stein F, Sauthoff G, Palm M (2002) J Phase Equilibria 23(6):480
2. McDeavitt SM, Abraham DP, Park JY (1998) J Nucl Mater 257:21–34
3. Granovsky MS, Arias D (1996) J Nucl Mater 229:29–35
4. Shklovskiy VA, Kuzmenko VM (1989) UFN 157(2):311–338
5. Glezer AM, Shuryigina NA (2013) Amorfno-nanokristallicheskie splavyi. Fiz.mat. lit, Moskva, p 452
6. Shpak AP, Lysov VI, Kunitskij YuA, Tsaregradskaya TL (2002) Crystallization and amorphization of metallic systems Akademperiodika, Kyiv, p 207 (in Ukrainian)
7. Lysov VI, Tsaregradskaya TL, Turkov OV, Saenko GV (2007) Rus J Phys 81(10):1765
8. Abrosimova GE, Aronin AS (1998) Fizika tverdogo tela 40(10):1768–1772 (in Russian)
9. Lysov VI, Tsaregradskaya TL, Turkov OV, Saenko GV (2013) J Phys Stud 7(2):2701
10. Diagrammy sostoyaniya dvoynyih metallicheskih sistem: Spravochnik: V 3-h t.: T.3. Kn.1./ Pod obsch. red. N.P. Lyakisheva. – M.: Mashinostroenie, 872 (2001)
11. Perdew JP, Burke K, Ernzerhof M (1996) Phys Rev Lett 77:3865

12. Gonzea X, Amadond B et al (2009) Comput Phys Comm 180:2582
13. Schlegel HB (1982) J Comput Chem 3:214
14. Ashok KV, Modak P, Svane A (2011) Christensen N.E. Phys Rev B 83:134205
15. Nagel SR, Fisher GB, Tauc G, Bardley BG (1976) Phys Rev B 13:3284
16. Nakonechnaya OI, Plyushchai IV, Semen'ko MP, Zakharenko NI (2000) Phys Metals Metallogr 90(5):439 (in Russian)

Modeling Multilayer Pyramidal-Like Adsorbate Structures Growth During Deposition at Homoepitaxy

Alina V. Dvornichenko and Olga M. Shchokotova

1 Introduction

Molecular-beam epitaxy (MBE) serves as a technique that is frequently used to fabricate single crystals which inherit atomic structure of the substrate. During the epitaxial growth, the deposited material forms a beam of neutral atoms. These atoms attach the substrate and can be adsorbed according to chemical bonding and become adatoms. Adsorbed atoms can diffuse on a substrate by probing energetically more favorable positions and interact with each other by making clusters of adsorbate. Isolated adatoms with nonzero probability can desorb back in vapor phase. Usually, the layer-by-layer growth during epitaxy provides formation of a new monolayer of adatoms. In the case of incomplete atomic layer, such process results in a formation of two-dimensional adsorbate structures on a substrate. In such a case, the interface roughness takes small values and the growing surface can be considered as atomically flat. At the same time the process of the smooth epitaxial growth is realized at initial stages of exposing (transient) and according to the statistical mechanics could not be considered as a stationary one. As it was shown before, during the layer-by-layer growth, the roughness of the surface grows in time for large substrate and long exposing time [1–4]. This effect becomes a result of incident flux fluctuations, which leads to accelerating the lateral diffusion of adatoms after some initial transient. Such processes will lead to monotonical increase in the surface roughness and at large exposing time it attains the stationary value. Hence, during epitaxial growth, interacting adatoms will self-organize into three-dimensional pyramidal islands. Here the morphology of a growing surface will be determined by the following processes: adsorption, desorption, and nucleation on top of islands; edge diffusion, diffusion down step, and dimer diffusion; attachment/detachment from island boundaries and island coalescence [1, 5, 6].

A. V. Dvornichenko (✉)
Sumy State University, 2 Rimskii -Korsakov St., Sumy 40007, Ukraine
e-mail: alina.dvorni4enko@ukr.net

O. M. Shchokotova
Institute of Applied Physics, National Academy of Sciences of Ukraine,
58 Petropavlovskaya St., Sumy 40000, Ukraine

O. Fesenko and L. Yatsenko (eds.), *Nanomaterials and Nanocomposites, Nanostructure Surfaces, and Their Applications*, Springer Proceedings in Physics 246, https://doi.org/10.1007/978-3-030-51905-6_31

In order to provide a theoretical study with a change in the morphology of the growing surface during adsorption/desorption processes nowadays, several approaches are used. These methods include atomistic models [4, 7–10] and citation therein, reaction-diffusion models [11–13], approaches, considering formation of separated islands [14], and devices with steps [15]. These models allow one to study evolution of the surface morphology [16–18]. The continuous approaches allow one to study evolution of the height of the growing surface. They are based on partial differential equations for the height field. Among these models, one can issue: a model of Kardar, Parisi and Zhang (KPZ) with nonconserved dynamics [19]; the Edwards–Wilkinson (EW) model for conserved solid-on-solid growth [20]; the Mullins–Herring (MH) model for conserved dynamics with significant surface diffusion [21, 22]; the model for molecular-beam epitaxy, known as conserved Villain–Lai–Das Sarma model [10, 23]. At the same time, these continuous models do not take into account the formation of step edges and, hence, could not be used to describe formation of pyramidal-like surface structures with terraces.

One of the theoretical methods for the step flow, allowing one to describe formation of steps in adsorbate structures is the mesoscopic Burton–Cabrera–Franck (BCF) model [24]. It allows one to represent island boundaries as steps, and is widely used in the modeling of islands formation during epitaxial growth. This model is continuous in the horizontal direction (along a substrate) and discrete in the growth (vertical) direction. Here the concentration of adatoms on any terrace is described by a continuous function and the location of the step is given by a continuous curve. According to this approach, the boundaries of islands are modeled by using diffusion-like equation for the concentration of adatoms. The boundary conditions are taken in accordance with the microscopic details of atomic processes on islands edges. One of the modifications of the BCF model is the phase-field approach. This method can be considered as an extension of the known Ginzburg–Landau theory for phase transitions. According to this approach, one simultaneously considers coupled partially differential equations for the adsorbate concentration and the phase field, which plays a role of the order parameter. This method was used by Liu, Matiu and Karma, Plapp (see [25, 26]) to perform theoretical study of adsorbate islands formation during epitaxial growth.

The BCF model [24] considers nucleation processes but does not take into account the interaction of adsorbed atoms. In real situations, adatoms are mobile interactive spices, which can diffuse along a substrate and form adsorbate clusters due to their interactions. Moreover, lattice mismatch between adatom and substrate and elastic tensions are able to induce additional interactions of adatoms. These interactions provide strong local bond, which is able to correct the rate for desorption and produce the lateral flux which is opposite to the standard diffusion flux. In theoretical studies of the surface structures formation during adsorption/desorption processes, the temperature of the growing surface is assumed to remain a constant. At the same time, from a physical point of view one should take into account that adsorption processes lead to local increase in the surface temperature; the desorption processes result to local cooling of the surface. Moreover, local changes in the surface temperature can be realized during processes of adatoms clustering and formation of islands of adsorbate [27]. Hence, a simultaneous influence of these processes will lead to local change in the surface temperature even on mesoscopic level of consideration. It was shown before, that local variations in temperature have an important influence onto dynamics of crystal/aggregates formation [28]. It is known that spatially extended dynamical systems are very sensitive to influence of fluctuating sources representing multiplicative noise. An influence of external and internal fluctuations onto dynamics of pattern formation was studied previously in systems described by reaction-diffusion models with quasi-chemical reactions [29–33], by studying adsorbate islands formation at condensation from gas phase [34], at modeling surface evolution at ion-beam sputtering [17]. It was shown that problems of pattern formation fluctuations can play an important and/or crucial role. At the same time by performing theoretical studies of surface morphology change, during epitaxial growth one usually assumes that fluctuations are negligible.

In this chapter, we perform a theoretical study of adsorbate islands formation at epitaxial growth processes in the framework of phase-field modeling by taking into account the interaction of adatoms. We will show that at small values of interaction energy the Gaussian-like surface will grow whereas well-organized pyramidal-like adsorbate structures can be realized at elevated values of the interaction energy of adatoms only. We will generalize the phase-field model by taking into account fluctuations in the total lateral flux of adsorbate and local changes in the surface temperature in time at adsorption/desorption processes. We will consider the system dynamics in the framework of mean field analysis and by using numerical simulations. The role of fluctuations in process of adsorbate islands formation will be analyzed. We will set a window for the system parameters (interaction strength of adsorbate and deposition flux) when oscillatory regime of surface growth is realized. We will discuss an influence of the deposition flux, interaction strength of adsorbate, and relaxation time of the temperature of the growing surface onto the morphology of the growing surface, dynamics of the number of pyramidal islands and their mean size and distribution of adsorbate islands over sizes.

The chapter is organized in the following manner. In the next section, we derive the mathematical model, describing the formation of pyramidal-like adsorbate islands on the substrate during deposition, by taking into account the interaction of adsorbate and local changes in the surface temperature. In Sect. 3, we perform numerical simulations of pattern formation in the system, characterized by the constant temperature. In Sect. 4, we introduce fluctuations of the total lateral flux of adsorbate and discuss noise-induced effects at deposition. In Sect. 5, we study effects of the local variations in the surface temperature caused by adsorption/desorption processes. Section 6 contains results of the scaling properties of the dynamics of pyramidal islands formation. Main conclusions are collected in the last section.

2 Phase-Field Model of Epitaxial Growth

In the framework phase-field modeling, we will monitor the spatio-temporal evolution of the dimensionless concentration field $x = \Omega(c - c_{eq}^0)$, where c is the concentration of adatoms, Ω is the atomic volume (area) of the substrate, and c_{eq}^0 represents the equilibrium concentration of adsorbate at a straight step. According to the classical BCF model, an evolution of the surface morphology is described by the following equations [24]

$$\partial_t x = F - \frac{x}{\tau_x} + \nabla \cdot D\nabla x,$$
$$v_n = D\left[\left(\frac{\partial x}{\partial n}\right)_+ - \left(\frac{\partial x}{\partial n}\right)_-\right], \quad x = d_0\kappa. \tag{1}$$

Here, $F = F_d - c_{eq}\Omega/\tau_x$ stands for the effective deposition rate which is defined through the actual deposition flux per atomic area F_d and the mean time for the desorption τ_x. The second term describes desorption processes. The last term in the first equation of (1) describes the lateral diffusion of adatoms with the diffusion coefficient D. The second equation in (1) describes evolution of the normal velocity of the step v_n, where $(\partial x/\partial n)_\pm$ are the normal gradients of concentration on the lower (+) and upper (−) side of the step. Here $d_0 = \Omega^2 c_{eq}^0 \gamma/T$ is the microscopic dimensionless capillarity length, which is defined through the local step curvature κ, the step stiffness γ, and the temperature T. In studies of adsorption/desorption processes, the desorption rate usually is defined in accordance with the Arrhenius relation $\tau_x = \nu_0^{-1}\exp(E_{\text{des}}/T)$, where ν_0 is the constant of desorption rate and E_{des} is the activation energy for desorption. The diffusion coefficient is defined in the standard manner $D = a^2\nu_d \exp(-E_{\text{diff}}/T)$, where a is the lattice constant of the substrate, ν_d is the attempt frequency, and E_{diff} is the activation energy for diffusion, $E_{\text{des}} < E_{\text{diff}}$.

By considering adatoms as interacting spices, following [35–37] we assume that the lateral (substratum mediated) interactions of the adsorbate can affect the desorption rate τ_x^{-1}. Hence, interaction potential $U(\mathbf{r})$ of adatoms leads to correction of the desorption rate in the following form: $\tau_x^{-1} = \tau_{x0}^{-1}\exp(U/T)$. As was pointed out before, the attractive interactions generate an adsorbate flux opposing the diffusive one [35–37]. Therefore, the interaction potential induces the thermodynamics (chemical) force $\mathbf{f} = -\nabla(U/T)$ acting adsorbed particles. This force induces speed $\mathbf{v} = D\mathbf{f}$ according to the Einstein's relation and the associated particle flux $\mathbf{v}x$ is possible on the $(1 - x)$ free sites. By taking into account both diffusion and induced fluxes the evolution equation for the concentration field takes the form

$$\partial_t x = F - \frac{x}{\tau_{x0}} e^{U/T} - \nabla \cdot \mathbf{J}_{\text{tot}};$$
$$\mathbf{J}_{\text{tot}} = -D\left[\nabla x + x(1 - x)\nabla(U/T)\right]. \tag{2}$$

In the framework of self-consistent approximation, an interaction potential for adsorbate can be represented in the following way: $U(x(\mathbf{r})) = -\int u(\mathbf{r}-\mathbf{r}')x(\mathbf{r}')d\mathbf{r}'$, where $u(r)$ is a symmetric binary attractive interaction potential for two adatoms separated by a distance r. By assuming that an interaction length of adatoms r_0 is small comparing to a diffusion length $L_d = \sqrt{D\tau_{x0}}$ ($r_0 \sim 1nm$, $L_d \sim 1\mu m$ for metals), i.e., $r_0 \ll \ell$, one can write $U(x(r)) \simeq -\epsilon x(r)$, where $\epsilon = \int u(r)dr$ stands for the interaction strength and depends on concrete material properties.

It is known that the existence of the steps in the direct numerical simulations of the sharp interface problem are not simple tasks [38]. Previously it was shown that the phase-field model is convenient to simplify the analysis of the processes of the surface growth and perform numerical simulations [25, 26].

The main idea of the phase-field approach is in the introduction of an effective order parameter $\phi(\mathbf{r}, t)$ which indicates phase at a particular position (see, for example, [26, 39–41]). In case under consideration, the phase field ϕ corresponds to the surface height in units of the lattice constant a of the adsorbed atoms and defines the number of layers. Following [25], local minima of the order parameter correspond to terraces of the pyramidal-like structures, whereas rapid spatial variation of ϕ relates to position of the step. According to [25, 26], the evolution equation for the phase field ϕ is of the form

$$\tau_\phi \partial_t \phi = -\frac{\delta H}{\delta \phi}, \quad H = \int d\mathbf{r}[\varpi^2 (\nabla\phi)^2/2 + h(\phi, x)], \tag{3}$$

with the free energy density

$$h(\phi, x) = \frac{1}{2\pi}\cos(2\pi[\phi - \phi_s]) - \lambda x\left(\phi + \frac{1}{2\pi}\sin(2\pi[\phi - \phi_s])\right). \tag{4}$$

Here ϖ is the width of the step, λ stands for the dimensionless coupling constant, $\phi_s/2$ is the initial height of the substrate. Here the term $1 + \cos(\pi[\phi - \phi_s])$ sets that minima of the free energy H are possible at $\phi - \phi_s = 2n + 1$, independently of the concentration of adatoms [26]. Therefore, the phase-field model for the adsorbate islands growth during epitaxy becomes the form:

$$\begin{aligned} \frac{\partial x}{\partial t} &= F - \frac{x}{\tau_{x0}} e^{-\varepsilon x} + D\nabla\cdot\{[1 - \varepsilon x(1-x)]\nabla x\} \\ \tau_\phi \frac{\partial \phi}{\partial t} &= -\frac{\delta H}{\delta \phi} = \varpi^2 \Delta\phi + \sin(\pi[\phi - \phi_s]) + \lambda x(1 + \cos(\pi[\phi - \phi_s])), \end{aligned} \tag{5}$$

where we have introduced the dimensionless interaction strength $\varepsilon = \epsilon/T$; τ_ϕ is the relaxation time of the phase field ϕ.

In order to take into account local variations in the surface temperature during adsorption/desorption processes, we will generalize the derived phase-field model. According to the linear Onsager theory, one can represent the total lateral flux of adsorbate in the following form $\mathbf{J}_{\rm tot} = -M\nabla(\mu/T)$, defined through the kinetic coefficient M and chemical potential μ with $\nabla\mu = (\partial\mu/\partial x)_T \nabla x + (\partial\mu/\partial T)_x \nabla T$. This allows us to rewrite the total flux in the form

$$\mathbf{J}_{\text{tot}} = -\frac{L}{T}\left(\left.\frac{\partial\mu}{\partial x}\right|_T \nabla x + \left.\frac{\partial\mu}{\partial T}\right|_x \nabla T - \frac{\mu}{T}\nabla T\right),$$

and write the free energy functional $\mathcal{F}$ according to the chemical potential $\mu = \delta\mathcal{F}/\delta x$ in the form:

$$\mathcal{F} = \int \mathrm{d}\mathbf{r}\left\{T\left[x\ln x + (1-x)\ln(1-x)\right] - \frac{\epsilon}{2}x^2\right\}. \tag{6}$$

Here the first term relates to the entropic contribution and the second term corresponds to the interaction of adatoms. By setting the kinetic coefficient M in the form $M \equiv Dx(1-x)$ the total flux of adsorbate becomes the form:

$$\mathbf{J}_{\text{tot}} = -\frac{D}{T}x(1-x)\nabla\frac{\delta\mathcal{F}}{\delta x}. \tag{7}$$

One knows that even in equilibrium internal processes in the system result in fluctuations of thermodynamic quantities (temperature, pressure, volume, and thermodynamic potentials) [42]. Usually, the thermal equilibration is considered as a fast process. From theoretical point of view, any physical process is characterized by its own time scale. For the actual case under consideration, the temperature field should satisfy the Fourier law $\partial_t T = \nabla \cdot \int_0^t \mathcal{M}(t, t'; \tau_\chi)\nabla T(\mathbf{r}, t')\mathrm{d}t'$. In order to avoid the situation with infinite propagation speed for the temperature field the correlation time τ_χ should be considered as small but nonzero parameter [43]. This leads to the heat waves emerging [43], pattern selection processes [44, 45] and oscillatory dynamics of pattern formation [11]. Such oscillatory behavior is characterized by small amplitude and can be realized at initial stages. From the physical viewpoint the following relations are realized: $\tau_{x0}/\tau_\chi \ggg 1$, $\tau_T \gg \tau_\chi$, where τ_T relates to the relaxation time of the bath temperature T_0. Next, we put $\tau_T/\tau_{x0} \equiv \nu \lesssim 1$. By using the memory kernel in a form $\mathcal{M}(t, t'; \tau_\chi) = \chi\tau_\chi e^{-|t-t'|/\tau_\chi}$, where χ stands for the thermal diffusivity and by taking into account local changes if the surface temperature during adsorption/desorption processes the evolution equation for the temperature of the growing surface attains the form:

$$\partial_t T = \frac{T_0 - T}{\tau_T} + \chi\Delta T + \alpha F x + \beta\partial_t x. \tag{8}$$

The third term corresponds to the re-heat of the surface with an intensity α. It is caused by deposition flux F promoting the energy exchange with the reservoir. This assumption is widely used by studying temperature instabilities in chemical reactions (see, for example [46] or [47]. In such systems temperature instability is caused by defects leading to reconstruction of the surface. Local variation in surface temperature for curved steps is described by the phase field that is taken into account by the last term in (8), that also relates to local heating ($\partial x/\partial t > 0$) or cooling ($\partial x/\partial t < 0$) the surface during adsorption/desorption processes with intensity β.

For further consideration, let us move to dimensionless quantities and measure time and space in units τ_{x0} and diffusion length L_d. The phase-field model describing processes of adsorbate islands formation during epitaxial growth and local variations in the surface temperature becomes the form

$$\begin{cases} \partial_t x = F - x e^{-\varepsilon x/\theta} - \nabla \cdot \mathbf{J}_{\text{tot}} - \frac{1}{2}\partial_t \phi + \xi(t), \\ \partial_t \phi = \varpi^2 \Delta \phi - \partial_\phi h(x, \phi), \\ \nu \partial_t \theta = 1 - \theta + \chi \Delta \theta + \alpha F x + \beta \partial_t x. \end{cases} \tag{9}$$

Here we have introduced dimensionless quantities $F \equiv F\tau_{x0}$, $\varepsilon \equiv \epsilon/T_0$, $\theta \equiv T/T_0$, $\chi \equiv \chi\tau_T$, $\alpha \equiv \alpha\tau_T/T_0$, $\beta \equiv \beta\tau_T/T_0$ and put $\tau_\phi/\tau_{x0} = 1$ The last term in the evolution equation for the adsorbate concentration represents additive Gaussian zero-mean and delta-correlated noise.

3 Pyramidal Islands Growth at Constant Temperature

In this section, we will discuss a possibility to control dynamics of pattern formation at epitaxial growth assuming the constant temperature of the growing surface. In order to provide a detailed study of the processes of the surface growth at epitaxy, we will exploit numerical simulations. To that end, we will solve numerically the dynamical system (5) on two-dimensional quadratic lattice with the linear size $L = 256$, mesh size $\Delta l = 1.0$, time step $\delta t = 0.005$ and periodical boundary conditions. Here and thereafter we put $\varpi^2 = 2.0$, $\lambda = 10.0$.

It is known that at epitaxial growth processing the morphology of the growing layer will repeat the morphology of the substrate. Moreover additional monolayers of adsorbate will appear as additional terraces during epitaxial processes. For the actual phase-field model at $\phi_s = 0$ no surface structures can be realized during exposing. An edge defect (one edge dislocation) will promote formation of terraces of the pyramid in parallel directions to the edge defect. A dot defect will induce formation of "kid-like pyramid," Pyramidal-like holes will realize on a vacancy on the substrate. The screw dislocation located on the substrate will stimulate the formation of pyramidal-like structure [26].

In order to provide a general description, we will consider the initial substrate with a lot of surface defects, by setting initial conditions of the order parameter field in a form of Gaussian distribution with small dispersion, i.e., $\langle\phi(\mathbf{r}, t = 0)\rangle = \langle\phi_s\rangle = 0$, dispersion $\langle(\delta\phi(\mathbf{r}, t = 0))^2\rangle = 0.1$. For the adsorbate concentration, we put $x(\mathbf{r}, t = 0) = 0$. Snapshots of the system evolution at different values of the deposition rate F and the interaction strength ε are shown in Fig. 1.

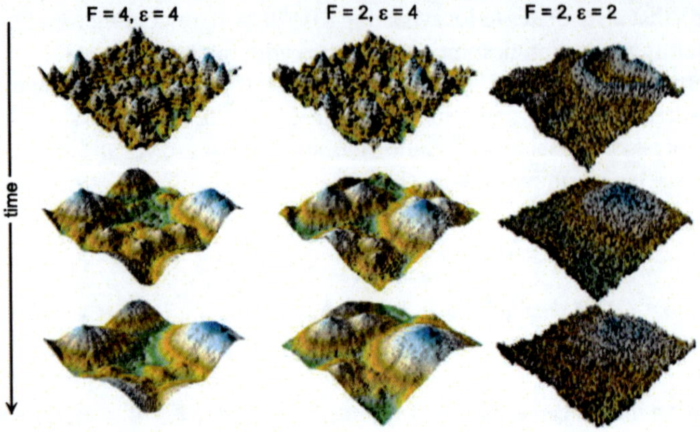

Fig. 1 3d snapshots of the order parameter ϕ evolution (from top to bottom) at different values of the interaction strength ε and deposition rate F

It follows that during the system evolution the pyramidal-like adsorbate structures are formed starting from the Gaussian surface at elevated interaction energy (see first and second columns in Fig. 1). These structures grow and interact, and at the large time only large pyramids are realized. Here a change in the deposition flux does not have crucial influence onto surface morphology. In the case of small values of the interaction strength ε patterns start to develop at early stages of the epitaxy, but light interaction between adsorbate could not provide coarsening (see third column in Fig. 1). In such a case the diffusion processes of adsorbate are more preferable than the agglomeration ones (accumulation due to the flux induced by the interaction potential). Therefore, a competition between diffusion flux and with interaction flux plays an important role in process of adsorbate islands formation during epitaxial growth.

In order to provide the quantitative description of the process of surface growth, we will study the following three statistical quantities: interface width W, skewness m_3, and kurtosis m_4, defined in the standard manner

$$m_3 = \frac{\langle(\phi(\mathbf{r}) - \langle\phi(\mathbf{r})\rangle)^3\rangle}{W^3}, \; m_4 = \frac{\langle(\phi(\mathbf{r}) - \langle\phi(\mathbf{r})\rangle)^4\rangle}{W^4}, \; W^2 = \langle(\phi(\mathbf{r}) - \langle\phi(\mathbf{r})\rangle)^2\rangle,$$

where $\langle\phi(\mathbf{r})\rangle \equiv L^{-2}\sum_{\mathbf{r}}\phi(\mathbf{r}, t)$ is the mean value of the height field. Skewness measures the symmetry of a profile of the distribution of heights over the reference surface level. The sign of skewness defines whether the points of the height are proportionately above ($m_3 > 0$) or below ($m_3 < 0$) average surface level. Kurtosis measures the sharpness of the height distribution and describes randomness of the surface relative to that of a perfectly random (Gaussian) surface. The Gaussian distribution is characterized by the value $m_4 = 3.0$. From the general description, for the surface with the most of the features located nearby the mean surface level the kurtosis will be less than one for the surface characterized by the larger portion of the surface features far from the mean surface level. Evolution of interface width, skewness, and kurtosis at different values of the interaction strength ε and deposition flux F is shown in Fig. 2.

It follows that at $\varepsilon = 4$ (see dash curve in Fig. 2a) the interface width increases toward its stationary value; whereas at small ε the quantity $W^2(t)$ increases at early stages only and with further exposing it decreases toward small quasi-stationary value (see dash curve in Fig. 2a). It means that at small values of the interaction strength, it is not enough to provide growth of well-defined patterns. Here diffusion processes result in a Gaussian-like envelope of the height distribution. The deposition flux F does not affect sufficient dynamics of W^2 at large ε (compare dash and dash-dot curves in Fig. 2a).

From results for the dynamics of the skewness and the kurtosis (see Fig. 2b, c, respectively), one gets that at elevated values of ε both m_3 and m_4 deviate from values related to the Gaussian distribution with $m_3 = 0$, $m_4 = 3.0$ at later stages. It means that the height distribution will differ from the Gaussian one: it will be asymmetric and non-centered around the mean value $\langle\phi\rangle$. For the small values of the interaction Strength, both skewness and kurtosis attain Gaussian values and fluctuate around them. Hence, the height distribution has the Gaussian properties, to be symmetric and centered around the mean value. Therefore, small values of the interaction strength ε can not induce well-developed patterns, and the surface is characterized by the Gaussian properties.

The height distributions over values at different values for ε and F are shown in Fig. 3. It follows that in the case of strong interactions (see Fig. 3a, b at $\varepsilon = 4$ and $F = 4$, $F = 2$, respectively) the distribution of heights is characterized by the large amount of peaks corresponding to the number of terraces: the smallest peaks relate to the lowest level of the pyramidal structure. Here the thick solid curves correspond to

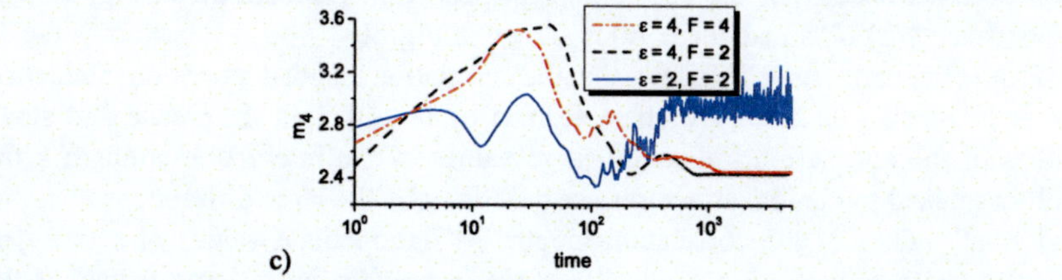

Fig. 2 Evolution of interface width W^2 (**a**), skewness m_3 (**b**) and kurtosis m_4 (**c**) at different values for deposition flux F and interaction energy ε

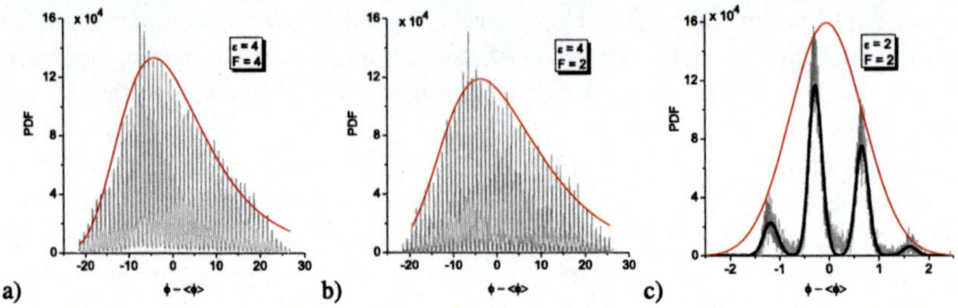

Fig. 3 Height distribution at $t = 3000$ and: **a** $\varepsilon = 4$, $F = 4$; **b** $\varepsilon = 4$, $F = 2$; **c** $\varepsilon = 2$, $F = 2$

the approximation of peaks by the Generalized Extreme Value Distribution (GEV) of the form $f_{GEV}(z; \mu, \kappa) = \frac{1}{\kappa} \exp\left(-\frac{z-\mu}{\kappa}\right) \exp\left\{-\exp\left(-\frac{z-\mu}{\kappa}\right)\right\}$, where μ and κ are approximation parameters. Similar distributions were observed for the distribution of adsorbate islands over sizes in the problem of condensation from gas phase [12]. The height distribution at small values of the interaction strength ε is shown in Fig. 3c. Here the small number of peaks means, that during deposition, when the concentration of adsorbate attains some supersaturation, desorption and diffusion processes start to play the major role and destroy terraces. Here the multi-peaks curve relates to the approximation of peaks by multi-peaks Gaussian; bell-shaped curve corresponds to the approximation of the main peaks by single Gaussian.

In order to characterize the surface order, one can exploit the correlation function defined in the following form: $C_0(r) = \langle(\phi(r_0 + r) - \langle\phi\rangle)(\phi(r_0) - \langle\phi\rangle)\rangle$. In the case of a surface with a self-affine behavior at short length scale and a smooth one at long scales, one can use the following approximation $C_0(r) = C_{\max} \exp(-(r/\xi)^{2\alpha})$, where ξ stands for the lateral correlation length. This quantity denotes the average length when the heights between two surface points are correlated. Hence, if ξ takes large values, then two points on terraces are correlated. It means that the height difference between two points is small, and no well-defined pyramidal patterns can be possible. In the opposite case, if ξ is small then two points on the terrace are less correlated and the corresponding average height difference is large. In such a case, different terraces are possible. The quantity ξ also is a measure of lateral roughness fluctuations.

To calculate ξ, we select the point $r = \xi$ corresponding to $C(r = \xi) = C_{\max}/e$. Typical dependencies of the correlation function $C_0(r)$ are shown in Fig. 4a. The dependence $\xi(\varepsilon)$ obtained at $t = 40$ is shown in Fig. 4b).

From obtained results, it follows that the correlation length depends on interaction strength. Small ε patterns are characterized by large height difference and small values of the correlation radius. At large values of the interaction strength ε the well-organized pyramidal structures are described by the large difference in heights and small values of the correlation length. At intermediate values of ε one gets patterns characterized by a small difference in heights and a large values of the correlation length (see snapshots in Fig. 4). The observed re-entrant dependence of the correlation length on the interaction energy ε indicates a transition from a

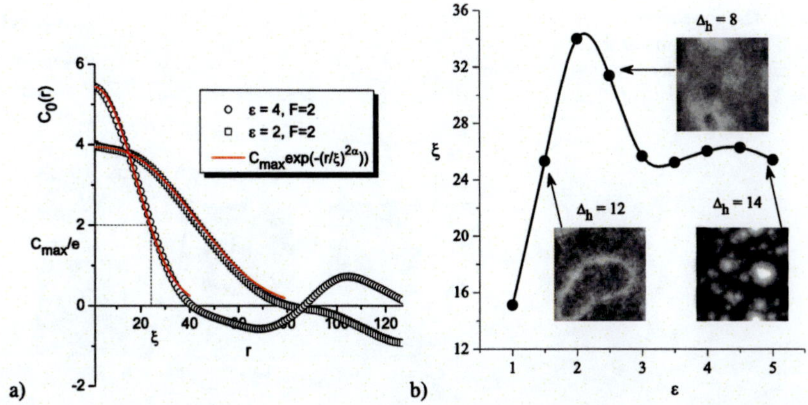

Fig. 4 Correlation function $C_0(r)$ at different values of the deposition rate F and interaction strength ε (**a**); and the correlation length ξ as a function of the interaction strength ε at $F = 2$. Here $\Delta_h = \langle \phi_{\max} - \phi_{\min} \rangle$ measures the averaged height difference between lower $\phi_{\min}$ and upper $\phi_{\max}$ points on the surface

disordered surface toward pyramidal patterns through the Gaussian surface. Hence, an increase in the interaction energy of adsorbate results in a competition between adsorption and desorption/diffusion processes: it leads to a formation of the surface with Gaussian-like surface and finally to the surface with pyramidal-like structures [48].

4 Effects of the Adsorbate Flux Fluctuations

By considering the process of adsorbate islands growth at epitaxy in real conditions, one should take into account the fluctuation source. Previously, it was shown that the multiplicative noise can be introduced into the system dynamics in an *ad hoc* form [29, 36, 37]. In our case in order to study stochastic system next we assume that the total lateral flux of adsorbate $\mathbf{J}_{\text{tot}}$ is composed by the regular part and stochastic component. In such a case, we can write

$$\begin{aligned} \frac{\partial x}{\partial t} &= F - \frac{x}{\tau_{s0}} e^{-\varepsilon x} - \nabla \cdot \mathbf{J}_{\text{tot}}, \\ \mathbf{J}_{\text{tot}} &= -D_{ef}(x)\nabla x + \boldsymbol{\zeta}(x; \mathbf{r}, t), \end{aligned} \tag{10}$$

where $D_{ef}(x) = D(1 - \varepsilon x(1 - x))$; for the stochastic source $\boldsymbol{\zeta}$ we assume Gaussian properties: $\langle \boldsymbol{\zeta}(x, \mathbf{r}, t) \rangle = 0$, $\langle \boldsymbol{\zeta}(x, \mathbf{r}, t)\boldsymbol{\zeta}(x, \mathbf{r}', t') \rangle = 2\Sigma D_{ef}(x)\delta(\mathbf{r} - \mathbf{r}')\delta(t - t')$, Σ is the noise intensity. By using the Stratonovich interpretation of the stochas-

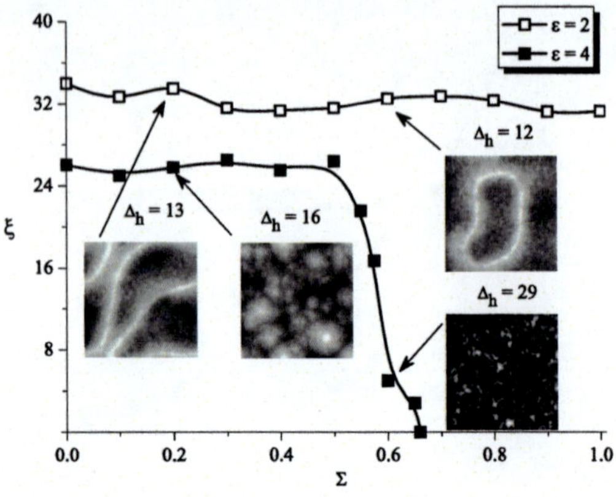

Fig. 5 The correlation length ξ as a function of the fluctuations intensity Σ at $F = 2$

tic source, the reduced equation for the evolution of the concentration of adatoms becomes the form [49]:

$$\begin{aligned}\frac{\partial x}{\partial t} =& F - xe^{-\varepsilon x} + \nabla \cdot \mathscr{D}_{ef}(x)\nabla x \\ & - \frac{\Sigma}{2}\nabla \cdot \left(\nabla \frac{\partial \mathscr{D}_{ef}(x)}{\partial x}\right) + \nabla \cdot \sqrt{\mathscr{D}_{ef}(x)}\boldsymbol{\xi}(\mathbf{r}, t),\end{aligned} \tag{11}$$

where $\mathscr{D}_{ef}(x) = 1 - \varepsilon x(1 - x)$, $\langle \boldsymbol{\xi}(\mathbf{r}, t)\rangle = 0$, $\langle \boldsymbol{\xi}(\mathbf{r}, t)\boldsymbol{\xi}(\mathbf{r}', t')\rangle = 2\Sigma\delta(\mathbf{r} - \mathbf{r}')$ $\delta(t - t')$.

In order to analyze the stability of the system, one can consider dynamics of the mean concentration $\langle x\rangle$. By averaging (11) over noise, one gets a dynamical equation for the first statistical moment in the following form

$$\frac{\partial \langle x\rangle}{\partial t} = F - \langle xe^{-\varepsilon x}\rangle + \nabla \cdot \langle \mathscr{D}_{ef}(x)\nabla x\rangle - \frac{\Sigma}{2}\nabla \cdot \left(\nabla \left\langle \frac{\partial \mathscr{D}_{ef}(x)}{\partial x}\right\rangle\right). \tag{12}$$

It is seen that in linear regime, the fluctuations of the lateral flux of adsorbate stabilize the state $\langle x\rangle = 0$. Indeed, by putting $\varepsilon \ll 1$ one can expand all functions in the right-hand side of (12) up to the first order in $\langle x\rangle$. Finally, for the Fourier transform $\langle x(t)_{\mathbf{k}}\rangle = \int \langle x(\mathbf{r}, t)\rangle e^{i\mathbf{kr}}d\mathbf{r}$ one gets

$$\frac{\partial \langle x_k\rangle}{\partial t} = F - \langle x_k\rangle - k^2(1 + \Sigma\varepsilon)\langle x_k\rangle. \tag{13}$$

It follows that in the linear regime, the fluctuations of the adsorbate flux can not destabilize the state $\langle x\rangle = 0$; they lead to re-normalization of the coupling parameter. While the deposition flux F leads to an increase in the concentration field x the noise action promotes stabilization of the system.

Next, we perform numerical simulations by solving coupling equations (3) and (11) on the quadratic grid according to the mechanism, presented above. In order to discuss an influence of the introduced fluctuations onto growing surface morphology let us analyze the dependencies of the correlation length ξ on the noise intensity Σ, shown in Fig. 5.

From obtained results, it follows that small ε fluctuations cannot induce formation of well-organized pyramidal patterns with terraces. In this case, the correlation length does not depend significantly on the noise intensity Σ (see empty squares in Fig. 5). For the case of large values of interaction strength, the noise action has no essential influence onto pattern formation processes at small intensity Σ (see filled squares in Fig. 5). At large values of the noise intensity, the introduced fluctuations lead to formation of needle-like patterns. Here two nearest points are spaced far from each other [49].

5 Oscillatory Dynamics in a Generalized Model

In this section, we will pay our attention onto dynamics of pyramidal-like adsorbate islands formation during epitaxial growth by taking into local changes in the temperature of the growing surface, caused by adsorption/desorption processes. To that end, we will solve the system (9) by setting $\chi = 10$.

5.1 Stability Analysis of the Two-Component Model

By analyzing the evolution of the phase field ϕ, one can describe different regimes of the system dynamics and the corresponding change in the surface morphology [25]. Indeed, if $\partial_t\phi = 0$ then the height of the surface remains constant and, as a result, surface does not evolve in time. In such a case adsorption and desorption processes are characterized by the same probability.

If the speed of the surface growth is positive, i.e., $\partial_t\phi \equiv v_\phi > 0$ then the height of the surface grows in time. In such a case one can distinguish two different regimes characterized by $v_\phi = \text{const}$ and $v_\phi = v_\phi(t)$. In the regime, when the surface grows with the constant speed one gets the layer-by-layer growth, similar to the Frank-van der Merwe mechanism of the surface growth. For the more complicated situation of the time-dependent speed of the surface growth, $v_\phi = v_\phi(t)$, dynamics of the phase field ϕ can govern processes of adsorbate islands formation according to the Volmer–Weber or Stranski–Krastanov mechanisms. In the case of negative values of the speed $\partial_t\phi < 0$ desorption and/or evaporation processes play a major role in surface height dynamics, and as a result, the height of the surface decreases. In this section, we will consider the case $\partial_t\phi > 0$, only, assuming $\phi \simeq v_\phi t$. The averaging of the second equation of the system (9) gives the simple relation $\langle v_\phi \rangle \simeq \lambda x_0$, where $x_0 \equiv \langle x(t \to \infty) \rangle = \text{const}$ relates to the stationary value of the concentration field.

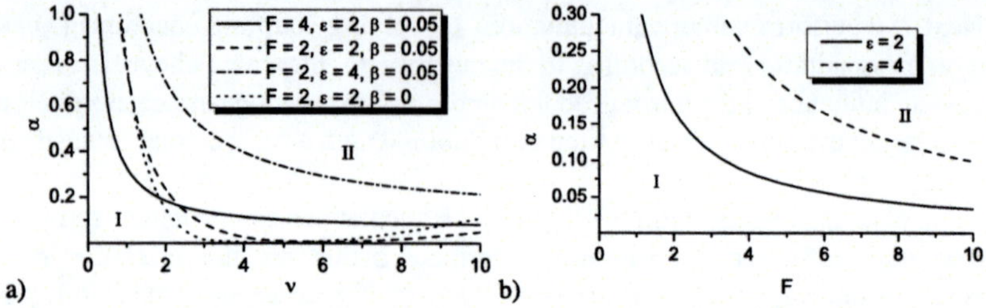

Fig. 6 Stability diagram of oscillatory regime realization. In panel **b** $\nu = 1.0, \beta = 0.05$. Oscillatory dynamics is realized in the domain II

Hence, macroscopic approximation dynamics of averages can be described by the system of two equations of the form [50]:

$$\begin{cases} \partial_t \langle x \rangle = F - \langle x \rangle e^{-\varepsilon\langle x \rangle/\langle \theta \rangle} - \frac{1}{2}\lambda x_0, \\ \langle v_\phi \rangle, \\ \nu \partial_t \langle \theta \rangle = 1 - \langle \theta \rangle + \alpha F \langle x \rangle + \beta \partial_t \langle x \rangle. \end{cases} \tag{14}$$

This system is characterized by the unique stationary homogeneous state, where the stationary value of the temperature field is defined as $\theta_0 = 1 + \alpha F x_0$ and the stationary value of the concentration field can be found from the equation $F - \lambda x_0/2 = x_0 e^{-\varepsilon x_0/(1+\alpha F x_0)}$. Next, proceeding in the standard scheme of the linear stability analysis one can define the corresponding stability exponents and find a set of main system parameters when the stability exponents have both real and imaginary parts: real part defines the stability of the stationary state and nonzero imaginary part determines the oscillation frequency. The stability diagrams shown in Fig. 6a, b indicate values of α, ν, F, and β bounding domain of the oscillatory regime (domain II). From the obtained diagrams, it follows that a decrease in the deposition flux F results to an increase in the critical values of ν at elevated α when oscillations are possible. Elevated values of interaction energy ε require larger values of both ν and α for oscillatory regime realization. The dot curve in Fig. 6a indicates that the oscillatory regime is also possible for the case $\beta = 0$. Hence, in the case of the constant positive value of the speed of the surface growth oscillatory dynamics of both averaged quantities $\langle x \rangle$ and $\langle \theta \rangle$ can be realized.

5.2 *Numerical Results*

In this section, we will study dynamics of pattern formation numerically by solving the system (9) on two-dimensional lattice $L \times L$ with $L = N\ell$ and $N = 512$; $\ell = 1$. As initial conditions for the temperature field we choose $\theta(\mathbf{r}, 0) = 1$.

Typical snapshots of the system evolution at different values of ν are shown in Fig. 7. By comparing results for the system with constant temperature (see snapshots at $\nu = 0$) and for the actual system when temperature field evolves in time ($\nu \neq 0$), one can see a principal difference in the number of terraces in pyramidal structures and their width; in the case $\nu \neq 0$ pyramidal structures are characterized by a small number of well-defined terraces, which are characterized by the large areas, comparing to the case $\nu = 0$. At the same time there are no principal differences in both systems with small and large values of ν.

Next, let us discuss dynamics of averages $\langle x \rangle$ and $\langle \theta \rangle$, shown in Fig. 8 for different set of the system parameters. It follows that during initial stages of the system evolution both the mean adsorbate concentration and the mean temperature evolve in time in an oscillating manner. Moreover, both quantities manifest single-phase oscillations meaning that if the concentration of adsorbate increases (deposition processes) then the temperature grows; temperature decreases when desorption/evaporation is

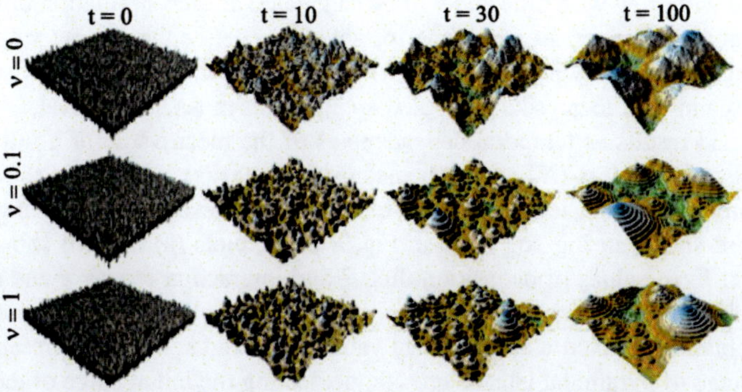

Fig. 7 Snapshots of the phase field ϕ evolution at $F = 4$, $\varepsilon = 4$ and different values of the relaxation time of the temperature field ν

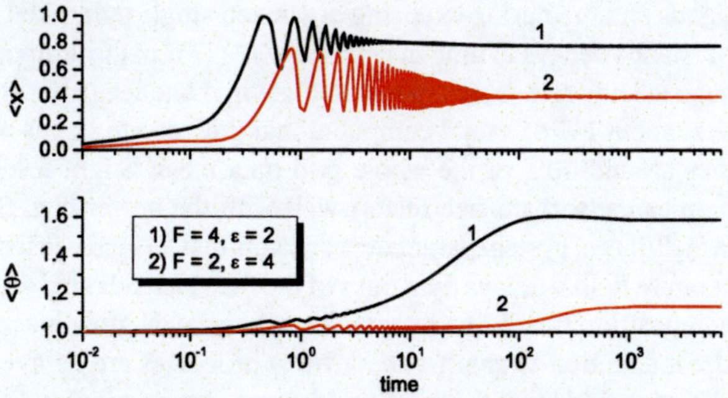

Fig. 8 Evolution of mean adsorbate concentration field, and mean temperature at different values of ε and F and (1) $F = 4, \varepsilon = 2$; (2) $F = 2, \varepsilon = 4$. Other parameters are $\nu = 10, \alpha = 0.01, \beta = 0.05$

realized. By comparing results for two different sets of parameters one can conclude that the decaying rate and the frequency of oscillations depends on the system parameters.

6 Scaling and Universality

6.1 Scaling Properties of the Islands Growth Processes

Next, we aim to discuss the dynamics of adsorbate islands growth by monitoring evolution of the mean size of adsorbate islands and their number. In order to calculate the area occupied by each pyramidal island on any height level, we will exploit the percolating clusters method. In the model under consideration, the height of the surface constantly grows according to the law $\langle\phi\rangle = Ft$. Hence, the mean height level also will grow with time and will be calculated at each simulation time step t_i in the standard manner. At each time instant t_i, we define the current value of the mean height of the growing surface $\langle\phi\rangle(t_i)$ and calculate the full number of adsorbate structures and their mean size (occupied area) on this mean height level.

Obtained results as temporal dependencies of the mean value of a full number of adsorbate structures $\langle N\rangle$ and their mean size (area) $\langle s\rangle$, counted in the value of the square of diffusion length L_d on the mean height level of the growing surface $\langle\phi\rangle(t_i)$ are shown in log–log plots in Fig. 9. Here, plots (a) and (b) show dependencies at fixed values of deposition flux F and interaction energy ε and different values of the relaxation time of the temperature field ν; plots (c) and (d) correspond to data obtained at fixed ν and different values of F and ε. Here we present results for the stage of pyramidal islands growth, neglecting the initial stage of the islands formation. In all plots, symbols correspond to numerical data and curves relate to power-law approximations, $\langle N\rangle \propto t^{-\gamma}$ and $\langle s\rangle \propto t^{\delta}$ with $\gamma, \delta > 0$. From the dependencies $\langle N\rangle(t)$ shown in Fig. 9a, c it is followed that initially formed ~50 small pyramidal islands emerge during exposing organized single pyramidal structure and the number of islands decays in time in power law $t^{-\gamma}$. From the temporal dependencies of the mean islands size $\langle s\rangle(t)$ shown in Fig. 9b, d one sees, that at initial stages the mean size is about 1% of a total computational grid; at late stages one pyramidal island occupies around 40% of the whole grid on a mean height level. A decrease in time of the mean adsorbate size relates well with the power law. By comparing obtained data at different system parameters one gets that an increase in the relaxation time of temperature field ν delays dynamics of the mean islands size growth. At large values of the deposition flux F the dynamics of the islands growth accelerate. With increase in the interaction strength ε the growth processes are delayed. According to obtained data one finds that the relation δ/γ remains invariant to the variation in main system parameters

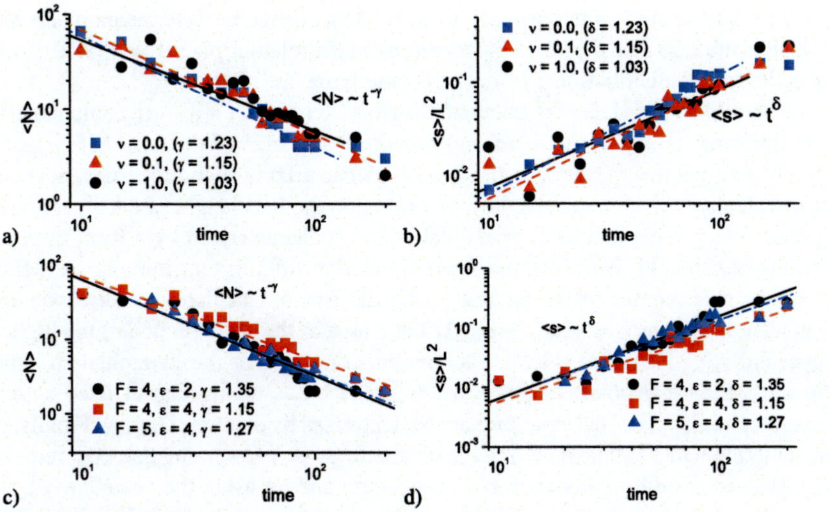

Fig. 9 Evolution of the mean number of pyramidal islands (a,c) and the mean area of structures on the half-height of the growing surface (b,d) at $F = 4$, $\varepsilon = 4$ and different values of the relaxation time for the temperature field ν (plots a and b); at $\nu = 0.1$ and different values of the deposition flux F and interaction strength ε (plots c and d)

6.2 Universality of the Islands Size Distribution

Finally, let us consider properties of the distribution of islands over sizes. As it was shown before, both the mean area occupied by island adsorbate islands and the mean number of islands on the mean height level of the growing surface evolve in time in power-law form, $\langle s \rangle \propto t^{\gamma}$ and $\langle N \rangle \propto t^{-\gamma}$. Hence, for the dependence of the number of islands on the mean islands size one gets the algebraic dependence in the form of the Zipf law, $N(s) \propto s^{-1}$. In such a case, the conservation law for the total area of all islands S_0 on a mean height level of the growing surface, $S_0 = \int_0^{\infty} sN(s)\mathrm{d}s = \text{const}$, must be satisfied, while mean height of the growing surface grows in time. Typical snapshots of areas occupied by adsorbate on the mean height level of the growing surface at $F = 4$, $\varepsilon = 4$ and $\nu = 0.1$ at different time instants are shown in Fig. 10a. Evolution of the total area occupied by adsorbate is shown in Fig. 10b. It follows that the total area S_0, occupied by adsorbate remains constant in time. Moreover, it remains invariant to the change in main system parameters.

Figure 10c shows the dependence of the number of adsorbate islands *versus* islands size (occupied area on the mean height level), calculated at different time instants. It follows that numerical data correspond well to the theoretical prediction $\langle N \rangle \propto \langle s \rangle^{-1}$, independently on deposition flux and interaction energy of adsorbate. In the insertion in Fig. 10c, we show the distribution of adsorbate islands over sizes $f(s)$ obtained at $\varepsilon = 4$, $F = 4$, $\nu = 0.1$ and $t = 50$ by using standard numerical differentiating scheme. Hence, the distribution of adsorbate islands over sizes on the mean height of the growing surface is universal, invariant to a change in system parameters and satisfies the Zipf law $p(s) \propto s^{-1}$. According to the relation $p(s)\mathrm{d}s = f(R)\mathrm{d}R$ with $s \propto R^2$ for the distribution $f(R)$ one gets the same law $f(R) \propto R^{-1}$.

It should be noted that the obtained distribution in a Zipf law form could not give any information about the mean and most probable size of the adsorbate islands. In order to get the appropriate distribution characterizing such properties, next, we will calculate the area occupied by adsorbate on the half-height of each pyramidal island. We assume that each pyramidal island is characterized by a form of circle on the half-height. Next, for the fixed simulation time instant initially we define the height difference of the highest pyramidal island, calculate the corresponding half-height of it as $h = (\phi_{\max} - \phi_{\min})/2$ and area of the cross section of the highest pyramid $s_{\max}$. Next, we calculate the inclination angle of the pyramidal structure in the standard manner: $\tan(\gamma) = h/\sqrt{s_{\max}/\pi}$, by assuming that at fixed system parameters all pyramidal structures are characterized by the same angle γ. Finally, by calculating half-heights of all pyramidal structures $\{h_i\}$ (assuming that $\phi_{\min}$ remains constant at fixed time instant for all structures) and by using the value $\tan(\gamma)$, we calculate the distribution of radii of pyramidal islands on their own half-heights [51]. Obtained results for the distributions $f(R/\langle R\rangle)$ are shown in Fig. 11a, b, at fixed system parameters and different time instants; and at fixed time instant and different system parameters, respectively. Here numerical data are shown by symbols; solid lines correspond to the approximation of numerical data by log-normal distribution for the quantity $x \equiv R/\langle R\rangle$: in the following form $f(x; \mu, \sigma) = \frac{1}{x\sigma\sqrt{2\pi}} e^{-\frac{(\ln x - \mu)^2}{2\sigma^2}}$, where μ and σ correspond to the shape and log-scale, accordingly. It follows that this distribution is universal (does not depend on simulation time) and is invariant to a change in the main system parameters.

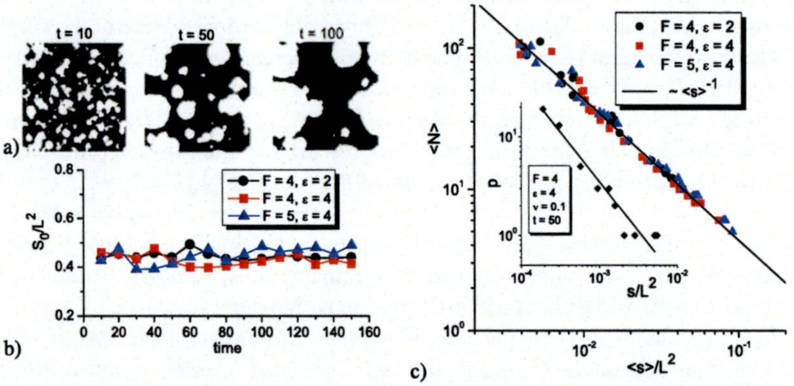

Fig. 10 Snapshots of the cross section of pyramidal structures on the mean height level of the growing surface at $F = 4$, $\varepsilon = 4$, $\nu = 0.1$ at different time instants (**a**). Dependencies of the total area of all islands area on the mean height level at $\nu = 0.1$ (**b**). Dependencies of the mean number of islands on the mean island area at different time instants at $\nu = 0.1$ (**c**). The distribution of adsorbate islands over sizes $f(s)$ at $\varepsilon = 4$, $F = 4$, $\nu = 0.1$ and $t = 50$ is shown in the insertion in plot c

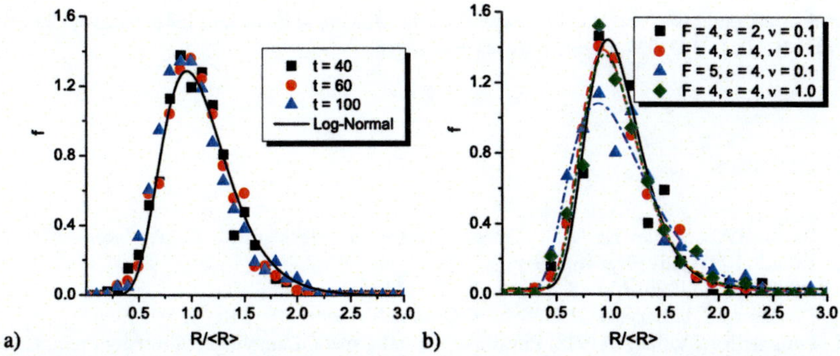

Fig. 11 Distributions of pyramidal islands over sizes calculated at a half-height of each pyramidal structure: **a** at $F = 4$, $\varepsilon = 4$, $\nu = 0.1$ and different time instants; **b** at $t = 100$ and different system parameters

7 Conclusions

In this chapter, we present results of theoretical and numerical studies of the process of pyramidal adsorbate islands formation during deposition at epitaxial growth in the framework of the phase-field modeling. We took into account stochastic nature of the lateral adsorbate flux and local variations in the temperature field during adsorption/desorption processes.

It is shown that well-defined pyramidal islands can be formed at large values of interaction strength of adsorbate. It is found that the fluctuations of the lateral flux decrease the lateral correlation length of adsorbate at large interaction strength, leading to formation of needle-like patterns.

It is shown that early stages of exposing a competition of adsorption/desorption processes leads to the realization of oscillatory regime of island formation.

We have found that the mean number of pyramidal islands decreases in time in a power-law form; whereas the mean area occupied by adsorbate on the mean height of the growing surface grows in time in the power-law form independently on main system parameters governing scaling exponent.

It is shown that on the mean level of the height of the growing surface, the distribution of adsorbate islands of sizes corresponds to the Zipf law. By calculating sizes of pyramidal islands on the half-height level of each pyramidal island, the distribution of adsorbate islands over sizes is characterized by the log-normal distribution. Both distributions are universal and invariant to a change in the main system parameters. An estimation of the obtained results for the mean linear size of pyramidal adsorbate islands on the half-height gives $\langle R \rangle \sim (2 \div 8)L_d$ with $L_d \sim (10^{-7} \div 10^{-6})m$.

Obtained results can be used to control dynamics of pyramidal-like adsorbate islands growth and to correct the technological conditions for the growth of nano-sized adsorbate islands during epitaxial growth in real experimental studies.

Acknowledgements Support of this research by the Ministry of Education and Science of Ukraine, project No. 0117U003927, is gratefully acknowledged.

References

1. Barabasi AL, Stanley HE (1995) Fractal Concepts in surface growth. Cambridge, New York
2. Krug J (1997) Adv Phys 46:139
3. Das Sarma S (1990) J Vac Sci Technol A 8:2174
4. Das Sarma S, Lanczycki CJ, Kotlyar R, Ghaisas SV (1996) Phys Rev E 53:359
5. Pimpinelli A, Villain J (1998) Physics of crystal growth. Cambridge University Press, Cambridge
6. Krug J (2002) Phys A 318:47
7. Ratsch C, Puggerone P, Scheffler M (1997) Surface diffusion: atomistic and collective processes. In: Tringides MC (ed). Plenum, New York
8. Tamborenea PI, Das Sarma S (1993) Phys Rev E 48:2575
9. Wilby D, Vvedensky DD, Zangwill A (1992) Phys Rev B 46:12896
10. Das Sarma S (1999) Dynamic scaling in epitaxial growth. In: Zhang Z, Lagally MG (ed) Morphological organization in epitaxial growth and removal, vol 14. World Scientific, Singapore, p 94
11. Kharchenko VO, Kharchenko DO, Kokhan SV et al (2012) Phys Srcipta 86:055401
12. Kharchenko VO, Kharchenko DO, Dvornichenko AV (2014) Surf Sci 630:158
13. Kharchenko VO, Kharchenko DO (2015) Surf Sci 637–638:90
14. Metiu H, Lu Y-T, Zhang ZY (1992) Sci 255:1088
15. Ratsch C, Gyure MF, Caflisch RE et al (2002) Phys Rev B 65: 195403
16. Park S, Jeong H, Kahng B (1999) Phys Rev E 59:6184
17. Kharchenko DO, Kharchenko VO, Lysenko IO, Kokhan SV (2010) Phys Rev E 82:061108
18. Kharchenko VO, Kharchenko DO (2011) Condens Matter Phys 14(2):23602
19. Kardar M, Parisi G, Zhang YC (1986) Phys Rev Lett 56:889
20. Edwards SF, Wilkinson DR (1982) Proc R Soc London Ser A 381:17
21. Wolf DE, Villain J (1990) Europhys Lett 13:389
22. Das Sarma S, Tamborenea PI (1991) Phys Rev Lett 66:325
23. Lai ZW, Das Sarma S (1991) Phys Rev Lett 66:2348
24. Burton WK, Cabrera N, Frank FC (1951) Philos Trans R Soc London A 243:299
25. Liu F, Metiu H (1994) Phys Rev E 49:2601
26. Karma A, Plapp M (1998) Phys Rev Lett 81:4444
27. Kharchenkoa VO, Kharchenkoa DO, Dvornichenko AV (2016) Phys A 444:689
28. Santamaria-Holek I, Gadomski A, Rubi JM (2011) J Phys Condens Matter 23:235101
29. Kharchenko DO, Kokhan SV, Dvornichenko AV (2009) Phys D 238:2251
30. Kharchenko VO (2009) Phys A 388:268
31. Kharchenko VO, Kharchenko DO (2012) Eur Phys J B 85:383
32. Kharchenko VO, Kharchenko DO (2014) Phys Rev E 89:042133
33. Kharchenko D, Lysenko I, Kharchenko V (2010) Phys A 389:3356
34. Kharchenko VO, Kharchenko DO (2012) Phys Rev E 86:041143
35. Batogkh D, Hildebrant M, Krischer F, Mikhailov A (1997) Phys Rep 288:435
36. Mangioni SE, Wio HS (2005) Phys Rev E 71:056203
37. Mangioni SE (2010) Phys A 389:1799
38. Emmerich H (2003) Continuum Mech Thermodyn 15:197
39. Fix G (1983) Free boundary porblems. In: Fasano A, Primicario M (ed). London, Pittman
40. Collins JB, Levine H (1985) Pys Rev B 31:6119
41. Langer JS (1986) Directions in condensed matter physics. In: Grinstein G, Mazenko G (ed). World Scientific, Singapore

42. Landau LD, Lifshitz EM (1980) Statistical physics, part 1, vol 5, 3rd edn. Butterworth-Heinemann
43. Joseph DD, Preziosi L (1989) Rev Mod Phys 61:41
44. Galenko PK, Kharchenko D, Lysenko I (2010) Phys A 389:3443
45. Kharchenko D, Kharchenko V, Lysenko I (2011) Cent Eur J Phys 9:698
46. Horsthemke W, Lefever R (1984) Noise induced transitions. Springer-Verlag
47. Mirzoev FKh, Panchenko VYa, Shelepin LA (1996) Physics-Uspekhi 39(1)
48. Kharchenko VO, Kharchenko DO, Lysenko IO (2015) Nanosistemi. Nanomateriali Nanotehnologii 13(4):577
49. Kharchenko DO, Kharchenko VO, Lysenko IO (2011) Phys Scripta 83:045802
50. Kharchenko DO, Kharchenko VO, Zhylenko TI et al (2013) Eur Phys J B 86:175
51. Kharchenko VO, Kharchenko DO, Dvornichenko AV (2015) Eur Phys J B 88:3

Dispersion Kinetics of Thin Double Chromium–Silver Films Deposited Onto Ceramic Materials and Annealed in Vacuum

I. I. Gab, T. V. Stetsyuk, B. D. Kostyuk, O. M. Fesenko, and D. B. Shakhnin

1 Introduction

Joining ceramic materials with each other and with metals is carried out by two main methods:

(1) brazing with molten metallic solders [1–3];
(2) solid-phase pressure welding [4–6].

Since ceramic materials, particularly oxide ones, are usually poorly wetted by molten metals, metal coatings are often used and applied in various ways (electron beam sputtering, magnetron sputtering, chemical deposition, etc.). Adhesive-active metals such as titanium, chromium, niobium, and others are most commonly used to make such coatings. Then, thus metallized ceramic parts are brazed in vacuum or an inert medium (argon, helium, etc.) by molten metallic solders on the basis of tin, silver, copper, nickel, etc. In this case, the thickness of the solder seam is from 50 to 100 μm up to several millimeters. Sometimes, multilayer metal coatings are used, but the thickness of the layers and brazed seams remains high [7–9].

At the same time, there is information that the reduction in the thickness of the brazed seam leads to a significant increase in the strength of the brazed joint [10–12]. Obtaining brazed or welded joints of metallized ceramics is possible through

I. I. Gab (✉) · T. V. Stetsyuk · B. D. Kostyuk
Frantsevich Institute for Problems of Materials Science of National Academy Sciences of Ukraine, 3, Krzhyzhanovskogo st., Kyiv 03142, Ukraine
e-mail: tvst@ukr.net

O. M. Fesenko
Institute of Physics of National Academy Sciences of Ukraine, 46, Nauky ave., Kyiv 03680, Ukraine

D. B. Shakhnin
University « Ukraine», 23 Lvivska st., Kyiv 03115, Ukraine

O. Fesenko and L. Yatsenko (eds.), *Nanomaterials and Nanocomposites, Nanostructure Surfaces, and Their Applications*, Springer Proceedings in Physics 246, https://doi.org/10.1007/978-3-030-51905-6_32

reducing the thickness of both the metallization coating on the ceramics and the brazing layer by itself. This can be achieved if the soldering metal or alloy is also applied in the form of a rather thin film, the thickness of which does not exceed several μm. In addition, a thin soldering seam allows you to obtain precision welded or brazed ceramic and ceramic-metal units which can be used in microelectronics, radio engineering, microwave engineering, etc.

This objective can be achieved by application onto the ceramic surfaces of double metal films, one of which is 100–200 nm thickness and consists of an adhesion-active metal such as Ti, Cr, etc., and the other is slightly thicker (2–3 μm) and serves as a solder, e.g., Cu, Ag, etc., which will ensure joining of metallized ceramic materials during brazing or welding with fine (2–4 μm thickness) solder seam.

In literature, there is information on the use of the above-mentioned two applications—and multilayer metal coatings in production of aircrafts, in structures to be implanted into the human body, in devices for flat displays, light-emitting devices and solar panels, in friction units, as sensitive elements of strain gauges, in erosion-resistant coatings of blades for engines of various purposes, as coatings for cutting tools and machine parts [13–18].

The structure of such two applications—or multiphase coatings, determination of the optimal ratio of the thickness of each layer, processes of interaction at the phases interface, including the ceramic phase, is also an important area of research.

The study of dispersion kinetics two-layer chromium–silver coatings (films) on oxide materials during annealing in vacuum and the creation of brazed and welded oxide ceramics joints based on them with super-thin brazed seam, the thickness of which does not exceed 5 μm, is the main task of the present work.

2 Materials and Experimental Methods

In this paper, an electron beam method for sputtering of metal films was used.

The thickness of the deposited nanofilms was measured by two methods:

(1) the metal of the given weight (calculated for films of the required thickness) is completely evaporated, then the thickness of the film can be easily calculated according to the law of Lambert [19, 20];
(2) with the help of a special quartz sensor located in the vacuum sputtering chamber near the sample on which the film is deposited.

Solid non-metallic substrates were made of sapphire, alumina, and zirconia ceramics as small thin plates 4 × 3 × 2 mm in size. One of the flat surfaces of each specimen was well polished to a roughness $R_z = 0.03-0.05$ μm. After polishing, all specimens were thoroughly defatted and burned in air at 1100 °C for 1 h.

The double metal films onto oxides (leucosapphire, alumina, and zirconia ceramics) consisted of two layers of different thicknesses. The first layer was a 150 nm thickness chromium nanofilm sprayed directly onto the polished oxide surfaces and served as a metallization coating for the deposition of second thicker layer onto them. This second layer of 1.5 μm was a silver film served as a solder material for joining metallized oxides by brazing or welding by pressure. The quality of the deposited films was monitored using an XJL-17 metallographic microscope.

Since the solder material is silver, in this case, specimens with metallic films deposited onto them were annealed in a vacuum chamber at different temperatures (from 900 to 1000 °C) in a vacuum no worse than 2×10^{-3} Pa for different times (from 5 up to 20 min).

Annealed specimens were investigated using scanning electron microscopes JSM-6700 F and Superprobe 733 with microphotographs storing. Using these microphotographs, the areas of metal islets on the surface of non-metallic samples were determined by the planimetric weighting method, i.e., by weighing the elements of the metallized surfaces of samples carved out from the paper microphotograph copy [11]. The experimental data obtained were processed in the form of graphs showing the dependence of the surface area of the samples covered with metal nanofilms on the annealing parameters (temperature, time).

3 Results and Discussion

The initial double films at all three oxide materials surfaces were continuous, without cracks, pores, and other defects.

The annealing of the initial double chromium–silver film at the leukosapphire surface at 900 °C made rather small changes in the film morphology after 20-min of exposure. The film annealing at 950 °C for 5 and 10 min also did not lead to the film morphology change. After 20-min of exposure at 950 °C, the film had already broken up into many interconnected fragments of different shapes covering about 85% of the chromium metallization coating surface (Figs. 1a and 2). In this state, the film is well suited for the formation of joints of metallized leukosapphire samples by pressure welding. Annealing the film at 1000 °C for 5 min resulted in the slight film disintegration (Fig. 1b). Further 10-min exposure led to the silver layer melting and disintegration into separate droplets covering about 75% of the chromium coating surface (Figs. 1c and 2). After 20-min of annealing, the molten silver droplets occupied only about 70% of the chromium surface area (Figs. 1d and 2). In this state, the film is also suitable for brazing leukosapphire samples with low load (up to 0.1 MPa).

In Fig. 2, dependence of the leucosapphire surface area covered with double chromium–silver film on the annealing time at different temperatures (900–1000 °C) is shown.

The appearance and character of the morphology change of the double chromium–silver film deposited onto alumina ceramics were initially almost the same as for films at the leucosapphire surface. Thus, annealing at 900 °C slightly impaired the integrity of the original continuous film, even after 20-min of exposure, when the area of its breaks and voids did not exceed 5–10% of the total film area (Figs. 3a and 4). Increase of the annealing temperature up to 950 °C intensified somehow the film dispersion process, and yet after 20-min of exposure, the film covered about 90% of the sample surface area (Figs. 3b and 4) which makes it possible to use this double film for

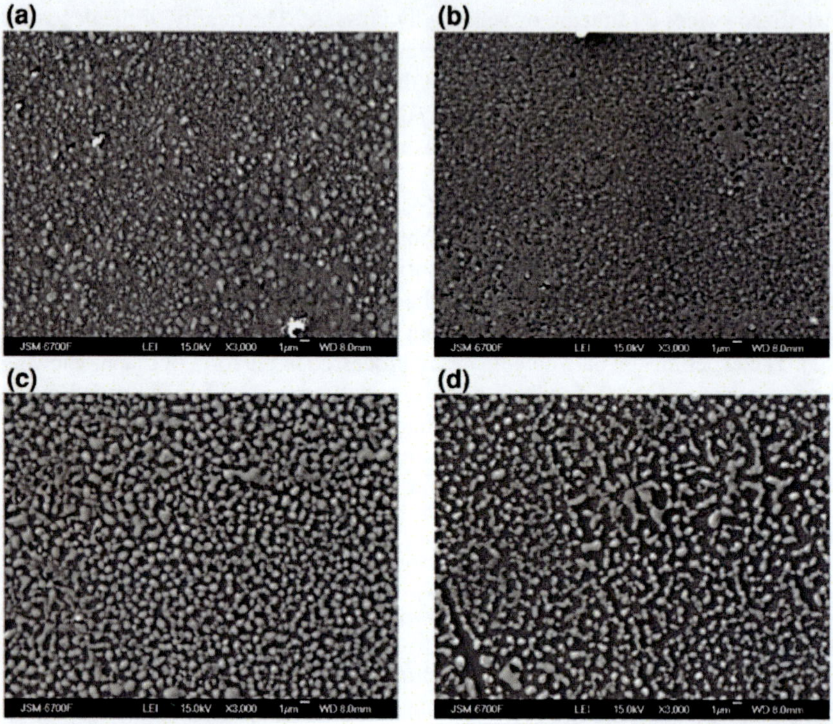

Fig. 1 SEM image (×3000) double chromium-silver film onto leucosapphire which was annealed in vacuum: **a** 950 °C, 20 min; **b** 1000 °C, 5 min; **c** 1000 °C, 10 min; **d** 1000 °C, 20 min

pressure welding of ceramics samples metallized with it under these conditions, i.e., at a temperature lower than the silver melting point (962 °C). Annealing at 1000 °C immediately led to the silver layer melting after 5 min exposure; and after 10 and 20 min annealing, the silver film started to disintegrate intensively into separate round droplets covering about 75% of the chromium metallization coating at the ceramic sample surface (Figs. 3c and 4). This double film can be used to join metallized ceramic samples by brazing only with low load of up to 0.1 MPa to prevent the silver layer from disintegration and the droplets from growing tall.

In Fig. 4, dependence of the alumina ceramics surface area covered with a double chromium–silver film on the annealing time at different temperatures (900–1000 °C) is shown.

The behavior of the double chromium–silver film deposited onto zirconia ceramics was almost exactly identical to its behavior at the alumina ceramics and leukosapphire surfaces, i.e., the annealing of the original solid film at 900 °C led to minor changes in the film as well (Fig. 5a) which intensified a bit after annealing at 950 °C (Fig. 5b). Therefore, this film is also suitable for zirconia ceramics joints manufacturing by pressure welding.

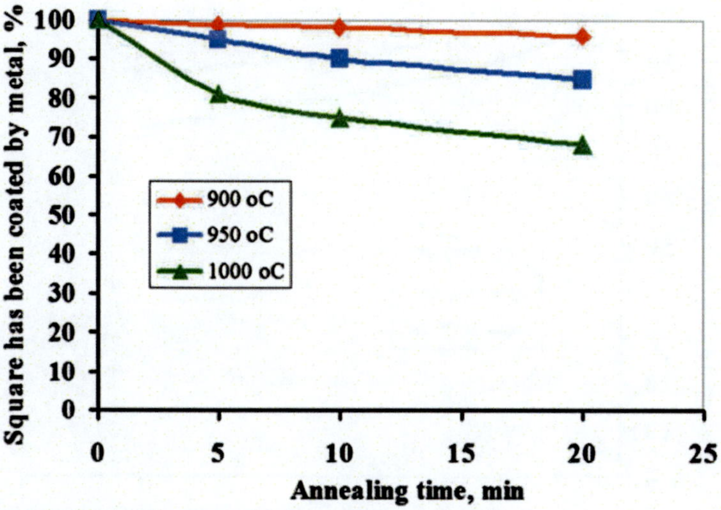

Fig. 2 Dependence of the leucosapphire surface area covered with a double chromium-silver film on the annealing time at different temperatures (900–1100 °C)

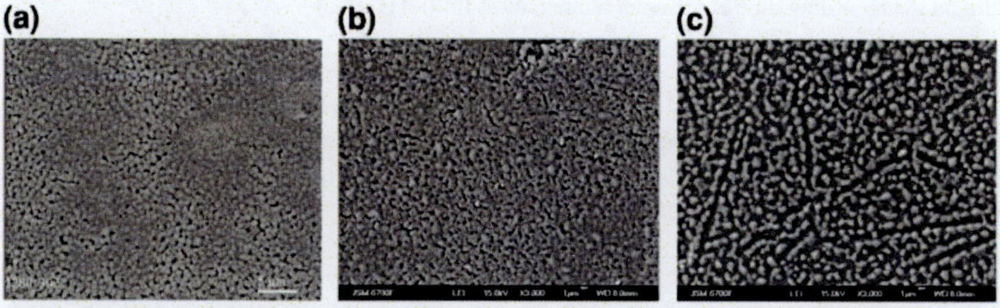

Fig. 3 SEM image (×3000) double chromium-silver film onto alumina ceramics which was annealed during 20 min in vacuum: **a** 900 °C; **b** 950 °C; **c** 1000 °C

After annealing at 1000 °C the silver layer, like the two previous oxide materials, started to disintegrate after 5-min of exposure. With the further annealing, the disintegration of the silver film increases; and, after 10 min of annealing, it covers only about 70% of the chromium coating (Figs. 5c and 6), and after 20-min of exposure—almost 60% of it (Figs. 5d and 6). Thus, it can be concluded that double chromium–silver film onto the zirconia surface can also be used at 1000 °C to join specimens of this ceramic by brazing with low load (about 0.1 MPa).

In Fig. 6, dependence of the zirconia ceramics surface area covered with a double chromium–silver film on the annealing time at different temperatures (900–1100 °C) is shown.

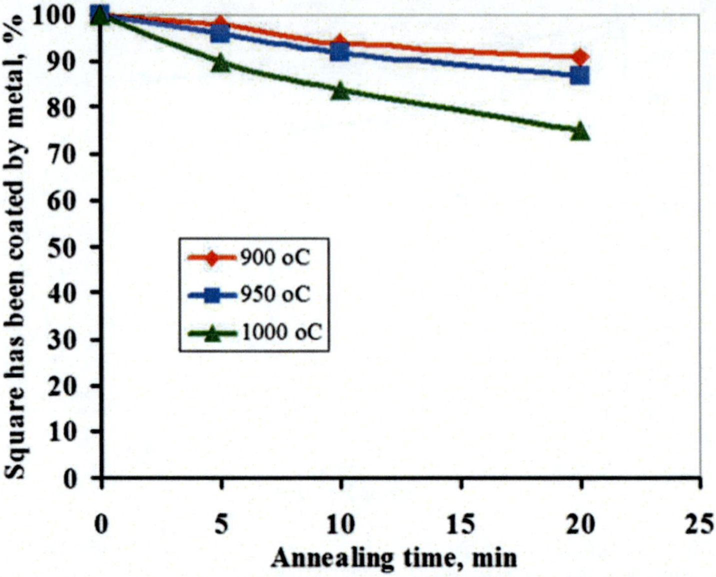

Fig. 4 Dependence of the alumina ceramics surface area covered with a double chromium-silver film on the annealing time at different temperatures (900–1100 °C)

On the basis of the studies conducted, experimental technologies for pressure welding and brazing of oxide materials with double chromium–silver films deposited onto their surfaces were developed, which ensured the production of ultra-thin brazing and welding seams 2 ÷ 4 μm thickness (Fig. 7). The strength of such seams reached 100 MPa.

4 Conclusions

Investigation of the kinetics of dispersion of chromium–silver films deposited onto leukosapphire, alumina, and zirconia ceramics revealed that the dispersion process of films at all the investigated oxides during their vacuum annealing proceeds in a similar manner. Thus, they remained rather solid when annealed at up to 900 °C; and with annealing temperature increase up to 950 °C, their disintegration intensity increases, especially with 20 min exposure at this temperature; nevertheless, in such case, the film fragments were still covering about 85–90% of the oxide surfaces. With the annealing temperature increase up to 1000 °C, the molten silver films disintegrate into separate droplets, which were still covering 65–80% of the oxide surfaces and may be suitable for joining oxide materials by brazing with low load.

Plotted kinetic dispersion curves of double films at oxide surfaces during annealing them in vacuum within the temperature range 900–1000 °C allow selection of the

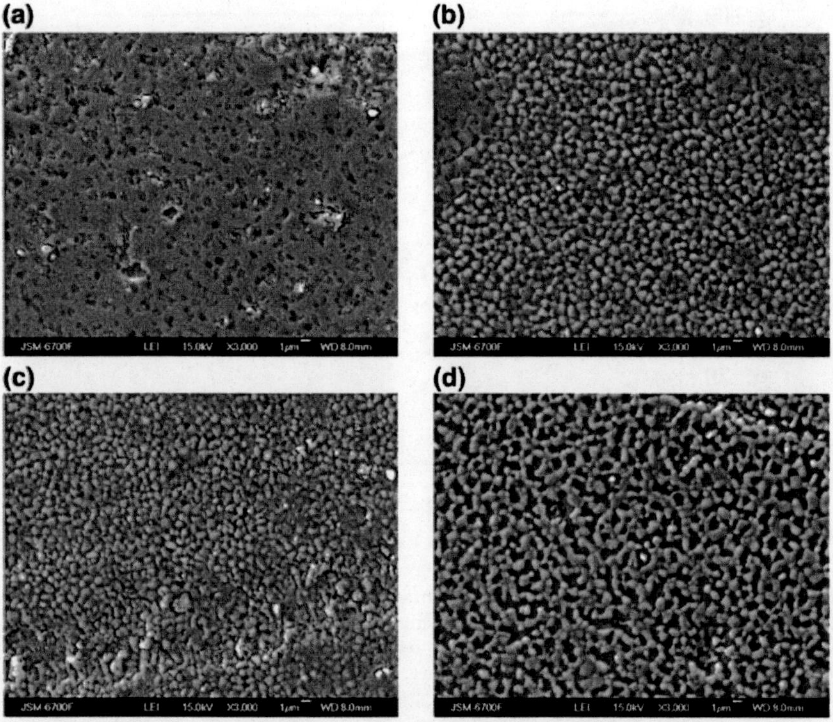

Fig. 5 SEM image (×3000) double chromium-silver film onto zirconia ceramics which was annealed in vacuum: **a** 900 °C, 20 min; **b** 950 °C, 20 min; **c** 1000 °C, 10 min; **d** 1000 °C, 20 min

optimal values of the joining processes parameters of metallized oxide materials by brazing or welding by pressure, temperature of the process, exposure time, etc.

According to the developed experimental technologies for the joining of oxides metallized with double chromium–silver films, we have manufactured the joints of these materials with ultra-thin solder seam 2 ÷ 4 μm thickness whose strength reaches 100 MPa.

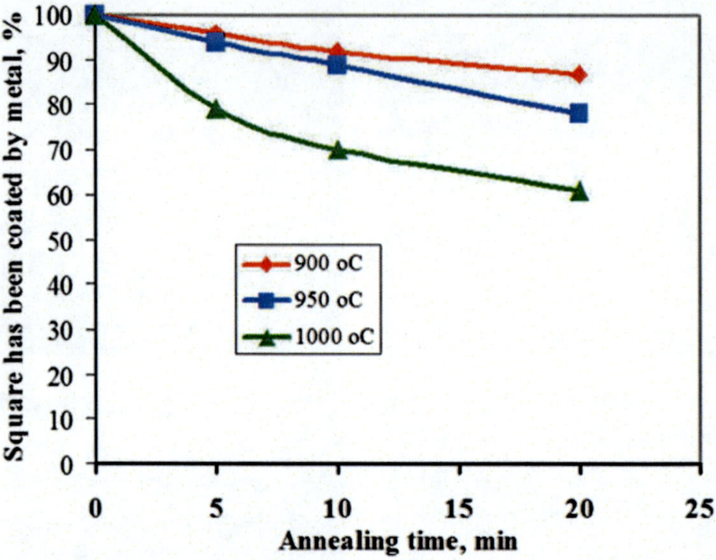

Fig. 6 Dependence of the zirconia ceramics surface area covered with a double chromium-silver film on the annealing time at different temperatures (900–1100 °C)

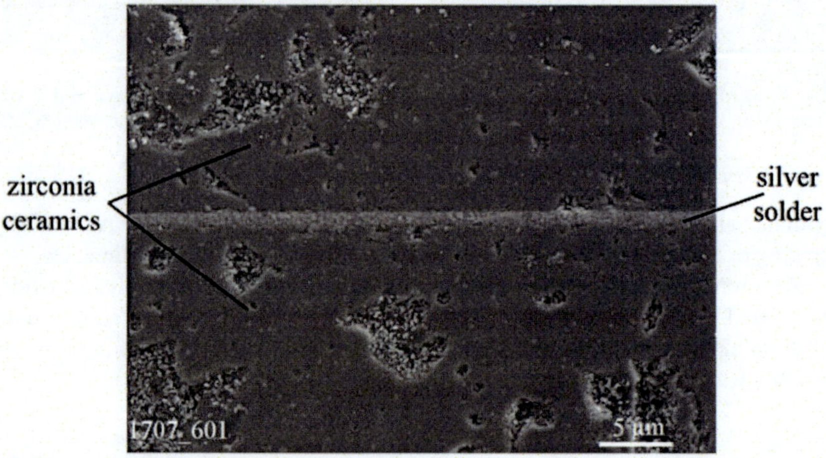

Fig. 7 The joint of zirconia ceramics, through a double film of Cr–Ag, with ultrathin welding seam which was made by pressure welding, ×600

References

1. Rubashev MA, Berdov VN, Gavrilov GI et al (1980) Termostoykiye dielektriki i ikh spai s metallami v novoy tekhnike. Atomizdat, Moscow (in Russian)
2. Naydich YV (1972) Contact phenomena in metallic melts. Naukova dumka, Kiev (in Ukranian)
3. Lashko NF, Lashko SV (1967) Payka metallov. Mashinostroyeniye, Moscow (in Russian)
4. Kazakov NF (1976) Diffuzionnaya svarka materialov. Mashinostroyeniye, Moscow (in Russian)
5. Bachin VA (1991) Teoriya, tekhnologiya i oborudovaniye diffuzi-onnoy svarki Mashinostroyeniye, Moscow. (in Russian)
6. Metelkin II, Pavlov MA, Pozdeeva NV (1977) Svarka keramiki s metallami. Metallurgia, Moscow (in Russian)
7. Yevdokimov VR, Kashtanov SL, Lado LN, Shubin SN (1995) Diffuzionnaya svarka okhvatyvayushchego soyedi-neniya alyumooksidnoy keramiki s korrozionnostoykoy stal'yu. J Svarochnoye proizvodstvo 8:2 (in Russian)
8. Request 60-239373, C04B 37/02, B23K 20/00, 28 Nov. 1985
9. Request 380163, C04B 37/02, B23K 3/00, 04 Apr. 1991
10. Naidich Y (1999) High strength ceramics brazed joints: scientific and technological bases. J Ind Ceram 19(3):162
11. Naydich YuV, Gab II, Kostyuk BD, Stetsyuk TV, Kurkova DI, Dukarov SV (2007) Investigation of the ceramic materials connection processes (soldering) using metal nanofilms. J Rep Natl Acad Sci Ukr 35:97 (in Ukranian)
12. Naydich YV, Gab II, Kostyuk BD, Stetsyuk TV, Kurkova DI, Dukarov SV (2006) Issledovaniye struktury, morfologii i smachiva-niya metallami metallicheskikh nanoplenok, nanesennykh na po-verkhnost' oksida alyuminiya s tsel'yu polucheniya svarnykh i paya-nykh soyedineniy materialov na yego osnove. J Tekhnika mashinostroyeniya 1:28 (in Russian)
13. Yakovich VM, Komarovskaya VM (2016) Mnogosloynyye pokrytiya. Inzhenerno-pedagogicheskoye obrazovaniye v XXI veke: materialy XII Respublikanskoy nauchno-prakticheskoy konferentsii molodykh uchenykh i studentov BNTU, Minsk, pp 215–218. (in Russian)
14. Andreyev AA, Kostyuk GI, Minayev NA (2012) Effektivnyye nanostrukturnyye mnogosloynyye pokrytiya dlya rezhushchikh instrumentov i detaley meditsinskogo naznacheniya, rabotayushchikh v dinamicheskom rezhime. J Aviatsionno-kosmicheskaya tekhnika i tekhnologiya 2(89):28 (in Russian)
15. Zaytsev SV, Gerasimenko YuV, Lobanov MV, Khoviv AM (2014) ssledovaniye morfologii poverkhnosti oksidiro-vannykh plenok sistem Ti-Nb. J Kondensirovannyye sredy i mezhfaznyye granitsy 16(2):153 (in Russian)
16. Lobanova TA, Volkhonskiy AO, Blinkov IV (2013). J Innovatsionnyye tekhnologii polucheniya iznoso-toykikh i triboadaptiruyemykh pokrytiy v industrii nanosistem 1(10):76. (in Russian)
17. Velikodnyy DV, Protsenko CI, Protsenko IYE (2008) Tenzoeffekt v dvukhsloynykh plenkakh Cu/Cr i Fe/Cr. J Fizicheskaya inzheneriya poverkhnosti. 6(1–2):37 (in Russian)
18. Naveed M, Obrosov A, Weib S (2015) Investigation of the wear resistance properties of Cr/CrN multilayer coatings against sand erosion. Conference Papers in Science. Hindawi Publishing Corporation. Conf. Papers in Sci. Volume. Article ID 873543. http://dx.doi.org/10.1155/2015/873543
19. Metfessel S (1963) Tonkie plenki, sh szgotovlenie I izmerenie. Gosenergoizdat, Moscow-Liningrad (in Russian)
20. Khaas G, Tun PE (1968) Thin film physics. Mir, Moskow (in Russian)

Nanocomposites and Nanomaterials

Epoxy Molecular Structure Alteration in Graphene-Epoxy Nanocomposites: Loading Effects

Borys M. Gorelov, Alla M. Gorb, Zbigniew Czapla, Sylwester Wacke, A. Nadtochiy, Vasyl V. Kuryliuk, Marek Kostrzewa, Adam Ingram, Oleksiy I. Polovina, and Nadia V. Sigareva

1 Introduction

Today, nanostructured graphene-filled polymer composites (G-PNCs) are promising candidates for advanced materials with improved multifunctional properties such as thermal-interface materials [1–5] or electromagnetic-interference shielding materials [6–10]. Among various nanofillers, graphene and its derivatives (namely, graphene oxide, functionalized graphene, functionalized graphene oxide, graphene nanoplatelets, etc.) are one of the most actively studied nanofillers for different types of polymers [11, 12].

In recent years, physical properties of graphene-based epoxy nanocomposites are in a scope of various studies. The incorporation of graphene or its derivatives in epoxy matrix has been extensively studied for lowering thermal expansion coefficient [13], improving glass transition temperature (T_g) [14–16], thermal stability [17], and thermal conductivity [18–24].

On the whole, it should be noted that results obtained by various authors are contradictory due to using various epoxy systems cured with different hardeners. For example, both increasing T_g [14–16] and its decreasing [25, 26] have been observed.

Today, aliphatic, cycloaliphatic, and aromatic amines are all employed widely as hardeners for epoxy resins [27–29]. The low equivalent weights of the aliphatic polyamines such as polyethylene polyamines (PEPAs) give tightly cross-linked

B. M. Gorelov · N. V. Sigareva
Chuiko Institute of Surface Chemistry of the National Academy of Sciences of Ukraine, Kyiv, Ukraine

A. M. Gorb (✉) · A. Nadtochiy · V. V. Kuryliuk · O. I. Polovina
Departments of Physics, Taras Shevchenko National University of Kyiv, Kyiv, Ukraine
e-mail: g_alla@ukr.net

Z. Czapla · S. Wacke · M. Kostrzewa · A. Ingram
Department of Physics, Opole University of Technology, Opole, Poland

O. Fesenko and L. Yatsenko (eds.), *Nanomaterials and Nanocomposites, Nanostructure Surfaces, and Their Applications*, Springer Proceedings in Physics 246, https://doi.org/10.1007/978-3-030-51905-6_33

networks with DGEBA-type resins providing good physical properties, including excellent chemical and solvent resistance. Other advantages include low viscosity, ambient cure temperature, a little color, and low cost [28]. Although PEPA-DGEBA-epoxies possess limited flexibility and toughness (and thermal resistivity) as compared to resins cured with aromatic amines, the latter have now largely been replaced by aliphatic or cycloaliphatic alternatives taking health concerns with handling epoxy materials into account [29, 30]. However, epoxy resins are brittle and have poor resistance to crack propagation [31, 32]. Consequently, certain efforts have been undertaken toward remedying some shortcomings from amine-cured epoxies.

From the other hand, physical properties of G-PNCs are affected by properties of graphene itself, namely, type, size (aspect ratio) and shape, also filler morphology (namely, dispersion/agglomeration, orientation, waviness), loading level, and the thermal resistance of an interface between graphene and epoxy matrix [16, 20, 33].

Generalizing the results of earlier studies, it can be concluded that they revealed two opposite effects of graphenic filler on molecular structure of a host epoxy matrix:

(1) restricting macromolecular chain mobility [5, 34] due to strong interfacial interactions;
(2) lowering the cross-linking degree (CLD) of epoxy [5, 35–37].

Thus, the above-mentioned circumstances—a great variety of epoxy systems studied and a spread in characteristics of filling graphenic nanoparticles hamper our understanding of graphene-epoxy interfacial interaction mechanisms that impact strongly the overall properties of graphene-epoxy nanocomposites.

Also, it is necessary to underline that utilizing the pristine graphene as nanofiller for polymers is strongly limited by its high-cost production and low yields. In contrast, multilayered graphene nanoplatelets (MLG), which may contain tens of graphene layers, have emerged as one of the attractive nanofillers for polymer composites providing an excellent balance between multifunctional properties and cost [38–40]. Another advantage of filing epoxy with MLG is a possibility to tailor physical parameters of a nanocomposite material by either varying morphological characteristics of MLG (such as aspect ratio, orientation, number of layers) or modifying its surface chemically.

Thus, the problem to treat molecular structure alterations in graphene-based epoxy nanocomposites and thus to predict their performance remains a big challenge for polymer material science. Therefore, this study has been undertaken in order to obtain better awareness of 3D-molecular network alterations in MLG-based DGEBA-epoxy nanocomposites from comparing loading effects in structure-sensitive properties, namely dielectric, thermal stability, and structural integrity.

2 Experimental Technique

2.1 Materials and Samples

Multilayered graphene nanoplatelets (MLGNPs) for our experiments have been prepared from graphite flakes by using the electrochemical technique described by Xia et al. [41]. The details of our preparation technique and structural properties of the platelets synthesized are described in [24].

The commercially available ED-20 (Ukraine) DGEBA-epoxy with molecular weight (M_n) 400 g/mol and the epoxy groups content (*epoxy number*) ~20.4% [42, 43] was used as the host polymer for preparing nanocomposites. The structural formulae of a repeating molecular unit of DGEBA-oligomer and molecules of hardening polyamines playing as amine cross-links (ACLs) for the oligomer's units are shown in Fig. 1.

The Polyethylene-polyamine (PEPA) was used as a curing agent. PEPA is a mix of primary amines of various molecular masses; its general structural formula is $H_2N[-CH_2CH_2NH-]_nH$, where $n = 3$ corresponds to triethylenetetramine (TETA), $n = 4$—to tetraethylenepentamine (TEPA), and $n = 5$ to pentaethylenesextamine (PESA). The PEPA is widely used as a hardener of epoxy resins due to its explosion-proof and hard ignitable properties [43]. Polyfunctional primary amines form an important class of epoxy hardeners. Primary amines undergo an addition reaction with the epoxide group to form a hydroxyl group and a secondary amine. The secondary amine can further react with an epoxide to form a tertiary amine and an additional hydroxyl group [27].

Given that the most important polar groups (N, OH) are located close to the cross-linking nodes, the increase of the CLD leads to an increase of the polar moiety density [44]. Under curing, the oligomer units combine with each other into long epoxy macromolecular chains (EMCs). The EMCs, in turn, undergo numerous entanglements and cross-links by amine molecules, thus forming 3D-network molecular

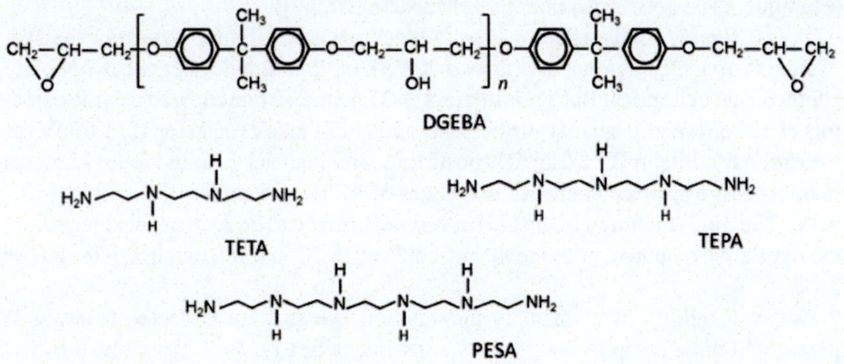

Fig. 1 The structural formulae of DGEBA-epoxy and hardening polyamines (TETA, TEPA, PESA)

structure. The structural formula of the resin can be found elsewhere [27]. The resin underwent no additional purification procedure before using. The epoxide to the hardener ratio was kept to be constant at 5:1.

The multiplication of the dipoles may result from breaking amine cross-links (ACLs) and shortening EMCs.

No efforts to orient MLG-NPs within the nanocomposites have been undertaken in the present study because the orientation-related anisotropy may hinder to unveil peculiarities of filler-epoxy interactions. The filler-mass-loadings (C_f) were 0.5, 1, 2, and 5% for both the nanocomposites. As prepared liquid composites of both the types were mixed manually until homogeneous suspensions were obtained, and their further polymerization occurred at room temperature during 72 h.

2.2 Experimental Techniques

Dielectric properties were obtained by measuring complex permittivity using the computer-controlled impedance meter Hioki IM3536 General Purpose LCR Meter. Details of the measuring technique can be found elsewhere [45]. The procedure of preparation of samples is described in [46].

An impact of MLGNPs on thermal stability of the epoxy has been determined by using a programmable thermal desorption technique combined with a mass-spectroscopic detection (TDMS) via the single-pole mass-analyzer MX 7304A. Details of the measuring technique can be found elsewhere [47, 48].

The experimental values of the dynamic elastic modules of the epoxy and its MLG-nanocomposites (namely, the Young's modulus E, the compression modulus M, the Lame constants λ and μ, and the Poisson's ratio ν) have been calculated via values of longitudinal (V_L) and shear (V_S) wave phase velocities (measured within the frequency range of 1.5–2.0 MHz) and calculated values of the nanocomposite's density ρ by using a procedure described in [48]. The theoretical values of the modules have been calculated by using the molecular dynamic (MD) simulation technique. (The details can be found elsewhere [24, 49]).

A fast timing configuration of the "Ortec" positron lifetime system was used. "Scionix" modules equipped with two XP 2020Q photomultiplier tubes optically coupled with cylindrical BaF_2 scintillators (Ø25.4 mm, 10 mm) operated as the detectors of two different γ-quanta attributed to start (1274.6 keV) and stop (511.0 keV) of a positron lifetime. A 0.1 MBq ^{22}Na positron source sealed by kapton foils (12.5 μm in thickness) was placed between two plates of the same sample (~1.5 mm in thickness). The time resolution of the device was determined to be 230 ps from the prompt curve of a ^{60}Co source, at the temperature $T = 18$ °C and relative humidity RH = 35%.

About a million of elementary annihilation events were collected to evaluate positron lifetime components τ_1, τ_2, τ_3, and intensities I_1, I_2, I_3 from row positron annihilation lifetime spectra fitted by three components. The mathematical treatment of PAL spectra was performed with LT 9.0 program [50] without constraints imposed on lifetime components. Source contribution was evidenced at the level of 12%, allowing compensation of input from positrons annihilated in the Kaptons foil ($\tau_s = 0.352$ ns). Other details of the measuring technique can be found elsewhere [51, 52].

3 Results and Discussion

3.1 Data Obtained by Dielectric Spectroscopy

Dynamic dielectric properties represent the response of the material under study, the disturbance caused by an ac-electric field applied externally. Quantitatively, the dielectric properties can be described in the terms of the complex dielectric permittivity $\varepsilon * (\omega, T) = \varepsilon_1(\omega, T) + j\varepsilon_2(\omega, T)$, where ε_1 is the dielectric permittivity and ε_2 is the dielectric loss factor. In the general case, both ambient temperature (T) and frequency of an external electric field (ω) have the strongest impact on the dielectric properties of materials. Therefore, studying frequency-temperature dependencies of both the dielectric constant and dielectric loss factor is one of the most convenient and sensitive methods of studying material's inner structure [45].

It is advisable to begin studying MLG-loading impact on epoxy dielectric properties at low-frequency region where dipole-related relaxation processes do not contribute to the dielectric response of the materials studied. In the neat resin, such response may originate from either absorbed water molecules or short-scale local motions of lateral and terminal moieties the polymer macromolecules, whereas in the nanocomposites—from both the polymer itself and the nanoplatelets (namely, from local motion of the residual surface hydroxyls and other dipole contaminant molecules populating the surfaces).

From the physical point of view, complex dielectric permittivity of the neat epoxy over the frequency range under study is governed by two basic polarization mechanisms, namely the orientation and the hopping ones [53], and also by the related effects of the dielectric relaxation, because, in accordance with the classic Debye theory of polarization [53], the orientation (thermal) mechanism gives its contribution into the permittivity in the frequency region ($\omega/2\pi$) ranging from 10^5 to 10^9 Hz, whereas the hopping mechanism contributes at the frequencies below 10^5 Hz.

Frequency dependences of ε_1 for the nanocomposites are similar to those for the unfilled resin and ε_1 decrease monotonously with increasing frequency in the range from ~40 Hz to 1 MHz regardless of filler's loading. The decreasing occurs due to delay of hopping and reorientation displacement of less movable dipole moieties [53].

Frequency-temperature dependences of both ε_1 and ε_2 are essentially dependent on the loading. Figure 2 shows low-temperature ($T = 95$ K) loading dependences of the dielectric permittivity ε_1 (Fig. 2a) and the dielectric loss factor ε_2 (Fig. 2b) at fixed frequencies.

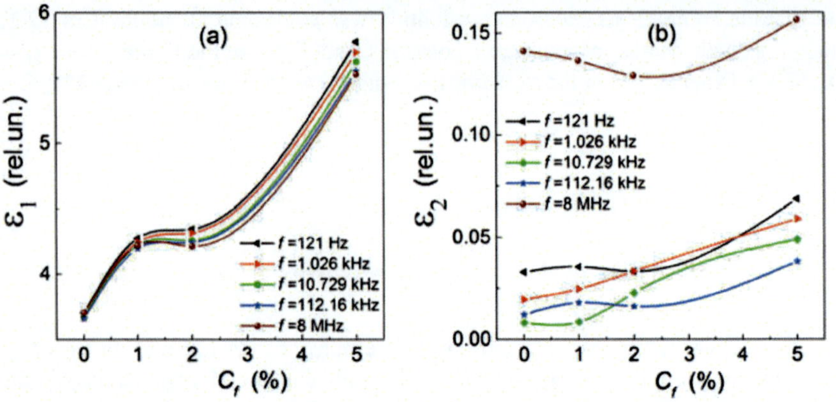

Fig. 2 Low-temperature ($T = 95$ K) loading dependences of the dielectric permittivity ε_1 (**a**) and the dielectric loss factor ε_2 (**b**) at fixed frequencies of 121 Hz, 1.026 kHz, 10.729 kHz, 112.16 kHz, and 8 MHz

It is remarkable that both $\varepsilon_1(C_f)$ and $\varepsilon_2(C_f)$ increases with increasing filler's loading in an unmonotonous way. This fact may be explained by supposing that several mechanisms of interaction occurring between epoxy and MLGNPs compete with each other in the nanocomposites. The first one is increasing a number of dipole moieties and plays for increasing ε_1. The multiplication of the dipoles may results from breaking amine cross-links (ACLs) and shortening EMCs. Breaking the cross-links means lowering the CLD. The effect is a result of the interactions. It have been observed by various authors [34–37, 54] and manifests itself as decreasing the glass-transition temperature T_g.

Both the above-mentioned processes (breaking ACLs and shortening EMCs) result in increasing number of short EMCs, free (non-cured) epoxy groups, and free amine molecules. The free moieties together with the terminal (OH- and CH_3-) and the lateral (OH- and CH_3-) groups of EMCs, all bear non-zero dipole moments and thus will contribute in increasing both ε_1 and ε_2. (Hereafter, these dipole moieties will be referred to as γ-dipoles and their density will be denoted as N_γ). Not numerous hydroxyl groups originating from absorbed and dissociated water molecules can interact with the dangling carbon bonds on the lateral surface of MLGNPs and can form dipolar COH-moieties playing also as γ-relaxators.

Another mechanism of interaction between EMCs and MLGNPs is restricting chain's mobility [5, 34, 54]. First of all, it comes from electrostatic binding of γ-dipoles with active surface sites (ASSs) of the nanoplatelets. It is clear that the contribution into both ε_1 and ε_2 from γ-dipoles bounded by either ASSs will be reduced due to their restricted mobility.

From the other hand, when the density of ASSs occupied with γ-dipoles becomes high enough, a process of forming double electric layers (DELs) develops in the vicinity of the nanoparticles. The physical processes responsible for the formation of a DEL on a charged surface embedded into an ionic liquid together with analytical models for such layers are widely discussed in west literature on electrochemistry [55, 56]. It has earlier been pointed out [57] that interfacial electric effects related to DEL-formation may give their contribution into the overall dielectric properties nanocomposites.

In our nanocomposites, such DELs consist of free γ-dipoles and lateral moieties of EMCs and will be localized along both lateral facets and defect areas of basal planes of the nanoparticles. It is clear, that γ-dipoles involved in DELs will give negligible contribution in ε_2.

Taking the above-mentioned mechanisms of epoxy–graphene interaction into account, loading dependences of ε_1 can be explained in the assumption that the first mechanism (breeding γ-dipoles) dominates at $0\% < C_f < 1\%$, whereas at $1\% < C_f < 2\%$ an intensity of the second mechanism (binding γ-dipoles) enhances as much as both the processes compensate each other. As a consequence, ε_1 undergoes negligible variations. At $C_f > 2\%$ ASSs having occupied and DELs having formed, the first process prevails again.

In contrast to $\varepsilon_1(C_f)$, $\varepsilon_2(C_f)$ varies with increasing f. At $0\% < C_f < 2\%$ $\varepsilon_2(C_f)$ varies in a non-regular way with increasing f. When $C_f > 2\%$ ε_2 increases monotonously with increasing f. Such behavior can be understood as follows. Obviously that small and movable γ-dipoles contribute into ε_2 over the entire frequency region studied. As mentioned above, γ-dipoles involved in DELs do not contribute in ε_2. However, the single DEL can behave as a spontaneously polarized domain, which is capable to dissipate the energy of an electric field applied. Its contribution will depend on its size and spatial orientation relative to the field.

To penetrate into physical background of the effect of increasing ε_1 and ε_2 of epoxy with increasing its MLGNP-loading, loading-related evolution of minor dielectric relaxation (MDR) processes has been studied. It is known that a number of the minor relaxation (as compared to α-relaxation) processes take place in the vitreous state of thermosets (i.e., at $T < T_g$). These are referred to as "the dipole-group relaxation," labeled by β, γ, δ, and so on (beginning at the high-temperature end of $\varepsilon_2(T)$), and originates mainly from local vibration modes of the molecules [53].

In pure DGEBA-epoxy resins, MDR-processes have been widely studied earlier [58–64]. So, Mikolajczak et al. [58] have studied dynamic mechanical behavior of an epoxy resin made with the diglicidyl ether of bisphenol A (DGEBA) and 4,4-diaminodiphenylmethane (DDM) and found both γ-and β-relaxation processes in the glassy region. They suggested that the existence of γ- and β-relaxations indicate merely two types of local regions of mobility in the epoxy. Namely, the γ-relaxation may indicate molecular motions in those regions where polymer chains are relatively loosely packed, whereas β-relaxation deals with the motions in those regions where loose packing is caused by unreacted molecular segments and/or by subtantially lower cross-linking densities. Another possible mechanism for the γ-relaxation is due to the crankshaft rotation [65] below the glass transition temperature.

Indeed, recent studies which have been made by using the atomic force microscope at fracture interfaces [66–70], IR-spectroscopy [66, 67, 69, 70], and differential scanning microscopy [66, 67] confirmed that epoxy-amine resins possess a heterogeneous, nodular structure, which is composed of regions of relatively higher cross-linking surrounded by an interstitial phase of a relatively lower cross-link density [66, 68].

Also, It has previously been shown [59–62] that the dielectric β-relaxation in epoxy resins cured with the amine-contained hardeners is affected mainly by the chemical structure of the curing agents. Namely, the β-peak was tentatively assigned to the local motion of the dipolar groups that are created during the cross-linking reactions. These were the hydroxyl-ether groups and the secondary or tertiary amines.

Later, Johari and Mangion [63] have studied dielectric relaxation of bisphenol-A-based thermosets cured with diaminodiphenyl methane and diaminodiphenyl sulfone. They found that the γ-relaxation process is more prominent than the β-relaxation process and supposed the γ-peaks to be partly due to local motions of those dipoles that remain as the unreacted components during the curing of the thermosets. Also, their results show that the β-relaxation peak becomes increasingly more prominent as the number of cross-links and the chain length increases. Not recently, Hassan et al. [64] reported substantial suppression of β-peaks as compared to γ-peaks for DGEBA-epoxy networks prepared using 3,3- and 4,4-diaminodiphenyl sulfone isomer cross-linkers.

On the whole, there is ambiguity in treating MDR-peaks in various epoxy systems, i.e., in ascribing various dipole moieties to β-or γ-relaxators. In our experiments, the MDR-processes manifest themselves in frequency dependences of ε_2 as rather broad single peaks. In the neat epoxy, the peaks have been observed at $T > 175$ K over the entire frequency region studied. Therefore, we assumed that β-and γ-peaks overlap due to various relaxators possessing close values of the activation energy are present in both the neat resin and its nanocomposites.

Analyzing shifts in γ-peak temperature position with increasing frequency for every nanocomposite, the loading dependence for effective activation energy (E_γ) of γ-dipoles can be determined. Within a frame of the classic Debye theory [52], in the temperature position (T_γ) of the peak in $\varepsilon_2(T)-$ curve obtained at the certain frequency ω_γ is related to the correspondent macroscopic relaxation time τ_γ by

$$\omega_\gamma \cdot \tau_\gamma\left(T_\gamma\right) = 1. \qquad (1)$$

Assuming that γ-relaxation process in both the resin and its nanocomposites obeys the Arrhenius law [53]:

$$1/\omega_\gamma = \tau_{\gamma 0} \exp(E_A/kT_\gamma) \ or \ T_\gamma\left(C_f\right) = E_A\left(C_f\right) / \left\{k \cdot \ln\left(1/\omega_\gamma \, \tau_{\gamma 0}\left(C_f\right)\right)\right\}, \qquad (2)$$

where k is the Boltzmann constant. Then, supposing the time constant $\tau_{\gamma 0}$ to be independent of frequency, the effective activation energy (E_A) for dipoles involved in the γ-process can be evaluated as an inclination of straight lines of $\ln \tau_{\gamma 0}$ verses $1/T_\gamma$.

The calculated loading dependence of E_A is shown on Fig. 3. It is close to linear one. The value of $E_A(0) = 0.525$ eV $= 50.6$ kJ/mol obtained for the neat epoxy

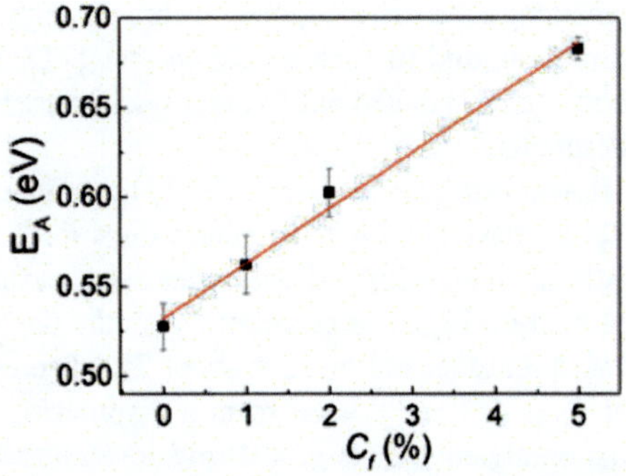

Fig. 3 Loading dependence of the activation energy for γ-relaxation process

is close to one given in literature −56.2 kJ/mol [64]. Obviously, that increasing E_A means an enhancement in strength of interaction between γ-dipoles and their neighbouring environment. On the other hand, it means a restriction in mobility of the γ-dipoles. However, such restriction is not accompanied with lowering ε_1. Simultaneous increasing of both ε_1 and E_A can be explained assuming electrical ordering of the γ-dipole ensemble due to forming DELs. As it is mentioned above, a DEL may have spontaneous polarization and therefore higher local values of ε_1 in the vicinity of the nanoplatelets.

The density N_γ of relaxators contributing to the γ-process (γ-dipoles) can be evaluated by using the Frohlich model from the relation [53]:

$$\sum_i N_{\gamma i} g_{\gamma i} p_{0i}^2 = 9\varepsilon_0 k_B T_\gamma F_\varepsilon(T_\gamma), \tag{3}$$

where p_{0i} is the dipole moment of the γ-relaxator of i-th type, and the correlation factor g_i is a measure of the local electrical ordering in the material for the i-th γ-relaxators. The correlation factors are supposed to characterize the short-range molecular interactions [53]. The constant $F_\varepsilon(T_\gamma)$ is determined by [53]

$$F_\varepsilon(T_\gamma) = \frac{[\varepsilon_{1S}(T_\gamma) - \varepsilon_{1\infty}(T_\gamma)] \cdot [2\varepsilon_{1S}(T_\gamma) + \varepsilon_{1\infty}(T_\gamma)]}{\varepsilon_{1S}(T_\gamma) \cdot [\varepsilon_{1\infty}(T_\gamma) + 2]^2}, \tag{4}$$

where ε_{1S} denotes the static dielectric constant and $\varepsilon_{1\infty}$ denotes the dielectric constant at high-frequency limit: $\omega \to \infty$. Hereafter, the left-side term of the (4) will be referred to as "the Frohlich Relaxation Term" $FRT(T_\gamma)$.

Analyzing (4), it may be assumed that effective (i.e., averaged over the volume) values of p_{0i} undergo negligible variations with increasing C_f. Such the assumption is confirmed by IR-absorption data where any line reveals no shift with increasing C_f. At the same time, we suggest that the correlation factors g_i may vary in the vicinity of the nanoplatelets, where a portion of γ-dipoles appeared bounded either by ASSs or in DELs. Here, if the neighboring dipoles line up in a parallel direction, then $g_i > 1$ [71]. On the other hand, in areas remote from the nanoparticles it is advantageous for γ-dipoles to line up with the neighboring ones in an antiparallel direction, and in such a case $g_i < 1$ [71]. This situation is likely to take place in both tightly packed and loosely packed areas where different types of γ-dipoles are dominant.

Averaging the sets of $FRT\left(T_\gamma\right)$ calculated by using (4) at different temperatures T_γ, we have obtained the mean values $FRT\left(C_f\right)$ which are a generalized characteristic of an ensemble of γ-dipoles. The normalized dependence of *FRT* versus C_f is plotted on Fig. 4. It correlates with the dependence $\varepsilon_{2\gamma}\left(C_f\right)$ at $C_f \leq 2\%$. At $C_f > 2\%$, *FRT* goes out on saturation. The discrepancy between $FRT\left(C_f\right)$ and $\varepsilon_{2\gamma}\left(C_f\right)$ at $C_f > 2\%$ may come from a disordering of the γ-dipoles in areas remote from nanoplatelets where $g_i < 1$ and lowers with increasing. Therefore, FRT remains nearly constant in spite of the fact that $N_{\gamma i}$ may increase.

A generalized picture of MLGNP's impact on both polarization and relaxation processes in epoxy can be obtained by analyzing frequency-loading dependences of the complex dielectric permittivity by using the Cole–Cole plots [53]. Figure 5 shows $\varepsilon_2(\varepsilon_1)$-plots for the neat epoxy (the curve "1") and its MLG-nanocomposites (the curves "2," "3," and "4") obtained for $T = 300$ K. The experimental locuses are plotted via correspondent dependences $\varepsilon_1(\omega, 300\text{ K})$ and $\varepsilon_2(\omega, 300\text{ K})$. The locuses manifest the effect of simultaneous increasing $\Delta\varepsilon_{1\gamma} = \varepsilon_{1S} - \varepsilon_{1\infty}$, and $\varepsilon_{2\gamma} = \varepsilon_{2\gamma}\left(T_\gamma\right)$. Along with increasing $\Delta\varepsilon_{1\gamma}$, increasing $\varepsilon_{2\gamma}$ with increasing loading is likely to evidence on increasing N_γ.

Fig. 4 Loading dependence of the normalized Frolich's relaxation term

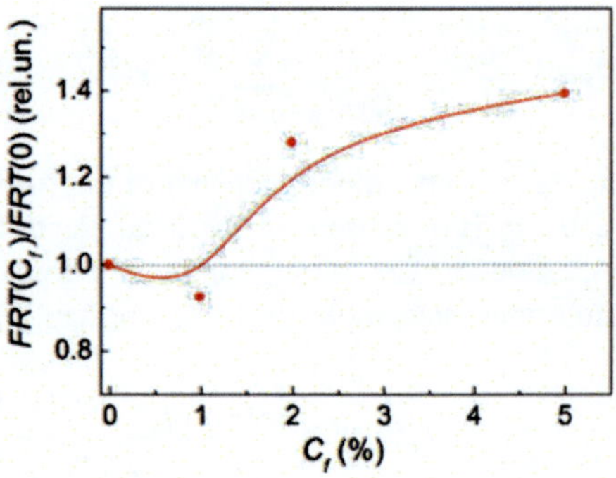

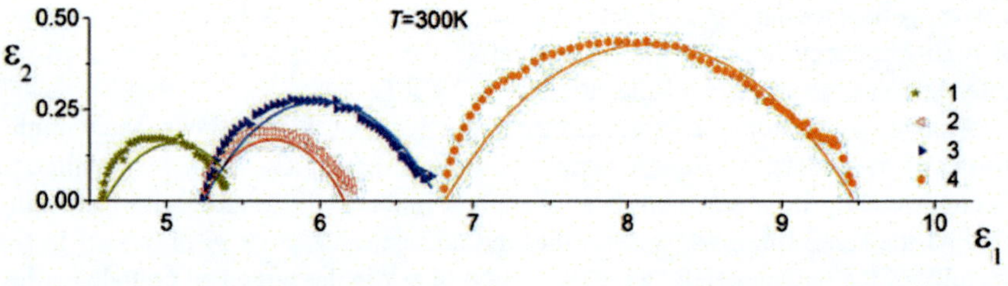

Fig. 5 $\varepsilon_2(\varepsilon_1)$-plots for the neat epoxy (the curve "1") and its MLG-nanocomposites (the curves "2", "3", and "4") obtained for $T = 300$ K

The physical mechanisms underlying loading-induced evolution of an ensemble of the γ-dipoles populating 3D-molecular architecture of the epoxy resin can be ascertained by using theoretical modeling of dependences of ε_2 versus ε_1. It is clearly seen from Fig. 5, that the experimental locuses are very close to circular ones. A careful geometric analysis of the locuses by using the Havriliak–Negami fitting procedure [72] shows that the Havriliak–Negami model's parameter β has proven to be greater than unity. In other words, the Havriliak–Negami model fails to comprehend the experimental data. Therefore, the Cole–Cole model has been used. The solid lines given on Fig. 5 have been calculated by using the relation [53]:

$$\frac{\varepsilon^* - \varepsilon_{1\infty}}{\varepsilon_{1S} - \varepsilon_{1\infty}} = \frac{1}{\left[1 + i\omega\tau_\gamma\right]^{1-\alpha}}, \tag{5}$$

where $\varepsilon* = \varepsilon_1 - i\varepsilon_2$, and the parameter α describes a distribution if γ-dipoles over its relaxation times, i.e., takes presence of various γ-dipoles into account. The theoretical curves have been calculated into approximation that maximal values ($\varepsilon_{2\gamma}$) of ε_2 coincide with the values ε_{2C} which correspond to values of $\varepsilon_1 = \varepsilon_{1C} = (\varepsilon_{1S} - \varepsilon_{1\infty})/2$.

The most essential differences between the experiment and the Cole–Cole model take place at higher frequencies (at $f > 10$ kHz). It can be explained in an assumption that various γ-dipoles are present in both the neat epoxy and its nanocomposites. Moreover, for every type of dipole, its contribution into ε^* will vary with loading and varying temperatures. In other words, a more accurate fitting between the experiment and the theory can be reached when the single term in the right hand of (5) will be replaced by the sum additives of individual values of α_i and $\tau_{\gamma i}$.

A detailed analysis shows that ε_{1S} increases linearly with increasing loading, whereas the loading dependence of $\varepsilon_{1\infty}$ is similar to that of ε_1 (see Fig. 2). Also, the dependences $\alpha(C_f)$ measured at various frequencies are unmonotonous. It is known that if α increases then a number of short-relaxation-time dipoles increases (i.e., $\tau_\gamma(C_f) > \tau_\gamma(0)$ and vice versa [72]. A case of decreasing α can be treated as the case of binding short (and, hence, light) γ-dipoles to ASSs. It may be assumed that when a number of such dipoles increases more rapidly with increasing C_f than a number of ASSs does, then α begin to increase.

A certain approach to interpreting the broad γ-relaxation peaks is to consider the broadening to result from a distribution of relaxation times τ, which can be described in terms of the distribution function $F(\tau/\tau_\gamma)$ [72]:

$$F_3(C_f) = \frac{\sin(\pi\alpha)}{2\pi\left\{\cosh\left[(1-\alpha)\ln\left(\frac{\tau}{\tau_\gamma}\right)\right] - \cos(\pi\alpha)\right\}}. \tag{6}$$

Havriliak and Negami emphasized [72] that although the distribution function is empirical, it is accurate, since the calculation of F from the relaxation function $(\varepsilon^* - \varepsilon_{1\infty})/(\varepsilon_{1S} - \varepsilon_{1\infty})^{-1}$ is exact and therefore the reliability of F is as good as the

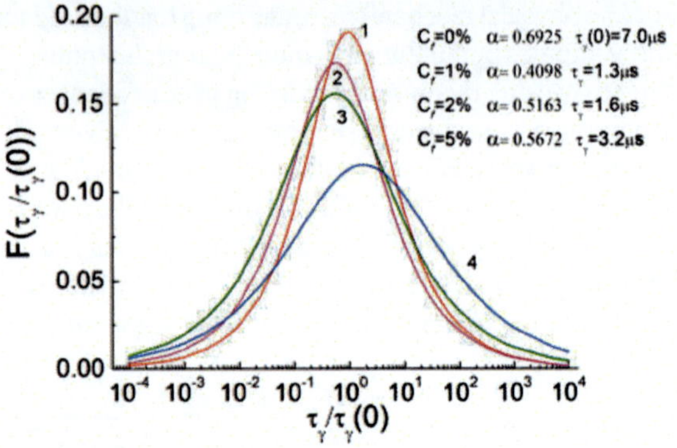

Fig. 6 Evolution of the shape and position of the distribution function F with increasing the loading for $T = 300$ K

reliability of the relaxation function. In our experiments, a dependence of $F(\tau/\tau_\gamma)$ on C_f comes from the dependences of both α and τ_γ on C_f.

Evolution of the shape and position of the distribution function with increasing the loading is shown on Fig. 6 for $T = 300$ K. In our experiments, a dependence of $F[\tau/\tau_\gamma(0)]$ on C_f comes from the dependences of both α_i and τ_γ on C_f. As shown in Fig. 6, for $C_f \leq 2\%$ $F[\tau/\tau_\gamma(0)]$ curves shifts to lower values of τ. It means that more light moieties are dominant among the γ-dipole ensemble. When C_f increases $F[\tau/\tau_\gamma(0)]$ curves shifts to higher τ, i.e., heavy moieties bring main contribution in both ε_1 and ε_2.

On the whole, BDS data evidenced on the effect of increasing dielectric permittivity of the epoxy induced by embedded MLGNPs. It may be supposed that the effect is due to 3D-molecular network alteration, which includes processes of breaking EMCs and binding dipole moieties by ASSs of the nanoplatelets. The processes impact on MDR is mainly caused by the γ-dipoles. MDR is satisfactorily described by well-known Cole–Cole model. Loading effects in MDR confirm a competition between above-mentioned structure-alteration processes. Non-monotonous evolution of the relaxation-time-distribution function with loading shows that light (heavy) dipoles are dominant at low (high) loadings.

3.2 *Data Obtained by Programmable Thermal Desorption Mass-Spectroscopy (TDMS)*

BDS provides general information about structural alteration of 3D-molecular network of the epoxy induced by embedded MLGNPs. In contrast, TDMS enables to study peculiarities of interactions occurring between individual moieties and the nanoparticles.

A detailed analysis of the thermal-desorption mass-spectra of the positively charged ($z = +1$) volatile products (m/z) of thermally induced destruction of the neat resin and its MLGNP-filled nanocomposites can be found elsewhere [24]. In this work, we studied the products which gave the double-peaked thermal desorption curves (TDCs).

Table 1 contains the list of such products of the neat resin along with the values of T_{max} and I_{max} obtained for both low temperature (LT) and main peaks from the correspondent TDCs.

An additional (wide, low temperature, LT) peak appears at $T < 300$ °C and is much softer than the main one for most of the moieties, except hydroxyl groups (m/z = 17) and water molecules (18). The small LT peaks indicate that some portions of correspondent moieties are weakly coupled with their molecular environment. It may be suggested that such moieties are constituent of EMCs located in loosely packed areas.

Loading-induced evolution of TDCs for some individual double-peaked moieties ($m/z = 7, 18, 28, 31, 43, 56$) described in Table 1, and is shown in Fig. 7a–f. One can see that introducing MLGNPs into the epoxy results in suppression of LT-peaks for most fragments. The suppression evidences binding moieties to the nanoplatelets. Suppression of the LT-peak for hydroxyl groups (see Fig. 7a) testifies that their contribution into the effect of simultaneous increasing $\Delta\varepsilon_{1\gamma}$ and $\varepsilon_{2\gamma}$ is insignificant.

Table 1 The most intensive volatile products of the neat resin along with temperature positions (T_{max}) and intensities ($I_{max} = I(T_{max})$) of LT- and main peaks

m/z	Chemical formulae	Definition	T_{max1}	I_{max1}	T_{max2}	I_{max2}
14	CH_2	Methylene	242	59.2	330	293.7
15	CH_3	Methyl	242	1013.1	330	4041.6
16	CH_4	Primary amine	220	1163.8	330	3513.9
17	OH	Hydroxyl, ammonia	220	2560.1	308	4415.7
18	H_2O	Water	110	4979.8	330	4659.5
28	CO	Carbon monoxide	244	1307.2	332	4540.2
30	OCH_2	Formaldehyde	244	1031.4	332	3044.3
31	NH_2CH_3,CH_2OH	Methylamine			332	894.8
40	$CHCH_2CH$, COC		–	–	333	2407.8
41	$CHCH_2CH_2$, COCH		245	146.1	333	1109.6
42	H_2CCO		245	89.9	333	3474.1
43	CH_2NHCH_2, H_3CCHNH, H_2CCHNH_2	Aziridine, Vynilamine, Ethanamine	245	573.0	311	3543.95
56	H_2COCH	Epoxy group	225	17.3	335	811.4
57	CH_2OCHCH_2, $(CH_2)_2CHNH_2$		225	27	335	386.2
58	CH_2OCHCH_3		225	74.9	335	574.5
59	$CH_3(CH_2)_2NH_2$, $CH_3CHOHCH_2$		225	43.1	313	225.7

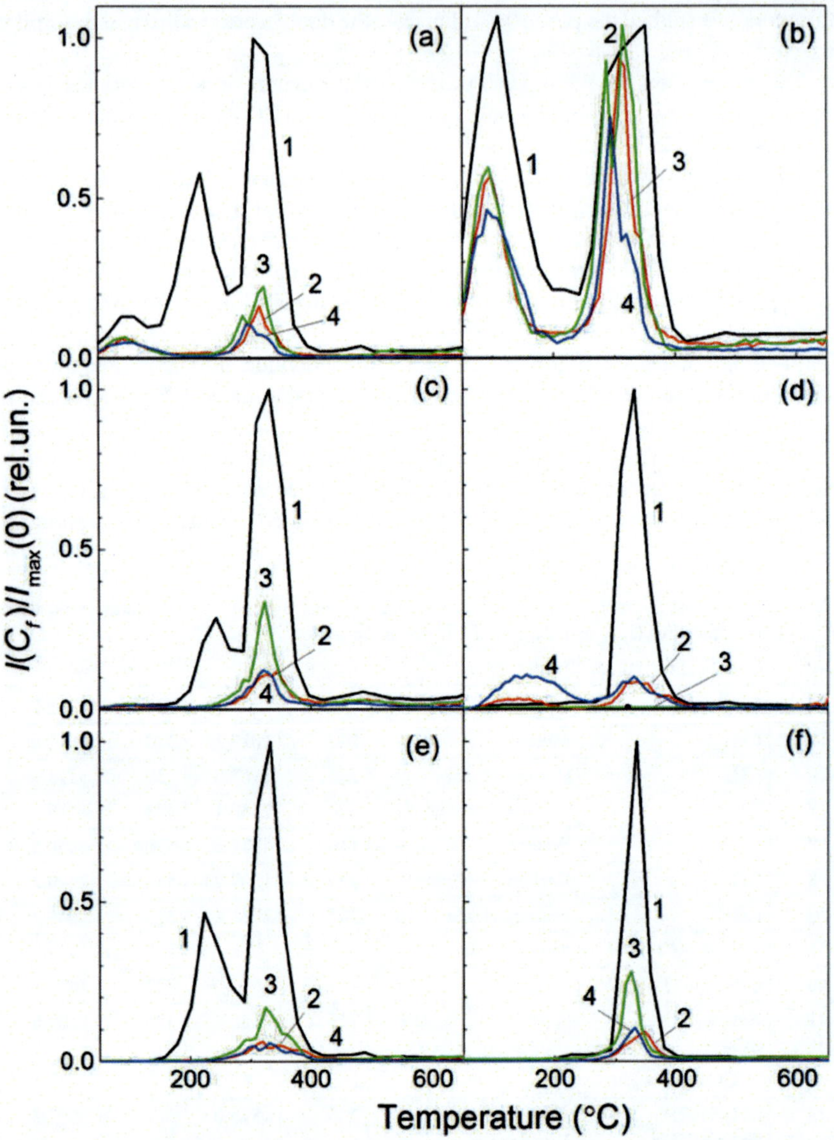

Fig. 7 Normalized thermal desorption curves $I(C_f,T)/I_{max}(0,T)$ related to the 17-th (**a**), 18-th (**b**), 28-th (**c**), 31-st (**d**), 43-rd (**e**), and 56-th (**f**) m/z-moieties for the unfilled epoxy resin (the black curves « 1») and its MLGNPs-nanocomposites of C_f= 1% (the red curves « 2»), 2% (the green curves « 3»), and 5% (the blue curves « 4»)

However, TDCs for water (see Fig. 7b) reveal two bounded state for water molecules inside the epoxy and its nanocomposites. It is remarkable that the water-related LT peak proved to be partially suppressed with increases the loading. Thus, loosely bounded water molecules can be considered as the relaxators contributing into the MDR-relaxation.

Also, different from the others, the behavior is demonstrated by the nitrogen-containing moiety of $m/z = 31$ (methylamine), which terminates polyamine's molecules of the PEPA-hardener. Its LT peak increases with increasing the loadings. This fact provides the direct evidence for breaking the cross-links in the epoxy.

Also, it's clear seen from Fig. 7 that different moieties show non-monotonous loading dependences of maximal value $I_{\max}$ of the intensity, namely: $I_{\max}$ decreases at $C_f \le 1\%$, then increases at $1\% < C_f \le 2\%$, and then shows a weak decreasing at $2\% < C_f \le 5\%$. It's remarkable that $I_{\max}(5\%) \approx I_{\max}(1\%)$. On the whole, the effect of lowering $I_{\max}$ is a result of binding the moieties to ASSs of MLGNPs.

From the other hand, non-monotonous behavior of $I_{\max}(C_f)$ means a similar non-monotonous behavior for outputs of the correspondent moieties. The amount or output $Q_n(C_f, T_1)$ of the n-th moiety emanated from a sample under heating it has been determined from the obvious expression

$$Q_n(C_f, T_1) = A_0 \int_{T_0}^{T_1} I_n(C_f, \tau)\, d\tau \tag{7}$$

where A_0 is a dimension parameter, $I_n(C_f, \tau)$ is the temperature dependence of the thermal desorption intensity of the n-th m/z-moiety, T_0 and T_1 are the temperature limits for integration.

Figure 8 shows loading dependences of normalized outputs $Q_n(C_f, T_1)/Q_n(0, T_1)$ evaluated at $T_1 = 600$ °C, where all the $Q_n(C_f, T_1)$ reach its saturated values. These dependences reveal non-monotonous increasing in thermal stability for MLG-nanocomposites (enhancements at $C_f \le 1\%$ and $2\% < C_f \le 5\%$).

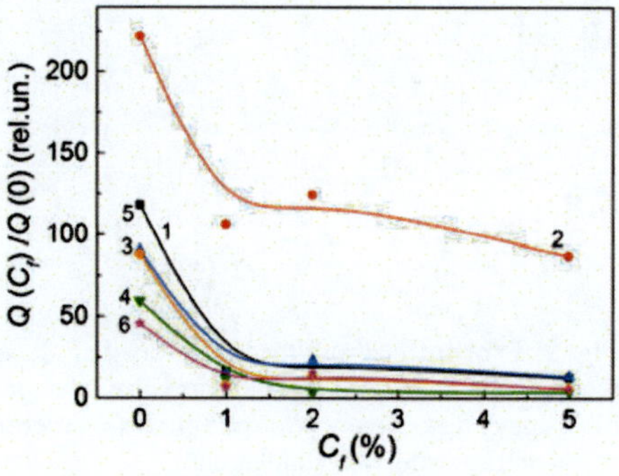

Fig. 8 Normalized loading dependences of normalized outputs $Q_n(C_f,T_1)/Q_n(0,T_1)$ evaluated at $T_1 = 600$ °C for the 17-th (curves "1"), 18-th ("2"), 28-th ("3"), 31-st ("4"), 43-rd ("5"), and 56-th ("6") m/z-moieties for the unfilled epoxy resin and its MLGNPs-nanocomposites

Thus, TDMS-data shows that embedding MLGNPs into the epoxy result in stabilizing its 3D-molecular network structure [73] due to binding of both free and loosely-coupled dipole moieties to ASSs and in its electric ordering in the vicinity of the nanoparticles.

3.3 Data Obtained by Ultrasound Probing and Static Mechanical Loading

3.3.1 Ultrasonic Probing

Both experimental and theoretical loading dependences of the normalized Young's module $E(C_f)/E(0)$ (a) and the normalized Poisson's ratio $\nu(C_f)/\nu(0)$ (b) for the MLG-filled epoxy nanocomposites are shown on Fig. 9 Experimental data show that the E decreases at $C_f \leq 1\%$ and then increases slowly with increasing the loading. The parameter ν and other modules show similar behavior. Decreasing dynamic elastic modules at low loadings can be ascribed to destruction of the amine cross-links and shortening of the epoxy macromolecular chains. When the loading increases, the graphene nanoplatelets form a supramolecular frame and it stabilizes 3D-molecular network of the epoxy matrix.

MD-calculations revealed that E increases at low loadings ($C_f \leq 0.5\%$) and decreases when C_f increases further, whereas ν increases gradually with increasing the loading. It should be pointed out that discrepancies between experimental and theoretical values of the elastic modules are likely to come from that the ultrasound wavelength are much higher than linear dimensions of these stress-anisotropy

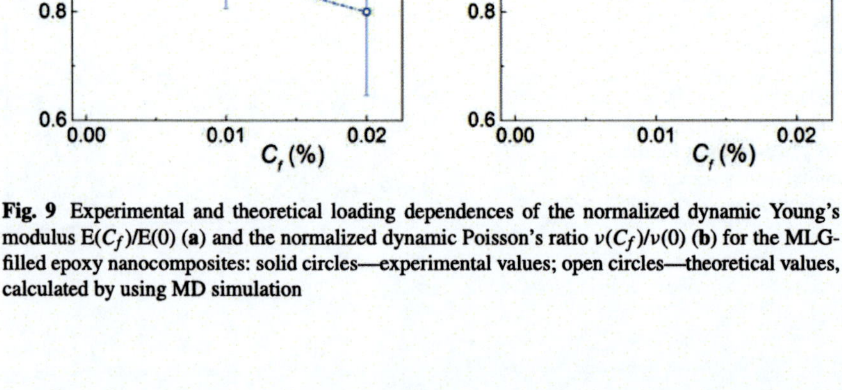

Fig. 9 Experimental and theoretical loading dependences of the normalized dynamic Young's modulus $E(C_f)/E(0)$ (**a**) and the normalized dynamic Poisson's ratio $\nu(C_f)/\nu(0)$ (**b**) for the MLG-filled epoxy nanocomposites: solid circles—experimental values; open circles—theoretical values, calculated by using MD simulation

domains. Also, both variability of geometric shape and non-uniformity in spatial distribution of the nanoplatelets over the sample have been ignored in the simulation.

3.3.2 Static Mechanical Loading

The "Shopper" testing machine has been used to compress the samples. The destruction strength (P_0) and the static Young's modules (E_0) have been determined from the strtess-strain curves as five-samples-averaged values. The static Poisson's ratio (ν_0) have been calculated by [74]:

$$(1 + \nu) = E_0(1 - 2\nu)/6\sigma_F, \tag{8}$$

where σ_F is the fluidity limit.

Experimental and theoretical loading dependences of the normalized static Young's modulus $E(C_f)/E(0)$ and the normalized static Poisson's ratio $\nu(C_f)/\nu(0)$ for the MLG-filled epoxy nanocomposites are shown on Fig. 10. In contrast to UP-data, the static values of both the Young's modulus and the Poisson's ratio increase slowly with increasing the loading. Such the discrepancy can be related with that residual mechanical stresses and micropores relax during compression of the samples, whereas ultrasonic waves are insensitive to small-scale non-uniformities and the latter remain unchanged under relatively low dynamic deformations induced by the waves.

From the other hand, one can see that there is a good agreement between measured and calculated values of $\nu_0(C_f)$ at $0\% < C_f < 2\%$. As to $E_0(C_f)$, the agreement takes place only at $0\% < C_f < 0.5\%$. The reasons for such a discrepancy is now not thoroughly understood.

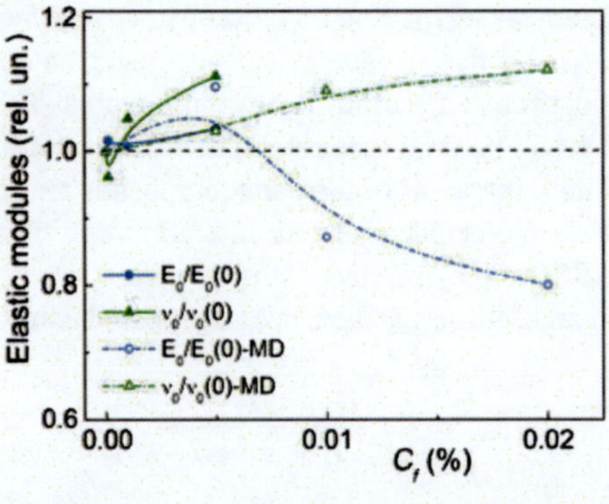

Fig. 10 Experimental and theoretical loading dependences of the normalized static Young's modulus $E_0(C_f)/E_0(0)$ (**a**) and the normalized static Poisson's ratio $\nu_0(C_f)/\nu_0(0)$ (**b**) for the MLG-filled epoxy nanocomposites: solid circles—experimental values; open circles—theoretical values, calculated by using MD simulation

3.4 Data Obtained by Positron Annihilation Life-Time Spectroscopy

It has been established in our PALS-study that there are three channels of positron annihilation take place in both the unfilled epoxy resin and its MLG-filled nanocomposites. The channels are characterized with intensities I_1, I_2, I_3 and the corresponding lifetimes τ_1, τ_2, and τ_3. The lifetimes correspond to the self-annihilation of the para-positronium atoms and the annihilation of the free positrons (τ_1), to the annihilation of positrons trapped by the structural defects (τ_2), and to the pick-off annihilation of the ortho-positronium atoms (τ_3). Due to the dominant interpretation of the lifetime-components in the literature [51], the ortho-positronium atoms diffuse into the so-called free volume areas (FVA) or holes, which are characterized by the lowered electron charge density (ρ_e) and then annihilate there. It should be noted that in amorphous glassy polymers (like epoxy resins) FVA can be located in either ordered or disordered regions [53, 75].

The resulted annihilation parameters (both measured and calculated) are summarized in Table 2.

It's known that τ_1 is inversely proportional to electron charge density in FVA (ρ_e), whereas τ_2 and τ_3 are inversely proportional to electron charge densities (ρ_+ and ρ_{Ps}, correspondently) on walls of the potential holes trapping positron and ortho-positronium [51, 76]. Thus, we can write

$$\tau_1 \sim 1/n_e, \quad \tau_2 \sim 1/\rho_+, \quad \tau_3 \sim 1/\rho_{Ps}, \tag{9}$$

Here, it should be noted that ρ_e can be regarded to be roughly equal to the average electron density along the epoxy chains.

It can be seen from Table 2, that loading dependences of τ_1 show decreasing within the annihilation volume of the nanocomposites at $C_f \leq 2\%$. Also, alteration of resin's chains with increasing loadings is accompanied by synchronous decrements in both $\tau_2(C_f)$ and $\tau_3(C_f)$ due to increasing local electron density in the vicinity of defects trapping positrons and ortho-positronium atoms. The similarity between $\tau_1(C_f)$, $\tau_2(C_f)$ and $\tau_3(C_f)$ shows that electron density variations occurred in both free and annihilation-bearing areas of the nanocomposites are governed by the same process, namely resin's molecular structure alteration due to interactions between polar moieties and ASSs of the graphene nanoplatelets.

The average positron lifetime τ_{av} have been calculated using the expression [77]:

$$\tau_{av} = \left(\frac{\tau_1 I_1 + \tau_2 I_2 + \tau_3 I_3}{I_1 + I_2 + I_3} \right). \tag{10}$$

Although, considerable uncertainties in the measurements are present, the tendency of decreasing of (τ_{av}) with increasing the loading. Such the behavior evidences on decreasing average electron density in the nanocomposite's bulk with increasing the loading.

Table 2 The positron lifetimes for *i*-th component (τ_i), the intensities of the *i*-th component (I_i), average positron lifetime (τ_{av}), free volume radius (R) and free volume fraction (f_v) for investigated samples

C, %	I_1, %	I_2, %	I_3, %	τ_1, ps	τ_2, ps	τ_3, ps	τ_{av}, ps	R, Å	f_v %
0	53.23 (0.51)[a]	31.96 (0.49)	14.82 (0.24)	199.6 (2.2)	527.0 (9.5)	1670 (12)	522.1 (3.7)	2.522 (0.026)	1.792 (0.047)
1	52.86 (0.63)	31.82 (0.63)	15.32 (0.25)	199.4 (2.5)	503 (10)	1627 (8,3)	514.6 (3.6)	2.476 (0.018)	1.754 (0.038)
2	49.66 (0.71)	34.52 (0.70)	15.82 (0.29)	190.7 (2.8)	473.5 (8.8)	1581 (10)	508.3 (3.7)	2.426 (0.023)	1.703 (0.045)
5	52.68 (0.72)	32.45 (0.71)	14.88 (0.29)	199.9 (2.8)	489 (10)	1597 (12)	501.6 (3.9)	2.443 (0.027)	1.635 (0.048)

[a]Standard deviations are given in round brackets

The relationship between the lifetime (τ_3) and the free volume size (R) was developed by Tao [78] and Eldrup et al. [79]. According to this model Ps is predicted to be localized in a spherical infinite potential barrier of radius R_0 with the electron layer in the region $R < r < R_0$. The free volume radius R can be evaluated using the expression:

$$\tau_3 = \frac{1}{2}\left[1 - \frac{R}{R_0} + \frac{1}{2\pi} sin\left(\frac{2\pi R}{R_0}\right)\right]^{-1} \tag{11}$$

The values τ_3 and R (hole radius) are expressed in ns and Å, respectively. R is equal to $R_0 + \Delta R$, where $\Delta R = 1.66$ Å.

The value of I_3 and the free volume radius (R) can be used for determination of the percentage value of the free volume fraction (f_v) according to the equation:

$$f_v = A \cdot I_3(\%) \cdot \left(\frac{4}{3}\pi R^3\right) \tag{12}$$

The value of the empirical parameter A for the polymers is predicted to be ~0.0018/Å^3 [80, 81]. The loading dependence $f_v(C_f)$ proved to be descending, that evidences tightening 3D-molecular network of the nanocomposites as compared to the neat epoxy.

4 Conclusions

1. Filling epoxy resin with MGNPs is accompanied with certain 3D-molecular structure alteration of the resin. From one hand, the alteration manifest itself results in non-monotonous increasing of both the dielectric permittivity and the dielectric loss factor. Both the effect reveal themselves in the minor dielectric relaxation process (MDR), which can satisfactorily be described by the well-known Cole-Cole relaxation model.
2. Thermal desorption data evidence on non-monotonous increasing in thermal stability for the MLG-nanocomposites (namely, enhancements at $C_f \leq 1\%$ and $2\% < C_f \leq 5\%$). By other words, MGNPs have a stabilizing impact on epoxy 3D-molecular network due to binding both free and loosely-coupled dipole moieties to ASSs and in its electric ordering in the vicinity of the nanoparticles. The ensemble of these moieties are seemed to include terminal (OH- and CH_3-) and the lateral (OH- and CH_3-) molecular groups of EMCs and water molecules loosely-coupled with EMCs.
3. Enhancement in static Young's modulus for the MLGNP-filled epoxy nanocomposites with increasing the loading provides another consequence of the stabilizing effect.

4. Alteration of 3D-molecular network of the epoxy, induced by embedded MLGNPs, accompanied with decreasing free-volume portion in the nanocomposites and increasing electric charge density along EMCs.

References

1. Balandin AA (2011) Thermal properties of graphene and nano-structured carbon materials. Nat Mater 10:569–581. https://doi.org/10.1038/nmat3064
2. Shahil KMF, Balandin AA (2012) Graphene-multilayer graphene nanocomposites as highly efficient thermal interface materials. Nano Lett 12(2):861–867. https://doi.org/10.1021/nl203906r
3. Shahil KMF, Balandin AA (2012) Thermal properties of graphene and multilayer graphene: applications in thermal interface materials. Solid State Commun 152(15):1331–1340. https://doi.org/10.1016/j.ssc.2012.04.034
4. Atif R, Shyha I, Inam F (2016) Modeling and experimentation of multi-layered nanostructured graphene-epoxy nanocomposites for enhanced thermal and mechanical properties. J Compos Mater 51(2):209–220. https://doi.org/10.1177/0021998316640060
5. Burger N, Laachachi A, Ferriol M, Lutz M, Toniazzo V, Ruch D (2016) Review of thermal conductivity in composites: mechanisms, parameters and theory. Prog Polym Sci 61:1–28. https://doi.org/10.1016/j.progpolymsci.2016.05.001
6. Zhu Z, Sun X, Xue H, Guo H, Fan X, Pan X, He J (2014) Graphene-carbonyl iron cross-linked composites with excellent electromagnetic wave absorption properties. J Mater Chem C 2:6582–6591. https://doi.org/10.1039/C4TC00757C
7. Kong L, Yin X, Yuan X, Zhang Y, Liu X, Cheng L, Zhang L (2014) Electromagnetic wave absorption properties of graphene modified with carbon nanotube/poly(dimethyl siloxane) composites. Carbon 73:185–193. https://doi.org/10.1016/j.carbon.2014.02.054
8. Kargar F, Barani Z, Balinskiy M, Magana AS, Lewis JS, Balandin AA (2018) Dual-functional graphene composites for electromagnetic shielding and thermal management fariborz. Adv Electron Mater 5(1):1800558. https://doi.org/10.1002/aelm.201800558
9. Azizi F, Jahangiri H (2016) Radar absorbing nanocomposites based multiLayered graphene platelets/epoxy. J Nanostruct 5(4):345–349. magiran.com/p 1499526
10. Saini P (2019) Historical review of advanced materials for electromagnetic interference (EMI) shielding: conjugated polymers, carbon nanotubes, graphene based compoisites. Indian J Pure Ap Phy 57:338–351. http://nopr.niscair.res.in/handle/123456789/47317
11. Atif R, Shyha I, Inam F (2016) Mechanical, thermal, and electrical properties of graphene-epoxy nanocomposites—A review. Polymers 8(8):281–318. https://doi.org/10.3390/polym8080281
12. Li J, Huang H, Zhou Y, Zhang C, Li Z (2017) Research progress of graphene-based microwave absorbing materials in the last decade. J Mater Res 32(7):1213–1230. https://doi.org/10.1557/jmr.2017.80
13. Seong M, Kim DS (2015) Effects of facile amine-functionalization on the physical properties of epoxy/graphene nanoplatelets nanocomposites. J Appl Polym Sci 132(28):42269. https://doi.org/10.1002/app.42269
14. Ribeiro H, Silva WM, Rodrigues M-TF, Neves JC, Paniago R, Fantini C, Calado HDR, Seara LM, Silva GG (2013) Glass transition improvement in epoxy-graphene composites. J Mater Sci 48(22):883–7892. https://doi.org/10.1007/s10853-013-7478-3
15. Zhao S, Chang H, Chen S, Cui J, Yan Y (2016) High-performance and multifunctional epoxy composites filled with epoxide-functionalized graphene. Eur Polym J 84:300–312. https://doi.org/10.1016/j.eurpolymj.2016.09.036

16. Olowojoba GB, Kopsidas S, Eslava S, Gutierrez ES, Kinloch AJ, Mattevi C, Rocha VG, Taylor AC (2017) A facile way to produce epoxy nanocomposites having excellent thermal conductivity with low contents of reduced graphene oxide. J Mater Sci 52(12):7323–7344. https://doi.org/10.1007/s10853-017-0969-x
17. Starkova O, Chandrasekaran S, Prado LASA, Tolle F, Mulhaupt R, Schulte K (2013) Hydrothermally resistant thermally reduced graphene oxide and multi-wall carbon nanotube based epoxy nanocomposites. Polym Degrad Stab 98(2):519–526. https://doi.org/10.1016/j.polymdegradstab.2012.12.005
18. Fu Y-X, He Z-X, Mo D-C, Lu S-S (2014) Thermal conductivity enhancement of epoxy adhesive using graphene sheets as additives. Int J Therm Sci 86:276–283. https://doi.org/10.1016/j.ijthermalsci.2014.07.011
19. Tang B, Hu G, Gao H, Ha L (2015) Application of graphene as filler to improve thermal transport property of epoxy resin for thermal interface materials. Int J Heat Mass Transf 85:420–429. https://doi.org/10.1016/j.ijheatmasstransfer.2015.01.141
20. Wu Sh, Ladani RB, Zhang J, Bafekrpour E, Ghorbani E, Mouritz AP, Kinloch AJ, Wang CH (2015) Aligning multilayer graphene flakes with an external electric field to improve multifunctional properties of epoxy nanocomposites. Carbon 94:607–618. https://doi.org/10.1016/j.carbon.2015.07.026
21. Wang Y, Yu J, Dai W, Song Y, Wang D, Zeng L, Jiang N (2014) Enhanced thermal and electrical properties of epoxy composites reinforced with graphene nanoplatelets. Polym Compos 36(3):556–565. https://doi.org/10.1002/pc.22972
22. Park W, Guo Y, Li X, Hu J, Liu L, Ruan X, Chen YP (2015) High-performance thermal interface material based on few-layer graphene composite. J Phys Chem C 119:26753–26759. https://doi.org/10.1021/acs.jpcc.5b08816
23. Shen X, Wang Z, Wu Y, Liu X, He Y, Kim J (2016) Multilayer graphene enables higher efficiency in improving thermal conductivities of graphene/epoxy composites. Nano Lett 16(6):3585–3593. https://doi.org/10.1021/acs.nanolett.6b00722
24. Gorelov BM, Gorb AM, Nadtochiy AB, Starokadomskiy DL, Kuryliuk VV, Sigareva NV, Shulga SV, Ogenko VM, Korotchenkov OO, Polovina OI (2019) Epoxy filled with bare and oxidized multi-layered graphene nanoplate-lets: a comparative study of filler loading impact on thermal properties. J Mater Sci 54(12):9247–9266. https://doi.org/10.1007/s10853-019-03523-7
25. Hu X, Qi R, Zhu J, Lu J, Luo Yu, Jin J, Jiang P (2014) Preparation and properties of dopamine reduced graphene oxide and its composites of epoxy. J Appl Polym Sci 131(2):39754. https://doi.org/10.1002/app.39754
26. Liu W, Koh KL, Lu J, Yang L, Phua S, Kong J, Chen Z, Lu X (2012) Simultaneous catalyzing and reinforcing effects of imidazole-functionalized graphene in anhydride-cured epoxies. J Mater Chem 22:18395–18402. https://doi.org/10.1039/C2JM32708B
27. May C (2018) Epoxy resins: chemistry and technology (second ed). CRC Press, p 65
28. Mohan P (2013) A Critical review: The modification, properties, and applications of epoxy resins. Polym Plast Technol Eng 52(2):107–125. https://doi.org/10.1080/03602559.2012.727057
29. Ding Y, Jia Y, Yang J, He H, Sun S, Shi T (2011) Correlation analysis of epoxy/amine resin cure, structure and chemorheological behavior in RTM processes. Polym Compos 32(4):648–656. https://doi.org/10.1002/pc.21060
30. Pham HQ, Marks MJ (2004) In Encyclopedia of Polymer Science and Technology. In: Kroschwitz JI (ed). Wiley, Hoboken, NJ 9:678 – 804
31. Monteserın C, Blanco M, Aranzabe E, Aranzabe A, Vilas JL (2017) Effects of graphene oxide and chemically reduced graphene oxide on the curing kinetics of epoxy amine composites. J Appl Polym Sci 134(19):44803. https://doi.org/10.1002/app.44803
32. Zaman I, Manshoor B, Khalid A, Meng Q, Araby S (2014) Interface modification of clay and graphene platelets reinforced epoxy nanocomposites: a comparative study. J Mater Sci 49(17):5856–5865. https://doi.org/10.1007/s10853-014-8296-y

33. Prolongo SG, Jiménez-Suárez A, Moriche R, Ureña A (2014) Graphene nanoplatelets thickness and lateral size influence on the morphology and behavior of epoxy composites. Eur Polym J 53:292–301. https://doi.org/10.1016/j.eurpolymj.2014.01.019
34. Ribeiro H, Silva WM, Rodrigues M-TF, Neves JC, Paniago R, Fantini C, Calado HDR, Seara LM, Silva GG (2013) Glass transition improvement in epoxy-graphene composites. J Mater Sci 48:7883–7892. https://doi.org/10.1007/s10853-013-7478-3
35. Gresil M, Wang Z, Poutrel Q-A, Soutis C (2017) Thermal diffusivity mapping of graphene based polymer nanocomposites. Sci Rep 7:5536. https://doi.org/10.1038/s41598-017-05866-0
36. Zaman I, Phan TT, Kuan H, Meng Q, La LTB, Luong L, Youssf O, Ma J (2011) Epoxy/graphene platelets nanocomposites with two levels of interface strength. Polymer 52(7):1603–1611. https://doi.org/10.1016/j.polymer.2011.02.003
37. Ma J, Meng Q, Michelmore A, Kawashima N, Izzuddin Z, Bengtsson C, Kuan H (2013) Covalently bonded interfaces for polymer/graphene composites. J Mater Chem A 1(13):4255–4264. https://doi.org/10.1039/C3TA01277H
38. Edwards RS, Coleman KS (2013) Graphene synthesis: relationship to applications. Nanoscale 5:38–51. https://doi.org/10.1039/C2NR32629A
39. Mao S, Pu H, Chen J (2012) Graphene oxide and its reduction: modeling and experimental progress. RSC Adv 2:2643–2662. https://doi.org/10.1039/C2RA00663D
40. Chang DW, Choi H-J, Jeon I-Y, Baek J-B (2013) Edge-selectively functionalized graphene nanoplatelets. Chem Rec 13(2):224–238. https://doi.org/10.1002/tcr.201200032
41. Xia ZY, Pezzini S, Treossi E, Giambastiani G, Corticelli F, Morandi V, Zanelli A, Bellani V, Palermo V (2013) The exfoliation of graphene in liquids by electrochemical, chemical, and sonication-assisted techniques: a nanoscale study. Adv Funct Mater 23(37):4684–4693. https://doi.org/10.1002/adfm.201370188
42. Bratychak M, Chopyk N (2013) Synthesis and application of oligoesteric mixtures of ED-20 epoxy resin peroxide derivative during cross-linking. Chem Chem Technol 7(4):435–440. https://doi.org/10.23939/chcht07.04.435
43. Bratychak M, Ivashkiv O, Astakhova O (2014) Chemical modification of ED-20 epoxy oligomer by 1,4-butanediol. Dopovidi Nats Acad Nauk Ukrainy 8:97–102. https://doi.org/10.15407/dopovidi2014.08.097
44. Bouvet G, Nguyen Dang S, Cohendoz X, Feaugas S, Mallarino S Touzain (2016) Impact of polar groups concentration and free volume on water sorption in model epoxy free films and coatings. Prog Org Coat 96:32–41. https://doi.org/10.1016/j.porgcoat.2015.12.011
45. Kremer F, Schonhals A, Luck W (2003) Broadband dielectric spectroscopy. Springer-Verlag Berlin Heidelberg. https://doi.org/10.1007/978-3-642-56120-7
46. Gorelov BM, Gorb AM, Polovina OI, Wacke S, Czapla Z, Kosrtzewa M, Ingram A (2018) Filler's impact on structure and physical properties in polyester resin–oxide nanocomposites. Adsorpt Sci Technol 36(1–2):549–570. https://doi.org/10.1177/0263617417706797
47. Pokrovsky VA (2000) Temperature-programmed desorption mass spectrometry. J Therm Anal Calorim 62(2):407–415. https://doi.org/10.1023/A:1010177813557
48. Gorelov B, Gorb A, Korotchenkov O, Nadtochiy A, Polovina O, Sigareva N (2015) Impact of nanosized oxide fillers on elastic properties and thermal destruction of a styrene cross-linked polyester resin. J Appl Polym Sci 132(22):42010/1-10. https://doi.org/10.1002/app.42010
49. Gorelov BM, Gorb AM, Czapla Z, Wacke S, Nadtochiy AB, Kuryliuk VV, Kostrzewa M, Ingram A, Starokadomskiy DL, Korotchenkov OO, Polovina OI, Sigareva NV (2019) Impact of multi-layered graphene nanoplatelets on 3D-molecular network of an epoxy resin. In: Proceedings of the 2019 IEEE 9th International Conference on Nanomaterials: Applications and Properties (NAP-2019), pp 01SSAN07-1–01SSAN07-5
50. Kansy J (1996) Microcomputer program for analysis of positron annihilation lifetime spectra. Nucl Instrum Methods A374(2):235–244. https://doi.org/10.1016/0168-9002(96)00075-7
51. Schrader DM, Jean YC (1988) Positron and positron chemistry. Study in physical and theoretical chemistry. Elsevier, Amsterdam-Oxford-New York-Tokyo, p 120
52. Jean YC, Van Horn JD, Wei-Song H (2013) Perspective of positron annihilation spectroscopy in polymers. Macromolecules 46(18):7133–7145. https://doi.org/10.1021/ma401309x

53. Kao K (2004) Dielectric phenomena in solids, chapt 2, 1st ed. Academic Press, p 579
54. Putz KW, Palmeri MJ, Cohn RB, Andrews R, Brinson LC (2008) Effect of cross-link density on interphase creation in polymer nanocomposites. Macromolecules 41(18):6752–6756. https://doi.org/10.1021/ma800830p
55. Oldham KB (2008) A Guoy-Chapman-Stern model of a double layer at a (metal)/(ionic liquid) interface. J Electroanal Chem 613(2):131–138. https://doi.org/10.1016/j.jelechem.2007.10.017
56. Zhang H, Hassanali AA, Shin Y, Knight C, Singer SJ (2011) The water–amorphous silica interface: analysis of the Stern layer and surface conduction. J Chem Phys 134(2):024705. https://doi.org/10.1063/1.3510536
57. Roy M, Nelson JK, MacCrone RK, Schadler LS, Reed CW, Keefe R (2005) Polymer nanocomposite dielectrics-the role of the interface. IEEE Trans DEI 12(4):629–643. https://doi.org/10.1109/TDEI.2005.1511089
58. Mikolajczak G, Cavaille JY, Johari GP (1987) Dynamic mechanical behaviour and its dependence on preparation method of structural epoxide resin. Polymer 28(12):2023–2031. https://doi.org/10.1016/0032-3861(87)90036-X
59. Ochi M, Okazaki M, Shimbo M (1982) Mechanical relaxation mechanism of epoxide resins cured with aliphatic diamines. J Polym Sci Part B Polym Phys 20(4):689–699. https://doi.org/10.1002/pol.1982.180200411
60. Ochi M, Iesako H, Shimbo M (1986) Relaxation mechanism of epoxide resin cured with acid anhydrides. III. Effect of alkyl side chains on mechanical and dielectric β relaxations. J Polym Sci Part B Polym Phys 24(6):1271–1282. https://doi.org/10.1002/polb.1986.090240607
61. Ochi M, Shimbo M, Saga M, Takashima N (1986) Mechanical and dielectric relaxations of epoxide resins containing spiro-ring structure. J Polym Sci Part B Polym Phys 24(10):2185–2195. https://doi.org/10.1002/polb.1986.090241003
62. Ochi M, Yoshizumi M, Shimbo M (1987) Mechanical and dielectric relaxations of epoxide resins containing the spiro-ring structure. II. Effect of the introduction of methoxy branches on low-temperature relaxations of epoxide resins. J Polym Sci Part B Polym Phys 25(9):1817–1827. https://doi.org/10.1002/polb.1987.090250903
63. Mangion MBM, Johari GP (1990) Relaxations of thermosets. III. Sub-Tg dielectric relaxations of bisphenol-A–based epoxide cured with different cross-linking agents. J Polym Sci Part B Polym Phys 28(1):71–83. https://doi.org/10.1002/polb.1990.090280106
64. Hassan M, Tucker S, Abukmail A, Wiggins J, Mauritz K (2016) Polymer chain dynamics in epoxy based composites as investigated by broadband dielectric spectroscopy. Arabian J Chem 9(2):305–315. https://doi.org/10.1016/j.arabjc.2015.07.016
65. Schatzki TF (1962) Statistical computation of distribution functions of dimensions of macromolecules. J Poly Sci 57(165):337–356. https://doi.org/10.1002/pol.1962.1205716526
66. Sahagun C, Morgan S (2012) Thermal control of nanostructure and molecular network development in epoxy-amine thermosets. ACS Appl Mater Interfaces 4(2):564–572. https://doi.org/10.1021/am201515y
67. Sahagun CM, Knauer KM, Morgan SE (2012) Molecular network development and evolution of nanoscale morphology in an epoxy-amine thermoset polymer. J Appl Polym Sci 126(4):1394–1405. https://doi.org/10.1002/app.36763
68. Haba D, Kaufmann J, Brunner AJ, Resch K, Teichert C (2014) Observation of elastic modulus inhomogeneities in thermosetting epoxies using AFM and discerning facts and artifacts. Polymer 55(16):4032–4040. https://doi.org/10.1016/j.polymer.2014.06.030
69. Morsch S, LiuY Lyon S, Gibbon S (2016) Insights into epoxy network nanostructural heterogeneity using AFM-IR. ACS Appl Mater Interfaces 8(1):959–966. https://doi.org/10.1021/acsami.5b10767
70. Morsch S, LiuY Greensmith P, Lyon SB, Gibbon SR (2017) Molecularly controlled epoxy network nanostructures. Polymer 108:146–153. https://doi.org/10.1016/j.polymer.2016.11.050
71. Hill NE (1969) Theoretical treatment of permittivity and loss. In Dielectric properties and molecular behavior. Van Nostrand series in physical chemistry Series in Physical Chemistry, Van Nostrand Reinhold, p 480

72. Havriliak S, Negami S (1967) A complex plane representation of dielectric and mechanical relaxation processes in some polymers. Polymer 8:161–210. https://doi.org/10.1016/0032-3861(67)90021-3
73. Grassie N, Scott G (1988) Polymer degradation and stabilization. Cambridge University Press, New York, p 222
74. Kozlov GV, Sanditov DS(1994) Anharmonic effects and physical-mechanical properties of polymers. Novosibirsk, Nauka, p 261 (Rus)
75. Engbrecht J, Green D, Hillmyer MA, Olson D, Todd EM (2013) Positron lifetime spectroscopy in ordered nanoporous polymers. J Polym Sci Part B Polym Phys 51(15). DOI: https://doi.org/1157–1161. https://doi.org/10.1002/polb.23314
76. Brandt W, Berko S, Walker WW (1960) Positronium decay in molecular substances. Phys Rev 120:1289–1295. https://doi.org/10.1103/PhysRev.120.1289
77. Seeger A, Banhart F, Bauer W (1989) Positron Annihilation. In: Dorikens-Vanpraet L, Dorikens M, Segers D (eds). World Scientific Publishers, Singapore, p 275
78. Tao SJ (1972) Positronium annihilation in molecular substances. J Chem Phys 56(11):5499–5510. https://doi.org/10.1063/1.1677067
79. Eldrup M, Lighbody D, Sherwood JN (1981) The temperature dependence of positron lifetimes in solid pivalic acid. Chem Phys 63(1–2):51–58. https://doi.org/10.1016/0301-0104(81)80307-2
80. Li Y, Zhang R, Jean YC (2003) Free-volume data in polymeric materials. World Scientific, Singapore, p 373. https://www.worldscientific.com/doi/pdf/10.1142/9789812775610_fmatter
81. Wang YY, Nakanishi H, Jean YC, Sandreczki TC (1990) Positron annihilation in amine-cured epoxy polymers-pressure dependence. J Polym Sci Part B Polym Phys 28(9):1431–1441. https://doi.org/10.1002/polb.1990.090280902

Structure and Properties of the POSS-Containing Nanocomposites Based on Polyurethane Matrix

L. V. Karabanova, L. A. Honcharova, and V. I. Shtompel

1 Introduction

Nanocomposites are a unique class of materials where a little quantity of nanofillers can significantly improve their properties [1–5]. Polyurethanes (PUs) are often used as polymer matrices for nanocomposites preparation due to their wide variety of physical and chemical properties [6–9]. Polyurethane materials are extensively used in technical and biomedical applications, whether in blood-contacting application and organ reconstruction [10], or in skin patches, coatings, and catheter application [11]. PU is seen as an attractive material with complicated structure (randomized hard and soft nanodomains) [12]; therefore, PU should highlight changes in its properties at the inclusion of functionalized nanofillers.

The incorporation of polyhedral oligomeric silsesquioxane (POSS) macromer into the polyurethane matrix results in improving its various properties [13–19]. The POSS nanoparticles have attracted considerable attention for creation of the nanocomposites because the organic substituents in the outer layer of POSS molecules make them compatible with polymers, biological systems, or surfaces [20, 21]. The polyhedral oligomeric silsesquioxanes are the silicon-containing nanoparticles with size from 1 to 3 nm, with cubic structure, and they could contain reactive groups, which allowed POSS to be imbedded into polymer systems [14–19]. Reactive groups in the outer layer of POSS molecules could be the hydroxyl groups,

L. V. Karabanova (✉) · L. A. Honcharova · V. I. Shtompel
Institute of Macromolecular Chemistry, National Academy of Sciences of Ukraine of, Kyiv, Ukraine
e-mail: lyudmyla_karaban@ukr.net

L. A. Honcharova
e-mail: glove@meta.ua

V. I. Shtompel
e-mail: vishtomp@bigmir.net

O. Fesenko and L. Yatsenko (eds.), *Nanomaterials and Nanocomposites, Nanostructure Surfaces, and Their Applications*, Springer Proceedings in Physics 246, https://doi.org/10.1007/978-3-030-51905-6_34

amine groups, carboxylic acid groups, acrylic groups, epoxy groups [22–24]. Due to variety in reactive groups in the outer layer of POSS, it becomes possible to create the nanocomposites and hybrids using such POSS for different applications. These could be the multifunction materials from polymer systems to ceramics.

Depending on how much reactive functional groups are there in the POSS, the POSS particle could be included as part of main polymer chain [25], could create cross-linked system [26], or could be a cage in backbone [25, 27]. Compared with traditional polymers, materials contained POSS have significant advantages: thermal conductivity, increased hydrophobicity, chemical resistance, mechanical strength, hardness, increased temperature of thermal decomposition. The incorporation of polyhedral oligomeric silsesquioxane (POSS) macromer into the polyurethane matrix results in the permeability of gas transport [10], increasing of the conduction and the permittivity [11]. The addition of POSS into the PU's polymers leads to enhancement of their thermal stability [12, 13], also to improving the mechanical properties [14, 15] due to the reinforcement effect of the nanoparticles. The increasing of resistance of the nanocomposites against oxidation was also detected by authors [14, 16] due to incorporation of POSS to the systems. These favorable effects are evident mainly when the functionalized POSS particles are integrated into polymer chains via a chemical reaction.

In our study, 1, 2-propanediolisobutyl-POSS (POSS) was used as a functionalized nanofiller for PU matrix. The presence of two reactive hydroxyl groups in peripheral substituent of POSS allowed it to react with diisocyanates, so resulting in the embedding of the POSS particles into the main polymer chain of the nanocomposites [20–22, 25].

The objective of the present study was the creation of the nanocomposites based on polyurethane matrix and 1,2-propanediolisobutyl polyhedral oligomeric silsesquioxane (POSS), and investigation of the effects induced by the POSS content on the structure, morphology, thermal, and dynamic mechanical properties of the nanocomposites.

2 Materials and Testing Methods

2.1 Materials

1,2-propanediolisobutyl polyhedral oligomeric silsesquioxane (POSS) was purchased from Sigma-Aldrich Inc. Trimethylolpropane (TMP) and toluene 2,4-diisocyanate (TDI) were obtained from Merck. The reagents were of high purity and used as received. The oligoetherglycol with $M\text{w} = 5.000$ g/mol^{-1} (Laprol 5003) was obtained from commercial sources and dried with liquid nitrogen under vacuum at 75–80 °C for 5 h before use. Ethyl acetate was used as a solvent and its purification was done by distillation in vacuum.

The molecular structures of neat components constituents of the nanocomposites are presented below.

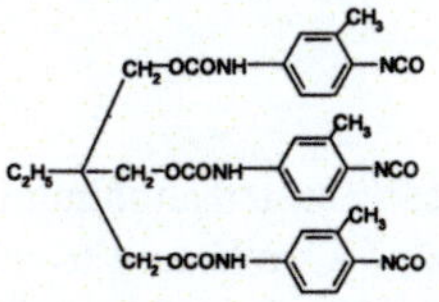

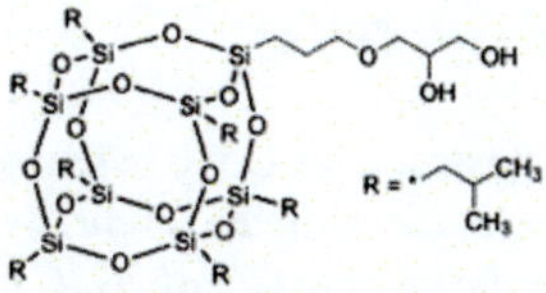

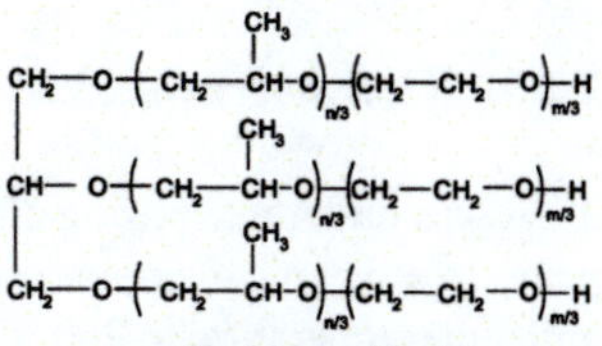

n=69÷81, m=10÷12

Adduct TMP-TDI POSS Laprol 5003

PU network used as polymer matrix was synthesized by two-step polymerization. In the first stage, the adduct of trimethylolpropane with toluene 2,4-diisocyanate (the adduct TMP-TDI) was obtained as described before [28]. The second step was the synthesis of three-dimensional polyurethane. Polyurethane network was obtained from a mixture of Laprol 5003 and adduct TMP/TDI (ratio 1:2 g-eq.) at 80 °C in nitrogen atmosphere. POSS (1–10 wt%) was added as a functionalized nanofiller at the second stage of PU network synthesis. The stoichiometric ratio NCO/OH = 1.5/1 was in all cases. The mixture was poured out on Petri dishes and cured at 70 °C for 2 h and post-cured at 80 °C for 24 h. The obtained nanocomposite films with ~1 mm thickness were post-cured for 2 h at 100 °C and then were held at 80 °C for 36 h in vacuum 10^{-5} Pa.

The nanocomposites were named PUX, where X denotes the content of POSS in the samples. For example, PU1 indicates that the POSS content in the nanocomposite is 1% by weight.

Neat PU also was synthesized for investigation and comparison of the results.

2.2 *Testing Methods*

2.2.1 Fourier Transform Infrared Spectroscopy (FTIR)

FTIR spectra of the neat POSS, neat adduct TMP-TDI and model adduct + POSS were recorded using a TENSOR 37 FT-IR spectrometer (BRUKER, USA, Germany) in the spectral range of 4000–400 cm^{-1} (with resolution of 0.8 cm^{-1}). The samples (thin films) of PU, PU1, and PU10 were presented in the ATR module.

2.2.2 Differential Scanning Calorimetry (DSC)

DSC was undertaken using a TA Instruments Q1000, on samples (cut thin film samples) of 10–15 mg in hermetically sealed aluminum pans. Experiments were conducted at the heating rate of 20 °C min^{-1} in the temperature range –90–200 °C under nitrogen. The glass transition temperature (T_g) was taken as the midpoint of the curve at the change in heat capacity.

2.2.3 Thermogravimetric Analysis (TGA)

TGA was undertaken using a TA Instruments Q500 in hermetically sealed aluminum pans. In a dry nitrogen atmosphere, the samples (about 5–6 mg) were heated from ambient temperature to 700 °C at the heating rate of 20 °C min^{-1}. The initial thermal degradation temperature was taken as the onset temperature at which a mass loss of 5 wt% occurs.

2.2.4 Dynamic Mechanical Analyses (DMA)

Dynamic mechanical measurements were carried out using a Dynamic Mechanical Thermal Analyzer Type DMA Q800 from TA Instruments over the temperature range from −100 to +220 °C and at fixed frequency 10 Hz with a heating rate of 3 °C/min. The experiments were performed under tension mode on rectangular specimens (35 mm × 5 mm × 1 mm). The samples were subsequently subjected to the following thermal cycle during DMA measurements: a first run from 20 °C up to 100 °C and then second run from –100 °C up to +220 °C. The second run was used for analysis of the results.

2.2.5 Wide-Angle X-Ray Scattering (WAXS)

The peculiarities of the amorphous structure of the polyurethane (PU) and the layered structure of 2-propanediolisobutyl polyhedral oligomeric silsesquioxane (POSS) were investigated by the wide-angle X-ray scattering (WAXS) method using an X-ray diffractometer DRON-4-07, whose optical scheme was performed using the Debye–Scherrer method (the "passing through" the sample of the primary X-ray beam).

2.2.6 Small-Angle X-Ray Scattering (SAXS)

The microphase structure of the native PU and the nanocomposites contained from 1 to 10 wt% of POSS was investigated by the method of small-angle X-ray scattering (SAXS) using a small-angle X-ray camera KRM-1, the slit collimation (primary beam) of which was performed by the Kratky method [29]. The geometrical parameters of the camera's collimation system satisfied the conditions of the infinite height of the collimation slit. The experimental values of the scattering intensity were normalized using the size of the sample's volume and the factor of weakening of the primary beam. The recalculation of the normalized values of the scattering intensity to the punctual collimation was performed by the Schmidt method [30]. All WAXS/SAXS measurements were performed using CuK_{α} radiation source (λ = 0,154 nm), monochromated Ni-filter, at a temperature of 20 ± 2 °C.

2.2.7 Scanning Electron Microscopy (SEM)

Scanning electron microscopy (SEM) was performed on a JEOL JSM 6060 LA (Tokyo, Japan) at the accelerating voltage of 30 kV and using a detector of secondary electrons. The samples were cut into strips, before being submerged in liquid nitrogen for 5 min and fractured as quickly as possible. Then the samples were warmed to room temperature and fixed to an SEM stub. The fracture surface of samples was coated with gold in vacuum to prevent accumulation of static charge and to increase the resolution. All measurements were done at 20 °C and at magnification of × 15000.

3 Results and Discussion

3.1 FTIR Analysis of Incorporation of 1,2-Propanediolisobutyl-POSS into PU Matrix

A model system adduct + POSS was created by mixing of the adduct TDI-TMP and POSS in the molar ratio NCO/OH = 1.5/1 for the FTIR study of the incorporation of POSS into PU matrix. A similar ratio was used for the synthesis of neat PU and POSS-containing nanocomposites. A given amount of POSS (10 wt%) was dissolved in ethyl acetate and added to the adduct. The composition was carefully mixed and coated on glass NaCl. The mixture was cured (and post-cured) at the 80 °C within 3 days.

In Fig. 1 the FTIR spectra of neat POSS, neat adduct, and model adduct + POSS before curing and after curing are shown.

In the FTIR spectrum of POSS the strong absorption band at 1108 cm^{-1} is observed (Fig. 1, curve 1) which could be attributed to the asymmetric stretching mode of ν_{as}(Si–O–Si) in caged structure [31–33] and it is always accompanied with the corresponding symmetric stretching counterpart mode ν_s(Si–O–Si) in the spectral region from 400 to 480 cm^{-1} (at 481 cm^{-1}) [31, 32, 34, 35] and corresponding bending vibrations δ_s(O–Si–O) at 565 cm^{-1} [34, 35]. The band at 742 cm^{-1} is attributed to the stretching vibration in the Si–C(H_2) isobutyl groups of POSS [34, 36] and accompanied by CH_2 (2871 and 2926 cm^{-1}) and CH_3 (2908 and 2955 cm^{-1}) stretching [37] and corresponding deformational modes (1465, 1402, 1384, 1230, and 839 cm^{-1}) [32, 34, 38]. The broadband at 3420 cm^{-1} is attributed to the stretching mode of ν(OH) [39].

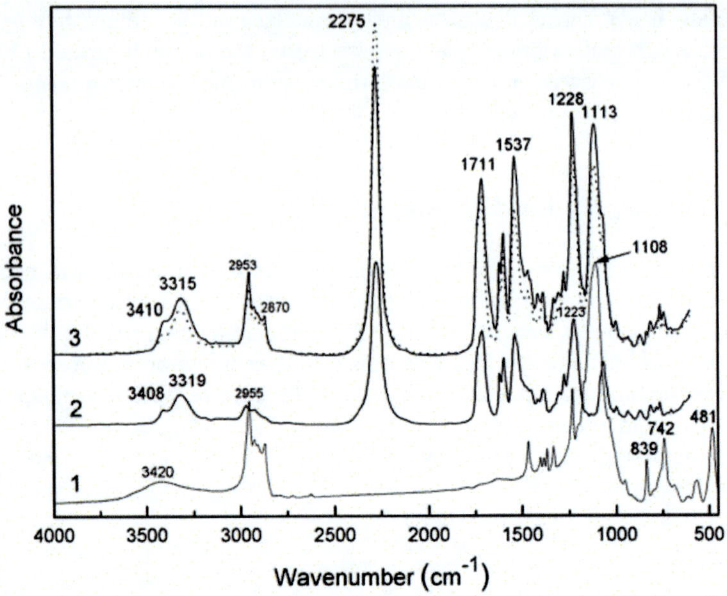

Fig. 1 FTIR spectra of 1—neat POSS; 2—neat adduct TDI-TMP; 3—model system adduct + POSS before (dot) and after curing (straight line)

The characteristics of absorption bands of the neat adduct (urethane units peaks) (Fig. 1, curve 2) are at 3408 (free N–H stretching), 3319 (hydrogen-bonded N–H stretching), 1730 (free C–O stretching), 1711 (H–bonded C–O stretching), 1537 (N–H bending), and 1233 cm^{-1} (N–CO–O asymmetric stretching) [38, 40]. The broadbands in the range from 2856 to 2951 cm^{-1} are assigned to the saturated C–H symmetric and asymmetric stretching vibrations of methylene groups [38].

In Fig. 1. (curve 3), the spectrum of model system adduct + POSS before (dot) and after curing (straight line) are presented. We could observe that the band of isocyanate group (–NCO stretching, 2275 cm^{-1}) of the model adduct + POSS decreased with the curing process. Conversely, urethane units bands increased and redistributed. So, we can conclude that the additional urethane linkages are formed in the model adduct + POSS that means the hydroxyl groups of POSS are capable to react with diisocyanates of adduct TDI-TMP in our system.

The proposed structure of cured model adduct of TDI-TMP + POSS is shown in Fig. 2.

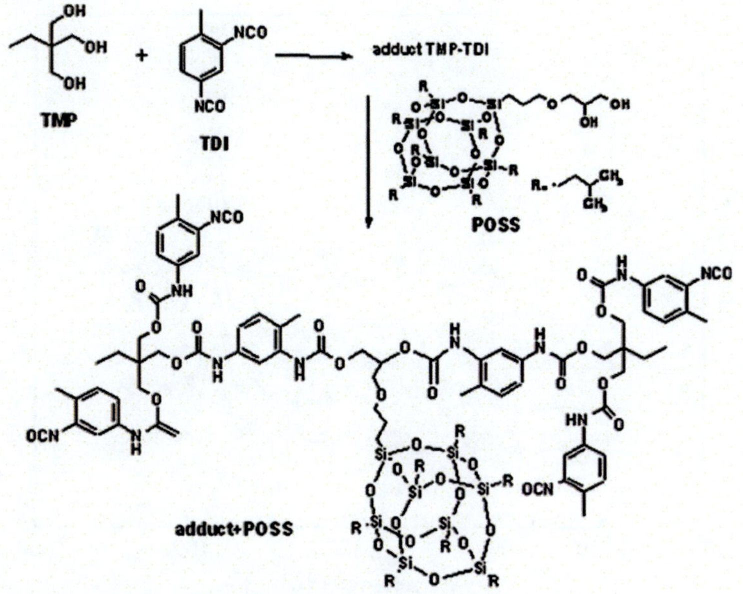

Fig. 2 The structure of cured model adduct + POSS

3.2 *Chemical Structure of the POSS-Containing Nanocomposites by FTIR-ATR Analysis*

In Fig. 3, the FTIR-ATR spectra of the nanocomposite films with different POSS content after curing are presented. FTIR spectrum of neat POSS is also presented for comparison in Fig. 3. The isocyanate band at 2275 cm^{-1} corresponding to the—NCO stretching vibrations has completely disappeared when the curing process completed in the neat PU and also in the nanocomposites (Fig. 3, curves 2–6).

For the POSS-containing PU nanocomposites not only the characteristic bands of urethane units' peaks, but also the absorption bands at 1108 and 743 cm^{-1} for Si–O–Si and Si–C stretching vibrations and corresponding bending modes at 839 and 1465 (1332) cm^{-1}, respectively, were observed. Furthermore, it can be seen that the characteristic silsesquioxane peaks increase their intensity with the increasing POSS content. Such results show that the POSS has been successfully incorporated into PU network. It is the confirmation that the POSS cube structure is chemically incorporated into the PU matrix rather than there being a physical mixture [41, 42].

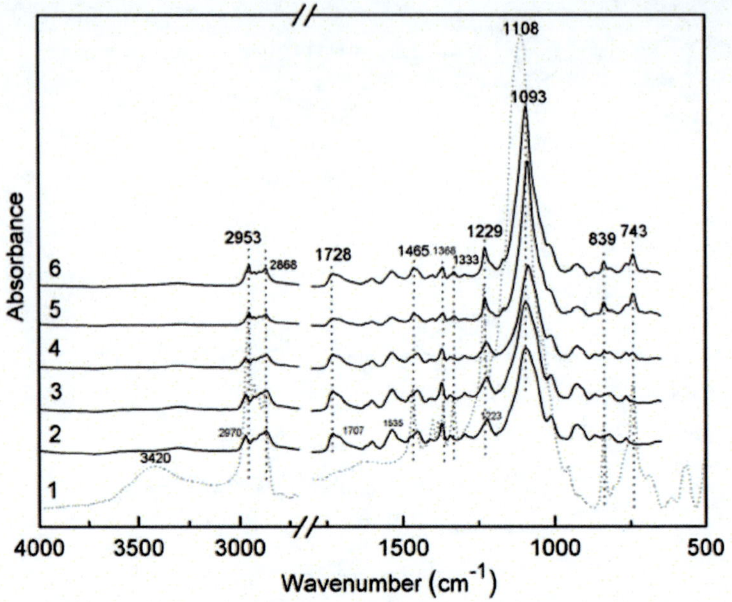

Fig. 3 FTIR spectra of 1—POSS and FTIR-ATR spectra of 2—neat PU, 3—PU1, 4—PU3, 5—PU5, and 6—PU10

3.3 *The Structure of Neat PU, POSS, and Nanocomposites by WAXS and SAXS*

From the analysis of WAXS curves (Fig. 4), it was found that the native PU demonstrates the presence of one asymmetric diffraction peak of a diffuse type (amorphous halo), indicating that this polymer has an amorphous structure (Fig. 4, curve 1). Based

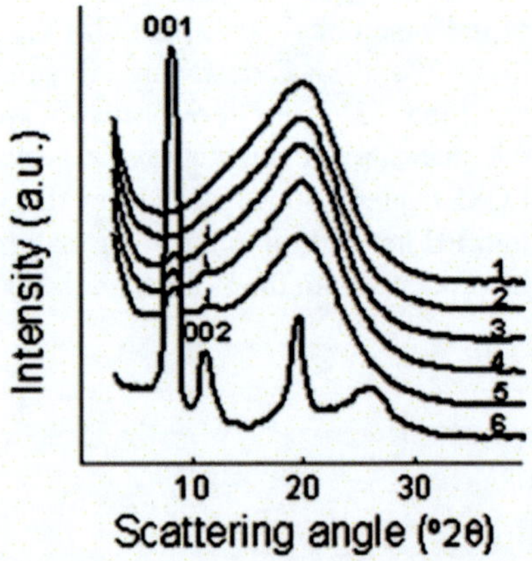

Fig. 4 WAXS profiles for native PU (1), POSS (6), and for the PU's nanocomposites with the POSS content: 1 (2), 3 (3), 5 (4), and 10 (5) wt%

on the angular position ($2\theta_m$) of the apex of the amorphous halo ($2\theta_m \approx$ 19,6°), the period (d) of the near ordering of molecular chain fragments between the cross-links in the volume of PU, according to the Bragg equation [43]:

$$d = \lambda(2\sin q_m)^{-1},$$

is equal to 0.45 nm.

Investigation of the native POSS by WAXS have shown (Fig. 4, curve 6) the existence of two discrete diffraction maxima which are significantly different in intensity, with the diffraction maximum of significant intensity having Miller's indices (001) and low intensity having Miller's indices (002). These two diffraction maxima indicate the existence of a two-level layered structure in the volume of the native POSS sample. The formation of such a structure of POSS is due to the bicyclic structure of its molecules, and the grafted aliphatic chain with two hydroxyl groups at the end. The Bragg's distance between the centers of molecular layers of the first level (d_{001}), according to the angular position of the diffraction maximum of greater intensity ($2\theta_{001} = 8.2°$), is 1.07 nm, and the distance between the centers of the layers (d_{002}), according to the less intensity diffraction maximum position ($2\theta_{002} = 11.2°$) is 0.79 nm. In addition to the layered structure, the POSS sample has also a crystalline structure. This is indicated by the presence of a singlet diffraction maximum at the curve of POSS (Fig. 4, curve 6) at $2\theta_m = 19.4°$ and the multiplet maximum at $2\theta_m$ ~ 26.8 °.

Comparing the WAXS X-ray diffractograms of the native PU and the nanocomposites containing POSS, it was found that the apex of the amorphous halo of the polyurethane matrix at 1 wt% of POSS insignificantly rises above the part of the amorphous halo below 17.2° (Fig. 4, curves 1, 2). With an increase of POSS content in the volume of PU up to 3 wt%, the apex of the amorphous halo becomes even more (than at 1 wt% POSS) raised above the part of the amorphous halo (Fig. 4, curves 2, 3). Also at this concentration of POSS the low-intensity discrete maximum (001) (at $2\theta_m = 8.2°$) and barely noticeable the maximum (002), reflecting the two-level layered structure of the nanofiller, could be observed.

With a further increase of POSS content up to 5 and 10 wt% in the volume of PU, unlike PU samples containing 1 and 3 wt% of POSS, the peak of the amorphous halo becomes less diffuse, that is, more sharpened (Fig. 4, curves 1–3 and 4, 5). At the same time, the intensity of the diffraction peaks (001) and (002), which characterize the layered structure of POSS, gradually increases.

The detected changes in the apex of the amorphous halo of the PU matrix at the introduction of a different amount of POSS into the reaction mixture during the PU synthesis may indicate that at least one terminal hydroxyl group of the aliphatic chain of POSS may participate in the urethane formation reaction, and therefore the POSS may be implanted into the chain between cross-links of PU. Consequently, the detected changes in the apex of the amorphous halo of PU matrix in the POSS-containing nanocomposites with different POSS content can be explained precisely by the grafting of a certain number of POSS molecules to the molecular chains between cross-links of PU.

Fig. 5 SAXS profiles for native PU (1), POSS (6), and for the PU's nanocomposites with the POSS content: 1 (2), 3 (3), 5 (4), and 10 (5) wt%

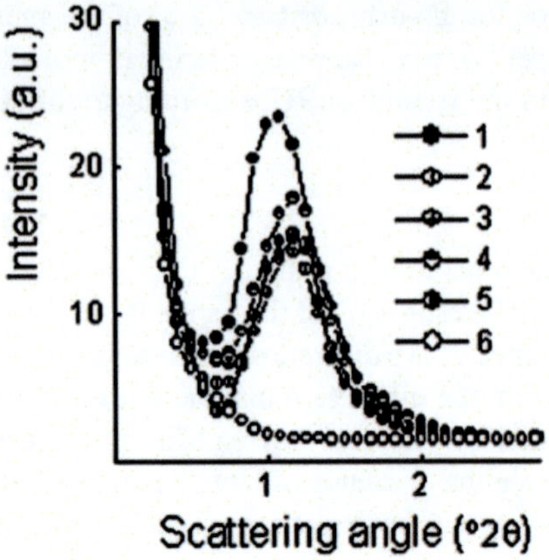

From the SAXS analysis of PU nanocomposites with a different amount of POSS, significant influence of the nanofiller on the microphase structure of PU was revealed (Fig. 5, curves 2–5). Thus, the manifestation of one clear interference maximum on the intensity profile of the native PU (Fig. 5, curve 1) indicates the existence of periodicity in the arrangement of rigid and flexible domains of the molecular chain.

Based on the position of the interference maximum peak on the intensity profile of the PU ($2\theta_m = \sim 1.05°$), the Bragg's period (D) of the appearance of the rigid or flexible domains in the volume of the PU is 8,4 нм. However, the introduction of only 1 wt% of POSS into the volume of PU (Fig. 5, curve 2) leads to a decrease in the intensity of the interference maximum and, accordingly, to the decrease of contrast in the electronic density ($\Delta\rho$) between the rigid and flexible domains ($\Delta\rho = \rho - <\rho>$, where ρ and $<\rho>$ are the local and average values of the electronic density in the two-phase system, respectively). The introduction of 1 wt% of POSS into the volume of PU leads also to the changes in interference maximum angular position ($2\theta_m = 1.12$ nm), that is, to the decrease of the period D of the appearance of rigid and flexible domains in the volume of PU (Table 1).

Table 1 Parameters of the microphase structure of native PU and PU's nanocomposites with different POSS contents

Samples	POSS content, wt%	$2\theta_m$, ° (degree)	D, nm	l_p, nm
PU	0	1.05	8.4	5.7
PU1	1	1.12	7.8	4.9
PU3	3	1.17	7.5	4.3
PU5	5	1.17	7.5	4.5
PU10	10	1.17	7.5	5.5
POSS	100	–	–	0

It is noteworthy that the increase of the amount of POSS up to 3 wt% in the volume of PU (Fig. 2, curve 3) results in the further shift of the interference maximum on the intensity profile to the higher values of 2θ ($2\theta_m = 1.7\ ^{\circ}$), and accordingly in the decrease of the D value (Table 1). At this POSS content in the PU, the intensity of the interference peak is slightly higher than the maximum intensity of the PU's nanocomposite containing 1 wt% of POSS (Fig. 2, curves 2, 3). With an increase of the POSS amount up to 5 wt% in the volume of PU, it was detected, that the intensity of the interference peak is significantly higher compared to the intensity of the peak of the PU's nanocomposite contained 3 wt% of POSS. It should be noted that at 5 wt% of POSS in the PU, the angular position of the interference peak, and accordingly the value of D, remain the same as for PU nanocomposite, which contains 3 wt% of the nanofilller (Table 1).

However, with an increase of the POSS content up to 10 wt% in the PU, the angular position and, accordingly, the value of the structural parameter D, is similar to the PU's nanocomposites, containing 3 and 5 wt% of POSS (Table 1). But the intensity of the interference maximum is slightly higher than the intensity of maximum of PU's nanocomposite, containing 3 wt% of POSS.

In Fig. 6, the dependence of the intensity of PU's interference maximum on the concentration of POSS nanofiller is presented. From the SAXS profile analysis, it is clear that the nanofiller POSS has a homogeneous structure at the nanoscale level.

For the purpose of evaluation of the effective size of the hard (or flexible) domains of the native PU and of the PU's nanocomposites, a structural parameter such as the range of inhomogeneity l_p, which is directly related to the average size of phase ($< l_1 >$ and $< l_2 >$), was determined in a two-phase system using the following equation:

$$l_p = \varphi_2 < l_1 >= \varphi_1 < l_2 >,$$

where φ_1, φ_2 are the volume fractions of the phases ($\varphi_1 + \varphi_2 = 1$).

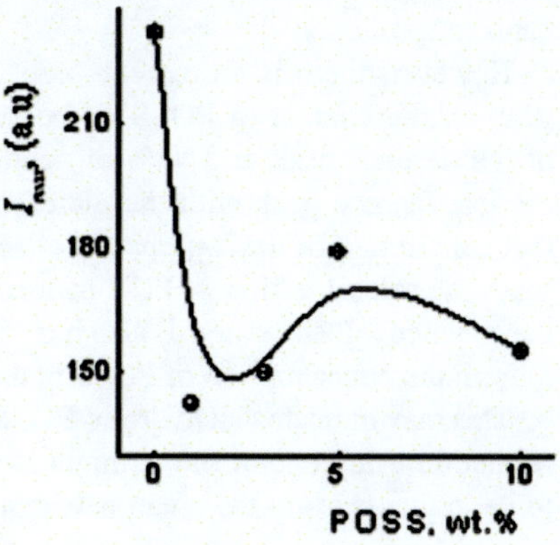

Fig. 6 Dependence of PU's interference maximum intensity on the concentration of POSS in the nanocomposites at SAXS measurements

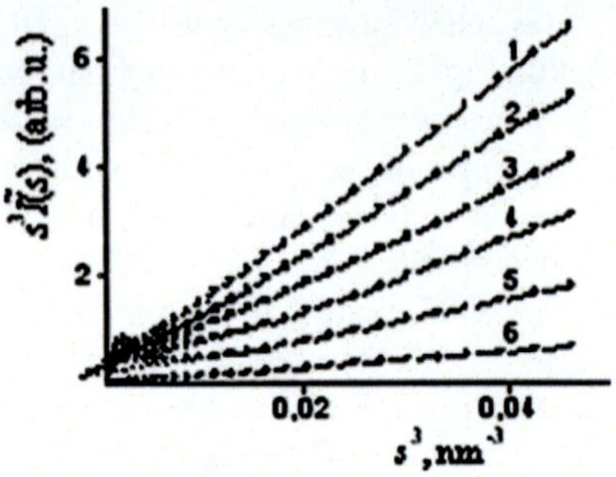

Fig. 7 SAXS profiles by Ruland's method for PU (1), for nanocomposites with 1% (2), 3% (3), 5% (4), 10% (5) of POSS and for native POSS (6)

The evaluation of $2\theta_m$ and D was performed from the analysis of the dependence of I from 2θ (Fig. 5), the estimation of the structural parameter l_p was carried out by the Ruland method [44, 45] from the analysis of the dependence $s^3\tilde{I}(s)$ from s^3 (Fig. 7), where I and $\tilde{I}$—the intensity of SAXS at point and slit collimation of the primary irradiation beam, respectively, and s is reciprocal space, with $s = \lambda^{-1}(2\sin\theta)$.

As shown by the calculations of the parameter $l_{p,}$ the largest value of the effective size of rigid (or flexible) domains has the native PU, and the smallest—the PU's nanocomposites, containing 3 and 5 wt% of POSS. While at 10 wt% of POSS in the volume of PU—the effective size of the same type of the domains is somewhat larger and close to the size of domains in the native PU (Table 1).

Thus, on the basis of the results of investigation of native PU and PU's nanocomposites contained different amount of nanofiller 2-propanediolisobutyl polyhedral oligomeric silsesquioxane (POSS) by WAXS/SAXS methods we can conclude that nanofiller POSS slightly affects the amorphous structure of PU. However, to a greater extent, the nanofiller POSS, containing two hydroxyl groups in aliphatic molecular chain affects the microphase structure of PU. This is manifested in the shifting of the interference peak to the larger values of the scattering angles (2θ), and accordingly in the decreasing of the D value with the increase of the POSS content in the volume of PU, as well as in reducing of the intensity of interference peak, and, respectively, in reducing of contrast in the electron density $\Delta\rho$ in the volume of POSS-containing nanocomposites.

However, there is an extreme dependence of the intensity of the interference peak on the content of POSS in the volume of PU, with the maximum intensity of interference peak at 5 wt% of nanofiller. The next decrease in the intensity of the interference peak with the increase of POSS content up to 10 wt% indicates that certain part of the 2-propanediolisobutyl polyhedral oligomeric silsesquioxane plays the role of a filler in PU's nanocomposites. It is noteworthy that the extreme dependence (albeit reversed) also has an effective size of rigid (or flexible) domains l_p from the concentration of POSS in the PU.

This may indicate that hydroxy-POSS could participate in the reaction of urethane formation using one of the terminal hydroxyl groups, which allow hydroxy-POSS to be implanted into the chain between cross-links of PU. But with an increase of

POSS amount, a certain part of the POSS is not implanted but plays a role of the nanofiller in the system.

3.4 *The Thermal Behavior of POSS-Containing Nanocomposites*

Thermal properties of PU and PU/POSS nanocomposites are investigated by TGA and DSC under nitrogen atmosphere. The related DSC and TGA thermographs of the PU/POSS nanocomposites are presented in Fig. 8 with the corresponding plot of the neat PU.

Figure 8a shows the DSC scans of the neat PU and PU/POSS nanocomposites with POSS content 1, 3, 5, and 10 wt%. From the DSC data (second scans) the glass transition temperature (T_g) of the nanocomposites was found to be increased with increasing of POSS content: from $T_g = -61.13$ °C (neat PU) to $T_g = -59.76$ °C and $T_g = -58.76$ °C (PU1 and PU10, respectively).

The TGA data of PU and PU/POSS nanocomposites are given in Figs. 8b, c and Table 2.

As could be seen, the TGA curves for all the samples display similar degradation profiles; for all the samples degradation was observed in two steps: around 280–300 °C and above 380 °C, indicating the soft and hard segments in the polyurethane (Fig. 8b). Maximum values of the thermal decomposition temperatures, $T_{d(\max)}$, defined as the second maximum of the TGA curves, are shifted toward higher temperatures with increasing POSS content: from 376 °C for neat PU to 379 and 382 °C for PU1 and PU10, respectively. According to the TGA data (Fig. 8c), addition of 1 wt% of POSS results in the increase in onset degradation temperature (T_{onset}) from 270 °C for native PU to 294 °C for the nanocomposite.

Thermal decomposition of PU occurs through double step with maximum rates at 288 and 376 °C with leaving 0.82 wt% carbonaceous char. The first and second

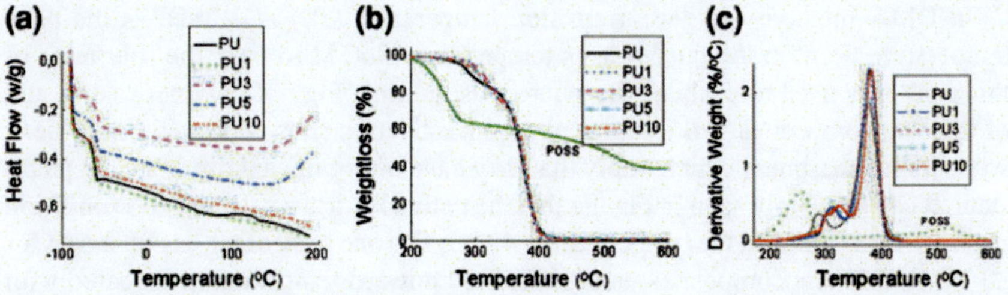

Fig. 8 **a** DSC thermograms of neat PU and PU/POSS nanocomposites under inert atmosphere (N_2) at the heating rate of 20 °C min^{-1}. **b** TGA thermograms of neat PU and PU/POSS nanocomposites under inert atmosphere (N_2) at the heating rate of 20 °C min^{-1}. **c** TGA thermograms (derivative weight) of neat PU and PU/POSS nanocomposites under inert atmosphere (N_2) at the heating rate of 20 °C min^{-1}

Table 2 TGA data of neat PU, and PU/POSS nanocomposites

Sample	$T_{5\%}$ (°C)[a]	T_{max1} (°C)[b]	T_{max2} (°C)[c]
PU	270	288	376
PU 1	294	313	379
PU 3	297	314	379
PU 5	291	310	379
PU 10	291	325	382
POSS	219	257	523

[a]Temperature at 5% weight loss
[b]The maximum 1 degradation rate temperature
[c]The maximum 2 degradation rate temperature

degradation stages are related to the fragmentation of polyols in the soft segments and the urethane bond decomposition in the hard segments, respectively [46]. POSS decomposes in double steps with maximum rates at 257 and 523 °C with leaving 38.64% inorganic residue mainly based on Si–O–Si cage structure (consisting of an inorganic core of silicon and oxygen and outer organic attachments of alkyl groups).

Overall, the results of thermal properties investigation of POSS-containing nanocomposites demonstrate that the created nanocomposites are significantly more thermally stable in comparison with native polymer matrix (PU network). The thermal stability of the nanocomposites increases with the amount of POSS in the systems.

3.5 Dynamic Mechanical Analysis of the Native PU and PU/POSS Nanocomposites

The results of DMA analysis of PU/POSS nanocomposites are presented in Figs. 9a, b, and compared with that of neat PU.

In DMA method, the glass transition temperature (T_g) is defined as the peak temperature point in the tan δ versus temperature plot. Moreover, the sharpness of this peak was used to evaluate the microphase morphology of polymer matrix and of the nanocomposites. All samples under investigation showed a single tan δ peak with similar sharpness which imply that they have no significant difference in phase state. It can be clearly seen in Fig. 9a that the native PU has a narrow glass transition peak with maximum at the temperature −42 °C. The one peak of tan δ is observed for all PU/POSS nanocomposites, but it is shifted upward temperature compared with neat PU network. In Fig. 9a, the shift of T_g of PU/POSS nanocomposites to higher temperature from −41 to −38 °C for PU1 and PU10, respectively, could be observed, which implies that the segmental mobility of PU decreases with the increasing of POSS content. At the same time, the amplitude of PU's tan δ peak significantly

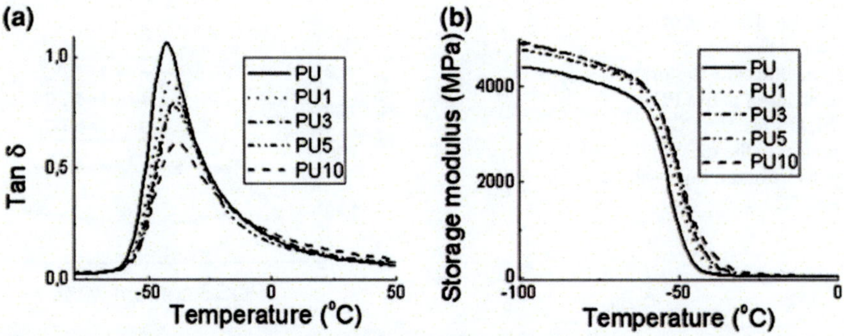

Fig. 9 **a** The temperature dependences of loss factor tan δ for the native PU and PU/POSS nanocomposites with different POSS content (1–10 wt%). **b** The temperature dependences of storage modulus (E') for native PU and PU/POSS nanocomposites with different POSS content (1–10 wt%)

decreases for PU/POSS nanocomposites compared with native PU (Fig. 9a). This proves the restriction of PU chains segmental motion due to the depressing effect of POSS nanoparticles in the system. As shown in previous studies [6, 46] POSS is capable to be chemically embedded into PU network structure. Obviously, it creates steric restriction for PU segmental motion, thus the intensity of PU tan δ peaks decreases with amount of POSS in the system.

The temperature dependence of PU/POSS nanocomposite's storage moduli E' is shown in Fig. 9b. The storage moduli of PU/POSS nanocomposites increase with increasing of POSS content, especially in the temperature range below T_g. At low temperature of −95 °C the PU matrix is in the glassy state and thus the storage moduli are significantly high. At this condition, E' value of PU was measured to be 4375 MPa, mean the values for PU1, PU3, PU5, and PU10 were 4716, 4870, 4729, and 4853 MPa, respectively. As seen the storage moduli of PU/POSS nanocomposites increases non-monotonically with POSS content.

When temperature was increased to a value which is well above the T_g, the PUs chains became flexible and thus the storage modulus became much lower. In the case of PU matrix, the E' at −20 °C is only 11 Mpa. The E' of PU/POSS nanocomposites at this condition were 30, 54, 56, and 107 MPa which is much higher than the E' of PU matrix, demonstrating higher rigidity of PU chains with increasing the POSS content in the nanocomposites. As a result, the PU/POSS nanocomposites with more POSS in the chain would show higher storage modulus.

The correlation of the storage moduli E' of PU/POSS nanocomposites relative to the E' of native PU (M_n/M_o) versus POSS content is shown in Fig. 10. It could be seen that the storage moduli of PU/POSS nanocomposites (M_n) exceed the storage modulus of PU (M_o). But M_n/M_o changes non-monotonically with POSS content. When the POSS content increases from 0 to 1 and 5 wt% the storage moduli of the nanocomposites increase in 1.5 times compared to the native PU. Non-monotonic changes of M_n/M_o may be explained by the results of SAXS investigation which shows that at 1–5 wt% of POSS, the nanoparticles are chemically implanted into

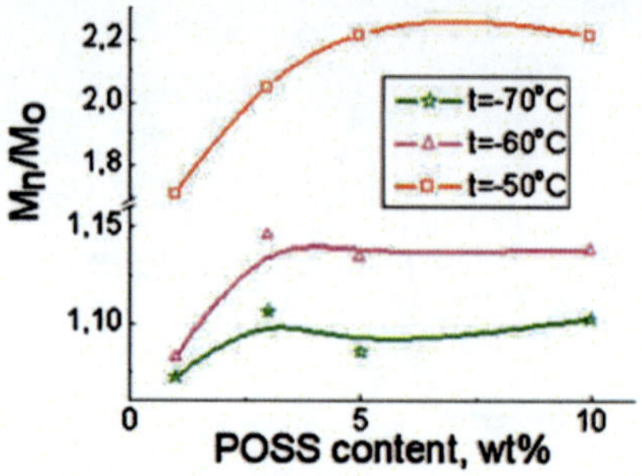

Fig. 10 The relation of storage modulus of nanocomposite to storage modulus of native PU (M_n/M_o) versus POSS content in the range of temperature from −80 to −50 °C

the polymer chain. But with an increase of POSS amount up to 10% a certain part of the POSS is not implanted, but plays a role of the nanofiller in the system.

3.6 Morphology of the POSS-Containing Nanocomposites

Figure 11 shows the SEM images of PU and PU/POSS nanocomposites with different POSS content. It was found that homogeneous at this level structure of PU samples becomes more segregated with increasing of POSS content. We observed POSS particles aggregation in the form of rectangular microdomains for nanocomposite with the maximum POSS content. Obviously, the POSS introduced in PU matrix acts as a nanostructuring agent. As a result, the nanocomposites with more ordered structure are formed thus leading to materials with improved thermal stability.

4 Conclusions

Nanocomposites based on PU matrix and 1,2-propanediolisobutyl-POSS (POSS) nanoparticles were prepared and investigated. POSS nanoparticles were incorporated into PU matrix in the process of PU synthesis. The structure peculiarities, the dynamic mechanical and thermal properties, the morphology of the nanocomposites have been investigated. Overall, it was found that POSS nanoparticles are capable to be incorporated into PU polymer chain by a chemical reaction between hydroxyl groups of POSS and isocyanate groups of PU. The incorporation of the POSS nanoparticles into PU matrix leads to the formation of more ordered structure, and the POSS acts as a nanostructuring agent in the system.

The introduction of the POSS nanoparticles into PU matrix significantly affects the thermal stability of the nanocomposites. The increase in thermal stability of the nanocomposites with the amount of POSS in the systems has been detected.

The investigation by WAXS/SAXS methods has shown that nanofiller POSS slightly affects the amorphous structure of PU. However, to a greater extent, the

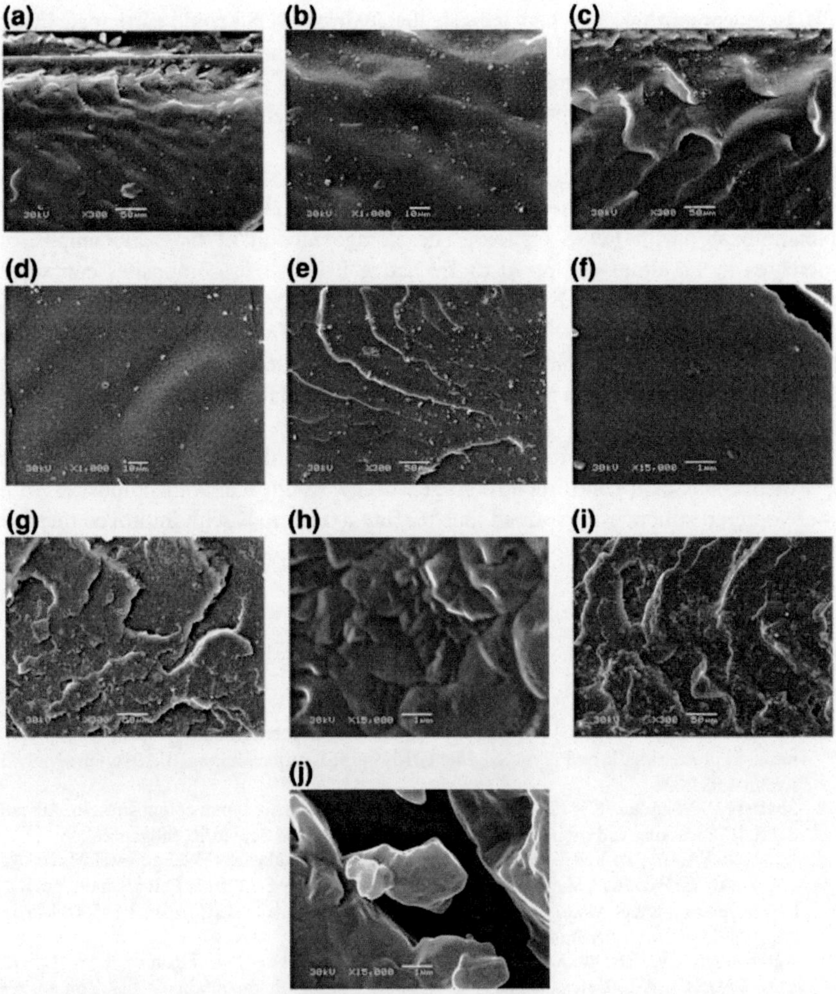

Fig. 11 SEM images of PU (a, b) and PU/POSS nanocomposites with POSS content of 1% (c, d), 3% (e, f), 5% (g, h), 10% (i, j)

nanofiller POSS affects the microphase structure of PU. This is manifested in the shifting of the interference peak to the larger values of the scattering angles (2θ), and accordingly in the decreasing of the Bragg's period D with the increase of the POSS content in the volume of PU, as well as in reducing of the intensity of interference peak. However, there is an extreme dependence of the intensity of the interference peak on the content of POSS in the volume of PU, with the maximum intensity of interference peak at 5 wt% of nanofiller. The next decrease in the intensity of the interference peak with the increase of POSS content up to 10 wt% indicates that a certain part of the 1, 2-propanediolisobutyl-POSS plays the role of a nanofiller in

PU's nanocomposites. This may indicate that hydroxy-POSS could participate in the reaction of urethane formation using one of the terminal hydroxyl groups, which allow to be implanted into the chain between cross-links of PU. But with an increase of POSS amount a certain part of the POSS is not implanted, but plays a role of the nanofiller in the system.

The dynamic mechanical analysis has shown that the storage moduli of PU/POSS nanocomposites exceed the storage modulus of PU. But M_n/M_0 changes non-monotonically with POSS content. The storage moduli of the nanocomposites increases in 1,5 times compared to the native PU for nanocomposites contained from 1 to 5 wt% of POSS. Non-monotonic changes of M_n/M_0 may be explained by the fact (have shown by SAXS investigation) that at 1–5 wt% of POSS, the nanoparticles are chemically implanted into polymer chain. But with an increase of POSS amount up to 10% a certain part of the POSS is not implanted, but plays a role of the nanofiller in the system.

From the morphology investigation, it is obvious that POSS introduced into PU matrix acts as a nanostructuring agent. As a result, the nanocomposites with more ordered structure are formed thus leading to materials with improved thermal stability.

References

1. Ray SS, Okamoto M (2003) Polymer/layered silicate nanocomposites: a review from preparation to processing. Prog Polym Sci 28(11):1539–1561. https://doi.org/10.1016/j.progpolymsci.2003.08.002
2. Shaffer MSP, Sandler JKW (2006) Carbon nanotube/nanofibre polymer composites. In: Advani SG (ed) Processing and properties of nanocomposites. World Scientific, Singapore
3. Bershtein VA, Gun'ko VM, Karabanova LV, Sukhanova TE, Yakushev PN, Egorova LM, Turova AA, Zarko VI, Pakhlov EM, Vylegzhanina ME, Mikhalovsky SV (2013) Polyurethane-poly(2-hydroxyethyl methacrylate) semi-IPN-nanooxide composites. RSC Adv 3:14560–14570. https://doi.org/10.1039/c3ra40295a
4. Karabanova LV, Bershtein VA, Sukhanova TE, Yakushev PN, Egorova LM, Lutsyk ED, Svyatyna AV, Vylegzhanina ME (2008) 3D diamond-containing nanocomposites based on hybrid polyurethane–poly(2-hydroxyethyl methacrylate) semi-IPNs: composition-nanostructure-segmental dynamics-elastic properties relationships. J Pol Sci B 46(16):1696–1712. https://doi.org/10.1002/polb.21506
5. Moniruzzaman M, Winey KI (2006) Polymer nanocomposites containing carbon nanotubes. Macromolecules 39(16):5194–5205. https://doi.org/10.1021/ma060733p
6. Wolinska-Grabczyk A, Jankowski A (2007) Gas transport properties of segmented polyurethanes varying in the kind of soft segments. Sep Pur Tech 57(3):413–417. https://doi.org/10.1016/j.seppur.2006.03.025
7. Gumenna MA, Shevchuk AV, Klimenko NS, Shevchenko VV (2007) Polyurethanes on the base of polyhedral oligosilsesquioxanes (POSS). Polym J 29(3):177–185 [in Russian]
8. Karabanova LV, Whitby RLD, Bershtein VA, Korobeinyk AV, Yakushev PN, Bondaruk OM, Lloyd AW, Mikhalovsky SV (2013) The role of interfacial chemistry and interactions in the dynamics of thermosetting polyurethane-multi-walled carbon nanotube composites with low filler content. Colloid Polym Sci 291(3):573–583. https://doi.org/10.1007/s00396-012-2745-4
9. Karabanova LV, Whitby RL, Bershtein VA, Korobeinyk AV, Yakushev PN, Bondaruk OM, Lloyd AW, Mikhalovsky SV (2012) Microstructure changes of polyurethane by inclusion of

chemically modified carbon nanotubes at low filler contents. Comp Sci Tech 72(8):865–872. https://doi.org/10.1016/j.compscitech.2012.02.008
10. Lelah MD, Cooper SL (1986) Polyurethane in medicine and surgery. CRC Press, Boca Raton, pp 158–167
11. Lloyd AW, Faragher RG, Denyer SP (2001) Ocular biomaterials and implants. Biomaterials 22(8):769–785. https://doi.org/10.1016/S0142-9612(00)00237-4
12. Karabanova LV, Lloyd AW, Mikhalovsky SV, Helias M, Philips GJ, Rose SF, Mikhalovska L et al (2006) Polyurethane/Poly(hydroxyethyl methacrylate) semi-interpenetrating polymer networks for biomedical applications. J Mater Sci Matter Med 17:1283–1296. https://doi.org/10.1007/s10856-006-0603-y
13. Madhavan K, Reddy BSR (2009) Structure–gas transport property relationships of poly(dimethylsiloxane–urethane) nanocomposite membranes. J Mem Sci 342(1–2):291–299. https://doi.org/10.1016/j.memsci.2009.07.002
14. Fomenko AA, Gomza YuP, Klepko VV, Gumenna MA, Klimenko NS, Shevchenko VV (2009) Dielectric properties, conductivity and structure of urethane composites based on polyethylene glycol and polyhedral silsesquioxane. Polym J 31(2):137–143 [in Ukrainian]
15. Mahapatra SS, Yadav SK, Cho JW (2012) Nanostructured hyperbranched polyurethane elastomer hybrids that incorporate polyhedral oligosilsesquioxane. React Funct Polym 72(4):227–232. https://doi.org/10.1016/j.reactfunctpolym.2012.02.001
16. Lewicki JP, Pielichowski K, Jancia M, Hebda E, Albo RLF, Maxwell RS (2014) Degradative and morphological characterization of POSS modified nanohybrid polyurethane elastomers. Polym Degrad Stab 104:50–56. https://doi.org/10.1016/j.polymdegradstab.2014.03.025
17. Wei K, Wang L, Zheng S (2013) Organic–inorganic polyurethanes with 3, 13-dihydroxypropyloctaphenyl double-decker silsesquioxane chain extender. Polym Chem 4:1491–1501. https://doi.org/10.1039/c2py20930f
18. Bourbigot S, Turf T, Bellayer S, Duquesne S (2009) Polyhedral oligomeric silsesquioxane as flame retardant for thermoplastic polyurethane. Polym Degrad Stab 94:1230–1237. https://doi.org/10.1016/j.polymdegradstab.2009.04.016
19. Huang J, Jiang P, Li X, Huang Y (2016) Synthesis and characterization of sustainable polyurethane based on epoxy soybean oil and modified by double-decker silsesquioxane. J Mater Sci 51(5):2443–2452. https://doi.org/10.1007/s10853-015-9557-0
20. Wang W, Guo Y, Otaigbe JU (2009) The synthesis, characterization and biocompatibility of poly(ester urethane)/polyhedral oligomeric silesquioxane nanocomposites. Polymer 50(24):5749–5757. https://doi.org/10.1016/j.polymer.2009.05.037
21. Lai YS, Tsai CW, Yang HW, Wang GP, Wu KH (2009) Structural and electrochemical properties of polyurethanes/polyhedral oligomeric silsesquioxanes (PU/POSS) hybrid coatings on aluminum alloys. Mater Chem Phys 117(1):91–98. https://doi.org/10.1016/j.matchemphys.2009.05.006
22. Kuo SW, Chang FC (2011) POSS related polymer nanocomposites. Prog Polym Sci 36(12):1649–1696. https://doi.org/10.1016/j.progpolymsci.2011.05.002
23. Hebda E, Pielichowski K (2018) Polyurethane/POSS hybrid materials. In: Kalia S, Pielichowski K (eds) Polymer/POSS nanocomposites and hybrid materials: preparation, properties, applications. Springer, Switzerland, pp 167–204. https://doi.org/10.1007/978-3-030-02327-0_5
24. Zhou H, Chua MH, Xu J (2019) Functionalized POSS-based hybrid composites. In: Pielichowski K, Majka TM (eds) Polymer composites with functionalized nanoparticles. Synthesis, properties, and applications. Elsevier, pp 179–210. https://doi.org/10.1016/B978-0-12-814064-2.00006-8
25. Ghermezcheshme H, Mohseni M, Yahiaei H (2015) Use of nanoidentation and nanoscratch experiments to reveal the mechanical behaviour of POSS containing polyurethane nanocomposite coatings: The role of functionality. Tribol Int 88:66–75. https://dx.doi.org//10.1016//j.triboint.2015.02.023
26. Mir Mohamad Sadeghi G (2018) Synthesis and evaluation of the effect of structural parameters on recovery rate of shape memory Polyurethane-POSS nanocomposites. Eur Polym J https://doi.org//10.1016//j.eurpolymj.2018.12.041

27. Blanko I (2018) Decomposition and ageing of hybrid materials with POSS. In: Kalia S, Pielichowski K (eds) Polymer/POSS nanocomposites and hybrid materials. Preparation, properties, applications. Springer, Switzerland, pp 415–462. https://doi.org/10.1007/978-3-030-02327-0_13
28. Karabanova LV, Boiteux G, Gain O, Seytre G, Sergeeva LM, Lutsyk ED (2004) Miscibility and thermal and dynamic mechanical behaviour of semi-interpenetrating polymer networks based on polyurethane and poly(hydroxyethyl methacrylate). Polym Int 53(12):2051–2058. https://doi.org/10.1002/pi.1627
29. Kratky O, Leopold H (1964) Messung und unterdruckung der blendenstreuung am kolimationsystem fur rontgenkleinwinkeluntersuchungen. Die Macromol Chemie 75(1):69–74
30. Schmidt PW, Hight RJ (1960) Slit height corrections in small angle X-ray scattering. Acta Cryst 13:480–483. https://doi.org/10.1107/S0365110X60001138
31. Wamke A, Dopierała R, Prochaska K, Maciejewski H, Biadasz A, Dudkowiak A (2015) Characterization of Langmuir monolayer, Langmuir-Blodgett and Langmuir-Schaefer films formed by POSS compounds. Col Surf A 464:110–120. https://doi.org/10.1016/j.colsurfa.2014.10.022
32. Jerman I, Koželj M, Orel B (2010) The effect of polyhedral oligomeric silsesquioxane dispersant and low surface energy additives on spectrally selective paint coatings with self-cleaning properties. Sol Energy Mater Sol Cells 94(2):232–245. https://doi.org/10.1016/j.solmat.2009.09.008
33. Jerman I, Mihelčič M, Verhovšek D, Kovač J, Orel B (2011) Polyhedral oligomeric silsesquioxane trisilanols as pigment surface modifiers for fluoropolymer based thickness sensitive spectrally selective (TSSS) paint coatings. Sol Energy Mater Sol Cells 95(2):423–431. https://doi.org/10.1016/j.solmat.2010.08.005
34. Kraus-Ophir S, Jerman I, Orel B, Mandler D (2011) Symmetrical thiol functionalized polyhedral oligomeric silsesquioxanes as building blocks for LB films. Soft Matter 7(19):8862–8869. https://doi.org/10.1039/c1sm05443k
35. Dittmar U, Hendan BJ, Florke U, Marsmann HC (1995) Funktionalisierte octa-(propylsilsesquioxane) (3-$XC_3H_6)_8(Si_8O_{12}$) modellverbindungen für oberflächenmodifizierte kieselgele. J Organomet Chem 489(1–2):185–194. https://doi.org/10.1016/0022-328X(94)05100-P
36. Bärtsch M, Bornhauser P, Calzaferri G, Imhof R (1994) $H_8Si_8O_{12}$: A model for the vibrational structure of zeolite A. J Phys Chem 98(11):2817–2831. https://doi.org/10.1021/j100062a016
37. Xue M, Zhang X, Wu Z, Wang H, Ding X, Tian X (2013) Preparation and flame retardancy of polyurethane/POSS nanocomposites. Chin J Chem Phys 26(4):445–450. https://doi.org/10.1063/1674-0068/26/04/445-450
38. Maoz R, Sagiv J, Degenhardt J, Möhwald H, Quint P (1995) Hydrogen-bonded multilayers of self-assembling silanes: structure elucidation by combined Fourier transform infra-red spectroscopy and X-ray scattering techniques. Supramol Sci 2(1):9–24. https://doi.org/10.1016/0968-5677(96)85635-5
39. Bellamy LJ (1980) The infrared spectra of complex molecules, v. II. Advances in infrared group frequencies. 2nd ed. Chapman and Hall, London, pp 1–299. https://trove.nla.gov.au/version/12064222
40. Moon JH, Seo JS, Xu.Y, Yang S (2009) Direct fabrication of 3D silica-like microstructures from epoxy-functionalized polyhedral oligomeric silsesquioxane (POSS). J Mater Chem 19(27):4687–4691. https://doi.org/10.1039/b901226e
41. Smetankina NP, Angelova AV, Lukas SD (1967) Polyurethane coatings based on polyoxypropylene glycols. Synthesis and physical chemistry of polymers (polyurethanes) 5:49–56 [in Russian]
42. Zhang S, Zou Q, Wu L (2006) Preparation and characterization of polyurethane hybrids from reactive polyhedral oligomeric silsesquioxanes. Macromol Mater Eng 291(7):895–901. https://doi.org/10.1002/mame.200600144
43. Guinier A (1956) Théorie et technique de la Radiocristallographic. Dunod, Paris, p 601
44. Ruland W (1971) Small-angle scattering of two-phase systems: determination and significance of systematic deviations from Porod's law. J Appl Cryst 4(1):70–73. https://doi.org/10.1107/S0021889871006265

45. Perret R, Ruland W (1971) Eine verbesserte Auswertungsmethode für die Röntgen-kleinewinkelstreuung von Hochpolymeren. Kolloid Z Z Polymere 247(1–2):835–843. https://doi.org/10.1007/BF01500257
46. Kuznetsova VP, Laskovenko NN, Zapunnaya KV (1984) Organosilicon polyurethanes. Naukova dumka, Kyiv, p 224. [in Russian]

Structure and Properties of Polyaniline Micro- and NanoComposites with Noble Metals

O. Aksimentyeva, Yu. Horbenko, and P. Demchenko

1 Introduction

Modern science is moving toward the creation of effective heterogeneous organo-inorganic systems that provide the appearance of a number of new physicochemical properties that are not inherent in the individual components that form such systems. Inorganic–organic hybrid nanocomposites are a class of materials defined as inorganic nanostructures included in an organic matrix, where both the constituents of the hybrid system show characteristics different from the ones they would have in the absence of the other [1–3]. Hybrid polymer nanocomposites are considered to be promising materials for a variety of technological applications due to their unique electronic, optical, and sensory properties. Interest in polymer–metal nanosystems is conditioned by the possibility of obtaining materials with new, enhanced functional properties that can be applied in sensor and medical chemistry, electrocatalysis, and biotechnology [4–7].

Understanding the physicochemical processes occurring in such systems is key to creating a new generation of functional devices and, at the same time, makes their manufacturing technology reliable and manageable. Collaborative studies by physicists, chemists, and material scientists have shown that reducing the particle size of metals (especially noble ones—Ag, Pt, Au, Pd) to the nanometer scale leads to new crystalline modifications or even new physical phenomena not peculiar to materials with micron-sized particles. The prevention of aggregation of nanoparticles with excess surface energy and the preservation of their unique properties is possible by the creation of composites with polymer matrices by the mechanism of encapsulation

O. Aksimentyeva (✉) · Yu. Horbenko · P. Demchenko
Ivan Franko National University of Lviv, 6/8 Kyryla&Mefodia, Lviv 79005, Ukraine
e-mail: aksimen@ukr.net

Yu. Horbenko
e-mail: y_bilka@ukr.net

O. Fesenko and L. Yatsenko (eds.), *Nanomaterials and Nanocomposites, Nanostructure Surfaces, and Their Applications*, Springer Proceedings in Physics 246, https://doi.org/10.1007/978-3-030-51905-6_35

("core-shell") or the insertion, intercalation, or other mechanisms of embedding of inorganic component into a polymer matrix with the formation of coordination, covalent, hydrogen, or ionic bonds [8–10].

For the creation of composites of semiconductor or metallic nanoparticles with polymeric matrices, organic polymers with their own electronic conductivity, of which polyaniline is a prime representative, are increasingly being considered. Due to the high stability, uniqueness, and variety of electronic properties, this polymer is attracting increasing attention of scientists around the world [11–15]. Control over the formation of the properties of metal-polyaniline composites consists of the ability to establish the relationship between the conditions of synthesis, structure, and physicochemical characteristics of the resulting hybrid systems [16–18]. Various approaches, including both single-step (one-step) and multistage synthesis, are used for the synthesis of nanocomposites [19, 20]. The multistage synthesis includes a three-step pathway: the synthesis of charged inorganic nanoparticles, the modification of their surface ("passivation-compatibilization"), and their introduction into the organic matrix [21].

One-step synthesis is based on the direct oxidation of the monomer by metal ions in solution in the absence of additional oxidants. The latter involves the formation of nanoparticles directly in the presence of the polymer, so that the nanocomposite is the product of the one-step reaction. The direct method is faster, but it is not able to produce contact-free identical nanoparticles. With such techniques, the inorganic precursor is usually introduced in the polymeric matrix and, by consequent chemical reduction, the desired product is obtained [19].

We used this approach to obtain PAn composites with noble metals. Studies show that such synthesis, in the absence of additional stabilizers, in most cases does not lead to the formation of nanoparticles. However, at a metal to polymer ratio of 0.1, it was possible to obtain 7 nm spherical particles [17]. The use of substances with bulk anions such as dodecylbenzenesulfonate, polystyrenesulfonate, or camphorsulfonic acid, which act both as dopant and stabilizer, results in the production of nanoparticles [20].

Attempts have also been made to use organic media for the one-step synthesis of PAn-Au [21], PAn in the form of nanoballs-Au [18], and palladium-poly (3,5-dimethylaniline) [19] composites. Interesting is the phase-boundary synthesis, which used aniline dissolved in chloroform as the lower phase and chloroplatinic acid with polystyrenesulfonate dissolved in water as the upper phase [20]. In the obtained composite, the particles were uniformly distributed in the polymer matrix.

In most works about single-stage synthesis, attention is focused on the reaction characteristics, the size and distribution of particles in the polymer matrix. Only in some cases, the conductivity or thermal stability of the composites is indicated. Regarding conductivity, the incorporation of metal into the polymer does not produce a definite result and causes both its growth [21–24] and decrease [25, 26] compared to the pure polymer. Conductivity depends not only on the amount of metal in the composite, but also on the state of the structure and the degree of oxidation of the polymer.

The purpose of the work was to study the influence of the nature of the noble metal (Ag, Au, Pt, Pd) on the patterns of synthesis and structure of composites with polyaniline (PAn) obtained by one-step synthesis in the conditions of emulsion polymerization.

2 Experimental

2.1 Synthesis of Composites

For the synthesis of metal-containing composites, the method of microemulsion polymerization [24] was chosen, which was realized by the next procedure. 0.64 g of $Na_2S_2O_8$, 3 ml of butyl alcohol C_4H_9OH, 40 ml of hexane C_6H_{14} were added to the solution containing 4.5 g of sodium dodecyl sulfate $C_{12}H_{25}SO_4Na$ in 15 ml of 1 M H_2SO_4 and stirred for about 20 min to a homogeneous state. After that, 0.5 ml of distilled aniline was added dropwise with stirring and the calculated amount of a solution of the metal compound ($HAuCl_4$/H_2PtCl_6/K_2PdCl_4/$AgNO_3$) to achieve a metal content of 15% by weight of the composite. The reaction was carried out with constant stirring at room temperature for 24 h. The precipitate was separated, washed by water and acetone, and dried at 80 °C to constant mass according to traditional procedure [11, 12]. Samples obtained by microemulsion polymerization at the ratio of oxidant: monomer = 1:2 became the basis for the synthesis of four composites of polyaniline-metal: PAn-gold, PAn-platinum, PAn-palladium, PAn-silver in 1 M H_2SO_4 organic-water solution. All reagents and materials used were of high quality, purchased from Sigma-Aldrich. The metal content was selected on the basis of literature data and previous experiments, since with smaller quantities of metal it does not have a significant effect on the properties of the composite, and at larger metal aggregation into clusters occurs.

2.2 Methods

Powder specimens were prepared for X-ray studies by careful grinding in an agate mortar of polyaniline or its composites. The powder of the sample was deposited in a uniform layer to a special amorphous film for X-ray experiments and was fixed with another film in the cuvette. The arrays of experimental intensities and reflected angles from the samples were obtained using an automatic STOE STADI P diffractometer (manufactured by STOE & Cie GmbH, Germany) with a PSD linear positional precision detector according to the modified Guignet geometry ($CuK\alpha_1$-radiation); concave Ge-monochromator (111) of Johann type; $2\theta/\omega$-scanning, the interval of angles 2ϑ $0.500 \leq 2\theta \leq 120.185$ °2θ with step 0.015 °2θ. Initial processing of experimental diffraction arrays, calculation of theoretical diffraction patterns of known compounds for the purpose of phase identification, refinement of the parameters of elementary cells was performed using a software package STOE WinXPOW 3.03 [27] та PowderCell 2.4 [28]. The calculation of the microstructural parameters was carried out in an isotropic approximation by the methods of the integral width of the diffraction peaks using a software package WinPLOTR [29] according to [30].

Structural studies of the morphology and content of the synthesized composites were performed by scanning electron microscopy (SEM) and energy-dispersive spectroscopy (EDAX) on a microscope microanalyzer, PЭMMA-102-02. Samples for study were prepared by applying polyaniline or a composite from solution of N-methylpyrrolidone on a polished platinum plate. The distance from the electron source to the sample surface was 300 mm. The electron energy was scanned in the range of 0–10 kV. Statistical processing of the results was performed by software MAGELLAN v1.2.

For investigation by atomic force microscopy, atom-force microscope Nanoscope IIIa Dimension 3000TM Digital instruments were used. Polymer–metal composite was dissolved in N-methylpyrrolidone and sonicated for 15 min. Then a drop of the solution was taken with a micro-syringe and deposited on a silvered glass, after which the solvent was dried in vacuum.

Molecular spectra of PAn and its composites were obtained by FT-IR spectroscopy on an IFS 66 spectrometer. For this, the sample PAn or composite was triturated in an agate mortar with pre-dried and crushed KBr and formed into a tablet under pressure.

The electrical conductivity of the samples was measured by the van der Pauw method [30]. Samples were formed in tablets with a diameter of 2×10^{-2} m using a mold, acting at a constant pressure of 100 kg/cm^2 for 5 min. They were placed between four paired silver contacts. To measure the temperature dependence of the resistivity, the sample was placed in a special quartz cell between the nickel flat contacts and the soldered thermocouple. The change in temperature at a constant speed in the range 273–333 K was controlled and kept of 5 K/min. During measurements, constant load of 10 N/cm^2 acted at the sample. The resistance of the tablet was recorded with a digital ohmmeter Щ-306-1. The specific volumetric resistance was calculated according to the formula:

$$R = \frac{\rho \cdot l}{S},$$

where S—cross-sectional area of cylindrical sample (m), l—height (m), R—measured resistance value (Ohms). The relative error of determination for the parallel measurement series did not exceed 5%.

3 Results and Discussion

The use of precious metals as components to create polymer–metal composites is due to their high performance, the relative stability of their nanoparticles compared to other metals, and the high chemical activity of precursor compounds, which allows to obtain filled materials during one-step synthesis without the use of additional reagents. Typically, precious metal particles incorporated into the polymer matrix exhibit increased activity in the catalysis of various chemical transformations and are therefore of practical interest.

The idea of synthesis was that due to the change in the oxidation degree of the newly formed polymer chains, the ions of the metal are reduced, forming nanosized particles. In the basis of the synthesis of polyaniline composites with noble metals the next chemical reactions lie:

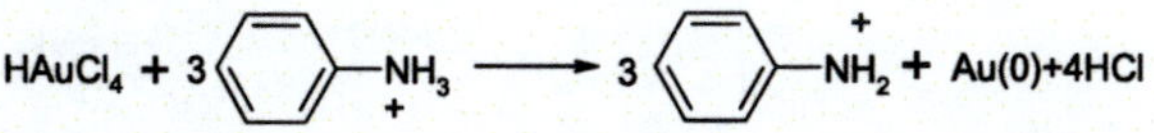

(1)

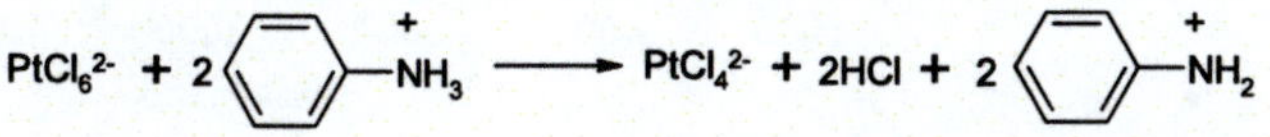

(2)

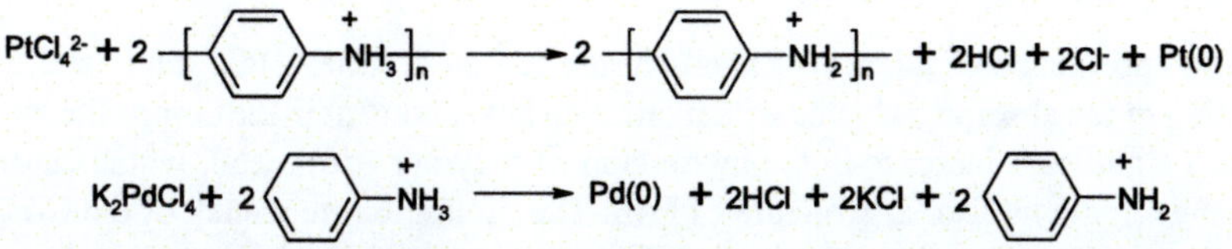

(3)

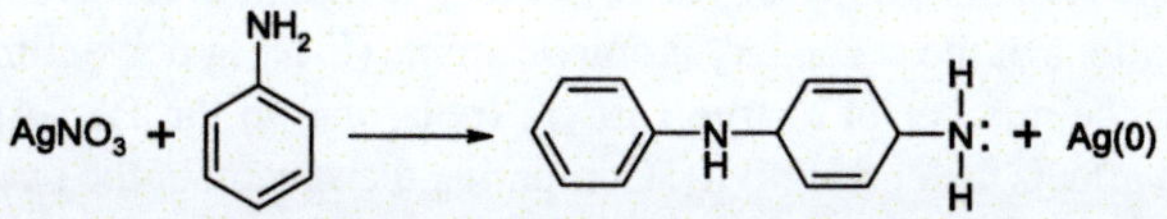

(4)

For polyaniline–metal composites, there is an increase in the yield of the product in comparison with the pure polymer from 3 to 24%, with a direct relationship between the increase in the yield of the product and the metal reduction potential. For example, $E(Au^{3+}/Au^0) = 1.691$ V responds to 24,2% and $E(Ag^+/Ag^0) = 0.799$ B—3.2%, indicating the direct involvement of the metal salt in the aniline polymerization process. The established dependence between the metal reduction potential and the increase in product yield compared to pure PAn indicates the involvement of metal ions (atoms) in the process of composite synthesis and formation.

Samples with low (2.5%) and high (50%) silver content were used to detect the effect of metal on the structure of composites (see Fig. 1). The wide absorption band with a maximum at 3400 cm^{-1} indicates the presence of adsorbed water molecules and is characteristic of hydrogen bonds formed by NH and NH^+ groups. The adjacent weakly expressed peak of 2923 cm^{-1} corresponds to hydrogen bonds between the hydrogen atoms of the main chain. The 1640 cm^{-1} band corresponds to the quinoid ring and the 1492 cm^{-1} band to the benzoid ring. Absorption bands between 1300 and 1100 cm^{-1} correspond to the basic and protonated form of polyaniline. Bands lower than 800 cm^{-1} are typical for C-H bonds [31].

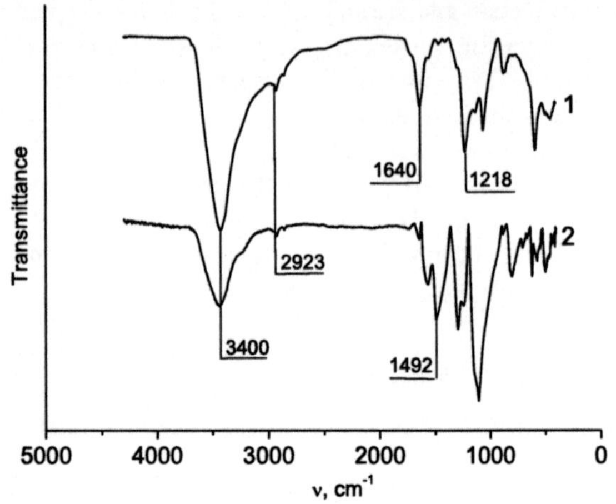

Fig. 1 FTIR spectra of polyaniline–silver composites with metal content 2,5 (1) and 50 (2) wt%

The appearance of additional bands in the region of 1500–1000 cm^{-1} and the splitting of the absorption bands with increasing silver content in the composite indicates a significant increase in the interaction of polymer with metal, which causes a change in the molecular structure of PAn. The oscillation frequency of the C-N^{+} fragment in the PAn polaron lattice at 1275 cm^{-1} is modified relative to the characteristic oscillations of the secondary aromatic amine (C-N) in polyaniline, indicating a decrease in the number of charge carriers (polarons) in the PAn-Ag composite, which in turn affects the conductivity. Comparing the spectra of the two composites, it can be seen that as the content of the filler increases, the intensity of the band corresponding to the adsorbed water (~3400 cm^{-1}) decreases. This can be explained by the fact that the metal displaces water in a competitive adsorption reaction due to the free hydrogen bonds of the polymer chain. The increase in the intensity of the absorption band corresponding to the quinoid ring indicates an increase in the proportion of quinoid fragments.

Therefore, as the concentration of silver increases, part of the emeraldine is spent on its reduction while passing to the peroxidized (pernigraniline) non-conducting form of polyaniline. The incorporation of metal in the polymer matrix in the case of gold and platinum led to an increase in the specific conductivity of about 1.5–2 times, and in the case of palladium and silver—to its decrease. It is worth noting that the mass content of metal of 15% corresponds to only a few percent of its volume.

Measurement of the resistivity versus temperature showed the semiconductor nature of the conductivity of composites with activation energies of charge transfer within 0.08–0.19 eV (see Fig. 2). This indicates the main role of the polymer matrix in the conductivity mechanism of the obtained composites.

The process of charge transfer in hybrid composites can occur in different ways. Due to the presence of a conjugated π-electron bonding system in the polymer matrix both one-dimensional (1D) carrier transport and inter-chain transitions can occur. On the other hand, the metal filler may form its own conduction channels inside the

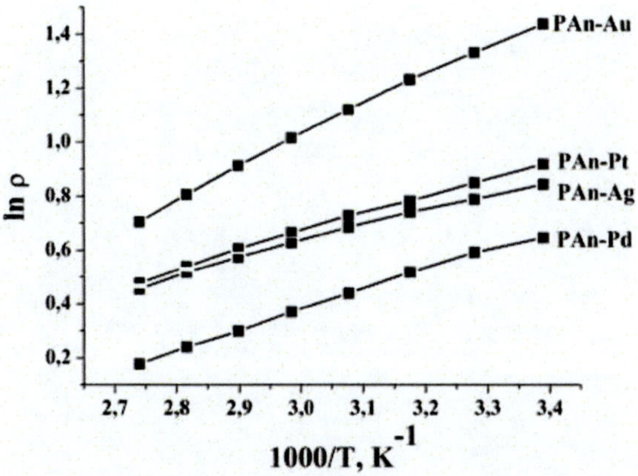

Fig. 2 Temperature dependence of specific resistance of polyaniline–metal composites

polymer matrix and on the surface of the composite [33]. An important aspect of providing the ability of electronic transport in the composite is its morphological features, which will affect the nature of the dependence of the conductivity of the composites on the metal content. These particularities were studied by scanning electron and atomic force microscopy of samples obtained as thin films of composites deposited from a solution of N-methylpyrrolidone on glass.

During the formation of a new phase during polymerization, processes of coagulation (when the length of the polymer chains exceeds some critical), aggregation, and crystallization occur. If inorganic (metallic) components are involved in the formation of a new polymer phase, they will in some way affect the structure of the synthesized composites. There are various types of supramolecular formations—lamellae, packs, strips, dendrites, fibrils, but the most common for polyaniline are globular structures, which are characteristic of polymers obtained by polycondensation. Under certain conditions, globules can be straightened to form fibrillar structures, such as when convolution into the sphere is complicated by the presence of charges or impurities. In this case, the macromolecules are arranged in packs.

Crystalline formations of conjugated polymers are usually embedded in an amorphous polymer matrix, which has a much lower conductivity than crystalline that determines the activation barrier for charge transfer. The formation of fibril packs results in a linear ordering of the chains in a certain area, which should lower the activation barrier of conductivity due to the possibility of one-dimensional (1D) movement of the carriers along the conjugated polymer chain system.

According to SEM images for the PAn-Au composite, a large number of low-dispersive particles are observed on the surface, which according to the EDAX results are gold (see Fig. 3). Image analysis showed that the average particle diameter is 4 μm and the average area is 16 μm^2, and the size distribution demonstrated that 99% of the particles have a diameter of up to 10 μm.

For the composite polyaniline–platinum, a combination of elongated particles and wires is observed on the surface, the EDAX confirmed that it consists of platinum (see Fig. 4). The average particle diameter is 35 μm and the average area is 415 μm^2,

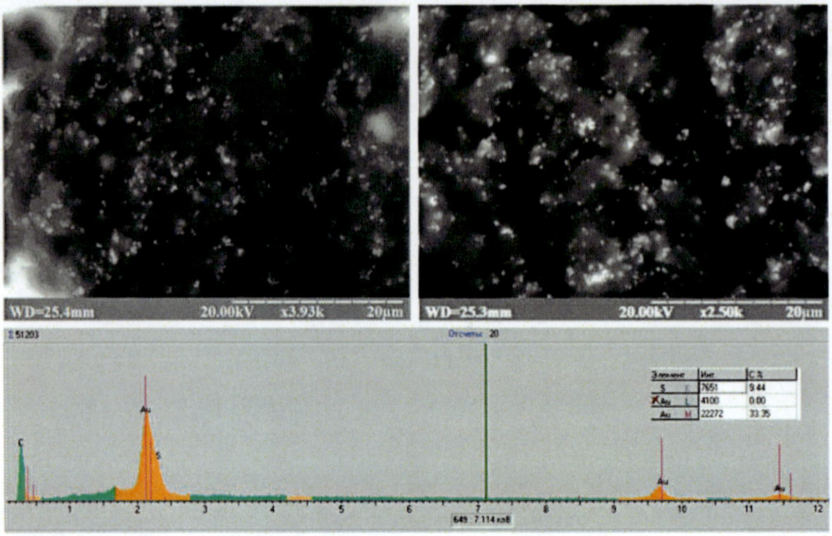

Fig. 3 SEM images and EDAX results for the PAn-Au composite

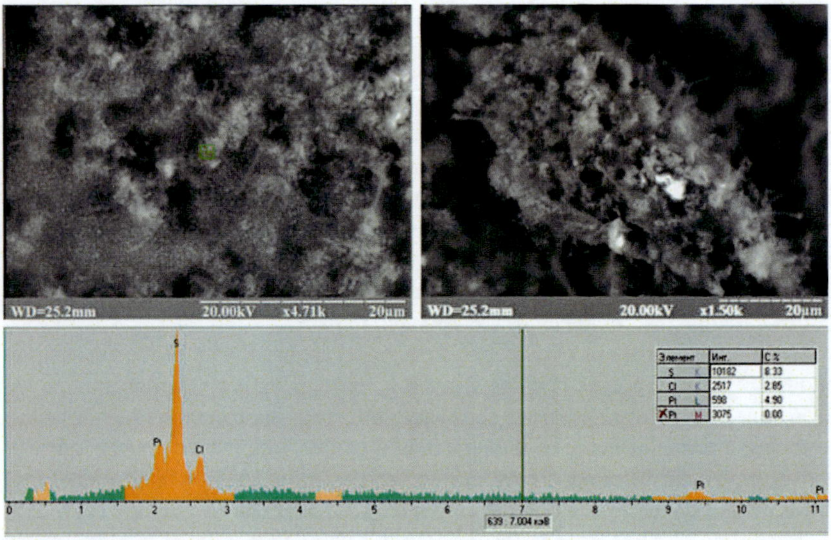

Fig. 4 SEM images and EDAX results for the PAn-Pt composite

the diameter distribution shows that size of 90% objects is 30–60 μm.

For polyaniline–palladium composite the surface has a morphology similar to pure polyaniline, with only slightly more elongated spheres (see Fig. 5). The EDAX indicated the presence of palladium, but no particles are visible on the surface. Based

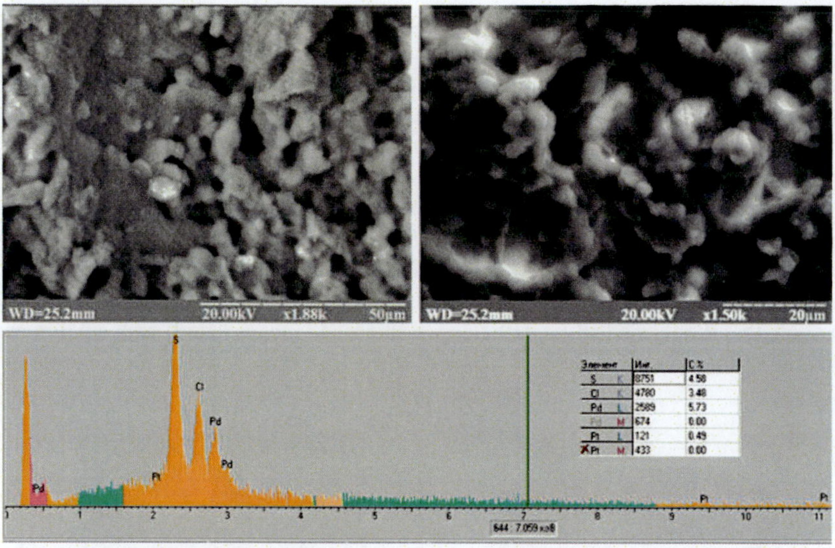

Fig. 5 SEM images and EDAX results for the PAn-Pd composite

on this fact, we can conclude that palladium particles are incorporated directly into the polymer matrix. In this case, the granulometric analysis failed due to the difficulty of setting the boundaries of individual objects.

On the example of PAn-Ag composites, we traced the influence of metal content on the structural features of the obtained materials. The surface morphology of composites with low silver content (2.5 wt%) is rather uneven and has the appearance of pyramidal structures similar to polymer crystals up to 150 nm in height (see Fig. 6).

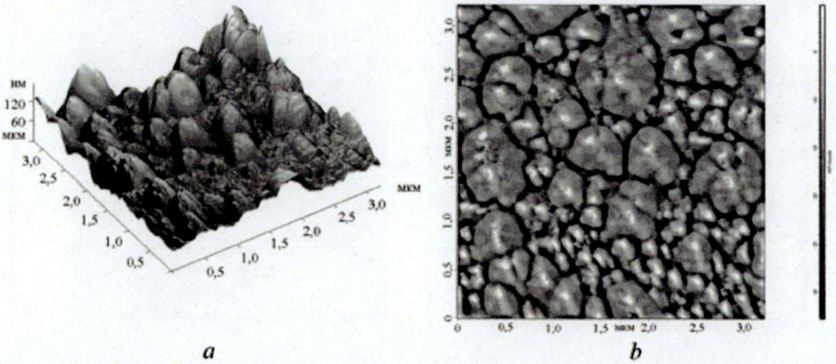

Fig. 6 AFM image of the surface of the PAn-Ag composite with a silver content of 2.5 wt%: **a** 3D spatial; **b** 2D image (scan area 1.5 × 1.5 μm²)

The surface of the composite with a metal content of 5 wt% is more ordered than the previous one. The formation of packs of macromolecules fibrils is observed, among which is the incorporation of metal particles of almost identical spherical shape with a diameter of 500–600 nm (see Fig. 7).

For composites with 10 and 20 wt% metal content, the tendency of change of surface morphology from broad striped to ordered chain persists (see Fig. 8).

The maximum degree of ordering (the average crystallite size of 0.4 μm^3) is achieved at 30 wt% silver in the composite (see Fig. 9). It can be seen that the polymer chains are being arranged in the oblong packs, which contributes to their maximum unwinding. Analyzing the appearance of the inhomogeneities of the surface, we can assume that the formed silver particles are enclosed (encapsulated) with a polymer. The morphology of the obtained samples is the transition between the amorphous

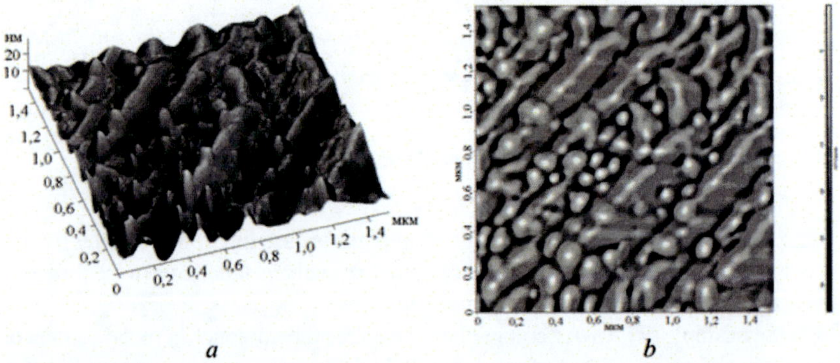

Fig. 7 AFM image of the surface of the PAn-Ag composite with a silver content of 5 wt%: **a** 3D spatial; **b** 2D image (scan area 1.5 ×1.5 μm^2)

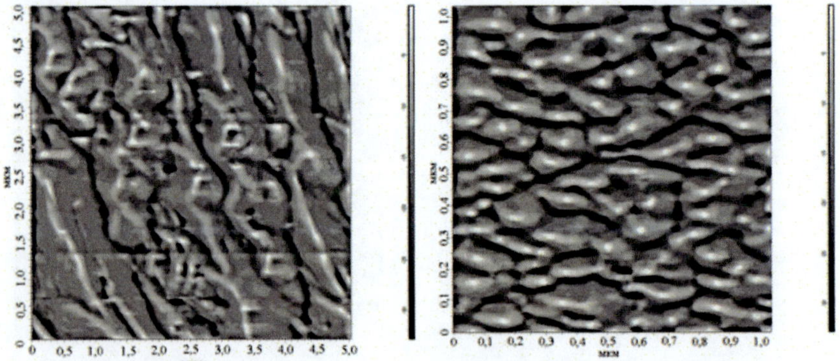

Fig. 8 AFM 2D image of a PAn-Ag composite surface with a silver content of 10 (**a**) and 20 (**b**) % by weight (scan area 5 × 5 μm^2)

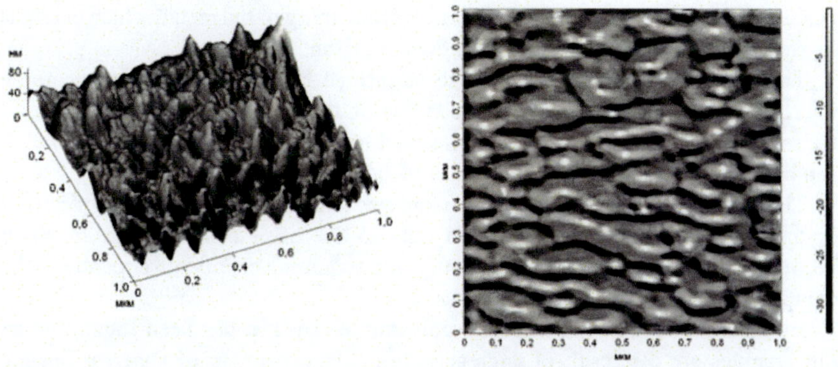

a *b*

Fig. 9 AFM image of the surface of the PAn-Ag composite with a silver content of 30 wt%: **a** – 3D spatial; **b** – 2D image (scan area 1.5×1.5 μm^2)

Fig. 10 AFM 2D image of a PAn-Ag composite surface with a silver content of 40% by weight (scan area 5× 5 μm^2)

morphology of the polymer and crystallized metal, which is characterized by clear faces (see Fig. 9). Obviously, the silver particles act as additional channels of current transfer between the individual polymer chains, which in combination with the ordering of the structure causes the high electrical conductivity of the composites.

The morphology of the PAn-Ag composite with a 40 wt% silver content is characterized by the presence of crystalline formations with distinct faces (see Fig. 10), which are significantly different from composites with low filler content, in which the structural elements have a fuzzy (amorphous) shape. That is, with increasing silver content in the composite it has happened the aggregation of metal particles with the formation of microcrystalline regions separated by polymer chains. When the percolation threshold is reached, between these regions contacts appeared, and the conductivity is already determined by the properties of the inorganic phase, that is, the conductivity is due to the electronic conductivity of the metal, which explains its weak change with increasing silver concentration.

The fusion of metal regions is also confirmed by the results of atomic force microscopy (see Fig. 11). It is well seen that on the surface there is an increase in the number of large and small interconnected formations with distinct faces, which can be interpreted as microcrystallites of metal.

That is, the composite PAn-Ag has an ordered structure that combines the features of organic and inorganic components only in the silver content range of about 30 wt%. Outside of this interval, the composite acquires a disordered structure of the component whose content is larger.

By means of SEM for composite polyaniline–silver, it has been found that its big granules are the result of aggregation of a large number of spheres. Despite the significant metal content according to the EDAX results, only a few particles are observed on the surface, which can indicate encapsulation, as in the case of palladium. The average particle diameter on the surface is 50 μm.

The encapsulation of the metal into the polymer matrix may result in a decrease in the specific conductivity of the composites, since in this case, the number of contacts between the metal particles separated by the organic phase goes to zero. This conclusion is confirmed by the results of electrical conductivity measurements (see Table 1).

To clarify the chemical structure and to identify possible interactions between the organic and inorganic component X-ray phase analysis of composites was carried out. The diffraction pattern of PAn-Au sample clearly identifies the gold phase (Fig. 12a): space group *Fm-3m*, structural type of copper, specified elementary cell parameter a = 4,0781(4) Å, elementary cell volume $V = 67{,}825(11)$ Å^3 (according to accepted standard literary data a = 4,0782 Å for Au). The average size of coherent scatter domains, in the approximation—the average grain size of the phase, is 126±34 Å, average maximum internal stress in grains—5.6×10^{-3}. In addition, you can see

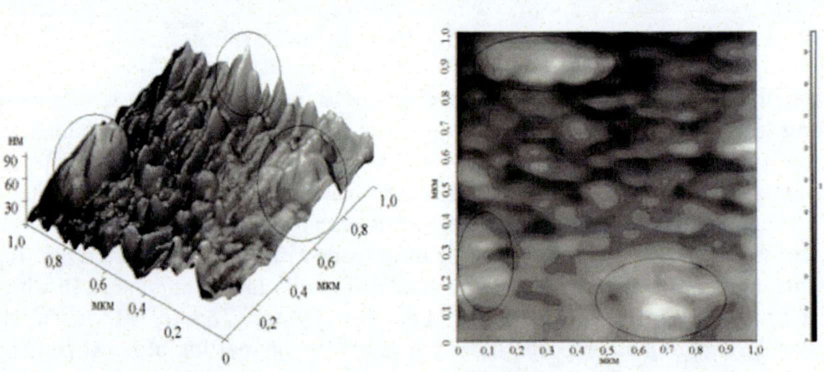

a *b*

Fig. 11 AFM image of the surface of the PAn-Ag composite with a silver content of 30 wt%: **a** – 3D spatial; **b** – 2D image (scan area 1.5×1.5 μm^2)

Table 1 Specific conductivity and polymer yield by weight for polyaniline–metal composites

Content of composite	Polymer yield, η, wt%	Specific conductivity, $\sigma \times 10^{-3}$, S/cm	The size of the globules, μm (according to SEM)
PAn-Au	79.5	61 ± 3.2	16
PAn-Pt	64.2	75 ± 4.1	35
PAn-Pd	67.3	13 ± 0.9	–
PAn-Ag	58.5	16 ± 0.8	50
Pan	55.2	36 ± 2.1	–

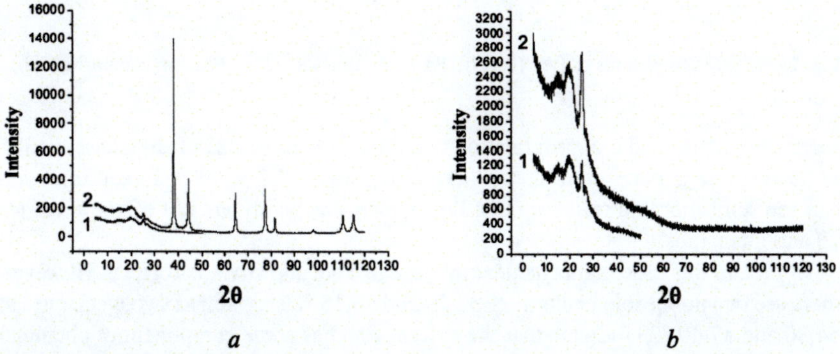

Fig. 12 Diffraction patterns of PAn (1, *a*, *b*), PAn-Au composite (2, *a*), PAn-Pt composite (2, *b*)

semi-amorphic reflexes from PAn (which practically are not offset in angles 2θ and intensities). There are no other additional peaks in the case of the PAn-Au samples. Namely, we can assume that a mechanical mixture of polyaniline with gold formed.

The diffraction pattern of PAn-Pt sample (Fig. 12*b*) identifies halo from polyaniline. There are no reflexes of the cubic face-centered lattice of platinum. Obviously, the platinum in it is in a completely amorphous state (particle size of the grains of its phase <~ 20 Å.

The diffraction pattern of PAn-Pd sample (Fig. 13*a*) shows well-formed reflexes, which obviously belong to the individual phase of $Pd_x((C_6H_7N)_n)_y$. The degree of crystallinity of the sample is good, although the high background line and its nature (coincidence with the halo for pure PAn) indicate the presence of a semi-amorphic component in the sample. Analytical indexing of the diffraction maxima indicated that the new compound crystallized in monoclinic synonymy, with the parameters of the unit cell $a = 23.066(9)$, $b = 4.757(2)$, $c = 6.063(2)$ Å, $\beta = 95.04(3)°$, $V = 662.77$ Å^3. Possible space groups are *C*2, *C*m, *C*2/*m* або *P*2$_1$/*a*.

The diffraction pattern of PAn-Ag sample (see Fig. 13*b*) clearly identifies the silver phase: the space group *Fm*-3*m*, structural type of copper, specified elementary cell parameter $a = 4{,}0858(4)$ Å, elementary cell volume $V = 68{,}207(12)$ Å^3 (according to accepted standard literary data $a = 4{,}0857$ Å for Ag). The average size of coherent

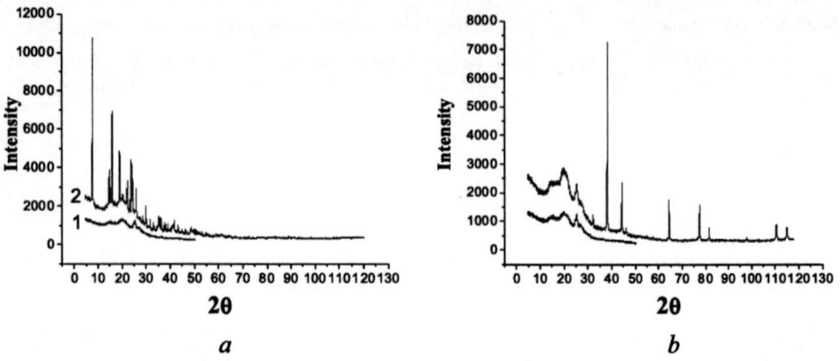

Fig. 13 Diffraction patterns of PAn (1, a, b), PAn-Pd composite (2, a), PAn-Au composite (2, b)

scatter domains, in the approximation—the average grain size of the phase, is 305 $\pm$ 61 Å, average maximum internal stress in grains—2.2×10^{-3}. In addition, you can see semi-morphic reflexes from PAn which practically are not offset in angles 2θ and intensities.

In fact, in this case, a mechanical mixture of PAn and silver is formed. However, there are two noticeable peaks at 2θ 32,20 and 46,15 °2θ, as well as two weak ones at 54,80 and 57,40 °2θ indicate that the system also has some component of chemical interaction with the formation of the composite phase PAn-Ag in a small amount, which is manifested in the existence of such reflexes.

4 Conclusion

Thus, the obtained polyaniline composites with noble metals have a semiconductor character of the conductivity; their morphology depends largely on the nature of the metal compound used for synthesis.

The X-ray structural studies and a scanning electron microscopy with energy-dispersive spectroscopy showed that the PAn-Au composite is characterized by the formation of uniformly distributed metallic grains in the polymer with the size of 12.6 $\pm$ 3.4 nm. In the PAn-Ag composite the main mass of metal with an average particle size of 30.5 $\pm$ 3.5 nm is included in the polymer matrix. The diffractogram of the PAn-Pt composite has only an amorphous halo, which is inherent to the polymer, while the energy-dispersion spectroscopy indicates the presence of metal in the composite. Obviously, platinum is in a completely amorphous state (the size of the particles of its phase grains is ~2 nm). The most interesting was the PAn-Pd composite, for which there are well-formed reflexes that may belong to the individual phase of $Pd_x((C_6H_7N)_n)_y$. Separate sections of the metallic phase have not been recorded on the SEM images, which suggests the formation of chemical compounds. Consequently, the metal fragment is included in the polymer chain or covalently, or in coordination with the polymer ligands (Pt, Pd). Part of the metal enters the composite in the form of nanoparticles (Au, Ag) uniformly distributed in the polymer matrix. Depending on the nature of the inclusion of the metal in the polymer matrix, and the metal center affects both the type of bond and the degree of polymerization.

References

1. Arancis R, Joy N, Aparna P, Vijayan R (2014) Polymer grafted inorganic nanoparticles, preparation, properties, and applications: a Review. Polym Rev 54:268–347
2. Erokhin V, Ram MK, Yavuz Ö (2008) The new frontiers of organic and composite nanotechnology. Elsevier, Amsterdam, Boston
3. Hu C-W, Yamada Y, Yoshimura K (2017) A new type of gasochromic material: conducting polymers with catalytic nanoparticles. Chem Commun 53(22):3242–3245
4. Kumar CSSR (eds) (2007) Nanomaterials for medical diagnosis and therapy. Nanotechnologies for the life sciences, vol 10. Wiley-VCH Verlag GmbH
5. Olenych IB, Aksimentyeva OI, Tsizh B, Horbenko Yu (2018) Transport and relaxation of charge in organic-inorganic nanocomposites. Acta Phys Pol, A 133(4):851–855
6. Pandey S (2016) Highly sensitive and selective chemiresistor gas/vapor sensors based on polyaniline nanocomposite: a comprehensive review. J Sci Adv Mater Devices 1:431–453
7. Dong X, Zhang X, Wu X, Cui H, Chen D (2016) Investigation of gas-sensing property of acid-deposited polyaniline thin-film sensors for detecting H_2S and SO_2. Sensors 16:1888–2003
8. Aksimentyeva OI, Savchyn VP, Dyakonov VP, Piechota S, Horbenko YY, Opainych IY (2014) Modification of polymer-magnetic nanoparticles by luminescent and conducting substances. Mol Cryst Liq Cryst 590:35–42
9. Reshenhyak O, Zaitcev V (eds) (2017) Computational and experimental analysis of functional materials. Apple Academic Press, Toronto
10. Sudha JD, Sivakala S, Prasanth R et al (2009) Development of electromagnetic shielding materials from the conductive blends of polyaniline and polyaniline-clay nanocomposite-EVA: preparation and properties. Compos Sci Technol 69:358–364
11. Stejskal J, Gilbert RG (2002) Polyaniline. Preparation of a conducting polymer. Pure Appl Chem 74:857–867
12. Chiang JC, Mac Diarmid AG (1986) Polyaniline: protonic acid doping of the emeraldine form to the metallic regime. Synth Met 13:193–205
13. Stejskal J, Exnerova M, Moravkova Z et al (2012) Oxidative stability of polyaniline. Polym Degrad Stab 97:1026–1033
14. Le T-H, Kim Y, Yoon H (2017) Electrical and electrochemical properties of conducting polymers. Polymers 9(4):150
15. Posudievsky OY, Kurys YI, Pokhodenko VD (2004) 2-Phosphormolibdic acid doped polyaniline-V_2O_5 composite. Synth Metals 144:107–111
16. Mallick K, Witcomb MJ, Dinsmore A et al (2005) Polymerization of aniline by auric acid: formation of gold decorated polyaniline nanoballs. Macromol Rapid Commun 26:232–235
17. Huang K, Zhang Y, Huang K, Long Y et al (2006) Preparation of highly conductive, self-assembled gold/polyaniline nanocables and polyaniline nanotubes. Chem A Eur J 12:5314–5319
18. Guoand L, Peng Z (2008) One-pot synthesis of carbon nanotube-polyaniline-gold nanoparticle and carbon nanotube-gold nanoparticle composites by using aromatic amine chemistry. Langmuir 24:8971–8975

19. Mallick K, Witcomb MJ, Dinsmore A et al (2005) Fabrication of a metal nanoparticles and polymer nanofibers composite material by insitu chemical synthetic route. Langmuir 21:7964–7967
20. Liu FJ, Huang LM, Wen TC et al (2007) Interfacial synthesis of platinum loaded polyaniline nanowires in poly(sterenesulfonicacid). Mater Lett 61:4400–4405
21. Breimer MA, Yevgeny G, Sy S et al (2001) Incorporation of metal nanoparticles in photopolymerized organic conducting polymers: a mechanistic insight. Nano Lett 1:305–308
22. Sivakumar MA, Gedanken A (2005) Sonochemical method for the synthesis of polyaniline and Au-polyaniline composite susing H_2O_2 for enhancing rate and yield. Synth Met 148:301–306
23. Pillalamarri S, Blum FD, Touhiro AT et al (2005) One-pot synthesis of polyaniline metal nanocomposites. Chem Mater 17:5941–5944
24. Li L, Yan G, Wu J et al (2008) Preparation of polyaniline-metal composite nanospheres by in-situ microemulsion polymerization. J Colloid Interface Sci 326:72–75
25. Kinyanjui JM, Hanks J, Hatchettet DW et al (2004) Chemical and electrochemical synthesis of polyaniline/gold composites. J Electrochem Soc 151:113–120
26. Neelgund G, Hrehorova E, Joyce M et al (2008) Synthesis and characterization of polyaniline derivative and silver nanoparticle composites. Polym Int 57:1083–1089
27. STOE WinXPOW (2010) version 3.03. Stoe & Cie GmbH, Darmstadt, Germany
28. Kraus W, Nolze G (2000) PowderCell for Windows (version 2.4). Federal Institute for Materials Research and Testing, Berlin
29. Roisnel T, Rodriguez-Carvajal J (2001) WinPLOTR: a Windows tool for powder diffraction patterns analysis. Sci Forum 378–381:118–123
30. de Keijser ThH, Langford JI, Mittemeijer EJ et al (1982) Use of the Voigt function in a single-line method for the analysis of X-ray diffraction line broadening. J Appl Crystallogr 15:308–314
31. Smirnov VI (2012) Non-destructive methods of control of parameters of semiconductor materials and structures: a textbook. Ulyanovsk. In rus
32. Tarutyna LI, Pozdniakova FO (1986) Spectral analysis of polymers. Khimia, Saint Petersburg In rus
33. Sudha JD, Sivakala S, Prasanth R et al (2009) Development of electromagnetic shielding materials from the conductive blends of polyaniline and polyaniline-clay nanocomposite-EVA: preparation and properties. Compos Sci Technol 69:358–364

Role of Active Complexes of Selected Thiopurine Derivatives with Bi(III) Ions Related to Kinetics and Mechanism of the Electrode Process in the Surfactant Presence

Agnieszka Nosal-Wiercińska and Waldemar Kaliszczak

1 Introduction

The development of industry, environmental degradation, and specific lifestyle of man led to the emergence of a new phenomenon which is civilization diseases. Among them, cancer, which is often difficult to diagnose and treat, is essential. One of the methods of systemic treatment is chemotherapy, based on drugs that affect the cell cycle, which causes it to be toxic to the whole body. There is therefore a need to study the mechanisms of these drugs' actions and to search for new systems of controlled release of drugs.

Modern electrochemistry gives such possibilities, for example, by undertaking the problem of testing electrode mechanisms, including multi-electron complex processes, in which there are intermediate stages leading to the creation of the product. The kinetics of these processes is largely dependent on the structure of the electrical double layer, which is conditioned by the phenomenon of adsorption, among others [1–4].

Studies of electroreduction processes of many depolarizers in the presence of organic substances show significant differences in the kinetics and mechanism of electrode processes [5–9]. The criteria to be met by the depolarizer/basic electrolyte/organic substance system show that the electrode processes can be accelerated by these substances ("cap-pair" rule) and point out to the mechanism of these

A. Nosal-Wiercińska (✉) · W. Kaliszczak
Faculty of Chemistry, Department of Analytical Chemistry, Institute of Chemical Sciences, Maria Curie Skłodowska University, M. Curie-Sklodowska Sq. 3, 20-031 Lublin, Poland
e-mail: anosal@poczta.umcs.lublin.pl

O. Fesenko and L. Yatsenko (eds.), *Nanomaterials and Nanocomposites, Nanostructure Surfaces, and Their Applications*, Springer Proceedings in Physics 246, https://doi.org/10.1007/978-3-030-51905-6_36

processes. The attention is paid to the large role of complexes forming on the electrode surface and the essence of the location of depolarization reduction potential in the area of labile adsorption equilibrium of organic matter [9].

The further significant impact of changes in the water activity and applied polarographically active accelerating organic substances on the mechanism and kinetics of electroreduction of Bi(III) ions [10] extended the interpretation of the mechanism of the "cap-pair" effect.

The studies of mixed adsorption layers in the aspect of their effect on the kinetics of reduction of Zn(II) ions confirmed the increased dynamics of this process in the presence of an inhibitor (n-butanol) and accelerating substances (thiourea and selected toluidine isomers) [11, 12]. The dynamics of this two-stage electrode process is reflected in a much greater increase in the value of kinetic parameters with an increase in the concentration of the accelerating organic substance in the presence of an inhibitor in the base electrolyte solution compared to a similar increase without an inhibitor. This was explained by the weaker hydration of the electrode surface and the interactions of co-adsorbate molecules.

The use of various surfactants for electrochemical studies allowed describing the mechanism of the Zn(II) ion electroreduction process aplaying various models. The catalytic effect of 1-decanesulfonic acid sodium salt on the electroreduction of Zn(II) ions is closely related to the formed active complex and the orientation of the adsorbed surfactant molecule on the mercury surface [13–15]. Among the theoretical Fawcett models, the possibility of using the EE model to describe subsequent stages of the electrode reaction was excluded [16].

Based on of the above, the study of the kinetics and mechanism of the electrode process in the presence of 6-mercaptopurine (6MP) (Scheme 1), 6-thioguanine (6TG) (Scheme 2), azathioprine (AZA) (Scheme 3) drugs used in cancer chemotherapy, mainly in myeloid and lymphoblastic leukemia [17–21] seemed to be absolutely justified.

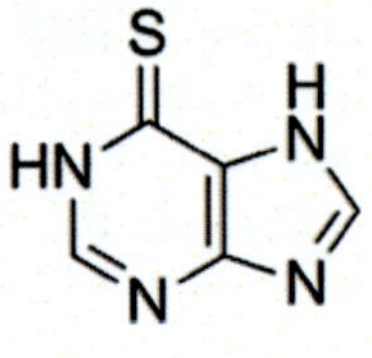

Scheme 1 6-mercaptopurine

Scheme 2 6-thioguanine

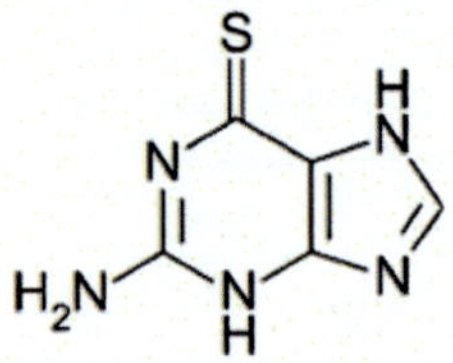

Scheme 3 Azathioprine

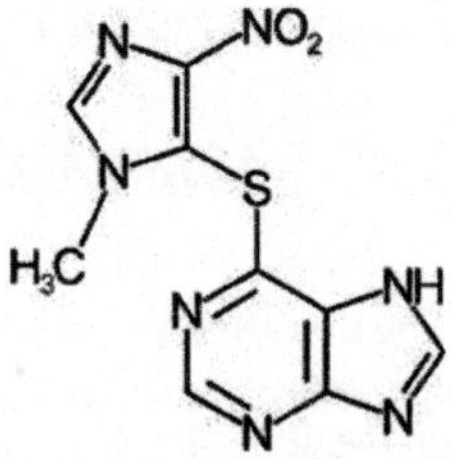

The use of Bi(III) ions as a depolarizer during the study gave the opportunity to observe changes in the kinetics of the reaction towards its acceleration. On the other hand, the introduction of basic surfactants to the electrolyte, Triton X-100 and Tween 80, changed the dynamics of the catalytic process of drugs on Bi(III) electroreduction as far as inhibition is concerned.

The research included: (i) adsorption of 6MP, 6TG, AZA, TX-100, Tween 80, and appropriate mixtures at the electrode/basic electrolyte interface; (ii) mechanism and kinetics of the Bi(III) ion electroreduction process in the presence of 6MP, 6TG, AZA, and mixed adsorption layers.

The results of the proposed research may contribute to the development of a controlled drug release system based on changing the dynamics of the electrode process kinetics. This, in turn, can be used to monitor the patient's health in order to create a relationship between the concentration of the drug in the blood and the achieved therapeutic effect which is so important in the treatment of cancer.

2 Materials and Instrumentation

$NaClO_4$, $HClO_4$, $Bi(NO_3)_3 \cdot 5H_2O$, and thiopurine derivatives such as 6- mercaptopurine, 6-thioguanine, azathioprine were purchased from Fluka. Water applied to prepare all solutions was purified in the Millipore system.

A solution of $1 \cdot 10^{-3}$ mol $\cdot$ dm^{-3} Bi(III) in the 2 mol $\cdot$ dm^{-3} chlorate(VII) was the supporting electrolyte.

There were used the following organic substance concentration ranges have been used, namely thiopurine derivatives od 0.1 do $10 \cdot 10^{-3}$ mol $\cdot$ dm^{-3}, Triton X-100, and Tween 80 from $1 \cdot 10^{-6}$ to $1 \cdot 10^{-3}$ mol $\cdot$ dm^{-3}. The solutions were prepared immediately before the measurements.

The electrochemical measurements were performed with an Autolab Fra 2/GPES (Version 4.9) frequency response analyzer (Eco Chemie, Utrecht, Netherlands). A three-electrode system consisting of Ag/AgCl/3M KCl electrode as a reference, a platinum wire as an auxiliary electrode and dropping or hanging mercury—electrode with a controlled increase rate and a constant drop surface (0.014740 cm^2), as a working electrode (MTM Poland) was used.

All electrochemical measurements were made in thermostated cells at 298 K.

3 Experimental Methods

3.1 Adsorption Measurements

The double-layer capacity (C_d) was measured using the electrochemical impedance spectroscopy (EIS). For the whole polarization range, the capacity dispersion was tested at different frequencies between 200 and 1000 Hz. To obtain proper equilibrium values of differential capacity, the linear dependence of capacity on the square element from frequency was extrapolated to the zero frequency [10, 22].

The potential of zero charge (E_z) was measured for each solution by the method of streaming the mercury electrode [10, 22]. The interfacial tension (γ_z) between mercury and the electrolyte solutions at E_z was measured with a conventional maximum bubble pressure capillary electrometer described earlier [10, 22].

3.2 Kinetic Studies

Research on the mechanism of the electrode process was associated with the need to determine kinetic parameters such as formal potential (E_f^0), reversible half-wave potential ($E_{1/2}^r$), transfer coefficients (α), standard rate constants (k_s), or apparent rate constants (k_f) of Bi(III) ion electroreduction as well as diffusion coefficients (D_{ox}). These parameters were determined in the way presented in the paper [23]. In the DC polarography, square-wave voltammetry (SWV), cyclic voltammetry (CV), and electrochemical impedance spectroscopy (EIS), the optimal experiment operating conditions were as given in the papers [24, 25].

4 Results and Discussion

As follows from the studies, 6-mercaptopurine, 6-thioguanine, azathioprine, and surfactants adsorbed on the mercury surface change the interface area structure. It was proved that thiopurine derivatives adsorb on the electrode specifically [29] whereas Tween 80 and Triton X-100 physically [26–29].

The differential capacity curves (Fig. 1) obtained in the studied systems for 6-mercaptopurine [29, 30], 6-thioguanine [31], azathioprine [32] point out to the changes in the capacity values compared with the supporting electrolyte.

In the region of the "hump" potentials, appearing in 2 mol · dm chlorate(VII) without the thiopurine derivatives (≈ −200 to −800 mV), after the addition of 6-mercaptopurine and 6-thioguanine to the solution, the height of the "hump" decreases. In turn after the introduction of azathioprine into the solution, the height

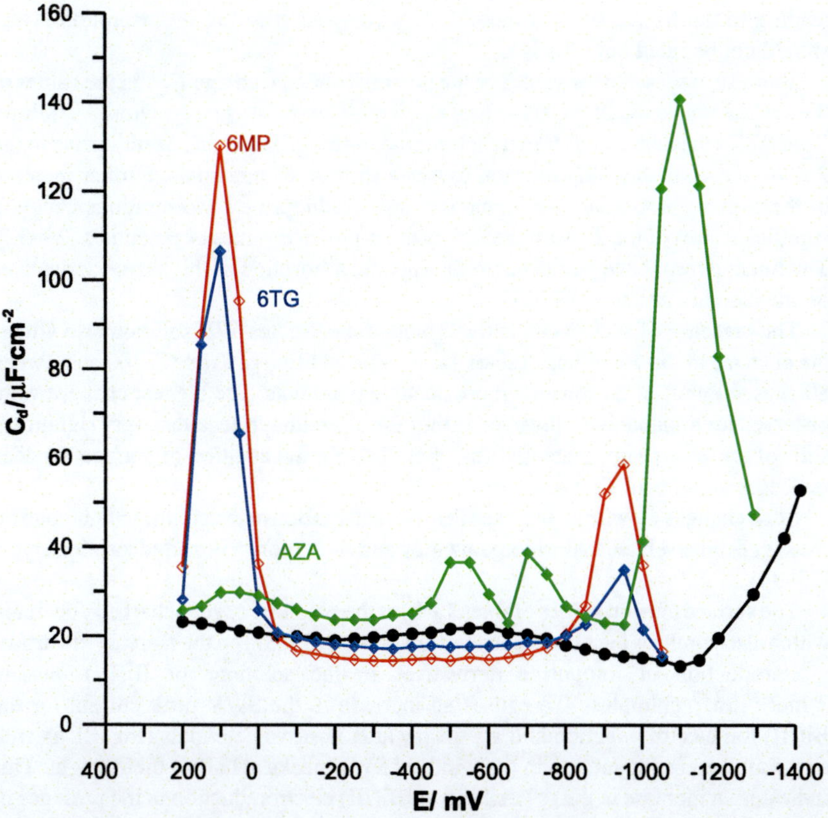

Fig. 1 Differential capacity-potential curves of double-layer interface Hg/2 mol · dm^{-3} chlorate(VII) (●) and with $1 \cdot 10^{-3}$ mol · dm^{-3} 6-mercaptopurine (◊), $1 \cdot 10^{-3}$ mol · dm^{-3} 6-thioguanine (◊) $1 \cdot 10^{-3}$ mol · dm^{-3} azathioprine (♦)

of the "hump" rises and takes the form of sharp educated peaks. In the region of higher potentials (≈100 mV) the adsorption peaks occur in the presence of 6-mercaptopurine, 6-thioguanine and azathioprine (Fig. 1), however in the area of the most negative potentials about ≈ −900 mV (in the case of 6MP and 6TG) or about ≈ −1100 mV the desorption peaks appear. It should be noted that the adsorption–desorption peaks in the case of 6MP are evidently higher which may indicate better adsorption properties. However, in the case of AZA the opposite situation, namely the desorption peak dominated the adsorption peak, is observed.

The addition of the surfactants to the basic electrolyte solution containing thiopurine derivatives results in the evident decrease of adsorption and desorption peaks [29–34]. This is associated with the influence of Triton X-100 or Tween 80 on adsorption equilibria. However, the interactions of thiopurine derivatives and surfactants leading to the formation of a more or less compact structure of adsorption layers should not be ruled out [29–34].

Table 1a, b presents the values of the potentials of zero charge E_z and the values of the surface tension γ_z at the zero charge potential for the studied systems. As follows from tables the addition of 6-mercaptopurine, 6-thioguanine, and azathioprine to the 2 mol · dm chlorates(VII) solution causes a shift of E_z potentials. Further increase in the 6-mercaptopurine, 6-thioguanine, and azathioprine concentrations causes a significant shift of the E_z values towards more positive values of potentials [29–32]. The linear dependence points out to the specific adsorption of thiopurine derivatives on the mercury surface [35–37].

The presence of surfactants in the 2 mol · dm chlorates(VII) solution also affects the changes in the E_z values (Tables 1a, b). The addition of TritonX-100 and Tween 80 causes a shift of E_z towards more positive potentials. The biggest changes were observed for thioguanine. However, in the case of azathioprine, after a very significant shift of the zero charge potential (by about 0.2 V), the addition of surfactants does not cause significant changes.

Such changes E_z with an increase in the concentration of tested surfactants confirm the suggestions related to the reorganization and formation of mixed adsorption layers [29–34].

The surface tension values (Tables 1a, b) at the potential of zero charge γ_z decrease which also confirms the phenomenon of adsorption [35–37] on the mercury electrode.

Introduction of thiopurine derivatives to the solutions of Bi(III) ions in 2 mol · dm^{-3} chlorates(VII) causes an increase in the SWV peak current for the Bi(III) ion electroreduction and a shift towards positive potentials (Fig. 2), as well as a simultaneous reduction in the width of SWV peaks at half of their height. This indicates an increase in the reversibility of Bi(III) electroreduction in the presence of the studied substances [10, 38, 39]. As can be seen from Fig. 2, the greatest catalytic effect was obtained in the presence of 6-thioguanine.

The effect of the studied surfactants on the picture of the SWV peaks Bi(III) ions electroreduction in 2 mol · dm^{-3} chlorates(VII) at the constant concentration of accelerating substance ($1 \cdot 10^{-3}$ mol · dm^{-3} 6TG, 6MP, AZA) depends on the change of electrode process reversibility towards inhibition (Figs. 2, 3, and 4). The pictures of peaks are still well defined and no drastic change of the width at half of their height is observed. That may point out compensation of the inhibiting process due to the presence of the catalyzing substance in the basic electrolyte solution [38, 39].

Only for azathioprine in the presence of Tween 80 the peak becomes not so well defined.

Similar changes in the reversibility of Bi(III) electroreduction due to the presence of thiopurine derivatives or surfactants are indicated by the CV voltammograms (Fig. 5).

Tables 1 a, b Potential of zero-charge E_Z versus Ag/AgCl electrode and surface tension γ_Z for E_Z of 2 mol · dm^{-3} chlorates(VII) solutions + 1 · 10^{-3}mol · dm^{-3} thiopurines + different surfactants concentration systems

a

$1 \cdot 10^{-3}$ mol · dm^{-3} thiopurine + $10^6 c_{TritonX\text{-}100}$/mol · dm^{-3}	0		6-mercaptopurine		6-thioguanine		Azathioprine	
	$-E_z$/V	γ_z/mN · m^{-1}	$-E_z$/V	γ_z/mN · m^{-1}	$-E_z$/V	γ_z/mN · m^{-1}	$-E_z$/V	γ_z/mN · m^{-1}
0	0.483	478.8	0.478	462.8	0.516	470.1	0.291	465.5
10	0.484	475.9	0.473	446.9	0.469	467.3	0.291	465.0
20	0.483	470.0	0.472	426.3	0.463	463.0	0.291	464.7
50	0.482	460.9	0.470	402.0	0.430	440.7	0.290	464.0
100	0.473	419.8	0.463	396.4	0.415	418.0	0.290	463.5

b

$1 \cdot 10^{-3}$ mol · dm^{-3} thiopurine + $10^6 c_{Tween80}$/mol · dm^{-3}	0		6-mercaptopurine		6-thioguanine		Azathioprine	
	$-E_z$/V	γ_z/mN · m^{-1}	$-E_z$/V	γ_z/mN · m^{-1}	$-E_z$/V	γ_z/mN · m^{-1}	$-E_z$/V	γ_z/mN · m^{-1}
0	0.483	478.8	0.478	462.8	0.516	470.1	0.291	465.5
10	0.484	475.9	0.469	463.6	0.438	450.8	0.290	464.0
20	0.483	470.0	0.466	458.0	0.425	447.8	0.285	463.0
50	0.482	460.9	0.466	434.7	0.388	430.6	0.282	462.0
100	0.473	419.8	0.460	399.2	0.379	425.0	0.273	460.0

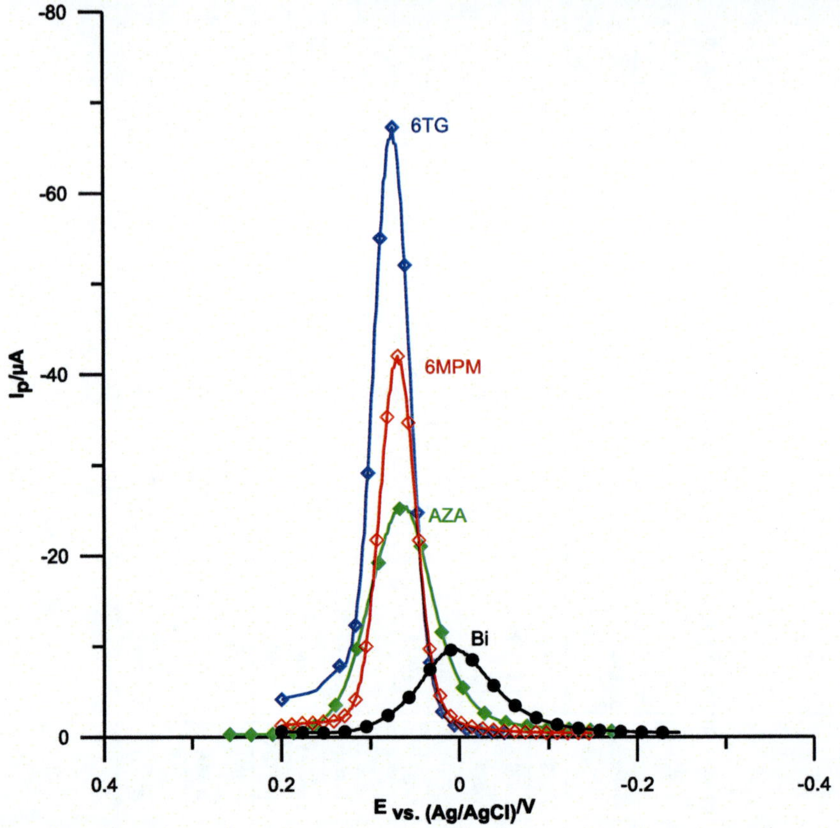

Fig. 2 The SWV peaks of $1 \cdot 10^{-3}$ mol · dm^{-3} Bi(III) (●) electroreduction in 2 mol · dm^{-3} chlorates(VII) and with $1 \cdot 10^{-3}$ mol · dm^{-3} 6-mercaptopurine (◇), $1 \cdot 10^{-3}$ mol · dm^{-3} 6-thioguanine (◇) $1 \cdot 10^{-3}$ mol · dm^{-3} azathioprine (◆)

The values ΔE_{a-c} decrease compared to those obtained for the basic electrolyte ($1 \cdot 10^{-3}$ mol · dm^{-3} Bi(III) in 2 mol · dm^{-3} chlorates(VII)). Thus, the electrode process becomes faster. This is particularly noticeable with 6TG and 6MP. The addition of surfactants to such a system affects the ΔE_{a-c} increase. There is a change in the dynamics of the catalytic action of thiopurine derivatives [14, 38, 39].

From the analysis of $\Delta E_{a-c} = f(\nu)$, dependence results that the step controlling the process of Bi(III) electroreduction in chlorates(VII) is the chemical reaction. This reaction is probably the formation of active Bi-thiopurine complexes on the electrode surface, evidently localized inside the adsorption layer (Fig. 6) which mediates in the electrons transfer [10, 38, 39]. Therefore, the aforementioned 6TG, 6MP, AZA adsorption [29–32] will not reduce the electrode surface, but will positively shift the balance of formation of these complexes.

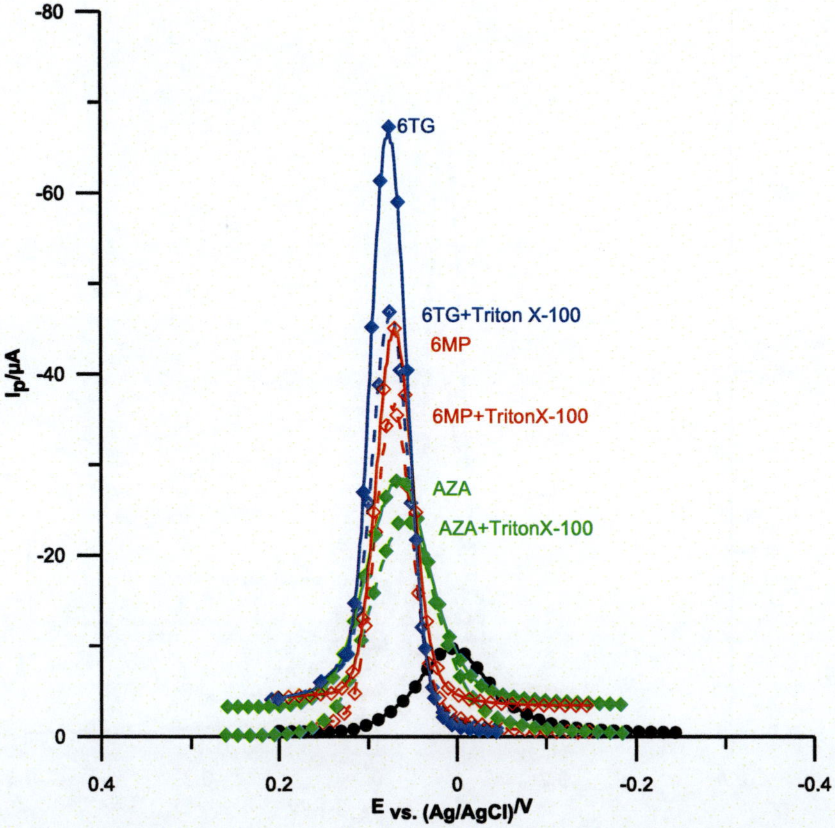

Fig. 3 The SWV peaks of $1 \cdot 10^{-3}$ mol · dm^{-3} Bi(III) (•) (—) electroreduction in 2 mol · dm^{-3} chlorates(VII) and with $1 \cdot 10^{-3}$ mol · dm^{-3} 6-mercaptopurine (—), $1{\cdot}10^{-3}$ mol · dm^{-3} 6-thioguanine (—), $1 \cdot 10^{-3}$ mol · dm^{-3} azathioprine (—) and in the system: $1 \cdot 10^{-3}$ mol · dm^{-3} 6MP + $1 \cdot 10^{-5}$ Triton X-100 (- -); $1 \cdot 10^{-3}$ mol · dm^{-3} 6TG + $1 \cdot 10^{-5}$ Triton X-100 (- -); $1 \cdot 10^{-3}$ mol · dm^{-3} AZA + $1 \cdot 10^{-5}$ Triton X-100 (- -)

However, the changes in the mechanism of Bi (III) ion electroreduction in the presence of a mixture of thiopurine derivatives and surfactants in the base electrolyte solution were associated with significant differences in ΔE_{a-c} with a change in the polarization rate, particularly above the CMC of TritonX-100 and Tween 80 [38, 39]. It is most likely that surfactant molecules block the electrode surface, pushing the previously formed Bi-thiopurine active complexes out of the adsorption layer (Fig. 6). The Bi-thiopurine type complexes have been shown to play a key role because they dominate in reaching the adsorption equilibria of the tested mixtures [38, 39].

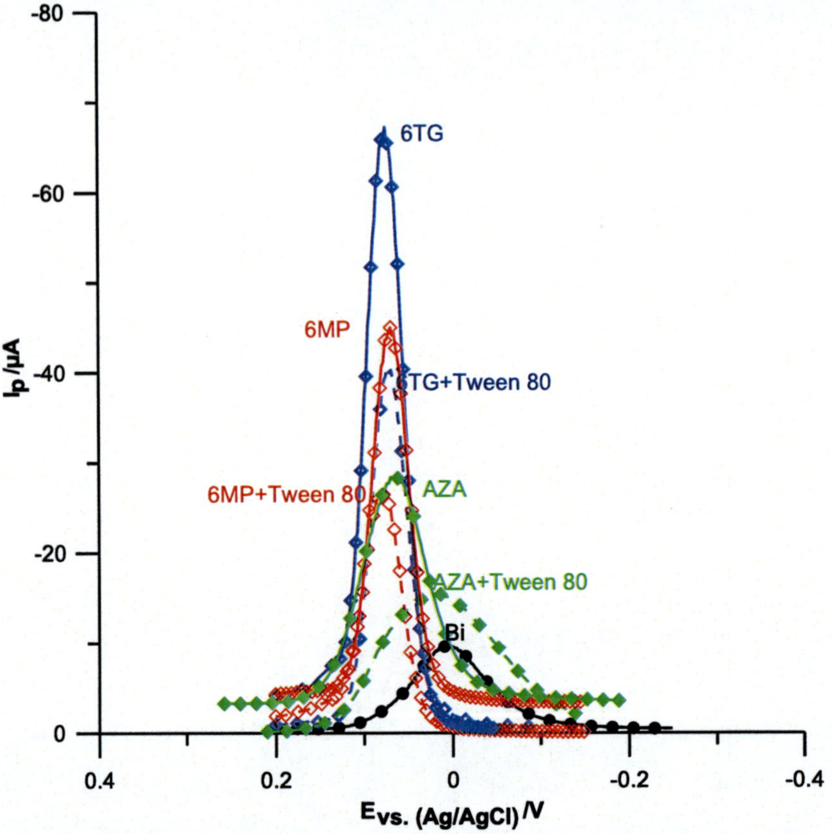

Fig. 4 The SWV peaks of $1 \cdot 10^{-3}$ mol · dm^{-3} Bi(III) (•) (—) electroreduction in 2 mol · dm^{-3} chlorates(VII) and with $1 \cdot 10^{-3}$ mol · dm^{-3} 6-mercaptopurine (—), $1 \cdot 10^{-3}$ mol · dm^{-3} 6-thioguanine (—), $1 \cdot 10^{-3}$ mol · dm^{-3} azathioprine (—) and in the system: $1 \cdot 10^{-3}$ mol · dm^{-3} 6MP + $1 \cdot 10^{-5}$ Tween 80 (- -); $1 \cdot 10^{-3}$ mol · dm^{-3} 6TG + $1 \cdot 10^{-5}$ Tween 80 (- -); $1 \cdot 10^{-3}$ mol · dm^{-3} AZA + $1 \cdot 10^{-5}$ Tween 80 (- -)

The $lnk_f = f(E)$ dependencies [10, 38, 39] (obtained by the electrochemical impedance spectroscopy) are not linear for all the studied systems which point out to the multi-stage process of Bi(III) ion electroreduction in 2 mol · dm^{-3} chlorates(VII).

The kinetic parameters αn_α and k_s determined using electrochemical techniques indicate the catalytic effect of thiopurine derivatives as well as changes of its amount due to the presence of the 6TG–TritonX-100 i 6TG–Tween 80, 6MP–TritonX-100 i 6MP–Tween 80, or AZA–TritonX-100 i AZA–Tween 80 mixtures (Tables 2a, b).

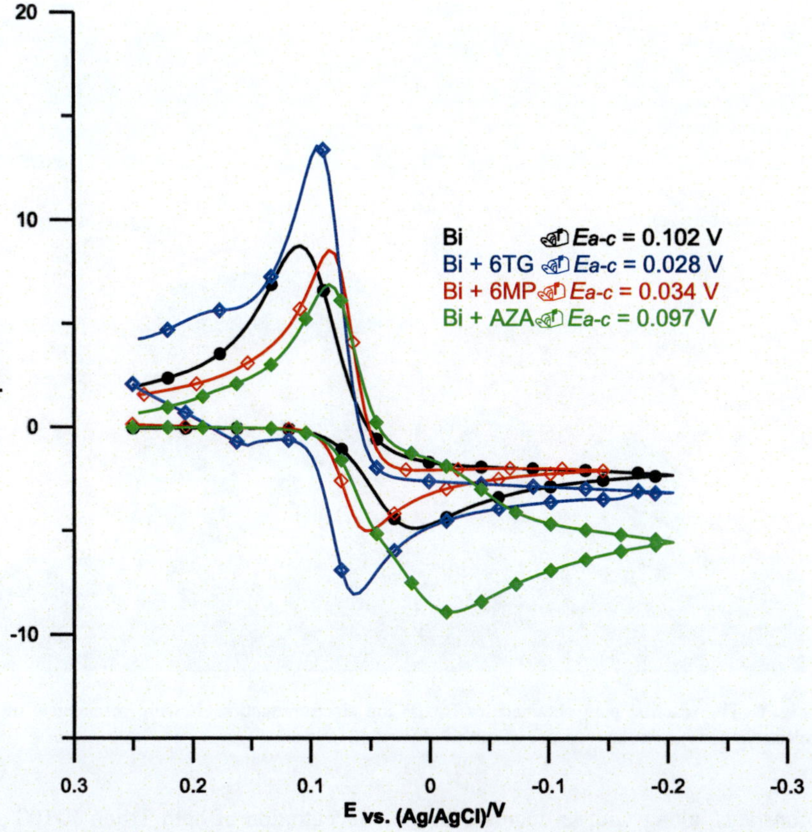

Fig. 5 The cyclic voltammograms of $1 \cdot 10^{-3}$ mol · dm^{-3} Bi(III) (●) electroreduction in 2 mol · dm^{-3} chlorates(VII) and with $1 \cdot 10^{-3}$ mol · dm^{-3} 6-mercaptopurine (◊), $1 \cdot 10^{-3}$ mol · dm^{-3} 6-thioguanine (◊) $1 \cdot 10^{-3}$ mol · dm^{-3} azathioprine (♦)

The increase in the value of transition coefficients, after introducing 6-thioguanine, 6-mercaptopurine, and azathioprine into the basic electrolyte solution, indicates a significant increase in the reversibility of the Bi(III) ion electroreduction process [38, 39]. The values of standard constant rates k_s confirm the catalytic activity of 6MP, 6TG, and AZA [10, 38, 39]. Probably the size of the effect is related to the equilibrium of the formation of the active complexes before passing more electrons [10]. On the other hand, the addition of surfactants to the tested system indicates the downward trend of α. This also contributes to a decrease in the value of rate

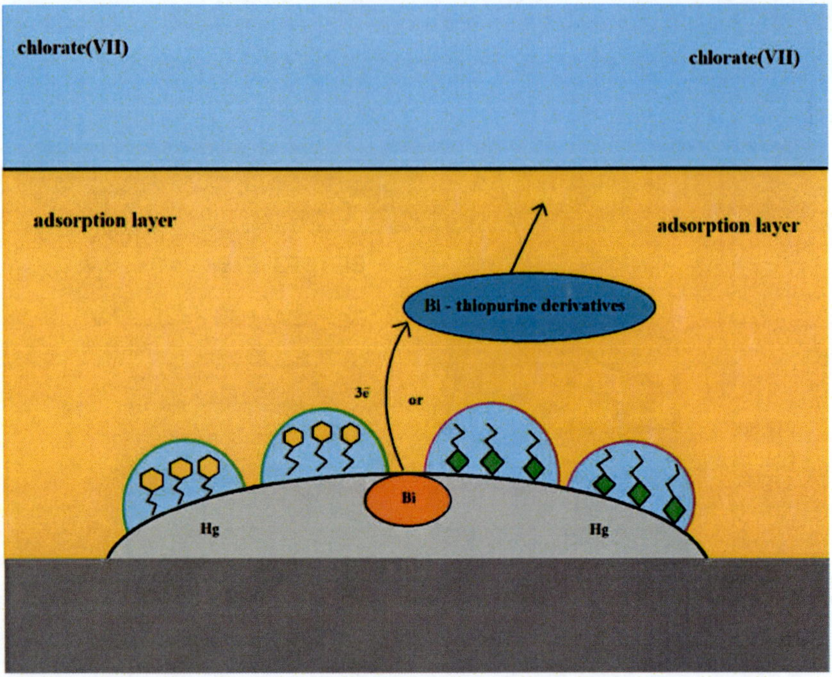

Fig. 6 The reaction path obtained for Bi(III) ion electroreduction showing mechanism of the electrode process in the presence of mixed adsorption layers

constants, along with an increase in the concentration of both Triton X-100 and Tween 80 in the chlorate(VII) solution containing $1 \cdot 10^{-3}$ mol $\cdot$ dm^{-3} thiopurine derivatives [38, 39]. Blocking the surface of the electrode, surfactant molecules push the above-mentioned active complexes out from the Bi-thiopurine adsorption layer. It should be also noted that for the same concentration of surfactants the values k_s are always lower in the presence Tween 80 (a larger molecule compared to TritonX-100), which proves that it is a more effective inhibitor of Bi(III) ions depolarization on a mercury surface.

The mechanism of catalytic action, for example, amino acids, proposed in the literature [10, 40] on electroreduction of Bi(III) ion with amalgam formation, taking into account the intermediate stages of the formation of active complexes, is most likely related to the examined systems [38, 39]. The presence of the surfactants changes the dynamics of the electrode process kinetics as a result of the aforementioned change in the mechanism of this process. This may indicate trends towards further research on the mechanisms of drug action, their understanding, modification, and control.

Tables 2 **a, b** The values of cathodic transition coefficients α and standard rate constants k_s [38, 39] of $1 \cdot 10^{-3}$ mol · dm^{-3} Bi(III) ion electroreduction in 2 mol · dm^{-3} chlorates(VII) and in the presence of $1 \cdot 10^{-3}$ mol · dm^{-3} thiopurines + different surfactants concentration systems

a

$1 \cdot 10^{-3}$ mol · dm^{-3} thiopurine + $10^6 c_{TritonX\text{-}100}$/mol · dm^{-3}	0		6-mercaptopurine		6-thioguanine		Azathioprine	
	α	10^3 k_s/cm · s^{-1}	α	$10^3 k_s$/cm · s^{-1}	α	10^3 k_s/cm · s^{-1}	α	10^3 k_s/cm · s^{-1}
0	0.28	0.197	0.68	9.180	0.68	9.760	0.51	7.360
10	0.28	0.197	0.68	8.850	0.68	9.755	0.46	7.300
20	0.28	0.197	0.67	7.190	0.67	9.680	0.43	7.150
50	0.28	0.197	0.64	5.082	0.67	9.540	0.42	7.000
100	0.28	0.197	0.60	0.925	0.66	6.320	0.38	5.250

b

$1 \cdot 10^{-3}$ mol · dm^{-3} thiopurine + $10^6 c_{Tween80}$/mol · dm^{-3}	0		6-mercaptopurine		6-thioguanine		Azathioprine	
	α	10^3 k_s/cm · s^{-1}	α	10^3 k_s/cm · s^{-1}	α	10^3 k_s/cm · s^{-1}	α	10^3 k_s/cm · s^{-1}
0	0.28	0.197	0.68	9.180	0.68	9.760	0.51	7.360
10	0.28	0.197	0.67	7.765	0.68	8.650	0.43	7.250
20	0.28	0.197	0.66	4.370	0.67	4.987	0.41	4.100
50	0.28	0.197	0.63	3.280	0.67	5.120	0.38	3.174
100	0.28	0.197	0.57	0.218	0.63	3.050	0.37	0.190

References

1. Alias K, Fawcett WR, Parsons R (1974) Effects of adsorbed ions on simple electrode reactions. Part 1. Reduction of periodate ion in the presence of toluene-p-sulphonate anions. J Chem Soc Faraday Trans 1(70):1046–1056
2. Parsons R (1980) Adsorption phenomena in electrochemistry. Surf Sci 101:316–326
3. Kalvoda R (2007) Is polarography still attractive? Chem Anal 52:869–873
4. Barek J (2013) Possibilities and limitations of mercury and mercury-based electrodes in practical electroanalysis of biologically active organic compounds. Port Electrochim Acta 31:291–295
5. Ikeda O, Watanabe K, Taniguchi Y, Tamura H (1984) Adsorption effect of highly polarizable organic compounds on electrode kinetics. Bull Chem Soc Jpn 57:3363–3367
6. Dalmata D (2005) Kinetics and mechanism of Zn(II) ions electroreduction catalyzed by organic compounds. Electroanalysis 17:789–793
7. Nosal-Wiercińska A (2010) Catalytic activity of thiourea and its selected derivatives on electroreduction of In(III) in chlorates(VII). Cent Europ J Chem 8:1–11
8. Sykut K, Dalmata G, Nieszporek J (1998) The catalysis of the reduction of Bi^{III} ions by methionine. Electroanalysis 10:458–461
9. Sykut K, Dalmata G, Nowicka B, Saba J (1978) Acceleration of electrode processes by organic compounds—"cap-pair" effect. J Electroanal Chem 90:291–302
10. Nosal-Wiercińska A (2014) Intermolecular interactions in systems containing Bi(III)–ClO_4^-–H_2O–selected amino acids in the aspect of catalysis of Bi(III) electroreduction. Electroanalysis 26:1013–1023
11. Saba J, Sykut K, Nieszporek J, Gugała D (2001) Influence of the mixed adsorption layer 1-butanol/thiourea on the two-step electroreduction of zinc(II) ions. Croat Chem Acta 74:75–89
12. Saba J, Nieszporek J, Gugala D, Sienko D, Szaran J (2003) Influence of the mixed adsorption layer of 1-butanol/toluidine isomers on the two step electroreduction of zinc(II) ions. Electroanalysis 15:33–39
13. Nieszporek J (2011) Influence of sodium 1-decanesulfonate on the enthalpy of activation of two-step Zn^{2+} ions electroreduction. J Electroanal Chem 662:407–414
14. Nieszporek J, Nieszporek K (2018) Experimental and theoretical studies of anionic surfactants activity at metal/solution interface: the influence of temperature and hydrocarbon chain length of surfactants on the zinc ions electroreduction rate. Bull Chem Soc Jpn 91:201–210
15. Perek-Dlugosz A, Socha A, Rynkowski J (2017) Electrochemical reactions of sodium 2-ethylhexyl sulfate salt. Electrocatalysis 8:270–278
16. Fawcett WR (2011) Fifty years of studies of double layer effects in electrode kinetics—a personal view. J Solid State Electrochem 15:1347–1358
17. Kowalska A (2009) Chloro and thiopurine derivatives with anticancer and immunosuppressive activity. Farm Przegl Nauk 5:15–20
18. Zochowska D, Zegarska J, Hryniewiecka E, Samborowska E, Jazwiec R, Tszyrsznic W, Borowiec A, Dadlez M, Paczek L (2016) Determination of concentrations of azathioprine metabolites 6-thioguanine and 6-methylmercaptopurine in whole blood with the use of liquid chromatography combined with mass spectrometry. Transplant Proc 48:1836–1839
19. Gowda JI, Mallappa M, Sharanappa T, Nandibewoor T (2017) CTAB functionalized multi-walled carbon nanotube composite modified electrode for the determination of 6- mercaptopurine. Sens Bio-Sens Res 12:1–7
20. Crawford DJK, Maddocks JL, Jones DN, Szawlowski P (1996) Rational design of novel immunosuppressive drugs: analogues of azathioprine lacking the 6-mercaptopurine substituent retain or have enhanced immunosuppressive effects. J Med Chem 39:2690–2695
21. Coulthard SA, Hogarth LA, Little M, Matheson EC, Redfern CP, Minto L, Hall AG (2002) The effect of thiopurine methyl-transferase expression on sensitivity to thiopurine drugs. Mol Pharmacol 62:102–109
22. Galus Z (1979) Electroanalytical methods of determination of physicochemical constants. PWN, Warsaw (in Polish)

23. Nosal-Wiercińska A (2010) The kinetics and mechanism of the electroreduction of Bi(III) ions from chlorates (VII) with varied water activity. Electrochim Acta 55:5917–5921
24. Nosal-Wiercińska A (2013) The role active complexes in the multistep process of Bi(III) ion electroreduction in chlorate (VII) solutions with varied water activity in the presence of cysteine. Electrochim Acta 92:397–403
25. Grochowski M, Nosal-Wiercińska A, Wiśniewska M, Szabelska A, Gołębiowska B (2016) The effects of homocysteine protonation on double layer parameters at the electrode/chlorates(VII) interface, as well as the kinetics and the mechanism of Bi(III) ion electroreduction. Electrochim Acta 207:48–57
26. Avranas A, Papadopoulos N, Papoutsi D, Sotiropoulos S (2000) Adsorption of the neutral macromonomeric surfactant Tween-80 at the mercury/electrolyte solution interface as a function of electrode potential and time. Langmuir 16:6043–6053
27. Munoz E, Rodriguez-Amaro R, Ruiz JJ, Avila JL, Camacho L (1992) Inhibition of the electrode reduction of an adsorbed species by the competitive adsorption of surfactant. Study of the cefazolin—Triton X-100 system at the Hg–H_2O interface. J Electroanal Chem 324:359–374
28. Sotiropoulos S, Nikitas P, Papadopoulos N (1993) Interfacial micellization of cetyl-dimethylbenzylammonium chloride and Tween 80 at the Hg/electrolyte solution interphase. J Electroanal Chem 356:225–243
29. Nosal-Wiercińska A, Kaliszczak W, Grochowski M, Wiśniewska W, Klepka T (2018) Effects of mixed adsorption layers of 6-mercaptopurine—Triton X-100 and 6-mercaptopurine—Tween 80 on the double layer parameters at the mercury/chlorates(VII) interface. J Molec Liq 253:143–148
30. Nosal-Wiercińska A, Kaliszczak W, Drapsa A, Wiśniewska M, Yilmaz S, Yagmur S, Saglikoglu G (2019) Impact of water activity on double layer parameters at the mercury/chlorates(VII) interface in the presence of mixed adsorption layers of 6-mercaptopurine–Triton X-100 Adsorption 25:819–824
31. Nosal-Wiercińska A, Kaliszczak W, Drapsa A, Grochowski M, Wiśniewska M, Klepka T (2019) Influence of nonionic surfactants and water activity on to adsorption of 6-thioguanine at the mercury/chlorates(VII) interface. Adsorption 25:251–256
32. Kaliszczak W, Grochowski M, Nosal-Wiercińska A, Brycht M, Checinska-Majak D, Gołębiowska B (2019) Effect of azathioprine on the parameters of double Hg/chlorate(VII) interface layer in the presence of nonionic surfactants. Physicochem Probl Miner Process 55:1350–1356
33. Wiśniewska M, Chibowski S, Urban T (2015) Modification of the alumina surface properties by adsorbed anionic polyacrylamide—impact of polymer hydrolysis. J Ind Eng Chem 21:925–931
34. Wiśniewska M, Chibowski S, Urban T (2016) Adsorption properties of the nanozirconia/anionic polyacrylamide system—effects of surfactant presence, solution pH and polymer carboxyl groups content. Appl Surf Sci 370:351–356
35. Nosal-Wiercińska A, Dalmata G (2010) Adsorption of methonine at mercury/aqueous solution of chlorate(VII) interface: dependence on the supporting electrolyte concentration. Electroanalysis 207:198–202
36. Nosal-Wiercińska A, Grochowski M (2011) Adsorption of thiourea and its methyl derivatives from chlorate(VII) with varied water activity. Collect Czech Chem Commun 76:265–275
37. Nosal-Wiercińska A, Wiśniewska M, Grochowski M, Kaliszczak W, Skrzypek S, Brycht M, Guziejewski D, Franus W (2017) The effect of homocysteine and homocystine protonation on double-layer parameters at the electrode/chlorates(VII) interface. Adsorpt Sci Technol 35:396–402
38. Kaliszczak W, Nosal-Wiercińska A (2018) The importance of the active complexes of 6-mercaptopurine with Bi(III) with regards to kinetics and electrode mechanism changes in the presence of non-ionic surfactants. J Electroanal Chem 828:108–115
39. Kaliszczak W, Nosal-Wiercińska A (2019) Influence of mixed 6-thioguanine-nonionic surfactant adsorption layers on kinetics and mechanism of Bi(III) ion electroreduction. Electrocatalysis 10:621–627

40. Nosal-Wiercińska A, Grochowski M, Wiśniewska M (2018) Effects of amino acids protonation on double-layer parameters of the electrode/chlorates(VII) interface, as well as kinetics and mechanism of Bi(III) ion electroreduction in the aspect of the "Cap-Pair" effect. Springer Proc Phys 210:285–300

Physical–Chemical Properties of Magnetite Nanoparticles Doped with Ag(I) and Au(III) Cations

O. Lavrynenko, N. Dudchenko, O. Pavlenko, and A. Brik

1 Introduction

Development of the material science of nanostructures and nanotechnologies alongside with biomedicine makes necessary in the creation of new kind biocompatible materials that are characterized by relevant physical–chemical properties [1]. Due to the combination of magnetic, optical, and catalytic properties, the core and shell-type nanocomposites consisted from superparamagnetic iron-bearing particles and clusters of precious metals on their surface belong to the most suitable structures for the creation of numerous materials for biology and medicine [2, 3]. At the same time, gold and silver nanoparticles are in demand to their practical biomedical application due to innative surface plasmonic resonance in a wide frequency range of exciting irradiation [4], biocompatibility, and selectivity in respect to varies biological substrates, high electro-conductivity, and sorption activity [5–7]. The coating of magnetic nanoparticles by precious metals, especially gold and silver, leads to their stabilization in corrosive biological media [8]. Meanwhile, the iron oxide nanoparticles, especially magnetite, doped with metal cations, assume new unique properties that will provide the enhancing their application field. So, the presence of precious metal atoms in the magnetite structure may change the physical–chemical properties of iron oxides. For instance, when aurum was included in magnetite crystal lattice the magnetic properties of nanocomposite became lower values in comparison to

O. Lavrynenko (✉) · O. Pavlenko
Frantsevich Institute for Problems of Material Science, NAS of Ukraine, Kiev, Ukraine
e-mail: alena.lavrynenko@gmail.com

N. Dudchenko · A. Brik
M.P. Semenenko Institute of Geochemistry, Mineralogy and Ore Formation, NAS of Ukraine, Kiev, Ukraine

O. Fesenko and L. Yatsenko (eds.), *Nanomaterials and Nanocomposites, Nanostructure Surfaces, and Their Applications*, Springer Proceedings in Physics 246, https://doi.org/10.1007/978-3-030-51905-6_37

impurities free magnetite and, simultaneously, such nanoparticles showed semiconducting behavior [9]. Moreover, the combination of plasmonic properties alongside magnetic was notified for magnetite particles preliminarily doped by silver [10].

The interaction of aurum (III) aquaforms with iron oxyhydroxides in natural systems may occur via adsorption mechanisms [11] and accompanied by donating electron that supplied reduction of dissolved aurum complexes to the metallic state during the steps of oxidation of the sulfide or oxide minerals [12]. Ferrous cations of magnetite or Green Rust structures can reduce aurum concentration and precipitate gold on the oxidized mineral surface. By the second way the interaction of ferrous cations, as a product of the dissolution of corresponding minerals with aurum-containing complexes takes place in the subsurface waters and water solutions. Such kind of reactions leads to the simultaneous precipitation of gold particles and disperse iron oxides. Hence, the ferrous-containing minerals and mineraloids (colloid particles) play a big role in supergene systems for the secondary silver and gold deposits formation [13] and take place in biogeochemical cycles of iron, aurum, and argentum, as well [14]. As a rule, gold species adsorption on the surface of iron oxyhydroxides accompanied by the formation of varies inner-sphere surface complexes. Generally, their chemical composition strongly depends on the concentration of hydrogen ion, so, it is a function of pH [15].

The laboratory studies of red–ox interaction of hydroxysulfate Green Rust suspense with the solution of aurichlorohydric acid showed the formation of free aurum clusters (2.9–7.7 nm) that form more complicate structures up to tens and hundreds of nanometers [12]. The results of other investigation confirmed [16, 17] that introduction of $AuCl_n(OH)_{4-n}$ solution into $GR(SO_4{}^{2-})$ dispersion under nitrogen atmosphere leads to the formation of mixed Green Rust–aurum associates and their following destruction and appearance of the individual gold and magnetite 15–30 nm particles.

While the contact of the hydroxysulfate and hydroxycarbonate Fe(II)–Fe(III) LDHs with water $AuCl_4{}^-$ complexes can carry out via solid-state Green Rust oxidation accompanied with reductive sedimentation of gold nanoparticles on their surface. For example, the size of gold particles equaled 10–60 nm when the phase $GR(CO_3{}^{2-})$ was applied as a nanoreactor, but it increased up to 20–120 nm when the phase $GR(SO_4{}^{2-})$ was used for the performing out the synthesis procedure [18]. Hence, analysis of the referred data shows that interaction of aurum-containing species with iron (oxy)hydroxide particles, (such as Fe(II)–Fe(III) layered double hydroxides (LDHs), can be carried out via varies chemical mechanisms that caused by the physical–chemical conditions of the phase formation processes and the genesis of the ferrous-bearing phases as well.

Generally, the process of the development of ferric–ferrous LDHs on the iron and steel surface was taken as a basis of the formation nanosized ferrimagnetic iron oxide particles under the process called rotation–corrosion dispergation (RCD). The process of phase formation in our system is based on simple red–ox reactions (such as the anodic dissolution of an iron-bearing component in the steel composition and oxygen depolarization in a cathodic part of the steel surface). Contact of the present system with the air influence the oxidation condition in the reaction area and promote the phase transformation of the metastable Fe(II)–Fe(III) LDHs (Green Rust) structures into ferric oxyhydroxides (lepidocrocite and goethite) in the outer layer, and, correspondingly, ferric–ferrous oxide—magnetite in the inner layer of the corrosion products. The presence of the 3d-metal cations transition goes to the non-stoichiometric spinel ferrites or magnetite formation. These formed particles are dopped with corresponding metals. But when we add some noble metal aqua forms to the water solutions, it is clear that the core and shell-type nanoparticles appear into the dispersion medium to form different ionic-stabilized sols. In our methodic system, Green Rust has role of a strong reducing agent, and because of this, it may transform to the metal state all the precious metal ions it has [16]. Lepidocrocite or goethite can bond precious metal ions via adsorption mechanism without red–ox reactions because of having a well-developed surface area [19], whereas magnetite simultaneously takes part in the various aurum and argentum aquaforms reducing and sorption [20, 21].

Our previous research showed the appearance of a continuous Green Rust layer on the steel surface in contact with water medium 1–3 h and its full oxidation into ferric oxyhydroxides within the following 3–5 h. Hence, analysis of the phase formation processes in the presence of aurum aquaforms took place on the activated steel surface or on the preliminary formed Green Rust or lepidocrocite surface layers. The present work is directed to study the phase and chemical composition, thermal behavior, surface, and magnetic properties of the nanostructures formed on the steel surface in the presence of aurum- and argentum-bearing aqua forms under the rotation–corrosion dispergation conditions.

2 Experimental

The rotating disk electrode is made of steel 3 (St3), it's composition has, %: C—0.14–0.22; Si—0.05–0.15; Mn—0.4–0.5; Cr—0.3; Ni—0.3; P—0.04; S—0.05; N—0.01. It was used to perform the phase formation process. The carrying out the synthesis procedure was performed in the open-air system under galvanostatic conditions via the rotation–corrosion dispergation route. During the experiment, our electrode surface was contacting alternately with dispersion medium and air. The steel surface was exposed to mechanical treatment and activating procedure to remove an oxidized layer and supply equal conditions for the phase formation process. We used as the dispersion medium distilled water and aurum- and argentum-containing water solutions with the precious metal concentrations in the range from 0.05 to 25 mg/dm^3. The pH value of the initial solutions lies in the range from 2.5 to 6.5 depending on the chemical composition of the solutions. The phase formation process was performed in thermostat TS-1/80-SPU and it lasted up to 24 h at T 20 and 50 °C, afterward the disk electrode was dried in the air atmosphere.

To analyze nanosized powders, we applied X-ray diffraction (XRD) method, thermal analytical measurements (TG/DTG, DTA), X-ray fluorescence spectroscopy (XRFS), Fourier transform infrared spectroscopy (FTIR), chemical–analytical investigations, magnetometry, transmission (TEM), and scanning electron microscopy (SEM).

Computer-aided X-ray diffractometer (DRON–UM1) equipped with two Soller's slits and filtered radiation of cobalt anode CoK_α was used to determine the phase composition of the high disperse samples. The rate of the recording was set $1° \cdot min^{-1}$, and the interfacial Wulff–Bragg's angle made up 80°. Lattice parameters and particle sizes were calculated according to obtained XRD-date. Thermogravimetric and differential thermal analyses (TG-DTA) of the powders was performed in the static air atmosphere by derivatograph Q-1500D (Hungary). The record was made using computer data registration. The parameters of the pattern recording were the following: the samples of 150 mg were heated at the rate 10 $°C \cdot min^{-1}$ from 20 to 1000 °C; the sensitivity was 20 mg; TG—500, DTG—500, and DTA—250. The samples were placed into a corundum crucible and covered by a quartz beaker to create an equal temperature field. FTIR spectra were recorded for the mineral phases placed on the "mirror". Thermo Nicolet FTIR spectrometer in the range 4000–400 cm^{-1} with 50 scans collected was used for measurement. A scanning electron microscopy (SEM) using the JOEL-6700 microscope was applied to estimate the morphology of the surface structures. The weight ratio of iron to gold or silver in the as-prepared samples was evaluated using an XRFS carried out in the automatic spectrometer «ElvaX» equipped with a titanium anode. The magnetic properties of the powders were studied using magnetometry performed with the help of a magnetometer equipped by Hall sensor.

3 Results

3.1 The Nanoparticles Phase Composition Formed on the Steel Surface in the Presence of Au(III) and Ag(I)

The phase composition of argentum-bearing samples formed on the St3 surface was studied in the wide range of the $AgNO_3$ concentration when c(Ag^+) was varied from 0.5 to 25 mg/dm^3. According to obtain data (Fig. 1a, b), two phases are formed in the composition of the surface structures simultaneously—spinel ferrite (magnetite) Fe3O4 (JCPDS file No 19-0629) and lepidocrocite (JCPDS file No 08-0098). The

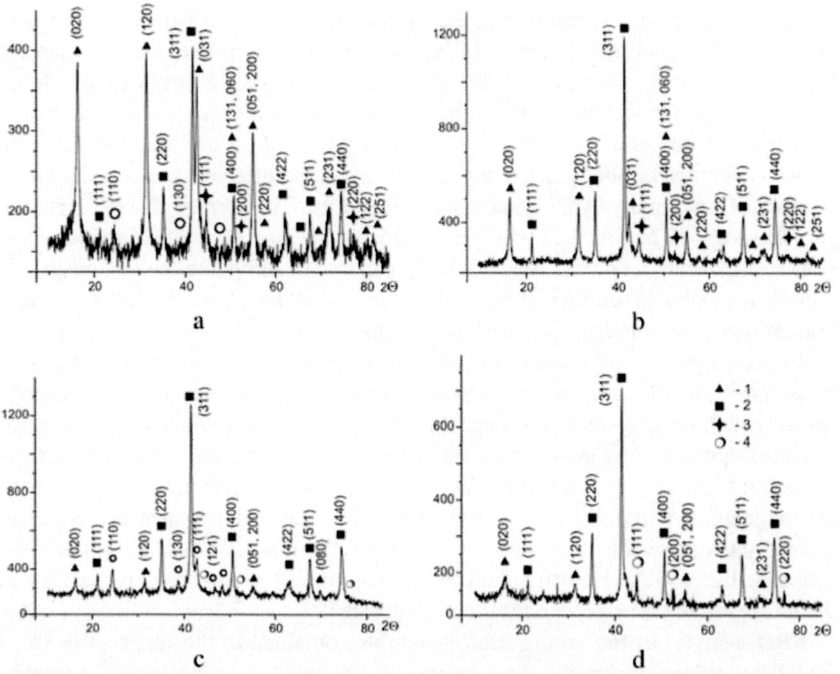

Fig. 1 XRD-patterns of the structures formed on the St3 surface contacted with the air and water solutions: **a** $AgNO_3$, $c(Ag^+) = 1$ mg/dm^3; **b** $AgNO_3$, $c(Ag^+) = 20$ mg/dm^3; **c** $HAuCl_4$, $c(Au^{3+}) = 1$ mg/dm^3; **d** $HAuCl_4$, $c(Au^{3+}) = 20$ mg/dm^3. Numbers correspond to 1—lepidocrocite; 2—magnetite; 3—silver; 4—gold

additional appearance of goethite (JCPDS file No 17-536) is closely connected with the phase transformation of lepidocrocite under oxidative conditions. The presence of small silver peaks on the XRD-patterns may be related to the reduction of Ag^0 clusters on the magnetite surface or sedimentation of silver grains as the separate phase. Overall, the intensity of the lepidocrocite reflexes is smaller than the intensity of the magnetite peaks. The relative amount of lepidocrocite peaks is gradually decreased by one third when the $c(Ag^+)$ is grown, but under the same conditions, such parameter increased threefold for magnetite. The intensity of the (111), (200), and (220) silver reflexes are growing when the argentum concentration in the solution is increased. Generally, the particle size of magnetite depends on the argentum concentration. Namely, the particles size gradually increased from 19.5 nm to 27 nm in the range of $c(Ag^+)$ from 1 to 20 mg/dm^3

The role of the chemical state of argentum species in the formation of the phase composition of the surface structures was studied using the X-ray diffraction method performed in situ. We chose pH = 6.5, when argentum was present in the solution as hydrated cations, and pH = 11, when argentum hydroxide was precipitated and dissolved in a water medium. According to the obtained data, in the presence of Ag^{+}, the primary magnetite particles appeared on the steel surface within 1 h. Lepidocrocite was formed on the St3 surface as a second phase. Development of the surface structures within 24 h led to an increase in the intensity of the reflexes in the XRD-patterns. The following phase transformation process was directed on the partial oxidation of magnetite and the appearance of the second oxyhydroxide phase—goethite. In addition, among the surface structures, we can see the small quantity of the primary hydroxycarbonate Green Rust. The average size of magnetite particles was ~22 nm. At the initial pH value 11, the typical structures formed on the steel surface were iron oxyhydroxides—lepidocrocite and goethite. The magnetite reflexes appeared on the XRD-pattern only after 24 h of the phase formation process. The average size of magnetite particles equaled 38 nm.

To determine the influence of the oxygen originate from air to the reaction area on the phase distribution, we analyzed three separate layers of particles: he outer (I) contacts with air and water solution, the inner (III) contacts with the steel surface and, corresponding, the middle one (II) between (I) and (III). The distribution of the surface phases includes decrease the lepidocrocite's part and, corresponding; increase the magnetite's part from layer (I) to layer (III). The particles size of magnetite gradually decreased from 24 nm in (I) layer to 21 nm in (II), and 19 nm in (III). Probably, the particle size can influence the processes of oxidation of magnetite and the formation of lepidocrocite shell on its surface [22].

XRD-analysis of the aurum-bearing samples obtained in the range of $c(Au^{3+})$ from 0.5 to 20 mg/dm^3 shows the formation of two-phase systems, where magnetite is estimated as a dominant component of the surface structures, while lepidocrocite is present as an admixture only. In all cases, the gold peaks (JCPDS file No 02-1095) are small because the aurum content in the nanocomposites lies on the detection limit of XRD method. Typical XRD-patterns of the surface structures are present in Figs. 1c, d.

The crystallite size of magnetite is insignificantly growing from 21 nm at $c(Au^{3+}) = 1$ mg/dm^3–24 nm at $c(Au^{3+}) = 20$ mg/dm^3. The CSR of lepidocrocite is found in the range 12.5–16 nm.

3.2 *The Thermal Behavior and Chemical Composition of the Nanoparticles Formed on the St3 Surface in the Presence of Au(III) and Ag(I)*

3.2.1 IR-FT Spectroscopy

The surface properties of magnetite nanoparticles doped with argentum cations were studied for the samples formed at $c(Ag^{+})$ 0.5–5 mg/dm^3 (Fig. 3). Analysis of infrared *spectrum (1)* corresponding to $c(Ag^{+}) = 0.5$ mg/dm^3 shows the band at 3509 cm^{-1} lied in the interval of the moisture vibration. The band at 1473–1430 cm^{-1} appears in the area of the CO_3^{2-} vibration. The band at 1369, 570, and 424 cm^{-1} can be related

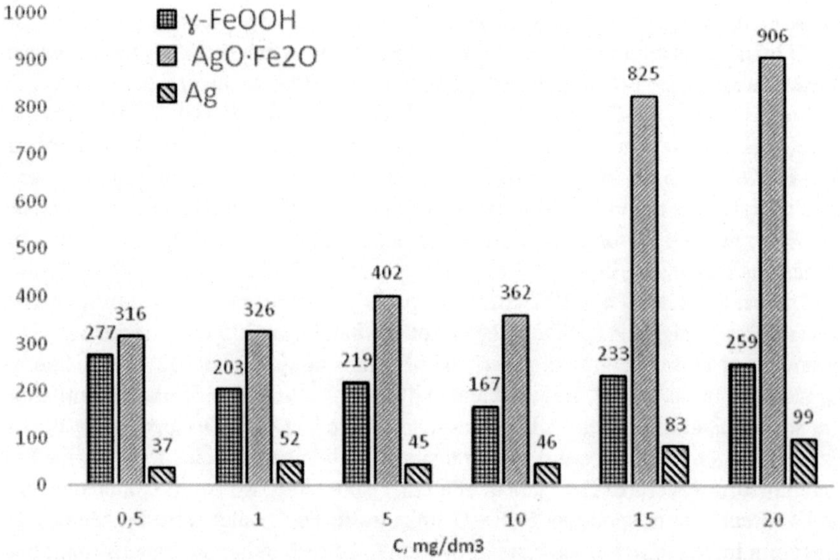

Fig. 2 Distribution (relative units) of the magnetite, lepidocrocite, and silver in the mineral composition of the surface structures formed on the St3 surface contacted with $AgNO_3$ solutions

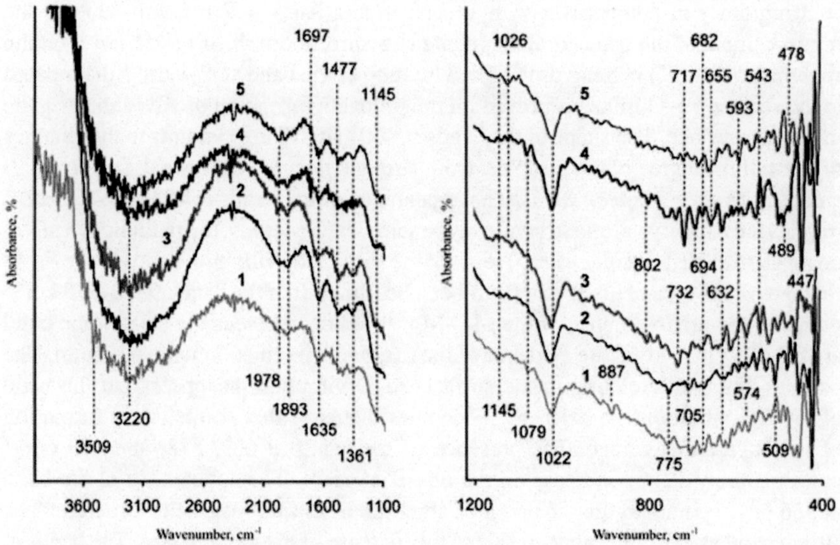

Fig. 3 FTIR spectra of magnetite nanoparticles doped with argentum cations. The initial argentum concentrations in the solutions contacted with St3 surface were, mg/dm^3: line 1–0.5; line 2–1; line 3–2; line 4–3; line 5–5

to the Fe–O linkages. The typical bands at 590 and 450 cm^{-1} corresponding to the Fe–O linkages in tetrahedral and octahedral positions of magnetite crystal lattice are present in *spectrum (1)* [23]. A band at 1149 cm^{-1} points to the presence of oxygen bridges as well as Fe–OH groups. The bands at 1029, 1164, 759 cm^{-1} correspond to γ-FeOOH and the band at 682 cm^{-1} can prove the presence of structural elements of γ-Fe_2O_3. The bands at 674, 628, and 586 cm^{-1} are typical for spinel structure and they characterize the uncoated ferrimagnetic surface. The band at 570 cm^{-1} can be related to Fe–O–Fe linkage and the bands at 555, 570, 647, and 655 cm^{-1} may be corresponded to γ-Fe_2O_3.

The bands at 3509 and 887 cm^{-1} disappear in the *spectrum (2)* corresponding to c(Ag^+) = 1 mg/dm^3 but the intensity of the band near 755 cm^{-1} increases. The bouncing of hydroxyl groups describes the bands at 1150 and 1029 cm^{-1} (the σ-OH linkage in the plane), and the band at 482 cm^{-1} (γ-OH out of plane). Probably, the strong band at 509 cm^{-1} is corresponded to α-FeOOH. Disappearance of the band at 474 cm^{-1} may confirm the transition of the nanoparticles from octahedral to tetrahedral structure. The band at 586 cm^{-1} splits into two bands centered at 593 and 582 cm^{-1} corresponding to Fe–O linkages in Fe_3O_4 that were broken due to argentum interaction, but the band at 674 cm^{-1} corresponding to spinel structure is not changed. At the same time, the band at 628 cm^{-1} related to Fe_3O_4 splits into three bands at 632, 624, and 613 cm^{-1}.

The intensity of the band at 447 cm^{-1} is growing in the *spectrum 4* (c(Ag^+) = 3 mg/dm^3) in comparison with *spectrum 3* (c(Ag^+) = 2 mg/dm^3) due to the reinforcement of the octahedral magnetite structure, and it shifts to 458 cm^{-1} on the *spectrum 5* (c(Ag^+) = 5 mg/dm^3). The vibration of the band at 474 cm^{-1} that related to the surface Fe–O linkages proved on the probable formation of silver shell on the magnetite surface. The shape of the bands at 460 and 437 cm^{-1} points to the changes of crystalline degree of the samples: from strong (*spectrum* 2) to weak (*spectrum 3*) and back to strong (*spectrum* 5). The appearance of the bands at 408 and 412 cm^{-1} implies the nanocrystalline structure of the samples. Generally, the reduction of silver on magnetite surface (*spectrum* 5) leads to the shift of the vibration bands of the Fe–O linkages to the area of great wave numbers. So, the shift of the band at 570–578 cm^{-1} points to the predominance of the M_{th}–M_{oh} structure, whereas the shift of the band at 586–593 cm^{-1} gives the evidence of the magnetite interaction with argentum. The band at 489 cm^{-1} relating to the surface Fe–O vibration disappears but the band of the OH vibrations at 1018 cm^{-1} becomes intensive that points to the formation of new hydrate coverage. The presence of the bands at 609, 578, and 478 cm^{-1} confirms the γ-Fe_2O_3 increase on the phase interface. The appearance of the band at 566 cm^{-1} points to the inclusion of argentum ions into magnetite structure. The strong band at 412 cm^{-1} appears due to the increase of the particle size. The bands at 794 and 632 cm^{-1} relate to the residual of α-FeOOH; the bands at 686 and 721 cm^{-1} correspond to γ-Fe_2O_3; the band at 740 and 663 cm^{-1} belong to γ-FeOOH; and, the bands at 505, 458, and 420 cm^{-1} can be related to α-Fe_2O_3.

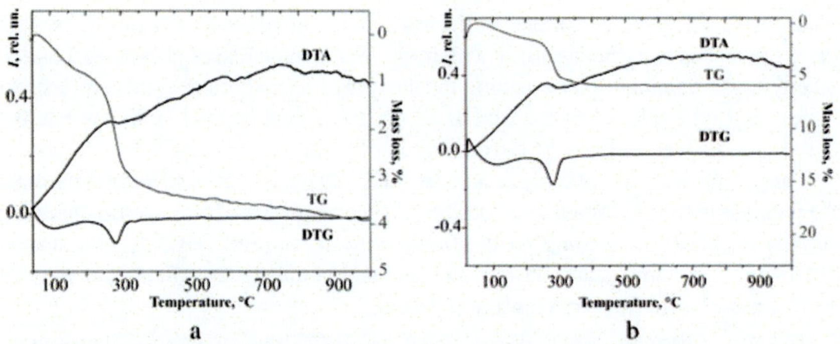

Fig. 4 TG-DTA curves of the structures formed on the St3 surface contacted with the air and water solutions: **a** $AgNO_3$, $c(Ag^+) = 5$ mg/dm^3; **b** $AgNO_3$, $c(Ag^+) = 10$ mg/dm^3

3.2.2 TG-DTA Analysis

DTA curves (Fig. 4) are sharply grown up to 280 °C. Two peaks on DTG curves indicate dehydrotation of samples at T 80–120 °C and dehydroxilation of lepidocrocite lattice at ~280 °C. The presence of silver clusters on the magnetite surface prevents the oxidation of magnetite surface accompanied by the appearance characteristic endotherm triplet at 230 °C [24]. But an exothermic hump in the range 320–340 °C may be corresponded to the phase transformation of magnetite into maghemite. The next reflexes at ~630 °C point to the polymorphic transformation of maghemite into hematite. Generally, the mass lost, % for the sample obtained at $c(Ag^+) = 5$ mg/dm^3 is 4.1 and, for the sample formed at $c(Ag^+) = 5$ mg/dm^3 it increased to 5.8%. Taking into account the same moisture of the samples, their mass loss is closely connected with the quantity of lepidocrocite part in the mineral composition of the surface structures.

3.2.3 X-ray Fluorescence Spectroscopy

The chemical composition of the nanoparticles was estimated using analytical investigation and XRF spectroscopy. Argentum-bearing samples were studied depending on the initial Ag^+ concentration and pH value of the $AgNO_3$ water solutions. Analysis of the samples was performed separately for the outer (I) and the inner (II) surface layers as well as for the precipitates in the solutions. So, the increase in the $c(Ag^+)$ from 0.5 to 20 mg/dm3 at pH value 6.5 leads to corresponding growth of the argentum part in the chemical composition of magnetite particles from 0.2 to 1.8 wt%. The part of silver in the solution precipitates was significantly bigger and reached ~40 wt% at $c(Ag^+) = 0.5$ mg/dm^3 and ~ 64 wt% at $c(Ag^+) = 5$ mg/dm^3.

Analysis of the samples formed in the wide range of pH values from 2.5 to 11.5 gives the reason to associate silver with magnetite in the range of pH values 5.6–9.5, where it was dominant phase among the structures formed on the steel surface. At lower and higher pH, when the iron oxyhydroxides were formed in the system, the correlation between silver and ferrum was not found.

The results of chemical analysis and data of X-ray fluorescence spectroscopy show the increase in the gold fraction in the chemical composition of the aurum-containing surface structures depending on the initial concentration of $HAuCl_4$. Whereas at $c(Au^{3+}) = 1$ mg/dm^3 near 0.05 wt% is contained in magnetite structure, at $c(Au^{3+}) = 20$ mg/dm^3 gold fraction grows to 0.9 wt%.

3.2.4 Chemical Composition of the Dispersion Medium Contacted with the Steel Surface

We perform the centrifugation of the solutions at 8 g within 10 min to sedimentation of nanosized argentum-containing particles in the water medium. The concentration of ferrous cations in the final solutions gradually decreased when the initial $c(Ag^+)$ was enhanced. So, depending on the pH value, the ferrous concentration varied 1.3–0.8 mg/dm^3 at $c(Ag^+) = 0.5$ mg/dm^3, it decreased to 0.04 mg/dm^3 at $c(Ag^+) = 3$ mg/dm^3, and it fully disappeared at $(Ag^+) > 5$ mg/dm^3. The average pH values of final solutions varied in the range of 6.5–7.2.

Kinetic regularities of the total iron concentration are present in Fig. 5. According to obtained data when the initial concentration of Au^{3+} is relatively low (0.5–1 mg/dm^3) the total iron concentrations are increasing during all experimental time (*curves* 1 and 2). Such dependences may point to the formation of iron–oxygen particles even when aurum cations are included in the composite structure. Under

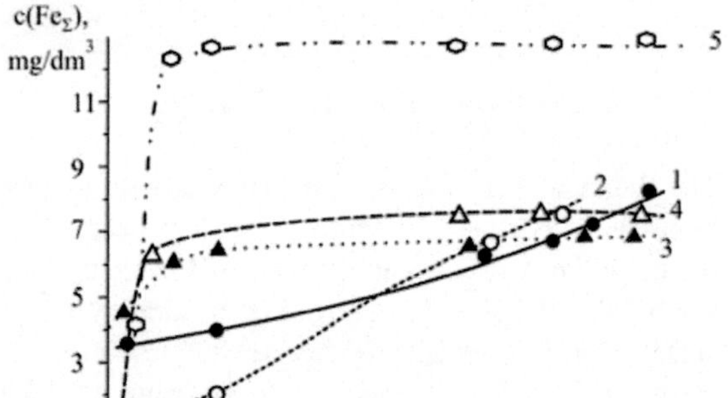

Fig. 5 Kinetic regularities of the total iron concentrations in the nanocomposite sols, where the initial $c(Au^{3+})$, mg/dm^3: 1–0.5; 2–1; 3–3; 4–5; 5–5; 5–10

the following conditions, the pH values are gradually increased from 4.8 to 5.8 and from 4.8 to 5.5 at the initial c(Au^{3+}) 0.5 and 1 mg/dm^3, respectively. At the same time, the increase in the initial c(Au^{3+}) up to 3–10 mg/dm^3 the total iron concentrations are increased to 6–12 mg/dm^3 and get stability (curves 3–5). Whereas the final pH values equal to 5.5 in all cases, but the increase in the time of phase formation process up to 24 h and more may increase the final values of hydrogen ion exponent. Generally, pH index insignificantly depended on the quantity of aurichlorohydric acid added into the work solutions. Whereas the pH value equaled 7.4 at the initial c(Au^{3+}) = 1 mg/dm^3, it decreased to 6.7at c(Au^{3+}) = 10 mg/dm^3, and the enhancement of the $HAuCl_4$ quantity did not give rise to the acidification of the work solutions. Such dependence well correlates with the property of the rotation–corrosion dispergation system steady (stationary) state to average out the pH of the dispersion medium to 7–9 in a wide range of initial values [15]. So, the formation of the nanoparticles runs out within a few hours of the contact of the steel surface with aurum-bearing water solutions. Hence, at the initial c(Au^{3+}) > 3 mg/dm^3, the system gets stability in 3–5 h.

3.3 The Morphology of the Nanostructures Formed on the Steel Surface Contacting with $AgNO_3$ and $HAuCl_4$ Solutions

The morphology of the composite structures formed on the steel surface that was contacted with water solutions containing noble metal aquaforms is present in Fig. 6. The surface structures formed at c(Ag^+) = 5 mg/dm^3 are shown in Fig. 6a, b, and at c(Ag^+) = 15 mg/dm^3, corresponding, in Fig. 6c, d. The structures obtained at c(Au^{3+}) = 20 mg/dm^3 are present in Fig. 6e, f. Independent of the precious metal concentrations, the obtained nanocomposites form aggregates that are distinguished by homogeneity and spherical or plate-like shapes. The size of individual aggregate lies in the range from 50 to 200 nm. But in all cases, the formation of big associates included more than ten aggregates clearly seen on the SEM images.

At the same time, the formation of the sore and shell type nanoparticles takes place in the water solution contacted with the steel surface. We can see the spherical particles of the following composites in Fig. 7a (Fe_3O_4 and Ag^0) and Fig. 7b (Fe_3O_4 and Au^0). Drying-out of the corresponding sol drops leads to obtain the film that included individual particles (Fig. 7c, d). In general, the particle size is 10–40 nm, but in the case of polynuclear growth of the nanocomposites, the particle size increased to 150–200 nm. Usually, the bi-phase system was formed in the sol composition included Fe_3O_4 and Ag^0 and lepidocrocite (Fig. 7e, f)) when the argentum concentration in the initial solution was relatively small (3–5 mg/dm^3).

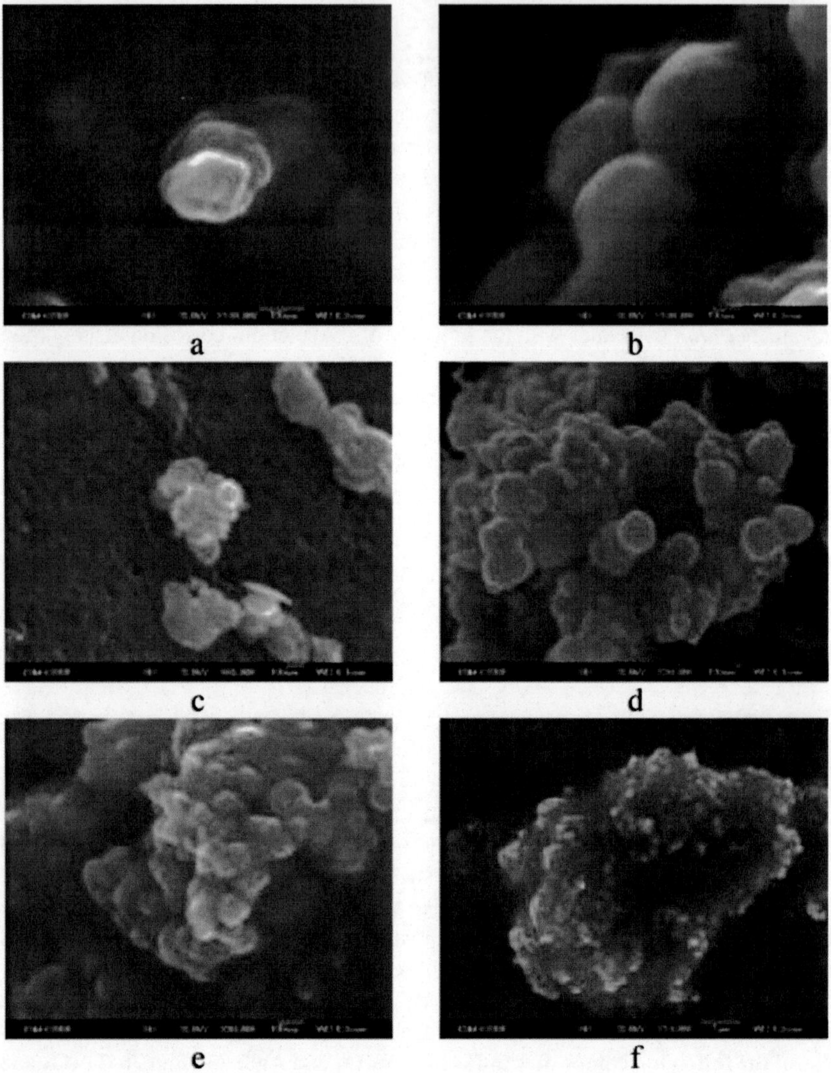

Fig. 6 The nanocomposite aggregates formed on the St3 surface contacting with noble metal solutions at **a**, **b** $c(Ag^{+}) = 5$ mg/dm^3; **c**, **d** $c(Ag^{+}) = 15$ mg/dm^3; **e**, **f** $c(Au^{3+}) = 20$ mg/dm^3

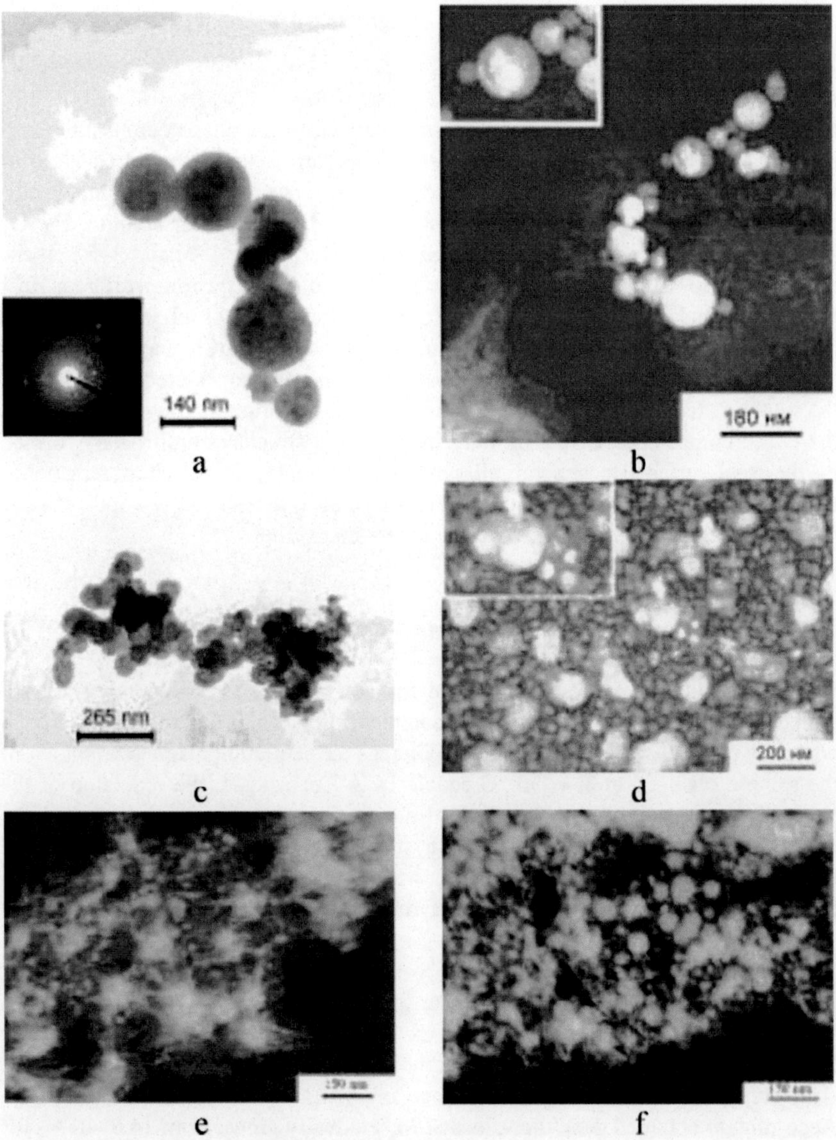

Fig. 7 TEM images of core and shell-type nanocomposites: **a** Fe_3O_4 and Ag^0; **b** Fe_3O_4 and Au^0; dehydrated drop of the sols; **c** Fe_3O_4 and Ag^0; **d** Fe_3O_4 and Au^0; bi-phases sols of Fe_3O_4 and Ag^0 and lepidocrocite formed at $c(Ag^+)$, mg/dm^3; **e** 3; **f** 5

3.4 Magnetic Properties of the Magnetite Doped with Ag^+ and Au^{3+}

The role of the oxidative state of the steel surface in the phase composition and magnetic properties of the surface nanostructures formed under the rotation–corrosion dispergation conditions were studied in the presence of aurum-containing solutions. Hence, to prepare the steel surface we used the activator solution H_2SO_4 and supply its following contact with distilled water within 1–3, and 3–5 h at the temperature 50 °C. The homogeneous surface layers of hydroxycarbonate Green Rust and, corresponding, lepidocrocite were formed as the result of such interaction. The as-prepared steel surface was moved into corresponding $HAuCl_4$ water solutions.

The magnetic characteristics of the nanopowders were considered depending on the synthesis temperature, the initial concentrations of aurum-bearing solutions, and the phase composition of the primary iron–oxygen structures preliminary formed on the steel surface. In general, all synthesized samples do not show remanence magnetization M_r and coercitivity H_c. The shape of hysteresis loops is indicative of superparamagnetic nanoparticles, in our case—magnetite.

3.4.1 The Role of the Synthesis Temperature

To compare the magnetic properties of the surface structures depending on the temperature conditions, we chose the samples formed at T 20 and 50 °C in 3 h when the $c(Au^{3+})$ was set 20 mg/dm^3. Whereas the saturation magnetization M_s of the sample obtained at $T = 50$ °C equals $14\ A \cdot m^2 \cdot kg^{-1}$, the reduction of the synthesis temperature up to $T = 20$ °C leads to decrease of M_s to $2\ A \cdot m^2 \cdot kg^{-1}$ and it results in the disappearance of a hysteresis loop. Hence, in the last case, the magnetite part of the surface structures was relatively small and the iron-oxygen particles were represented as ferric oxyhydroxides, probably, lepidocrocite.

3.4.2 The Influence of Aurum Concentration

The role of the aurum concentration on the saturation magnetization of the surface structures was studied in the wide range of $c(Au^{3+})$ from 20 to 1000 mg/dm^3. According to obtained data, the values of M_s gradually grown from 14 to $40\ A \cdot m^2 \cdot kg^{-1}$ when the initial $c(Au^{3+})$ concentration was, correspondingly, increased from 20 to 500 mg/dm^3. But the parameter M_s sharply fell to $4\ A \cdot m^2 \cdot kg^{-1}$ at $c(Au^{3+}) = 1000\ mg \cdot dm^{-3}$. In the latter case, the hysteresis loop had disturbed mode. So, we suppose the appearance among the surface structures of the second non-magnetic aurum-bearing phase, not excepting gold nanoparticles.

3.4.3 The Role of the Primary Phases Formed on the Steel Surface

We compare magnetic properties of the surface structures obtained when the aurum-bearing solutions were contacting with cleaned and activated St3 surface with the same characteristics of the powders obtained when Green Rust or lepidocrocite layers were formed preliminary on the steel surface, afterward the corroded steel was placed into contact with aurum-bearing solutions. The initial $c(Au^{3+})$ was set 50 and 500 mg/dm^3. So, at $c(Au^{3+}) = 50\ mg/dm^3$ the parameter M_s gradually increased from 18 $A \cdot m^2 \cdot kg^{-1}$ for the sample obtained on the non-oxidized St3 surface to 25 $A \cdot m^2 \cdot kg^{-1}$ for nanopowders formed on the surface Green Rust layer and to 30 $A \cdot m^2 \cdot kg^{-1}$ when the particles were formed on the lepidocrocite layer. But at $c(Au^{3+}) = 500\ mg/dm^{3,}$ the value of M_s equaled 40 $A \cdot m^2 \cdot kg^{-1}$ for the flimsy samples obtained on the cleaned steel surface and on the lepidocrocite layer. But the nanostructures formed on the Green Rust layer were characterized by lower value, where $M_s = 14\ A \cdot m^2 \cdot kg^{-1}$. Hence, we expect that the role of the primary species in the process of composite structure formation will be the most determinative, but the obtained data are ambiguous.

3.4.4 The Role of Noble Metal's Cations

The comparative study of magnetic properties of the samples formed on the steel surface contacting with aurum- and argentum-containing solutions was performed in the concentration range of the precious metals from 1 to 20 mg/dm^3. So, all samples, like a previous case, showed the absence of the coercivity and remanence magnetization. The highest values of the saturation magnetization are typical for the samples formed at the metal concentration 10 mg/gm^3. It got 49 $A \cdot m^2 \cdot kg^{-1}$ for aurum-bearing system and 24 $A \cdot m^2 \cdot kg^{-1}$ for the argentum-bearing system, respectively. The similar high $M_s = 40\ A \cdot m^2 \cdot kg^{-1}$ was found for the sample obtained at $c(Au^{3+}) = 1\ mg/dm^3$. The range of concentration 3–5 mg/dm^3 was not favorable for the formation of magnetic nanoparticles, the values of M_s did not exceed 5 (for Ag^+)–13 (for Au^{3+}) $A \cdot m^2 \cdot kg^{-1}$ But increase in the initial $c(Ag^+)$ up to 20 mg/dm^3 led to sharply decreasing of M_s to 1.3 $A \cdot m^2 \cdot kg^{-1}$.

3.4.5 Core and Shell Type Nanocomposites

Whereas the most common structures formed on the steel surface in the open-air system are magnetite and lepidocrocite, core and shell-type composite nanoparticles are typical products formed in the dispersion medium under the RCD process. Determination of the shell thickness precious metallic shell on the magnetic properties of the composite particles was studied for argentum-bearing structures. Precious metal aquaforms may be reduced on the magnetite cores as clusters, porous or compact layers depending on their initial concentration in the solution. We fixed a decrease in the saturation magnetization for Fe_3O_4 and Ag^0 nanocomposite particles from 40 to 25 $A \cdot m^2 \cdot kg^{-1}$ when the initial argentum's concentration was increased from 0.5 to 5.0 $mg \cdot dm^{-3}$ [8]. But such a trend is not confirmed for Fe_3O_4 and Au^0 nanocomposites.

So, analysis of the results of magnetic study let to confirm the association of relative high saturation magnetization for the nanopowders formed under conditions when magnetite was formed as a dominated phase. The increase in the lepidocrocite part in the phase composition of the surface structures caused a decrease in the values of M_s.

3.5 *Biomedical Application of Magnetite Nanoparticles Doped with Argentum and Aurum*

As prepared sols of nanocomposites based on magnetite doped with argentum obtained when the St3 surface was contacting with $AgNO_3$ water solutions at $c(Ag^+)$ 0.5 and 5 mg/dm^3 were used to study of the lipid peroxygenation in the lecithin-containing model system under the conditions of oxidative stress. The choice of such composites was conditional on the well-known antimicrobial and antifungal activity of silver and capability of argentum species to interact with cell membranes via toxin inactivation. So, the nanoparticles $FeFe_2O_4$ and Ag^0 were added into an incubating medium containing 1.5 mg/ml of lecithin dissolved in 20 mM Tris-HCl buffer at pH value 7.4. Incubation was performed in 1 h at $T = 37$ °C. The content of lipid peroxyl radicals was measured by chemoluminescence enhanced by luminol. The results show an insignificant decreasing of the chemoluminescence intensity of the composite particles in comparison with нульовою пробою due to the antioxidative activity of silver. Our preliminary investigation permits mare conclusion about the perspective of $FeFe_2O_4$ and Ag^0 nanocomposite sol's application to target supply of pharmaceuticals and therapy of various pathological states accompanied by the oxidative injuries of cells and bio-tissues.

The testing of antimicrobial activity of Fe_3O_4 and Ag^0 nanoparticles was performed for Gram-positive and Gram-negative bacterial cultures *S. Aureus* and *E. Coli* containing 4×10^5 CFU. Incubation lasted 24 h at T = 37 °C. The results showed the effectiveness of composite particles with respect to both kinds of species. Hence, the composite structures based on magnetite and silver were admitted as viable to the creation of new medical products directed on the medical treatment of infectious complications of the patients of resuscitation and intensive care units, especially at the development of nosocomial diseases.

At the same time, the core and shell type nanocomposite sols obtained via RCD method are capable to initiate the lipid peroxygenation in the phosphatide dispersions that accompanied by the burst of chemoluminescence even when EDTA overmeasure was added into water medium. Because the appending of chelating agents decreased the free radical generation, the obvious mechanism of the interaction in the system was provided not only with coming out of ferrous cations from composite structures but with catalytic activity of noble metal's shell.

Moreover, aurum-containing nanocomposites are characterized by the specific optical (bioplasmonic) properties. In particular, water sols of Fe_3O_4 and Au^0 exerted the fluorescence peaks in the area of 563 and 659 nm when they were exposed by the wideband light source at 500–650 nm. The following optical phenomenon appears as an effect of action on plasmons of individual particles and their associates that included a few particles as well. Such structures move of the peak to far-red—near-infrared light diapason that is the most permeable for biological tissues. So, the optical properties of nanocomposites open the perspectives for photodynamic and photothermal effects of the composite particles on tumorous tissues.

4 Conclusion

1. The contact of the steel 3 surface with the air and water solutions containing argentum and aurum aquaforms under the rotation–corrosion dispergation (RCD) conditions leads to the formation two-phase system that includes magnetite doped with Ag(I) or Au(III) cations and lepidocrocite. At the same time, the nanoparticles of core and shell-type nanocomposites are the typical products collected in the sols within the RCD process.
2. The general characteristics of the products of the phase formation process are the following. Whereas the particle size for both systems varies from 19 to 24 nm, the content of noble metals in the chemical composition of magnetite particles depends on their initial concentration and changes from 0.2 to 1.8 wt% for Ag and from 0.05 to 0.9 wt% for Au. According to a magnetic study, all samples do not show remanence magnetization and coercitivity. The shape of hysteresis loops is typical for superparamagnetics.
3. The preliminary investigation shows the row of the possible application of as-prepared nanoparticles to target the supply of pharmaceuticals and therapy of various pathological states accompanied by the oxidative injuries of cells and bio-tissues.

References

1. Zhichuan X, Yanglong H, Shouheng S (2007) Magnetic core/shell Fe_3O_4/Au and Fe_3O_4/Au/Ag nanoparticles with tunable plasmonic properties. J Am Chem Soc 129:8698–8699
2. Hoskins C, Min Y, Gueorguieva M et al (2012) Hybrid gold-iron oxide nanoparticles as a multifunctional platform for biomedical application. J Nanobiotechnol. 10: 27. https://doi.org/10.1186/1477-3155-10-27
3. Muniz-Miranda M, Gellini C, Giorgetti E, Margheri G (2017) Bifunctional Fe_3O_4/Ag nanoparticles obtained by two-step laser ablation in pure water. J Colloid Interf Sci 489:100–105

4. Haes AJ, Zou S, Schatz GC, Van Duyne RP (2004) *A Nanoscale* optical biosensor: the short range distance dependence of the localized surface plasmon resonance of silver and gold nanoparticles. J Phys Chem B 108:6961–6968
5. Ipe B, Yoosaf K, Thomas KG (2006) Functionalized gold nanoparticles as phosphorescent nanomaterials and sensors. J Am Chem Soc 128:1907–1913
6. Zhuo Y, Yuan PX, Yuan R et al (2008) Nanostructured conductive material containing ferrocenyl for reagentless amperometric immunosensors. Biomaterials 29:1501–1508
7. Daniel M-C, Astruc D (2004) Gold nanoparticles: assembly, supramolecular chemistry, quantum-size-related properties, and applications toward biology, catalysis, and nanotechnology. Chem Rev 104:293–346
8. Xu Z, Hou Y, Sun S (2007) Magnetic core/Shell Fe_3O_4/Au and Fe_3O_4/Au/Ag nanoparticles with tunable plasmonic properties J Am Chem Soc 129:8698–8699
9. Mahmood A, Ramay SM, Al-Zaghayer YS, Al Hazaa AN, Al Masary WA, Atiq S (2015) Au doping effect on the electrical and magnetic properties of Fe_3O_4 nanoparticles Modern Phys Lette B 29:33, 1550213 (9 p). https://doi.org/10.1142/s0217984915502139
10. Amendola V, Scaramuzza S, Agnoli S, Granozzi G, Meneghetti M, Campo G, Bonanni V, Pineider F, Sangregorio C, Ghigna P, Polizzi S, Riello P, Fiameni S, Nodari L (2015) Laser generation of iron-doped silver nanotruffles with and plasmonic properties nano research. https://doi.org/10.1007/s12274-015-0903-y
11. Berrodier I, Farges F, Benedetti M, Winterer M, Brown Jr GE, Deveughe'le M (2004) Adsorption mechanisms of trivalent gold on iron- and aluminum-(oxy)hydroxides. Part 1: X-ray absorption and Raman scattering spectroscopic studies of Au(III) adsorbed on ferrihydrite, goethite, and boehmite. Geochimica et Cosmochimica Acta 68(14):3019–3042
12. Heasman DM, Sherman DM, Ragnarsdottir KV (2003) The reduction of aqueous Au^{3+} by sulfide minerals and green rust phases. Am Mineralogist 88:725–738
13. Greffié C, Benedetti MF, Parron C, Amouric M (1996) Gold and iron oxide associations under supergene conditions: an experimental approach. Geochim Cosmochim Acta 60(9):1531–1542
14. Shuster JP (2013) The biogeochemical cycling of gold under surface and near-surface environmental conditions. Electronic Thesis and Dissertation Repository, p 1319. https://ir.lib.uwo.ca/etd/1319
15. Berrodier I, Farges F, Benedetti M, Winterer M, Brown Jr GE, Deveughe'le M (2004) X-ray absorption and Raman scattering spectroscopic studies of Au(III) adsorbed on ferrihydrite, goethite, and boehmite. Geochimica et Cosmochimica Acta 68(14):3019–3042
16. Ed O'Loughlin J, Kelly SD, Kemner KM, Csencsits R, Cook RE (2003) Reduction of AgI, AuIII, CuII, and HgII by FeII/FeIII hydroxysulphate green rust. Chemosphere 53:437–446
17. Ed O'Loughlin J, Kemner KM, Burris DR (2003) Effects of AgI, AuIII, and CuII on the reductive dechlorinatiom of carbon tetrachloride by green rust. Environ Sci Technol 37:2905–2912
18. Ayadi S, Perca C, Legrand L (2013) New one-pot synthesis of Au and Ag nanoparticles using green rust reactive particle as a micro-reactor. Nanoscale Res Lett 8:95. http://www.nanoscalereslett.com/content/8/1/95
19. Schoonen MAA, Fisher NS, Wente M (1992) Gold sorption onto pyrite and goethite: A radiotracer study. Geochim Cosmochim Acta 56:1801–1814
20. Ran Y, Fu J, Rate AW, Gilkes RJ (2002) Adsorption of Au(I, III) complexes on Fe, Mn oxides and humic acid. Chem Geol 185:33–49
21. Uchida A, Yokoyama T, Motomura Y (2002) Role of iron(III) and aluminum hydroxides in concentration/ reduction of Au(III) complexes. Resourc Geol 52(3):223–230
22. У.А. Дир, Р.А. Хауи, Дж. Зусман. Породообразующие минералы. Несиликатные минералы. 1966, **5**, М.: Мир, 1966
23. García Casillas PE, Rodriguez Gonzalez CA, Martínez Pérez CA (2012) Infrared spectroscopy of functionalized magnetic nanoparticles, infrared spectroscopy. Mater Sci Eng Technol. Prof. The ophanides Theophile (Ed). InTech. ISBN 978-953-51-0537-4
24. Mandal M, Kundu S, Ghosh SK, Panigrahi S, Sau TK, Yusuf SM, Pal T (2005) Magnetite nanoparticles with tunable gold or silver shell. J Colloid Interf Sci 286:187–194

Structure and Properties of Nanostructured Metallic Glass of the Fe–B–Co–Nb–Ni–Si High-Entropy Alloy System

O. I. Kushnerov, V. F. Bashev, and S. I. Ryabtsev

1 Introduction

Over the past decade, a new class of materials called multi-principal element alloys (MPEAs) or high-entropy alloys (HEAs) or complex concentrated alloys (CCAs) has received significant research attention. Such alloys usually contain from 5 to 13 elements in equiatomic or close to equiatomic concentrations [1]. The correct choice of the number of components and their concentrations makes it possible to obtain an alloy with high mixing entropy, the value of which is preserved both in the melt and after solidification. Simple solid solutions of substitution with face-centered or body-centered cubic crystal lattices are usually formed during the crystallization of such multicomponent alloys because of their high mixing entropy. On the other hand, it is possible to obtain the HEAs with a structure in which intermetallic compounds with high hardness values (Laves phases, σ phase) will be combined with a simple solid solution characterized by high ductility [1]. Many HEAs possess unique properties, such as wear resistance, resistance to corrosion and oxidation, radiation resistance, high hardness, and strength [1–6]. It should also be noted the superior biocompatibility demonstrated by some HEAs [5]. So, the HEAs may find use as materials for nuclear reactors applications, medicine, electronics devices, mechanical equipment, rocket casings, and engines, etc.

In contrast to HEAs, another type of modern material, the so-called metallic glass (MG) contains only one, occasionally two principal elements. The HEAs and MGs were explored separately until the amorphous HEAs have been successfully synthesized [1]. Such alloys were called high-entropy metallic glasses (HE-MGs). The development of HE-MGs alloys was the basis for new methods of development

O. I. Kushnerov (✉) · V. F. Bashev · S. I. Ryabtsev
Oles Honchar Dnipro National University, Gagarin Ave., 72, Dnipro 49010, Ukraine
e-mail: kushnrv@gmail.com

O. Fesenko and L. Yatsenko (eds.), *Nanomaterials and Nanocomposites, Nanostructure Surfaces, and Their Applications*, Springer Proceedings in Physics 246, https://doi.org/10.1007/978-3-030-51905-6_38

and manufacturing of MGs. The HE-MGs (as happens with substances that simultaneously belong to different types of materials [7, 8]) have excellent physical and mechanical properties combining the advantages of both metallic glasses and HEAs. So, HE-MGs exhibit considerable potential for various technical applications [9]. In particular, MGs usually are excellent soft magnetic materials. They have high saturation magnetization, low coercivity, and power loss. Till now, serials of HE-MG systems have been synthesized [1, 9–13].

This study is dedicated to developing a new nanostructured $Fe_{25}B_{17.5}Co_{21.35}Nb_{3.65}Ni_{25}Si_{7.5}$ HE-MG which combines good thermal stability with high mechanical and soft magnetic properties.

2 Materials and Methods

The as-cast sample of Fe–B–Co–Nb–Ni–Si alloy with a nominal composition of $Fe_{25}B_{17.5}Co_{21.35}Nb_{3.65}Ni_{25}Si_{7.5}$ (in at.%) was prepared with a Tamman high-temperature electric furnace in the argon gas flow using a copper mold. The mass losses during ingot preparation did not exceed 1% and the average rate of cooling was ~10^2 K/s. The as-cast ingot was thereafter remelted and the nanostructured films were obtained from the melt by splat quenching (SQ) technique. A technique for splat quenching used in the present work consisted of rapid cooling of melt drops upon their collision with the internal surface of a rapidly rotating hollow cylinder of copper. The cooling rate was estimated following the methodology proposed in [3]. We used the expression (1)

$$V = \frac{\alpha\vartheta}{c\rho\delta}, \tag{1}$$

where c is the heat capacity of film, ρ is the film density, α is the coefficient of heat transfer, ϑ is the excess temperature of the film, and δ is the thickness of film [3]. Considering the thickness of fabricated splat quenched films, the estimated rate of cooling was ~10^6 K/s. The X-ray diffraction analysis (XRD) was carried out using a DRON-2.0 diffractometer with monochromatized Cu $K\alpha$ radiation. The diffraction patterns were processed using QualX2 software [14]. The temperature dependencies of electrical resistivity of the film were measured by the four-point technique upon continuously heating and subsequently cooling in the high vacuum with a pressure of $4 \cdot 10^{-2}$ Pa. The heating rate from room temperature to ~1010 K was 0.2 K/s. Annealed films were used for structural studies by X-ray diffraction analysis (with a photographic registration, in a Debye camera on the URS-2.0 diffractometer in filtered Co K_α radiation). Debyegrams were digitally microphotometred and processed using a qualitative phase analysis software Qualx2. The thermal behavior of the nanostructured films was studied by a differential thermal analysis technique (DTA) at a heating rate of 0.2 K/s. The microhardness was examined using a tester PMT-3 at a load of 100 g. The coercive force (H_C) was measured at room temperature with a B–H curve tracer. The hysteresis loops of the samples were measured at room temperature via a vibrating sample magnetometer (VSM) with the field applied parallel to the plane of the film.

3 Results and Discussion

There are two main criteria by which the HEAs are usually characterized. This is the entropy of mixing ΔS_{mix} and the enthalpy of mixing ΔH_{mix}. However, to predict the phase composition of HEAs, some additional parameters were proposed [1, 3]. These parameters include in particular the valence electron concentration (VEC), the thermodynamic parameter Ω, which takes into account the melting temperature, mixing entropy and the mixing enthalpy. The important parameter is an atomic-size difference between alloy components which is denoted as δ.

Lately, it was found that the topological instability of atomic packing is also affected by the atomic-size difference [15]. In accordance with the results of [15], a decisive role in determining the stability of the atomic packaging in HEAs is played by atoms with minimum and maximum radii. To characterize quantitatively the effects of the atomic-size difference on the packing in HEAs it is necessary to determine the solid angle of packing ω_S for the smallest size atom

$$\omega_S = 1 - \sqrt{\frac{(r_S + \bar{r})^2 - \bar{r}^2}{(r_S + \bar{r})^2}}, \quad (2)$$

and the solid angle of packing ω_L for the largest size atom

$$\omega_L = 1 - \sqrt{\frac{(r_L + \bar{r})^2 - \bar{r}^2}{(r_L + \bar{r})^2}}. \quad (3)$$

Here, r_L is the atomic radius of the largest atom and r_S is the atomic radius of the smallest atom [15].

$$\bar{r} = \sum_{i=1}^{n} c_i r_i, \quad (4)$$

c_i—an atomic fraction of the component with number i,r_i—an atomic radius of the component with number i.

Thus, the normalized parameter γ characterizing the atomic packing state has been defined as follows:

Table 1 Electronic, thermodynamic and the atomic-size parameters of the $Fe_{25}B_{17.5}Co_{21.35}Nb_{3.65}Ni_{25}Si_{7.5}$ HEA

ΔS_{mix}, J/(mol · K)	ΔH_{mix}, kJ/mol	Ω	VEC	δ	γ
13.653	−26.295	0.987	7.439	14.41	1.545

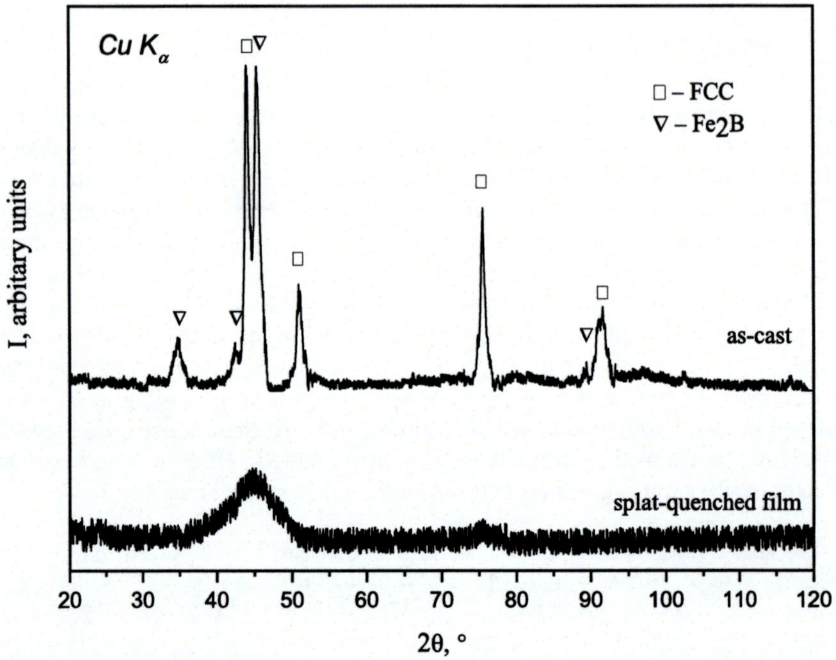

Fig. 1 XRD patterns of as-cast ingot and splat-quenched nanostructured film of $Fe_{25}B_{17.5}Co_{21.35}Nb_{3.65}Ni_{25}Si_{7.5}$ HEA

$$\gamma = \frac{\omega_S}{\omega_L}. \quad (5)$$

As pointed out in [15], the value of $\gamma = 1.175$ is a boundary between the metallic glasses or alloys with intermetallic compounds and the simple solid solution alloys. Using the data from [16–18], we calculated ΔS_{mix}, ΔH_{mix}, δ, *VEC, Ω,* and γ of the $Fe_{25}B_{17.5}Co_{21.35}Nb_{3.65}Ni_{25}Si_{7.5}$ HEA (Table 1).

According to [19], HE-MGs may form when ΔS_{mix}, ΔH_{mix} and δ simultaneously satisfy the following conditions: $7 \le \Delta S_{mix} \le 16$ J/(mol · K), $-49 \le \Delta H_{mix} \le -5.5$ kJ/mol and $\delta \ge 9$. Another condition indicating the possibility of forming metallic glass or intermetallic compounds is the value of the parameter $\Omega \le 1.1$. As we can see from the Table 1, the $Fe_{25}B_{17.5}Co_{21.35}Nb_{3.65}Ni_{25}Si_{7.5}$ alloy is fully satisfied with all of the mentioned above conditions and ΔS_{mix}, ΔH_{mix}, δ, Ω, γ are the appropriate parameters for forecasting the phase composition of HEAs. As pointed in [10] the

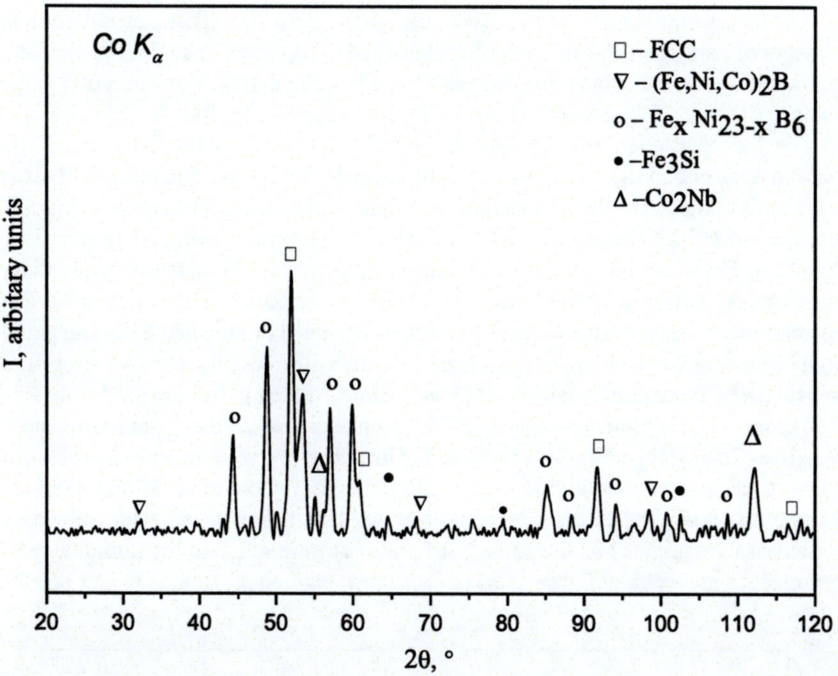

Fig. 2 XRD patterns of annealed nanostructured $Fe_{25}B_{17.5}Co_{21.35}Nb_{3.65}Ni_{25}Si_{7.5}$ HE-MG film

atomic-size difference δ is one of the most important parameters that determine the formation of a solid solution or amorphous phase in HEA. The value of δ should be large enough to provide sufficient stress at the atomic level, which is necessary for the destabilization of the solid solution phase [10]. Furthermore, to limit the long-range atomic diffusion and increase the efficiency of local packing, the negative value of ΔH_{mix} and large δ are also necessary. So, the fulfillment of these conditions leads to high glass-forming ability and ensures the suppression of the crystalline phase formation during the cooling process [10].

The parameters of the crystal lattice and the phase composition of the investigated alloy (Table 2) were determined from X-ray diffraction patterns (Figs. 1 and 2).

Table 2 The phase composition of $Fe_{25}B_{17.5}Co_{21.35}Nb_{3.65}Ni_{25}Si_{7.5}$ HEA

Alloy	Phase composition
As-cast $Fe_{25}B_{17.5}Co_{21.35}Nb_{3.65}Ni_{25}Si_{7.5}$	FCC ($a = 0.3574$ nm) + Fe_2B
Splat-quenched film $Fe_{25}B_{17.5}Co_{21.35}Nb_{3.65}Ni_{25}Si_{7.5}$	Fully glassy phase
Annealed film $Fe_{25}B_{17.5}Co_{21.35}Nb_{3.65}Ni_{25}Si_{7.5}$	FCC ($a = 0.3527$ nm) + $(Co, Ni, Fe)_2B$ + $Fe_xNi_{23-x}B_6$ + Fe_3Si + Co_2Nb

For the as-cast sample of $Fe_{25}B_{17.5}Co_{21.35}Nb_{3.65}Ni_{25}Si_{7.5}$ HEA, the structure is consists of a simple face-centered cubic phase and Fe_2B phase. The lattice parameters of the face-centered cubic phase suggest that it is formed based on γ-iron (in [3] the lattice parameter *a* of γ-iron was extrapolated at room temperature as 0.3572 nm).

For the splat-quenched $Fe_{25}B_{17.5}Co_{21.35}Nb_{3.65}Ni_{25}Si_{7.5}$ alloy films, the XRD pattern does not contain any clear crystalline peaks, only a broad maximum. This is a sign of a completely glassy structure. A coherently scattering domain size (crystallite size) of HE-MG estimated by the Scherrer formula was ~ 3 nm. After annealing to ~1010 K (above the crystallization temperature) and subsequent slow cooling, the films crystallize with the formation of several crystalline phases (Table 2). The much more complex phase composition that is observed in annealed films compared to as-cast samples is obviously explained by their low cooling rate. As a result, a substantially more equilibrium multiphase state is formed in the annealed films.

Figure 3 shows the DTA curve of the nanostructured $Fe_{25}B_{17.5}Co_{21.35}Nb_{3.65}Ni_{25}Si_{7.5}$ HE-MG film. The curve contains an obvious glass transition stage, with a supercooled liquid region after it and then an exothermic peak, indicating the crystallization. This behavior indicates the glassy nature of the sample. The temperature of glass transition T_g and the initial temperature of crystallization T_x are marked by arrows and equal to 800 K and 863 K,

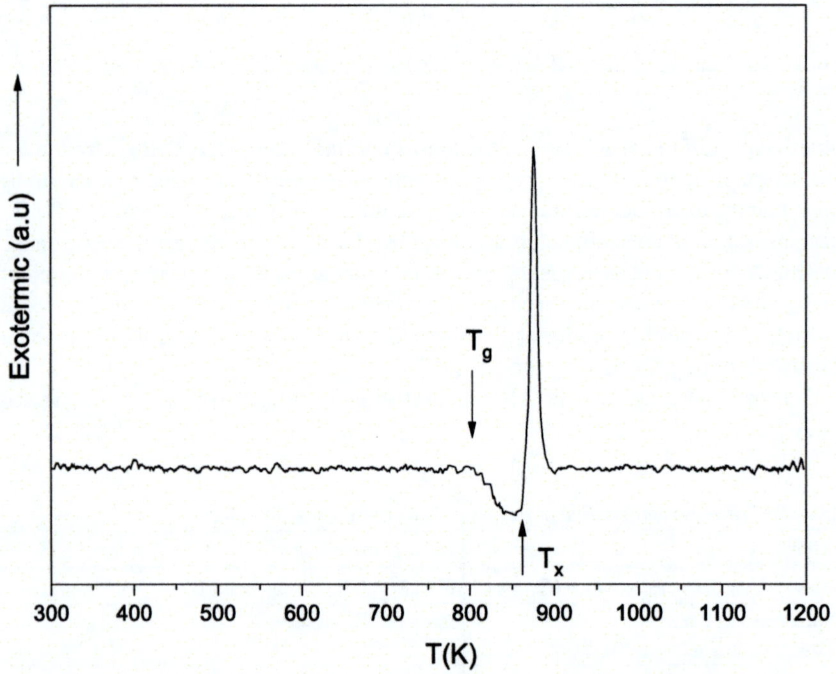

Fig. 3 DTA curve of the nanostructured $Fe_{25}B_{17.5}Co_{21.35}Nb_{3.65}Ni_{25}Si_{7.5}$ HE-MG film

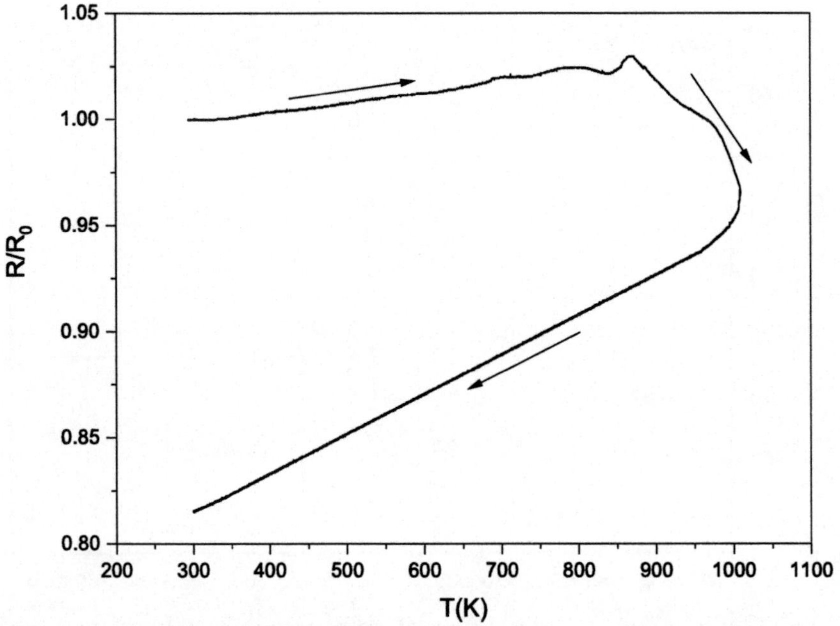

Fig. 4 The temperature dependencies of relative electrical resistivity of $Fe_{25}B_{17.5}Co_{21.35}Nb_{3.65}Ni_{25}Si_{7.5}$ HE-MG film

respectively. Considering that only one exothermic peak is visible on the DTA curve, crystallization occurs through one stage.

Let us consider the dependence of the relative electrical resistivity of the nanostructured films $R(T)/R_0$ (300 K). The beginning of an irreversible decrease of the resistivity during heating indicates the occurrence of some phase transformations in the film. As can be seen from Fig. 4, phase transformation temperature (temperature of crystallization) in $Fe_{25}B_{17.5}Co_{21.35}Nb_{3.65}Ni_{25}Si_{7.5}$ HE-MG film is ~869 K. This value agrees well with the value obtained from DTA. At the same time, the presence of bends on the curve in the phase transformation region indicates the formation of several crystalline phases. After completion of the crystallization process, the dependence of the resistance on temperature has the usual character, where the resistance decreases with decreasing temperature.

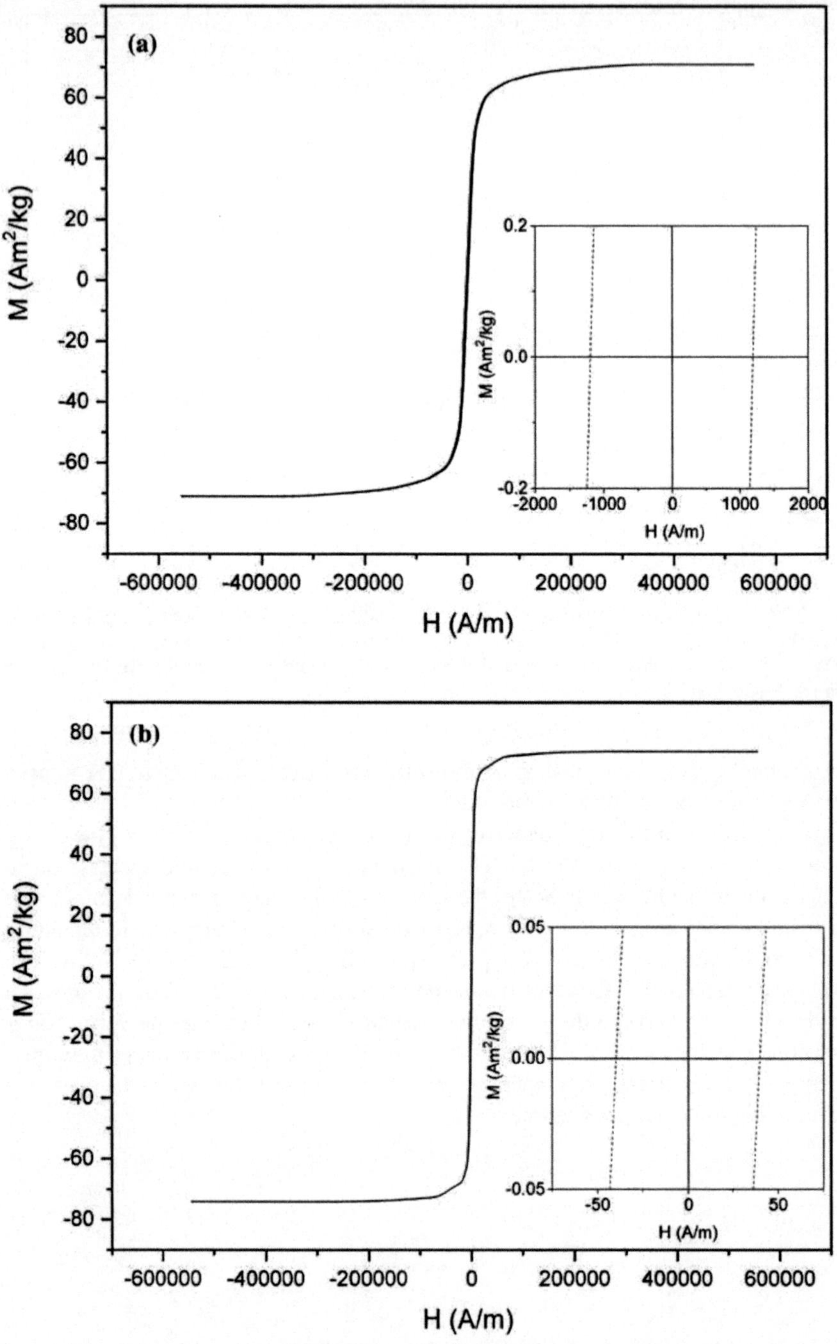

Fig. 5 Hysteresis loops of as-cast (**a**) and splat-quenched nanostructured (**b**) samples of $Fe_{25}B_{17.5}Co_{21.35}Nb_{3.65}Ni_{25}Si_{7.5}$ HEA

Table 3 The saturation magnetization (M_S), coercive force (H_C) and microhardness (H_μ) of $Fe_{25}B_{17.5}Co_{21.35}Nb_{3.65}Ni_{25}Si_{7.5}$ HEA

Alloy	M_S, $A \cdot m^2/kg$	H_c, A/m	H_μ, MPa
As-cast $Fe_{25}B_{17.5}Co_{21.35}Nb_{3.65}Ni_{25}Si_{7.5}$	71 ± 7	1200 ± 100	9000 ± 400
Splat-quenched nanostructured film $Fe_{25}B_{17.5}Co_{21.35}Nb_{3.65}Ni_{25}Si_{7.5}$	74 ± 7	40 ± 4	8500 ± 400

Figure 5 shows the hysteresis curves of the as-cast $Fe_{25}B_{17.5}Co_{21.35}Nb_{3.65}Ni_{25}Si_{7.5}$ HEA and splat-quenched nanostructured metallic glass. In accordance with the measured values of coercive force H_C (Table 3), the SQ film is a typical soft magnetic material, while coercivity of the as-cast HEA approximately 30 times larger.

Thus, it was shown that the transformation of the $Fe_{25}B_{17.5}Co_{21.35}Nb_{3.65}Ni_{25}Si_{7.5}$ alloy from crystalline to glassy state leads to a shift in the magnetic characteristics from the hard magnetic to the soft magnetic. The low value of the coercive force of HE-MG with a high degree of structural homogeneity and amorphicity can be explained by the insignificant density of the pinning sites for domain walls [10]. Such behavior is typical for alloys in the nanostructured amorphous state [20].

Both as-cast and SQ $Fe_{25}B_{17.5}Co_{21.35}Nb_{3.65}Ni_{25}Si_{7.5}$ samples exhibit high microhardness values (Table 3), which indicates good mechanical properties of the alloys. Obviously, the presence of brittle Fe_2B compound precipitates, which have a high hardness is the reason for the higher values of microhardness observed in the as-cast crystalline alloy. At the same time, a homogeneous splat quenched metallic glass without such precipitates is more ductile but has a lower microhardness.

4 Conclusions

In this study, a new nanostructured soft magnetic $Fe_{25}B_{17.5}Co_{21.35}Nb_{3.65}Ni_{25}Si_{7.5}$ HE-MG has been synthesized by a dint of splat quenching technique. The temperature of glass transition T_g and the initial temperature of crystallization T_x are equal to 800 K and 863 K, respectively. The HE-MG demonstrates the high value of saturation magnetization of 74 $A \cdot m^2/kg$, low H_C of 40 A/m, and a high value of microhardness $\geq$8000 MPa. So, the obtained soft magnetic HE-MG with good mechanical properties can be used for both engineering and functional purposes.

Acknowledgments The authors would like to thank Prof. O. V. Sukhova for kindly providing the starting Fe-B alloys.

References

1. Gao MC, Yeh J-W, Liaw PK, Zhang Y (2016) High-entropy alloys: fundamentals and applications. Springer
2. Firstov GS, Kosorukova TA, Koval YN, Odnosum VV (2015) High entropy shape memory alloys. Mater Today Proc 2:S499–S503. https://doi.org/10.1016/j.matpr.2015.07.335
3. Bashev VF, Kushnerov OI (2017) Structure and properties of cast and splat-quenched high-entropy Al–Cu–Fe–Ni–Si alloys. Phys Met Metallogr 118:39–47. https://doi.org/10.1134/S0031918X16100033
4. Miracle DB, Senkov ON (2017) A critical review of high entropy alloys and related concepts. Acta Mater 122:448–511. https://doi.org/10.1016/j.actamat.2016.08.081
5. Wang S, Xu J (2017) TiZrNbTaMo high-entropy alloy designed for orthopedic implants: As-cast microstructure and mechanical properties. Mater Sci Eng, C 73:80–89. https://doi.org/10.1016/j.msec.2016.12.057
6. Pogrebnjak AD, Yakushchenko IV, Bondar OV, Beresnev VM, Oyoshi K, Ivasishin OM, Amekura H, Takeda Y, Opielak M, Kozak C (2016) Irradiation resistance, microstructure and mechanical properties of nanostructured (TiZrHfVNbTa)N coatings. J Alloys Compd 679:155–163. https://doi.org/10.1016/j.jallcom.2016.04.064
7. Akimov SV, Duda VM, Dudnik EF, Kushnerev AI, Tomchakov AN (2006) Secondary ferroic properties of partial mixed ferroelectric ferroelastics. Phys Solid State 48:1073–1076. https://doi.org/10.1134/S1063783406060175
8. Dudnik EF, Duda VM, Kushnerov AI (2001) Second-order ferroic properties of a Pb5Ge3O11 uniaxial ferroelectric. Phys Solid State 43:2280–2283. https://doi.org/10.1134/1.1427957
9. Qi T, Li Y, Takeuchi A, Xie G, Miao H, Zhang W (2015) Soft magnetic Fe 25 Co 25 Ni 25 (B, Si) 25 high entropy bulk metallic glasses. Intermetallics 66:8–12. https://doi.org/10.1016/j.intermet.2015.06.015
10. Li Y, Zhang W, Qi T (2017) New soft magnetic Fe25Co25Ni25(P, C, B)25 high entropy bulk metallic glasses with large supercooled liquid region. J Alloys Compd 693:25–31. https://doi.org/10.1016/j.jallcom.2016.09.144
11. Xu Y, Li Y, Zhu Z, Zhang W (2018) Formation and properties of Fe25Co25Ni25(P, C, B, Si)25high-entropy bulk metallic glasses. J Non Cryst Solids 487:60–64. https://doi.org/10.1016/j.jnoncrysol.2018.02.021
12. Wei R, Tao J, Sun H, Chen C, Sun GW, Li FS (2017) Soft magnetic Fe 26.7 Co 26.7 Ni 26.6 Si 9 B 11 high entropy metallic glass with good bending ductility. Mater Lett 197:87–89. https://doi.org/10.1016/j.matlet.2017.03.159
13. Ding J, Inoue A, Han Y, Kong FL, Zhu SL, Wang Z, Shalaan E, Al-Marzouki F (2017) High entropy effect on structure and properties of (Fe Co, Ni, Cr)–B amorphous alloys. J Alloys Compd 696:345–352. https://doi.org/10.1016/j.jallcom.2016.11.223
14. Altomare A, Corriero N, Cuocci C, Falcicchio A, Moliterni A, Rizzi R (2017) Main features of QUALX2.0 software for qualitative phase analysis. Powder Diffr 32:S129–S134. https://doi.org/10.1017/S0885715617000240
15. Wang Z, Huang Y, Yang Y, Wang J, Liu CT (2015) Atomic-size effect and solid solubility of multicomponent alloys. Scr Mater 94:28–31. https://doi.org/10.1016/j.scriptamat.2014.09.010
16. Takeuchi A, Inoue A (2005) Classification of bulk metallic glasses by atomic size difference, heat of mixing and period of constituent elements and its application to characterization of the main alloying element. Mater Trans 46:2817–2829. https://doi.org/10.2320/matertrans.46.2817
17. Li W-K, Zhou G, Mak TCW (2008) Advanced structural inorganic chemistry. Oxford University Press, Oxford, New York
18. Troparevsky MC, Morris JR, Kent PRC, Lupini AR, Stocks GM (2015) Criteria for predicting the formation of single-phase high-entropy alloys. Phys Rev X 5:011041. https://doi.org/10.1103/PhysRevX.5.011041

19. Guo S, Liu CT (2011) Phase stability in high entropy alloys: formation of solid-solution phase or amorphous phase. Prog Nat Sci Mater Int 21:433–446. https://doi.org/10.1016/S1002-0071(12)60080-X
20. Gulivets AN, Zabludovsky VA, Shtapenko EP, Kushnerev AI, Dergachov MP, Baskevich AS (2002) Multilayer compound Co-P films with controlled magnetic properties. Trans IMF 80:154–156. https://doi.org/10.1080/00202967.2002.11871457

Predicting Phase Stability of Materials Based on $(Sc_{1-x}Ln_x)[(SiO_4)_{0.5}O_{0.5}]$, Ln = Tb – Lu and Y Solid Solutions

E. I. Get'man and S. V. Radio

1 Introduction

Solid solutions based on scandium oxyorthosilicate, Sc_2SiO_5, are promising materials for creating new efficient lasers for medicine, laser location [1], military equipment, telecommunications, metal cutting [2], environmental monitoring [3–5], and other purposes. They serve as matrices (host materials), while the triple-charged Ln^{3+} ions, which are contained in small amounts (up to 5 at.%), act as activators (dopants). In addition to oxyorthosilicates containing only scandium cations in their matrices [6, 7], we also studied materials based on "mixed" oxyorthosilicates with two different cations—scandium and rare-earth elements (REEs), which represent solid solutions of Sc_2SiO_5–Ln_2SiO_5 systems [8–11]. The use of "mixed" scandium and REE oxyorthosilicates is stipulated by the desire to obtain materials with better properties compared to Ln_2SiO_5, and which are less expensive compared to Sc_2SiO_5 since the cost of scandium is higher than one of REE.

No research of physical and chemical fundamentals for the synthesis of solid solutions—state diagrams and, in particular, solubility regions based on the components of the Sc_2SiO_5 and Ln_2SiO_5 systems—has been carried out, while this knowledge is necessary for choosing the composition of materials. As far as we know, only information about the $Lu_{2-x}Sc_xSiO_5$ [12] and $Er_xSc_{2-x}SiO_5$ [13] systems are available. In [12], it was reported that three compositions of polycrystalline solid solutions with $x = 0.5, 0.8, 1.0$ at a temperature of 1670 K were obtained, while in [13], the synthesis of $Er_xSc_{2-x}SiO_5$ solid solutions in the form of multinanolayer films within the temperature range 1173–1373 K was studied. In [13], it was also reported that $Er_xSc_{2-x}SiO_5$ films can be used to create a light source with high optical gain since they have a higher Er concentration compared to Si-based materials doped

E. I. Get'man · S. V. Radio (✉)
Vasyl' Stus Donetsk National University, Vinnytsia, Ukraine
e-mail: radio@donnu.edu.ua

O. Fesenko and L. Yatsenko (eds.), *Nanomaterials and Nanocomposites, Nanostructure Surfaces, and Their Applications*, Springer Proceedings in Physics 246, https://doi.org/10.1007/978-3-030-51905-6_39

with Er. However, no information on the substitution limits in the $Lu_{2-x}Sc_xSiO_5$ and $Er_xSc_{2-x}SiO_5$ systems is available.

It is important to determine experimentally solubility regions in the solid phase, which requires special equipment, additional expensive reagents, and more time. This forces researchers who study the properties of "mixed" REE oxyorthosilicates to choose the composition of host matrix materials and activators either by analogy with related systems or by trial and error.

It is often ignored that "mixed" REE oxyorthosilicates tend to decompose and change their phase composition and properties upon cooling. This can lead to the degradation of materials based on them if used in practical applications. Therefore, before carrying synthesis and studying it, it is recommended to evaluate the stability of solid solutions in the corresponding systems during their synthesis and intended use.

In view of this, the aim of this research is to predict the phase stability and substitution limits in materials based on solid solutions of oxyorthosilicate of scandium and REEs of Tb–Lu row, and Y.

Yttrium subgroup REEs and yttrium were chosen as second cations due to their isostructurality with Sc_2SiO_5, as well as the proximity of crystalline ionic radii of Sc^{3+} (0.885 Å) and triple-charged cations of yttrium subgroup REEs and yttrium (1.063–1.001 Å [14]), which suggests the wide presence of isomorphic substitution of scandium by these REEs. The radii of the cerium subgroup REE cations (1.172–1.078 Å) significantly differ from the ionic radius of scandium, and their oxyorthosilicates are not isostructural with Sc_2SiO_5 [15], which, according to the theory of isomorphic miscibility [16–18], should significantly limit the solubility of components in systems with scandium and REEs of the La–Gd series.

2 Calculation Procedure and Initial Data

The main task in calculating the substitution limits of solid solutions using the Urusov's crystal energy method [16–18] is to determine the mixing energy Q_{mix} (interaction parameter). As to isostructural components of the system and the possibility of their pseudobinary representation, there are two contributions to the mixing energy, which are caused by the difference in size of the substituting structural units Q_δ and the difference in the degree of ionicity of the chemical bond Q_ε:

$$Q_{\text{mix}} = Q_\delta + Q_\varepsilon = \text{Cmnz}_m\text{z}_x\delta^2 + 1390\text{mz}_m\text{z}_x\alpha(\Delta\varepsilon)^2/(2\text{D}),\ (\text{kJ/mol}),$$

where C = 112.6 kJ, a constant calculated from the equation $C = 20(2\Delta\chi + 1)$ [18] based on the difference in electronegativity χ of Ln^{3+} cations and anions [19]. The value $\chi(SiO_4^{4-})$, as recommended [20], was accepted equal to $\chi(O^{2-}) = 3.7$ [19]; $m = 2$, the number of formula units in the pseudobinary approximation of components. Since the anionic sublattice of the crystal structure of oxyorthosilicates contains SiO_4^{4-} and O^{2-} anions that are not bonded to the Si atom [15] and the substitution limits are calculated per 1 mol of the replaceable ion, the oxyorthosilicate formulas will be presented below as a pseudobinary compound $Ln[(SiO_4)_{0.5}O_{0.5}]$; $n = 6$, the coordination number of the replaceable structural unit in the pseudobinary approximation of the structure; $z_m = 3$, $z_x = 3$, the formal charges of the replaced and general structural units in the components: $z_m = 3$, $z_x = 4 \cdot 0.5 + 2 \cdot 0.5 = 3$; δ is a dimensional parameter, which for each system is characterized by the relative difference of cube roots of unit cell volumes taken from [15, 21, 22], calculated by the formula:

$$\delta = \left(V_{Ln}^{1/3} - V_{Sc}^{1/3}\right) / V_{Sc}^{1/3};$$

$\alpha = 1.73$, the reduced Madelung constant calculated by the Hoppe formula [23]: $(\alpha/n)^2 + \alpha = 1.81$, where $n = 6$ is the coordination number in the pseudobinary approximation of the structure; $\Delta\varepsilon$ is the difference in the degree of ionicity of the chemical bond in the components of the systems. D is the interatomic distance of the cation—tetrahedral anion, $D = 3.111$, calculated from the data [22].

3 Calculation Results

Some initial data and calculation results are summarized in Tables 1 and 2 and Fig. 1. The Table 1 shows that as the number of REE in the Tb–Lu row increases, the contributions of Q_δ values to the total mixing energy become smaller (from 34.5 to 10.8 kJ/mol), which is explained by smaller differences in the size of substitutable structural units—scandium and REE.

Contributions due to different degrees of ionicity of chemical bond in the components of the Q_ε systems are substantially smaller (in most cases, 2–3 times smaller) than in Q_δ, and they can be neglected in this case. This agrees with the recommendation not to take them into account, provided that $\Delta\varepsilon \leq 0.05$ [16–18] (in this case, $\Delta\varepsilon \leq 0.015$). Therefore, it is accepted that $Q_{mix} = Q_\delta$.

In all systems, the size parameter (δ) does not exceed 0.1 with its maximum value of 0.0535 (Table 1). This, according to [16–18], makes it possible to use the approximation of regular solutions when calculating the decomposition temperatures (stability) of solid solutions. In this case, the curve showing the dependence of the decomposition temperatures on the system composition will be nearly symmetric. Therefore, to calculate the maximum (critical) decomposition temperatures, the

Table 1 Data for the calculation of mixing energies and critical (maximum) decomposition temperatures of solid solutions $(Sc_{1-x}Ln_x)$ $[(SiO_4)_{0.5}O_{0.5}]$, Ln = Tb–Lu and Y (space group C2/c)

Ln	V, Å^3	δ^a	Q_δ, kJ/mol	χ_{Ln}	ε	Δε	T_{cr}, K
Tb	876.80 [21]	0.0535	34.5	1.410	0.708	0.001	2060
Dy	856.57 [15]	0.0453	24.8	1.426	0.706	0.003	1480
Ho	843.04 [15]	0.0398	19.1	1.433	0.704	0.005	1150
Er	836.70 [15]	0.0372	16.7	1.438	0.703	0.006	1010
Tm	828.59 [15]	0.0338	13.8	1.455	0.699	0.010	860
Yb	824.07 [15]	0.0319	12.3	1.479	0.694	0.015	810
Lu	819.31 [15]	0.0299	10.8	1.431	0.705	0.004	650
Sc	749.97 [22]	–	–	1.415	0.709	–	–
Y	852.25 [21]	0.0435	22.8	1.340	0.722	0.013	1400

[a]*Note* According to the recommendations in [17, 18] and considering the dependence of the interaction parameter on the difference in volumes of the unit cells of components [29], the calculation of the dimensional parameter δ was carried out according to the volumes of the unit cells

Table 2 Decomposition temperatures (stability) of $(Sc_{1-x}Ln_x)$ $[(SiO_4)_{0.5}O_{0.5}]$ solid solutions for $x = 0.01, 0.03, 0.05, 0.10$, and 0.20

x	Tb	Dy	Ho	Er	Tm	Yb	Lu
0.01	880	630	490	430	370	350	280
0.03	1110	800	620	550	469	440	350
0.05	1260	910	700	620	520	490	400
0.10	1500	1080	840	730	620	590	470
0.20	1780	1280	990	870	740	700	560

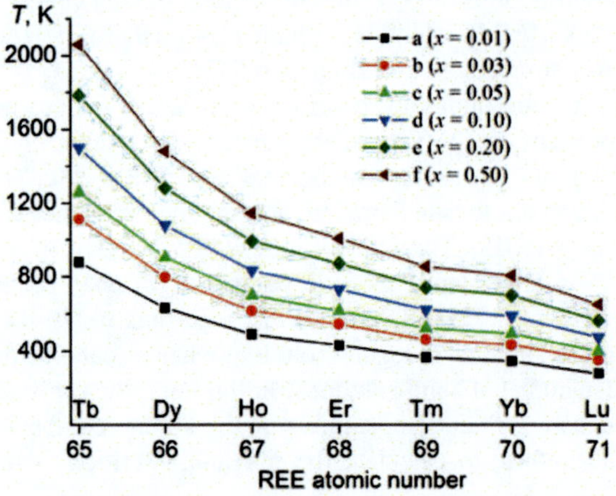

Fig. 1 The diagram of thermodynamic stability of solid solutions in the systems $(Sc_{1-x}Ln_x)$ $[(SiO_4)_{0.5}O_{0.5}]$, where $x = 0.01$ (**a**), 0.03 (**b**), 0.05 (**c**), 0.10 (**d**), 0.20 (**e**), and 0.50 (**f**)

following equation was used:

$$T_{cr} = Q_{mix}/2kN,$$

where k is the Boltzmann constant and N is the Avogadro number. In order to calculate the substitution limits for a given decomposition temperature of a solid solution (T_d) or the decomposition temperature for a given substitution limit [24], the Becker equation was used [24]:

$$-(1-2x)/\ln[x/(1-x)] = RT_d/Q_{mix},$$

where R is the universal gas constant, Q_{mix} is a mixing energy (or interaction parameter), and x is a substitution limit.

As can be seen from Table 1 and Fig. 1 (curve f), the values of maximum decomposition temperatures, as expected, become smaller as REE number increases. The Becker equation was also used to calculate the decomposition temperatures (stability) of solid solutions for the substitution limits $x = 0.01$, 0.03, 0.05, 0.10, and 0.20 (Table 2), and to build their dependencies (Fig. 1) on the REE number (curves a, b, c, d, and e, respectively). The latter can be used to determine the substitution limit of scandium for REE based on a given temperature or calculate the decomposition temperature based on the substitution limit. In the first case, it is necessary to draw an isotherm from a given temperature to the intersection with the vertical line for this REE. The intersection point makes it possible to estimate the range of x values within which the substitution limit lies. The substitution limit should be defined by interpolating the vertical segment between the closest to the intersection point dependencies of the substitution limit on the REE number. In the second case, based on the given composition the point is determined on the vertical line of the REE, and then the horizontal line is drawn until its intersection with the temperature axis. More precise results can be obtained if using the Becker equation.

It is commonly known that as the temperature decreases, the mobility of the structural units in solid solution becomes smaller due to a decrease in the diffusion rate, while the solubility regions become narrower [17]. This happens until the diffusion rate becomes so low that the decrease in the solubility regions practically ceases, i.e., spontaneous quenching occurs, and solid solutions become metastable. If we assume that the hardening temperature is close to the minimum temperature at which the interaction of the components in the solid phase begins that leads to the formation of a solid solution, we can estimate the temperature of spontaneous hardening and the region of metastability in the system. It was previously established (Table 3) that the temperature during the synthesis of solid solutions of $Sc_{2-x}Er_xSiO_5$ oxyorthosilicates as a part of the preparation of multinanolayer films is in the range of 1173–1373 K,

Table 3 Methods and temperatures for the synthesis of REE oxyorthosilicates

Method of synthesis	Composition	*T*, K
Calcination of multinanolayer films [13]	$Sc_{2-x}Er_xSiO_5$	1173–1373
Solution combustion synthesis (SCS) [25]	Lu_2SiO_5:Ce	1273
	Gd_2SiO_5:Ce	1273
	Y_2SiO_5:Ce	1273
Sol–gel method followed by calcination [26]	Y_2SiO_5	1323
	Tb_2SiO_5	1323
	Dy_2SiO_5	1323
	Ho_2SiO_5	1273
	Er_2SiO_5	1273
	Tm_2SiO_5	1273
	Yb_2SiO_5	1223
	Lu_2SiO5	1173

while the temperature during the synthesis of Gd, Lu, and Y oxyorthosilicates using the solution combustion synthesis method (SCS) is 1273 K, and the temperature during the solid-phase synthesis of REE oxyorthosilicates of the Tb–Lu row, and Y, using the sol–gel method is in the range of 1173–1323 K.

Thus, at a temperature of less than 1173 K, the diffusion rate of structural units is apparently insufficient for the synthesis of REE oxyorthosilicates and solid solutions based on them. Therefore, it can be assumed that the decomposition of solid solutions at temperatures below ~1173 K is unlikely to occur, therefore, the solid solution will be metastable.

The diagram also makes it possible to evaluate the regions of thermodynamic stability of solid solutions of scandium oxyorthosilicate and REEs of the Tb–Lu series. In the $(Sc_{1-x}Ln_x)[(SiO_4)_{0.5}O_{0.5}]$ systems with Ln = Tb, Dy, and Y, unlimited solid solutions are thermodynamically stable in the entire range of concentrations $0 < x < 1$ at temperatures above the critical one (2060–1400 K; Table 1, Fig. 1); when the temperature decreases to the range between T_{cr} and ~1173 K, they become thermodynamically unstable and can decay. At $T < 1173$ K, solid solutions will not decompose, i.e., spontaneous hardening will occur, and they will become metastable.

In the systems containing REE from Er to Lu, the maximum decomposition temperatures (1010–650 K) are lower than the spontaneous quenching temperature (~1173 K), unlimited solid solutions do not decompose upon cooling, and remain stable at temperatures higher than the critical one and metastable at temperatures lower than the critical one.

The difference between the critical temperature for a system containing Ho (1150 K) and the temperature of spontaneous quenching (~1173 K) is less than the calculation error (±100 K [17]); therefore, it is difficult to predict decomposition of an unlimited solid solution in this system.

Similarly, limited solid solutions with $x = 0.01, 0.03, 0.05, 0.10$, and 0.20 in the regions above the curves a, b, c, d, and e, respectively, are thermodynamically stable (Fig. 1), in the regions below the curves they are unstable and can decay, while at $T < 1173$ K they are metastable.

Although there are numerous publications, which study the laser properties of scandium oxyorthosilicate, doped with, for example, 0.5 at.% of holmium [3], 1 at.% of neodymium [9, 27], 4 at.% of thulium [7], 5 at.% of ytterbium [28], and others, and in some papers on the properties of "mixed" oxyorthosilicates $Y(Lu,Sc)_2SiO_5$ [2], $(Sc_{0.5}Y_{0.5})_2SiO_5$ [8, 10], $(Sc_{0.2}Y_{0.8})_2SiO_5$ [11], there is practically no data on the limits of isomorphous substitutions in the corresponding systems. This, of course, makes it difficult to assess the reliability of our calculations. However, they do not contradict the experimental data obtained previously for $Lu_{2-x}Sc_xSiO_5$ and $Er_xSc_{2-x}SiO_5$ solid solutions. For example, in the $Lu_{2-x}Sc_xSiO_5$ system, a mixture of Lu_2O_3, Sc_2O_3 and SiO_2 was calcinated to synthesize solid solutions for compositions with $x = 0.5, 0.8, 1.0$ at a temperature of 1670 K [12], i.e., in the region of continuous series of solid solutions, which are thermodynamically stable according to the results of our calculation (Fig. 2).

In [13], the $Er_xSc_{2-x}SiO_5$ solid solutions were synthesized by calcination of atomized multinanolayer films in the temperature range 1173–1373 K; i.e., also in the region of continuous series of solid solutions above the decomposition curve calculated by us (Fig. 3).

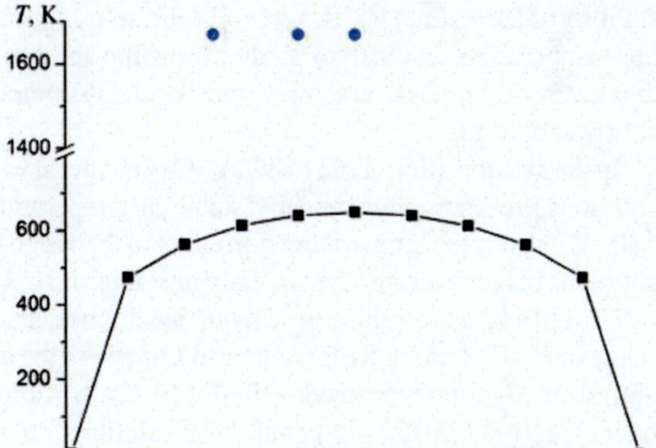

Fig. 2 Dependence of the calculated decomposition temperatures (stability) of solid solutions for the system $(Sc_{1-x}Lu_x)$ $[(SiO_4)_{0.5}O_{0.5}]$ (■) on the molar fraction of Lu, and experimental data on the synthesis temperatures (●) for the compositions with $x = 0.25, 0.4$, and 0.50 [12]

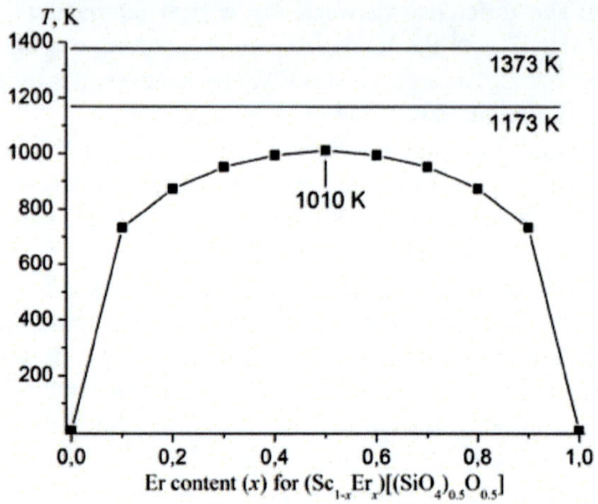

Fig. 3 Dependence of the calculated decomposition temperatures of solid solutions for the system $(Sc_{1-x}Er_x)$ $[(SiO_4)_{0.5}O_{0.5}]$ on the mole fraction of Er, as well as the temperature range (1173–1373 K) during the synthesis of $(Sc_{1-x}Er_x)$ $[(SiO_4)_{0.5}O_{0.5}]$ according to experimental data [13]

4 Conclusion

The crystal–chemical approach in the approximation of regular solutions was used to calculate the mixing energies (interaction parameters) of solid solutions based on scandium oxyorthosilicate $(Sc_{1-x}Ln_x)$ $[(SiO_4)_{0.5}O_{0.5}]$, modified with rare-earth elements with $x = 0.01, 0.03, 0.05, 0.10, 0.20$, and 0.5. With an increase in the REE number, the calculated mixing energies and critical decomposition temperatures of solid solutions become smaller, which is caused by the decrease in the REE ionic radii in the series from La to Yb.

The diagram of thermodynamic stability makes it possible to evaluate not only the stability of $(Sc_{1-x}Ln_x)$ $[(SiO_4)_{0.5}O_{0.5}]$ solid solutions in a wide range of compositions and temperatures but also to predict for some solid solutions the substitution limits at a given decomposition temperature, or the decomposition temperature at a given substitution limit.

In the systems $(Sc_{1-x}Ln_x)$ $[(SiO_4)_{0.5}O_{0.5}]$ with Ln = Tb, Dy, and Y, unlimited solid solutions are thermodynamically stable at temperatures above critical one (2060–1400 K), and if the temperature decreases to the range between T_{cr} and ~1173 K, the solutions become thermodynamically unstable and can decompose. At a temperature of $T < 1173$ K, solid solutions will not decompose, since they become metastable. In the systems containing REE from Er to Lu, where the critical decomposition temperatures are significantly lower (1010–650 K) than the temperature of spontaneous quenching (~1173 K), unlimited solid solutions do not decompose upon cooling, and they remain stable at temperatures higher than the critical one and metastable at temperatures lower than the critical one.

The calculation results obtained do not contradict the experimental data obtained previously for $(Sc_{1-x}Lu_x)[(SiO_4)_{0.5}O_{0.5}]$ and $(Sc_{1-x}Er_x)[(SiO_4)_{0.5}O_{0.5}]$ solid solutions since the temperatures of their synthesis are in the ranges attributed by us to thermodynamically stable ones.

Acknowledgments This work was carried out within the Fundamental Research Programme funded by the Ministry of Education and Science of Ukraine (grant ID 0119U100025).

References

1. Zavartsev Y, Zagumennyi AI, Kalachev Y, Kutovoi SA, Mikhailov VA, Shcherbakov IA (2013) Study of Tm: Sc_2SiO_5 laser pumped into the $^3H_6$ – $^3F_4$ transition of Tm^{3+} ions. Quantum Electron 43(11):989–993
2. Denoyer A (2007) Études optiques de nouveaux matériaux laser: des orthosilicates dopes à l'ytterbium Yb:Y(Lu,Sc)$_2$SiO$_5$. Thèse présentée au département de physique en vue de l'obtention du grade de docteur ès sciences (Ph.D.). Faculté des sciences, Université de Sherbrooke, Sherbrooke, Québec, 177 p. https://savoirs.usherbrooke.ca/handle/11143/5072
3. Yang X-t, Xie W-q, Liu L, Li L-j (2017) Compact intra-cavity pumped low-threshold passively Q-switched Ho:Sc_2SiO_5 laser by a LD-pumped Tm:YAP laser at room temperature. Laser Phys Lett 14(8):085806
4. Yang X-T, Liu L, Zhang P, Xie W-Q (2017) A resonantly pumped single-longitudinal mode Ho:Sc2SiO5 laser with two fabry-perot etalons. Appl Sci 7(5):434
5. Yang X-t, Yao B-q, Ding Y, Li X, Aka G, Zheng L-h, Xu J (2013) Spectral properties and laser performance of Ho:Sc_2SiO_5 crystal at room temperature. Opt Express 21(26):32566–32571
6. Zheng L, Su L, Xu J (2012) Growth and characterization of ytterbium doped silicate crystals for ultra-fast laser applications. In: Modern aspects of bulk crystal and thin film preparation. IntechOpen. https://www.intechopen.com/books/modern-aspects-of-bulk-crystal-and-thin-film-preparation/growth-and-characterization-of-ytterbium-doped-silicate-crystals-for-ultra-fast-laser-applications
7. Zheng L, Xu J, Su L, Li H, Ryba-Romanowski W, Lisiecki R, Solarz P (2010) Crystal structure and optical study of Tm:Sc_2SiO_5 single crystal. Appl Phys Lett 96(12):121908
8. Shi J, Liu B, Zheng L, Wang Q, Tang H, Liu J, Su L, Wu F, Zhao H, He N, Li N, Li Q, Guo C, Xu J, Yang K, Xu X, Ryba-Romanowski W, Lisiecki R, Solarz P (2018) Optical study of Tm-doped solid solution $(Sc_{0.5}Y_{0.5})_2SiO_5$ crystal. J Cryst Growth 487:83–86
9. Li L, Cao M, Liu S, Peng Y, Zheng L, Xu J (2017) High-peak-power Nd:$ScYSiO_5$ laser at 1355 nm pumped by a pulsed laser diode. Optik 140:248–252
10. Aleksandrov V, Iliev H, Zheng L, Su L, Xu J, Aka G, Buchvarov I (2013) Passive mode-locking of a diode pumped Nd:$ScYSiO_5$ laser. In: 2013 conference on lasers and electro-optics—international quantum electronics conference (Optical Society of America, 2013), paper CA_P_11
11. Aleksandrov V, Iliev H, Trifonov A, Zheng L, Xu J, Su L, Buchvarov I (2014) Passive mode locking of a diode pumped Nd:$Sc_{0.2}Y_{0.8}SiO_5$ laser. In: CLEO: 2014 OSA technical digest (online) (Optical Society of America, 2014), paper JTu4A.127
12. Ben Yahia H, Rodewald UC, Heying B, Balamurugan S, Pottgen R (2011) The solid solution $Lu_{2-x}Sc_xSiO_5$. Z Naturforsch B 66(11):1183–1187
13. Najar A, Omi H, Tawara T (2015) Effect of structure and composition on optical properties of Er–Sc silicates prepared from multi-nanolayer films. Opt Express 23(6):7021–7030
14. Shannon RD (1976) Revised effective ionic radii and systematic studies of interatomic distances in halides and chalcogenides. Acta Cryst A32:751–767
15. Felsche J (1973) The crystal chemistry of the rare-earth silicates. In: Rare earths. Structure and bonding, vol 13. Springer, Berlin, Heidelberg, pp 99–197
16. Urusov VS (1975) Energetic theory of miscibility gaps in mineral solid solutions. Fortschr Mineral 52:141–150
17. Urusov VS (1977) Teoriia izomorfnoi smesimosti (Theory of isomorphous miscibility). Nauka, Moscow, 251 p. (in Russian)

18. Urusov VS, Tauson VL, Akimov VV (1997) Geokhimiya tverdogo tela [Geochemistry of solid state]. GEOS, Moscow, 500 p. (in Russian)
19. Li K, Xue D (2006) Estimation of electronegativity values of elements in different valence states. J Phys Chem A 110(39):11332–11337
20. Batsanov SS (1968) The concept of electronegativity. Conclusions and prospects. Russ Chem Rev 37(5):332–351
21. Liu Y, Xu CN, Nonaka K, Tateyama H (2001) Strong blue, green and red light emission at elevated temperatures from Y_2SiO_5 doped by the rare-earth ions. J Mater Sci 36(18):4361–4364
22. Alba MD, Chain P, González-Carrascosa T (2009) Synthesis, rietveld analysis, and solid state nuclear magnetic resonance of X_2-Sc_2SiO_5. J Am Ceram Soc 92(2):487–490
23. Hoppe R (1970) Madelung constants as a new guide to the structural chemistry of solids. Adv Fluor Chem 6:387–438
24. Becker R (1937) Über den Aufbau binarer Legierungen. Z Metallkunde 29:245–249
25. Yukihara EG, Jacobsohn LG, Blair MW, Bennett BL, Tornga SC, Muenchausen RE (2010) Luminescence properties of Ce-doped oxyorthosilicate nanophosphors and single crystals. J Lumin 130(12):2309–2316
26. Wang J, Tian S, Li G, Liao F, Jing X (2001) Preparation and X-ray characterization of low-temperature phases of R_2SiO_5 (R = rare earth elements). Mater Res Bull 36(10):1855–1861
27. Zheng L, Xu J, Su L, Li H, Wang Q, Ryba-Romanowski W, Lisiecki R, Wu F (2009) Estimation of low-temperature spectra behavior in Nd-doped Sc_2SiO_5 single crystal. Opt Lett 34(22):3481–3483
28. De Tan W, Tang D, Xu X, Zhang J, Xu C, Xu F, Zheng L, Su L, Xu J (2010) Passive femtosecond mode-locking and cw laser performance of Yb^{3+}:Sc_2SiO_5. Opt Express 18(16):16739–16744
29. Li Y, Kowalski PM, Blanca-Romero A, Vinograd V, Bosbach D (2014) Ab initio calculation of excess properties of $La_{1-x}(Ln, An)_xPO_4$ solid solutions. J Solid State Chem 220:137–141

Influence of Different Type of Irradiation to Carbon Nanotubes

H. Yu. Mykhailova, I. Sydorchenko, V. Koda, and B. V. Kovalchuk

The study of radiation effects in CNT is due to the improvement and complementation of their properties during radiation exposure and high technological capabilities. Insightful radiation not only causes radiation damage in materials but can also serve as an effective technological tool for defects to improve the quality and obtain materials with fundamentally new properties. The electronic irradiation of MCNT at room temperature results from the formation of vacancies on their walls and, ultimately, by amorphization at high radiation doses, MCNTs are more stable to radiation than SCNTs, probably because the atoms are knocked out of the internal ones. Radiation effects and radiation dose on the properties of carbon nanotubes have been investigated in this article.

Carbon nanotubes (CNTs) attract the attention of scientists from many countries of the world for the last 30 years. This interest is due to their unusual properties: they combine high strength and elasticity, heat, and electrical conductivity. Mechanical deformation and defects influence the electronic structure, concentration of charge carriers, electrical, emission, and other properties of CNT. Defects arise both in the process of synthesis (growth and gas admixtures) of CNT and with external influences, especially in radiation. They reject the form of CNT from the rectilinear, change the conditions of the passage of current, affect the concentration of charge carriers, Fermi energy, and electrical conductivity.

The geometry of CNT varies under the action of elastic deformation, which can significantly deform the profile of the entire nanotube, or locally in the region of the intersection of contacting nanotubes, which affects the electronic structure at the deformation point, which controls the electrical conductivity of not only one

H. Yu. Mykhailova (✉) · I. Sydorchenko · V. Koda · B. V. Kovalchuk
G. V. Kurdyumov Institute for Metal Physics of the N.A.S. of Ukraine, 36 Academician Vernadsky Boulevard, 03142 Kiev, Ukraine
e-mail: mihajlova.halina@gmail.com

O. Fesenko and L. Yatsenko (eds.), *Nanomaterials and Nanocomposites, Nanostructure Surfaces, and Their Applications*, Springer Proceedings in Physics 246, https://doi.org/10.1007/978-3-030-51905-6_40

nanotube but also the entire array of nanostructures. Considerable attention to radiation defects has arisen in clarifying the possibility of using nanotubes in space and nuclear reactors.

Electrophysical properties of isolated CNT are investigated in many laboratories of the world for their application in nanoelectronics. The properties of an individual nanotube are known to some extent—along the axis, it is a good conductor of current and heat, but in the radial direction, the motion of charge carriers and phonons is limited to an outer cylindrical carbon layer. To expand the scope of CNT, it is necessary to increase the flow of heat and current, for which it is necessary to investigate their properties in the bulk or consolidated state. In a bulk state, the CNT, as a rule, does not conduct an electric current, but with a slight compression, the situation is changing.

The properties of an array of carbon nanotubes or a material containing CNT differ from the properties of individual nanotubes (single-layered or multilayered). An array of CNTs, unlike individual nanotubes, may not conduct an electric current. This is due to the fact that the connections between adjacent nanotubes are van der Waals.

Adding a small amount of CNT to a dielectric matrix facilitates the appearance of electrical conductivity [1–6], whose size has a wide range. This dispersion in the magnitude of electrical conductivity is reflected by the complex interaction of various factors, the most important of which are the own conductivity of CNTs and their contact resistance.

The electrical conductivity of the individual BWNT is 10^4–10^7 $(\Omega m)^{-1}$ [7–9], the measured contact resistance for the SCNT bundle is 100–400 kΩ for metal/metal or semiconductor/semiconductor compounds and twice that for metal/semiconductor systems [10].

Defects in CNT may occur during synthesis (growth defects) or during external influences, for example, high-energy irradiation.

Experimentally obtained nanostructures may contain defects that lead to the distortion of their structure and differences in properties. The formation of curved, colloid, swirling CNTs, as well as tubes of variable diameters occur only in the presence of topological defects associated with re-hybridization and unsaturated bonds [11].

Topological defects can disappear when heated to a temperature of up to 2773 K and higher in an inert medium.

In nanostructured materials, the peculiarity of the atomic structure is that each element of the system or nanocomposite has a local structure of the free volume, which leads to the appearance of potential barriers limiting the charge carriers in the 1-to-2D (interlayer intervals of MCNT) on the boundary, 2-to-1D (in CNT) or 3-to-0D (fullerenes in the matrix and isolated clusters) directions, and the electronic processes are predominantly quantum in nature.

The stability of materials and components of onboard equipment to the influence of the surrounding outer space plays an important role in ensuring the trouble-free operation of spacecraft. More than half of refusals in the operation of onboard equipment are due to the influence of the factors of outer space: electron fluxes, cold and hot cosmic plasma, solar electromagnetic radiation, meteoric matter, solid particles of artificial origin (space debris), etc. [12]. Analysis of the statistics of the conditions for the occurrence of anomalies in the work of spacecraft shows that in most cases failure and failure are due to the influence of cosmic radiation on materials and elements of space equipment [13].

The radiation impact on the spacecraft is due to electrons and ions with energy 10^3–10^6 eV capable of penetrating depths of more than several tens of micrometers, causing in the volume of matter ionization of atoms, the formation of local electrical charges, the formation of defects, nuclear transformations, etc.

Radiation effects are any changes in the structure, properties, state of a substance or material caused by the action of radiation. Such conversions may have a reverse and irreversible character. The most sensitive to the effects of cosmic radiation are semiconductor and optical materials, to a lesser extent polymeric materials, the most resistant to radiation exposure are metals [14].

Radiation effects arising from ionizing radiation in nanostructures have a number of peculiarities in comparison with similar effects in objects whose sizes lie in micro and macro ranges.

Most nanocomposites have a higher radiation resistance relative to the various components of cosmic ionizing radiation compared to traditional composite materials, but the amount of these data is currently insufficient.

The high radiation resistance of nanocomposites is due to the physical model, which is based on the notion of a less dense, compared with bulky bodies, anisotropic packaging of atoms in the CNT and the adsorption surface of the tube with carbon atoms that it is bulging [15]. Migration processes lead to the healing of radiation defects [16]. CNTs are able to shield electromagnetic radiation due to high strength characteristics and to remove thermal loads, static voltage charges due to high heat and electrical conductivity [17]. It is obvious that during the interaction of an electron or an ion with the sufficiently high energy characteristic of cosmic radiation, only very small energy of the incident particle is transmitted to nanoparticles [18]. In a nanosized object, there is a small number of additional charge carriers or structural defects. At the same time, with increasing energy of the flying particles, the number of created carriers and defects will decrease in line with the decrease of the linear transmission of energy. By contrast, in conventional bulk materials, the total number of charge carriers and structural defects increases with increasing energy of flying particles if their run is invested in the linear dimensions of the object.

Figure 1 shows electron microscopic images of CNTs containing inside the catalyst (*a*) and the microstructure of the catalyst Al_2O_3–Fe_2O_3–$MoO_{30,21}$ in the initial state (b)

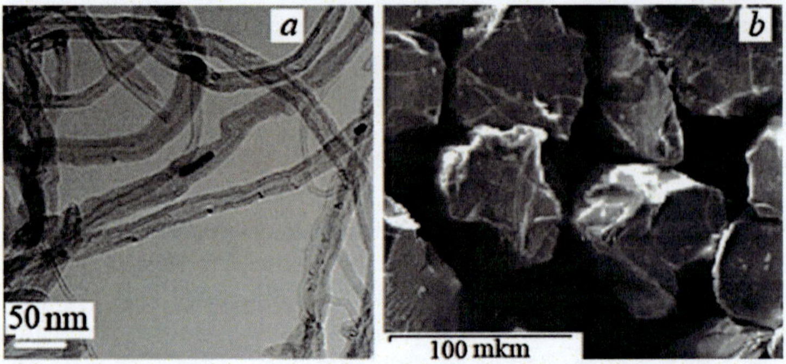

Fig. 1 Electron microscopic image of MCNTs containing inside the catalyst nanoparticles (**a**); microstructure of Al_2O_3–Fe_2O_3–$(MoO_3)_{0,21}$ catalyst in the initial state (**b**)

MCNTs are identified by transmission electron microscopy. The purity of the nanotubes was 99%. The observed narrow dark bands belong to the walls of the nanotubes.

The statistical distribution of CNTs in outer diameter is in the range of 10–25 nm. Inside some CNTs, dark catalyst particles are visible.

Spectroscopic analysis showed that the sample of CNTs contained 98.1 wt% carbon, 1.51 wt% oxygen and 0.39 wt% iron. This composition confirms that catalyst nanoparticles are present inside the CNT.

The Al_2O_3–Fe_2O_3–$(MoO_3)_{0,21}$ catalyst contains 25.83 wt% oxygen, 55 wt% iron. It also has carbon and other elements.

The effect of electron irradiation with energies 2 and 21 MeV and doses up to $2.21 \cdot 10^{17}$ e/cm^2 to the MCNT of their properties was studied.

The electrical conductivity of the CNT ensemble was measured by the developed method [19]: the sample of CNT was placed in a dielectric capsule under the piston, lowering the latter changes the density of the CNT and fixes the transition to a wired state.

This is possible to obtain a curve of electrical conductivity vs density ($\sigma(\rho)$).

The value of $\sigma(\rho)$ depends on the orientation of the nanotubes, their ordering in the process of compression, as well as on defects in the structure, the presence of which can lead to bending of CNTs. In the case of weak compression of the non-irradiated CNT to a density of $\rho_f = 0.08$ g/cm^3, the initial stage of the transition to the electrically conductive state is fixed at $\sigma = 1.5 \cdot 10^{-5}$ $(\Omega\ \text{cm})^{-1}$. Subsequent CNT compression leads to a jump $\sigma(\rho)$ of more than 4 orders of magnitude, then the maximum conductivity value $\sigma_{max} = 0.22$ $(\Omega\ \text{cm})^{-1}$ at the density $\rho = 0.25$ g/cm^3 is observed, after which σ starts smoothly decrease to a value of $\sigma = 0.06$ $(\Omega\ \text{cm})^{-1}$ with the highest compression. The rapid growth of $\sigma(\rho)$ at the initial stage is due to the compaction of the CNTs and the increase in the total contact area between

Table 1 Critical parameters for the initial and irradiated by 2 and 21 MeV samples of CNT

	Standard	2 MeV		21 MeV	
		$0.63 \cdot 10^{17}$ e/cm^2	$1.9 \cdot 10^{17}$ e/cm^2	$0.89 \cdot 10^{17}$ e/cm^2	$2.21 \cdot 10^{17}$ e/cm^2
ρ_f, g/cm^3	0.080	0.12	0.13	0.10	0.08
ρ_r, g/cm^3	0.24	0.27	0.23	0.34	0.19
ε, %	80	46	50	28	62
σ, (Ohm cm)$^{-1}$	0.23	0.22	0.19	0.30	0.43

σ—the conductivity of the CNT array in the process of compression deformation under the piston
ρ_f—the critical density of the compressed CNT array at which the electrical conductivity jump is recorded
ρ_{rel}—density at unloading at which the gap of an electric circuit is fixed
ε—coefficient of elasticity limit of the CNT array

them. The density of the transition to the conducting state ρ_f depends on the degree of deviation of the shape of the CNT in the array from the ideal rectilinear characteristic of a defect-free nanotube. When the pistons are reversed and the volume filled with the nanotubes increases, they are elastically relaxed, which keeps the nanotubes in contact with each other and with the electrodes. Upon reaching the value of $\rho_{rel} = 0.33$ g/cm^3, there is a significant decrease in the contact area, a sharp fall in σ and an electric circuit break, indicating the completion of the process of elastic relaxation.

Data of the density of the transition to the conductive state and electrical conductivity are presented in Table 1.

Electron irradiation with 2 MeV energy leads to an increase in ρ_f from 0.08 to 0.13 g/cm^3 (Table 1), which is related to the more dense packing of the CNT array as a result of the rectification of their shape by the interaction of growth and radiation defects. Under the action of high-energy electrons, radiation defects are formed, with the value of σ almost unchanged from the radiation dose. This behavior may be related to the presence of adsorbed gas impurities inside the MCNT that do not remove during irradiation and stabilize the electrical conductivity.

Irradiation of the array with electrons with energy of 21 MeV leads to an increase in σ_{max} and a slight increase in ρ_f (after irradiation with a smaller dose). This is explained by the influence of migration processes that lead to the interaction of growth and radiation defects or annihilation of them with each other if they lead to a positive or negative curvature of the graphene layer. Thus, the specimens have growth and radiation defects, as well as adsorbed gas impurities, which have different effects on their properties.

The calculation of the coefficient of elasticity limit ε showed that for the initial sample it is 80%. However, after irradiation with high-energy electrons, the elasticity (ε) of the CNT array decreases from 80 to 30% for a smaller dose.

This fact underscores the complex nature of the effect of radiation defects generated in CNTs under the action of electron irradiation on the elastic properties of the CNT array. Perhaps at some irradiation dose, the concentration of point defects (vacancies and adsorbed atoms) reached saturation in the CNT.

As the radiation dose increased, the subsequent accumulation of structural defects did not occur but the role of their "healing" effects and the reorganization of carbon–carbon bonds under radiation loading increased. Both effects could have contributed to an increase in the coefficient of elasticity ε at the maximum dose $D = 2.21 \cdot 10^{17}$ e/cm^2 compared to the value of ε at the dose $D = 0.89 \cdot 10^{17}$ e/cm^2.

For CNTs irradiated with electrons with E = 2 MeV and E = 21 MeV and γ-quanta (E = 1.2 MeV), there is an increase in the Seebeck coefficient α with increasing dose D up to 38 μV/K, 44 μV/K for 2 MeV and 21 MeV, respectively, and up to 62 μV/K for γ-irradiation. Such an increase in the coefficient α is associated with the appearance of radiation defects in the acceptor type under the action of high-energy electrons.

Research showed that irradiation of an array of multilayer carbon nanotubes by electrons with an energy of 2 and 21 MeV and a dose up to $2.21 \cdot 10^{17}$ e/cm^2 affects its electrical and elastic properties. The irradiation leads to a noticeable decrease in its elastic limit but contributes to a significant (approximately twofold) increase in the electrical conductivity of the material, as well as the Seebeck.

These results are due to the complex nature of the interaction of primary and radiation defects: primary defects compensate for the effects of radiation defects on the electrical properties of the array but do not change their effect on the elasticity of the material. A stronger decrease in the elastic limit of the CNT was observed at a lower dose of radiation, which was explained by the effects of the "healing" of structural defects generated in the CNT and the rearrangement of carbon–carbon bonds under radiation irradiation.

Therefore, controlled changes in the concentration of primary defects in CNTs by thermal and mechanical treatment can offset the influence of radiation on the conductivity of the CNT array, which is important for their practical application.

1 Conclusion

It is established that after irradiation of the MCNT array by high-energy (2 and 21 MeV) electrons with a dose up to $2 \cdot 10^{17}$ e/cm^2, their elastic properties decrease by 2–3 times, the conductivity changes significantly, and by two times the Seebeck coefficient increases the formation of acceptor-type defects that reduce the concentration of conduction electrons. It is shown that by a controlled change in the concentration of defects in thermal and mechanical treatment, it is possible to compensate for the influence of radiation irradiation on electrical conductivity, which is important for the use of the CNT array in the long term as sensitive sensors and energy converters in the conditions of radiation exposure: at cosmic, radioactive states, waste, etc.

Radiation irradiation with a dose of $D \approx 10^{17}$ r.p./cm^2 leads to a decrease in electrical conductivity and an increase in the Seebeck coefficient (α): when electrons are irradiated with $E = 21$ MeV, α increases from 26 to 36 μV/K, and from $E = 2$ MeV—up to 45 μV/K, when γ-irradiated with $E = 1.2$ MeV the highest increase of Seebeck coefficient up to 55 μV/K is observed. These results indicate the different nature of the defects that are formed during the growth of nanotubes and under the action of radiation. The prospect is to create defects in the CNTs that reduce the potential barriers between the contacting nanotubes through which charge carriers are tuned, and the phonons on them are scattered, which reduces unwanted heat dissipation and is important for increasing the thermoelectric figure of merit of the carbon nanotube array.

The density of the transition to the conductive state of the CNT array is determined by the concentration of topological defects (Stone–Wales, paired vacancies, adsorbed carbon atoms), which increase the degree of curvature of the shape relatively straightforward for defect-free nanotubes.

Each of the defects of different genesis: growth and radiation origin, separately leads to distortion of the nanotube shape, which indicates the creation of a topological disorder, which is accompanied by a decrease in electrical conductivity and the increase of thermo-EMF. But with their combined action, the degree of curvature of the nanotube shape does not increase, and on the contrary, it substantially decreases, which indicates the interaction of defects and the effective "healing" of both radiation and growth defects by the migration of broken carbon atoms along with graphene layers and their recombination.

References

1. Yao Z, Postma HWCh, Balents L, Dekker C (1999) Carbon nanotube intramolecular junctions. Nature 402:273–276
2. Seamus A, Curran SA, Talla J, Dias S, Zhang D (2009) Electrical transport measurements of highly conductive carbon nanotube/poly(bisphenol A carbonate) composite. Carroll D J Appl Phys 105:073711–073718
3. Ahmad K, Pan W, Shi SL (2006) Electrical conductivity and electrical properties of multiwall carbon nanotubes and alumina composities. Appl Phys Lett 89:133122–133128
4. Sandler JKW, Kirk JE, Kinloch IA, Shaffer MSP, Windle AH (2003) Ultra-low electrical percolation threshold in carbon-nanotube-epoxy composites. Polymer 44:5893–5899
5. Bryning MB, Islam MF, Kikkawa JM, Yodh AG (2005) Very low conductivity threshold in bulk isotropic single-walled carbon nanotube-epoxy composites. Adv Mater 17: 1186–1191
6. Kymakis E, Alexandou I, Amaratunga GA (2002) Single-walled carbon nanotube-polymer composites: electrical, optical and structural investigation. Synthetic Metals 127:59–62
7. Bryning MB, Milkie DE, Islam MF, Kikkawa JM, Yodh AG (2005) Thermal conductivity and interfacial resistance in single-wall carbon nanotube epoxy composites. Appl Phys Lett 87:161909–161909-3
8. DeHeer WA (1995) Aligned carbon nanotu be films: production and optical and electronic properties. DeHeer WA, Bacsa WS, Chatelain A, Gerfin T, Humphreybaker R, Forro L, Ugarte D. Science 268:845–847
9. Fischer JE, Dai H, Thess A, Lee R, Hanjani NM, Dehaas DL, Smalley RE (1997) Metallic resistivity in crystalline ropes of single-wall carbon nanotubes. Phys Rev B 55:R4921–R4924

10. Ebbesen TW, Lezec HJ, Hiura H, Bennett JW, Ghaemi HF, Thio T (1996) Electrical conductivity of individual carbon nanotubes. Nature London 382(4):54–56
11. Харрис П (2003) Углеродные нанотрубки и родственные структуры. Новые материалы XXI Века / П. Харрис. – М.:Техносфера, 335 с
12. Ахметов АО (2010) Система контроля работоспособности функционально сложных интегральных микросхем при проведении радиационных исследований / А.О. Ахметов, Д.В. Бобровский, П.В. Некрасов // Электроника, микро- и наноэлектроника. Сборник научных трудов. М.:НИЯУ МИФИ, С.258–265
13. Kotakoski J, Krasheninnikov AV, Nordlung K (2007) Atomistic simulations of irradiation effects in carbon nanotubes: an overview. Rad Effec Defect Solids 162(3–4):157–169
14. Wilkins R (2005) Ground-based space radiation effects studies on single-walled carbon nanotube materials. Wilkins R, Pulikkathara MX, Khabashesku VN et al. Mater Res Soc Sympos Proc 851:NN6.5
15. Новиков ЛС (2010) Радиационные воздействия на материалы космической техники / Л.С. Новиков. – М.: Университетская книга, 192 с
16. Макунин АВ (2011) Полимер-наноуглеродные композиты для космических технологий. Часть 1. Синтез и свойства наноуглеродных структур. Учебное пособие / А.В. Макунин, Н.Г. Чеченин. – М: Университетская книга, 150 с
17. de Jonge N, Bonard JM (2004) Carbon nanotube electron sources and applications. Philosoph Trans Royal Soc London. 362:2239–2266
18. Ackland G (2010) Controlling radiation damage. Science 327:1587–1588
19. Патент на корисну модель "Спосіб визначення поперечної електропровідності порошку з анізотропією форми частинок" / Нищенко М.М., Михайлова Г.Ю., Шевченко М.Я.; заявник та патентовласник Інститут металофізики ім.. Г.В. Кюрдюмова НАН України. – №94148; заяв. 16.06.2014, опубл.27.10.2014 р., Бюл.20

Conductance Length Dependence in Carbon-Conjugated Nanoscale Systems. Use of Extended Quasi-correlated Orbitals

Anatoliy V. Luzanov

Abbreviations

DFT	Density functional theory
EQC	Extended quasi-correlated (model)
EUE	Effectively unpaired electron
FCI	Full configuration interaction
GF	Green's function
HOMO	Highest occupied MO
HPHF	Half-projected Hartree-Fock
LUMO	Lowest unoccupied MO
MO	Molecular orbital
NBA	Normal bond alternation
QCTB	Quasi-correlated tight-binding (model)
RBA	Reverse bond alternation
RHF	Restricted Hartree-Fock
SMC	Single-molecule conductance
TB	Tight-binding (model)

1 Introduction

Despite the increasing research being undertaken in the area of single-molecule conductance (SMC) theory (e.g., see [1–10]) there is still a need for careful assessment of various existing approaches to single-molecule conductance (SMC). Really,

A. V. Luzanov (✉)
SSI "Institute of Single Crystals", NAS of Ukraine, 60 Nauky Ave., 61001 Kharkiv, Ukraine
e-mail: avluzanov@gmail.com

O. Fesenko and L. Yatsenko (eds.), *Nanomaterials and Nanocomposites, Nanostructure Surfaces, and Their Applications*, Springer Proceedings in Physics 246, https://doi.org/10.1007/978-3-030-51905-6_41

many interesting effects are revealed by very simple models such as the tight-binding (TB) approximation (see review [10]). However, not all results of TB models can be accepted with certainty even at the qualitative level. One of the issues that require to be scrutinized more closely is an assumed increase of SMC with increasing size of the chain-like molecules (usually SME is expected to be exponentially decaying with length). It is found in the interesting papers [11–13] that for some structural types TB predicts an increase of SMC with enlarging π-network, and these predictions are in qualitative agreement with density functional theory (DFT) results. However, note especially that generally DFT cannot be treated as fully reliable for SMC because the typically used energy functionals frequently come to underestimating HOMO-LUMO gaps [14–16]. As recognized in [17], "DFT tends to overestimate conductances by up to several orders of magnitude".

Thus, in investigating SMC for large systems we focus on the good-quality semiempirical models which take into proper account of π-electron interaction effects. Our previous studies [18, 19] also support use of relatively uncomplicated π-models which, nonetheless, are beyond the crude TB topological approximation. Therefore, the main goal of the paper is to elucidate, by using π-electron correlation models, whether it is indeed possible to predict an increase of single-molecule conductance in long π-conjugated systems.

2 The Used Models

We will follow notations and definitions as in [18, 19]; for general notions and terminology the reader refers to book [9]. In the ballistic regime, the T = 0 molecular conductance at the Fermi energy is determined by matrix elements of the one-electron Green's function (GF), G, evaluated for a given electronic model. The position of contacts (electrodes) predetermines which matrix elements of G must be computed. Within π-electron schemes we deal with a standard basis set $\{|a\rangle\}$ including the 2pπ atomic orbital of each carbon atom of the conjugated molecule in question ($1 \le a \le N$, and N is a number of the atoms). For the given pair (a, b) where a and b signify contacting carbon atoms, the corresponding molecular conductance g_{ab} can be estimated, within the conventional Landauer approach, by the relation

$$g_{ab}/g_0 = \Gamma_a \Gamma_b |G_{ab}|^2. \quad (1)$$

Here g_0 is the quantum conductance unit, $G_{ab} = \langle a|G|b\rangle$ is the GF matrix element, and Γ_a and Γ_b are broadening parameters expressing approximately coupling effects between electrodes and respective sites a and b of the molecule in question.

Now for the retarded GF at the Fermi energy let us consider the customary approximations along with the schemes elaborated recently (for more detail see Appendices). They all can be encompassed by a general one-electron resolvent of the form:

$$G_0^{\mathrm{X}} = (i0^+ - h^{\mathrm{X}})^{-1}, \tag{2}$$

where X specifies the used model and the respective effective Hamiltonian h^{X}. The occurrence of the imaginary infinitesimal $i0^+$ is not essential for our purposes here. More specifically, take the simplest known model, i.e. the standard TB approximation for π-electrons. As usual, the TB Hamiltonian matrix, h^{TB}, is identified (up to a factor) with an adjacency matrix of the corresponding graph (it is in fact the Hückel energy matrix). Then

$$G_0^{\mathrm{TB}} = (i0^+ - h^{\mathrm{TB}})^{-1} \tag{3}$$

is a suitable working formula having a huge body of applications. The merits and demerits of the Hückel π-electron approximation are well discussed in literature ([20] and many others).

The characteristic drawback of TB is neglecting direct long-distance electronic effects (see, e.g., [19, 21]). One can partially avoid this difficulty by using the restricted Hartree-Fock (RHF) scheme:

$$h^{\mathrm{RHF}} = f, \tag{4}$$

with f being the usual Fock matrix which is computed easily within the conventional Pariser-Parr-Pople approximation. Additionally, it is suitable to shift all orbital energy in (4) in such a way that the Fermi energy becomes a zero reference point. The shifted and renormalized h^{RHF} matrix is further used for evaluating G_0^{RHF} and for other models. For more detail see Appendices; in particular, the renormalization of h^{RHF} is made by factor $1/\beta_{\mathrm{eff}}$ where β_{eff} is written in (23), and the energy shift value is identified with the Fermi level given by (30).

However, do not forget that RHF can suffer from serious disadvantages (RHF instability and ignoring electron correlation) which are particularly significant for large-scale problems. At the same time, a very simple way to improve TB model has been was proposed [22, 23] in the form of a quasi-correlated TB (QCTB) scheme. As shown in [18], in QCTB we in fact replace h^{TB} by the formal Hamiltonian

$$h^{\mathrm{QCTB}} \Rightarrow [(h^{\mathrm{TB}})^2 + \delta^2]/h^{\mathrm{TB}}, \tag{5}$$

which makes sense only for computations of G_0^{QCTB} by (2). Here δ is an orbital splitting parameter by which different orbital for different spins are introduced into the TB scheme, thus taking into implicit account the π-electron correlation effects.

A further improvement is attained by combining QCTB and RHF models. The approach is suggested in [21] and termed the extended quasi-correlated (EQC) method. The preliminary results were given in [19] in terms of the quasi-correlated long-range interaction (QCLRI) scheme which is identical to EQC from [21]. In this case we construct the intermediate Fock matrix $f_{[1]}$ (also shifted by the Fermi energy). This matrix is normally computed by using the TB density matrix (bond-order matrix) P. Hence, we do not invoke any self-consistency procedure. The obtained intermediate Fockian $f_{[1]}$ is then renormalized [see (25)] and used straightforwardly in computing G_0^{EQC} by (2) with applying a formal Hamiltonian similar to (5):

$$h^{\mathrm{EQC}} \Rightarrow [(f_{[1]})^2 + \delta^2]/f_{[1]}$$

The explicit representation is provided by (26) and (27). The benefit of EQC is owing to accounting for long-range effects, so that the drawbacks inherent to QCTB (due to a short-range nature of TB) are removed. In doing so, π-correlation effects are also presented in EQC via the previous splitting parameter δ. Hence, this approach is more preferable, especially for nano-sized systems and singular π-structures.

It is also worth mentioning another recently developed model [19] based on the variatonal treatment of electron detachment/attachment processes within the half-projected Hartree-Fock (HPHF) method. The appropriate GF will be represented by G_0^{HPHF}. As seen from [19], among the models considered there, G_0^{HPHF} is the best but only for not too large systems because HPHF lacks a size-consistency.

At last, the complete account for all electronic effects is achieved by the π-electron full configuration interaction (FCI) method [24]. The related approach for GF was elaborated in [25]. The thus obtained G_0^{FCI} can be qualified as a "gold standard" in the molecular Green's function theory. Unfortunately, FCI is feasible for small systems only (e.g., in our examination below we could provide G_0^{FCI} for maximum 14-electron π-structures). Therefore, G_0^{FCI} serves us here for testing G_0^{X} (X= TB, RHF, EQC and HPHF) on small-size systems. The QCTB results are not included explicitly in the present paper because they are qualitatively close to those of EQC.

3 Polyene-like Toy Models

As our first concrete problem we consider model polyene-like π-systems in the same specific setting as in [13]. In the cited work the simple TB model of even cumulene chains is exploited in order to examine the size dependence of molecular conductivity. Recall that each cumulene molecule with even number of C atoms contains two, short and long, orthogonal π-subsystems, and for the conductivity problem one can take into consideration only the long one which will be symbolized by (-C = C- $)_n$. In the cumulenes a weak interaction between the subsystems can be safely neglected. In [13] the focus is on alternation of resonance integrals, and two types of their alternation were investigated, the normal bond alternation (NBA) and reverse bond alternation (RBA). In the NBA case 'double bond' resonance integral is $\beta_{c=c} = \beta_0$, and that of 'simple' bond is $\beta_{c-c} = 0.935\beta_0$ with $\beta_0 = -3.0$ eV; for RBA we must take $\beta_{c=c} = 0.935\beta_0$ and $\beta_{c-c} = \beta_0$. In this section we make using this TB parametric scheme of [13], and to the TB results of the cited work we will add appropriate data from the more advanced π-models (RHF, EQC, HPHF, and FCI). For linear n-polyene-like chains with $n = 1 \div 7$, the obtained results are displayed numerically in Tables 1 and 2, and graphically in Fig. 1.

It can be seen from the tables that, if excluding the TB scheme, the approximate (relative to FCI) GF results, particularly the HPHF ones, are very close to those of the FCI gold standard. The principal outcome from these calculations is that with increasing the polyene-like chain (n in the tables) no increase of GF (and thereby

Table 1 Comparison of π-electron GF element $(G_0^{\mathrm{X}})_{1,2n}$ for various models (X = FCI, HPHF, EQC, RHF, and TB) in n-polyene-like chains with NBA

n	FCI	HPHF	EQC	RHF	TB
2	−0.557	−0.575	−0.564	−0.606	−0.935
3	0.369	0.383	0.382	0.395	0.874
4	−0.256	−0.264	−0.270	−0.264	−0.817
5	0.182	0.186	0.196	0.180	0.764
6	−0.131	−0.132	−0.145	−0.123	−0.715
7	0.096	0.094	0.109	0.085	0.668

Table 2 Comparison of π-electron GF element $(G_0^{\mathrm{X}})_{1,2n}$ for various models (X = FCI, HPHF, EQC, RHF, and TB) in n-polyene-like chains with RBA

n	FCI	HPHF	EQC	RHF	TB
2	−0.632	−0.647	−0.645	−0.699	−1.144
3	0.470	0.481	0.489	0.516	1.223
4	−0.370	−0.374	−0.389	−0.396	−1.308
5	0.300	0.300	0.317	0.312	1.399
6	−0.249	−0.245	−0.262	−0.250	−1.497
7	0.209	0.202	0.216	0.202	1.601

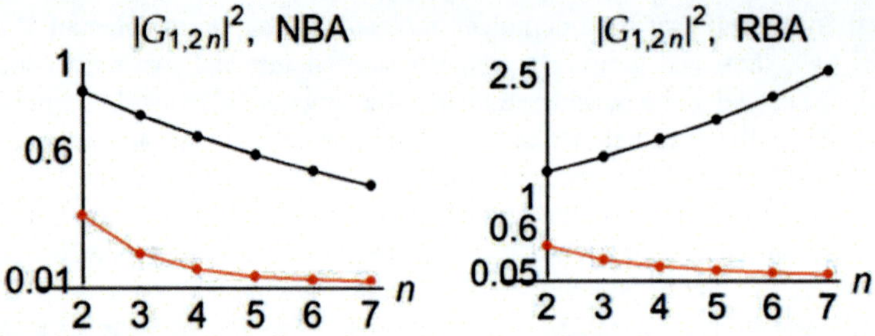

Fig. 1 Length dependence of squared TB (in black), and FCI (in red), elements for the model n-polyene-like chains with the normal bond alternation (left panel) and the reverse bond alternation (right panel)

no increase of the conductivity) is observed. In further analysis we will use more concise notations $G_{1,2n} = (G_0)_{1,2n}$, $G_{p,p'} = (G_0)_{p,p'}$ etc.

The difference between TB and the advanced π-theories is evident from Fig. 1 as well. The TB approximation provides the qualitatively correct results only for usual NBA polyene systems (the left panel in Fig. 1). At the same time, for the RBA structures (the right panel in Fig. 1) TB predicts a nonphysical effect of growing the electric conductance in the finite-size molecular conductors with nonzero orbital gap. In addition, it is worth bearing in mind that as a rule, SMC has basically a tunnel nature (with exponentially decaying tunnel transitions).

4 Polyacene and Nanographene Molecules

Now we examine more complex (acene-based) systems that are also discussed in the current literature in the context of size behavior of molecular conductance [12]. To study this we first consider exemplary cases of linear [n]acene series (Fig. 2). In the figure we show the contact carbon atom pairs of which the GF matrix elements are computed. These pairs are taken of two types: para positions to each other, or (p, p') pairs, and the farest-distance pairs, or shortly (f, f') pairs. In (p, p') pairs of the linear [n]acenes, the distance between sites p and p' is in fact a constant value (nearly twice length of aromatic bond), whereas in (f, f') pairs the distance increases almost linearly with increasing n (a number of rings). For the two mentioned types of contacts, we compare the GF elements at the FCI level with those at the approximate levels of theory (Table 3). In computations of acenes we employ the same π-electron parameters as in previous papers [18, 19, 21] (all resonance integrals are equal to $\beta_0 = -2.4$ eV etc.).

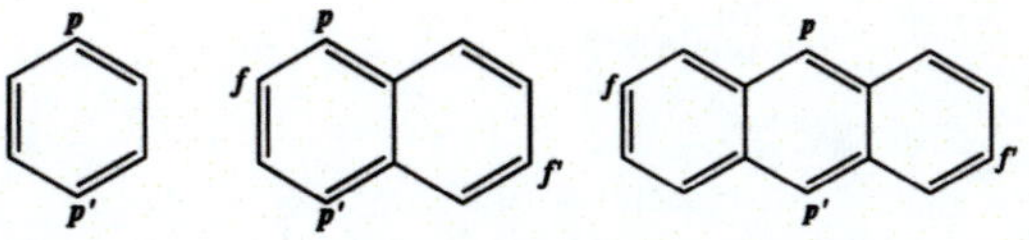

Fig. 2 Polyacenes with two types of connectivities to electrodes; labeling contact pairs: (p, p') for the para type and (f, f') for the farest-distance type

Table 3 Comparison of π-electron GF elements $|(G_0^{\mathrm{X}})_{p,p'}|$ (in bold face) and $|(G_0^{\mathrm{X}})_{f,f'}|$ (in plane face) in various models (X = FCI, HPHF, EQC, RHF, and TB) for the first three members of the acene series

Acene	FCI	HPHF	EQC	RHF	TB
Benzene	**0.447**	**0.458**	**0.439**	**0.496**	**0.500**
Naphthalene	**0.519**	**0.527**	**0.495**	**0.568**	**0.667**
	0.236	0.245	0.232	0.270	0.333
Anthracene	**0.640**	**0.626**	**0.587**	**0.708**	**1.000**
	0.125	0.131	0.126	0.153	0.250

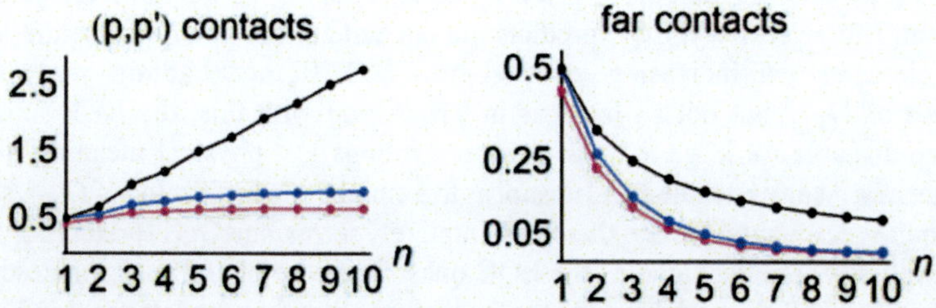

Fig. 3 Length dependence of $|(G_0)_{p,p'}|$ and $|(G_0)_{f,f'}|$ for TB (in black), RHF (in blue), and EQC (in magenta) in $[n]$acenes

We observe from Table 3 that for the small acenes from Fig. 2 all the models are qualitatively more or less similar (even despite marked quantitative differences between TB and more sophisticated models). With this, we see again that EQC is not worse than then HPHF, and for (f,f') pairs the EQC is even better. On this account and on other similar accounts (do not also forget about the lack of size-consistency in HPHF) we suggest EQC as the most reliable for large-scale $[n]$ acenes and graphene systems which are examined below.

Before analyzing the obtained data for the long acenes (Fig. 3) we briefly touch their specific behavior in the TB (Hückel) description. As known, for many classes of π-electron networks it is possible to present analytical solution to (3) (see, e.g., [10, 11, 26, 27]) via the graph-theoretic formula due to Heilbronner [28]. For the studied acenes the required GF elements turn out to be very simple. Let $G_{p,p'}^{\mathrm{TB}}[n]$ signify the absolute value of GF matrix element for a para-type contacts in $[n]$acene.Then, for odd $n = 2k + 1$ we obtain

$$G^{\mathrm{TB}}_{p,p'}[2k+1] = (k+1)/2, \tag{7}$$

and for even $n = 2k$

$$G^{\mathrm{TB}}_{p,p'}[2k] = k(k+1)/(2k+1). \tag{8}$$

Respectively, for a farest-distance type contacts one finds

$$G^{\mathrm{TB}}_{f,f'}[n] = 1/(n+1). \tag{9}$$

Of course, (7)–(9) reproduce the results of the last column in Table 3. No less interesting is the asymptotic behavior of these expressions when n goes to infinity:

$$G^{\mathrm{TB}}_{p,p'} \to n/4, \tag{10}$$

$$G^{\mathrm{TB}}_{f,f'} \to 0. \tag{11}$$

Now look at the plots in Fig. 3. We see that the TB curve for para contacts naturally follows (10) whereas the curves given by other methods are qualitatively perfectly different. More precisely, EQC predicts the opposite effect of a quick saturation of $G_{p,p'}$ elements with increasing acene length. The RHF model admits only a slight increase of $G_{p,p'}$ but not so large as in TB. Along with this, the RHF solutions become unstable for large n, so that these solutions lost physical meaning for too long acenes. At any rate, the EQC upshot as to a small finite limit value of $G_{p,p'}$ seems to be quite reasonable. Notice that in qualitatively terms, the QCTB data (not given here explicitly) are the same as the EQC ones. The case of (f,f') pairs is described realistically by all π- models used here for linear acenes (see the left panel in Fig. 3).

Next we consider the graphenic structures. Small and large graphene networks are typical for applications of the SMC theory [29], particularly in its TB version [3, 26]. The conductance length dependence for graphene-like molecules was examined recently in [12] from which we take representative examples of small systems comprising of the anthracene structure subunits. Here we extend the consideration and explore [n]perianthracene for sufficiently large number n of anthracenic units (Fig. 4). Notably, invoking electron-correlation π-models is of primary importance for such nanoscale systems.

Fig. 4 Structure and (p,p') contact pair for [n]perianthracene

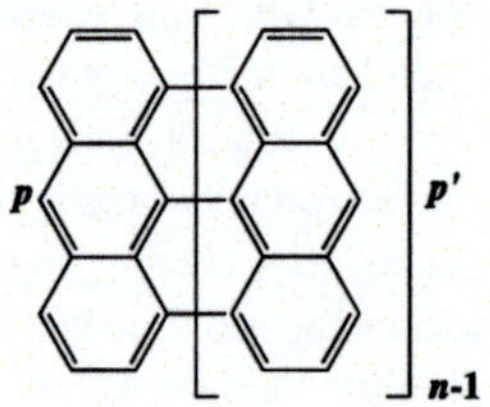

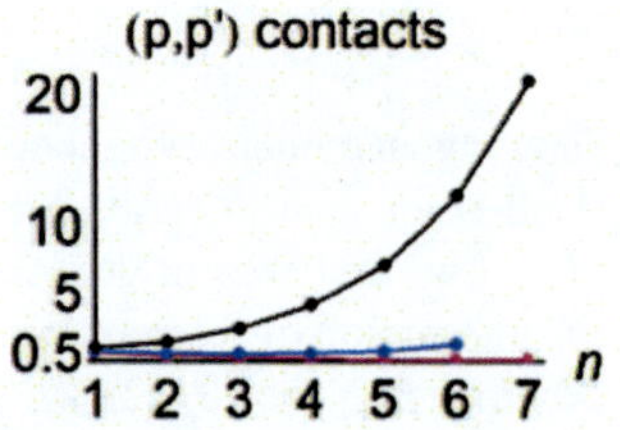

Fig. 5 [*n*]Perianthracenes: length (*n*) dependence of $|(G_0)_{p,p'}|$ in TB (in black), RHF (in blue), and EQC (in magenta)

Firstly, we discuss the (p,p') contact problem in [*n*]perianthracene in analytical terms at the TB level. The computations are performed by the above-mentioned Heilbronner formula [28]. The latter reduces the required GF elements to the ratio:

$$|G^{\mathrm{TB}}_{(p,p')}| = N^{\mathrm{K}}_{(p,p')}/N^{\mathrm{K}}_0, \tag{12}$$

where N^{K}_0 is a number of Kekulé structures of the whole molecule, and $N^{\mathrm{K}}_{(p,p')}$ is that of the molecule after removing sites p and p'. For [*n*]perianthracene $N^{\mathrm{K}}_0 = 4^n$, and $N^{\mathrm{K}}_{(p,p')}$ can be found in [30]. p. 213. It gives

$$|G^{\mathrm{TB}}_{p,p'}| = [(1 + 1/\sqrt{2})^n + (1 - 1/\sqrt{2})^n]/2 \tag{13}$$

which is asymptotically exponentially rising as a function of n:

$$|G^{\mathrm{TB}}_{p,p'}| \to [(1 + 1/\sqrt{2})^n/2$$

Let us now turn to the EQC results for the same problem. Again, in a sharp contrast to the TB picture, the $G^{\mathrm{EQC}}_{p,p'}$ systematically decreases with increasing n, as clearly seen from Fig. 5. And once more RHF do not provide reliable data (occurrence of symmetry breaking solutions etc.).

5 Polyquinoids

Polyquinoids are a specific example of the conjugated systems with fixed double bonds (that is having a single classical structure like polyenes; see Fig. 6). Various electronic properties of large polyquinoid systems are discussed in recent works [11, 18, 31, 32].

In particular, in [11] the molecular conductance within TB is considered, and it is shown that simple graph-theoretic consideration comes to the exponential dependence of $G^{\mathrm{TB}}_{p,p'}[n]$ as a function of the length parameter n. However, this TB result seems to be anomalous. A more sophisticated study is needed here.

Let us start with the appropriate analytical expression given in the cited paper [11]:

$$G^{\mathrm{TB}}_{p,p'} = 2^n, \tag{14}$$

showing an evident exponential growth with n. This expression is valid (within TB) if all resonance integrals are set equal, and as such (14) is simply derived from (12). The first step in our study was to use a more adequate TB model including the alternation of resonance integrals. We took from [18] the appropriate alternation scheme: $\beta_{\mathrm{c=c}} = 13/12\beta_0$ $\beta_{\mathrm{c-c}} = 11/12\beta_0$ with $\beta_0 = -2.4$ eV. Yet, it provided the better TB results which are different, but only quantitatively, not qualitatively: instead of (14), merely a more slow exponential growth,

$$G^{\mathrm{TB}}_{p,p'} = 0.923 \times 1.432^n,$$

was numerically obtained.

These results have induced us to invoke more refined models that include explicitly electron interaction effects. To this end we have employed RHF and EQC schemes. As in the above studies, for controlling accuracy of the used approximations the standard FCI model was restrictively applied (only for $n = 1$ and $n = 2$). Thus, we could see that EQC and RHF, but not TB, offer a reasonable estimation of $G^{\mathrm{FCI}}_{p,p'}$ in small [n]quinodimethanes:

$$n = 1 : 0.492, 0.563, 0.574, 1.322,$$

$$n = 2 : 0.347, 0.453, 0, 430, 1.892.$$

where the first number of each set is $G^{\mathrm{FCI}}_{p,p'}$, and the rest are related to EQC, RHF and TB, respectively.

The results of computing GF in the long quinoids are presented in Fig. 7. We see that in fact the poly(phenoquinodimethanes) exhibit the same behavior as the poly(perianthracenes) in Fig. 5. The main conclusion from Fig. 7 is also evident—there is no increasing trend in the electric conductance at the EQC levels (and actually at the RHF level too).

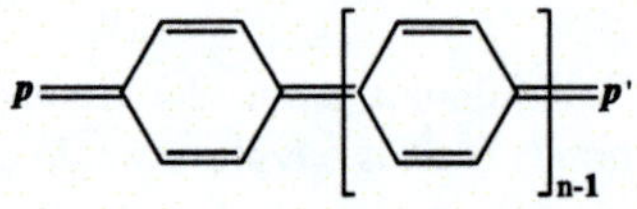

Fig. 6 Structure of [n]phenoquinodimethane with (p,p') farest contact pair

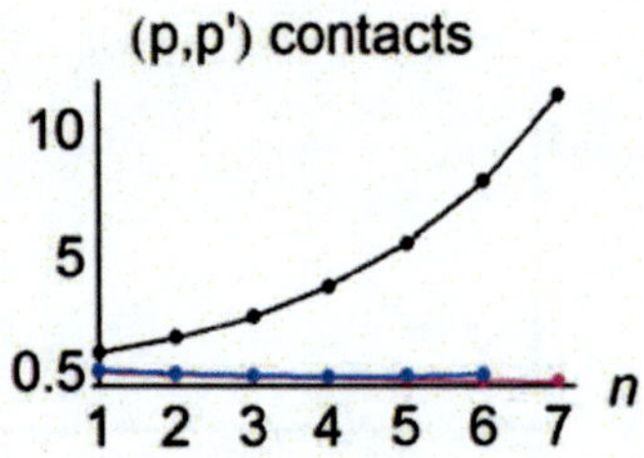

Fig. 7 [*n*]Phenoquinodimethanes: length (n) dependence of $G_{p,p'}$ in TB (in black), RHF (in blue), and EQC (in magenta)

6 Influence of Electron Unpairing

The considered systems are sufficiently representative to discuss π-electron conductance in a slightly more wide context of the electronic structure theory. The two known characteristics are relevant to be brought into play too. These are the conventional (Coulson-type) bond orders and local densities of effectively unpaired electrons (EUE). As to the bond orders, specifically $P_{p,p'}$, that is (*p*,*p*′) intersite elements of the one-electron density matrix, we only recall that they are some sort of distant cousins of the GF elements $G_{p,p'}$. The both quantities describe an electron coupling between the given sites, and moreover, it is not accidental that GF is sometimes called the influence function. In π-electron quantum chemistry there is a specific definition [33] of the π-bond order which is named the Pauling bond order. The latter is precisely identical to $G^{\mathrm{TB}}_{p,p'}$ [34], so that as a rule this quantity is not equivalent to the Coulson bond order $P_{p,p'}$. Of course, GF quantities being viewed as bond orders suffer from many deficiencies ([35–37] and others). Nevertheless, our experience says that the $G_{p,p'}$ values computed in better approaches to GF might be more adequate for treating π-electron bond orders. Thus, it is reasonable to expect that $P_{p,p'}$ and $G_{p,p'}$ behave more or less similarly. We explore this issue for bipartite (alternant) networks in Appendix B, where nondiagonal long-range matrix elements are crudely estimated, as folows:

$$G^{appr}_{p,p'} \approx 2P_{p,p'} / \Delta_c, \tag{15}$$

In above, Δ_c is a HOMO-LUMO energy gap, or more exactly, a fundamental gap (the latter will be discussed somewhat later).

The second of the above-mentioned quantities is related to the EUE atomic distribution which reflects a radical nature of molecules. It should be emphasized right away that the influence of the radical character on SMC is really not so simple as seems at first sight, and the results given below provide as well a critical look at current points of view on this issue. For a detailed account and relevant references on EUE see review [23].

In few words, the EUE theory [38, 39] affords a quantitative picture of the open-shell (frequently hidden) structure for any many-electron system (even for ground singlet states if allowing for electron correlation effects). In fact, EUE gives a measurement of molecular radical character, and it is interesting to apply EUE to

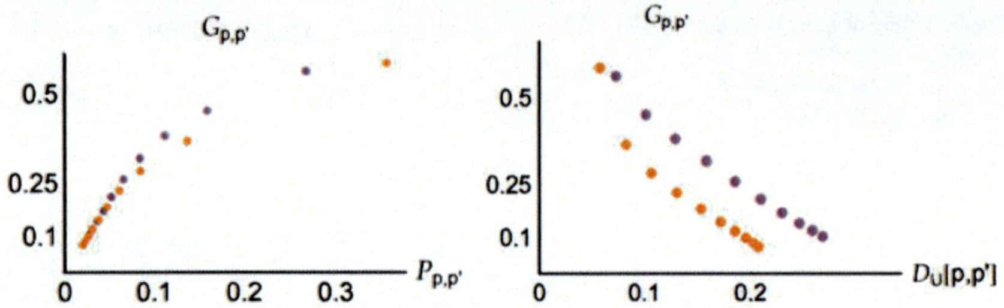

Fig. 8 Correlations for [*n*]perianthracenes (in red) and [*n*]phenoquinodimethanes (in violet). Left panel: $G_{p,p'}$ *vs* bond order $P_{p,p'}$; right panel: $G_{p,p'}$ *vs* $D^{\mathrm{U}}[p, p']$

the molecular conductivity problems. Notice that in [12] frontier natural orbital occupancies have been used for similar problems. At this point our approach is somewhat different from that of [12]. We believe that when analyzing conductance across a contact pair one must deal with the respective local EUE characteristics rather than the general EUE ones.

Let us introduce atomic EUE distribution $\{D_\mu^{\mathrm{U}}\}_{1\le\mu\le N}$ [39], following the definition of matrix D^{U} in [23], (6.42). Then, for the given highly correlated electron system the respective GF element, say $G_{p,p'}$, should be compared with an appropriate geometrical (not arithmetical) average of EUE indexes as follows:

$$D_{\mathrm{U}}[p, p'] = (D_p^{\mathrm{U}} D_{p'}^{\mathrm{U}})^{1/2} \tag{16}$$

etc. Turn to the results we display for specific examples in Fig. 8. From the plot in the right panel we see that, as expected from (15), GF elements at the Fermi level are positively correlated with long-range bond orders. Less trivial is the fact that the same quantities are negatively correlated with the averaged electron-unpairing densities, (16), related to contacting sites (the right panel in Fig. 8).

These observations are supported by direct computations of statistical correlation coefficients, r, for each structural series. Analyzing $G_{p,p'}$ data *vs* bond-order ones we obtained, in concordance with (15), the positive r values (for both perianthracene and phenoquinodimethanes series $r = 0.96$). Likewise, the statistical analysis of $G_{p,p'}$ *vs* $D^{\mathrm{U}}[p, p']$ leads to the negative values ($r = -0.96$ for perianthracene series, and $r = -0.99$ for phenoquinodimethane series).

The observed anticorrelation of GF *vs* electron unpairing can be elementarily clarified in terms of Mott's model of metal-insulator transition [40]. In this theory one usually deals with the so-called half-filled systems (one electron per site), just as in our π-electron problems. A rather simple consideration (see [41], Sect. 1.4.6) shows that in Mott insulators the Coulomb repulsion gives rise to the electron localization (Mott's localization mechanism) which blocks ability of electron to move freely between sites. Furthermore, in the Mott theory, the charge gap Δ_c, or the fundamental gap, is one of the basic energy parameters of electronic materials (for currently used terminology see, e.g., [42]). This Δ_c can be defined as

$$\Delta_c = I_0 - A_0, \tag{17}$$

where I_0 is ionization potential and A_0 electron affinity of the ground state of molecule in question. It is just this quantity that is poorly determined in the conventional DFT schemes [14]. Recently, important high-quality π-electron computations of Δ_c in polyacenes and graphene molecules have been accomplished using density matrix renormalization group method [43–46]. In all calculations of various nanoribbons with limited width, a finite extrapolated value of Δ_c was obtained; in particular, $\Delta_c = 2.45\,\text{eV}$, was given in [46] for a long poly(perinaphthalene). Hence, there is seemingly no possibility about the correctness of the TB prediction concerning a metal nature of finite-width graphene ribbons. Related to this topic are the results reported in [47] where it was shown experimentally that there exists an energy gap in nominally metallic carbon nanotubes. The same conclusion was reached theoretically in [48] for zigzag nanotubes and in [49] for graphenes.

In the light of these facts it seems also doubtful the conductance growth (at the Fermi level) with length of graphene molecules. To this reasoning we turn attention to the fact that the modified EUE localization indexes can reflect a partial antiferromagnetic ordering in molecules (see Fig. 6 in [23]). This ordering is in a direct relation to the Mott electron localization, which inevitably hinders electron transfer, so the statistical anticorrelations in the right panel of Fig. 8 become quite understandable.

7 Concluding Remarks

In this paper we have examined at the π-electron level the problem of length dependence of electron transport through single conjugated molecules. We employed several types of π-electron approximations, namely the very popular one-electron TB, the new advanced but simple approximation EQC, and more restrictively, the conventional reference FCI method; the two last are used to account for π-electron correlation effects.

It is found that TB, i.e., Hückel's model (even with taking into account bond alternation) cannot be considered as an adequate tool for treating such nontrivial issues as the conductance length dependence. Unlike the EQC and FCI theories, the TB schemes for polyene-like and polyquinoid structures produce abnormal increase of the molecular conductance (at the Fermi level). The SME problem was also examined here for periacene-type nanographene molecules, and again no increase of the conductance was observed at the EQC level, opposite to the TB and DFT predictions. This fact gives additional arguments to regard DFT as unreliable for SMC, as occasionally stated in literature [14–16]. Thus, the main result of our paper is that for large conjugated networks the TB estimations of long-range GF matrix elements are by no means reliable too, and no increasing molecular conductance takes a place in long oligomeric chains or typical π-conjugated nanoscale systems.

Stress once more that the previously assumed increase of conductance with length of graphene molecules (when electron correlation effects are strengthened!) is in a sharp contradiction with the main mechanisms in strongly correlated systems, in particular with the key Mott localization caused by sufficiently strong electron correlation. More than that, due to a general applicability of the Mott theory to finite-sized conjugated systems with half-filled bands, it is natural to suggest that conductance does not increase with length for any π-conjugated bipartite (alternant) network.

Turn also attention to the most recent paper [50] where the specifically improved TB scheme is proposed for computing molecular conductivity. This model is interesting per se, while the whole approach seems to be rather artificial from the viewpoint of the conventional quantum chemistry. Indeed, the needed long-range interactions elementarily follow from the usual PPP Fockian matrix, especially if computing the Fockian with TB (Hückel) bond orders (e.g., in the form of $f_{[1]}$ from (9) as was suggested in [19, 21]). Besides, π-electron correlation is a fundamental feature of nanoscale systems, and it cannot be handled if dealing with any Hückel-like or even RHF wave functions.

Appendices

Appendix A: Working Formulas for Green's Functions in Various π-Schemes

After performing the needed transformations described in Sect. 2, we obtain the working equations as follows. As known, the TB Hamiltonian of arbitrary bipartite network can be always taken in the block-matrix form:

$$h^{\mathrm{TB}} = -\begin{pmatrix} 0 & B \\ B^T & 0 \end{pmatrix}, \tag{18}$$

where the B matrix elements are ones for the chemically bonded pairs of sites only; other elements are zeros. The associated GF matrix G_0^{TB} is of the same skew-diagonal form as in (18). The same is true for any other π-models of bipartite structures [18, 21]. Thus, for the given model X we can specify G_0^{X} as follows:

$$G_0^{\mathrm{X}} = \begin{pmatrix} 0 & g_{\mathrm{X}} \\ g_{\mathrm{X}}^{\mathrm{T}} & 0 \end{pmatrix}. \tag{19}$$

Invoking the results of [18, 21] for X = TB and X = QCTB, we have explicitly

$$g_{\mathrm{TB}} = (BB^{\mathrm{T}})^{-1}B, \tag{20}$$

$$g_{\mathrm{QCTB}} = (BB^{\mathrm{T}} + \delta^2)^{-1}B. \tag{21}$$

where, as in previous works [18, 19, 21–23], splitting parameter $\delta = 7/24$.

In the case of RHF one must define the renormalized counterpart of the B-matrix, B_{RHF}. For representing the latter we need the corresponding block, P_{cc}, of the RHF bond-order matrix P and the block, γ_{cc}, of two-center Coulomb integral matrix γ, so

$$B_{\mathrm{RHF}} = (\beta_0 B - \gamma_{\mathrm{cc}} \circ P_{\mathrm{cc}}/2)/\beta_{\mathrm{eff}}. \tag{22}$$

Here β_0 is the standard resonance integral for CC π-bond. The renormalized resonance integral is

$$\beta_{\mathrm{eff}} = \beta_0 - \gamma_{12}/2, \tag{23}$$

where γ_{12} is the Coulomb two-center integral for the same bond. Symbol $\circ$ denotes usual element-wise (Shur's) multiplication: $(A \circ B)_{ij} = A_{ij}B_{ij}$. Then, similarly to (18) and (19),

$$h^{\mathrm{RHF}} = -\begin{pmatrix} 0 & B_{\mathrm{RHF}} \\ B_{RHF}^T & 0 \end{pmatrix}, \; G_0^{\mathrm{RHF}} = \begin{pmatrix} 0 & g_{\mathrm{RHF}} \\ g_{RHF}^T & 0 \end{pmatrix} \tag{24}$$

where $g_{\mathrm{RHF}} = (B_{\mathrm{RHF}} B_{\mathrm{RHF}}^{\mathrm{T}})^{-1} B_{\mathrm{RHF}}$. At last, within EQC we define

$$f_{[1]} = -\begin{pmatrix} 0 & B_{[1]} \\ B_{[1]}^T & 0 \end{pmatrix}, \; B_{[1]} = (\beta_0 B - \gamma_{\mathrm{cc}} \circ P_{\mathrm{cc}}^{[0]}/2)/\beta_{\mathrm{eff}}, \tag{25}$$

where block $P_{\mathrm{cc}}^{[0]}$ is taken from the TB solution for bond-order matrix in the Hall form [51]: $P_{\mathrm{cc}}^{[0]} = (BB^{\mathrm{T}})^{-1/2} B$. Then

$$G_0^{\mathrm{EQC}} = \begin{pmatrix} 0 & g_{\mathrm{EQC}} \\ B_{\mathrm{EQC}}^T & 0 \end{pmatrix}, \tag{26}$$

and the EQC block g_{EQC} is computed as simply as in (21):

$$g_{\mathrm{EQC}} = [B_{[1]}(B_{[1]})^{\mathrm{T}} + \delta^2]^{-1} B_{[1]}. \tag{27}$$

We also notice that rescaling the modified B-matrices in (22) and (22) by renormalization factor $1/\beta_{\mathrm{eff}}$ allows us to minimize the deviation norms $||G_0^{\mathrm{RHF}} - G_0^{\mathrm{TB}}||$ and $||G_0^{\mathrm{EQC}} - G_0^{\mathrm{TB}}||$, so that for the simplest reference two-center(ethylene) π-system these norms are zero.

Appendix B: Rough Estimations and Correlations Between Long-Range GF Elements and Bond Orders

Before discussing the estimations and statistical correlations let us consider the behavior of the full $G = G(E)$ near $E = E_F$. As usual, essential contributions arise due to frontier orbitals (HOMO and LUMO). But we first look at the problem being approached from a general GF theory. Let us take a spectral representation of GF (see (5.76) in [9] or (22.A.2) in [19]) and write down the contribution, ΔG, arising from lowest electron-detached and electron-attached states of the initial N-electron system with its singlet ground-state energy E_N:

$$\Delta G(E) = (E + I_0)^{-1}|\mathbf{I}\rangle\langle\mathbf{I}| + (E + A_0)^{-1}|\mathbf{A}\rangle\langle\mathbf{A}|\,.. \tag{28}$$

Here $I_0 = E_{N-1} - E_N$ and $A_0 = E_N - E_{N+1}$ are the lowest ionization potential and electron affinity; $|\mathbf{I}\rangle$, $|\mathbf{A}\rangle$ are the associated spin-free Dyson orbitals for ionization and attachment processes, respectively. In the case of TB these orbitals are exactly the frontier MOs. By analogy we name orbitals $|\mathbf{I}\rangle$ and $|\mathbf{A}\rangle$ as the frontier Dyson orbitals.

Now recall the Hush-Pople-McLachlan theorem [52, 53] for the chemical potential (in fact the Fermi energy) E_F. The theorem is valid for bipartite π-system within PPP theory [54]. It states that E_F which can be defined as

$$E_F = -(I_0 + A_0)/2, \tag{29}$$

is a constant for all bipartite:

$$E_F = W_C + \gamma_C/2. \tag{30}$$

Here W_C is the standard valence state ionization potential for the C atom, and γ_C is the corresponding Coulomb one-center integral (the both quantities are taken as semiempirical parameters). Then, combining (29) with (17) gives the following: $I_0 = \Delta_c/2 - E_F$ and $A_0 = -\Delta_c/2 - E_F$, so that (28) becomes

$$\Delta G(E_F) = 2\,(|\mathbf{I}\rangle\langle\mathbf{I}| - |\mathbf{A}\rangle\langle\mathbf{A}|)/\Delta_c, \tag{31}$$

where Δ_c is defined by (17). From (31) we can obtain

$$\Delta G_{p,p'} \equiv \Delta G_{p,p'}(E_F) = \langle p|\,\Delta G(E_F)\,|p'\rangle, \tag{32}$$

that is the contribution to $G_{p,p'}$ from the frontier Dyson orbitals. Here $|p\rangle$ and $|p'\rangle$ are the atomic π-orbitals of sites p and p', respectively. When simplifying (32) one must keep in mind that scalar products $\langle p \mid \mathbf{I}\rangle$,$\langle p' \mid \mathbf{I}\rangle$, etc. are expansion coefficients of the Dyson orbitals. For them the same sign rule as for the usual TB orbitals can be proven [25] in the case of bipartite π-structures. It comes to the equality of the **I** and **A** terms in (31) and (32), thus giving the preliminary result:

$$\Delta G_{p,p'} = 2\,\Delta P_{p,p'}/\Delta_{\mathrm{c}}, \tag{33}$$

with $\Delta P_{p,p'} = 2\langle p \mid \mathbf{I}\rangle\langle \mathbf{I} \mid p'\rangle$ being the contribution of the frontier Dyson orbitals to bond order $P_{p,p'}$. In practice we frequently observe that $P_{p,p'} \approx \Delta P_{p,p'}$ for long-range bond order $P_{p,p'}$, and one can expect the expression

$$G^{\mathrm{appr}}_{p,p'} = 2P_{p,p'}\big/\Delta_{\mathrm{c}} \tag{34}$$

to be reasonable for order-of-magnitude estimations of the respective long-range GF elements. This suggestion was verified numerically. In particular, ratio $G_{p,p'}\big/G^{\mathrm{appr}}_{p,p'}$ is about one for [*n*]periacene series with large *n*: 1.41 for $n = 2$; 1.21 for $n = 3$; 1.09 for $n = 4$; 1.04 for $n = 5$, etc.; the ratio is equal to 1.22 for polyquinoids from Fig. 6. At the same time, $G_{p,p'}\big/G^{\mathrm{appr}}_{p,p'}$ is approximately 1.56 for the para contacts in [*n*]phenacenes. At least, for all the cases studied in the paper, (34) has provided sensible estimates of long-range elements of π-electron GF.

References

1. Datta S (2005) Quantum transport: atom to transistor. Cambridge University Press, Cambridge
2. Nitzan A (2001) Electron transmission through molecules and molecular interfaces. Ann Rev Phys Chem 52:681–720
3. Chen F, Tao NJ (2009) Electron transport in single molecules: from benzene to graphene. Acc Chem Res 42:573
4. Zimbovskaya NA, Pederson MR (2011) Electron transport through molecular junctions. Phys Rep 509:1–87
5. Solomon GC, Herrmann C, Ratner MA (2012) Molecular electronic junction transport: some pathways and some ideas. Top Curr Chem 313:1–38
6. Lambert CJ (2015) Basic concepts of quantum interference and electron transport in single-molecule electronics. Chem Soc Rev 44:875–888
7. Moth-Poulsen K (ed) (2016) Handbook of single-molecule electronics. Pan Stanford Publishing, Singapore
8. Xiang D, Wang X, Jia C, Lee T, Guo X (2016) Molecular-scale electronics: from concept to function. Chem Rev 116:4318–4440
9. Cuevas JC, Scheer E (2017) Molecular electronics: an introduction to theory and experiment, 2nd edn. World Scientific, Singapore
10. Tsuji Y, Estrada E, Movassagh R, Hoffmann R (2018) Quantum Interference, graphs, walks, and polynomials. Chem Rev 118:4887–4911
11. Stuyver T, Fias S, De Proft F, Geerlings P (2015) The relation between delocalization, long bond order structure count and transmission: an application to molecular wires. Chem Phys Lett 630:51–56
12. Stuyver T, Zeng T, Tsuj Y, Geerlings P, De Proft F (2018) Diradical character as a guiding principle for the insightful design of molecular nanowires with an increasing conductance with length. Nano Lett 18:7298–7304

13. Garner MH, Bro-Jørgensen W, Pedersen PD, Solomon GC (2018) Reverse bond-length alternation in Cumulenes: candidates for increasing electronic transmission with length. J Phys Chem C 122:26777–26789
14. Refaely-Abramson S, Sharifzadeh S, Jain M, Baer R, Neaton JB, Kronik L (2013) Gap renormalization of molecular crystals from density-functional theory. Phys Rev B 88:081204-1-5
15. Chen Y, Tamblyn I, Quek SY (2017) Energy level alignment at hybridized organic-metal interfaces: the role of many-electron effects. J Phys Chem C 121:13125–13134
16. Xuan F, Chen Y, Quek SY (2019) Quasiparticle levels at large interface systems from many-body perturbation theory: the XAF-GW Method. J Chem Theory Comput 15:3824–3835
17. Hüser F, Solomon GC (2015) From chemistry to functionality: trends for the length dependence of the thermopower in molecular junctions. J Phys Chem C 119:14056–14062
18. Luzanov AV (2019) Single-molecule electronic materials: Conductance of π-conjugated oligomers within quasi-correlated tight-binding model. Funct Mater 26:152–163
19. Luzanov AV (2019) Single-molecule conductance theory using different orbitals for different spins: Applications to π-electrons in graphenes molecules. In: Fesenko O, Yatsenko L (eds) Nanophotonics, nanooptics, nanobiotechnology and their applications. Springer proceedings in physics, vol 222. Springer, Cham, pp 341–358
20. Kutzelnigg W (2007) What I like about Hückel theory. J Comput Chem 28:25–34
21. Luzanov AV (2019) Extended quasi-correlated orbitals with long-range effects: application to organic single-molecule electronics. Funct Mater 27:147–158
22. Luzanov AV (2014) Effectively unpaired electrons in bipartite lattices within the generalized tight-binding approximation: application to graphene nanoflakes. Funct Mater 21:437–447
23. Luzanov AV (2016) Effectively unpaired electrons for singlet states: from diatomics to graphene nanoclusters. In: Leszczynski J, Shukla MK (eds) Practical aspects of computational chemistry IV. Springer, Boston, pp 151–206
24. Amos T, Woodwars M (1969) Configuration-interaction calculations for small Pi systems. J Chem Phys 50:119–123
25. Pedersen KGL, Strange, M, Leijnse M, Hedegard P, Solomon GC, Paaske J (2014) Quantum interference in off-resonant transport through single molecules. Phys Rev B90:125413- 1-11
26. Fowler PW, Pickup BT, Todorova TZ, Myrvold W (2009) Conduction in graphenes. J Chem Phys 131:244110-1-8
27. Tsuji Y, Hoffmann R, Movassagh R, Datta S (2014) Quantum interference in polyenes. J Chem Phys 141:224311-1-13
28. Heilbronner E (1962) Über einen graphentheoretischen Zusammenhang zwischen dem Hückel'schen MO-Verfahren und dem Formalismus der Resonanztheorie. Helv Chim Acta 45:1722–1725
29. Seminario JM (ed) (2007) Molecular and nano electronics: analysis, design and simulation. Elsevier, Amsterdam
30. Cyvin SJ, Gutman I (1988) Lecture Notes in Chemistry, Kekulé Structures in Benzenoid Hydrocarbons. Springer, Berlin
31. Kobayashi M, Taketsugu T (2016) Divide-and-conquer hartree-fock-bogoliubov method and its application to conjugated diradical systems. Chem Lett 45:1268–1270
32. Tobe Y, Kubo T (eds) (2018) Physical organic chemistry of quinodimethanes. Springer, Cham
33. Pauling L, Brockway LO, Beach JY (1935) The dependence of interatomic distance on single bond—double bond resonance. J Am Chem Soc 57:2705–2709
34. Ham NS (1958) Mobile bond orders in the resonance and molecular orbital theories. J Chem Phys 29:1229–1231
35. Emri J, Lente G (2004) Use of an electron equivalent relationship between bond length and bond order to study chemical bonding. Part II. A study of bond orders, bond lengths and aromaticity in polycyclic aromatic hydrocarbons. J Mol Struct THEOCHEM 671:211–219
36. Radenković S, Gutman I, Antić M (2014) A case of breakdown of the Pauling bond order concept. Chem Phys Lett 614:104–109
37. Luzanov AV (2018) Cyclic aromaticity within Hückel and quasi-correlated Hückel-like models. Kharkov Univ Bull Chem Ser 31(54):6–18

38. Takatsuka K, Fueno T, Yamaguchi K (1978) Distribution of odd electrons in ground-state molecules. Theor Chim Acta 48:175–183
39. Head-Gordon M (2003) Characterizing unpaired electrons from the one-particle density matrix. Chem Phys Lett 372:508–511
40. Mott N (1974) Metal-insulator transitions. Taylor & Francis, London
41. Gebhard F (1997) The Mott metal-insulator transition: models and methods. Springer, New York
42. Brédas J-L (2014) Mind the gap! Materials Horizons 1:17–19
43. Raghu C, Pati YA, Ramasesha S (2002) A density matrix renormalization group study of low-lying excitations of polyacene within a Pariser-Parr-Pople model. Phys Rev B 66:035116-1-11
44. Das M (2014) Low-lying excited states in armchair polyacene within Pariser-Parr-Pople model: a density matrix renormalization group study. J Chem Phys 140:124317-1-6
45. Alfonsi J (2014) Size dependence of excited state properties in zigzag graphene nanoribbons. J Phys Conf Ser 566:012013-1-6
46. Goli VMLDP, Prodhan S, Mazumdar S, Ramasesha S (2016) Correlated electronic properties of some graphene nanoribbons: A DMRG study. Phys Rev B 94:035139-1-12
47. Deshpande VV, Chandra B, Caldwell R, Novikov DS, Hone J, Bockrath M (2009) Mott insulating state in ultraclean carbon nanotubes. Science 323:106–110
48. Mironov GI (2017) The energy spectrum of zigzag single-walled carbon nanotubes in the Hubbard model, in the static-fluctuation approximation. Low Temp Phys 43:719–723
49. Bazzanella M, Faccioli P, Lipparini E (2010) Effect of interactions on the conductance of graphene nanoribbons. Phys. Rev. B 82:205422-1-9
50. Tsuji Y, Estrada E (2019) Influence of long-range interactions on quantum interference in molecular conduction. A tight-binding (Hückel) approach. J. Chem. Phys. 150:204123-1-13
51. Hall GG (1955) The bond orders of alternant hydrocarbon molecules. Proc R Soc Lond 229:251–259
52. Pople JA, Hush NS (1955) Ionization potentials and electron affinities of conjugated hydrocarbon molecules and radicals. Trans Faraday Soc 51:600–605
53. MacLachlan AD (1961) Electrons and holes in alternant hydrocarbons. Mol Phys 4:49–56
54. Pople JA (1953) Electron interaction in unsaturated hydrocarbons. Trans araday Soc 49:1375–1385

Wear Resistance of Ti–Al–C MAX Phases-Based Materials for Pantographs Inserts of Electric Vehicles

V. Ya. Podhurska, O. P. Ostash, B. D. Vasyliv, T. O. Prikhna, V. B. Sverdun, M. V. Karpets, and T. B. Serbeniuk

1 Introduction

The pantograph is a device on top of trams, trains, or electric buses that collect electric current from the overhead wires (catenary). It pushes a contact strip up against the overhead to draw electricity. Above the pantograph and between the contact wire, there is a pantograph insert. Nowadays in Ukraine, the pantograph insert is made of aluminum alloy the lifetime of which is limited to 5,000 km [1, 2].

Composite materials based on titanium MAX phases are very promising [3, 4]. They combine the properties of both metals and ceramics such as high strength, corrosion and heat resistance, thermal and electrical conductivity, as well as low coefficients of friction and thermal expansion. The low coefficient of friction of the materials on the basis of the MAX phases is caused by their layered nanolaminate microstructure [5, 6]. Besides, such materials are easily processed by traditional mechanical methods [5–7]. Therefore, they are increasingly used in an electric vehicle, in particular for the manufacture of pantograph inserts [8, 9].

In order to increase the reliability and lifetime of electric vehicles, it is necessary to study and improve the wear and friction of contacting materials since the faults in the mobile connections account for about 95% of all failures [1]. Sliding electrical contacts are important moving connections in public vehicles. The materials of the pantograph inserts have to provide high electrical conductivity, durability, hardness, and wear resistance. The resistance to the action of the electric arc can serve as an important characteristic [1] since the decrease in the lifetime of the pantograph inserts

V. Ya. Podhurska (✉) · O. P. Ostash · B. D. Vasyliv
Karpenko Physico-Mechanical Institute of the NAS of Ukraine, 5 Naukova str., Lviv 79060, Ukraine
e-mail: podhurskavika@gmail.com

T. O. Prikhna · V. B. Sverdun · M. V. Karpets · T. B. Serbeniuk
Institute for Superhard Materials of the NAS of Ukraine, 2 Avtozavodska str., Kyiv 04074, Ukraine

O. Fesenko and L. Yatsenko (eds.), *Nanomaterials and Nanocomposites, Nanostructure Surfaces, and Their Applications*, Springer Proceedings in Physics 246, https://doi.org/10.1007/978-3-030-51905-6_42

is also caused by a pulsed electric arc that occurs due to insufficient contact between the insert and the wires of a catenary. As a result, hardened metal alloys appear on the surface of the insert, which intensifies the wear of contacting components during friction [2].

The aim of this work is to study composites based on the MAX phases of Ti–Al–C system as promising materials for the pantograph inserts in the public vehicles from the point of view of the microstructure and phase composition effects on their physical and mechanical properties.

2 Materials and Methods

Previous composites based on MAX phase were obtained using two-step technology, which consisted of synthesis in a vacuum (1350 °C) and hot pressing in the air (1350 °C, 30–35 MPa) [10]. However, such technology involves the use of a complex incorporating expensive equipment for high-temperature synthesis in vacuum, which is a disadvantage in industrial applications. Therefore, it is necessary to study the properties of such materials obtained by one-step technology, which is only hot pressing. We investigated titanium composites based on 312 (Ti_3AlC_2) and 211 (Ti_2AlC) MAX phases of structural modifications made using the following one-step technology, which consisted of hot pressing at pressures of 15 and 30 MPa at 1350 °C for 30 min of powders TiH_2, TiC, and Al. These powders were mixed and milled in a planetary MPF-1 activator in argon environment and then compacted at a pressure of 30 MPa. The phase composition of the obtained materials is shown in Table 1 (variants 1–6). These composite materials were compared with the optimum material obtained previously by two-step technology (variant 7) and with the aluminum alloy of the Al–Fe–Cu–Si system (variant 8) used for the production of tram pantograph inserts [10]. To determine the microhardness of the materials a PMT-3 M device with a load of 200 g (HV200) was used.

A series of prismatic specimens of 4 × 5 × 40 mm in size were subjected. The strength of the composite materials in the initial state (σ_f) was determined by the three-point bend test of the specimens in the air at 20 °C. The wear resistance of the materials was evaluated by the weight loss (Δm) of the body 1 (composite materials) and the counterbody 2 (M1 copper) under non-lubricating slip conditions (Fig. 1a). Taking into account the operating conditions of the friction pair, a specific contact load of 0.25 MPa and a running distance of 5 km were selected. The specific electrical conductivity of the composite material was determined in the air at 20 °C using a four-point scheme. The resistance of materials to the action of an electric arc was determined, whose power was regulated by the capacitance (300–600 μF) of the discharge capacitor ($W = CU^2/2$, where C is the capacitor capacitance and U is the maximum voltage in the capacitor discharge). To estimate the damage degree to the surface of specimens, the diameter (d) and depth (h) of the crater were used (Fig. 1b). These parameters were determined using the interference profilograph [10].

Table 1 Phase compositions and mechanical properties of obtained materials

Variant	Technological parameters	Phase composition, %	HV_{200}, MPa	σ_f, MPa
1	Hot pressing at 1350 °C, 30 MPa	$100Ti_3AlC_2$	6460	425
2	–	$96Ti_3AlC_2$ 4TiC	5140	361
3	–	$61Ti_3AlC_2$ $22Ti_2AlC$ 17TiC	5190	285
4	–	$50Ti_3AlC_2$ $7Ti_2AlC$ 43TiC	6820	450
5	Hot pressing at 1350 °C, 15 MPa	$100Ti_3AlC_2$	5450	285
6	–	$93Ti_3AlC_2$, 4TiC, $3Al_2O_3$	930	88
7	Two-step technology [10]	$95Ti_3AlC_2$ 5TiC	5790	545
8	Aluminum alloy [10]	–	1100	300

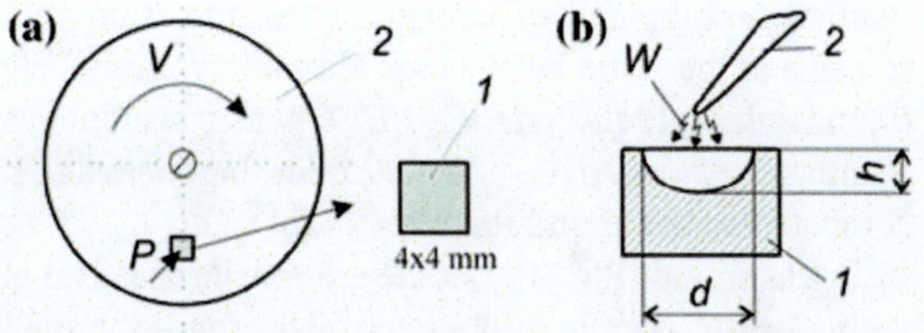

Fig. 1 Schematic representation of **a** the wear resistance test [(1) body, (2) counterbody] and **b** the resistance test to electric arc [(1) specimen; (2) electrode]

For microstructural and microfractographic analyses, we used a Carl Zeiss EVO-40XVP scanning electron microscope. X-ray diffraction analysis was performed on a DRON-4.0 diffractometer in CuKα monochromatic irradiation.

3 Results and Discussion

The investigated composite materials (Table 1) in their phase composition contained phases of titanium MAX phase of 312 structural modification or mixtures of phases 312 and 211 modifications, and some of them also had an additional amount of titanium carbide (TiC) and aluminum oxide (Al_2O_3). The conductivity of these materials is 2–5 MS/m, which corresponds to the requirements for the pantograph contact insert materials [1]. The microhardness of composites based on MAX phases obtained after hot pressing under 30 MPa (variants 1–4) is not inferior to the best material

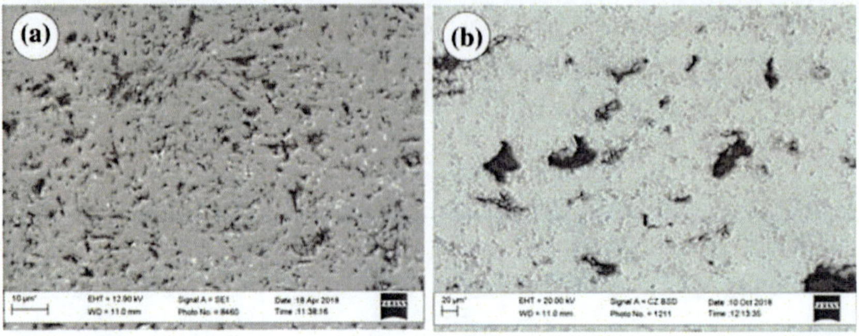

Fig. 2 SEM microstructures of composites based on MAX phase: (**a**, **b**) variants 1 and 5 (see Table 1)

made using two-step technology (variant 7), but they have lower strength. However, mechanical properties of such composites are higher in times than of the aluminum alloy (see Table 1).

When the hot pressing pressure decreases from 30 to 15 MPa, the porosity of the composite increases (about 10%) and the pore size increases to 20–30 μm (Fig. 2b vs. a). In the cases when the phase compositions are the same and 100% of the Ti_3AlC_2 MAX phase is fixed in the materials, their microhardness is practically unaffected (variant 5 compared to variant 1), but the strength level decreases by ~30% and remains at the same level as for aluminum alloy [11]. When the composite (variant 6), in addition to the Ti_3AlC_2 MAX phase, also included titanium carbide (TiC) and aluminum oxide (Al_2O_3), then its microhardness and strength are significantly lower compared to the aluminum alloy (Table 1).

It was found (Fig. 3) that the composite material of variant 1 fails in accordance with a brittle–ductile micromechanism (with the predominance of the micromechanism of destruction), and the surface of fracture is characterized by the presence of clearly visible nanolaminate-layered components of the MAX phase (see Fig. 3a) like in composite materials obtained by two-step technology [10].

The fractured structure of the material with higher porosity (variant 5), despite the same phase composition with variant 1, is markedly different (Fig. 3b): the low-energy dimple microrelief in combination with the facets of transgranular cleavage, which explains its lower strength, prevails here. The lowest strength of the material of variant 6 is caused by the similar low-energy dimple fracture micromechanism in combination with a pronounced intergranular fracture (Fig. 3c).

The wear resistance of the composites obtained by one-step (variants 1—6) and two-step (variant 7) technologies is almost the same (despite the difference in microhardness and strength) and high enough as compared to the aluminum alloy (Fig. 4a, Table 2). The lowest wear resistance is demonstrated by the material of variant 3, the phase composition of which is basically a mixture of the titanium MAX phase of modifications 312 and 211, similar to the materials obtained by two-step technology [10].

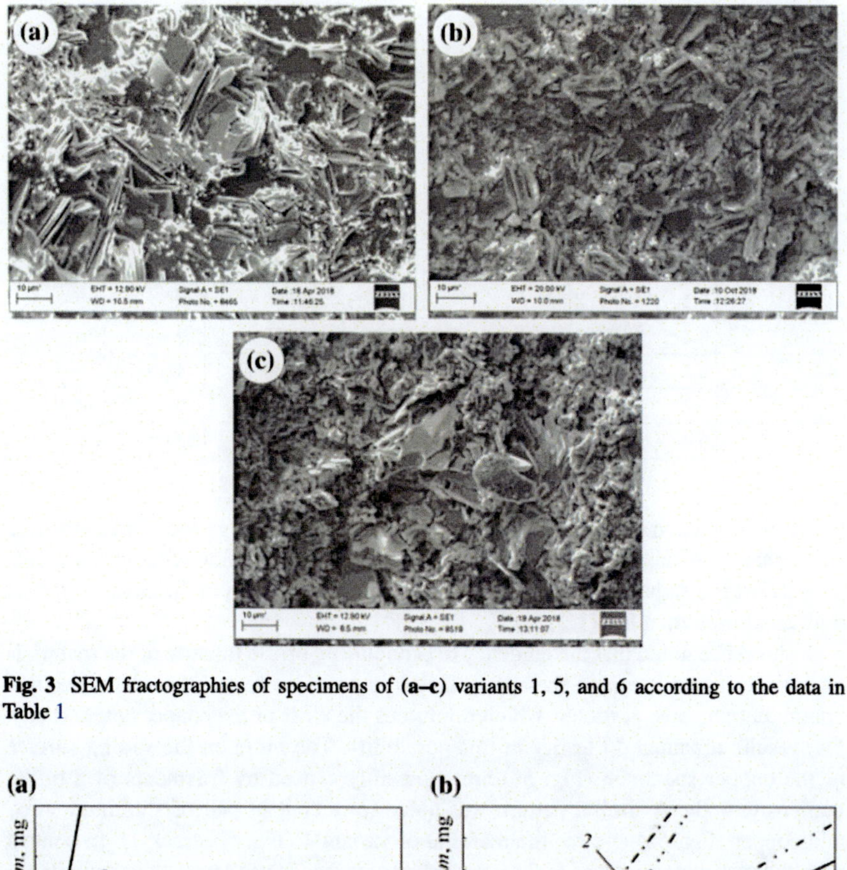

Fig. 3 SEM fractographies of specimens of (**a–c**) variants 1, 5, and 6 according to the data in Table 1

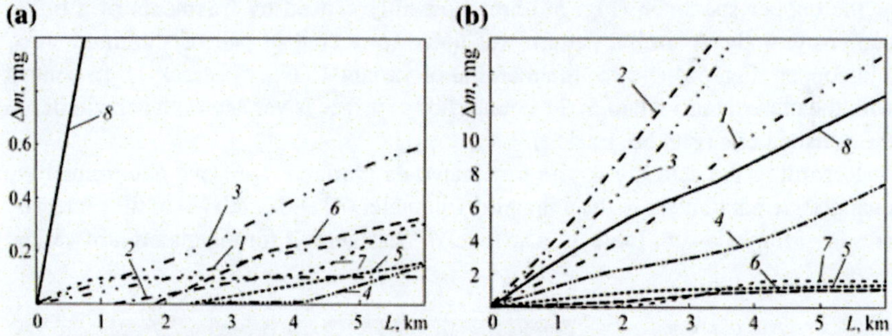

Fig. 4 Wear kinetics of the composites based on titanium MAX phase and aluminum alloy in pair with copper (**a**) and of copper in pair with the composites and aluminum alloy (**b**). The curve numbers correspond to the material variant number in Table 1

A more diverse picture is observed by the results of copper wear in contact with the studied composites (Fig. 4b, Table 2). Composite materials of variants 1–4 obtained by one-step technology: hot pressing at 30 MPa cause significant wear of copper, even more (variants 1–3) than aluminum alloy (variant 8). When the hot pressing pressure decreases from 30 to 15 MPa, sufficiently porous materials (variants 5 and

Table 2 Characteristics of wear and arc resistances of the studied materials

Variant	Weight losses Δm, mg		Defects parameters for arcs of various power			
			μF			
			d, μm		*h*, μm	
	Body	Counterbody	300	600	300	600
1	0.2	12.4	900	1430	11	13
2	0.1	21.1	1100	1500	14	8
3	0.5	22.8	960	1450	5	7
4	0.2	8.5	1050	1550	9	9
5	0.1	1.2	1000	1320	7	6
6	0.3	1.1	850	1300	4	8
7	0.3	1.7	814	1145	10	14
8	6.0	10.5	813	1200	42	76

6) are obtained and copper wear is reduced by an order of magnitude and more, like in the case of testing the material obtained by two-step technology (Table 2; variants 5, 6, 7). At the same time, copper wears out 9 times less than in the case of contact with an aluminum alloy [12].

It should be noted that the tendency to pronounced brittle fracture of the materials of variants 1 and 8 causes the appearance in a contact zone of the friction couple containing abrasive particles, which reinforces the wear of the copper counterbody. Our results obtained do testify to this possibility. The burrs on the sliding surface of the copper specimen (Fig. 5a) are apparently caused by fragments of a brittle composite body. A similar pattern was noted for a contact pair of aluminum alloy and copper (Fig. 5a) and for the material of variant 1 (Fig. 5b, track 1). In contact with the material of variant 5, the counterbody surface is very smooth and similar to the polished one (Fig. 5b, track 2).

Regarding the stability of the test materials (variants 1–6), we determined the quantitative parameters of the damaging (diameter (*d*) and depth (*h*) of the crater) by arcs of various power (Table 2). The level of parameter *d* for the material of variant

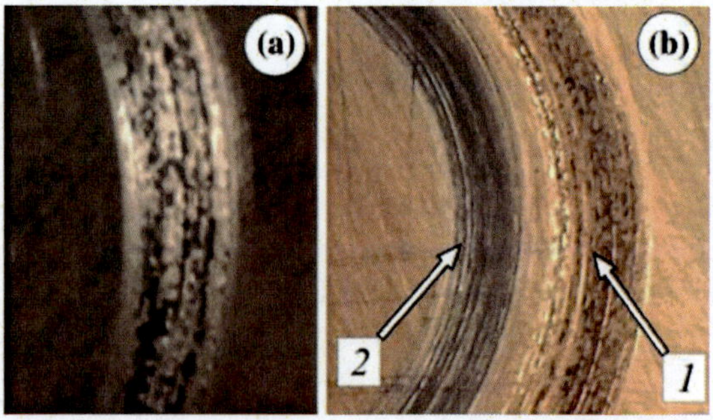

Fig. 5 Surfaces of copper specimens after wear test with the investigated materials: (**a**) variant 8; (**b**) variant 1 (track 1) and variant 5 (track 2) (see Table 2)

5 is the same as of variant 7 obtained by two-stage technology. However, it is shown that the level of parameter *h* for this material is lower than for the material of variant 7, and it is significantly lower than that of aluminum alloy (variant 8). According to this, the material of variant 5 has a higher resistance to arc damage.

4 Conclusions

It is found that materials based on the Ti_3AlC_2 MAX phase obtained only by hot pressing (one-step technology) practically are not inferior in microhardness, strength, and wear resistance to similar materials obtained by two-step technology (vacuum synthesis and hot pressing). However, they cause intense wear of the copper specimens in the "composite–M1 copper" friction pair. It is shown that a significant reduction in copper wear can be achieved due to the formation of the porous microstructure of the body material for such tribocouple by reducing the hot pressing pressure from 30 MPa to 15 MPa. For this case, wear resistance of the copper counterbody is nine times higher than for the case of the "aluminum alloy–copper" tribopair that models the interaction of the electrical contact inserts of trams pantographs with contact wires of the catenary used in Ukraine.

References

1. Gershman IS, Bol'shakov YuL, Sychenko VG (2008) Compatibility of different current-collecting materials in the same segment of a contact wire, Zaliznych Transp Ukrainy, No 5:52–56
2. Serhienko SA, Koval' VA, Current-collecting sliding element, Patent No. 77745 of Ukraine, Publ. On 25.02.2013, Bull. No. 4
3. Barsoum M (2000) The Mn+1AXn phases: a new class of solid thermodynamically stable nanolaminates. Progr Solid State Chem 28(1–4):201–281
4. Radovich M, Barsoum MW (2013) MAX phases: bridging the gap between metals and ceramics. Amer Ceram Soc Bull 92(3):20–27
5. Prikhna T, Ostash O, Basyuk T et al (2015) Thermal stability and mechanical characteristics of densified Ti3AlC2-based material. Solid State Phenom 230:140–143
6. Ivasyshyn AD, Ostash OP, Prikhna TO et al (2015) Influence of technological media on the mechanical and physical properties of materials for fuel cells. Mater Sci 51(2):149–157
7. Ivasyshyn A, Ostash O, Prikhna T et al (2016) Oxidation resistance of materials based on Ti_3AlC_2 nanolaminate at 600 °C in air. Nanoscale Res Lett 11:358. https://doi.org/10.1186/s11671-016-1571-x
8. Xiao QD, Lv ZL (2012) Current-carrying friction and wear characteristics of Ti3AlC2 by novel method of infiltration sintering. Adv Appl Ceram 111(4):202–207
9. Huang X, Feng Y, Qian G, Liu K (2017) Erosion behavior of Ti3AlC2 cathode under atmosphere air arc. J Alloys Comp 727:419–427
10. Prikhna TO, Podhurs'ka VYa, Ostash OP et al (2019) Influence of the technology of production of composites based on the Max phases of titanium on the process of wear in contact with copper. Part I. Two-stage technology. Mater Sci 54(4):589–595

11. Chen Q, Li DY, Cook B (2009) Is porosity always detrimental to the wear resistance of materials?—a computational study on the effect of porosity on erosive wear of TiC/Cu composites. Wear 267:1153–1159
12. Prikhna TO, Podhurs'ka VYa, Ostash OP et al (2019) Influence of the technology of production of composites based on the Max phases of titanium on the process of wear in contact with copper. Part II. Single-stage technology. Mater Sci 55(1):1–8

Polarization of Germanium Quantum Dots in Heterostructure Ge/Si Caused by Spatially Indirect Exciton Transitions: Theory

Sergey I. Pokutnyi

1 Introduction

Ge/Si heterostructures with germanium quantum dots (QDs) are of the second type of heterostructures. In them, the main electron level was located in the silicon matrix, and the main hole level was in the germanium QD. When studying the optical properties of Ge/Si nanosystems with germanium QDs, experimental work [1] was the first to reveal the spatial separation of electrons and holes, as a result of which electrons were localized above the QD surface, and holes moved into QDs. The transition between such states was indirect in space. In experimental studies [1, 2], it was established that in nanosystems consisting of germanium QDs located in silicon matrices, the excitation of spatially indirect excitons (SIE) is possible.

In [3], heterostructures, which are linear germanium QD chains on silicon substrates, were obtained using the method of electron beam lithography. The average radii of QD of germanium did not exceed 30 nm. In Ge/Si heterostructures with germanium QDs, it was established in experimental works [1, 2] that low-temperature optical absorption and photoluminescence spectra were caused by interband electron transitions from the valence band of germanium QD to the conduction band of the silicon matrix. The photoluminescence signal of nanostructures in the infrared spectral region (0.20–1.14) eV was observed up to room temperature [1, 2].

The theory predicts that exciton, containing a hole in QD volume and electron localized above outer spherical interface (QD—silicon matrix), has the binding energy two orders of magnitude higher than exciton in appropriate bulk semiconductors [4, 5]. Electron and the hole are separated by centrifugal potential, as well as polarization interaction with a curved interface (QD—silicon matrix) [5]. The energy

S. I. Pokutnyi (✉)
Chuiko Institute of Surface Chemistry of National Academy of Sciences of Ukraine, 17 General Naumov Street, Kyiv 03164, UA, Ukraine
e-mail: pokutnyi.serg@gmail.com

O. Fesenko and L. Yatsenko (eds.), *Nanomaterials and Nanocomposites, Nanostructure Surfaces, and Their Applications*, Springer Proceedings in Physics 246, https://doi.org/10.1007/978-3-030-51905-6_43

spectrum of this exciton depends on QD size [5]. Due to low electron and hole wave function overlap, an SIE shows the long lifetime [1, 2].

The use of nanosystems based on semiconductor and dielectric QDs as the active regions of nanolasers is hindered by a small binding energy of exciton $E_{ex}(a)$ (where a is the QD radius) in these nanosystems [6]. In this context, investigations aimed at finding nanostructures with significantly increased $E_{ex}(a)$ values are of considerable importance. The effect of a significant increase in the SIE binding energy (by almost two orders of magnitude) is found in nanosystems containing semiconductor (zinc selenide, germanium) and dielectric (aluminum oxide) QD, compared with the exciton binding energy in the corresponding single crystals [4–12]. Such an effect of the significant increase in the SIE binding energy opens the possibility for the use of nanosystems as an active field of nanolasers operating on exciton transitions at room temperatures.

The most promising applications for fiber-optic communication lines were Ge/Si heterostructures with germanium QDs, on the basis of which light-emitting diodes have already been fabricated that emit at room temperature in the near-infrared (200–960) meV wavelength range [13–17]. It was shown that interband and intraband transitions between SIE states caused significant emission in the infrared region of wavelengths in the energy range ($\leq$330 meV) in the studied Ge/Si heterostructure with germanium QDs [5]. Moreover, this energy range ($\leq$330 meV) was contained in the spectral region (200–960) meV, which was observed in experiments [13–17]. Such Ge/Si nanostructures with germanium QDs were promising for the creation of new elements of nanoelectronics in terms of realizing effective radiation sources of the visible and infrared ranges, and can also be used for new optoelectronic devices (in optical modulators, in optical switches, in filters, and converters) [18, 19].

An important role in exciton physics is played by SIE. A high lifetime makes it possible to use SIE in a wide area from Bose gas creation [20, 21] to instrumentation as for the development of excitonic devices with energy-efficient computation and seamless coupling to optical communication [22]. These devices include traps, lattices, conveyers, and ramps, which are used for studying basic properties of cold SIE, as well as excitonic transistors, routers, and photon storage devices, which hold the potential for creating excitonic signal processing devices and excitonic circuits [22].

However, the temperature range for exciton existence was limited by low temperatures for a long time, due to low its binding energy. Recently, the authors of [23, 24] have found the SIEs with high binding energy in van der Waals heterostructures. This gives the opportunity to study high-temperature quantum Bose gases and to create excitonic devices operating at room temperature [25–27].

In [1–5], it is shown that the formation of SIE can make a significant contribution to the optical absorption spectra in the visible and near-infrared regions and blur the absorption edge. The mentioned above mechanisms of absorption spectra formation and low-threshold optical nonlinearity are effective under the condition of large oscillator strength and large SIE electric dipole moment of transition between SIE states.

At present, the optical absorption of Ge/Si nanosystems with germanium QDs was not sufficiently clarified. Investigations in the theory of interaction of weak optical fields with SIE in Ge/Si nanosystem with germanium QDs have not been performed as of yet; to fill this gap, a theory of interaction of weak optical fields with SIE states emerging in Ge/Si nanosystem with germanium QDs is developed in this study.

In this paper, we use the Ge/Si nanosystems with germanium QDs, from works [1–3] to assess, within the framework of dipole approximation, the values of oscillator strength, transition dipole moments of the SIE ground state, QD polarizability, and size effect impact. The results show significantly higher values than for electron transitions in bulk semiconductors. It makes possible the appearance of optical nonlinearity even in weak optical fields when single transitions to exciton states are taking place.

2 Energy Spectrum of Exciton States in the Nanosystem

Quantitative estimation of the QD size corresponding to various exciton states can be carried out using the real nanosystem [1–3] containing the germanium QD with a radius a and permittivity $\varepsilon_2 = 16.3$, in silicon matrix with $\varepsilon_1 = 11.7$. In this nanosystem, a hole with effective mass $(m_h/m_0) = 0.39$ moves in QD volume, and electron with effective mass $(m_e/m_0) = 0.98$ moves in silicon matrix (m_0 is free electron mass), r_e and r_h—electron and hole distances from the QD center [4, 5]. In nanosystem, the main electronic level is located in a silicon matrix, and the main hole level is in the QD. The energy of electron–hole Coulomb interaction and energy polarization interaction causes localization of the electron in a potential well above the surface of the QD [4, 5]. For simplicity, we assume that a hole h with effective mass m_h is located at the QD center (with $r_h = 0$), and an electron e with effective mass m_e is localized over the spherical surface of the QD in the silicon matrix ($r_e = r$—distance of the electron e from the QD center).

In the model of quasi-zero-dimensional nanosystem in the framework of the effective mass approximation, the Hamiltonian of the SIE takes the form [5]:

$$H(r,a) = -\frac{\hbar^2}{2\mu_{ex}}\frac{1}{r^2}\frac{d}{dr}\left(r^2\frac{d}{dr}\right) + \frac{\hbar^2 l(l+1)}{2\mu_{ex} r^2} - \frac{e^2}{\tilde{\varepsilon} r} - \frac{e^2\beta a^3}{\varepsilon_1 r^2\left(r^2 - a^2\right)} + \frac{e^2\beta}{\varepsilon_2 a} + E_g \tag{1}$$

where the first term is the kinetic energy operator of the SIE, $\mu_{ex} = m_e m_h/(m_e + m_h)$—reduced mass of the SIE, the second term describes the centrifugal energy of the SIE (where $L^2 = l\,(l+1), l = 0, 1, 2, \ldots$– orbital quantum number of an electron), the third term describes the energy of electron–hole Coulomb interaction, the fourth and fifth terms describes the energy of the polarization interaction of an electron and a hole with a QD surface, $\tilde{\varepsilon} = 2\,\varepsilon_1\varepsilon_2/(\varepsilon_1 + \varepsilon_2)$—the permittivity of the nanosystem,

$\beta = (\varepsilon_2 - \varepsilon_1)/(\varepsilon_1 + \varepsilon_2)$—the nanosystem parameter, E_g is the bandgap energy of the silicon matrix.

To estimate the energy spectrum of the ground state of an exciton ($n = 1, l$), a variational method was used. The average value of the Hamiltonian (1) SIE on hydrogen-like wave functions is obtained in [5]:

$$E_{1,l}(a, j_l(a)) = R_{1,l}(r)|H(r, a)|R_{1,l}(r), \tag{2}$$

where

$$R_{1,l}(r) = A_l(r - a)^l exp(-j_l(r - a)/a),$$
$$A_{1,l}^2 = (2j_l/a)^{3+2l}(4(2l)!j_l^2 + 4(2l + 1)!j_l + (2l + 2)!)^{-1}, \tag{3}$$

are the hydrogen-like radial wave functions, $j_l(a)$—variational parameter. From the results of a variational calculation of the energy $E_{1,l}(a)$ (2) of the ground state of the SIE follows, that in contrast to excitons in a bulk material when the radius of QD a increased, then starting from the size

$$a_c^*(1, l) \le a \le a_c(1, l) \tag{4}$$

the band of exciton quasi-stationary states appears above the bottom of the conduction band of the silicon matrix. With a further increase of QD radius

$$a > a_c(1, l) > a_c^*(1, l), \tag{5}$$

the quasi-stationary states become the stationary states in the bandgap of the silicon matrix [5]. The critical QD radii for these states ($n = 1, l \le 3$) have corresponding values [5]: $a_c^*(1, l) = 8.04$ nm; 11.1 nm; 15.5 nm; $a_c(1, l) = 6.54$ nm; 8.35 nm; 11.95 nm; 17.34 nm.

The results have shown that the upper boundary of the quasi-stationary states band $E_{1,l}^{\max}(a)$ takes the maximum value for ($n = 1$, $l = 3$), and $E_{1,l=3}^{\max}\left(a = a_c^*(1, l = 3)\right) \cong 248.4$ meV. In the interval of QD radii a

$$a \ge \tilde{a}_c(1, l) = 20.8 \text{ nm} \tag{6}$$

given in (6), exciton stationary states $E_{1,l}(a)$ (2) approach to the states of 2D SIE localized close to a flat interface (germanium–silicon) [5]. The energy of this exciton [5] is equal to:

$$E_{1,l}(a) = -E_{ex}^{2D}, E_{ex}^{2D} = 2\hbar^2/\mu_{ex}(a_{ex}^{2D})^2 \tag{7}$$

where $E_{ex}^{2D} = 82$ meV is the binding energy of 2D SIE. The 2D SIE Bohr radius is defined as follows:

$$a_{ex}^{2D} = \tilde{\varepsilon}(m_0/\mu_{ex})(\hbar^2/m_0e^2), \tag{8}$$

where $a_{ex}^{2D} = 2.6$ nm.

Thus, the ground state ($n = 1$) of SIE energy spectrum $E_{1,l}(a)$ is formed by stationary states band with $\Delta E = E_{ex}^{2D} = \hbar\omega_{ex}^{2D} = 82$ meV width and quasi-stationary states band with $\Delta E = E^{max} = 248.4$meV width. It means that SIE energy spectrum is limited by (−82 meV and 248.4meV), relatively to silicon matrix conduction band bottom. It clearly depends on the size of the QD and contains a finite number of exciton states ($n = 1, l = 0, 1, 2, 3$) that contribute to the formation of the nanosystem light absorption spectrum discussed in [5].

3 Polarizability of Quantum Dots

The wavelengths of frequencies near $\omega_{1,l}(a) = (E_{1,l}(a)/h)$ are higher than SIE radius a_{ex}^{2D} (8); therefore, the polarizability of this SIE in the weak optical field can be described by dipole approximation [16]. If the kinetic energy of SIE is lower than its binding energy $k_BT < E_{1,l}(a)$ (where k_B—Boltzmann constant, T—temperature), and states $E_{1,l}(a)$ are narrow with a width $\Gamma_{1,l}(a) \ll \omega_{1,l}(a)$, then the polarizability of QD can be found, if we consider QD as one giant atom [28–30]:

$$A''(\omega, a) = \frac{e^2}{\mu_{ex}} \sum_l \frac{f_{1,l}^{1,l+1}(a)}{\omega_{1,l+1}^2(a) - \omega^2 - i\omega\Gamma_{1,l+1}(a)} \tag{9}$$

where

$$f_{1,l}^{1,l+1}(a) = \frac{2\mu_{ex}}{\hbar^2e^2}\left[E_{1,l+1}(a) - E_{1,l}(a)\right](D_{1,l}^{1,l+1}(a))^2 \tag{10}$$

is the oscillator strength of electron transition from ($n = 1, l$) to ($n = 1, l + 1$) SIE state. The oscillator strength depends on size, and shape of QD and expressed via matrix element of the transition dipole moment $D_{1,l}^{1,l+1}(a)$. Here, $E_{1,l}(a) = \hbar\,\omega_{1,l}(a)$ and $E_{1,l+1}(a) = \hbar\,\omega_{1,l+1}(a)$ are the energy of ground ($n = 1, l$) and ($n = 1, l + 1$) exciton states, respectively, and $\Gamma_{1,l+1}(a)$ is the SIE state ($n = 1, l + 1$) width. We introduce the notation

$$\left[E_{1,l+1}(a) - E_{1,l}(a)\right] = L_{1,l}^{1,l+1}(a)E_{ex}^{2D} \tag{11}$$

With allowance for (7) and (11), we can write the oscillator strength (10) of the transition of an electron in the form

$$f_{1,l}^{1,l+1}(a) = 4L_{1,l}^{1,l+1}(a)(D_{1,l}^{1,l+1}(a)/ea_{ex}^{2D})^2 \tag{12}$$

In the weak optical field, the main contribution to the polarizability $A''(\omega, a)$ (9) of QD is made by single transitions in the discrete spectrum of SIE states. Separating in polarizability (9), the contribution from only one resonant term corresponding to the transition between the ground ($n = 1, l$) and ($n = 1, l + 1$) SIE states, we can write polarizability $A''(\omega, a)$ (9) of the QD in the form [30]

$$A''(\omega, a) = f_{1,l}^{1,l+1}(a)\frac{e^2}{\mu_{ex}}\left[\frac{\omega_{1,l+1}^2(a) - \omega^2}{(\omega_{1,l+1}^2(a) - \omega^2)^2 + (\omega\Gamma_{1,l+1})^2} + i\frac{\omega\Gamma_{1,l+1}}{(\omega_{1,l+1}^2(a) - \omega^2)^2 + (\omega\Gamma_{1,l+1})^2}\right] \tag{13}$$

The restriction of these states does not affect the clarification of the behavior of the polarizability $A''(\omega, a)$ (9) of the nanosystem depending on the radius a QD when taking into account other states and for other resonant frequencies, since, as follows from (10)–(12), the dependence of the oscillator strengths of the $f_{1,l}^{1,l+1}(a)$ transitions between such states on the radius a QD is general.

Using (13), we obtain the expression for polarizability, provided that frequencies of the external field waves ω is lower and far from the resonant exciton state ($n = 1$, $l + 1$) frequency $\omega_{1,l+1}(a)$ (i.e. $\omega^2 << (\omega_{1,l+1}(a))^2$:

$$A''(a) \approx (\tilde{\varepsilon}/4)\left(E_{ex}^{2D}/E_{1,l+1}\right)^2 f_{1,l}^{1,l+1}(a)\left(a_{ex}^{2D}\right)^3 \tag{14}$$

For frequencies close to resonant $\omega \approx \omega_{1,l+1}(a)$, the real part of polarizability is small compared to the imaginary part and can be neglected:

$$A_2''(a) \approx i(\tilde{\varepsilon}/4)\left(E_{ex}^{2D}\right)^2/\left(E_{1,l+1}\hbar\Gamma_{1,l+1}\right) f_{1,l}^{1,l+1}(a)\left(a_{ex}^{2D}\right)^3 \tag{15}$$

In the case $\omega \approx \omega_{1,l+1}$, the real part of polarizability is negative and decreases with frequency increasing more slowly than the imaginary part:

$$A_3''(a) \approx -(\tilde{\varepsilon}/4)\left(E_{ex}^{2D}/E_{\omega}\right)^2 f_{1,l}^{1,l+1}(a)\left(a_{ex}^{2D}\right)^3 \tag{16}$$

where $E_\omega = \hbar\omega$.

The ratios of the polarizability absolute values are equal to:

$$K_1 = A_1''(a)\big/A_2''(a) = h\Gamma_{1,l+1}\big/E_{1,l+1} \ll 1 \tag{17}$$

$$K_2 = A_3''(a)/A_2''(a) = h\Gamma_{1,l+1}E_{1,l+1}/(E_\omega)^2 \ll 1 \tag{18}$$

$$K_3 = A_3''(a)/A_1''(a) = \left(E_{1,l+1}/E_\omega\right)^2 \ll 1 \tag{19}$$

The expressions (14–16) are applicable to an ensemble of noninteracting QD, i.e., to QD concentration N:

$$a_{ex}^{2D} N^{1/3} \ll 1 \tag{20}$$

4 Dipole Moments of Transitions Between SIE States in a Quantum Dot

Because the electron is out of QD, the transition dipole moment $D_{1,l}^{1,l+1}(a)$ consists of dipole moments (i) of polarization the QD by electron, (ii) of the electron relative to the hole localized in QD. Within the dipole approximation framework, when the electromagnetic field on the scale of a QD does not vary and can be considered as uniform, the operator of transition dipole moment to the SIE state takes the following form [31]:

$$\boldsymbol{D}(r) = \left(1 - K(a/r)^3\right)e\boldsymbol{r}, \tag{21}$$

where radius vector defines the distance between an electron and a hole localized in QD center, the coefficient $K = ((\varepsilon_2 - \varepsilon_1)/(\varepsilon_2 + 2\varepsilon_1)$.

For estimating the $D_{1,l}^{1,l+1}(a)$ value, we consider the electron transition between $(n = 1, l)$ and $(n = 1, l + 1)$ ground SIE states in nanosystem containing the QD with average radii a satisfying the condition (4) A transition between such states is allowed by the selection rules (in this case, the principal quantum number n changes arbitrarily, whereas the orbital quantum number l changes by unity). If the direction of the external radiation electric field vector $\boldsymbol{\varepsilon}(\omega,t)$ is along the Z-axis, then:

$$(a) = \Psi_{1,l+1}(r,\theta)|D_{1,l}^{1,l+1}1 - K(a/r)^3)e\mathbf{r}|\Psi_{1,l}(r,\theta) \tag{22}$$

The exciton wave functions $\Psi_{1,l+1}(r,\theta)$ and $\Psi_{1,l}(r,\theta)$ will be chosen as follows:

$$\Psi_{1,l}(r,\theta) = R_{1,l}(r)Y_{l,0}(\theta),\ \Psi_{1,l+1}(r,\theta) = R_{1,l+1}(r)Y_{l+1,0}(\theta) \tag{23}$$

where $Y_{l,0}(\theta)$ and $Y_{l+1,0}(\theta)$ are the exciton spherical wave functions, is the azimuthal angle defining the position of electron radius vector, $R_{1,l}(r)$ and $R_{1,l+1}(r)$ are the hydrogen-like radial wave functions defined by (3). The wave function $Y_{l,0}(\theta)$ can be expressed by the Legendre polynomial:

$$Y_{l,0}(\theta) = ((2l+1)/4\pi)^{1/2} P_l(\cos\theta) \tag{24}$$

It should be noted that only projection of transition dipole moment on the vector $\boldsymbol{\varepsilon}(\omega,t)$ direction is not equal to zero because of the QD spherical shape.

Using (22), (23), (24) expressions we can to write:

$$D_{1,l}^{1,l+1}(a) = (2l+1)^{1/2}(2l+3)^{1/2}2^{-1}e$$

$$\int_0^\pi P_l(\cos\theta)P_{l+1}(\cos\theta)sin\theta cos\theta d\theta \int_0^a (r^3 - Ka^3)R_{1,l}(r)R_{1,l+1}(r)dr \tag{25}$$

After integration in (25), taking into account (3), we obtain the expression for the transition dipole moment $D_{1,l}^{1,l+1}(a)$ [30]:

$$D_{1,l}^{1,l+1}(a) = (2l+1)^{1/2}(2l+3)^{1/2} j_l^{(2l+3)/2} j_{l+1}^{(2l+5)/2}(j_l + j_{l+1})^{-(2l+5)}$$

$$[(1-K)(2l+1)!(j_l + j_{l+1})^3 + 3(2l+2)!(j_l + j_{l+1})^2 + 3(2l+3)!(j_l + j_{l+1}) + (2l+4)!][4(2l)!j_l^2 + 4(2l+1)!j_l + (2l+2)!]^{-1/2}[4(2l+2)!j_{l+1}^2 + 4(2l+3)!j_{l+1} + (2l+4)!]^{-1/2} ea$$

$$\int_0^\pi P_l(\cos\theta)P_{l+1}(\cos\theta)sin\theta cos\theta d\theta \tag{26}$$

In the interval of QD radii a

$$(a_c^*(1, l=1) = 8.04\,\text{nm}) \leq a \leq 12.2\,\text{nm} \tag{27}$$

given in (6), using (26), we obtain the formula for the transition dipole moment $D_{1,0}^{1,1}(a)$ of an electron between the ground $(n = 1, l = 0)$ and $(n = 1, l = 1)$ SIE states

$$\begin{aligned} D_{1,0}^{1,1}(a) = {} & 2^2 3^{-1/2} j_0^{3/2} j_1^{5/2}(j_0 + j_1)^{-5}[(1-K)(j_0 + j_1)^3 \\ & + 6(j_0 + j_1)^2 + 18(j_0 + j_1) + 24] \\ & (2j_0^2 + 2j_0 + 1)^{-1/2}(j_1^2 + 3j_1 + 3)^{-1/2} ea \end{aligned} \tag{28}$$

Let us write the expression for the transition dipole moment $D_{1,1}^{1,2}(a)$ of an electron between the ground $(n = 1, l = 1)$ and $(n = 1, l = 2)$ SIE states in the interval of QD radii

$$(a_c^*(1, l=2) = 11.1\,\text{nm}) \leq a \leq 16.1\,\text{nm}, \tag{29}$$

using formula (26) in this form:

$$\begin{aligned} D_{1,1}^{1,2}a) = {} & 2^{9/2} 5^{-1/2} j_1^{5/2} j_2^{7/2}(j_1 + j_2)^{-7}[(1-K)(j_1 + j_2)^3 \\ & + 12(j_1 + j_2)^2 + 60(j_1 + j_2) + 120]\left(j_1^2 + 3j_1\right)^{-1/2} \\ & \left(2j_2^2 + 10j_2 + 15\right)^{-1/2} ea \end{aligned} \tag{30}$$

In the interval of QD radii a

$$(a_c^*(1, l=3) = 15.5\text{nm}) \leq a \leq \tilde{a}_c(1, l) = 20.84\text{nm} \tag{31}$$

given in (31), using (26), we obtain the expression for the transition dipole moment $D_{1,2}^{1,3}(a)$ of an electron between the ground ($n = 1$, $l = 2$) and ($n = 1$, $l = 3$) SIE states

$$D_{1,2}^{1,3}(a) = 2^6 3^{1/2} 7^{-1/2} j_2^{7/2} j_3^{9/2} (j_2 + j_3)^{-9} \left[(1-K)(j_2 + j_3)^3 + 18(j_2 + j_3)^2 + 126(j_2 + j_3) + 136\right] (2j_2^2 + 10j_2 + 15)^{-1/2} (j_3^2 + 28j_3 + 56)^{-1/2} ea \quad (32)$$

5 Calculation Results and Discussion

Using formulas (28) and (12), in the interval of radii a (27) QD for QDs with an average radius $a_1 = 8.4$ nm, we obtain the highest values of the transition dipole moment $D_{1,0}^{1,1}(a_1) = 14.9\ D_0$(where $D_0 = e\,\mathring{A}$ in Debye units) of an electron and oscillator strength of the transition $f_{1,0}^{1,1}(a_1) = 0.47$ electrons between the ground ($n = 1$, $l = 0$) and ($n = 1$, $l = 1$) by SIE states (taking into account the values $j_0 = 2.04$, $j_1 = 2.18$, $L_{1,0}^{1,1}(a_1) = 0.355$ taken from [5], and $a_{ex}^{2D} = 2.6$ nm (8)). In this case, the polarizability (14) of QDs with a radius a_1, caused by such a transition, takes the value $A''(a_1) = 2.2 \,.\, 10^{-19}\text{cm}^3$.

In the interval of radii a (29) QD for QDs with an average radius of $a_2 = 13.1$ nm, using (30) and (12), we find the largest values of the dipole transition moment $D_{1,1}^{1,2}(a_2) = 15.2\ D_0$ electrons and oscillator strengths of the transition $f_{1,1}^{1,2}(a_2) = 0.50$ electrons between the ground ($n = 1$, $l = 1$) and ($n = 1$, $l = 2$) SIE states (taking into account the values $j_1 = 2.24$, $j_2 = 2.36$, $L_{1,1}^{1,2}(a_2) = 0.363$ taken from [5]). Due to this transition, the polarizability (14) of QDs with radius a_2 takes the value $A''(a_2) = 2.3 \,.\, 10^{-19}\text{cm}^3$.

From formulas (32) and (12), it follows that in the interval of radii a (31) QD for QDs with an average radius $a_3 = 18.4$ nm, the maximum values of the dipole transition moment are $D_{1,2}^{1,3}(a_3) = 15.8\ D_0$ electron and oscillator strengths of the transition $f_{1,2}^{1,3}(a_3) = 0.56$ electron between the ground ($n = 1$, $l = 2$) and ($n = 1$, $l = 3$) SIE states (taking into account the values $j_2 = 2.5$, $j_3 = 2.62$,$L_{1,2}^{1,3}(a_3) = 0.376$ taken from [5]). The polarizability (14) of the QDs with a radius a_3, caused by such a transition, takes the value $A''(a_3) = 2.5 \,.\, 10{-}19\ \text{cm}^3$.

Quantitative values of the dipole moment of transitions $D_{1,l}^{1,l+1}(a)$ (28), (30), (32), oscillator strengths of transitions $f_{1,l}^{1,l+1}(a)$ (12) and polarizabilities $A''(a)$ (14) are listed in the table.

6 Conclusion

Table 1 shows the estimated values of $f_{1,l}^{1,l+1}(a)$,$D_{1,l}^{1,l+1}(a)$ and $A_1''(a)$ for transitioning between SIE states $(n = 1, l) \rightarrow (n = 1, l + 1)$ (where l = 0,1,2), for frequency $\omega_{1,l}(a)/\omega_{ex}^{2D} = 10^{-1}$. For frequencies $\omega_{1,l}(a)/\omega_{ex}^{2D} = 1$ and $\omega_{1,l}(a)/\omega_{ex}^{2D} \gg 1$, the values of $f_{1,l}^{1,l+1}(a)$,$D_{1,l}^{1,l+1}(a)$ remain constant, and polarizabilities $A_2''(a)$ and $A_3''(a)$ can be estimated using the expressions (17–19). The frequencies $\omega_{1,l}(a)$ are in the infrared region.

As follows from Table 1, the influence of QD size on QD transition dipole moment, oscillator strength, and polarizability is insignificant. For the frequency $\omega_{1,l}(a)/\omega_{ex}^{2D} = 10^{-1}$, value of $D_{1,l}^{1,l+1}(a)$(26) changes from 14.9 D_0 to 15.8 D_0. This is due to the fact that the dipole size is comparable to the Bohr radius a_{ex}^{2D} (8) of SIE, which is defined by permittivities of QD and matrix, and effective masses of electron and hole. It should be noted that because of this circumstance the use of variational wave functions (3), (23), (24) for an estimate the matrix element of transition dipole moments $D_{1,l}^{1,l+1}(a)$ (26) values does not introduce a significant error. The values of the oscillator strength and polarizability also slightly change. As follows from (10) and (26), the oscillator strength and transition dipole moments are independent of radiation frequency, therefore, it can be expected that size effect will have a slight effect on the QD polarizability at other radiation frequencies, but the polarizability frequency dependence is significant. When the resonance takes place, the polarizability reaches its maximum which is $E_{1,l+1}/\hbar\Gamma_{1,l+1}$ times higher than $A_1''(a)$ (14). Exciton lifetime is from nanoseconds to microseconds range [32, 33], that corresponds to the value of exciton state width $\hbar\Gamma_{1,l+1} = 10^{-4}$–$10^{-7}$ eV. Therefore, the QD polarizability $A_2''(a)$ (15) at resonance can be up to 10^7 times higher than at lower frequencies. For frequencies higher than the resonant one, the polarizability $A_3''(a)$(16) is $E_{1,l+1}^2/E_\omega^2$ and $\hbar\Gamma_{1,l+1}E_{1,l+1}/E_\omega^2$ times lower than $A_1''(a)$ (14) and $A_2''(a)$ (15), respectively, and inversely proportional to the square of frequency. This indicates that in weak fields the polarizability of semiconductors nanostructures with SIE formation possibility can be much greater than the one of bulk semiconductors. As follows from a_{ex}^{2D} (8), the polarizability (14), (15), (16) is inversely proportional to exciton effective mass and directly proportional to the permittivity of nanosystem. This allows underlining the exact nanosystem parameters which can be used to achieve the results expected.

Table 1 SIE states parameters in nanosystem. The estimated values of oscillator strength, transition dipole moments, and polarizability for $(n = 1, l) \rightarrow (n = 1, l + 1)$ transition (where l = 0, 1, 2)

a, nm	l	$f_{1,l}^{1,l+1}$	$D_{1,l}^{1,l+1}$, D_0	A'', 10^{-19} cm^3
8.4	0	0.47	14.9	2.2
13.1	1	0.50	15.2	2.3
18.4	2	0.56	15.8	2.5

A mechanism is proposed for the onset of the dipole moments of electron transitions in a nanosystem containing germanium QDs with average radii a not exceeding 25 nm, grown in a silicon matrix studied under experimental conditions [1–3].

The mechanism of QD polarizability photoexcitation described above does not take into account the exciton–phonon interaction, which mainly leads to a broadening of the kinetic energy spectrum and a decrease in the lifetime of the exciton. But this broadening will be small due to the small lifetime of the exciton, and the thermal equilibrium with the surrounding atoms does not take place. It is shown that electron transitions from SIE levels ($n = 1, l$) to higher excited exciton levels ($n = 1, l + 1$) belonging to the surface exciton zone, which was located in the bandgap of the silicon matrix, as well as in the conduction band of the silicon matrix [5] caused the dipole moments of the transitions $D_{1,l}^{1,l+1}(a)$ (26), the values of which are proportional to the radii a of QDs and exceeded the values typical for semiconductors by two orders [33]. The resonance optical absorption of the nanosystem in the infrared region of the spectrum (from 80 meV to 248 meV) can apparently be observed experimentally at room temperature.

Using the present model of a Ge/Si heterostructure with germanium QDs, it was shown [5] that interband and interband transitions between SIE states caused significant radiation in the infrared region of wavelengths in the energy range ($\leq$330 meV). This energy range ($\leq$330 meV) was contained in the spectral region (0.20–1.14) eV, which was observed in experiments [1–3]. Therefore, the present model of the Ge/Si heterostructure with germanium QDs partially explains the experimental results [1–3].

References

1. Yakimov AI, Dvurechensky AV, Nikiforov AI (2001) JETP Lett 73:529
2. Yakimov AI, Dvurechensky AV, Nikiforov AI (2001) JETP 119:574
3. Smagina JV, Dvurechenskiy AV, Seleznev VA (2015) Semiconductors 49:749
4. Pokutnyi SI (2016) Low Temp Phys 42:1151
5. Pokutnyi SI (2018) Low Temp Phys 44:819
6. Alferov ZhI (2002) Nobel lecture. Usp Fiz Nauk 172:1068
7. Pokutnyi SI (2013) Semiconductors 47:791
8. Pokutnyi SI, Culchin YuN, Dzyuba VP (2015) Semiconductors 49:1311
9. Pokutnyi SI (2013) Technical Phys Lett 39:233
10. Pokutnyi SI (2015) Technical Phys 60:1615
11. Pokutnyi SI (2007) Semiconductors 41:1323
12. Pokutnyi SI (2010) Semiconductors 44:507–514
13. Chang WH, Chou AT, Chen WY (2003) Appl Phys Lett 83:2958
14. Elcurdi M, Boucaud P, Sauvage S (2003) Physica E 16:523
15. Elfving A, Hansson GV, Ni WX (2003) Physica E 16:528
16. Dvurechensky AV, Yakimov AI, Kirienko VV (2018) Defect Diffuss. Forum. 386:68
17. Pakhanov N, Pchelyakov OP, Yakimov AI (2017) Optoelectron Instrum Data Process 53:190
18. Achtstein AW, Prudnikau AV, Ermolenko MV, Gurinovich LI, Gaponenko SV (2014) ACS Nano 8:7678
19. Gaponenko SV, Demir HV, Seassal C (2016) Optics Express 24(A):430

20. Lozovik YuE, Yudson VI (1976) JETP 44:389
21. Butov LV et al (2001) Phys Rev Lett 86:5608
22. Butov LV (2017) Superlattices Microstruct 108:2
23. Geim AK, Grigorieva IV (2013) Nature 499:419
24. Calman EV, Fogler MM, Butov LV, Hu S, Mishchenko A, Geim AK (2018) Nature Commun 9:1895
25. Fogler MM, Butov LV, Novoselov KS (2014) Nature Commun 5:4555
26. Ye Z et al (2014) Nature 513:214
27. Chernikov A et al (2014) Phys Rev Lett 113:076802
28. Pokutnyi SI (2006) Semiconductors 40:217
29. Pokutnyi SI (2018) J Nanophoton 12:012506
30. Dzyuba VP, Pokutnyi SI, Kulchin YN (2019) J Phys Chem C 123:26031
31. Landau LD, Lifshitz EM (1984) Course of theoretical physics, Vol 8: Electrodynamics of continuous media. (Pergamon Press, New York)
32. Gaisler AV, Yaroshevich AS, Derebezov IA, Kalagin AK, Bakarov AK, Toropov AI, Aseev AL (2013) JETP Lett 97:274
33. Cardona M, Guntherodt G (1982) Light scattering in solids. Recent results. (Springer, Berlin, Heidelberg, New York)

Activation Mechanism of the Cyclic Switchover Effect for Quantum Selective Detection with Dendritic Yanson Point Contacts

Alexander P. Pospelov, Gennadii V. Kamarchuk, Anna O. Herus, N. D. Sakhnenko, Maryna Ved, and Volodymyr L. Vakula

1 Introduction

The theoretical and experimental studies carried out in recent years have unequivocally demonstrated that the Yanson point contact [1, 2] placed in an ion-conducting medium can be considered as an electrochemical gapless electrode system (GES) [3, 4]. This effect is realized due to the ability of Yanson point contact to concentrate electric field, to keep its identity under a super high current density and to ensure an opposite directed electrochemical activity located at the ends of the point contact conductivity channel. One of the characteristic features of GES behavior is the transfer of charge along two parallel branches corresponding to the electronic and ionic conduction channels [5, 6]. The electronic channel is realized in the bulk of the conductor and is responsible for the so-called direct conduction. The ionic channel is localized in the electrolyte near the metal–electrolyte interface. The transfer processes in these two channels are conjugated which stimulates a series of physicochemical effects and transformations. One of the most interesting results of this interaction is the point-contact cyclic switchover effect [3]. The cyclic processes of synthesis and dissolution of dendritic point contacts, it consists of, occurring in accordance with the quantum shell effect [7] and are characterized by a number of features which explicitly demonstrate the quantum properties of Yanson point contacts. This makes it possible to apply the innovative operation principle of the

A. P. Pospelov · N. D. Sakhnenko · M. Ved
Department of Physical Chemistry, National Technical University "Kharkiv Polytechnic Institute", 2 Kyrpychov Str., Kharkiv 61002, Ukraine

G. V. Kamarchuk · A. O. Herus · V. L. Vakula (✉)
Department of Spectroscopy of Molecular Systems and Nanostructured Materials, B. Verkin Institute for Low Temperature Physics and Engineering of the National Academy of Sciences of Ukraine, 47 Nauky Ave., Kharkiv 61103, Ukraine
e-mail: vakula@ilt.kharkov.ua

O. Fesenko and L. Yatsenko (eds.), *Nanomaterials and Nanocomposites, Nanostructure Surfaces, and Their Applications*, Springer Proceedings in Physics 246, https://doi.org/10.1007/978-3-030-51905-6_44

new class of selective quantum sensing elements to analysis of liquid and gaseous media [4].

The key factor for the operation of the new selective quantum sensors is the influence exerted by the analyzed agent on the cyclic process of formation and destruction of point contacts in an electric field. Therefore the development of quantum sensorics requires a detailed study of nature and kinetic parameters of all the phases of the cyclic switchover effect: formation and further growth of a metallic dendrite, formation of a conduction channel and of a dendritic point contact, where the dendrite tip touches the counterelectrode, electrolytic dissolution of the conduction channel and interruption of the direct transfer of electrons between the electrodes.

This work presents the results of the study of a nanostructured point-contact system in the crucial phase of the cyclic switchover effect [3]—in the phase of electrochemical dissolution of the point-contact conduction channel. The purpose of this study is to reveal the peculiarities of the processes occurring in dendritic Yanson point contacts during the cyclic switchover effect and providing the ground for the realization of selective quantum detection mechanism and enhanced sensitivity of point-contact sensors to liquid and gaseous analytes. Analysis of the experimental data on the dynamics of the electric conduction and the lifetime of dendritic Yanson point contacts during the cyclic switchover effect allows us to consider a model of the process in which a copper Yanson point contact is destroyed and direct conduction is disrupted. The importance of understanding the nature of the processes taking place in this phase is dictated by the unique set of quantum physicochemical transformations a Yanson point contact undergoes which allows it to become the sensing element of a new type of quantum sensors [4].

The analytic opportunities provided by dendritic point-contact sensors are based on the interaction between atoms or molecules of the analyte and the atoms of the conduction channel surface, which impacts the parameters of the cyclic switchover effect and thereby the response function of the quantum sensor. The main elements of the system which shape the cyclic switchover effect and enable quantum detection are metallic nanostructures formed and destroyed by the electric field in every cycle. One of the key problems in studying the nature of this process is finding the peculiarities in the behavior of diachronically distributed sensing elements which may be crucial for the analytic procedures, since the knowledge of the physicochemical properties, structural and energetic features of the active surface of point contacts is an important prerequisite for optimization of the processes of improvement of the construction and metrological characteristics of quantum sensor devices.

2 Experimental Method

The object of the study was dendritic copper Yanson point contacts [3, 4]. The contacts were produced using the modified "needle-anvil" technique [2]. For this purpose, a special device was applied [8] to ensure a high-precision shift of needle-like and lamellar electrodes along with each other. The needle-like electrode was attached to a spring-damper in order to lower the probability of mechanically breaking the structure. The gap between the electrodes was filled with an electrolyte to enable the formation of a gapless electrode system which initiates the cyclic switchover effect in an electric field [3]. The electrolyte was a 0.05 mol/dm^3 aqueous solution of $CuSO_4 \cdot 5H_2O$ prepared with bidistilled water and a very high-purity salt. The electrode system was created in an airtight chamber blown through with argon for 30 min. The dendritic point contact was formed in an electric field generated by a stabilized DC source. The negative and positive poles of the source were connected to the needle and to the "anvil," respectively. A special multichannel device—point-contact spectrometer [9]—was used to perform the measurements, which allowed controlling parameters of the studied processes, recording electric characteristics of the point contacts, simultaneously measuring the current–voltage characteristics of the point contact along with its first and second derivatives, etc. The measurements were made using the four-probe scheme to exclude the influence of any current-carrying conductors.

3 Results and Discussion

Figure 1 shows a typical resistance R versus time t dependence registered in the process of the cyclic switchover effects taking place in the system of a dendritic point contact immersed in the electrolyte. It was recorded when the electrolyte was in contact with a human breath gas. Similar curves are also observed in other gaseous media [4]. The curve in Fig. 1 reflects the cyclic variations in the resistance of dendritic point contacts with time in the process of their synthesis and destruction in an electric field. The low values of electric resistance correspond to the contacts with

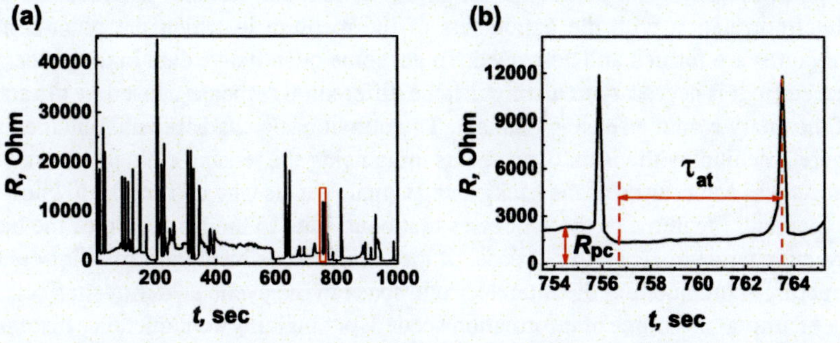

Fig. 1 A typical dependence of resistance R of dendritic point contacts in the electric field on time t (cyclic switchover effect). **a** General form of the curve $R(t)$; **b** Magnified section of the curve $R(t)$ inside the red box in "a." Monotonous horizontal sections of the curve in "b" correspond to the initial resistance of the dendritic point contact. τ_{at} is the period of dissolution of point contact down to an atomic size; R_{pc} is the point contact resistance. The current in the conduction channel is $I = 20\ \mu A$

large diameters of the conduction channel, while high resistances are those of the contacts with small diameters, down to the size of an atom [3]. The steps in the curve $R(t)$ (Fig. 1b) reflect the existence of metastable quantum energy states of contacts with certain diameters [10], which emerge in the process of the contacts' evolution.

The process of the cyclic switchover effect can be thought of as follows. As the current starts flowing through the needle tip, a dendrite starts growing along the lines of force of the electric field toward the "anvil" (precommutation phase) [3]. Sometime later the dendrite tip reaches the surface of the counterelectrode. In the point of contact, a dendritic Yanson point contact is formed (commutation phase). Further passage of the stabilized current through the system causes dissolution of the conduction channel of the Yanson point contact (decommutation phase). The dissolution phase is initiated by the gapless electrode system formed on the surface of the conduction channel upon the formation of the point contact. The process acquires a cyclic character with alternating phases of formation and dissolution of dendritic point contacts, which can be seen in Fig. 1a in the form of cyclic variations in the electric resistance of the system under study. This is the cyclic switchover effect [3]. The transformations of the Yanson point-contact nanostructure are shaped by the quantum shell effect [7] and characterized by the existence of metastable quantum states corresponding to certain diameters of conduction channel, which can be registered as steps (Fig. 1b) in the dependence electric resistance R versus time t [3]. A set of the metastable quantum states the system can be in during its evolution is a unique characteristic of the system. Its registration in the form of a histogram of the conductance of dendrite point contacts allows using it as an efficient tool for selective detection in gaseous and liquid media [4]. Thus, the conductance histogram of point contacts formed in the analyzed medium is the output characteristics of the selective quantum sensors based on the cyclic switchover effect [3, 4]. The conductance histogram characterizes the probability of formation of the metastable quantum states of the system during its evolution. The distribution of the probabilities of formation of point contacts with certain values of the conduction channel conductance is in direct correlation with the parameters of the medium in which the point-contact structures are formed and destroyed. To get some quantitative data in the process of detection, it is necessary to distinguish the differential response caused by the action of the analyte on the sensing element. The output basic characteristic obtained in a "pure" medium with virtually no agents influencing the sensing element or analyzed substances can be used as the background signal. That is why it is crucial to know the fundamental features of the processes that contribute to the formation of the background characteristics—in our case, of the conductance histogram of point-contact structures commutating the interelectrode space in the cyclic switchover effect.

An important source of information needed for obtaining a conductance histogram is the phase of the anode destruction of a Yanson point contact from the moment it is created to the complete interruption of direct conductance (decommutation phase). A key role in this process belongs to the gapless electrode system formed on the surface of the conduction channel of a point contact at the moment of its creation. Information about the nature of the GES operation can be gained by analyzing the lifetimes of Yanson point contacts during the evolution of the nanostructure from its initial state characterized by some initial diameter to the conduction channel of an atom-sized cross section.

Each cycle of the switchover process implies "birth" and "death" of a new point contact in an electric field. If we neglect the drift of the system parameters during the process, then each cycle can be seen as a parallel experiment. So, we can obtain a large database of characteristic properties of these nanostructures with reliably confirmed parameters for each cycle. The lifetime of a point contact as a function of its initial resistance is one of the parameters. By comparing the data on this dependence obtained from a model and from experiments, we will be able to arrive at a more adequate description of the process and thus reveal some important peculiarities in the behavior of the analyzed nanostructure. This is the task the present paper is dealing with below.

When considering the way the gapless electrode system functions, we believe that the intensity of the destruction of the analyzed nanostructure (decommutation phase) is directly dependent on the density of the electrochemical current at the positively polarized surface of the conduction channel. Let us suppose that the properties of the surface layer of the conductor are virtually the same as those in the bulk of the material. In this case, we neglect the impact of the energetically more saturated surface layers on conductance, as well as the influence of compounds that may be formed on the surface of the conductor as a result of its interaction with the electrolyte. The anode polarization is directly correlated with the rate of the anode dissolution. If we assume that the positive polarization in the analyzed medium is determined by the slowdown in the electron transfer, we can obtain an expression describing the relation between the potential drop responsible for the transfer of electrons through the metal–electrolyte interface and the kinetics of the relevant electrode process [11]:

$$j_{\text{el}} = j_0\left[\left(\exp\frac{\alpha zFU}{2RT}) - \exp(-\frac{(1-\alpha)zFU}{2RT}\right)\right] \tag{1}$$

where j_{el} is the current density reflecting the rate at which the material dissolves on a unit surface; j_0 is the exchange current; α is the transfer coefficient; F is the Faraday constant; z is the number of electrons transferred during a unit electrode process; RT *is* the gas constant and the thermodynamic temperature, respectively; U is the voltage applied to the conduction channel.

It follows from (1) that in order to find the specific rate of the conduction channel dissolution it is necessary to know the value of the anode polarization related to the voltage in the channel. Since the cyclic switchover effect can be achieved most easily in the galvanostatic regime [3, 12], we assume in our calculations that we have direct current flowing through the point contact. Therefore, the change in the voltage at the contact during the dissolution will be determined by the conduction channel resistance. Bearing this in mind, we analyzed the recurring sections of the contacts' resistance change in the decommutation phase of the cyclic process and performed some mathematical simulation in order to choose an appropriate function to describe them. Using the method of successive approximations, we managed to achieve a good agreement with the experimental data (Fig. 2b).

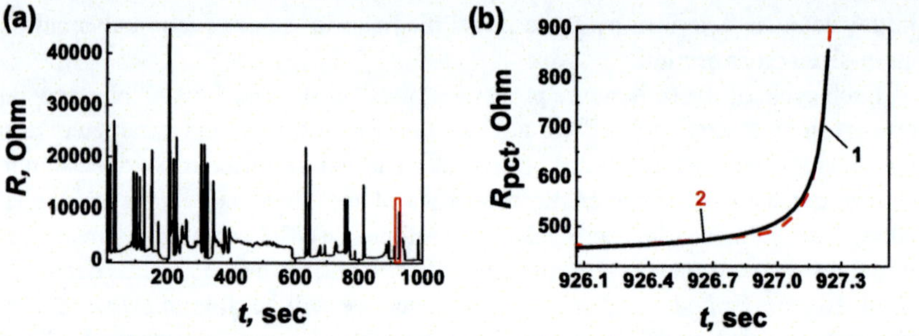

Fig. 2 Comparison of the model function used to describe the process of dissolution of a dendritic Yanson contact with the experimental data. **a** General cyclic dependence of resistance *R* of dendritic point contacts in an electric field on time *t*, the red box shows the equivalent section in "b"; **b** calculated (1) and experimental (2) curves of point-contact resistance R_{pct} as a function of exposure time *t* in an ion-conducting medium during the cyclic switchover effect, the current in the conduction channel is 20 μA

It turned out that the dynamics of the growth in the point-contact resistance during the cyclic switchover effect [3, 6, 13] allows considering the GES in this phase as a system with the following positive feedback:

$$\frac{dR_{pct}}{dt} = kR_{pct}^3, \tag{2}$$

where R_{pct} is the point-contact resistance as a function of time *t* and *k* is the control parameter.

The solution of (2) is an expression which reflects the dynamics of the variation in resistance of a point contact in an ion-conducting medium with a direct current of certain value flowing through the contact:

$$R_{pct} = \frac{R_{pc0}}{1 - \frac{t}{\tau_{at}}\left(1 - \frac{R_{pc0}}{1.664 \cdot 10^8}\right)}, \tag{3}$$

where R_{pc0} and R_{pct} are the initial and current values of the point-contact resistance, respectively; τ_{at} is the time of dissolution of the conduction channel from its initial diameter down to that of an atom-sized contact (Fig. 1b).

Using (3) in the case of a direct current flowing through the conduction channel, it is easy to find the potential drop over the point contact at any time during its dissolution. As a result of the formation of GES, the maximum polarization of the surface of the conduction channel to which potential *U* is applied is $|\Delta E_{max}| = U/2$. Let us assume that the maximum anode and cathode polarization have the same absolute values and are localized at the opposite ends of the channel. If the channel diameter is the same over the whole channel, then in the generatrix points lying in the plane which is perpendicular to the channel axis and crosses it in its geometric centre there is no polarization. These points belong to the polarization inversion boundary (PIB) [5, 6]. On both sides of the PIB at the "conduction channel surface–electrolyte" interface, oppositely directed electrochemical processes take place: the anode processes occur on the surface adjacent to the positive pole of the electric current source, while the cathode processes are observed on the opposite side. The electrolyte loops the ionic conduction channel. In this channel, the charge is transferred by both the autochthonous ions of the electrolyte and the ions coming to the solution from the crystalline lattice of the metal as a result of the exchange processes occurring at the interface. In an electric field, the area of the most intense passage of ions into the solution is the positively polarized surface which, by definition, is in the zone affected by the anode processes. Therefore, the chemical reduction processes occur on the negatively polarized surface [6]. The positively polarized section experiences a growing with time narrowing of the conduction channel which almost completely determines the point-contact resistance. It is obvious that the interruption of the direct conduction must occur in the area of the maximum density of the anode current. This area is in the zone of contact between the dendrite tip and the bulk counterelectrode ("anvil") characterized by the maximum anode polarization $\Delta E_{max} = U/2$.

The explicit equation (1) implies a relation between the parameters of the process of electrochemical dissolution and the material of the system. Parameters j_0 and α are sensitive to material nature. In our experiments, we studied dendritic copper point contacts during the cyclic switchover effect; therefore, we are going to use the physicochemical characteristics of copper in our subsequent calculations. Equation (3) allows calculation of temporal dependence of the maximum density of the anode current. Since the phenomenon under study is based on the Faraday processes, this dependence can be used to find a correlation between the duration of the anode reaction and the linear dimensions of the channel. Indeed, if we assume a 100% current yield, the following expression holds true [11]

$$t = \frac{\Delta\delta\gamma}{k_e j_{el}}, \tag{4}$$

where $\Delta\delta$ is the thickness of the layer that gets dissolved; k_e is the electrochemical equivalent; j_{el} is the current density; t is the duration of the process; γ is the density of the material.

According to (4), determination of the thickness of the layer dissolved in the anode process over a time τ with a monotonously changing current density ($\Delta\delta_{calc}$) can be reduced to calculating the following integral:

$$\Delta\delta_{calc} = \frac{k_e}{\gamma}\int_0^{\tau} j_{el}dt. \tag{5}$$

As a result of the step dependence of the point-contact resistance on time of its exposure to an electric field in the final phase of dissolution (Fig. 1), it is natural to assume that the time τ_{at}, i.e., the time of dissolution of a point contact to a size of an atom, is very much the same as the time of complete interruption of the direct conduction.

In our work, the control parameter of the kinetics of local dissolution of the conduction channel is the electric resistance of the point contact. This is due to the following reasons. It is known that resistance of a conductor with a nonuniform cross section is largely determined by the resistance of the section with the smallest diameter. It is more so for a higher degree of nonuniformity. A quantitative measure of the influence of a monotonous nonuniformity of cross section can be the length of a conductor with the resistance being half the total resistance. If we approximate a conductor with a nonuniform cross section by a truncated cone with a height L and diameters of the lower and upper bases d_1 and d_2, respectively ($d_1 > d_2$), we can easily find that the length of the section with a halved resistance adjacent to the basis with the smaller diameter from this equation:

$$L_{1/2} = \frac{L}{3\frac{d_1}{d_2} - 1}. \tag{6}$$

Apparently, if $d_1 = d_2$, that is for a uniform conductor $L_{1/2} = L/2$. But if $d_2 = 0.1d_1$, then the total conductor resistance is halved over its section with a length $L/29$. A growing conicity, which simulates the anode dissolution of the conduction channel in a GES, results in the point-contact resistance tending to the resistance of the area of contact between the dendrite tip and the counterelectrode. This means that with a GES we tend to describe the Yanson point contact within the model of an aperture rather than the model of a long channel [14].

Earlier, it was shown that point contacts formed in the area of contact of the growing copper dendrite and the copper counterelectrode are characterized by a ballistic regime of flight of electrons through the conduction channel with large electron mean free paths at room temperature and thus are the Yanson point contacts [3, 4]. To such contacts, we can apply Sharvin's formula [15, 16]:

$$R_0 = \frac{16}{3\pi}\frac{\rho l}{d^2}, \tag{7}$$

where R_0 is the point-contact resistance for a zero-voltage bias; ρl is a product of the electric resistivity ρ and the electron mean free path l, which is a characteristic constant of each metal.

The product ρl can be estimated using the following reasoning. It is known that the mean free path of a conduction electron can be found from this relation:

$$l = v_F \tau_r, \tag{8}$$

where v_F is the electron velocity at the Fermi surface; τ_r is the relaxation time. The velocity of electron at the Fermi surface can be estimated from this equation [17]:

$$v_F = \frac{\hbar}{m}(3\pi^2 n)^{1/3}, \tag{9}$$

where m is the electron mass, n is the concentration of Fermi electrons in the metal, and $\hbar$ is the Planck constant.

If we express the relaxation time as follows:

$$\tau_r = \frac{m}{\rho n e^2}, \tag{10}$$

where e is the elementary charge, and substitute (9) and (10) into (8), we can arrive at the following relation:

$$\rho l = \frac{\hbar (3\pi^2)^{1/3}}{e^2 n^{2/3}}. \tag{11}$$

Assuming that there is one Fermi electron, i.e., one conduction electron, for each of the copper atoms that form the conduction channel of the Yanson point contact, we can obtain $n = 8.49 \cdot 10^{28}$ m^{-3}. Then, according to (11), we have $\rho l = 0.658 \cdot 10^{-15}$ Ωm^2, which is in agreement with the published data for pure copper: $\rho l = 0.66 \cdot 10^{-15}$ Ωm^2 [18–20]. It follows from the available data that for copper at T = 293 K the resistivity is $\rho = 1.673 \cdot 10^{-8}$ Ωm [21], therefore, the electron mean free path at room temperature is $l = 394$ Å, which is in good agreement with the estimation reported in [22].

To destroy the conduction channel in a GES, it is enough to dissolve a layer of metal $\Delta\delta_{pc}$ which is half the channel diameter, that is $\Delta\delta_{pc} = d/2$. If we estimate the conduction channel diameter using Sharvin's equation (7), we can find that

$$\Delta\delta_{pc} = \frac{2}{\sqrt{3\pi}}\sqrt{\frac{\rho l}{R_0}}. \tag{12}$$

So, the difference $\Delta\delta_{pc} - \Delta\delta_{calc}$ can be used as a basic control function in establishing the dependence of the dissolution time τ_{at} of the point-contact conduction channel on the conduction channel resistance.

To find $\Delta\delta_{calc}$ from (5), we used Simpson's method of numerical integration [23]. Iteration procedures with the termination criterion $(\Delta\delta_{pc} - \Delta\delta_{calc})/\Delta\delta_{calc} \leq 0.01$ allowed us to calculate the time of complete interruption of direct conduction (contact's lifetime) as a function of the point-contact initial resistance. The obtained values were compared with the point-contact lifetimes experimentally observed in the $R(t)$ curves during the self-oscillating switchover process (Fig. 1b). The experimental lifetimes of point contacts were believed to be equal to the lengths of the sections corresponding to the local minima in the $R(t)$ curves, while the point-contact resistance was taken as the ordinate values of the sections. Processing of five experimental curves has yielded a dependence of the lifetime of the point-contacts on their resistance (Fig. 3) with confidence intervals indicated for a confidence probability of

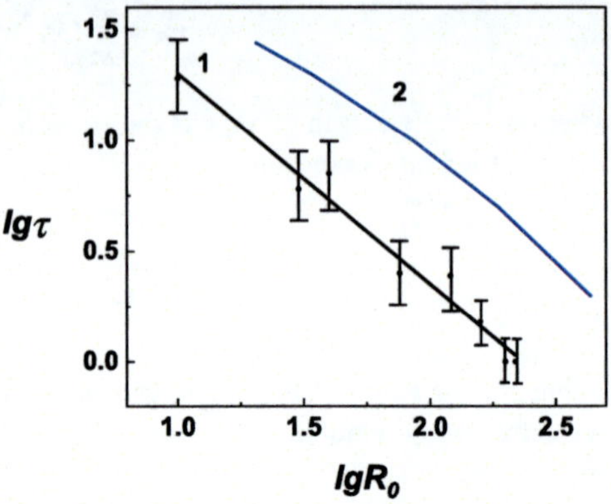

Fig. 3 Experimental (1) and calculated (2) dependences of the time of interruption of direct conduction of a point contact on its initial resistance. Current through the conduction channel is 20 μA

95%. Comparison of the experimental and calculated dependences in Fig. 3 shows that the calculated point-contact lifetimes are significantly higher than those obtained experimentally for the same resistance values. The reason can be measurement inaccuracy and imperfections of the model. The probability of a systematic error in our measurements is pretty low since we use a well-tested technique for studying resistance properties of Yanson point contacts [1, 2], which allows us to obtain reliable experimental results. The principles used for modelling our point-contact system are based on the fundamental laws of electrochemical kinetics. This guarantees high reliability and credibility of the analyzed correlations. Therefore, the problems may come from the parameters used for calculation.

Let us analyze the influence of the parameter values on the calculated lifetime. We assumed that the electric current yield of the process of anode dissolution of copper was close to 100%. Lower values of the yield can only result in a longer calculated time. It follows from the model that the lifetimes comparable with those experimentally observed can only be obtained with a tenfold decrease in the electron mean free path. But our earlier experiments clearly demonstrated that dendritic point contacts formed during the self-oscillating switchover process are Yanson point contacts characterized by a ballistic regime of electrons flow through the conduction channel, i.e., contacts with high values of the electron mean free path [3]. One of the model parameters which are most susceptible to the structure of the material and the state of its surface is the exchange current. However, even for the highest known exchange current obtained for the copper monocrystal face 110 ($j_0 = 20.0$ A/m^2) [24], which is more than an order of magnitude higher than the value normally used for calculations ($j_0 = 1.0$ A/m^2) [11], the calculated times of interruption of the direct conduction are significantly higher than those measured in our experiments.

Thus, we believe that the discrepancies between the model and experimental values may be due to the fact that the model of electrochemical kinetics used for calculations does not take into account the elevated free surface energy possessed by the analyzed nanostructure. This elevated free surface energy can be a result of the size effect [25, 26] or be due to the participation of the newly formed surface in the anode process [11, 24]. The experiments with distilled water as a liquid medium in which dendritic copper point contacts were successfully created [4, 12] confirm this assumption. The high chemical potential of the copper nanostructure caused an easy transition of copper atoms from the metallic lattice into the liquid phase followed by hydration. As a result, the near-surface layer of the conduction channel became enriched with charge carriers necessary for the efficient functioning of GES.

The obtained results shed light on one of the aspects of the high efficiency of the new mechanism of selective detection in gases and liquid media which is based on the quantized electric conductance of dendritic Yanson point contacts and the cycle switchover effect [3, 4]. Besides the quantized growth and quantized conductance displayed in the cyclic switchover effect, dendritic Yanson point contacts possess the whole set of properties typical of the surface of newly formed nanosized structures. The tip of the dendritic point contact in all phases of the cyclic process is a fresh nanostructure with elevated free surface energy. The high level of energy is caused by the two main factors: the size effect and the specific features of juvenile interfaces [27, 28]. The fundamental trend of a system to lower its free energy is responsible for the high adsorptive capacity of the surface which results in a higher probability of effective interaction between these structures and the analyte and eventually in the formation of new physical or chemical bonds. Since these processes occur simultaneously with the quantum transformations in the dendritic Yanson point contacts during their synthesis, they provide additional opportunities for the new quantum mechanism of selective detection [4]. As a result, quantum point-contact sensors are *a priori* characterized not only by the virtually absolute selectivity but also by an enhanced sensitivity to liquid and gaseous analytes.

4 Summary

The present paper analyzes the experimental data on the dynamics of electric conductance and lifetime of dendritic Yanson point contacts during the cyclic switchover effect and studies the peculiarities of the processes that can enable the realization of the quantum mechanism of selective detection and lead to an enhanced sensitivity of point-contact sensors to liquid and gaseous analytes. A mathematical model is proposed for the anode destruction of copper point contacts during the cyclic switchover process. The GES concept allowed us to logically correctly determine the localization of the sections where conduction channel destruction takes place and use the basic equation of electrochemical kinetics as the basis for calculations in the case of the rate-determining step of electronic transition. It is shown that a GES in an electric field function as a system with a positive feedback, which determines the dynamics of the point-contact resistance variation in an ion-conducting medium. Taking into account the key influence of the Faraday processes on the mechanism of the nanostructure destruction, we find a correlation between the duration of the anode reaction and the linear dimensions of the conduction channel. All calculations are performed for Yanson point contacts [1, 2], with their main fundamental properties and parameters taken into consideration. A comparative analysis of the calculated and measured times of interruption of the direct conduction of Yanson point contacts suggests an enhanced chemical activity of the conduction channel surface of these nanostructures. This peculiarity of dendritic point contacts may be due to the size effect and the presence of a fresh metallic solid surface at the interface. The approaches and results of the present paper can be used for the development of selective quantum sensors based on point-contact nanostructures.

Acknowledgments This work was partly supported by the NATO SPS Programme (Ref: SPS.MYP 985481), the Ministry of Education and Science of Ukraine and the National Academy of Sciences of Ukraine.

References

1. Naidyuk YuG, Yanson IK (2005) Point-contact spectroscopy. New York: Springer 300
2. Kamarchuk GV, Pospelov AP, Kamarchuk LV, Kushch IG (2015) Point-contact sensors and their medical applications for breath analysis. A review. In: Karachevtsev VA (ed) Nanobiophysics: Fundamentals and Applications, Pan Stanford Publishing Pte. Ltd. p Chapter 11, 327–379
3. Pospelov AP, Pilipenko AI, Kamarchuk GV, Fisun VV, Yanson IK, Faulques E (2015) A new method for controlling the quantized growth of dendritic nanoscale point contacts via switchover and shell effects. J Phys Chem C 119(1):632–639
4. Kamarchuk GV, Pospelov AP, Savytskyi AV, Herus AO, Doronin YuS, Vakula VL, Faulques E (2019) Conductance quantization as a new selective sensing mechanism in dendritic point contacts. SN Appl Sci 1:244
5. Pilipenko AI, Pospelov AP, Kamarchuk GV, Bondarenko IS, Shablo AA, Bondarenko SI (2011) Point-contact sensory nanostructure modeling. Funct Mater 18(3):324–327
6. Pospelov AP, Kamarchuk GV, Savytskyi AV, Sakhnenko MD, Ved MV, Vakula VL (2017) Macroscopic simulation of atom-sized structures of functional materials: phenomenology of the elongated electrode system. Funct Mater 24(3):463–468
7. Yanson AI, Yanson IK, Ruitenbeek JVM (1999) Observation of shell structure in sodium nanowires. Nature 400:144–146
8. Bobrov NL, Rybal'chenko LF, Khotkevich AV, Chubov PN, Yanson IK (1991) Device for creation of a cooled point contact between metal electrodes. In: Published in B. I. No. 8: p 168: USSR
9. Savytskyi AV, Zaika OS, Pletnev OM, Gudymenko VO, Pospelov AP, Kamarchuk GV (2014) Modern research complex for problems solution by Yanson point-contact spectroscopy. Scientific Herald of Uzhhorod University. Series Phys (35):189–195. (in Ukrainian)
10. Glazman LI, Lesovik GB, Khmel'nitskii DE, Shekhter RI (1988) Reflectionless quantum transport and fundamental ballistic-resistance steps in microscopic constrictions. JETP Lett 48(4):218–220
11. Damaskin BB, Petriy OA, Tsirlina GA (2006) Electrochemistry. Moscow: Kolos 672. (in Russian)
12. Kamarchuk GV, Pospelov AP, Savitsky AV, Koval' LV (2014) Nonlinear cyclic transport phenomena in copper point contacts. Low Temp Phys 40(10):1198–1205
13. Bofanova M, Pospelov A, Herus A, Kamarchuk G, Sakhnenko M, Ved M, Pletnev O (2018) Simulation of a dynamic point-contact system. In: VIII All-Ukrainian Congress on Electrochemistry Book of papers Part 2, p 263–266 (in Ukrainian)
14. Khotkevich AV, Yanson IK (1995) Atlas of point contact spectra of electron-phonon interactions in metals. Boston/Dordrecht/London: Kluwer Academic Publishers, p 168

15. Sharvin YV (1965) On one possible method for investigating the Fermi surface. Sov Phys JETP 48(3):984–985
16. Kulik IO, Omelyanchuk AN, Shekhter RI (1977) Electrical conductivity of point microcontacts and the spectroscopy of phonons and impurities in normal metals. Sov J Low Temp Phys 3(12):1543–1558 (1977)
17. Kittel C (2005) Introduction to solid state physics. New York: Wiley, p 680
18. Gniewek JJ, Moulder JC, Kropschot RH (1966) M.90. Electrical conductivity of high purity copper. In: Proceeding of the Xth international conference on low temperature physics. Moscow, USSR, August 31–September 6, pp 366–370
19. Aleksandrov BN (1978) Residual resistance as a criterion of metal purity In: Solid State Physics Kharkov ILTPE of AS of Ukrainian SSR, pp 52–101 (in Russian)
20. Yanson IK (1983) Microcontact spectroscopy of the electron-phonon interaction in clean metals (review article). Sov J Low Temp Phys 9(7):676–709
21. Matula RA (1979) Electrical resistivity of copper, gold, palladium and silver. J Phys Chem Ref Data 8(4):1147–1298
22. Yanson IK, Shalov YuN (1976) Electron-phonon interaction spectrum in copper. Sov Phys JETP 44(1):148–155
23. Davis PJ, Rabinowitz P (1984) Methods of numerical integration. Academic Press, p 612
24. Bockris JOM, Reddy AKN, Gamboa-Aldego M (2002) Modern Electrochemistry, v. 2A Fundamental of Electrodics. Vol. 2A. New York/Boston/Dordrecht/London/Moscow: Kluwer Academic Publishers, p 764
25. Kumar N, Rumbhat S (2016) Essentials in nanoscience and nanotechnology. Wiley, p 486
26. Roduner E (2006) Nanoscopic materials: size-dependent phenomena. Cambridge 286
27. Suzdalev IP (2008) Nanotechnology: physical chemistry of nanoclusters, nanostructures and nanomaterials. 2nd ed. Moscow: URSS, p 589
28. Natelson D (2015) Nanostructures and nanotechnology. Cambridge University Press, p 639

Influence of the Carbon Allotropes on Dilatometric Properties of the Fe–Cu Nanocomposites

M. C. Bouleklab, S. Hamamda, Y. Naoui, S. Nedilko, T. Avramenko, K. Ivanenko, S. Revo, O. Gomenyuk, V. Sheludko, V. Strelchuk, and A. Nikolenko

1 Introduction

The Fe–Cu nanoscale intermetallic composite systems have attracted considerable attention for a long time, despite problems with their manufacturing. This interest is related to attractiveness of the Fe–Cu systems for many applications due to their well-known high strength, thermal, and electrical properties [1–7]. Here, it is appropriate to say that such composites can be used not only in the traditional "metallurgical" direction. Thus, the Fe–Cu metastable material was studied as a mesoporous layer, which could be used as an alternative to the TiO_2 oxide used for increasing efficiency of the solar cells [8]. However, synthesis of the similar metastable intermetallic meets a number of obstacles, which are mainly related to the low solubility of components at equilibrium state if the temperature is below 700 °C [1, 3, 9, 10].

Mechanical alloying (milling) (MA) of the components of the intermetallic system allowed to increase their solubility comparing to equilibrium values and, at the same time, is a rather simple and effective way to produce a large amount of the composite. That is why manufacturing nanocomposite materials by the MA method is today a

M. C. Bouleklab · S. Hamamda · Y. Naoui
Laboratoire de thermodynamique et traitement de surfaces des Matériaux, Université Frères Mentouri Constantine1, B.P. 325 Route, Ain El Bey, Constantine 25017, Algeria

S. Nedilko (✉) · T. Avramenko · K. Ivanenko · S. Revo
Physics Faculty, Taras Shevchenko National University of Kyiv, 64/13, Volodymyrska Street, 01601 Kyiv, Ukraine
e-mail: SNedilko@univ.kiev.ua

O. Gomenyuk · V. Sheludko
Oleksandr Dovzhenko Hlukhiv National Pedagogical University, Hlukhiv, Ukraine

V. Strelchuk · A. Nikolenko
V. Lashkaryov Institute of Semiconductor Physics, the National Academy of Science of Ukraine, 41, Nauky pr-t, 03028 Kyiv, Ukraine

O. Fesenko and L. Yatsenko (eds.), *Nanomaterials and Nanocomposites, Nanostructure Surfaces, and Their Applications*, Springer Proceedings in Physics 246, https://doi.org/10.1007/978-3-030-51905-6_45

recognized method of the "cold" treatment of solid powder composites [11–16]. The materials thus obtained reveal new, interesting properties, which determine the long-term interest in their manufacture and research in comparison with the materials obtained in another way [17–20]. A quantitative thermodynamic explanation for the formation of solid solutions of immiscible elements by mechanical alloying has also been proposed [21].

The mechanical alloying of copper is also a way of developing high-strength alloys maintaining high electrical and thermal conductivity. Some examples are Cu–Fe, Cu–Ta, Cu–V, and Cu–W [20] composites. The Fe–Cu system is nearly immiscible in equilibrium at room temperature and up to 600 °C and it has a positive heat of mixing [20]. That is why the creation of metastable single-phase solid solutions of nanocrystalline sizes of grains by mechanically alloying the Fe–Cu powders has been extensively studied [20]. Interestingly, the Fe–Cu composites with layered microstructure were also produced, where nanoscaled Fe clusters were dissolved in the Cu matrix [20]. The results of a large number of studies have confirmed the possibility to make metastable Fe–Cu composites by the MA method [22–24]. The potential to extend the mutual solubility of Fe and Cu due to mechanical alloying has been widely reported. The formation of face-centered cubic (fcc) single-phase solid solutions was observed for Fe_xCu_{1-x} alloys with a Fe content $x < 60$ at.%, two-phase alloys consisting of an fcc and a body-centered cubic (bcc) phase were formed for an Fe content of $60 < x < 80$ at.%, and bcc single-phase solid solutions have been obtained for an Fe content of $x > 80$ at.%. It was observed that the mixing occurs on the atomic level in the milled alloys [20] and the crystallite size in the milled powders was reported to be ~20 nm or smaller. Decomposition of the unstable Fe–Cu solid solutions occurred due to annealing treatments was also reported [20]. It was found that many structural defects were incorporated in a material during severe plastic deformation) processes, which was accompanied with a higher dislocation density, a higher vacancy concentration, and the continuous formation of new grain boundaries [20].

Various additives can influence both properties and temperature behavior of the Fe–Cu composites. Thus, the dimensional changes in ferrous sintered compacts are influenced by several factors, including particle size, density, composition, sintering time and temperature, cooling rate, and microstructure. Reduced particle size and increased sintering time and temperature can be used to reduce dimensional change. Given a set base iron and sintering conditions, factors such as composition and cooling rate play an important role on microstructure, dimensional change, and mechanical properties. The most widely used additives to P/M iron are graphite and copper. Separately, copper and graphite additions cause growth in the sintered compact [25].

Additives of various carbon forms cause different effects on the intermetallic composites. Uncontrolled impurities of carbon materials could be various carbon forms aroused under milling due to the by-products of the destruction of the organic compounds. The latter could be used in the process of milling and specifically as lubricants [26, 27]. That is why the role of carbon materials in the MA processes and in determining the properties of the resulting composites was a special subject of research. As a result, it was shown that the kinetics of solid-state reactions, as well as the phase composition of the made materials, depend on the type of used allotropic carbons, their content in the initial mixture, and on the degree of its degradation under mechano-chemical synthesis [11–13, 28–30].

In this work, we briefly consider, on the basis of literary data, the role of carbon forms in determining the characteristics of some intermetallic composites Further, we present data on their temperature behavior, mainly on dilatometric characteristics, and then describe the results of our own studies of the effect of temperature on the characteristics of Fe–Cu nanocomposites with the addition of multiwall carbon nanotubes (MWCNT). There, as above, the focus will be on the study of dilatometric characteristics. The conditions and details of the procedure and the compositions those ensure the control of characteristics of the resulting composites were defined early by us, as we were the first who successfully used MWCNT for Fe–Cu composites production [31–33].

2 Impact of Impurities on Properties of the Fe-Based Composite Materials

It is known that the composition of P/M steels plays an important role in the microstructure and physical properties of sintered parts [25]. The different levels of alloying elements prealloyed into base powders change the hardenability of the material. Copper and graphite are very important additives. Reasonably, start considering with a simple system, namely Cu–C.

Studies of the various carbon allotropic forms influence on the formation of the structural-phase composition of copper-based composites have particularly been performed in the work [30]. The composites Cu–$C_{60/70}$ and Cu–C_g (in proportion Cu-25 at.% C, 30 g by weight) were prepared by MA in planetary ball mill AGO-2S in inert atmosphere $P_{Ar} =$ 0.1 MPa, with drums made of hardened stainless steel, and balls 8 mm in diameter made of ball-bearing steel. The rotation speed of disc and vial were 890 and 2050 rpm, respectively. It has been found there that the kinetics of solid-state reactions in the copper–fullerite (Cu–$C_{60/70}$) and copper–graphite (Cu–C_g) systems obtained in process of the MA depend on deformational stability and oxidation–reduction properties of fullerite and graphite. It has been also shown that partial destruction of fullerene molecules in the Cu–$C_{60/70}$ sample results in the formation of an amorphous fullerite-like phase, copper oxide Cu_2O, and supersaturated solid solution Cu(C, O). Total destruction of fullerene molecules in the system Cu–$C_{60/70}$ results in the formation of supersaturated solid solution Cu(C), just like in the case of the composite Cu–C_g. Main obtained results can be understood from the grafice abstract published in the paper (Fig. 1) [30].

Main attention in that study was focused on the copper–carbon composites because of good conductivity of both constituents, which makes it possible to use Cu–C_g materials as current-collecting elements of sliding electrical contacts. At the same time, the search was directed on for the improvement of the characteristics such as

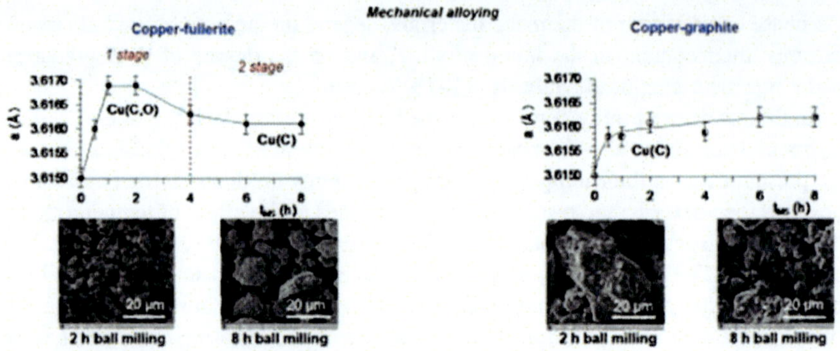

Fig. 1 Graphic abstract showing the content of the paper 30 [30]

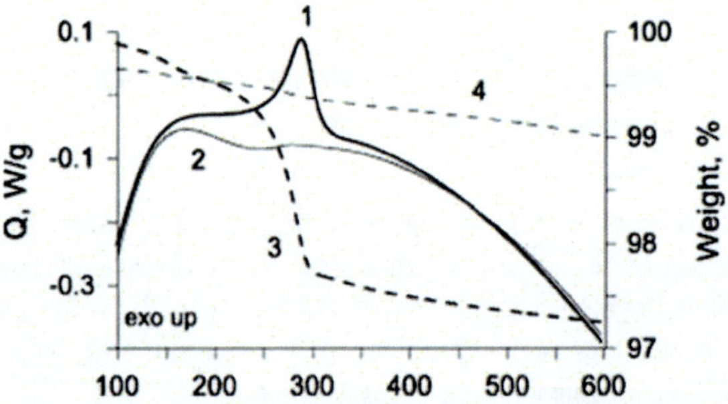

Fig. 2 Heat patterns obtained after heating the samples mechanically alloyed during 1 h: 1, 3—Cu–C60/70; 2, 4—Cu–Cg; DSC, TGA. Heating rate 10 °C/min, Ar [30]

wear resistance and tracking resistance, with a simultaneous reduction of wear rate of the copper counter body of the contact wire. An exothermic effect was observed at heating the system Cu–$C_{60/70}$ (1 h MS) on the DSC curve in the temperature range 230–330 °C (Fig. 2). In addition, the sample loses ~2% of its weight. Because of the copper oxide formation in Cu–$C_{60/70}$ powders, it is difficult to entirely exclude the influence of oxygen on the sequence of solid-phase reactions [30].

Experimental confirmation of different reactivity of fullerite and graphite with respect to oxygen has been obtained by differential scanning calorimetry. It was shown (Fig. 2) that the kinetics of oxidation in mechanocomposites Cu–C_g and Cu–$C_{60/70}$ in process of their heating in an oxidizing medium depends on the allotropic form of carbon and its structural state. The formation of the amorphous carbon with a disordered defect structure promotes reducing reactions.

As for simple Fe–C system it could be interesting to show some data about the unusual side of their possible application. The synthesis by MA was made using a Retsch PM 100 planetary ball mill in a 25-mL stainless steel grinding bowl to mechanically alloy a starting amount of 9 g of high-purity copper (<425 μm, 99.5%) and iron (≥99%) powders, used as received from the supplier (Sigma-Aldrich) [8]. Differential scanning calorimetry (DSC) is performed on the milled powders to provide insight into the energy of formation and mixing of the resulting compounds (Fig. 3).

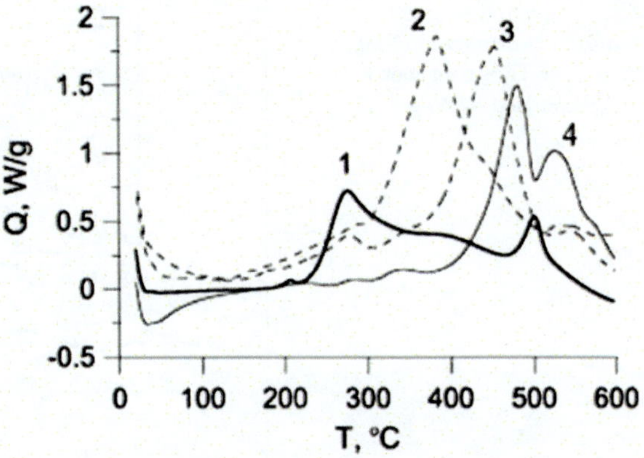

Fig. 3 The DSC curves obtained after heating the samples in oxidizing medium: 1—initial Cu; 2 and 3—Cu–$C_{60/70}$ after 1 and 8 h of MA, respectively; 4—Cu–C_g after 1 h of MA [30]

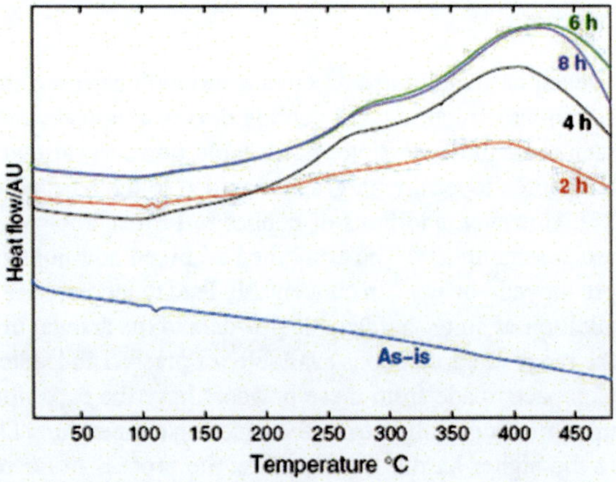

Fig. 4 Differential scanning calorimetry heat flow plots versus milling times [8]

The results taken from the DSC are shown in Fig. 4 for all milling times and also for the as-is specimen [8]. While the latter shows no phase reaction and its thermal response remains flat, the rest of the samples show a trend that indicates more thermal stability as milling time increases. The results for 6 and 8 h showed the least difference between the obtained peaks, indicating a limiting time for the milling process. The data were inverted to show a positive scale and they correspond to those mentioned in the work [30]. Some peaks also take place at higher temperatures that reflects a better phase stability.

The evolution of the microstructure with milling has impacted not only mechanical and thermal properties of the composite. Optical absorption of the samples was also influenced there (Fig. 5).

As for milling times of 2, 6, and 8 h, the onset of absorptivity is seen to be an abrupt step increase at progressive wavelengths of 341, 396, and 434 nm, respectively.

Adding simultaneously both copper and carbon leads to new and complex effects. Thus, the sinter-hardenable powders have been used as replacements for traditional

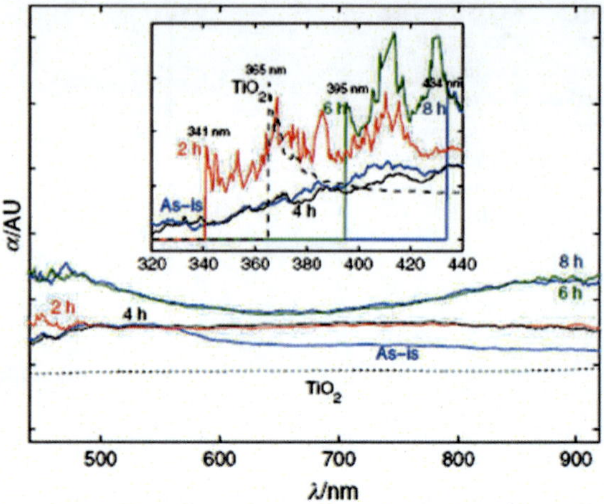

Fig. 5 Absorption of the compacts at different milling times. The TiO_2 data shown for comparison by dotted line [8]

quench-hardened and tempered materials due to their ability to transform to a martensitic microstructure upon cooling during a typical sintering operation. In the manufacture of the base sinter-hardenable powders, alloying additions are made to the melt (prealloyed), with graphite added to the base powder as the carbon source [8, 25]. However, additions of copper to further improve the hardenability of the mix are commonplace. The combined admixed additions of graphite and copper to the prealloyed powder can conceivably lead to increased retained austenite contents The addition of sinter-hardenable powders to the arsenal of P/M materials has expanded the range of applications available for pressed and sintered powder metallurgy parts. Compacts made from these powders have the capability to transform to martensite upon furnace cooling from the sintering temperature. Due to this ability to transform to the higher hardness martensite, the process of quench-hardening and tempering can be eliminated from the manufacturing process sequence where surface or through hardening is required. This capacity to form martensite during the sintering process is developed by an increase in hardenability, accomplished through both prealloyed, and admixed elemental additions. The use of both alloying techniques results in an increase in the alloy content of the final mix. The base materials used for sinter-hardening applications are prealloyed. In prealloying, the alloy additions are made to the molten metal bath prior to atomizing. This creates uniformity in alloy content. And, as the particulates are formed, the chemical composition within each particle is the same. In most sinter-hardening situations, two additional materials are used as admixed powder additions. They are carbon, in the form of graphite, and copper. These elements are almost always admixed because prealloying them in the melt has deleterious effects on both compressibility and green strength. The additions of copper and graphite provide further enhancement of the hardenability. During the sintering process, carbon is diffused into the base powder before the copper becomes liquid. The consequence of this alloying sequence is twofold. The diffusion of copper is inhibited in mixes with high carbon contents, and as the copper diffuses, high concentrations are often located along particle and grain boundaries [8, 25]. The evidence from these two studies appears to show that the carbon content of the sinter-hardened compact has a strong effect on the diffusion of the added copper, and, that at higher carbon contents, the diffusion of the copper is inhibited by the diffused carbon. This situation causes a higher local hardenability at the particle/pore surfaces and promotes the formation of retained austenite in the high-copper regions [8, 25].

The hard metallic balls used at the MA are contaminated by powders under milling. So, in order to avoid excessive sticking of the powdered sample to the balls some liquid organic (e.g., ethyl or methyl alcohol) are added to the powder mixes. The effect of those organic on the product of MA (Fe–Cu system) was particularly studied in the paper [21].

The sintering behavior and sintered microstructures of mechanically alloyed Fe-9.5 mass%Cu–0.5 mass% P system have been examined in the work [34]. Started materials for MA were reduced pure iron powder, water atomized pure copper powder, and crushed Cu-14 mass% P alloy powder. MA was performed in argon gas atmosphere by using a planetary ball mill. The change in the crystallite size of the powders with MA time almost finished for 36 ks and the yielded size was estimated to be about 10 nm. Sintering of MA powders at 1073 K for 36 ks gave a consolidated alloy, the density of which is comparative to that of an atomized powder alloy sintered at 1473 K. The microstructure of the sintered MA powder alloy is composed of copper particles of a few micrometers in diameter distributed in the matrix the grain size of which is less than about 5 μm [34].

The characteristics of the Fe–Cu alloys prepared by ball milling nominally pure (99%) Fe and Cu powders and compacted (at 300 °C) were measured in the [35]. Surprisingly, however, calorimetric measurements indicate the presence of a large endothermic peak for these nanocrystalline composites on heating near 600 °C and an exothermic peak near 400 °C on cooling. Furthermore, heating to temperatures above 600 °C results in the dissolution of Cu into the iron oxides which does not reprecipitate on cooling. The differential scanning calorimetry data for this sample, shown in Fig. 6, further support the likelihood that the large features in the M versus T data are due to phase transformations. In this figure, endothermic events are indicated by "upward" deviations, while exothermic events are the reverse. It is worth to note the endothermic reaction near 600 °C and the exothermic reaction near 425 °C, which roughly repeat with thermal cycling. The similarity in the temperatures of these two arrests to the temperatures of the largest magnetic effects is suspiciously coincidental. For a first-order phase transformation between equilibrium crystallographic phases (the common type), an endothermic heat effect will occur during heating and the opposite effect will occur during cooling, just as observed for this sample [35].

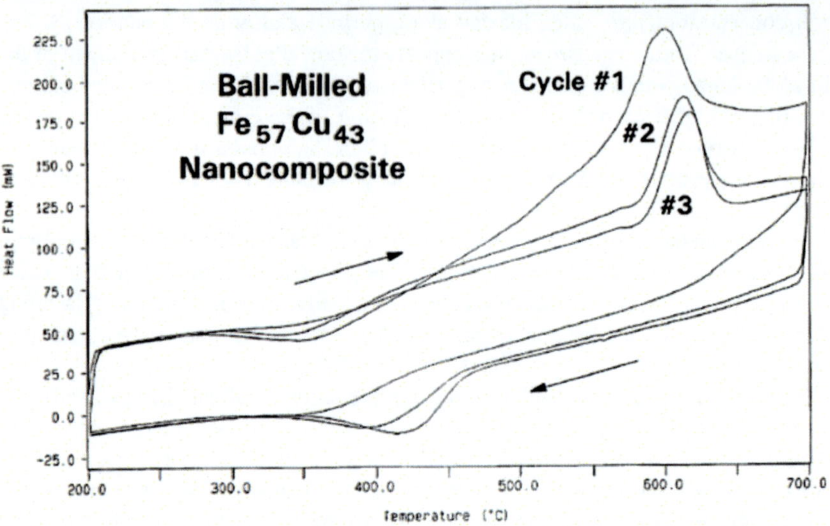

Fig. 6 DSC data versus temperature for the ball milled Fe–Cu nanocomposite during heating and cooling (indicated by arrows) for three successive thermal cycles [35]

3 Dilatometry Behavior of the Fe-Based Composite Materials

Early performed dilatometric studies have shown that the change of dimension is a direct function of Cu content up to levels of 10% [25]. The works have also shown that upon heating graphite quickly goes into solution in iron after the alpha to gamma transformation [25]. During this solutioning, the sample exhibits considerable growth beyond that associated with the thermal expansion of iron. The dimensional change (shrinkage) that occurs at the sintering temperature is smaller than the carbon solutioning growth [25]. When both copper and carbon are added to a base alloy, there is an interesting interaction between the two alloying elements. The carbon inhibits Cu diffusion in the compact. The amount of growth observed in a dilatometer at 1083 °C is significantly reduced in high carbon alloys. Therefore, growth and dimensional change in a high-carbon–high-copper mix can be less than if either additive was made alone [25].

The effect of Cu and C on the dimensional change (DC) of three different base alloys is shown in Fig. 7. The overall shape of the curves is the same for all three alloys. With no added Cu, the 0.6% graphite addition had the smallest dimensional change of all compositions and the addition of further graphite to 1% increased the dimensional change (causes growth). At 1% Cu, the addition of graphite between 0.6% and 1.0% has little effect on dimensional change and, at 2% Cu, the addition of C decreases the dimensional change. The difference in dimensional change is greatest at 0.6% graphite for the different levels of Cu, and further increases in C

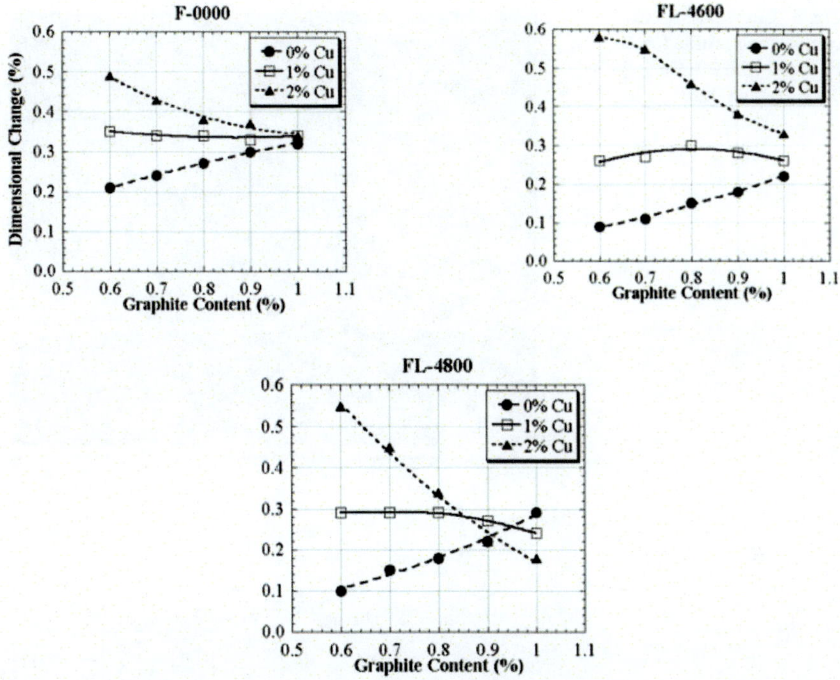

Fig. 7 Effect of the C and Cu on dimensional change of three different base alloys [25]

content reduces this difference as the curves tend to merge at high graphite levels. The interaction between the Cu and C has been shown by other authors in the literature [25].

Incorporation of the graphite to the base alloy is an effective way to minimize dimensional change variations in these compositions. The shape of the curves clearly displays the copper–carbon interaction. It is possible that the effect of carbon on copper diffusion throughout the compact causes different slopes. As carbon increases, copper diffusion decreases to the point where the 0% and 2% copper curves behave similarly (i.e., at 1% graphite in Fig. 7a). Literature data suggest that at high C levels, copper no longer diffuses down grain boundaries and is limited to the pore/prior particle boundary regions [25].

To know the copper distribution the line scans using electron dispersive spectroscopy (EDS) in an SEM were made across prior particles in sintered compacts of FL-4800 base alloy with 2% Cu and different carbon levels. A typical trace and copper profile is shown in Fig. 8. The Cu levels are highest near the pores and decrease toward the center of the base iron particle. Increase of the Cu levels was found to a depth of nominally 10 μm from the edge of the particle. It can be seen that the copper content near the pores is higher for the high-carbon alloys, while the average value at the center of the particles is higher for the low-carbon sample. The large range in Cu levels in the center of the particles at 0.6% graphite may be the result of Cu diffusion

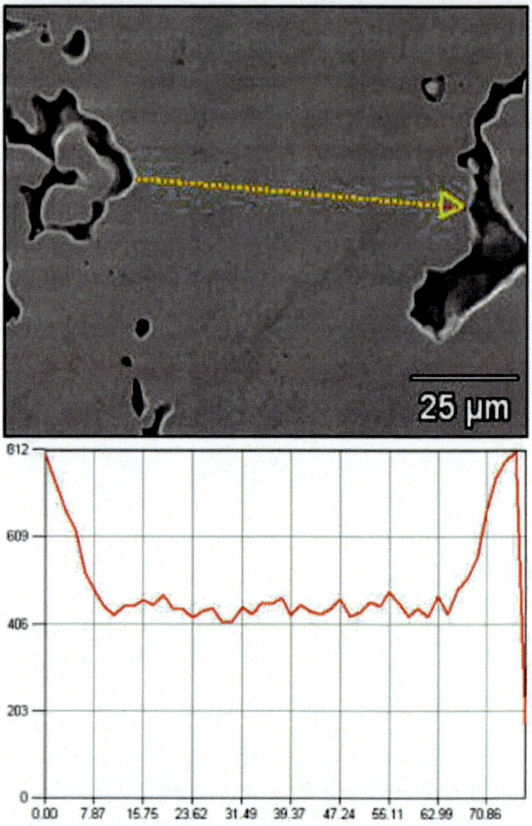

Fig. 8 Line scan and Cu composition profile for FL-4800 with 2% Cu and 0.8% graphite [25]

along grain boundaries. The carbon content clearly affects the Cu distribution in the samples. No difference in C content was found along with the same line scans.

The wide study of the dimensional change characteristics of PM steels was performed for the Fe–Cu composites containing various carbon forms [27]. Briefly, the authors concluded that the dimensional change is generally not determined by the effects of the processes that occurred at the sintering temperature. Instead, dilatometry showed that the largest contributions are typically due to the combined effects of various lower temperature processes. Given the emphasis on time and temperature in discussions of sintering, it is perhaps natural to conclude that the dimensional change during the process is mainly, if not entirely, a consequence of the processes that occur at the sintering temperature. In fact, numerous dilatometric studies have shown that while sintering conditions are important, they generally account for less than half and frequently less than a third of the dimensional change that was observed in the most common P/M compositions. Indeed, the balance of the dimensional change and often the greatest part of it are determined by various processes and effects that occur during the heating and cooling stages of the process. These include the behavior of

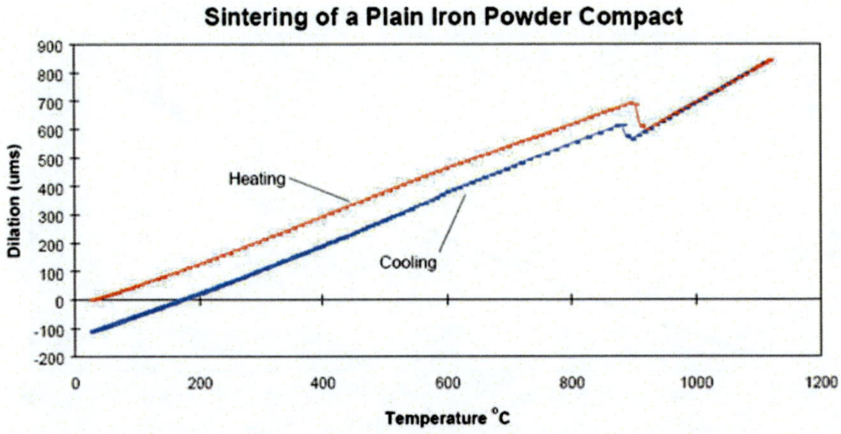

Fig. 9 Dilation versus temperature during sintering of a plain iron powder compact [27]

the lubricant as it affects the early thermal expansion of the compact and the amount of low-temperature sintering that occurs as well as contributions associated with the heating and cooling transformation behaviors and the thermal expansion and contraction characteristics which otherwise typify the process [27].

Thus, the series included iron, iron plus lubricant, iron plus graphite, iron plus lubricant plus graphite, and, lastly, iron plus lubricant plus graphite plus copper were examined. It may be useful to begin by pointing out that solid-state sintering may occur by any one of four mechanisms and in the general case is a cumulative effect of all four. These include vapor transport and the surface, volume, and grain boundary diffusion mechanisms. On the other hand, there are also mechanisms other than sintering that can cause dilatometric change [27]. The dilatometric results of the sintering a plain iron powder specimen as compacted in the die wall lubricated condition are shown below in Fig. 9 [27].

The figure shows that most of this change occurred in the alpha-stability range and it was, at least, partially connected with differences in the dilations that accompanied the phase changes. In addition, although it is not as obvious as the foregoing, there was also a thermal effect due to the differences in the expansion/contraction coefficients of the two phases and the temperatures at which the phase changes occurred. The important point is that the net effect of the phase transformations may be positive or negative and is largely unpredictable and therefore, uncontrollable. The deviation in the heating curves with increasing temperature confirms the existence of the indicated sintering effect. The reason, of course, for the greater sintering in the α phase than at the higher temperatures in the γ phase is that the iron generally diffuses faster in the α than it does in the γ [27].

The role of the admixed lubricant can be seen in Fig. 5 (see the work [27]), where the dilatometric results of sintering a compact of the same grade iron powder but with lubricant are shown.

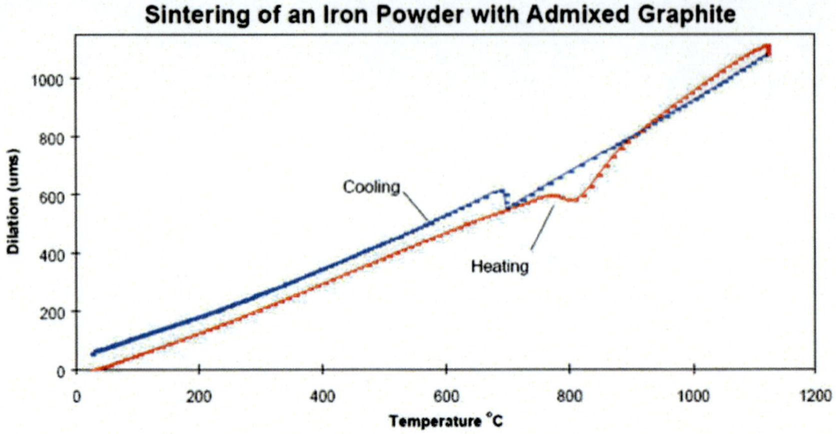

Fig. 10 Dilation versus temperature during sintering of an iron powder compact with admixed graphite [27]

The dilatometric data of sintering the compact of the same grade iron powder but with admixed graphite are shown below in Fig. 10 [27]. As in the case of the plain iron specimen, die wall lubrication was used to facilitate the compaction step.

The growth, in this case, is mostly the result of the presence of the admixed graphite and its effects on the specimen's behavior during α to γ transformation and subsequently during heating in the γ range. There are several conflicting processes that determine the specimen's dimensions and the observable dilation reflects the occurrence of all of them rather than anyone [27]. Those include in order of increasing importance: (1) sintering; (2) thermal expansion; (3) graphite solution; (4) transformation; and, (5) the response of the specimen to the stress state that arises in consequence of the thermal gradients that are present [27].

The dilatometric results of sintering the compact of the same grade iron powder but with admixed graphite, copper and lubricant are shown in Fig. 10 [27].

If to compare the data in Fig. 11 with the dilatometric record of the previous figures, it is easy to see only two obvious differences, both of which are tracible to the presence of the copper. One is that the cooling curve, in this case, intersects the heating curve five times instead of twice as in the earlier case and subsequently descends above it to indicate a small net growth. The other is a rapid expansion in the heating curve just before the maxima at 1120 °C. As explained below, this was apparently due to the melting of the copper and the appearance of a liquid phase. Copper can only penetrate the grain boundaries at their intersections, (e.g., at triple points), and combined with its poor solubility for iron will typically provide little if any shrinkage by grain boundary transport [27].

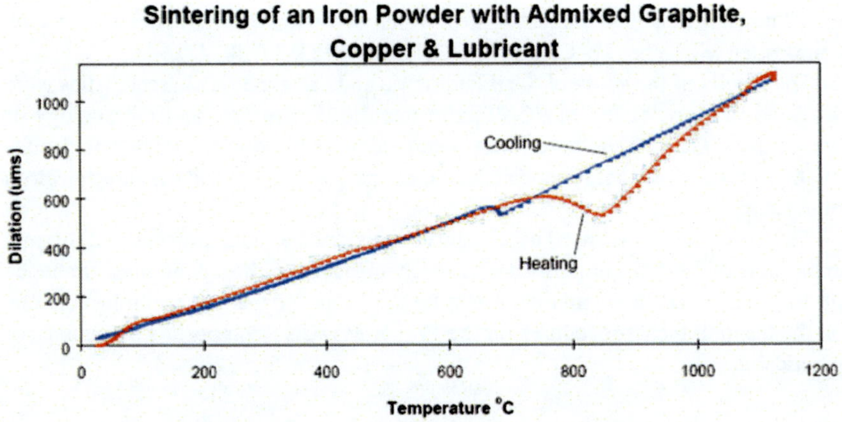

Fig. 11 Dilation versus temperature during sintering of an iron powder compact with admixed graphite, copper, and lubricant [27]

4 Effect of the MWCNT on Dilatometric Behavior of the Fe–Cu Nanocomposites

The CNTs are quasi-one-dimensional nano-objects with nanometric diameters (0.7–50) nm and micrometric length (1–100 μm or more). Therefore, they have a huge form factor and a very high specific surface area (up to 1300 m^2/g). The strong C–C covalent bonds and the structure of the CNTs walls provide exceptional mechanical properties such as high the Young's modulus (in excess of 1 TPa) coupled to strong resistance to breaking (more than 100 GPa). They also exhibit enhanced thermal conductivity and electrical behavior of the semi-conductor or metallic type; their electrical conductivity is high (10^{-3}–10^{-4} S cm^{-1}) and their capacitance allows the flow of high current densities (>10^{-9} A cm^{-2}) without internal damage [36].

The published results on the Al–CNT and Cu–Ti–CNT systems studies have confirmed the improvement of their characteristics under MWCNT addition, notably the low thermal expansion of those nanocomposites [37–39]. Not only these characteristics; in particular, thermal behavior features have been used to create materials with new characteristics.

Thus, the aim of work described in this section was to study the characteristics of thermal expansion and calorimetric data of Fe–Cu composites doped with MWCNT; elucidation of mechanisms that determine the temperature behavior of the noted characteristics and to clarify the carbon nanotubes role in these processes.

The application of mechano-chemical activation (mechanical alloying) allowed us to make a concentration range of the Fe–Cu nanocomposite materials (Fe:Cu = 4:1 ratio) doped with MWCNT, Fe–Cu-MWCNT, where the MWCNT concentration was equal to 0.5, 1.0, and 2.0% vol.). The indicated Fe:Cu and MWCNT contents were selected to study the effect of temperature on their properties, taking into account their known mechanical characteristics, which exceeded the characteristics of such composites with other Fe–Cu and MWCNT ratios [33, 37, 40, 41].

The results of the study of the dilatometric and calorimetric characteristics were analyzed in terms of the MWCNT influence on the structure and morphology of the samples. Data on the latter were obtained by monitoring the made composites by the X-ray diffraction, Raman and IR spectroscopy, optical and scanning electron microscopy.

The obtained data allowed us to draw conclusions about the processes of thermal expansion of Fe–Cu composites and mechanisms of carbon nanotubes influence on them. The results of this work can be the basis for making recommendations on the technological procedures for the Fe–Cu-MWCNT composites producing for practical use.

4.1 Sample Preparation and Methods

Multi-walled carbon nanotubes (TU U24.1-03291669-009:2009) were synthesized by us by means of the chemical vapor deposition procedure in a rotating reactor [42]. Their resulting characteristics were: average diameter—between 10 and 20 nm; specific surface area (determined through Ar—adsorption)—between 200 and 400 m^2/g; bulk density—between 20 and 40 g/dm^3 [43, 44]. The charge was made of IP-1 (GOST 9849–86) iron and PMS-1 (GOST 4960–2009) copper powders in a 4:1 weight ratio. These powders were mixed with MWCNT of concentrations 0.5, 1.0, and 2.0% in volume. After processing in the ball mill the composites were subjected to a 40% compression followed by annealing at 950 °C for 30 min in Argon flow. Then, the powder samples were subjected to cold rolling followed with annealing at 900 °C in Argon flow. The described cycle was repeated three times; thus, the samples were ultimately undergone to 80% thinning. Finally, the thickness of the obtained ribbons was near of 0.3 mm. The temperature behavior of some physical characteristics of the made composites, which were named here and before as Fe–Cu, Fe–Cu-CNT(0.5), Fe–Cu-CNT(1.0), and Fe–Cu-CNT(2.0), was studied using dilatometry and calorimetry technique. Structural characteristics and composition of the samples were monitored by X-ray diffraction, Raman scattering, and IR. The next equipment was used to perform mentioned studies: the NETZSCH 402C dilatometer (NETZSCH, Selb, Germany) with 3% of accuracy. The heating rate was near 10 °C/min. The thermal expansion coefficients were measured in the temperature range from 30 to 800 °C. Differential scanning calorimetry (DSC) test was performed using the Jupiter STA 449 F3 NETZSCH calorimeter (NETZSCH, Selb, Germany). The same heating rate was applied as under dilatometric measurements. The X-ray diffraction patterns of composites under study were obtained using the automated DRON-4.0 diffractometer with the filtered cobalt radiation $K\alpha = 1.7909$ Å under the following scanning parameters: monitoring range 2 h = (40–130), step scan of 0.05, and counting time per step at 3 s. The collecting of the diffraction data was performed by a full-profile analysis using the program providing the experimental diffraction peaks interpolation by the Lorentz function. The average grain size formed after mechanical alloying was estimated by the well-known Debye–Scherrer approach. The TEM images of the CNTs and composites after milling were obtained using transmission electron microscope SELMI PEM125 K operated at 100 kV. The IR absorption and Raman (macro-Raman monitoring) spectra were measured using a Jasco FT/130 IR-6300 (Jasco Analytical Instruments, Easton, MD, USA), Bruker SENTERRA (Bruker, Billerica, MA, USA). The triple T64000 Horiba Jobin-Yvon spectrometer equipped with quasi-confocal scanning microscope was used for the micro-Raman scattering measurements. The scanning and optical systems allowed the movement of the object at XYZ-coordinates with 100 nm accuracy and collection information about light scattering with submicron spatial resolution. The Ar–Kr Spectra-Physics 2018 laser with a wavelength of the incident light, $\lambda_{inc} = 488$ nm was used for these measurements. Optical microscopy of the sample was performed simultaneously with micro-Raman measurements.

4.2 Results and Discussion

The variations of the relative dimensions (sample length) of the samples ($\Delta L/L_0$) and the variations of their coefficients of thermal expansion (CTE) dependently on the type of the samples and temperature are shown in the Figs. 12 and 13, respectively. In addition, data on the dilatometric behavior of these samples are given in Table 1.

The Table 1 data give a quantitative assessment of the results shown in the figures. In particular, there are the estimated average values of the CTE at certain selected temperature intervals, the limits of the intervals themselves, and the position of special points on the curves of temperature dependences. As we will see below, there are several reasons for the observed temperature behavior of dilatometric characteristics, and therefore the table data are useful for the analysis of the contribution of impacts

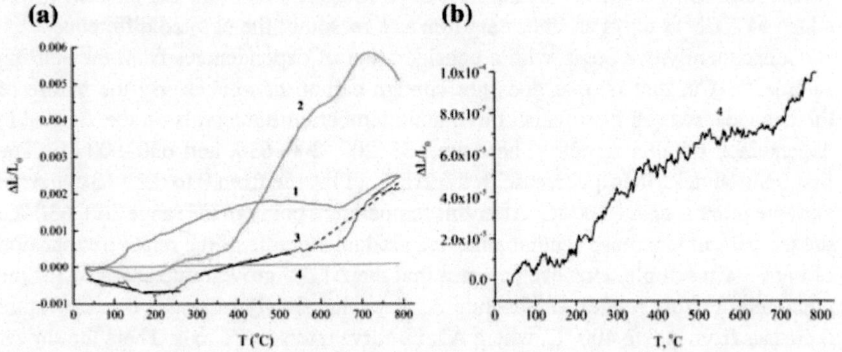

Fig. 12 The relative dimensional variations with temperature, $\Delta L/L_0(T)$, for the Fe–Cu (1), Fe–Cu-MWCNT(0.5) (2), Fe–Cu-MWCNT(1.0) (3), and Fe–Cu-MWCNT(2.0) nanocomposites (4)

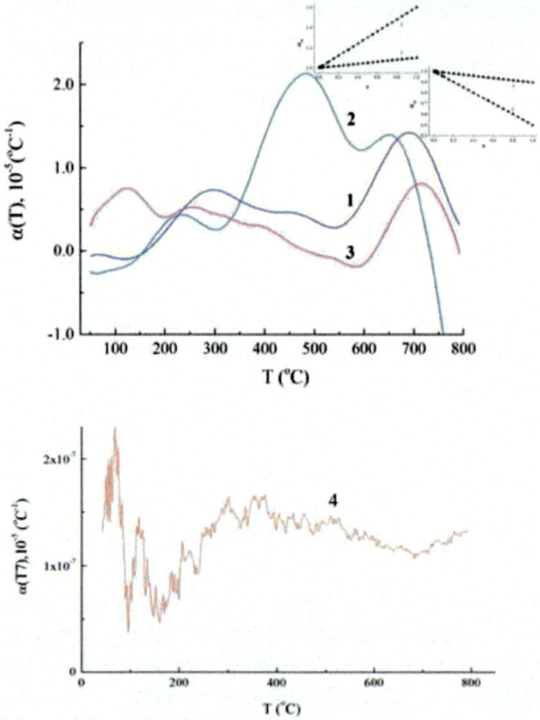

Fig. 13 Dependences of the thermal expansion coefficient on temperature, α(T), for the samples under study: Fe–Cu (1), Fe–Cu-MWCNT(0.5) (2), Fe–Cu-MWCNT(1.0) (3) and Fe–Cu-MWCNT(2.0) (4). The inserts illustrate the α_m dependence on v_2 according to the formula (3): to the left—for the case $\alpha_2 > \alpha_1$ and to the right—for the case $\alpha_2 < \alpha_1$; lines 1: α_2 differs from α_1 by 10%, lines 2: α_2 differs from α_1 by 50%

of different nature. First, we have to note that if the temperature is in the range from room temperature up to ~125 °C, the $\Delta L/L_0$ are of similar values for all the samples, as these values are close to the zero starting point. The maximal difference is realized for the Fe–Cu sample at T ~ 90 °C and it is near to $0.52 \cdot 10^{-3}$. Second note is the next. When the temperature is higher than 125 °C, then the behavior of the $\Delta L/L_0$ values noticeably depends on the MWCNT concentration, and the amount of the added MWCNTs plays an important role as a reason of the pointed difference.

Consequently, we begin with a consideration of dependencies from the starting sample Fe–Cu, that is, that does not contain carbon nanotubes. By the nature of the changes, we can distinguish three main temperature intervals on the $\Delta L/L_0(T)$ dependence for this sample. There are ~35–200, 200–630, and 630–800 °C. The first interval describes a decrease in the $\Delta L/L_0(T)$ value from 0 to $0.8 \cdot 10^{-3}$, so, an extreme point is near ~200 °C. After this temperature point, in the range 200–630 °C, the temperature increase accompanies the gradual increase of the relative expansion of the Fe–Cu sample. However, we see that the $\Delta L/L_0$ growth rate changed for the noted temperature range. An absolute $\Delta L/L_0$ value slowly increases if temperature increases from 200 to 400 °C, where $\Delta L/L_0$ curve intersects T-axis. Then, the slower increase of the $\Delta L/L_0$ quantity takes place up to temperature T = 630 °C. The rate of the $\Delta L/L_0$ changing sharply increases after the temperature reaches 650–670 °C

Table 1 Position of the singularities (T ± 3 °C) and their character—local maximum (max) or minimum (min); Δ_iT ($i = (1 \div 5)$)—temperature ranges chosen for the $\Delta L/L_0$ and α(T) dependences calculation analysis; Average values of the linear thermal expansion coefficients, <α_i(T)> $\cdot 10^5$, and normalized starting values of the Fe–Cu sample length, L^*_{oi} 10^{-4}, for each temperature range which has been chosen; temperature of the extremum points (T_{extr}) on the DSC curves

	Singularities number													
Sample	1	2	3	4	5	6	7	8	9	10	11	12	13	14
Fe–Cu	65 (max)		118 (min)			295 (max)			419 (min)	453 (max)	542 (min)			690 (max)
Fe–Cu-CNT(0.5)	63 (min)	90 (max)	120 (min)		232 (max)	300 (min)				484 (max)		593 (min)	653 (max)	
Fe–Cu-CNT(1.0)	65 (max)	88 (min)	122 (max)	200 (min)	255 (max)	300 (min)	322 (max)	366 (min)	390 (max)	482 (min)	540 (max)	583 (min)		715 (max)
Fe–Cu-CNT(2.0)	65 (max)	95 (min)	120 (max)	160 (min)	215 (max)	300 (max)		370 (max)			520 (max)			690 (min)
	Temperature ranges													
Δ_iT	65–120			165–260			310–525					575–650		725–790
<α_i(T)>	−0.26			0.31			0.27					0.82		0.79
L^*_{oi}	3.23			6.86			1.9					4.99		19.42
T_{extr}	68–75	120–125		155	265			435–470		565–590			675	

and the $\Delta L/L_0$ shows the maximum value near of $2.1 \cdot 10^{-3}$ at T ~ 770 °C (Fig. 12, curve 1).

If small amount (0.5 vol. %) of the MWCNTs is added to the Fe–Cu sample, the tendency of the $\Delta L/L_0(T)$ dependence is practically the same as for the described before Fe–Cu sample, but the absolute $\Delta L/L_0$ values are higher for overall temperature range that was applied. It is also important to note, that this increase and its temperature behavior are different for the above-mentioned temperature ranges. So, the $\Delta L/L_0$ is on average higher than $0.5 \cdot 10^{-3}$ and its dependence is similar to that for Fe–Cu sample in the range of 35–375 °C. Further temperature increase leads to a sharp change of the $\Delta L/L_0$ value for the Fe–Cu-MWCNT(0.5) composite from the value of ~$0.6 \cdot 10^{-3}$ up to ~$4.5 \cdot 10^{-3}$ at 600–630 °C. A further increase in temperature to 710–715 °C leads to a change in the $\Delta L/L_0$ value in the same range ΔT, as it was found for the Fe–Cu sample in this temperature range and that is $\Delta = 1.3 \cdot 10^{-3}$. Unlike the case of the Fe–Cu sample, the further increase in temperature in the case of the Fe–Cu-MWCNT(0.5) sample leads to a decrease in the $\Delta L/L_0$ value (Fig. 12, curve 2). As a result, the absolute maximum of the $\Delta L/L_0$ temperature dependence occurs at a temperature near 716 °C and $\Delta L/L_0$maximum is $5.85 \cdot 10^{-3}$. Consequently, we can state that the effect of the MWCNT additive is different for different ranges of the temperature behavior of the relative linear expansion of the Fe–Cu-MWCNT(0.5) composite. Namely, a) MWCNT only affect the value without changing the character of the $\Delta L/L_0$ temperature dependence in the range 35–375 °C; b) the MWCNT addition highlights the presence of two parts of the $\Delta L/L_0$ dependence in the range 200–630 °C (it concerns 200–375 and 400–630 °C ranges) which were hidden in the case of Fe–Cu composite; c) the plateau of the $\Delta L/L_0$, which took place for the Fe–Cu sample for temperatures after ~720 °C, was changed to the decrease of the $\Delta L/L_0$ when MWCNTs were introduced into the composite (the sample Fe–Cu-MWCNT(0.5)).

Two peculiarities, in our opinion, should be emphasized when considering changes in the dilatometric behavior of the Fe–Cu-MWCNT composites with a further increase in the content of carbon nanotubes (please compare curves 1 and 2 in Fig. 12 for the Fe–Cu-MWCNT(0.5) and Fe–Cu-MWCNT(1.0) composites). It is easy to see that the increase of the $\Delta L/L_0$ value is continued in the 35–375 °C range, whereas in the temperature range 400–800 °C the $\Delta L/L_0$ values decrease markedly—approximately by 2.5 times. At the same time, we even have a decrease in the relative linear expansion $\Delta L/L_0$ with increasing temperature in the range 400–630 °C. As a result, the $\Delta L/L_0$ values and their temperature dependences in the range 650–800 °C are close to ones measured for the samples without MWCNTs. The $\Delta L/L_0(T)$ riches the peak ($2.9 \cdot 10^{-3}$) at $T \approx 790$ °C for the Fe–Cu-MWCNT(1.0) composite.

When the MWCNT concentration reaches 2%, the $\Delta L/L_0(T)$ values do not practically depend on temperature if to regard the $\Delta L/L_0(T)$ curve in the same scale as other curves (see Fig. 12a). In fact, mentioned dependence is similar to measure for the Fe–Cu-MWCNT(1.0) composites, but if it is observed in other scales (see Fig. 1b). So, the maximal $\Delta L/L_0$ value for the Fe–Cu-MWCNT(2.0) is near to $1 \cdot 10^{-4}$ at $T = 790$ °C. Thus, the set of the maximal $\Delta L/L_0$ values is $2.1 \cdot 10^{-3}$, $5.85 \cdot 10^{-3}$,

$2.9 \cdot 10^{-3}$, and $0.1 \cdot 10^{-3}$ for the Fe–Cu, Fe–Cu-MWCNT(0.5), Fe–Cu-MWCNT(1.0), and Fe–Cu-MWCNT(2.0), respectively.

Some additive details of the dilatometry behavior of the samples under study become better visible from the temperature variations of the thermal expansion coefficients, CTE = α(T), (Fig. 13).

The shape of the dilatometric curves, α(T), for all four materials is different. At the same time, there are some common features, those allowed us to select the temperature ranges similar to those noted above for the $\Delta L/L_0$ curves. If regarding the Fe–Cu sample, this statement concerns the next of the temperature ranges: ~35–150, 200–325, 325–550, 575–700, and 700–800 °C (Fig. 13, curve 1). Moreover, if compare the data for the Fe–Cu, Fe–Cu-MWCNT(1.0), and Fe–Cu-MWCNT(2.0) samples we see similar behavior of the α(T) curves in the ranges ~150–300, 300–600 and 600–800 °C. However, the α(T) values for Fe–Cu and Fe–Cu-MWCNT(1.0) samples are fairly close, while for the Fe–Cu-MWCNT(2.0) sample with the largest content of the MWCNTs, they are smaller by two orders of magnitude. The values and behavior of the α(T) curves for the samples Fe–Cu and Fe–Cu-MWCNT(0.5) are similar in the ranges 0–125–200°C, but α(T) dependence for the Fe–Cu-MWCNT(0.5) sample differs radically in the range 300–600 °C from other three curves noted above. Thus, the described situation is similar to that observed for the $\Delta L/L_0$ data. At the same time, we have to emphasize that despite the similarity in general terms, there are a number of differences within each of the selected ranges. This refers to both of the number of features (singularities) in each of the ranges, the position of the features, and their nature—it is a local maximum or a minimum. The data about the position and character of the singularities have been accumulated in Table 1. Taking into account the significant, in some cases, inaccuracy of the α(T) measurement it is easy to see that the same features are inherent in various curves. We found four singularities in the ranges 0–125–200 °C, five singularities in the range 200–400 °C, three singularities in the range 400–600 °C, and at least two singularities in the range 600–800 °C. Consequently, we can assume that even in those cases where there are no distinct peaks on the wide bands of the both $\Delta L/L_0$ and α(T) curves, those bands are in fact a superposition of several strongly overlapping components. Besides, it worth to note that number of singularities is larger for the composite samples those contain a higher amount of the MWCNTs. We see only two features for the Fe–Cu sample in the range 0–400 °C, while there are five singularities for the Fe–Cu-MWCNT(0.5) composite, and 8 details for the Fe–Cu-MWCNT(1.0) composite. At the same time, the number of singularities is more stable (4–5) for the temperature range of 400–800 °C. Some of the singularities are well pronounced, while other ones are poorly selected (like to shoulders, see Fig. 13 and Table 1, e.g., columns 5–11). (The α(T) curve for the Fe–Cu-MWCNT(2.0) composite is out of consideration here as measured α(T) changes are small and close to experimental inaccuracy.) Thus, we have to state the rather complicated temperature behavior of the $\Delta L/L_0$ and α(T) characteristics. The data on the dilatometric characteristics of such composite materials are very limited in the literature, even for the Fe–Cu samples [25, 27, 34, 45–48]. The data for Fe–Cu composites where carbon was introduced as carbon nanotubes, according to our information, are just given for the first time in published literature.

The performed description of dilatometric characteristics, without a doubt, shows that their complex behavior should be associated with the multicomponent composition of Fe–Cu-MWCNTs composites. Therefore, further analysis of the dilatometric data, of course, should start with the simplest case—the Fe–Cu composite.

Analysis of the data on the temperature dependence of the CTE, α(T), for the Fe–Cu composite allows us to select at least five ranges with different temperature behavior of the α(T) value ($i = 1 \div 5$): $\Delta_i T = \sim 65$–120, 165–260, 310–525, 575–650, and 725–790 °C, respectively. The average values of α(T) in these ranges, $<\alpha_i(T)>$ were determined by us according to the usual formula for calculating averages values:

$$\langle \alpha_i(T) \rangle = \frac{1}{\Delta_i(T)} \int_{T_i}^{T_{i+1}} \alpha(T) dT \tag{1}$$

The resulting values of the $<\alpha_i(T)>$ thus obtained for each temperature range are given in Table 1. These $<\alpha_i(T)>$ values can be compared to the values of previously known for similar systems. First of all, we note that for iron and copper, the components of such systems, the coefficients of thermal expansion for bulk samples lie in rather wide limits and depend on the composition (brand) and therefore on the structure of the lattice and the state of the metal (temperature, etc.). Thus, in the temperatures range 100–300 °C for α-Fe it is possible to give for α the values (12.8–14.4)·10^{-6} K^{-1} and (12.4–14.1)·10^{-6} K^{-1} with a carbon content of 0.05 and 0.5%, respectively. The α value for Cu in the temperature range 0–100 °C is 17·10^{-6} K^{-1}. It is evident that these values are significantly (in 4.3 and 5.5 times) more than the value we estimated for the CTE coefficient or the investigated Fe–Cu composite in the same range of temperatures (~100–300 °C). Approximately the same, these relationships remain at higher temperatures (300–550 °C (Table 1). If to compare that with powder compact compositions of iron or Fe–Cu composites, then these ratios, according to our estimations of the data (taken from [25, 27, 34, 45–48]), may vary widely, and they may be characteristic for massive samples of Iron, that is, they vary greatly from our data (4.0 and 5.5 [46], 3.0–4.0 [34], 5.0 [25]), and to be close to them: 1.4 (see, for example, [41]). So, we must conclude that a comparative analysis of own and literature data regarding the absolute values of the coefficients of thermal expansion is not constructive due to significant differences in the conditions of obtaining and characteristics of the studied samples, while the consideration of the $\Delta L/L_0$ and α(T) values temperature changes is more attractive from this point of view.

Obviously, there is a significant number of reasons for the complex temperature behavior of these dilatometric characteristics for composite materials. The mechanisms of their thermal expansion/reduction can be considered both from a macroscale and from a microscale point of view. Those possible mechanisms are thermal expansion of the crystalline solid state lattice; transformation of the crystal lattice structure and phase transformation; reduction/shrinkage due to the temperature-activated release of the initial adsorbed gases and gases formed as a result of the adsorbed compounds thermal decomposition, the processes of sintering; and, conversely, the degradation of the composite grains in rolled materials, etc. From the point of view of the micro/nanoscale phenomena, it should be noted that a large number of possible combinations where these mechanisms can be actual. There can be the interaction of grains/particles: iron–iron, copper–copper, and iron–copper, as well as some influence of uncontrolled factors (pores, defects, impurities, etc.). Besides, the effect of high local temperatures under mechano-chemical activation can result in the presence in investigated composites of iron particles, where the surface layer is a solid copper solution in iron, and vice versa,copper particles, where the surface layer is a solid solution of iron in copper [31]. The interaction between iron and copper particles and their interaction with the particles of the above-mentioned core–shell type are also a factor that complicates the dilatometric behavior of the materials under study. It should be also emphasized that the activity of these processes can essentially depend on the size of the grains/particles of the material. More precisely, it has to say, that the ratio of the volume of the surface layer to the volume of the central core of the particle changes depending on the particles size, and therefore the role of the corresponding zones of the particles in the dilatometric processes also varies. It is clear that a detailed analysis of all these factors, at this moment, cannot be made especially in the case of so multicomponent samples. Therefore, we will try to distinguish the most significant of them and analyze their role initially on the simplest of the composition of the studied systems: Fe–Cu.

"Pure" thermal expansion effects are those caused by an increase in the equilibrium distance between atoms of solid state due to the increase of their vibrations amplitude when the temperature rises. For most of solids, their expansion occurs if temperature increases. Obviously, the temperature ranges, where the contribution of the "pure" thermal expansion of the samples studied by us is preferable, should be those where the $\Delta L/L_0(T)$ dependence is close to the linear and, at the same time, the CTE in this range does not change. An approximation of the $\Delta L/L_0(T)$ experimental dependence for the Fe–Cu composite (Fig. 12, curve 1) was made by assuming a linear expansion of this material. To do this, the above calculated $<\alpha_i(T)>$ values were used (Table 1). In this case, the initial values of the sample length L_{0i} on each of the temperature ranges $(\Delta T)_i$ ($i = 1 \div 5$) were selected as the parameter corresponding to the actual sample length at the initial temperature of the corresponding range, T_{0i}. The normalized values of those parameters, $L^*_{oi} = L_{0i}/L_0$, are also given in Table 1. It was assumed for such approximation that the L(T) value is an additive sum of the straight lines $L_i(T) = L^*_{oi} + L^*_{oi} <\alpha_i(T)> \cdot (T - T_{0i})$, and each of which is relevant in the appropriate range of temperatures $(\Delta T)_i$:

$$L(T) = SUM\left[L^*_{oi} + L^*_{oi} \cdot < \alpha_i(T) \cdot (T - TL_{0i})\right]. \quad (2)$$

The line segments $(L_i(T) - L_{0i}(T))/L_{0i}(T))$ constructed in this way are shown in Fig. 12. It is evident that they well describe the behavior of the $\Delta L/L_0(T)$ experimental dependence on the first three low-temperature ranges (35–550 °C), while in the high-temperature range (575–800 °C), the calculated $<\alpha_i(T)>$ values using and hence the consideration of the Fe–Cu composite expansion in a linear approximation does not give a satisfactory result. (The constructed lines on this temperature range are given by dots.)

Taking into account the above-made analysis of the $\Delta L/L_0(T)$ and $\alpha(T)$ experimental dependences on temperature and approximation of the $\Delta L/L_0(T)$ dependence on the calculated lines as well, we go to the conclusion that the mechanism of "pure" thermal expansion of the Fe–Cu composite dominates in the range of temperatures 375–550 °C.

There are cases in the solids dilatometry when the size of the solid body decreases with temperature increasing. That dependence may be caused by a sharp increasing the repulsion forces if compare with attraction forces increasing. It is obvious that such cases are more likely for solids which include three or more types of atoms [49]. It is clear that the same behavior is quite possible to observe for multicomponent and multiphase solids: alloys, compacts, composites, etc. where the role of atoms is already played by separated grains/particles of such composite materials.

Among the results presented in this paper, the decrease in the $\Delta L/L_0$ value with temperature increasing also takes place at some temperature intervals. These are the ranges 35–75, 135–200 °C for the Fe–Cu samples; the range 50–150 °C for the Fe–Cu-MWCNT (0.5), and the range 500–625 °C for the Fe–Cu-MWCNT (1.0) samples (Fig. 12). However, it should be emphasized that other mechanisms of the complex dilatometric behavior, which are caused by the multicomponent composition of the investigated composites, can give a stronger temperature effect than described above.

In general, the temperature behavior of the dimension of compacted powders, composite materials, etc. can be largely due not so much to "pure" thermal effects, but as a manifestation of the structural and phase transformations of such materials components, that is, it may be the result of their multi-component composition. So, if we consider the simplest model case of a composite material containing only two components that do not interact with each other, for example, powders of two metals, then for such material, at a certain temperature range, the thermal expansion coefficient α_m can be considered by so-called "rule of mixture":

$$\alpha_m = \alpha_1(1 - v_2) + \alpha_2 v_2, \tag{3}$$

where α_1 is the CTE of the main phase 1 of the material, and α_2 and v_2 are the CTE and the volume contribution of phase 2 to which the first phase is transformed. From the following form of formula (3):

$$\alpha_m = \alpha_1 - (\alpha_1 - \alpha_2)v_2 \tag{4}$$

It is easy to see that for a case $\alpha_1 > \alpha_2$, the v_2 increasing leads to a decrease in the value of α_m at a rate $(\alpha_1 - \alpha_2)$, whereas in the case of $\alpha_1 < \alpha_2$, the v_2 increasing is accompanied by the α_m increase with the same rate: $(\alpha_1 - \alpha_2)$. The inserts in Fig. 2 illustrate the described influence. Thus, the above-mentioned transitions to the ranges of increase (decrease) of the CTE on the $\alpha(T)$ curves may be due to phase transformations in the compositions of the samples. Taking into account the diffusion between phases and the formation of phase interfaces (e.g., by using instead of the ratio

(3), the so-called Turner formula [49]) can only increase the speed of the indicated temperature changes. Existence of more than two components in the composite can lead to complication of temperature dependencies, in general. The result obtained leads us to the conclusion that in the temperature range of 575–800 °C the expansion behavior, $\Delta L/L_0$, for the Fe–Cu composite can be mainly due to the manifestation of a phase transformation of two components of the composite, at least. The behavior of the $\Delta L/L_0$ curve for the Fe–Cu sample in the ranges 150–250 and 650–750 °C can also be the consequence of such transformations. (This question will be discussed below.) Thus, the increase of CTE for Fe–Cu sample with temperature increasing in the ranges 150–275 and 575–690 °C both as CTE decreases with further temperature increasing, as well as the corresponding temperature changes of these samples length, $\Delta L/L_0$, (Fig. 12) can be due to transformations of their phase composition. The answer which phase components are involved in these and other transformations of the structure of the sample we derived from studies of their composition, structure and morphology. To do this, we used procedures such as infrared absorption (scattering) of light, XRD, and Raman scattering.

The IR vibration spectra of the samples under study are presented in Fig. 14. In the spectrum of the Fe–Cu composite, as well as in the spectra of all other samples, the same set of vibration lines was observed in the range of 1700–3500 cm^{-1}, the maxima of which are located at the same frequencies for all the samples, but intensity changes somewhat from sample to sample (Fig. 14). At the same time, a series of additional lines appears in the spectra of composites, where carbon nanotubes were introduced, in the lower frequencies range: 1400–1700 cm^{-1}) (Fig. 14, curves 2–4). The intensity of these lines, as well as of the lines in the vicinity of 2920, 2950, and 2345 cm^{-1}, increases markedly with the increase of the nanotubes contents in the composite. The appearance of these lines and their concentration behavior confirm that their relation to carbon nanotubes. An analysis of the possible origin of the observed IR spectral lines allowed us to conclude that they can be identified as belonging to different functional molecular groups (carboxylic, $-CO_2H$, carbonyl,

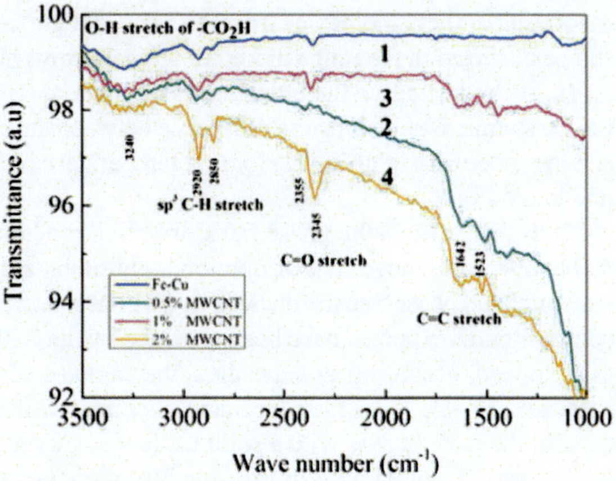

Fig. 14 Infrared vibration spectra of the Fe–Cu (1), Fe–Cu-MWCNT(1.0) (2), Fe–Cu-MWCNT(0.5) (3), and Fe–Cu-MWCNT(2.0) (4) nanocomposites

> CO, hydroxyl, –OH, etc.). Thus, the line at 3240 cm^{-1} is associated with the O–H bond of the CO_2H group [50]. The doublet at 2850 and 2920 cm^{-1} corresponds to the symmetric and antisymmetric valence vibrations of the CH_2 group [51]. The lines at 2345 and 2355 cm^{-1} are a manifestation of the > C=O bonds. Two peaks at 1642 and 1523 cm^{-1} are due to C=C bonds that confirm their origin from carbon nanotubes.

In our opinion, the appearance of the mentioned molecular groups in the content of the Fe–Cu composite should be considered as a consequence of certain participation of the environment atmosphere in the processes of mechano-chemical activation and subsequent heat treatment and rolling of the investigated materials. Increase in the IR absorption intensity caused by such groups in the composites containing carbon nanotubes (Fig. 14, curves 2–4) clearly shows that tubes and products of their destruction can adsorb this type of molecular groups and their radicals from the environment in conditions of high local temperatures, which are realized on the surface of iron and copper grains under milling. It is obvious that under the heating of the Fe–Cu composite in the temperature range of 35–200 °C, desorption of the mentioned molecular groups occurs, that is accompanied by the displacement and shrinkage of the composite. The further increase in temperature, as already shown above, leads to the thermal expansion of the composite, which begins to dominate the effect of shrinkage, and as a result, in the range of temperatures from 350 to 500 °C, the dimension of the sample varies, mainly due to the "pure" thermal expansion.

The shrinkage effect in the region of 35–200 °C is weakened for the Fe–Cu-MWCNT(0.5) composite containing the smallest amount of the MWCNT (Fig. 12, curve 2), and for samples with a higher content of nanotubes, the reduction of samples at these temperatures is not observed at all. This result, in our opinion, is due to the fact that in the presence of a large number of the tubes in the samples, adsorption of the mentioned molecular groups occurs predominantly on carbon nanotubes. The binding energy of such molecules with MWCNT is quite significant and the desorption of molecules is reduced at temperatures below 200 °C, and therefore the associated shrinkage is minimized. (In particular, the binding energy is 0.92–1.72–3.0 eV for the $-CO_2H$, = CO and –OH groups, respectively, according to various data [53–55]. A reduction of the Fe–Cu and Fe–Cu-MWCNT(0.5) samples can result from the desorption of these molecules from the MWCNT surface, but this desorption already can be occurred in the range of higher temperatures: 500–650 °C (Fig. 12, curves 1, 3). If for the last curve the reduction is obvious, then the effect of shrinkage of the Fe–Cu sample is shown by the difference between the experimental curve 1 (Fig. 12) and the model straight line (2) for this temperature range, which is a dashed–dotted line in the Fig. 12.

At the same time, the phase transformation processes are already competitive in this temperature range, in our opinion. Additional arguments regarding the role of structural and phase transformations in the dilatometric behavior of Fe–Cu-MWCNT composites, as expected have been provided by the XRD data (Fig. 15) (see also [31, 41]). Indeed, according to these data, the samples of the iron and copper powders compacts, Fe–Cu, after their mechano-chemical activation are multicomponent and contain the α-Fe phases with a solid Cu (α-Fe, Cu) solution and the composition of which varies from 95 wt% of iron and 5% of copper to 83 wt% of iron and 17% of

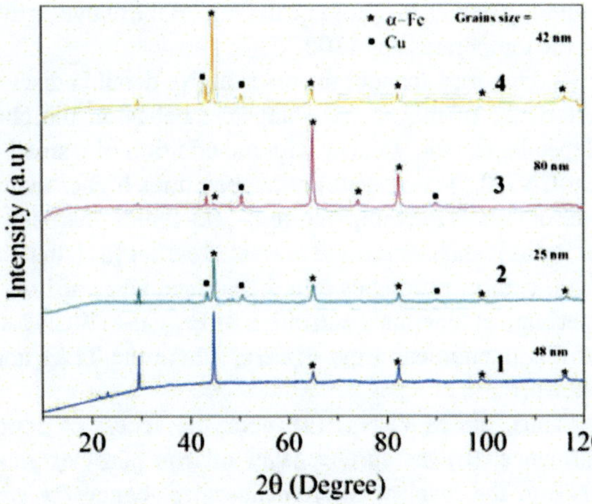

Fig. 15 The XRD patterns of the rolled Fe–Cu (1), Fe–Cu-MWCNT(0.5) (2), Fe–Cu-MWCNT(1.0) (3), and Fe–Cu-MWCNT(2.0) samples (4)

copper phase. Subsequent rolling and annealing of these powdered composites, in which the samples were obtained in the form of tapes, obviously, only intensified the sintering processes. Thus, the XRD pattern of the tape obtained from the above mentioned Fe–Cu compact does not detect the peaks characteristic for the copper grains (Fig. 15, curve 1), but this fact, however, does not mean that the Cu grains are not there. Most likely, a main part of the copper forms a solid solution in the iron, and the size of the remaining copper grains is too small to be identified by the XRD method. This view was confirmed by the fact that for the tapes made from the Fe–Cu-MWCNT compacts the copper peaks on the XRD patterns are reliably observed (Fig. 15, curves 2–4). This fact should be considered as a result of the carbon atoms blocking effect following with the placement of copper in the interstitial space, agglomeration of the particles and their partial sintering. Thus, the copper particles become "visible' in the XRD spectra [31].

Some similarity of the data on the structure, composition and dilatometric characteristics of the samples studied by us, in particular in the range of temperatures from 200 to 800 °C, to the literature data on sintering and temperature influence on the size of such powder composites testify to the similarity of thermal processes [46, 47]. The increase in the content of a copper solid solution in α-Fe with temperature increasing is the most important among them, in our opinion, and on the opinion of other researchers of the Fe–Cu powders, as well [27, 34, 40, 46, 47]. It is this process and the α-Fe → γ-Fe structural transformation provides, as we consider, the rapid increase of the sample length and the increase of CTE in the temperature ranges of 650–750 and 550–700 °C, respectively (see, respectively, Fig. 12, curve 1, and Fig. 13, curve 1). The output of the $\Delta L/L_0(T)$ dependence on saturation at a temperature of ~800 °C and a decrease in the value of the CTE after 700 °C for the Fe–Cu sample should be related to sintering, which can be started from low temperatures for powder compact compositions of iron–copper (~500–530 °C), if compare with bulk iron–copper samples [46, 47]. It is obvious that the reduction of the sample due to sintering begins to dominate the role of the above-mentioned expansion mechanisms at the temperature of ~800 °C.

Taking into account the made above detailed discussion of the dilatometric characteristics of the Fe–Cu composite behavior, the changes in the $\Delta L/L_0(T)$ α (T) dependences those occur with the addition of a small amount of the MWCNT (Fe–Cu-CNT(0.5) composite) at temperatures higher than 350 °C become clearer. (The role of the CNT at temperature <~300–350 °C has been observed above.) We suppose, as it was just observed and evaluated earlier [31], that the penetration of carbon atoms from carbon nanotubes into the surface layer of iron grains is primarily due to the mechanical–chemical activation of Fe–Cu-MWCNT samples. (see as an illustration of this phenomenon the Fig. 16, where the TEM images of some composites are shown.)

Thus, the MWCNT: (a) block the access of copper to iron and neutralize its influence; (b) the surface layer of iron particles/grains is diluted, changing their CTE at the level of micro/nanoscale; change the condition of the α-Fe $\rightarrow$ γ-Fe transformation and the sintering conditions too. The combined effect of these factors leads to the observed temperature behavior of the dilatometric characteristics of the Fe–Cu-MWCNT(0.5) sample. In particular, the low-temperature shift of the peak on

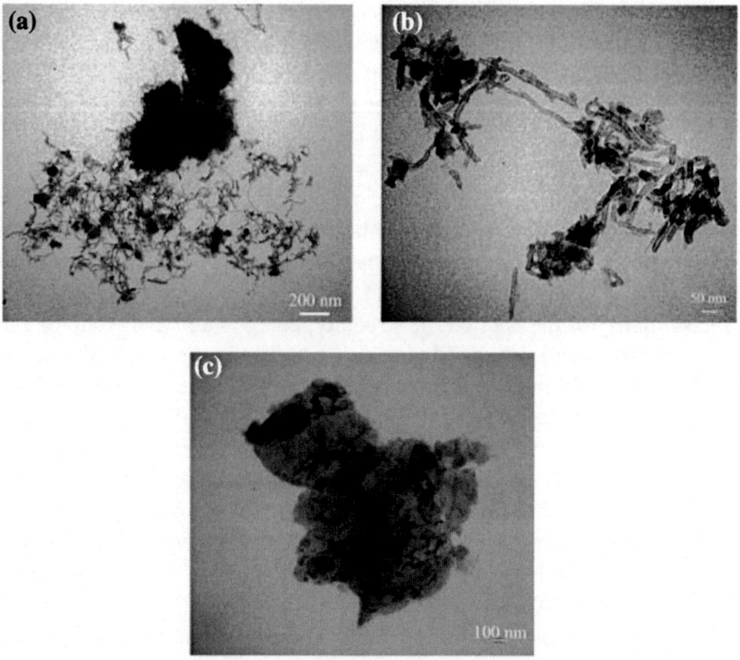

Fig. 16 The TEM images of the Fe–MWCNTs(30.0) (**a**), Fe–Cu-MWCNT(1.0) (b, c) nanocomposites milled for 20 (**b**) and 60 min (a, c) [31, 42, 58]. These images illustrate the penetration of the CNTs into iron particles near surface layer

curves 2 (Figs. 12 and 13) is in accordance with the diagram of Fe–C system states, as it is consistent with the shift to lower values of temperature of the α-Fe → γ-Fe transformation with carbon content increasing [27].

The further increase of the MWCNT content (Fe–Cu-MWCNT(1.0) composite) suppresses the described processes, first of all, as we believe due to the formation of carbides in the surface layers of a large number of iron grains, and leads to a certain separation of the carbides, iron and coppers grains and carbon nanotubes. As a result, the temperature behavior of the $\Delta L/L_0(T)$ and $\alpha(T)$ characteristics for samples Fe–Cu and Fe–Cu-MWCNT(1.0) becomes quite similar.

The incorporation into the composite of twice concentration of carbon nanotubes (sample Fe–Cu-MWCNT(2.0)) enhances, in our opinion, the processes described above lead to the formation of the iron carbide and the products of their decomposition layer between metallic grains. It is known that CNTs in the axial and radial directions have the CTE of opposite sign: $-1.2 \cdot 10^{-5}\ K^{-1}$ and $(1.6–2.6)\cdot 10^{-5}\ K^{-1}$ [56, 57]). This means that in their disordered state, and that is, obviously, characteristic of the nanotubes state in the investigated composites, carbon nanotubes should have a low, or close to zero, CTE. Thus, the above-mentioned MWCNT's layers in the significant amount of the MWCNT (2 vol.%) actually damp an expansion of metal fractions of composites, which is illustrated by the behavior of the curve 4 in Figs. 12 and 13. The results discussed are in agreement with data on some other similar CNT nanomaterials, where their CTEs become practically equal to the CTE of carbon nanotubes [38, 39, 51, 58, 59]. Such a strong decrease in CTE is associated with the formation of iron oxides and the formation of M–O-CNT bonds [60, 61] or with the formation of iron carbides due to carbon tube atoms [55, 62]. Both types of formations have relatively low CTEs at the grains boundaries and thus, they block the possible expansion of the metal matrix of the composite.

Our study of macro-Raman light scattering in near-surface layers of the samples actually revealed a number of lines in the frequency range 150–800 cm^{-1}, which correspond to internal vibrations of various types of iron and copper oxides (see Fig. 17). (In fact, the lines at this spectral range are located on a wide background of the excitation line wing.) The goethite (α-FeO_2H), hematite (α-Fe_2O_3), hydrohematite ($H_2Fe_4O_7$), protogemate (Fe_2O_3), and other its derivatives [63] can be such iron oxides, for example Their phases are unstable to mechanical influences, and therefore their content can considerably vary for composite samples with the same set but with different content of components. The data give in Fig. 16 from the frequency range lower than 500 cm^{-1} are consistent with this assertion. The described results, obviously, confirm the possibility of implementation of the above noted "oxide" mechanism of influence on the temperature behavior of the nanocomposites under study.

As for the "carbide" mechanism of influence, the MWCNT state is important for its implementation, as well as for the activity of all other processes that determine the dilatometric behavior of Fe–Cu-MWCNT nanocomposites. Obviously, the formation of carbides, as well as the procedure of mechanical–chemical activation, is detrimental to CNT. Thus, the loss of carbon atoms that diffuse to grains and then combine with iron forming carbides, at the same time, is the source of defects, and

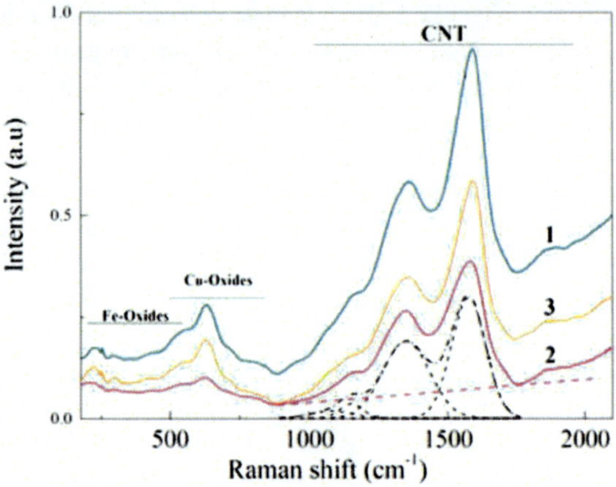

Fig. 17 The macro-Raman spectra of the Fe–Cu-MWCNT(0.5) (1), Fe–Cu-MWCNT(2.0) (2), and Fe–Cu-MWCNT(1.0) (3) nanocomposites

in the process of milling and further processing of the samples, the MWCNT should be destroyed. We think the different types of defected and destroyed MWCNTs fractions have to interact differently with other components of composites, namely to form carbides, to counter agglomeration of metal particles, to influence their size, to change the distribution of grain in size, which, in turn, will affect on the efficiency of the formation of the carbide [31]. The role of the size factor in the dilatometry of composites, including those such as Fe–Cu and Fe–Cu–Carbon, has already been noted earlier [27]. Regarding this factor, we note that the average grain size evaluated by us from the Williamson–Hall analysis of the shape of the lines in the X-ray diffraction spectra for the various studied composites is in the range from 25 to 80 nm (Fig. 15). However, no direct relationship was found between the content of the MWCNT, grain size and dilatometric manifestations. So, the grain size is the smallest for the Fe–Cu-MWCNT(0.5) sample; it is larger (42 nm) for the Fe–Cu-MWCNT(2.0) sample, and the highest on mean grain sizes values (80 nm) are formed in the Fe–Cu-MWCNT(1.0) composite. The absence of a correlation between the contents of the nanotubes and the average grain size once again demonstrates the impossibility of allocating only one factor that would determine the behavior of the dilatometric characteristics of the studied composites. That is, we must assert the cumulative nature of the action of various mechanisms. On the other hand, we see that here, the defect of nanotubes and their destruction are manifested as a very important factor in determining the Fe–Cu-MWCNT composites thermal characteristics.

Valuable information on the structure, morphology, and defectiveness of carbon nanotubes can be obtained from their Raman spectroscopy data. It is known, only one so-called *G* band is observed in the spectra of the ideal CNT structures, due to the vibrations of the double bond C = C in the aromatic cycle of carbon nanotubes (Raman shift ~1580 cm^{-1}, half-width ~10–15 cm^{-1}). In the presence of structural violations (defects, impurities, etc.) the *D* band arises (Raman shift ~1250–1450 cm^{-1}, half-width ~20–200 cm^{-1}). The *D* band is due to the vibration mode associated with boundary effects and it is in Raman—the spectra of any carbon material indicating the presence and number of violations of its structure [67, 68]. The increased intensity and the considerable width of this band may be due to the high content of defective tubes or a large number of shortened nanotubes in which the boundary effects are more pronounced, high concentration of the amorphous phase, etc. The intensity of this band is determined by defect, that is, the degree of violation of sp^2—the hybridization of carbon atoms (e.g., the carbon in the state of sp^3 hybridization). The ratio of the values of the intensity of the bands *D* and *G* (I_D/I_G) to a certain degree characterizes the relationship between the number of materials in the sample with a disordered and ordered structure [69]. The widths of the *D* and *G* bands also increase with structure disorder increasing, and the positions of these bands in a complicated manner depend on the structure and content of the carbon material components [63–68].

The range of the *D* and *G* bands in the macro-Raman spectra of the studied nanocomposites is shown in Fig. 17. It should be noted that the intensity of the *D* and *G* bands on these spectra (indicated by wide arrows) are located on a broad continuous background, which is probably due to the luminescence of functionalized, in particular oxidized, carbon forms [69]. (Its contribution to the Fe–Cu-CNT(1.0) spectrum is separated by a direct dashed line.) The wing of the observed *D* and *G* bands in the range 900–1200 cm^{-1} partially overlap with bands of lower intensity. Up to now, their origin remains under question. It is believed that these bands can be reflected by the vibration of the bonds of the methyl $C–CH_3$ (preferably) and methylene ($-CH_2-$) groups on the CNT surface [63, 67, 69]. At the same time, the mention is that, in particular, the band in the vicinity of 1100–1160 cm^{-1} (half-width is 50–130 cm^{-1}), should be attributed to the manifestation of sp^3 states in amorphous and glassy-like carbon and in other forms containing deformed carbon layers [70, 71]. Therefore, the manifestation of this so-called T band, and characteristics of its contour (intensity, half-width) can also be an indicator of the CNT disordered structure.

Analysis of the macro-Raman spectra of Fe–Cu-CNT composites in the range of frequencies of the CNT vibrations was carried out by removing the contribution of the above-noted background and the breakdown of the remainder to the components of the Gaussian form. The breakdown variant for the Fe–Cu-CNT composite (1.0), e.g., is shown in Fig. 17 and the breakdown parameters for all spectra are given in Table 2. It is easy to see that the basic, in intensity, bands have characteristics of the *D* and *G* bands, and the characteristics of less intensive bands in the vicinity ~960, ~1080 and ~1160 cm^{-1} correspond to the bands of the methyl groups and *T* bands of the CNT, respectively. It should be noted here that usually as I_D and I_G values are either intensity of the bands at their maximum (I_{peak}) or their integral intensity, which is proportional to the area of the spectrum under the corresponding band: I_{Integr}. Therefore, the values of the ratio of intensity, found both in the first, and in the second definition are presented in Table 2.

The manifestation of the bands in the ~960 and ~1080 cm^{-1} vicinity once again points to the role of different types of molecular functional groups in the discussed phenomena of composites expanding in this paper, while the reevaluation of the band

Table 2 Parameters for macro- and micro-Raman spectra breakdown; ν and δ—peak position and half-width, respectively (in cm^{-1})

Sample and area			D	Wide		G	D′		
	ν/δ	ν/δ	ν/δ			ν/δ		I_D/I_G (peak)	I_D/I_G (Integral)
MWCNT Micro-Raman			1350/41	–		1574/36		0.86	0.75
Fe–Cu-CNT (0.5)									
Macro-Raman	965 1086/99 1155/55		1358/204			1583/106	1591	0.58	1.11
Area p1(0.5)			1353/39	1383/356		1585/61		0.45	0.29
Area p2(0.5)		1278/59	1350/48	–	1431/77	1602/42		1.4	1.6
Area p3(0.5)		1300	1359		1432; 1487			1591	
Fe–Cu-CNT (1.0) Macro-Raman	958 1075/148 1148/84		1345/170			1572/127	1583	0.64	0.86
Fe–Cu-CNT (2.0)									
Macro-Raman	967/54 1082/87 1151/59		1355/199			1584/104	1592	0.55	1.06
Area p1(2.0)			1370/97		1466/44	1604/58		1.7	2.9
Area p2(2.0)			1346/80		1449/29	1599/57		0.74	1.39
Area p3(2.0)			1353/47		1496	1588/83			
Area p4(2.0)			1359/470	1383/223		1589/83			

at ~1160 cm^{-1} is a confirmation of the MWCNT destruction. However, the main confirmation of the last fact is the data on the D and G bands characteristics. We see that these characteristics are close to magnitude for the Fe–Cu-MWCNT(0.5) and Fe–Cu-MWCNT(2.0) nanocomposites. As for their values (half-width in the range of 104–204 cm^{-1}, and the ratio of I_D/I_G (Integral)—within the range of 0.86–1.11), they indicate a significant defect and destruction of the MWCNT. We used integral intensity values here since the D and G bands for our samples are a superposition of several components (see below), and therefore, in our opinion, in this case, we should operate precisely with the integral values of intensity. Indeed, the macro-Raman spectra, obviously, reflect the state of MWCNT and Fe–Cu matrix in areas with a cross section up to ~500 microns. This is quite larger in comparison to the size of metal grains and MWCNT sizes. Consequently, the analyzed macro-Raman data are a certain averaging. The temperature behavior of the characteristics of the sample is the result of a set of processes at the micro- and nanoscale levels. In order to highlight this issue, we conducted a Raman scattering study on a microlevel, when the Raman scattering cross section was only ~1 μm. This type of measurement was performed for different areas of the sample under study. The typical micro-Raman spectroscopy data for the areas (points) p1(0.5), p2(0.5), and p3(0.5) (Fe–Cu-CNT(0.5) sample) and for the areas (points) p1(2.0), p2(2.0), p3(2.0), and p4(2.0) (Fe–Cu-CNT(2.0) sample) are shown in Figs. 18 and 19. Quantitative data from this study are presented in the Table 2. The data on micro-Raman spectroscopy of the initial MWCNT are shown on the same Figures and given in Table 2, for comparison.

The main result of these measurements is the conclusion that there is significant heterogeneity in the distribution of MWCNTs on the surface of the samples. This follows, firstly, with a significant Raman intensity difference of the nanotubes spectra measured at different areas (points) of the samples, as well as the different ratio of the intensity of the spectrum associated with MWCNT (spectral range 900–1800 cm^{-1}) and the wing of Rayleigh scattering (range 100–250 cm^{-1}) due to the reflection of light from the metal components of the sample. It is obvious, that the higher values of this ratio correspond to those sample areas where the carbon material content is higher (see curves p1(0.5), p3(2.0), and p4(2.0) in Fig. 18). In addition, the composition of the bands, their shape, the values of width and position of the maxima are also evidence of the diversity of carbon forms in the investigated composites. The detailed view and characteristics of the breakdown for the Gaussian components of bands from the MWCNT-scattering spectral range are the illustration of this statement (Fig. 19 and Table 2).

Comparing the spectra of nanocomposites with the spectrum of initial carbon nanotubes (Fig. 10, MWCNT curve), we see that there are areas of composites (p1(0.5) and p4 (2.0)), where the spectra are quite similar to the MWCNT spectrum. In these areas, as noted above, the contents of the tubes are higher. However, in the case of nanocomposites, the D and G bands have considerably larger half-widths (especially the band D) and, unlike the MWCNT spectrum, they lie on the background of a very wide scattering band (band W in Table 2). In addition, the G band for these areas of the composites is shifted toward higher frequencies by ~10 cm^{-1}.

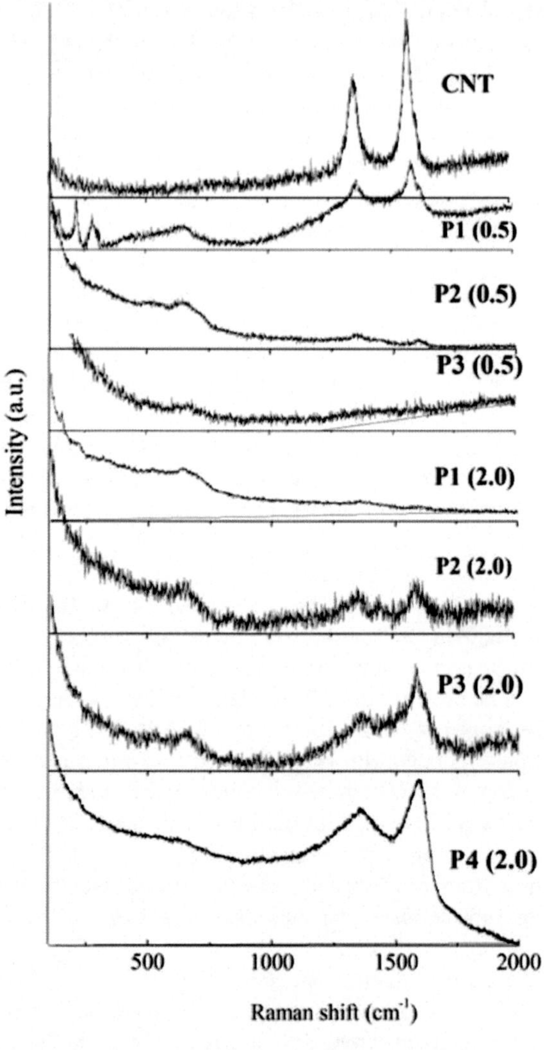

Fig. 18 The micro-Raman spectra of the studied composites monitored at various points (p1(0.5), p2(0.5), p3(0.5)) of the sample Fe–Cu-MWCNT(0.5) and at the points (p1(2.0), p2(2.0), p3((2.0), p4((2.0)) of the sample Fe–Cu-MWCNT (2.0)

The positions of the G bands are slightly different for other areas (Table 2). Consequently, we must state that in the considered areas of composites in addition to the nanotubes there are products of their destruction, which cause the appearance of a wide scattering band. The spectral components related to different carbon forms can be distinguished at the sample areas, where the concentration of tubes is lower. So, we can already see an additional band in the vicinity of 1496 cm^{-1} in the spectrum of the Fe–Cu-MWCNT(2.0) sample obtained on the site p3(2.0). Additional bands are also observed at 1431 and 1449 cm^{-1} for other areas (p2(0.5) and p2(2.0)) (Fig. 8, Table 2). Besides, in the spectra of sites p2 (0.5) and p3 (0.5), the new bands were identified at lower frequencies at 1278 and 1300 cm^{-1}t than they are for the *D* band.

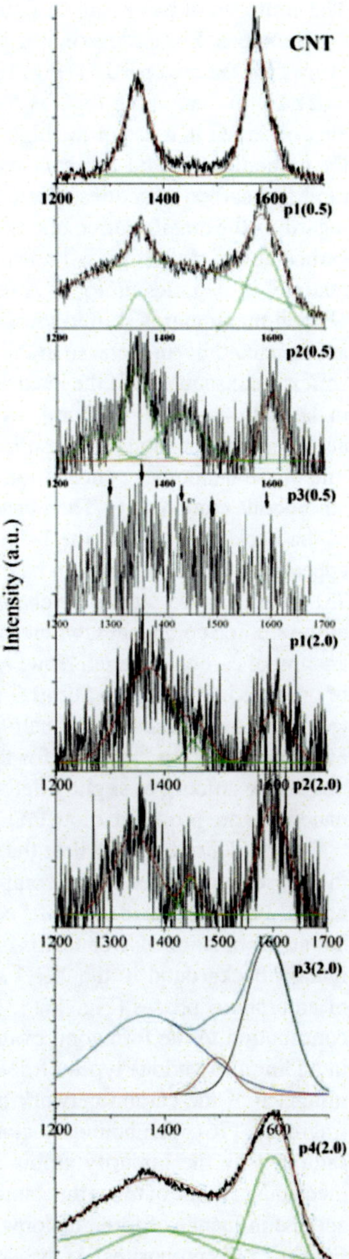

Fig. 19 The detailed view of micro-Raman spectra of the studied composites monitored at various points (p1(0.5), p2(0.5), p3(0.5)) of the sample Fe–Cu-CMWNT(0.5) and at the points (p1(2.0), p2(2.0), p3((2.0), p4((2.0)) of the sample Fe–Cu-MWCNT (2.0)

The entire set of bands can be found on the areas with a significantly lower content of nanotubes. So, the maximum positions of all the above bands are marked with arrows for the area p3(0.5) (Fig. 19).

Thus, we can argue that, as a wide background W band, the *D* and *G* bands, etc., observed in areas of the high MWCNT concentration, in fact, are the result of the numerous spectral components superposition due to the presence of different or modified carbon structures. The described results are consistent with existing ideas regarding the manifestation of defects and structures in the Raman spectra of various carbon forms. Indeed, it is known that at the gradual transition from macroscopic material to particles up to 20 Å (e.g., from mono to nanocrystalline graphite), the *G* band maximum is shifted toward high frequencies: up to 1600 cm^{-1}, and this is accompanied by an increase in the I_D/I_G ratio f to values greater than 1. In part, this shift is a consequence of the intensity increase of another "defective" band, D′, that is in the vicinity ~1610–1620 cm^{-1}. The CNT grinding due to the process of mechanical and chemical activation of composites can be accompanied by an increase of the sub-fine phase nanotubes contribution to the carbon material composition (such as the amorphous component). The *D* and *G* band widening, the appearance and increase of the background band, the I_D/I_G ratio increase up to value more than 1, and the appearance of the T band can be the consequence of such "amorphization" [70, 72, 73]. These manifestations are characteristic for the areas p1(0.5), p3(2.0) and p4(2.0) of the studied composites, whereas it has been explained above, there is a significant amount of carbon material. However, the obtained data show that, on the background of the noted *W* band, the *D* and *G* strips can be located, the characteristics of which (positions of the maxima and half-width) are close to the characteristics of the initial MWCNT (see, e.g., the data for the area p1(0.5)). This observation showed that a noticeable amount of slightly deformed CNT may be in the interstitial space These nanotubes are provided a low CTE for a composite containing 2% MWCNT.

It should also be noted here that, due to the specifics of the milling processes and the subsequent stages of the composite samples preparation, amorphization occurs not only with respect to the "soft" components of these composites, such as MWCNT, but amorphization also concerns of the metal constituents. The presence of a rather intense background in the XRD spectra should be considered as a manifestation of amorphous phases (Fig. 15) [31, 74]. Small-sized crystallites make a dominant contribution to the formation of the *D* band in the Raman spectra [75]. Moreover, in addition to various types of defects (vacancies, doubling lines, etc.), there is an influence of the carbon network breaking at the edges of the nanotubes and their fragments [76]. The boundary atoms have unsaturated valences those can become saturated by the impurity atoms or molecular groups binding [70, 72, 73]. This mechanism is one of those that binds one together "oxide" and "carbide" mechanisms and the influence of foreign atoms and groups on the dilatometric characteristics of Fe–Cu-CNT composites. At the same time, such a linkage significantly complicates dilatometric behavior and the analysis of the results of dilatometric studies.

In addition to defects, disordering of the bonds in the sp^2 rings, the small length of the nanotubes, and the presence of sp^2 chains as a result of the nanotubes destruction can influence the characteristics of Raman bands [74]. The reduction in the length

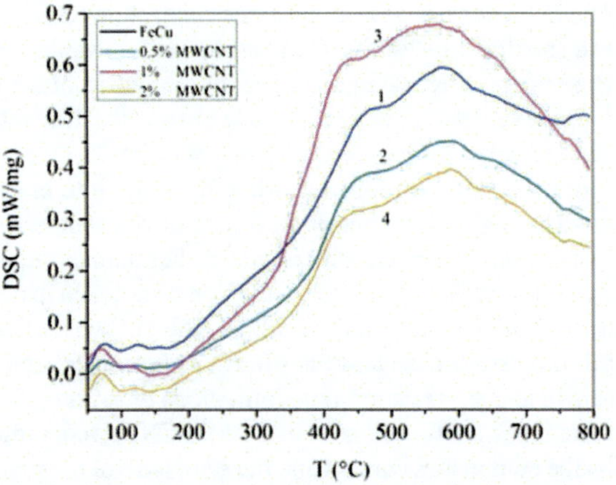

Fig. 20 Differential scanning calorimetry data for the samples Fe–Cu (1), Fe–Cu-MWCNT(0.5) (2), Fe–Cu-MWCNT(1.0) (3), and Fe–Cu-MWCNT(2.0) (4)

of the chain to 25 carbon atoms can also lead to high-frequency shift of the *G* band up to 1600 cm^{-1} [72, 76]. Consequently, we have the right to assume that nanosized MWCNT fragments and conjugate chain structures dominate in the carbon material content on the p2(0.5), p1(2.0), and p2(2.0) areas of the studied composites (see data in Table 2). In general, the diversity of carbon forms in the Fe–Cu-MWCNT composites complicates the temperature dependences of their dilatometric characteristics, reduces their features by "smearing" and enhancing the superposition of the components (see Figs. 12 and 13). The cumulative effect of these mechanisms is manifested in the behavior of other characteristics of Fe–Cu-MWCNT composites, in particular, in their calorimetry.

The measured calorimetric curves are shown on the Fig. 20. We see, that the shapes of the curves describing the variations of the heat capacities with temperature are similar to each other for all the materials under study. Each of them reveals the exothermic band with the peak position of the main component in the range 565–590 °C. The position and the intensity (0.40–0.68 mW/mg) of the band depend on the MWCNT content in the composite. The highest intensity of this band is for the Fe–Cu-MWCNT(1.0) nanocomposite, and therefore the ratio of its intensity, in comparison with other composites, is as follows: $I^{0\%}/I^{1\%} = 87\%$, $I^{0.5\%}/I^{1\%} = 66\%$, and $I^{2\%}/I^{1\%} = 58\%$. In addition to the main intensive peak, you can see three peaks of insignificant intensity in the range of 50–175 °C; the peak lying on the background of the main left of it and the "shoulders" in the ranges of 200–335 and 635–755 °C. The position of all peaks and conditional mid-shoulders, T_{extr}, are given in Table 1. The literature data regarding the DSC of Fe–Cu and Fe–Cu–Carbon composites are rather limited, and in relation to Fe–Cu-MWCNT composites, such data are completely unknown. Nevertheless, we can see that both temperature dependences and their numerical characteristics, in spite of the significantly different composition and conditions for obtaining composites from the described in the literature, to some extent, are similar to those already published [27, 74, 75]. In particular, the heat release reached the maximum (~2.25 mW/mg) at ~500 °C, for the Fe–Cu composite of 50:50 composition [24].

The DSC dependence, at the same time, showed another blurred peak near 700 °C and shoulder at ~200 °C. Comparing the T_{extr} values with the data about the temperature ranges selected by us above for estimation of the thermal expansion coefficient average values $<\alpha_i(T)>$, it is easy to see that the T_{extr} values correspond to the temperature ranges in which there are changes in the $<\alpha_i(T)>$ values. This result confirms the above made conclusions regarding the main role in the dilatometric manifestations of changing their composition (range of low temperatures up to 200 °C) and phase transformations of the structure at higher temperatures. The origin of the described wide band is directly related to the iron component in the composite matrix. This band is quite wide, and its half-width is ~350 °C, which lies from 400 to 750 °C. Within this temperature range, iron is in α-Fe allotropic form. However, as we have already shown above, structural transformations of α-Fe → γ-Fe occur in the same temperature range for Fe–Cu and Fe–Cu-MWCNT nanocomposites and this range depends on the composites content and the state (size of the particles, etc.) and the interaction of their components, which is reflected in the position and intensity of the main peak.

The position of other high-temperature features (T > 200 °C) on the DSC curves also correspond to the structural transformations discussed above and to the role of carbon material (starting MWCNT) in the Fe–Cu-MWCNT composites thermal characteristics determination. So, we have already accentuated, above, on the role of particle size and grain components of investigated composites in their dilatometry. In addition, we note that the small size and homogeneity of particles and grains contribute to the increase of surface energy and energy of interface layers, which are the main sources of storage energy accumulated under milling. The release of this energy during heating, as our data show, should increase the activity of composite components by increasing the speed and ways of their atoms transport and, thus, affecting the processes of phase transformation, sintering, etc. Here, we must add that the influence of carbon forms on the size of the structural components and on the oxidation processes of the metallic components of composites is also important. As it had been shown earlier, the kinetics of oxidation on the stage of such composites heating depends on the allotropic carbon and their structure, and the formation of an amorphous form with a disordered defective structure weakens such reactions [30]. The various carbon forms and their structure appear at peaks of varying position and intensity on the DSC curves [30]. Consequently, the carbon component, in addition to the "blocking" role described above, through its influence on oxidation processes determines the relation between the "carbide" and "oxide" mechanisms of influence on the temperature behavior of the characteristics of the investigated Fe–Cu-MWCNT composites. The cumulative effect of all processes described in this work is reflected, in particular, in the temperature dependences described herein for dilatometric and calorimetric characteristics.

5 Conclusions

Dilatometric characteristics (relative expansion, $\Delta L/L_0(T)$, and linear thermal expansion coefficient, CTE or $\alpha(T)$), and calorimetric data of the concentration set of composite materials Fe–Cu-MWCNT (concentration of the MWCNT = 0, 0.5, 1.0, and 2.0 vol.%; Fe: Cu ratio was 4:1) were discussed together with data on the composition, structure and morphology of the samples, for that the studies of their X-ray diffraction, Raman spectroscopy, IR absorption, optical and electron microscopy were carried out.

Temperature dependences of dilatometric characteristics revealed a number of specific components that can not be accurately interpreted up to now due to their significant superposition. Several temperature ranges were distinguished for the Fe–Cu composite by the character of the $\Delta L/L_0(T)$ and $\alpha(T)$ dependences. The average values of the coefficient of thermal expansion $<\alpha_i(T)>$ calculated for each of the temperature ranges were of ~4–6 times smaller than the known literature data for Fe–Cu and Fe–Cu-CNT composites.

The addition of the MWCNTs to the Fe–Cu composite results in changes in the $\Delta L/L_0(T)$ and $\alpha(T)$ dependencies), and the effect of MWCNTs being different for different temperature ranges; moreover, this effect depends on the content of the nanotubes. This is evidence of the manifestation of various mechanisms of thermal expansion of the investigated composites and, at the same time, of the various effects of carbon nanotubes on these mechanisms.

Temperature changes of crystalline lattice constants; multivariate interaction of components; transformation of the lattice structure; phase transformation of components; reduction/shrinkage of the sample due to the temperature-activated release of adsorbed gases and gases of adsorbed compounds formed as a result of the thermal decomposition; sintering; and, conversely, the disintegration of grains and particles, as well as the destruction of MWCNTs, were discussed as possible mechanisms of thermal expansion of the studied composites.

In particular, it was shown that mechanism of "pure" thermal expansion of Fe–Cu composite dominates in the temperatures range 375–550 °C, if to approximate the experimental dependence of $\Delta L/L_0(T)$ on Fe–Cu composites with the assumption of its linear expansion and to use the calculated $<\alpha_i(T)>$ values.

The processes related to the change in composition and structure of composites can be conventionally divided into two groups, such as "oxidation" and "carbidization," and carbon atoms of nanotubes play an important role both for each of them and in ensuring their interconnection. Thus, the analysis of the IR and Raman spectroscopy data allowed us to reveal carboxyl, carbonyl, and hydroxyl groups and the appearance of which in the samples is due to the impossibility of avoiding contact with the atmosphere in the processes of mechano-chemical activation and subsequent heat treatment and rolling of composites. In the presence of nanotubes in the samples, the number of molecular groups increases due to their adsorption by tubes and products of their destruction. The mentioned molecular groups desorption occurs during heating, which is accompanied by displacement and shrinkage of the composite grains. The temperature ranges from 35 to 75 and 135 to 200 °C for the Fe–Cu specimens, a range of 50–150 °C for the Fe–Cu-MWCNT(0.5), and a range of 500–625 °C for the Fe–Cu-MWCNT(1.0) are highlighted by us as such, where there may be a significant manifestation of shrinkage.

Both structural and phase transformations determine the dilatometric behavior of the Fe–Cu-MWCNT composites in the temperature range above 500 °C. The main ones among them at higher concentrations of CNT are increase of the copper solid solution content in α-Fe and increase in the content of γ-Fe due to the α-Fe → γ-Fe phase transformation with temperature increasing; the formation of iron oxides following formation the bonds such as M–O-CNT and the formation of carbides in the surface layer of iron grains.

The output of the $\Delta L/L_0(T)$ dependences at the maximum when the temperature reaches ≥800 °C and the decrease of the CTE value at $T \geq 650$ °C must be associated with sintering, which is known to begin for iron–copper–carbon powder compacts at temperatures of ~500–530 °C.

Increasing of the MWCNT content up to 2% suppresses the effects of thermal expansion and radically (two orders of magnitude) reduces the amount of CTE. This fact should be considered as a consequence of the cumulative effect of the formation of the carbide in the surface layer of iron grains, the blocking of copper to iron particles incorporation by carbon atoms and the leveling of its influence, the placement of copper in the intergranular space, changes in the α-Fe → γ-Fe transformation processes, separation of iron, carbides, and carbon nanotubes fractions, and the formation of MWCNTs layer of and the products of their decomposition between metallic grains and changes in the sintering conditions.

Milling and further processing of samples lead to the destruction of the MWCNTs, and the loss of carbon atoms in the formation of oxides and carbides is a source of the MWCNT defects that was confirmed by the data of the macro- and micro-Raman scattering.

Significant heterogeneity of the of carbon material distribution over the samples and the dispersion of the particle size in the range of 25–80 nm are the sources of the variety of MWCNT's influence on the above-mentioned processes of interactions and transformations.

The results obtained in this paper, in our opinion, are very interesting on the scientific side and are an important contribution to nanotechnology. They determine the important role of MWCNTs in the formation of Fe–Cu metallic nanocomposites and confirmed that mechano-chemical synthesis is an effective method for the development of new types of nanoscaled composite materials.

Acknowledgments The work was financially supported by the State budgets of Ukraine and Algeria via Ministry of Education and Science of Ukraine and Ministry of Higher Education and Scientific Research of Algeria, respectively.

References

1. Ying Yu C, Rainer S, Austin CY (1984) Thermodynamic analysis of the iron-copper system I: the stable and metastable phase equilibria. MTA A 15:1921–1930, https://doi.org/10.1007/BF02664905
2. Mazzone G, Antisari MV (1996) Structural and magnetic properties of metastable fcc Cu–Fe alloys. Phys Rev B 54:441–446. https://doi.org/10.1103/PhysRevB.54.441
3. Sumiyama K, Yoshitake T, Nakamura Y (1984) Magnetic properties of metastable bcc and fcc Fe–Cu alloys produced by vapor quenching. J Phys Soc Jpn 53(9):3160–3165. https://doi.org/10.1143/JPSJ.53.3160
4. Ma E, Atzmon M, Pinkerton FE (1993) Thermodynamic and magnetic properties of metastable Fe_xCu_{100-x} solid solutions formed by mechanical alloying. J Appl Phys 74:955–962. https://doi.org/10.1063/1.354837
5. Huang X, Mashimo T (1999) Metastable bcc and fcc alloy bulk bodies in Fe–Cu system prepared by mechanical alloying and shock compression. J Alloy Compd 288:299–305. https://doi.org/10.1016/S0925-8388(99)00108-5
6. Liu JZ, van de Walle A, Ghosh G, Asta M (2005) Structure, energetics, and mechanical stability of Fe–Cu bcc alloys from first-principles calculations. Phys Rev B 72:144109. https://doi.org/10.1103/PhysRevB.72.144109
7. He J, Zhao JZ, Ratke L (2006) Solidification microstructure and dynamics of metastable phase transformation in undercooled liquid Cu–Fe alloys. Acta Mater 54:1749–1757. https://doi.org/10.1016/j.actamat.2005.12.023
8. Alami AH, Abed J, Almheiri M, Alketbi A, Aokal C (2016) Fe–Cu metastable material as a mesoporous layer for dye-sensitized solar cells. Energy Sci Eng 4:166–179. https://doi.org/10.1002/ese3.114
9. Ravi C, Wolverton C, Ozolin V (2006) Predicting metastable phase boundaries in Al–Cu alloys from first-principles calculations of free energies: the role of atomic vibrations. Europhys Lett 73:719. https://doi.org/10.1209/epl/i2005-10462-x
10. Das N, Mittra J, Murty B, Pabi S, Kulkarni U, Dey G (2013) Miedema model based methodology to predict amorphous-forming-composition range in binary and ternary systems. J Alloy Compd 550:483–495. https://doi.org/10.1016/j.jallcom.2012.10.124
11. Suryanarayana C (ed) (1999) Non-equilibrium processing of materials. Pergamon, Oxford, UK
12. Suryanarayana C (2001) Mechanical alloying and milling. Progr Mater Sci 46:1–184. https://doi.org/10.1016/S0079-6425(99)00010-9
13. Suryanarayana C, Al-Aqeeli N (2013) Mechanically alloyed nanocomposites. Progr. Mater Sci 58:383–502. https://doi.org/10.1016/j.pmatsci.2012.10.001
14. Hamzaoui R, Elkedim O, Gaffet E (2004) Milling Condition effect on structure and magnetic properties for mechanically alloyed nanocrystalline Fe—10% Ni and Fe—20% Ni alloys. Mater Sci Eng, A 381:363–371. https://doi.org/10.1016/j.msea.2004.05.008
15. Le Brun P, Gaffet E, Froyen L, Delaey L (1992) Structure and properties of Cu, Ni and Fe powders milled in planetary ball mill. Scr Metall Mater 26:1743–1748. https://doi.org/10.1016/0956-716X(92)90545-P
16. Raviathul BM, Roy RK, Pramanick AK, Srivastava VC, ukhopadhyay NK (2016) Inverse Hall-Petch like behavior in a mechanically milled nanocrystalline Al_5Fe_2 intermetallic phase. Philosophical. Magazine 96:2445–2456, https://doi.org/10.1080/14786435.2016.1204474
17. Di Maggio R, Ischia G, Rossi F, Molinari A, Bortolotti M (2007) The microstructure and mechanical properties of Fe–Cu materials fabricated by pressure-less-shaping of nanocrystalline powders. J Mater Sci 42:9284–9292. https://doi.org/10.1007/s10853-007-1892-3
18. Kakisawa H, Minagawa K, Halada K (2003) Tensil behavior change depending on the microstructure of a Fe–Cu alloy produced from rapidly solidified powder. Mater Sci Eng A 340:175–180. https://doi.org/10.1016/S0921-5093(02)00171-5
19. Ma E (2005) Alloys created between immiscible elements. Prog Mater Sci 50:413–509. https://doi.org/10.1016/j.pmatsci.2004.07.001

20. Bachmaier M, Kerber D, Setman R (2012) Pippan, The formation of supersaturated solid solutions in Fe–Cu alloys deformed by high-pressure torsion. Acta Mater 60:860–871. https://doi.org/10.1016/j.actamat.2011.10.044
21. Faudot F, Gaffet E, Harmelin M (1993) Identification by DSC and DTA of the oxygen and carbon contamination due to the use of ethanol during mechanical alloying of Cu–Fe powder. J Mater Sci 28:2669–2676
22. Jiang JZ, Gente C, Bormann R (1998) Mechanical alloying in the Fe–Cu system. Mat Sci Eng A 242:268–277. https://doi.org/10.1016/S0921-5093(97)00522-4
23. Uenishi K, Kobayashi KF, Nash S (1992) Mechanical alloying in the Fe–Cu system. Z für Metallkunde 83:132–135
24. Sun J, Wang M, Li X, He Z, Zhao Y (2007) Mechanical alloying influence on the sintering of Cu–Fe compound powders. Key Eng Mater 353–358:1350–1353. https://doi.org/10.4028/www.scientific.net/KEM.353-358.1350
25. Lindsley B, Fillari G, Murphy T (2005) Effect of composition and cooling rate on physical properties and microstructure of prealloyed P/M Steels. Advances in Powder Metallurgy & Particulate Materials, compiled by C. Ruas and T. A. Tomlin, Metal Powder Industries Federation, Princeton, NJ, part 10, p 10–353
26. Eremina MA, Lomaeva SF, Elsukov EP, Ulyanov AL, Chulkina AA (2013) The Cu–Fe_3C nanocomposites mechano-synthesis using liquid hydrocarbon. Chem Steady-state Dev 21:639–646 (in Russian)
27. Semel FJ (2001) Processes determining the dimensional change of PM Steels, Advances in Powder Metallurgy & Particulate Materials, compiled by W. G. Eisen and S. Kassam. Metal Powder Industries Federation, Princeton, NJ part 5:5–113
28. Saji S, Kadokura T, Anada H, Notoya K, Takanoю N (1998) Solid solubility of carbon in copper during mechanical alloying. Mater Trans 39:778–781
29. Liu X, Liu Y, Ran X, An J, Cao Z (2007) Fabrication of the supersaturated solid solution of carbon in copper by mechanical alloying. Mater Charact 58:504–508. https://doi.org/10.1016/j.matchar.2006.06.022
30. Larionova NS, Nikonova RM, Ladyanov VI (2018) Mechanosynthesis of nanostructured composites copper-fullerite, copper-graphite. Adv Powder Technol 29:399–406. https://doi.org/10.1016/j.apt.2017.11.027
31. Boshko O, Dashevskyi M, Mykhaliuk O, Ivanenko K, Hamamda S, Revo S (2016) Structure and strength of Iron–Copper carbon nanotube nanocomposites. Nanoscale Res Lett 11:78. https://doi.org/10.1186/s11671-016-1298-8
32. Revo SL, Boshko OI, Dashevskyi MM, Ivanenko KO (2016) Structural relaxation of the Iron–Copper–Carbon nanotubes materials after mechanochemical activation. Nanosyst Nanomater Nanotechnol Ukraine 14:169–180
33. Revo SL, Melnichenko MM, Dashevskyi MM, Belyavina NN, Nakonechna OI, Ivanenko KO, Boshko OI, Avramenko TG (2017) Structure features, strength, and Microhardness of Nanocomposites obtained from Fe, Cu, and Carbon Nanotubes, O. Fesenko, L. Yatsenko (eds.), Nanophysics, Nanomaterials, Interface Studies, and Applications, Springer Proceedings in Physics, 195 (2017). Springer, p 799. https://doi.org/10.1007/978-3-319-56422-7_613
34. Lawcock RL, Davies TJ (1990) Effect of carbon on dimensional and microstructural characteristics of Fe–Cu compacts during sintering. Powder Metall 33:147–150. https://doi.org/10.1179/pom.1990.33.2.147
35. Shull RD, Cline JP, Baker I, Liu F (1996) Identification of a high-temperature magnetic phase transition in ball-milled and compacted nanocrystalline Fe–Cu alloys. J Appl Phys 79:6028–6030
36. Kasperski A (2013) Nanocomposites nanotubes de carbone—céramique à microstructure contrôlée: préparation et propriétés. Université de Toulouse, Thèse de doctorat
37. Lozovyi F, Ivanenko K, Nedilko S, Revo S, Hamamda S (2016) Thermal analysis of polyethylene + X% carbon nanotubes. Nanoscale Res Lett 11:97. https://doi.org/10.1186/s11671-016-1315-y

38. Revo S, Hamamda S, Ivanenko K, Boshko O, Djarri A, Boubertakh A (2015) Thermal analysis of Al + 0.1% CNT ribbon, Nanoscale Res Lett 10:170. https://doi.org/10.1186/s11671-015-0878-3
39. Hamamda S, Jari A, Revo S, Ivanenko K, Jari Y, Avramenko T (2017) Thermal analysis of copper-titanium-multiwall carbon nanotube composites. Nanoscale Res Lett 12:251. https://doi.org/10.1186/s11671-017-2025-9
40. Trudel Y, Angers R (1975) Properties of iron copper alloys made from elemental or prealloyed powders. Int J Powder Metal Powder Technol 11:5–16
41. Boshko O, Nakonechna O, Dashevsky M, Ivanenko R, Belyavina N, Revo S (2016) Effect of the carbon nanotubes on structure and magnetic properties of the Fe–Cu (4:1) composites. Adv Powder Technol 27:1101–1108. https://doi.org/10.1016/j.apt.2016.03.019
42. Kovalska EO, Sementsov YuI, Prykhodko HP, Kartel' MT (2012) Process for the preparation of catalysts for chemical deposition of carbon nanotubes from gas phase (in Ukrainian), (2012) Patent # 70847 Ukraine, C01B11/00 D01F9/00
43. Melezhyk AV, Yanchenko VV, Sementsov YuI (2007, Nanocarbon materials, NATO security through science series a: chemistry and biology, 529–537
44. Sementsov YuI, Gavrilyuk NA, Prikhod'ko GP, Melezhyk AV, Pyatkovsky ML, Yanchenko VV, Revo SL, Ivanenko EA, Senkevich AI (2007) Properties of PTFE-MWNT composite materials, NATO Security through Science Series A: Chemistry and Biology, 757–763
45. Chagnon F, Gagne M (2001) Dimensional control of sinter hardened P/M components. Advances in Powder Metallurgy & Particulate Materials, compiled by W. G. Eisen and S. Kassam, Metal Powder Industries Federation, Princeton, NJ. 5:5–31
46. Ohashi T, Tsutsui Sh (2000) Solid phase sintering of mechanically alloyed Fe–Cu powder alloy. J Jpn Soc Powder Powder Metall 47:474–479. https://doi.org/10.2497/jjspm.47.474
47. Kin YCh (2005) The effect of Cu addition method on dimensional change of Fe–Cu compacts. J Jpn Soc Powder Powder Metall 52:89–96. https://doi.org/10.2497/jjspm.52.89
48. Baran MC, Graham AH, Davala AB, Causton RJ, Schade C (1999) A superior sinter-hardenable materials, advances in powder metallurgy & particulate materials, compiled by C. L. Rose and M. H. Thibodeau, Metal Powder Industries Federation, Princeton, NJ. 7:185
49. Novikova CI (1974) Coefficient of thermal expansion of solids cores, Nauka Edition (in Russian)
50. Osswald S, Havel M, Gogotsi Y (2007) Monitoring oxidation of multiwalled carbon nanotubes by Raman spectroscopy. J Raman Spectrosc 38:728–736. https://doi.org/10.1002/jrs.1686
51. Kouklin N, Tzolov M, Straus D, Yin A, Xu JM (2004) Infrared absorption properties of carbon nanotubes synthesized by chemical vapor deposition. Appl Phys Lett 85:4463–4465. https://doi.org/10.1063/1.1812837
52. Veloso MV, Souza Filho AG, Mendes Filho J, Fagan SB, Mota R (2006) Ab initio study of covalently functionalized carbon nanotubes. Chem Phys Lett 430:71–74. https://doi.org/10.1016/j.cplett.2006.08.082
53. Park Hyoungki, Zhao Jijun, Jianping Lu (2005) Distinct properties of single-wall carbon nanotubes with monovalent sidewall additions. Nanotechnology 16:635–638. https://doi.org/10.1088/0957-4484/16/6/003
54. Wang C, Zhou G, Liu H, Wu J, Qiu Y, Lin Gu B, Duan W (2006) Chemical functionalization of carbon nanotubes by carboxyl groups on stone-wales defects: a density functional theory study. J Phys Chem B 110:10266–10271. https://doi.org/10.1021/jp060412f
55. Boshko O, Nakonechna O, Belyavina N, Dashevskyi M, Revo S (2017) Nanocrystalline Fe–C composites obtained by mechanical alloying of iron and carbon nanotubes. Adv Powder Technol 28:964–972. https://doi.org/10.1016/j.apt.2016.12.026
56. Shirasu Keiichi, Yamamoto Go, Tamaki Itaru, Ogasawara Toshio, Shimamura Yoshinobu, Inoue Yoku, Hashida Toshiyuki (2015) Negative axial thermal expansion coefficient of carbon nanotubes: experimental determination based on measurements of coefficient of thermal expansion for aligned carbon nanotube reinforced epoxy composites. Carbon 95:904–909. https://doi.org/10.1016/j.carbon.2015.09.026

57. Maniwa Y, Fujiwara R, Kira H, Tou H, Nishibori E, Takata M et al (2001) Multiwalled carbon nanotubes grown in hydrogen atmosphere: an x-ray diffraction study. Phys Rev B 64:073105. https://doi.org/10.1103/PhysRevB.64.073105
58. Alamusi N, Hu N, Jia B, Arai M, Yan C, Li J, Liu Y, Atobe S, Fukunaga H (2012) Prediction of thermal expansion properties of carbon nanotubes using molecular dynamics simulations. Comp Mat Sci 54:249–254
59. Shirasu K, Nakamura A, Yamamoto G, Ogasawara T, Shimamura Y, Inoue Y, Hashida T (2017) Potential use of CNTs for production of zero thermal expansion coefficient composite materials: an experimental evaluation of axial thermal expansion coefficient of CNTs using a combination of thermal expansion and uniaxial tensile tests. Composites: Part A 95:152–160. https://doi.org/10.1016/j.compositesa.2016.12.027
60. Bittencourt C, Hecq M, Felten A, Pireaux JJ, Ghijsen J, Felicissimo MP, Rudolf P, Drube W, Ke X, Van Tendeloo G (2008) Platinium-carbon nanotubes interaction. Chem Phys Lett 462:260–264. https://doi.org/10.1016/j.cplett.2008.07.082
61. Felten A, Ghijsen J, Pireaux JJ, Drube W, Johnson R, Liang D, Hecq M, Van Tendeloo G, Bittencourt C (2009) Electronic structure of Pd nanoparticles on carbon nanotubes Micron. Micron. 40:74–79. https://doi.org/10.1016/j.micron.2008.01.013
62. Ponosov YuS, Uymin MA, Ermakov AE, Scheholeva NN, Mysyk AA (2013) Raman scattering of light and elecnron microscopy of nanocomposites with metal nucleous-carbon shell structure (in Russian). Fiz Tverdogo tela 55:1425–1432 (in Russian)
63. Alekseev AD, Ulyanov EB, Trachevskii VV, Ivaschuk LI, Zymina SV (2010) Application of NMR and Raman spectroscopy for study of natural nature carbon nanomaterials genesys. Physica Techn High Pressure 20:126–139
64. Keszler AM, Nemes L, Ahmad SR, Fang X (2004) Characterisation of carbon nanotube materials by Raman spectroscopy and microscopy—A case study of multiwalled and singlewalled samples. J Optoelec Adv Mater 6:1269–1274
65. Delhaes P, Couzi M, Trinquecoste M, Dentzer J, Hamidou H, Vix-Guterl C (2006) A comparison between Raman spectroscopy and surface characterizations of multiwall carbon nanotubes. Carbon 44:3005–3013. https://doi.org/10.1016/j.carbon.2006.05.021
66. Crossley BL, Glauvitz NE, Quinton BT, Coutu RA, Collins PJ (2011) Carbon nanotubes applications on electron devices, Prof. Jose Mauricio Marulanda (Ed.), ISBN: 978-953-307-496-2, In Tech
67. Trovati S et al (2006) Human exposure to space radiation: role of primary and secondary particles. Radiat Prot Dosim 122:362–366. https://doi.org/10.1093/rpd/ncl438
68. Li X, Hayashi JI, Li CZ (2006) FT-Raman spectroscopic study of the evolution of char structure during the pyrolysis of a Victorian brown coal. Fuel 85:1700–1707. https://doi.org/10.1016/j.fuel.2006.03.008
69. Strauss V, Roth A, Sekita M, Guldi DM (2016) Efficient energy-conversion materials for the future: understanding and tailoring charge-transfer processes in carbon nanostructures. Chem 1:531–556. https://doi.org/10.1016/j.chempr.2016.09.001
70. Ferrari RO, Robertson J (2001) Resonant Raman spectroscopy of disordered, amorphous, and diamondlike carbon. Phys Rev B 64:075414. https://doi.org/10.1103/PhysRevB.64.075414
71. Tan PH, Dimovski S, Gogotsi Y (2004) Raman scattering of non-planar graphite: arched edges, polyhedral crystals, whiskers and cones. Phil Trans Royal Soc Lond A 362:2289–2310. https://doi.org/10.1098/rsta.2004.1442
72. Ferrari AC, Robertson J (2000) Interpretation of Raman spectra of disordered and amorphous carbon. Phys Rev B 61:14095. https://doi.org/10.1103/PhysRevB.61.14095
73. Bukalov SS, Mikhalitsyn LA, Zubavitchus YaV et al (2006) Study of the graphite and some other sp^2 carbon naterials by means of micro Raman spectroscopy and X-Ray diffraction (in Russian). Russian Chem J 1:83–91
74. Guérault H (2000) Propériétés structurales et magnétiques de poudres de fluorures nanostructurées MF_3 (M = Fe, Ga). Université du Maine, Thèse de doctorat

75. Cuesta A, Dhamelincourt P, Laureyns J, Martínez-Alonso A, Tascón JMD (1998) Comparative performance of X-ray diffraction and Raman microprobe techniques for the study of carbon materials. J Mater Chem 8:2875–2879. https://doi.org/10.1039/A805841E
76. Carry P (1985) Raman and Resonance Raman scattering application in biochemistry, (in Russian)—Moscow: Mir, 272

Influence of Nanofillers Concentration on Physical and Mechanical Characteristics of Their Polymer Composites

K. Ivanenko, L. M. Ushakova, T. Avramenko, S. Revo, M. T. Kartel, and Yu. I. Sementsov

1 Introduction

Physical and mechanical characteristics (strength, creep, toughness, etc.) are among the most important characteristics of the material that determine its performance regardless of functional purpose.

Due to the combination of mechanical, thermal, and electrophysical properties that are inherent in carbon nanotubes (CNTs), there is a constant expansion of the spectrum not only of their application but also of the possibilities of use for fundamental research.

The complex physical properties of nanotubes make them ideal fillers for polymer composite materials [1–10]. Today, it is possible to significantly increase the thermal conductivity of matrices and obtain leading polymers with a low concentration of MWCNTs. Such materials are used to create conductive elements, battery electrodes, sensors, protective screens, antistatic, and anticorrosive coatings and the like. At the same time, mechanical properties, namely: the average modulus of elasticity of MWCNTs is more than 1.8 TPa, (measured value—1.3 TPa) [11], tensile strength—63 GPa [12] put them among the most promising reinforcing fillers in the creation of composite materials with increased mechanical characteristics.

However, the tendency of CNTs to form agglomerates due to van der Waals forces (0.5 eV/nm) is a limiting factor in realizing the potential of these materials [1, 4–12]. For effective use of CNTs, it is important to ensure high homogeneity of their

K. Ivanenko · T. Avramenko · S. Revo (✉)
Research Laboratory of Physics of Metals and Ceramics, Taras Shevchenko National University of Kyiv, 64/13, Volodymyrska Street, Kyiv 01601, Ukraine
e-mail: s_revo@i.ua

L. M. Ushakova · M. T. Kartel · Yu. I. Sementsov
Department of Physics and Chemistry of Carbon Nanomaterials, O. Chuiko Institute of Surface Chemistry National Academy of Sciences of Ukraine, 17, General Naumov Street, Kyiv 03164, Ukraine

O. Fesenko and L. Yatsenko (eds.), *Nanomaterials and Nanocomposites, Nanostructure Surfaces, and Their Applications*, Springer Proceedings in Physics 246, https://doi.org/10.1007/978-3-030-51905-6_46

distribution in the polymer matrix, which is relevant to the efforts of many researchers [1–4, 7–10]. The potential for CNT-filled polymers is much higher because they have higher mechanical characteristics than unfilled material.

In this paper, the regularities of the influence of carbon nanotubes are considered and another nanoscale system of pyrogenic silica A-300 on the electrical and mechanical characteristics of a composite material based on polymers: polyethylene (PE), polypropylene (PP), and polytetrafluoroethylene (PTFE) is selected for comparison.

2 Methods

Low-density PE and FP as matrixes and MWCNTs and ground EG as fillers were chosen for preparation of NCM samples for testing. The multi-walled carbon nanotubes (Fig. 1) (TU U 24.1-03291669-009:2009 (ISC NAS of Ukraine)) were synthesized using chemical vapour deposition procedure in a rotating reactor. Characteristics of the produced MWCNTs were: average diameter—between 10 and 20 nm; specific surface area (determined through Ar—adsorption)—between 200 and 400 m^2/g; bulk density—between 20 and 40 g/dm^3 [13, 14].

The MWCNTs synthesized by CVD method are obtained in the form of agglomerates. Therefore, the powders of the initial components (PE-MWCNT) of the composite material were mechanically mixed in a drum (1, Fig. 2) with a speed of 120 min^{-1}. The drum has membranes with holes (2), forcing the mixture to move along a complex trajectory. The drum is placed in a revolver-type cassette (3), in which four mixtures can be mixed simultaneously in the same mode. The cassette is driven by a motor (4). Mixing of mixtures lasts 4 h. At this time, mixing—the mixture is completely isotropic.

After mixing, the mixtures were placed to a vacuum mold, where they were heated to a temperature (377 ± 5) K and kept at a pressure of ~30 MPa for 20 min, after which they were cooled without removing the pressure. The heating time was 60 min the total time was 140 min.

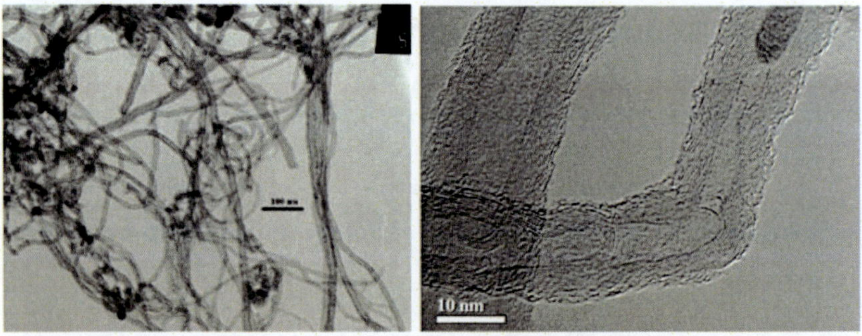

Fig. 1 TEM images of MWCNTs obtained on Fe–Al–Mo–O catalyst

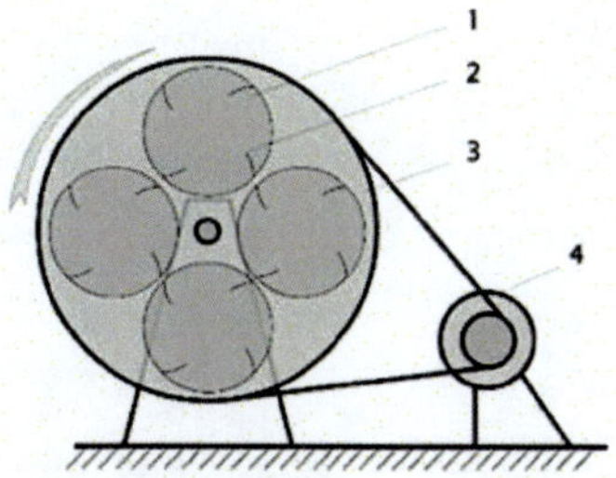

Fig. 2 Rotary stirrer revolver type for four mixture. 1—drum, 2—membrane, 3—cassette for drums, 4—engine

Polypropylene composites with MWCNT were obtained in the form of granules by mixing in a twin-screw extruder, from which samples were also made in the form of films and volumetric cylinders. The concentration of nanotubes was 0.1, 0.5, 1.0, 3.0, and 5.0% wt. [15].

In the case of PTFE, which is insoluble and non-fusible, coagulation of the aqueous dispersion of PTFE was carried out in a mixture with the addition of CNTs ethanol and mechanical stirring.

The deagglomeration of CNTs was carried out in a device operating on a cavitation effect, with a capacity of 4–7 kW in an aqueous medium. In the system, working on the cavitation principle, poured 10 liters of water and added 124 g of initial MWCNTs, i.e., about 100 g of purified CNTs. The treatment was carried out for 4 min. After that, the CNTs was cleaned in the traditional way—with a solution of hydrofluoric acid. After cleaning, CNTs were washed to pH = 6–7.

The PTFE–MWCNTs system was prepared by mixing an aqueous emulsion of PTFE with an aqueous dispersion of MWCNTs, which was obtained by homogenization in a cavitation device and without such homogenization. The PTFE–A300 system was prepared by introducing the A300 powder into the PTFE suspension and mechanical stirring [16, 17].

X-ray studies were carried out on an automated X-ray diffractometer DRON-3 M with radiation $\lambda_{Co} = 0{,}17902$ nm.

Compression or tension tests of the polymeric materials and their composites were performed using tensile machine 2167-R50 with automatic recording of the deformation diagram. Measurements were carried out at a load speed of 5 mm/min.

The conductivity measurements were carried out by a two-contact method at low frequencies using an E7-14 imitansmeter at room temperature.

3 Results and Discussion

3.1 Structural Features of Polymer Nanocomposites

Figure 3 shows the data of X-ray diffraction, and in Table 1, the crystallinity degree of PE–MWCNTs, PP–MWCNTs, PTFE–MWCNTs, and PTFE–A300 composites

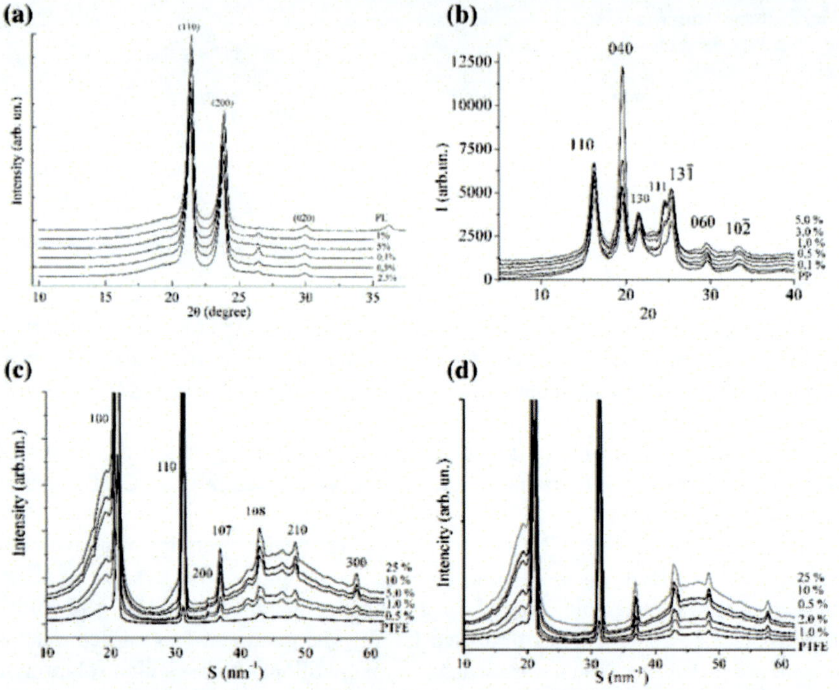

Fig. 3 X-ray diffraction of PE–MWCNTs **a** PP–MWCNTs **b** PTFE–MWCNTs **c** PTFE–A300 **d** composites with different nanofillers concentration

and the crystallite size depending on the CNTs concentration were calculated from the X-ray reflex profile. The introduction of CNTs into the polymer matrix demonstrates the structure-forming properties of CNTs. This is observed from a non-monotonic change in the degree of crystallinity depending on the concentration of CNTs (Table 1). The result is in good agreement with the data [7, 15–19], where it was shown that the crystallization of the polymer under shear deformation in the presence of single, five, and multi-wall CNTs leads to a change in the structure of the matrix. For the PE–CNTs system, the distances between the crystal planes (110) and (200) are smaller than the interplane distances in the PE control sample, and the polymer chains are oriented along the CNTs axis.

It can be considered that the nanofiller creates the effectiveness of crystallization centers. For example, for PTFE, small concentrations of MWCNTs or pyrogenic silica A-300 lead to an increase in the degree of crystallinity. When the concentration of the filler increases, the crystallization centers become too much, they “interfere” with the growth of the crystal phase, which leads to an increase in the number of globules, their “tails”, i.e., the concentration of the disordered phase PTFE.

The same result (increase in the degree of crystallinity) is observed for the PP–MWCNTs system, however, in a very narrow range of MWCNTs concentration up

Table 1 The crystallinity degree (χ), size of X-ray coherent scattering blocks (D) of PE–MWCNTs, PP–MWCNTs, PTFE–MWCNTs, PTFE–A300 system from nanofillers concentration

PE–MWCNTs							
Concentration CNT's, % wt.	0	0.1	0.25	0.5	1	2.5	5
Degree of crystallinity (χ), %	85.6	85.1	79.5	82.8	83.8	81.4	82.3
D, nm	21.92	19.28	20.5	21.04	21.46	20.12	21.35
PP–MWCNTs							
Concentration CNT's, % wt.	0	0.05	0.1	0.5	1	3	5
Degree of crystallinity (χ), %	70.9	71.8	60.9	63.5	63.8	64.0	68. 2
D, nm	11.97	13.01	11.24	10.85	10.40	10.32	10.27
PTFE–MWCNTs							
Concentration CNT's, % wt.	0	0.5	1	5	10	25	
Degree of crystallinity (χ), %	28.1	40.0	31.8	28.9	34.3	30.5	
D, nm	25.04	23.05	21.58	21.00	20.11	19.28	
PTFE–A300							
Concentration CNT's, % wt.	0	0.5	1	2	10	25	
Degree of crystallinity (χ), %	28.1	47.6	37.6	32.7	34.8	26.7	
D, nm	25.04	24.78	24.77	24.86	24.29	18.18	

to 0.05% wt. from 70.0 to 71.8% wt. Introduction of 0.1% wt. MWCNTs reduces the degree of crystallinity to 61%, followed by an increase in the concentration of MWCNTs to 5% wt.

3.2 *Electrical Conductivity of Filled Polymers*

One of the most common areas of research of polymer–MWCNTs composite materials is the creation of the conductive compositions [7, 18, 20, 21]. The value of the percolation threshold in PE–MWCNTs systems in different sources will differ by several orders of magnitude. Thus, in [20], it is noted that in the composition of PE with MWCNTs, which were obtained by mixing the melt in a twin-screw extruder, the percolation threshold is observed at a concentration of 7.5% wt., and the conductivity jump is 16 orders of magnitude from 10^{-20} to 10^{-4} Cm/cm.

Percolation theory is most often used for the analytical description of conductivity dependences on filler concentration [22, 23]. The framework of the probability of the formation of clusters of particles in contact with each other is considered. The description of the critical electric current flow in composite materials (CM) is given by means of the percolation problem formulated for a continuous medium. According to this problem, each point of space with probability $p = \theta_F$ corresponds to conductivity $\sigma = \sigma_F$ and with probability (*1–p*)—conductivity $\sigma = \sigma_m$. The threshold in the percolation terms means the filler concentration at which the phase transition of the second kind of dielectric-conductor occurs. In this case, in the composite system, the regions with high conductivity occupy a minimum fraction of the space—θ_F. At small *p*, all conductive elements are contained in the isolated clusters of finite size. With increasing *p*, the average size of clusters increases and at $p = \theta_F$, a through conduction channel, i.e., a continuous grid of conducting clusters, appears in the system for the first time. At high *p* values, nonconducting regions may already be isolated from each other.

Based on the percolation theory, the following expressions are obtained to describe the dependence of electrical conductivity on the filler concentration [22, 23]:

$$\begin{aligned} \sigma &\sim \sigma_F(\theta - \theta_F)^t, \theta > \theta_F, \\ \sigma &\sim \sigma_m(\theta_F - \theta)^q, \theta < \theta_F, \end{aligned} \tag{1}$$

where *t, q*—critical indices of percolation theory.

For a three-dimensional model of a composite material with the spherical filler particles in the percolation theory, the following values of the threshold and critical indices are obtained: $\theta_F = 17\%$ vol., $t = 1.6\ldots1.9$, $q = 1$. In practice, significant deviations from the theoretically calculated values are possible. For example, for the graphite–polystyrene system [24] $\theta_F = 2\%$ vol., $t = 0.35$; for the carbon black–polyvinyl chloride system, the authors [25] obtained $\theta_F = 11\%$ wt., and the authors [26]—9…9.5% wt. at $t = 1.9 \pm 0.2$, and the authors [21] for the PE–CNTs system obtained the value $\theta_F = 0.07\%$ wt. at $t = 2.1$.

The dependence of the electrical conductivity on the concentration of MWCNTs for the studied systems is shown in Fig. 4. The jump in conductivity in the transition

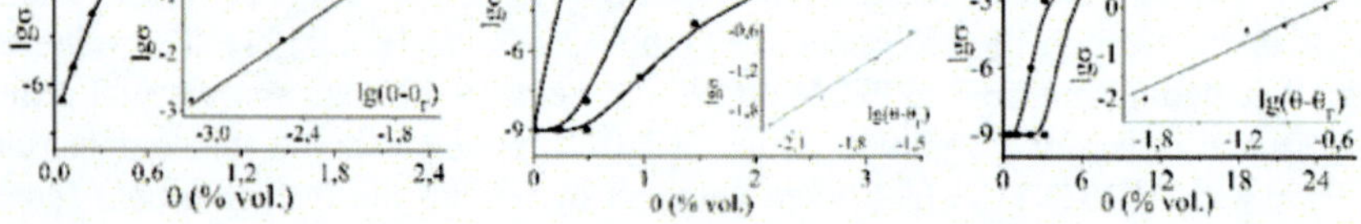

Fig. 4 Dependence of electrical conductivity of PE–MWCNTs **a** PP–MWCNTs **b** PTFE–MWCNTs **c** systems on the concentration of MWCNTs and linear dependence *lgσ ~ lg(θ–θ$_F$)* (insert)

to a composition with a concentration of 5% wt. is almost seven orders of magnitude for PE–MWCNTs (Fig. 4a) and nine orders for PP–MWCNTs (b), PTFE–MWCNTs (c). Of course, to create a continuous conductive grid in the polymer matrix, the size of the filler (the length of the MWCNTs agglomerates) and its distribution in the matrix are of great importance.

Approximating the experimental curves (Fig. 4) and given the linear dependence of the $lg\sigma \sim lg(\theta - \theta_F)$, percolation parameters for the three systems were determined. The percolation threshold for the PE–MWCNTs system was (0.45 ± 0.02) % vol. In this case, the value of electrical conductivity is $\approx 1.1 \times 10^3$ $(\Omega m \cdot cm)^{-1}$, and the critical index is $t \sim 1.8$, corresponding to the three-dimensional system. The calculation of critical indices for the system PP–MWCNTs gives the following values: $\theta_F = 0.345\%$ vol., $t = 1.83$. Deterioration of the homogeneity of the distribution of MWCNTs (curves *2, 1* Fig. 4b) corresponds to an increase in the critical concentration (percolation threshold) in the range of 0.345% vol., 0.92% vol., 1.28% vol. For the PTFE–MWCNTs system (Fig. 4c), we also observe a shift of the percolation threshold to the region of lower values from $\theta_F = 6.68\%$ vol. for initial MWCNTs to $\theta_F = 3.89\%$ vol. ($t = 1.64$) for MWCNTs pre-deaglomerated in a rotary hydrodynamic homogenizer (RHDH) using the simultaneous action of the shear deformation and a cavitation mixing. Therefore, the value of the critical MWCNTs concentration can be used as a parameter to determine the degree of the homogeneity of filled polymer systems.

Analysis of the results on the electrical conductivity of the obtained systems indicates a significant role of the polymer particle size in the formation of the conductive clusters. This is due to the MWCNTs distribution in the polymer matrix provided that the polymer particles have a much larger size than MWCNTs agglomerates [7]. Due to mechanical mixing, the filler covers the surface of the polymer particles, and during hot pressing, this structure changes little and the filler remains at the boundary of the polymer particles. Therefore, a cluster conductive structure with a concentration of MWCNTs above the average and a system of cells with a lower concentration of MWCNTs or their absence can be formed. In this case, the PE–MWCNTs system has a sufficiently low percolation threshold and a relatively low influence of the MWCNTs agglomerate size on the percolation properties of the systems [7].

3.3 The Mechanical Characteristics of the Nanocomposites

The tests of tensile (Fig. 5) and compression (Fig. 6) showed that the introduction of MWCNTs changes the mechanical characteristics of the considered composite systems. The dependence of elongation relative to tension stress (Fig. 5) for the PE–MWCNTs system can be divided into four regions. The region with positive curvature in the range up to 5% of deformation is used for composites. It may be due to the presence of pores (nanoscale), which works as elastic deformation elements [27], the elastic, plastic regions and fracture of the samples. Thus, the introduction of nanotubes in the polymer matrix leads to a change in the limit of the tensile strength (increase), modulus of elasticity (120–213 MPa), the conditional

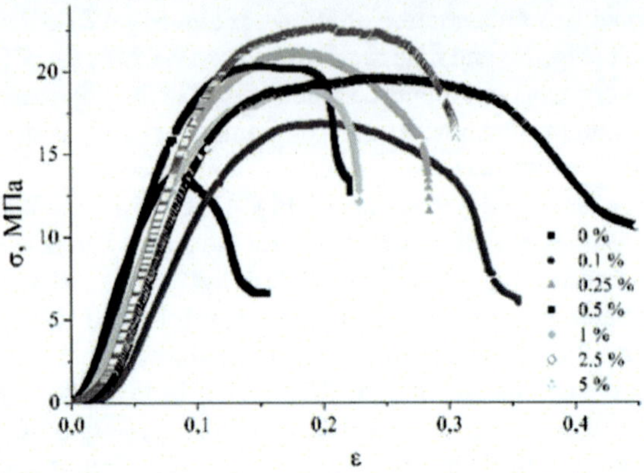

Fig. 5 The dependence of the relative elongation from tensile stress for the samples of PE and system PE–MWCNTs

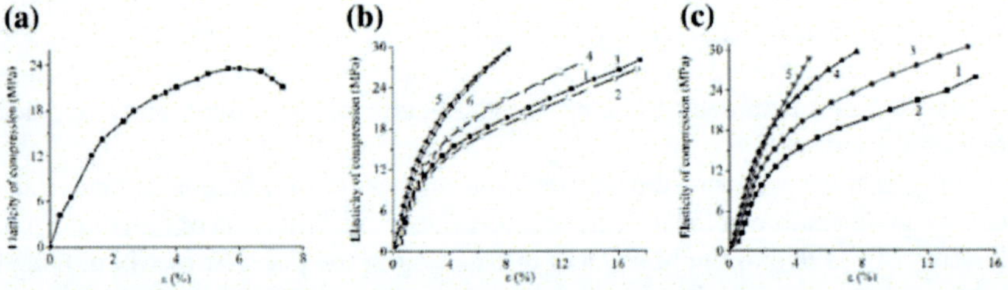

Fig. 6 Dependence of relative deformation vs compression stress for samples: **a** PP–MWCNT (0.1% wt. CNTs), **b** PTFE–MWCNT: 1—PTFE, 2—0.05% MWCNT, 3—0.5% MWCNT, 4—5% MWCNT, 5—15% MWCNT cavitation, 6—25% MWCNT cavitation; **c** PTFE–A300: 1—PTFE, 2—0.5% A300, 3—5% A300, 4—15% A300, 5—25% A300

yield strength (increase by ~50–60%), and to a significant expansion of the area of plastic deformation of the material. It increases the destruction deformation from 8% to almost 40%), it means the increases of the work of fracture or fracture energy (area under the curve of deformation).

The stress–strain relationships of the relative strain with respect to compression for the PP–MWCNT, PTFE–MWCNT, and PTFE–A300 systems are shown in Fig. 6. As can be seen from the above compression diagrams, the introduction of MWCNTs into the composition of the composite does not change the nature of the deformation—compression stress curve, only the quantitative values of the characteristics change. The curves are typical for solid polymers [28]. However, with the addition of MWCNTs, the fracture stresses increase and the fracture strain decreases. Reduction of fracture deformation under compression indicates an increase in the brittleness of the polymer in the presence of MWCNTs.

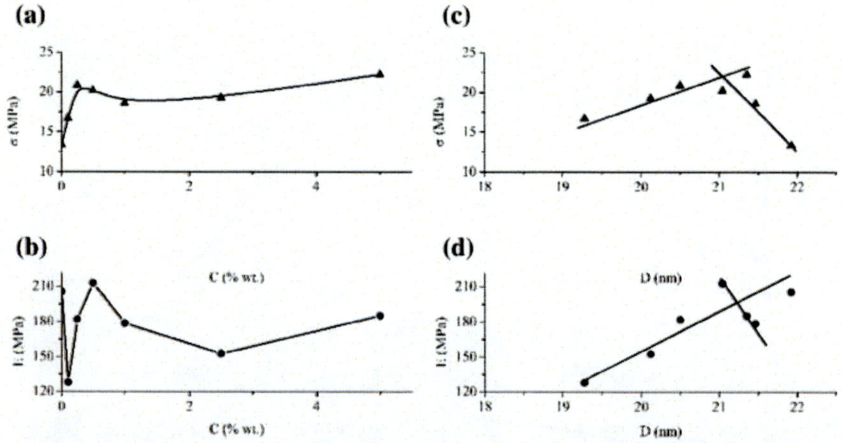

Fig. 7 Dependences of the conditional yield strength of composites PE–MWCNT (**a**, **c**) and the modulus of elasticity (**b**, **d**) on the concentration of CNTs (**a**, **b**) and the size of the coherent scattering blocks (**c**, **d**)

Note that the dependences of strength characteristics on the concentration of nanotubes are not monotonous (Figs. 7, 8, 9 and 10a, b). Theoretical analysis carried out in various models, e.g., [5, 6, 29–31] shows that such a change in properties is due to the characteristics of the different phases formed at the interface of the nanofiller and bulk polymer. The molecular dynamics modelling [5] demonstrates the formation of an ordered polymer matrix layer around the CNTs. This layer, known as the interfacial, plays a central role in the overall mechanical response of the composite. In case of a bad transfer of loading from a matrix to CNT, the effect of amplification concerning CNT is insignificant. Therefore, the presence of an interfacial surface is considered, as the only reason for the enhancement of the composite characteristics. If this approach is correct, it is possible to determine experimentally some structural parameter of the system that would characterize the interfacial surface and show a monotonic change in the strength characteristics of the composite from such a parameter.

For polymeric materials, the parameter that reflects the influence of CNTs on the structural hierarchy of the matrix can be, e.g., the size of the coherent X-ray scattering unit or the degree of crystallinity [5, 6, 31]. The average crystallite size is calculated by Scherer's formula [32].

In Fig. 7c, d, it is shown that the dependences of the conditional yield strength and elasticity modulus are depending on the value of the size of the coherent X-ray scattering unit for PE–MWCNTs systems. In Fig. 8c, d, it is shown that the dependences of the stress of destruction at compression and modulus of elasticity depending on the value of XRCSU for systems PP–MWCNTs have resulted. In Figs. 9 and 10c, d, it is shown that the dependencies of conditional yield strength and modulus of elasticity are depending on the value of XRCSU for PTFE–MWCNTs and PTFE–A300 systems.

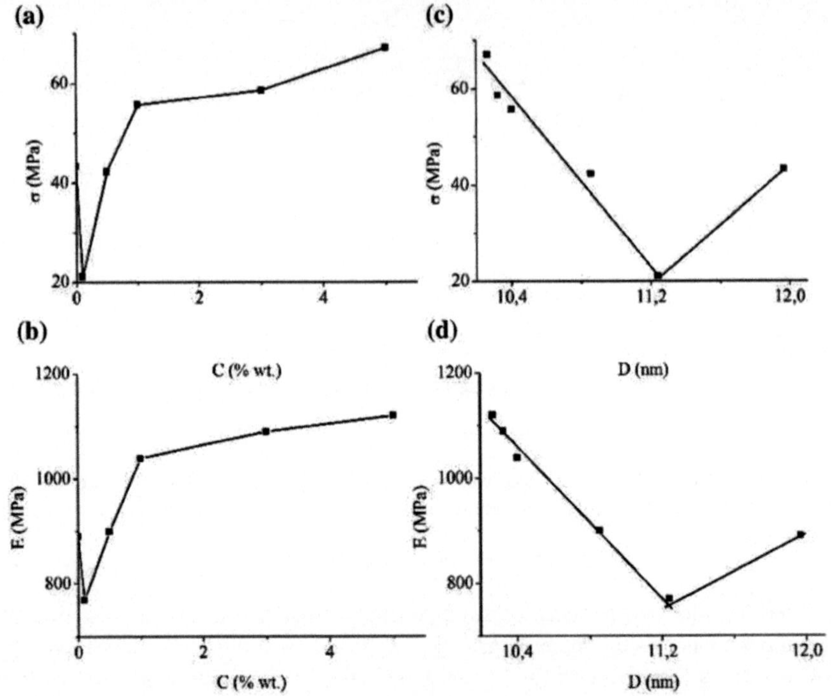

Fig. 8 Dependencies of fracture stress by compression of PP–MWCNTs composites (**a, c**) and modulus of elasticity (**b, d**) on CNT concentration (**a, b**) and size of coherent scattering blocks (**c, d**)

For all the above, the strength characteristics depending on the size of the CXRSU are linear and are divided into two regions. The boundary between which is obviously the percolation threshold. For example, for a PE–MWCNTs system, the CXRSU size is D ~ 21 nm, which corresponds to a CNT concentration of ~0.5% vol. and is in good agreement with the percolation threshold determined from the dependence of the conductivity on the MWCNTs concentration. Similar results are observed for PP–MWCNT, PTFE–MWCNT, and PTFE–A300 systems. For the PP–MWCNT system, the CXRSU size is approximately 11.2 nm, PTFE–MWCNT is ~21 nm, and for PTFE–A300 is ~24 nm.

The proposed approach is valid in the field of small MWCNTs concentration. For the PTFE–MWCNTs system, a continuous grid is created by the concentration of MWCNTs ~3% wt. and for this area of small concentrations of MWCNTs, and large values of CXRSU, there is a monotonic (almost linear) dependence of the strength characteristic on the size of the zones of structural inhomogeneity created by the filler.

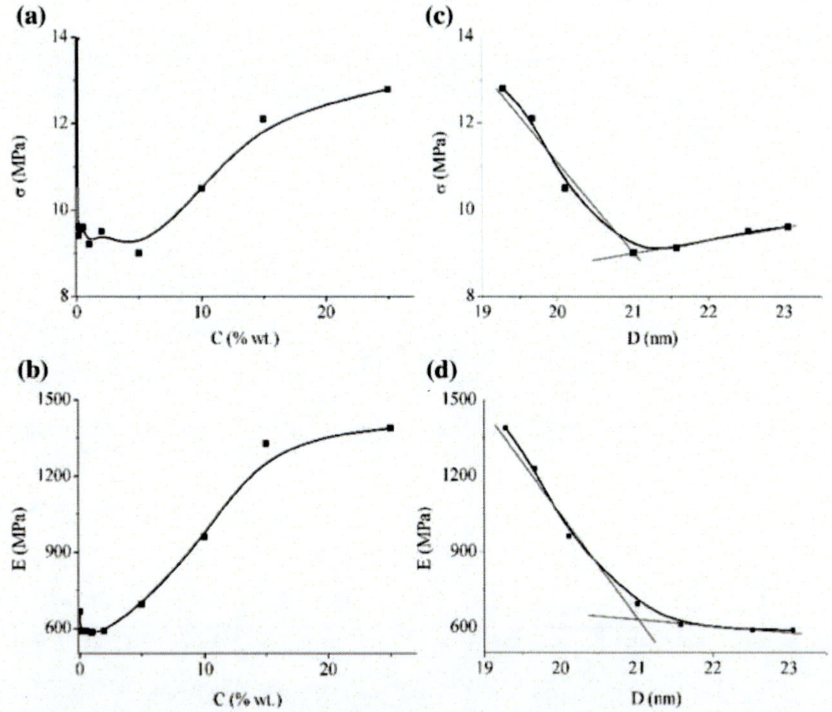

Fig. 9 Dependencies of the conditional yield strength of PTFE–MWCNTs composites (**a**, **c**) and the modulus of elasticity (**b**, **d**) on the concentration of CNTs (**a**, **b**) and the size of the coherent scattering blocks (**c**, **d**)

The XRCSU size of the PP–MWCNTs system is almost half that of the PE–MWCNTs and PTFE–MWCNTs systems. In [33] the authors conducted a quantum chemical study of the interaction energy of carbon nanotube fragments with polyethylene and polypropylene oligomers and optimized the most probable structures of their intermolecular complexes, where it is shown that the value of the interaction energy of a carbon nanotube fragment with a polypropylene oligomer is greater than that of polyethylene, which may explain the smaller size of coherent scattering blocks for the PP–CNT system. Since the polymer with the outer surface of the carbon nanotube forms a micromolecular complex, which is not covalently bound and is held by intermolecular dispersion forces [33], i.e., the oligomers of polymers and nanotube surfaces in the formed nanocomposites are placed closer to each other than separate ones.

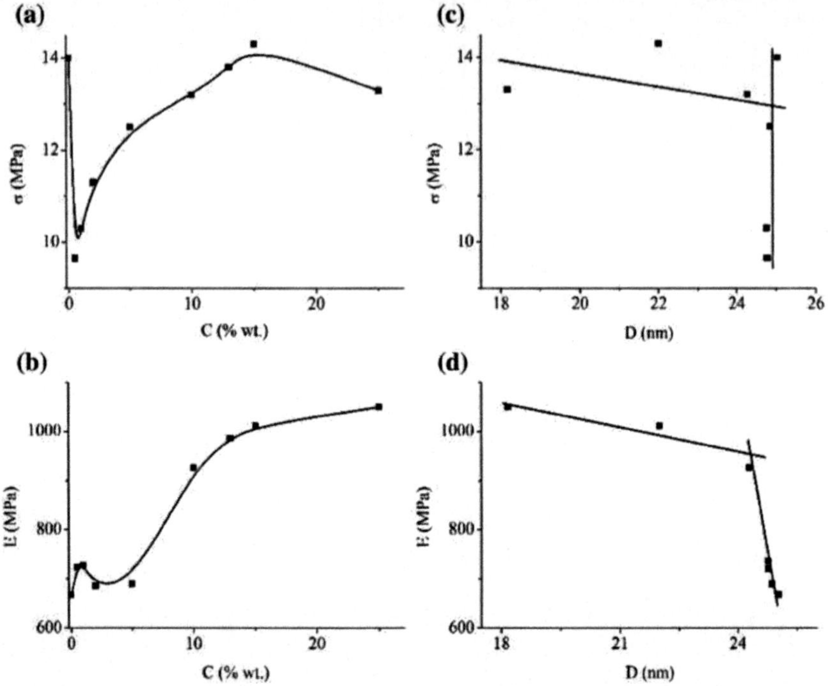

Fig. 10 Dependencies of the conditional yield strength of PTFE–A300 composites (**a**, **c**) and the modulus of elasticity (**b**, **d**) on the concentration of CNTs (**a**, **b**) and the size of the coherent scattering blocks (**c**, **d**)

4 Conclusions

It is experimentally shown that the introduction of a small concentration of nanofillers (MWCNTs and pyrogenic silica) into the polymer matrix (PE, PP, PTFE) significantly changes the structural characteristics of composites, namely the degree of crystallinity and the size of the X-ray coherent scattering blocks.

The dependence of electrical conductivity on MWCNTs concentration is percolation. At a critical concentration, a phase transition of the second kind dielectric-conductor occurs. Critical indices showing the three-dimensionality of these systems are calculated for the studied systems. The critical concentration at which a continuous volumetric current grid is formed depends on the degree of homogeneity of the MWCNTs distribution in the matrix and can be a parameter that characterizes it.

The strength characteristics of the studied composites are non-monotonically dependent on the filler concentration but linearly depend on the size of the X-ray coherent scattering blocks. These dependencies have two sections the inflection point for which corresponds to the critical concentration of the percolation transition. This confirms the theoretical calculations of various approximations that quantitatively the response of the matrix is proportional to the surface area of the interface.

References

1. Bokobza L (2007) Multiwall carbon nanotube elastomeric composites: a review. Polymer 48(17):4907–4920. https://doi.org/10.1016/j.polymer.2007.06.046
2. Bauhofer W, Kovacs JZ (2009) A review and analysis of electrical percolation in carbon nanotube polymer composites. Comp Sci Technol 69(10):1486–1498. https://doi.org/10.1016/j.compscitech.2008.06.018
3. Sementsov YI, Melezhyk AV, Pyatkovsky ML, Yanchenko VV, Gavrilyuk NA et al (2007) Hydrogen materials science and chemistry of carbon nanomaterials. In: Veziroglu TN, Zaginaichenko SYu, Schur DV, Baranowski B, Shpak AP, Skorokhod VV, Kale A (eds) NATO security through science series a: chemistry and biology. Springer, Dordrecht, pp 757–763. https://doi.org/10.1007/978-1-4020-5514-0_95
4. Zare Y (2016) Study of nanoparticles aggregation/agglomeration in polymer particulate nanocomposites by mechanical properties. Compos Part A 84:158–164. https://doi.org/10.1016/j.compositesa.2016.01.020
5. Malagù M, Goudarzi M, Lyulin A, Benvenuti E, Simone A (2017) Diameter-dependent elastic properties of carbon nanotube-polymer composites: Emergence of size effects from atomistic-scale simulations. Compos Part B 131:260–281. https://doi.org/10.1016/j.compositesb.2017.07.029
6. Cen-Puca M, Oliva-Avilés AI, Avilés F (2018) Thermoresistive mechanisms of carbon nanotube/polymer composites. Physica E 95:41–50. https://doi.org/10.1016/j.physe.2017.09.001
7. Sementsov YI, Makhno SN, Zhuravsky SV, Kartel MT (2017) Properties of polyethylene–carbon nanotubes composites. Chem Phys Technol Surface 8(2):107–119. https://doi.org/10.15407/hftp08.02.107
8. Sementsov YI, Kartel NT (2019) The influence of small concentrations of carbon nanotubes on the structuralization in matrices of different nature. Chem Phys Technol Surface 10(2):174–189. https://doi.org/10.15407/hftp10.02.174
9. Lozovyi F, Ivanenko K, Nedilko S, Revo S, Hamamda S (2016) Thermal analysis of polyethylene + X% carbon nanotubes. Nanoscale Res Lett 11(1):97. https://doi.org/10.1186/s11671-016-1315-y
10. Avramenko TG, Khutoryanskaya NV, Naumenko SM, Ivanenko KO, Hamamda S, Revo SL (2019) Effect of carbon nanofillers on processes of structural relaxation in the polymer matrixes. Springer Proc Phys 221:293–305. https://doi.org/10.1007/978-3-030-17759-1_20
11. Treacy MMJ, Ebbesen TW, Gibson JM (1996) Exceptionally high Young's modulus observed for individual carbon nanotubes. Nature 381:678–680. https://doi.org/10.1038/381678a0
12. Barber AH, Cohen SR, Wagner HD (2003) Measurement of carbon nanotube-polymer interfacial strength. Appl Phys Lett 82(23):4140–4142. https://doi.org/10.1063/1.1579568
13. Melezhik AV, Yanchenko VV, Sementsov YI (2007) in Nanocarbon materials. Hydrogen materials science and chemistry of carbon nanomaterials. In: Veziroglu TN, Zaginaichenko SY, Schur DV, Baranowski B, Shpak AP, Skorokhod VV, Kale A (eds) NATO security through science series A: chemistry and biology. Springer, Dordrecht, pp 529–537. https://doi.org/10.1007/978-1-4020-5514-0_67
14. Melezhik AV, Sementsov YI, Yanchenko VV (2005) Synthesis of fine carbon nano-tubes on coprecipitated metal oxide catalysts. Russ J Appl Chem 78(6):917–923. https://doi.org/10.1007/s11167-005-0420-y
15. Pinchuk-Rugal TM, Dmytrenko OP, Kulish MP, Grabovskyy YY, Nychyporenko OS, Sementsov YI, Shlapatskaya VV (2015) Radiation damages of isotactic polypropylene nanocomposites with multi-walled carbon nanotubes. Probl Atomic Sci Technol 96(2):10–17

16. Nychyporenko OS, Dmytrenko OP, Kulish MP, Pinchuk-Rugal TM, Grabovskyy YY, Zabolotnyy AM, Strelchuk VV, Nikolenko AS, Sementsov YI (2015) Defects of structure of nanocomposites of polytetrafluorethylene with multiwalled carbon nanotube. Nanosyst Nanomater Nanotechnol 13(4):673–686
17. Sementsov YI, Piatkovsky ML, Gavriliuk NA, Prikhodko GP, Kartel MT, Grabovsky YE (2009) Nanocomposites are fluoroplast 4/carbon nanotubes. Production, structure and mechanical properties. Chem Ind Ukr 5:59–64 [in Uktainian]
18. Minus ML, Chae HG, Kumar S (2012) Polyethylene crystallization nucleated by carbon nanotubes under shear. ACS Appl Mater Interfaces 4(1):326–330. https://doi.org/10.1021/am2013757
19. McNally T, Potschke P, Halley P, Murphy M, Martin D, Bell SEJ, Brennan GP, Bein D, Lemoine P, Quinn JP (2005) Polyethylene multiwalled carbon nanotube composites. Polymer 46(19):8222–8232. https://doi.org/10.1016/j.polymer.2005.06.094
20. Novak DS, Bereznenko NM, Shostak TS et al (2011) Strumipriva on the nanocomposites of polyethylene. Rock destruction and metal-working tools—technology of manufacture and applications (Collection of scientific papers. The ISM NAS of Ukraine.) 14:394. (in Russian)
21. Lisunova MO, Mamunya YP, Lebovka NI, Melezhyk AV (2007) Percolation behaviour of ultra-high molecular weight polyethylene/multi-walled carbon nanotubes composites. Eur Polym J 43:949–958. https://doi.org/10.1016/j.eurpolymj.2006.12.015
22. Kirkpatrick S (1973) Percolation and conduction. Rev Modern Phys 45:574–588. https://doi.org/10.1103/RevModPhys.45.574
23. Efros AL (1982) Physics and geometry of disorder. Nauka, Moscow [in Russian]
24. Mamunya YP (2000) Electrical and thermal conductivity of polymer composites with dispersed fillers. Ukr Chem J 66(3):55 [in Ukrainian]
25. Quivy A, Deltour R, Jasen AG et al (1989) Transport phenomena in polymer-graphite composite materials. Phys Rev B 39(2):1026–1030. https://doi.org/10.1103/PhysRevB.39.1026
26. Balberg I, Binenbaum N, Bozovsky S (1983) Anisotropic percolation in carbon black-polyvinylchloride composites. Sol St Comm 47(12):989–992. https://doi.org/10.1016/0038-1098(83)90984-5
27. Sementsov YI, Revo SL, Ivanenko KO (2016) Exfoliated graphite. Kyiv, Interservice, pp 78–90. (in Ukrainian)
28. Encyclopedia of Polymers (1977) Ed. VA Kabanova, vol 3. Soviet Encyclopedia, Moscow. [in Russian]
29. Halpin JC, Kardos JL (1976) The Halpin-Tsai equations: a review. Polym Eng Sci 16(5):344–352. https://onlinelibrary.wiley.com/doi/pdf/10.1002/pen.760160512
30. Haque A, Ramasetty A (2005) Theoretical study of stress transfer in carbon nanotubes reinforced polymer matrix composites. Compos Struct 71(1):68–77. https://doi.org/10.1016/j.compstruct.2004.09.029
31. Kartel M, Sementsov Y, Mahno S, Trachevskiy V, Bo W (2016) Polymer composites filled with multiwall carbon nanotubes. Univ J Mater Sci 4(2):23–31. https://doi.org/10.13189/ujms.2016.040202
32. Patterson A (1939) The scherrer formula for X-ray particle size determination. Phys Rev 56(10):978–982. https://doi.org/10.1103/PhysRev.56.978
33. Terets MI, Demianenko EM, Zhuravsky SV, Chernyuk OA et al (2019) Quantum chemical study on the interaction of carbon nanotube with polyethylene and polypropylene oligomers. Chem Phys Technol Surface 10(1):75–86. https://doi.org/10.15407/hftp10.01.075

The Influence of Carbon, Carbon, and Boron on the Formation of Diffusion Nanocomposite Hardened Layers on the Surfaces of Steel Parts

Andrew E. Stetsko and Yaryna T. Stetsko

1 Introduction

The main task of chemical–thermal treatment of steel parts is to increase their resource. Surface hardening gives a significant reinforcing effect on conventional structural steels: increasing strength, hardness, and durability. The core of such parts remains soft—resistant to dynamic influences. Also, the price of such parts made of structural steel is low.

The method of chemical deposition and diffusion of boron and carbon forms is done on the parts made of morphology steel having a reinforced layer of thickness 500 microns and microhardness of 14 GPa. The stability of these parts after chemical deposition and diffusion boron and carbon increases by 2–3 times.

Objective: Method for strengthening steel surfaces of friction pairs by chemical sedimentation and diffusion of boron and carbon that provides strength to manufacture or restore detailed machines.

2 Materials and Methods

Conducted of investigating the diffusion of carbon in the iron sheet during the decarburization of the gas in a solid atmosphere in a weak oxidizing atmosphere ($Ar + H_2 + H_2O$), the microstructures of the iron sheets after bleaching were discovered through SEM [1]. This discovery found a carbon gradient after the discoloration of iron sheet

A. E. Stetsko (✉) · Y. T. Stetsko
Mendeleeva Str., 4, Apt. 6, Lviv 79005, Ukraine
e-mail: andrew73@ukr.net

Y. T. Stetsko
e-mail: yaryna75@ukr.net

O. Fesenko and L. Yatsenko (eds.), *Nanomaterials and Nanocomposites, Nanostructure Surfaces, and Their Applications*, Springer Proceedings in Physics 246, https://doi.org/10.1007/978-3-030-51905-6_47

in the thickness direction. The results showed that the sheet of iron in the direction of thickness after discoloration formed is of three layers: layer (ferrite phase) that was near the surface; layer (cementite phase), which was located between (ferrite phase + cementite phase + ferrite phase) + (graphite phase) layers; and layer between (cementite phase + ferrite phase) + (graphite phase), which was at the center of the sample. The thickness of decarburization layer (ferrite phase + cementite phase) showed a nice linear correlation to the square root of the decarburization time. A staircase type during discoloration was the carbon gradient in the iron sheet in the thickness direction. The carbon migration in the iron sheet is contained of stable carbon decomposition and free carbon diffusion. The carbon content in the center of iron sheet was influenced by the time or temperature of decarburization. The carbon migration at S_2 iron stage was the control stage of the entire bleaching process.

Researchers conducted studies [2] of gas dispersion of high-alloy stainless steel to the growth surface hardness as general mechanical surface characteristics. It has been observed that the growth of chromium-rich carbides during the transfer of carbon to steel causes precipitation on the surface but reduces the chromium content of the solid solution. For this reason to keep a good corrosion resistance in the carbonated layer, it is obvious to optimize the stainless steel composition and the mintage process. Thermodynamic calculations are first used to find optimal parameters to maintain the highest Cr and Mo contents in the austenitic solid solution. A good correspondence between the calculated and experimental values was observed for martensitic stainless steel at 955 and 980 °C.

Published data [3] on carbon diffusion in austenite is calculated by the estimation in full austenitic range and then to be extrapolated to the temperature range of phase transformation. Data are limited to temperatures above 750 °C. In this study, new experiments are conducted to determine the carbon diffusion coefficient in austenite at temperatures up to 500 °C. Carburization experiments are completed in the austenitic range for a Fe–Mn–C and a Fe–Ni alloy. This method is first validated on the Fe–C–Mn steel. Carburization experiments are implemented on a Fe–Ni alloy at a temperature from 900 to 500 °C with step 100 °C.

The boronization and carburization of fine-grained superplastic stainless steel is studied in [4]. New experimental consequences on fine-grained Ti88.5Al4.5V3Fe2Mo2 were obtained. The diffusion of boron into steel superplastic alloy was scrutinized.

In [5], we can see prestraining of plastic accelerates the diffusion of carbon and boron. We apply boron-cemented layers with a homogeneous structure and enriched physicochemical properties hardened with significantly grained boron carbides.

In [6], the technological facet of boriding processe was done taking into account the pros and cons of each method. The influences of the boriding technique on the microstructure of borided materials have been indicated. Consideration in this part was given to the powder-pack processes, electrochemical boronizing in borax to the gas boronizing with the use of boron halides or boranes.

In [7], it has been shown that the diffusion front of the boride layer has a sawtooth form, while that of the RE-borosulfide layer is flat. Another from the BL layer, the RBSL layer is thin, entire, and plane. The results of FeS, Fe_2B, and FeB phases on the substrates were obtained by Auger electron spectroscopy analysis. The wear resistance test reflected that within a certain range, the abrasion resistance of the RBSL layer is finer than that of the BL layer, precisely under the high-load circumstance. The corrosion antiperistalsis test using the weight loss method has proven that the corrosion resistance of the RBSL layer is finer but declines faster with time extension than that of the BL layer.

3 Discussion

Diffusion saturation is carried out on parts made of steel (1.2% carbon). The complex method consists of several stages (Fig. 1). The main stages are chemical deposition and diffusion saturation on the surfaces of the machine parts [8–10].

First, machining of the surfaces of parts is carried out to a size that takes into account the surface build-up during chemical deposition and diffusion saturation. Mechanical cleaning, degreasing, and then decapitating are required, which significantly activate the surface prior to deposition. Decapitation is carried out in acids, followed by washing. During the chemical deposition, care should be taken to ensure that suspended specimens do not come into contact with each other and that the medium is maintained at a pH of 9–10 (Fig. 2).

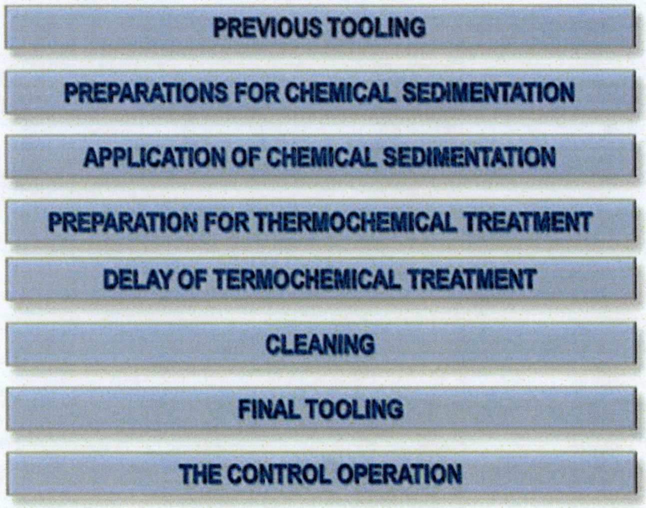

Fig. 1 Stages of the complex method

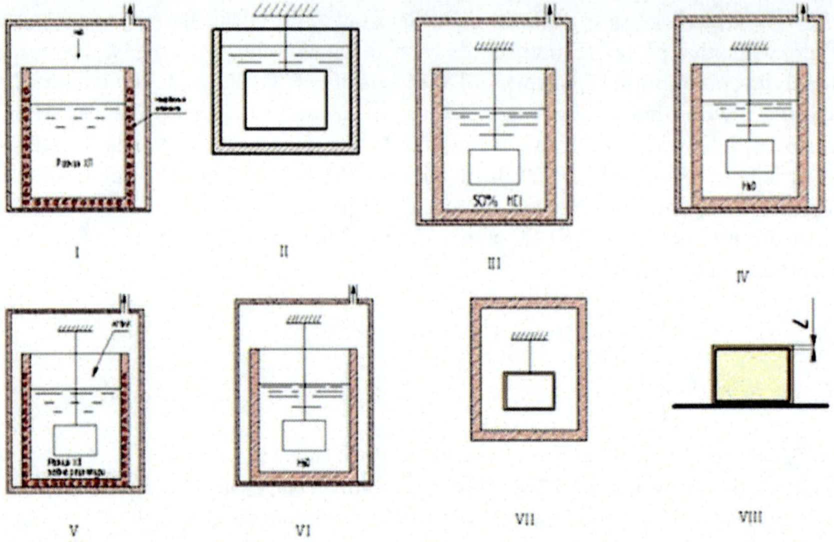

Fig. 2 Stages of chemical deposition

Chemical–thermal treatment was carried out in saturated powder mixtures. Retorts with the parts placed in them in powder mixtures have fusible closures (fusible caps) to seal saturated media (Fig. 3). Such retorts provide quality diffusion coatings on the surfaces.

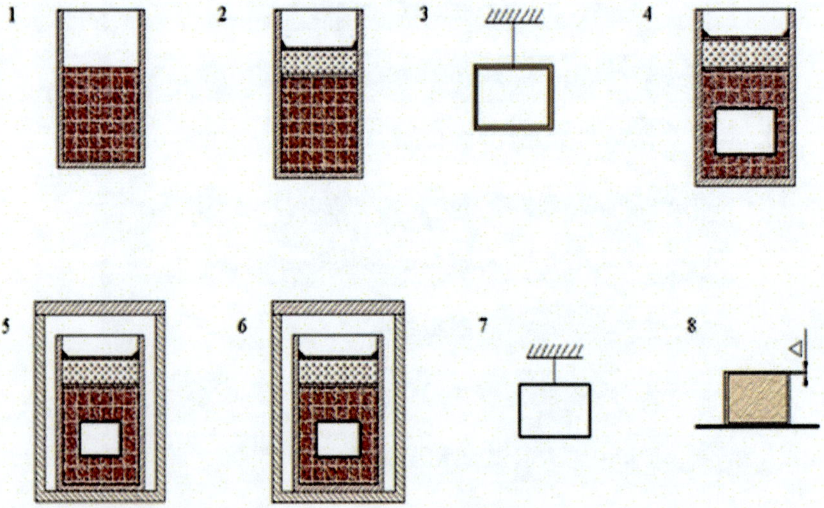

Fig. 3 Retorts with fusible closures

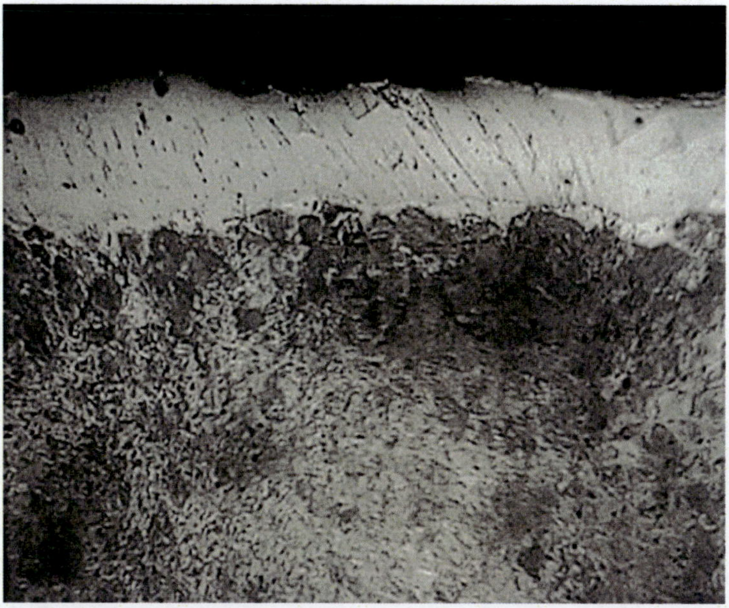

Fig. 4 Microstructure of hardening layer by chemical sedimentation and diffusion of carbon. Magnification × 600

Chemical element	#1	#2	#3	#4
$CoCl_2$, (g/l)	15	30	–	–
$NiCl_2$, (g/l)	30	30	–	–
$Na_3C_6H_5O_7$, (g/l)	90	–	–	–
NaH_2PO_2, (g/l)	25	25	25	30
NH_4Cl, (g/l)	–	50	50	45
$CoCO_3$, (g/l)	–	–	–	15
$NiCO_3$, (g/l)	–	–	–	30
$CoSO_4$, (g/l)	–	–	20	–
$NiSO_4$, (g/l)	–	–	30	–
CH_3COONa, (g/l)	–	90	–	90
$(CHCOONa)_2$, (g/l)	–	–	90	–
$C_4H_6O_5$, (g/l)	15	–	–	–
CH_2NH_2COOH, (g/l)	25	–	–	–
NH_4Cl, (g/l)	50	50	50	50

Fig. 5 Receptions of chemical deposition

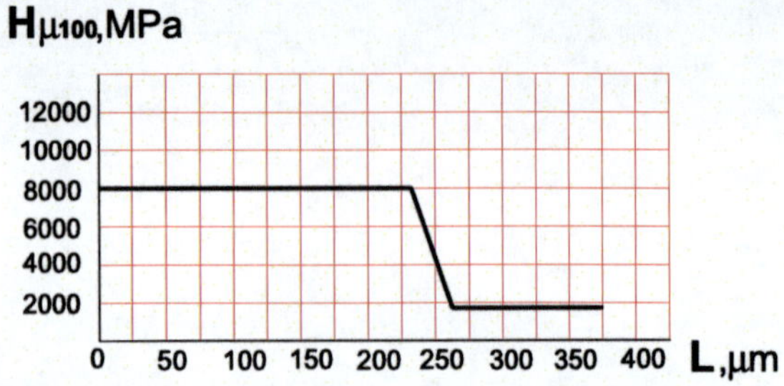

Fig. 6 Microhardness of hardening layer by chemical Ni–Co–P sedimentation and diffusion of carbon

Fig. 7 Microstructure of hardening layer by chemical Ni–Co–P sedimentation and diffusion of carbon. Magnification × 600

The thickness of the precipitated chemical coating is approximately 8–10 microns. The deposited chemical layer has a pseudo-amorphous structure. It has been experimentally found that the melting point of this chemically deposited layer reaches 720–740 °C. Due to low melting point, the chemical layer significantly accelerates diffusion processes during the chemical–thermal treatment. When heated, the chemical layer is first melted by spots (at a temperature of 700–720 °C), which are already sources of rapid diffusion of elements that saturate the surface layer.

The precipitated chemical layer is also a source of saturating elements that diffuse during chemical–thermal treatment. For example, diffusion of a chemical element Ni from a chemically deposited layer leads to a significant positive effect since Ni has the

Chemical element	#6	#7	#8	#9	#10
$CoCl_2$, (g/l)	15	30	–	–	–
$NiCl_2$, (g/l)	30	30	–	–	–
$Na_3C_6H_5O_7$, (g/l)	90	–	–	–	–
NaH_2PO_2, (g/l)	25	25	25	30	–
$K_2H_2PO_2$ (g/l)	–	–	–	–	30
NH_4Cl, (g/l)	–	50	50	45	–
$CoCO_3$, (g/l)	–	–	–	15	–
$NiCO_3$, (г/л)	–	–	–	30	–
$CoSO_4$, (g/l)	–	–	20	–	–
$NiSO_4$, (g/l)	–	–	30	–	–
$Co(NO_3)_2$ (g/l)	–	–	–	–	18
$Ni(NO_3)_2$ (g/l)	–	–	–	–	30
CH_3COONa, (g/l)	–	90	–	90	–
$(CHCOONa)_2$,(g/l)	–	–	90	–	–
$C_4H_6O_5$ (g/l)	15	–	–	–	–
CH_2NH_2COOH, (g/l)	25	–	–	–	–
$Na_2C_4H_4O_4$ (g/l)	–	–	–	–	100
NH_3 (ml)	50	50	50	50	60

Fig. 8 Receptions of chemical deposition

property to simultaneously increase the hardness and plasticity. Nickel diffuses into the surface and is placed at the boundaries of carbide and boride grains—envelopes them (Fig. 4). Such a structure of reinforced material has a very good wear resistance and a considerable resource.

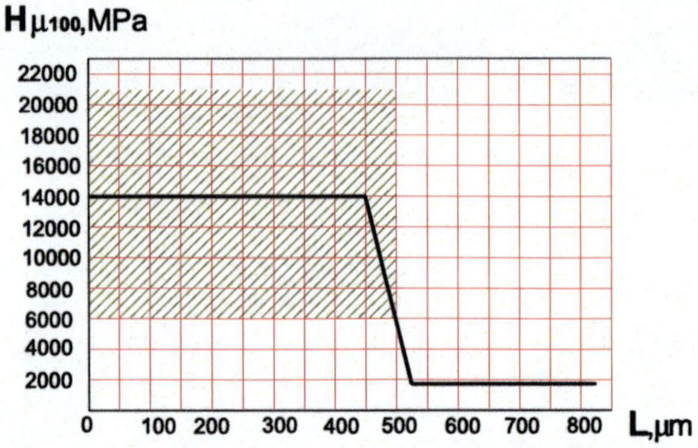

Fig. 9 Microhardness of hardening layer by diffusion boron and carbon

Fig. 10 Microstructure of hardening layer by chemical Ni–Co–P sedimentation and diffusion boron and carbon. Magnification × 450

For parts made of structural steel (containing 1.2 °C) and chemically deposited according to the formulations (Fig. 5), diffused carbon saturation was applied according to the modes: saturation temperature 950 °C, saturation time 5 h, isothermal holding temperature 727 °C, and isothermal holding time 1 h.

After chemical deposition and chemical thermal saturation with carbon, machine parts receive reinforced layers of considerable thickness—up to 250 microns. The structure of the hardened layers is cementite with a hardness of not less than 8000 MPa (Fig. 6). Such reinforced layers have significant advantages: low surface roughness and waviness, considerable thickness, and no carbon-free surface area under the diffusion layer (Fig. 7). In the vast majority of machine parts thus reinforced, there will be no need for final machining. The disadvantage of such reinforced layers is their fragility, so they cannot be applied to parts that operate at high dynamic loads.

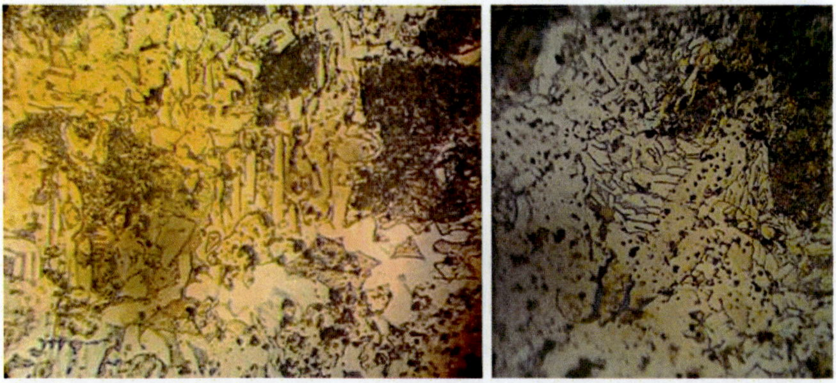

Fig. 11 Microstructure of hardening layer by chemical Ni–Co–P sedimentation and diffusion boron and carbon. Magnification × 600

To eliminate the aforementioned disadvantage, the following reinforcement of machine parts was carried out: chemical precipitation by formulations (Fig. 8) and compatible diffusion saturation with boron and carbon.

The chemical precipitation was completed in an aqueous solution over 45 min and had a thickness of 8–10 μm. Chemical–thermal treatment was making at a temperature of 950 °C for a time of 5 h. During heating, an isothermal holding at 727 °C was carried out for 1 h. Hardening under these modes forms diffusion layers of considerable thickness (up to 500 μm) and hardness (up to 14,000 MPa of integral hardness) on the surfaces of the details (Fig. 9). The morphology depends on the chemical-deposited layer (Figs. 10 and 11).

4 Conclusions

Chemical deposition applied to the surface of the reinforcing part before chemical–thermal treatment (diffusion carbon saturation, diffusion saturation with boron and carbon) significantly changes the morphology of the diffusion layers.

The results of the studies show that chemical precipitation and carbon diffusion should be used to strengthen the parts that operate under low dynamic loads and specific pressures. There is no need for finishing after machining.

For heavy-duty friction and wear parts, chemical precipitation and diffusion saturation with boron and carbon should be used. The composite structure of such reinforced layers greatly increases the life of the machine parts.

References

1. Hong L, Cheng R, Ai L et al (2019) Trans Indian Inst Met 72:335. https://doi.org/10.1007/s12666-018-1484-8
2. Turpin T, Dulcy J, Gantois M (2005) Metall and Mat Trans A 36:2751. https://doi.org/10.1007/s11661-005-0271-4
3. Thibaux P, Méténier A, Xhoffer C (2007) Metall and Mat Trans A 38:1169. https://doi.org/10.1007/s11661-007-9150-5
4. Matsushita M (2011) Boronization and carburization of superplastic stainless steel and titanium-based alloys. Materials 4(7):1309–1320. https://doi.org/10.3390/ma4071309
5. Filonenko NY, Bereza OY, Pilyaeva SB (2015) Mater Sci 51:172. https://doi.org/10.1007/s11003-015-9825-9
6. Kulka M. (2019) Trends in thermochemical techniques of boriding. In: Current trends in boriding. Engineering materials. Springer, Cham
7. Wang D, Zhang X, Zhao H et al (2013) Appl Phys A 113:115. https://doi.org/10.1007/s00339-013-7714-3
8. Stetsko AE (2017) Chapter 46. Effect of chemical vapor deposition on the morphology of the nanocomposite layer in the complex method. In: Fesenko O, Yatsenko L (eds) Nanophysics, nanomaterials, interface studies, and applications, NANO 2016. Springer proceedings in physics, vol 195. Springer, Cham, Switzerland, pp 611–622. https://doi.org/10.1007/978-3-319-56422-7_46
9. Stetsko A (2018) The nanocomposite diffusion coating of details prepared of boriding. In: Fesenko O, Yatsenko L (eds) Nanochemistry, biotechnology, nanomaterials, and their applications, NANO 2017. Springer proceedings in physics, vol 214. Springer, Cham, Switzerland, pp 345–356. https://doi.org/10.1007/978-3-319-92567-7_22
10. Stetsko AE, Stetsko YT (2019) Complex method of the composite nanocoatings formation. In: Pogrebnjak A, Novosad V (eds) Advances in thin films, nanostructured materials, and coatings. Lecture notes in mechanical engineering. Springer, Singapore, pp 179–188. https://doi.org/10.1007/978-981-13-6133-3_18

Synthesis of HA–Collagen and HA–Collagen–Alginate Nanocomposites

Zoriana Voitko, Alla Serhiienko, Tetiana Dontsova, Svitlana Nahirniak, and Andrii Lapinskyi

1 General

1.1 Introduction

Hydroxyapatite (HA) is a crystal–chemical analogue of the mineral component of skeletal tissues of animals and humans, which is part of the tooth enamel (95–97%), dentin (70–75%), and bone tissue (60–70%). The crystalline structure of HA belongs to the spatial group P63/m of hexagonal syngony. Stoichiometric HA can also be attributed to the spatial group P21/b of monoclinic syngony. The reduction in symmetry to monoclinic results from the ordering of the OH groups arrangement in calcium channels, as well as the mutual ordering of these channels, in the way that there is a twofold increase in the parameter b of the unit cell, whose parameters are $a = 9432$ Å, $c = 6881$ Å [1].

The main structural element of HA is phosphate tetrahedra PO_4. It forms a rigid three-dimensional framework with axial channels along the crystallographic direction (001). Ca^{2+} cations occupy two crystallographically different positions in the

Z. Voitko (✉) · A. Serhiienko · T. Dontsova · S. Nahirniak · A. Lapinskyi
Igor Sikorsky Kyiv Polytechnic Institute, Kyiv, Ukraine
e-mail: zoriana1007l@gmail.com

A. Serhiienko
e-mail: allasergienko706@gmail.com

T. Dontsova
e-mail: dontsova@ua.fm

S. Nahirniak
e-mail: nagirnyak_sv@ukr.net

A. Lapinskyi
e-mail: andlapinskiy@gmail.com

O. Fesenko and L. Yatsenko (eds.), *Nanomaterials and Nanocomposites, Nanostructure Surfaces, and Their Applications*, Springer Proceedings in Physics 246, https://doi.org/10.1007/978-3-030-51905-6_48

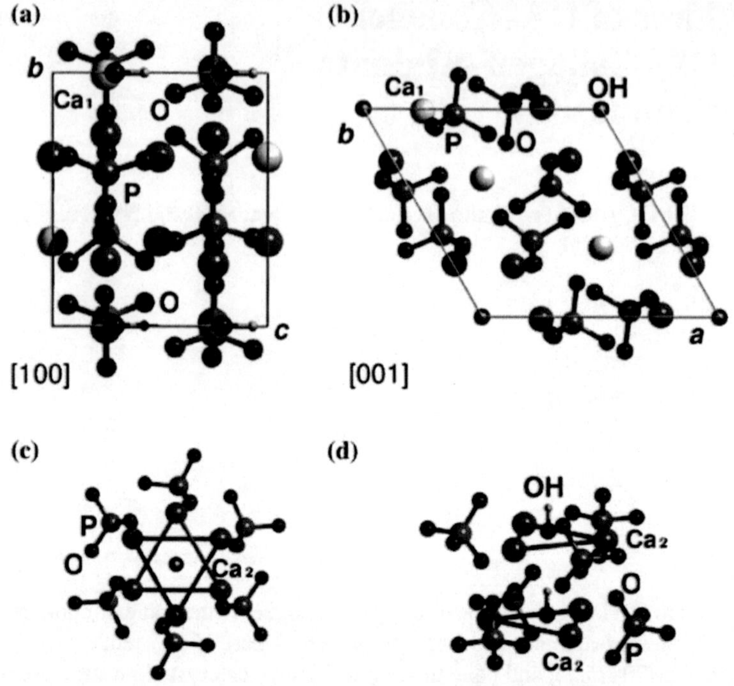

Fig. 1 Structure of the HA elementary cell: **a** Ca_1 coordination surrounding taking into account the crystallographic direction (100), **b** Ca_1 coordination surrounding taking into account the crystallographic direction (001), **c, d** coordination surrounding of the Ca_2 ions

HA structure (Fig. 1a, b) [2]. Ca_1 ions are bound to PO_4 tetrahedra. In this case, Ca_1 has surrounding close to the octahedral structure and is surrounded by nine oxygen ions, also belonging to the PO_4-groups. Ca_2 ions form an axial channel and are surrounded by oxygen ions from PO_4-groups and OH-groups. (Fig. 1c, d). The Ca_2 atoms form a triangle in a plane perpendicular to the *c*-axis, the triangles rotated 60° relative to each other along that axis. Thus, the refined formula for the HA structure can be represented as $Ca_{(1)4}Ca_{(2)6}(PO_4)_6(OH)_2$ [2].

Biomaterials synthesis based on HA has long become a priority direction in modern medical materials science. HA is widely used: in clinical practice because it has high biocompatibility, devoid of immunogenic, and allergenic activity; as a major component of therapeutic and prophylactic pastes and gels; for the repair of defects and treatment of bone pathologies, etc. Since natural bone tissue is a composite material, the use of HA is the most promising in the form of composite materials. Bone tissue is known to consist of HA, collagen, water and other components (proteins, salts, etc.). By introducing appropriate additives into HA-based composites, their mechanical and biological characteristics and, first their biocompatibility with the tissue of a living organism can be improved. In addition to collagen, such additives include alginate, which is a polysaccharide extracted from red, brown, and some green algae.

1.2 Synthesis of HA and Its Composites

There are many synthesis methods of hydroxyapatite: precipitation from aqueous solutions of salts (liquid-phase methods) [3–5], sol–gel [6], solid-phase synthesis [7], hydrothermal method [8], and some others [9]. Particular attention should be paid to methods that are capable of providing a synthesis of materials with specified and constant characteristics and which are easy to implement. In our view, such methods include the saccharine method of nanocrystalline HA powder producing. This method is realized by the interaction of ammonium hydrophosphate with calcium saccharate solutions [10]. The synthesis is described by the following chemical equation:

$$10CaC_{12}H_{20}O_{11} + 6(NH_4)_2HPO_4 + 3NH_4OH \rightarrow$$
$$\rightarrow Ca_{10}(PO_4)_6(OH)_2 + H_2O + 15NH_3 + 10C_{12}H_{22}O_{11}.$$

This method makes it easy to obtain biocompatible nanoscale powders with homogeneous morphology without the use of specialized equipment.

HA-based composites are obtained either in the form of ceramics reinforced with HA particles or in the form of composites with polymers and HA particles [11]. Synthesis of composites with polymers (in particular, collagen) has recently become increasingly in demand as they are able to get as close as possible to the chemical and morphological composition of bone tissue. Currently, intensive research is being conducted on the choice of collagen type, the conditions of mineralization, the porosity of the material, the parameters of HA synthesis, and its cross-linking in the composite. Usage of sodium alginate as a polymer allows obtaining composites in the form of capsules. Also, it was found that biocomposite materials based on alginate and HA have high biocompatibility, good hydrophilicity, have no immunogenic properties, and capable of stimulating osteoplastic processes. The presence of sodium alginate contributes to the formation of stoichiometric hydroxyapatite with controlled nanosizes of crystallites [12].

Therefore, the purpose of this work was to create of the HA–collagen and HA–collagen–alginate composites using the saccharine method and their characterization by physical and chemical methods of analysis.

2 Experimental

2.1 Synthesis of HA

Synthesis of the HA was implemented as follows. Initially, a sucrose solution with the concentration of 0.1 mol/dm^3 was prepared, to which 2 g of calcium oxide was added and left for 7 days to form the solution of calcium saccharate. A stoichiometrically calculated amount of ammonium hydrophosphate was added to the obtained calcium saccharate solution and stirred. After that, ammonia water (25%) was added dropwise under stirring to pH 10. The procedure was carried out at 313–323 K. After the addition of all reagents, the mixture was stirred for another 1 h and left for 7 days aging. Thereafter, the suspension was filtered, and the resulting precipitate was dried at 353 K.

2.2 Synthesis of HA–Collagen Composite

Synthesis of the HAP–collagen composite was provided in the same way as HA synthesis, but collagen was also added together with ammonium hydrophosphate.

2.3 Synthesis of HA–Collagen–Alginate Composite

For the synthesis of the HA–collagen–alginate composite, 1%wt alginate solution was first prepared. After obtaining a viscous solution of alginate, an HA–collagen composite was added to it. The obtained suspension was added dropwise to the 5%wt $CaCl_2$ solution. As a result, the balls which contained HA–collagen–alginate were formed. The resulting balls were separated from the solution and air-dried.

2.4 X-Ray Analysis

XRD (X-ray diffraction) measurements of the HA and HA–collagen samples were conducted using X-ray diffractometer Ultima IV Rigaku with CuKa radiation [13, 14]. Samples were automatically analyzed by the PDXL software package using ICDD/PDF-2 database (standard card of HA is ICDD 9002214).

The diffraction patterns of the pure HA and HA–collagen composite are shown in Fig. 2. Both samples contained the crystalline phase of 100% hydroxyapatite. The crystallite size was 5.3 nm for the HA sample and 2 nm for the HA–collagen sample. Therefore, in both cases, nanocrystalline HA was obtained, but in the presence of

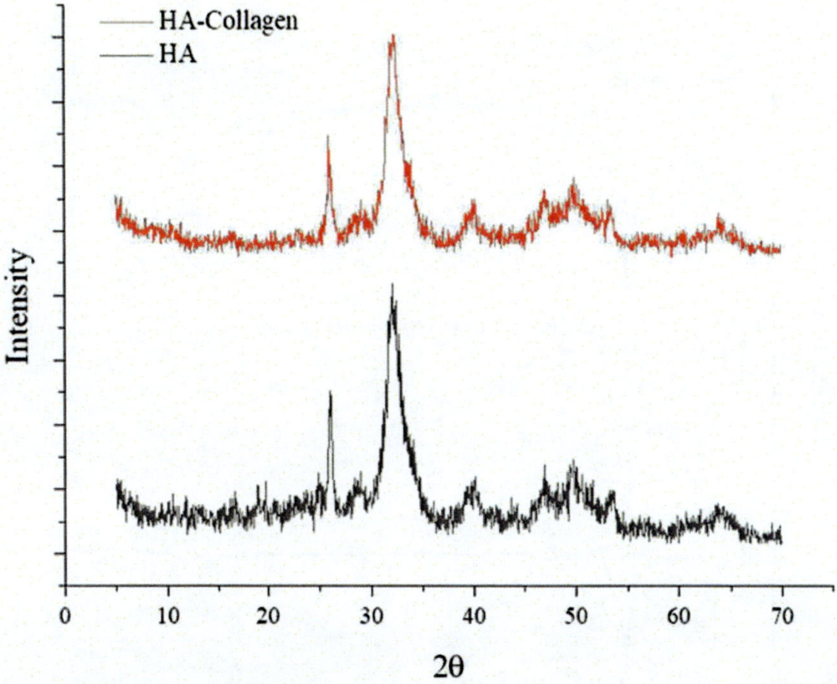

Fig. 2 The diffraction patterns of the samples

collagen, the crystallite size was 2.6 times smaller than the sample of pure HA, indicating the stabilizing effect of collagen during HA crystallization.

2.5 *IR Spectroscopy*

IR transmittance spectra were obtained on the Bruker vertex 70 V spectrometer in the range of 4000–400 cm^{-1}.

The IR spectra of samples of pure HA, composite HA–collagen, and pure collagen (for comparison) are shown in Fig. 3. The assignment of the characteristic peaks for all samples is shown in Table 1. As can be seen from Fig. 3 and Table 1, all characteristic PO_4^{3-} peaks are observed for samples containing HA. It should be noted that the presence of the absorption bands at 873–876 cm^{-1} and 1418–1461 cm^{-1} in the obtained spectra of hydroxyapatite-containing samples indicates the substitution of phosphate ions in the apatite by CO_3^{2-} groups, which is testifying the formation of carbonate-containing hydroxyapatite [15]. The characteristic peak at 3740 cm^{-1} for the OH group in HA was not found, which may indicate a low crystallization degree of the hydroxyapatite obtained by this method.

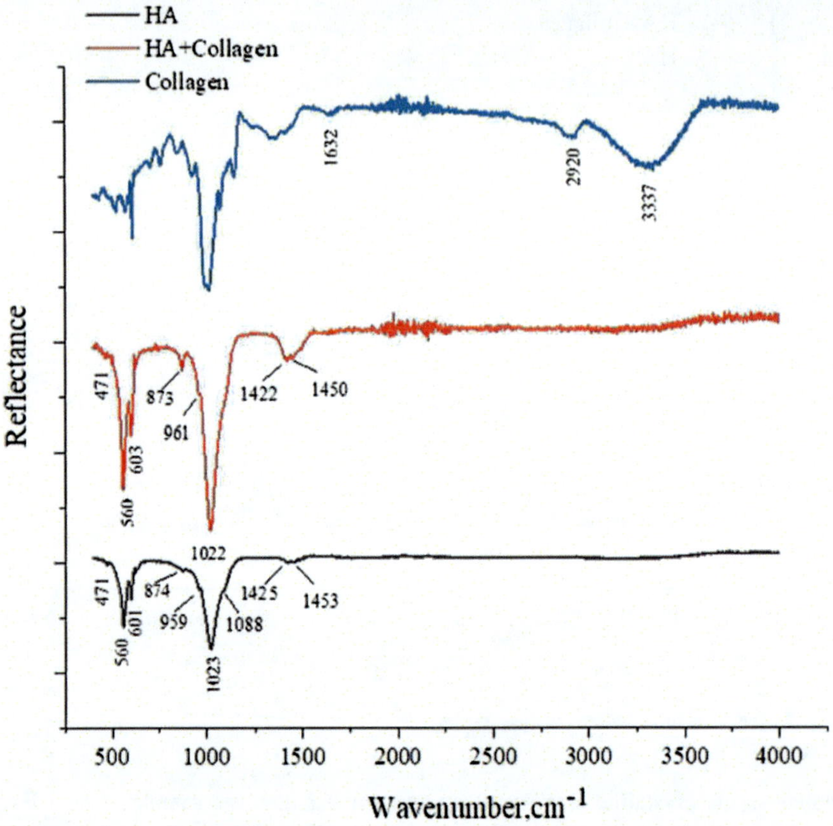

Fig. 3 IR spectra of the HA, HA–collagen composite samples, and pure collagen

Table 1 IR absorption bands for all tested samples [15, 16]

Adsorption band (cm^{-1})	Functional groups
HA	
471, 560, 601, 959, 1023, 1088	PO_4^{3-}
874	HPO_4^{2-}
1425, 1453	CO_3^{2-}
HA–collagen composite	
471, 560, 603, 961, 1022	PO_4^{3-}
873	HPO_4^{2-}
1422, 1450	CO_3^{2-}
Collagen	
1632	– CO (amide I)
2920	– CH_2–
3337	– NH (amide A)

Peaks 1632, 2920, and 333 cm^{-1} in the spectrum of the collagen sample indicate the presence of the characteristic for collagen functional groups –CO (amide I), $-CH_2-$, –NH (amide A), respectively [16]. However, in the HA–collagen sample, there are no peaks corresponding to collagen, which is, in our opinion, due to the chemical binding of hydroxyapatite with these groups.

2.6 *Raman*

Raman spectra for the determination of phase components were recorded in the range 100–1200 cm^{-1} using a three-stage Raman spectrometer T64000 (Horiba, Japan) with an Ar–Kr laser (Stabilite 2018-RM Spectra-Physics, USA). The excitation laser was operated at 488 nm (1 mW).

The Raman spectra of selected HA, HA–collagen composite samples, and pure collagen are shown in Fig. 4. According to the obtained spectra, the HA and HA–collagen samples have a clearly defined vibration band at 960 cm^{-1}, which corresponds to the symmetric stretching (P–O) mode. Oscillation bands characteristic of bending (O–P–O) mode at about 430–450 cm^{-1}, antisymmetric stretching (P–O) mode at 1020–1080 cm^{-1}, and bending (O–P–O) mode at 585–610 cm^{-1} were not

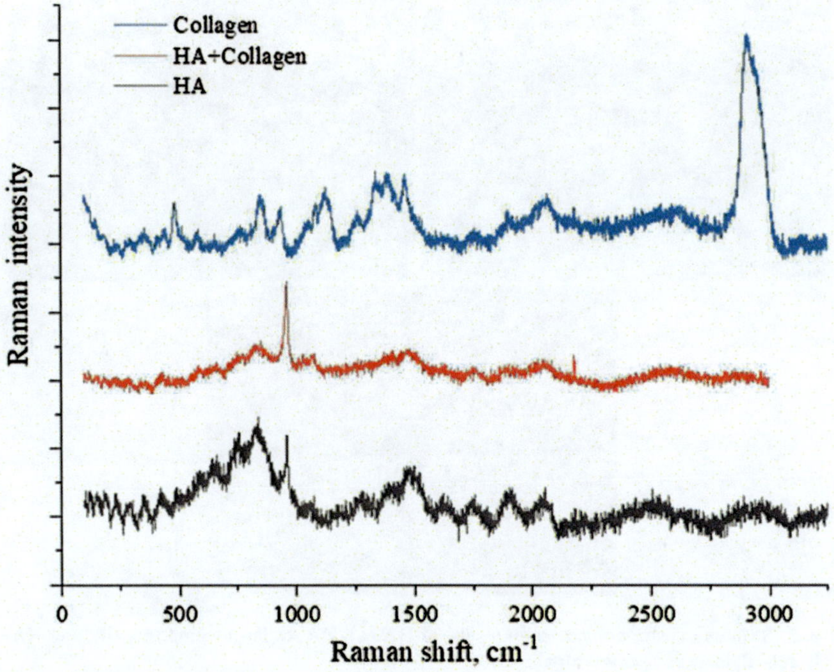

Fig. 4 Raman spectra of the HA, HA–collagen composite samples, and pure collagen

detected [17]. The main peak of pure collagen is 2856 cm^{-1}, which corresponds to the asymmetric and symmetric CH stretching modes [18].

Raman spectroscopy results are in a good agreement with X-ray diffraction and IR spectroscopy results.

2.7 Microscopy

Scanning electron microscopy studies were performed for samples of HA, HA–collagen, and HA–collagen–alginate composite. The latter was obtained in the form of balls to increase overall biocompatibility, facilitate storage, and use.

SEM images of HA, HA–collagen, and HA–collagen–alginate samples were obtained on the REM-106I scanning electron microscope. The obtained SEM images of the samples are presented in Fig. 5. It can be seen from the figure that pure HA has aggregates about 100–300 μm in size. HA–collagen is finer and has smaller particles than pure HA. The dimensions of the dual composite aggregates range from 50 to 250 μm. The morphology of the composite HA–collagen–alginate is characterized by a spherical shape with the ball's diameter of 50–250 μm. In addition, the balls have

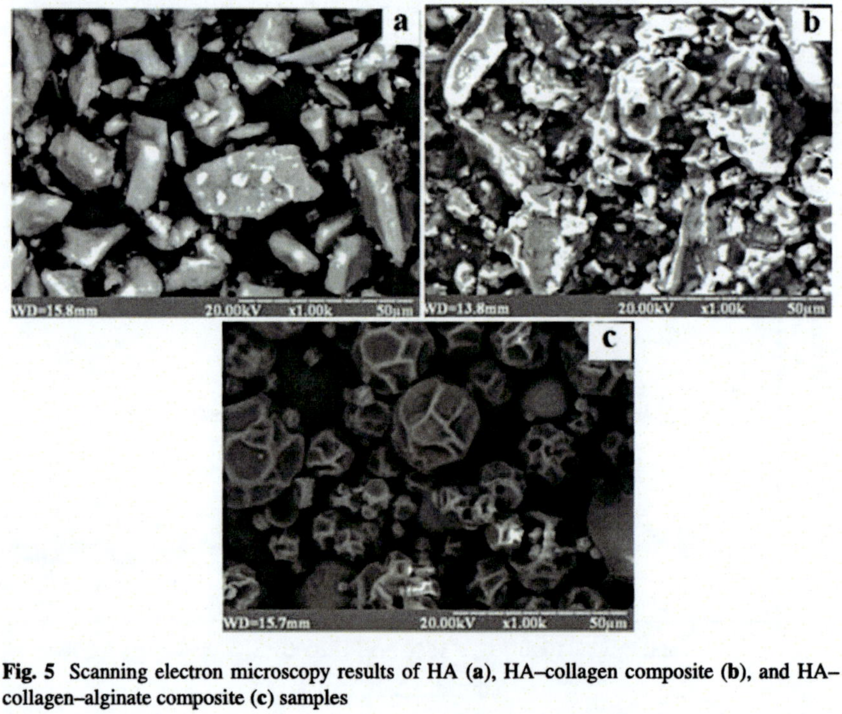

Fig. 5 Scanning electron microscopy results of HA (**a**), HA–collagen composite (**b**), and HA–collagen–alginate composite (**c**) samples

pores that will play a positive role in the processes of dissolution and regeneration when the balls get into bone tissue defects.

3 Conclusions

Synthesis of HA and its composites is carried out in a simple way by the saccharine method. Pure hydroxyapatite and two types of nanocomposites based on it have been synthesized—HA–collagen and HA–collagen–alginate. X-ray analysis established the pure phase of hydroxyapatite in all cases. IR spectroscopy indicates the chemical bond between collagen and HA. The presence of the carbonate ions in the HA structure was also found, which are characteristic of B-type hydroxyapatite. Scanning electron microscopy for HA, HA–collagen, and HA–collagen–alginate showed that aggregates with size 100–300 μm are characteristic for HA, 50–250 μm for HA–collagen, and aggregates of the HA–collagen–alginate composite have pores and are characterized by diameter 50–250 μm. This indicates a smaller size of HA aggregates in the composite, and alginate usage allows the production of composites that will better dissolve when they reach bone defects. All results are in good agreement with each other.

Acknowledgments The authors thank the Faculty of Physical Engineering of National Technical University of Ukraine «Igor Sikorsky Kyiv Polytechnic Institute» for their support in conducting this research.

References

1. Corno M, Orlando R, Civalleri B et al (2007) Periodic B3LYP study of hydroxyapatite (001) surface modelled by thin layer slabs. Eur J Mineral 5(19):757–767
2. Kay MI, Young RA, Posner AS (1964) Crystal structure of hydroxyapatite. Nature 204:1050–1052
3. Kramer E, Podurgiel J, Wei M (2014) Control of hydroxyapatite nanoparticle morphology using wet synthesis techniques: reactant addition rate effects. J Mater Lett 131:145–147
4. Abidi SA, Murtaza Q (2013) Synthesis and characterization of nano-hydroxyapatite powder using wet chemical precipitation reaction. J Mater Sci Technol 30:307–310
5. Orlovskii VP, Komlev VS, Barinov SM (2002) Hydroxyapatite and hydroxyapatite-based ceramics. Inorg Mater 38(10):973–984
6. Liu D-M, Troczynski T, Tseng WJ (2001) Water-based sol–gel synthesis of hydroxyapatite: process development. Biomaterials 22(13):1721–1730
7. Pramanik S, Agarwal AK, Rai KN (2007) Development of high strength hydroxyapatite by solid-state-sintering process. Ceram Int 33(3):419–426
8. Riman RE, Suchanek WL, Byrappa K et al (2002) Solution synthesis of hydroxyapatite designer particulates. Solid State Ionics 151(1–4):393–402
9. Varma HK, Kalkura SN, Sivakumar R (1998) Polymeric precursor route for the preparation of calcium phosphate compounds. Ceram Int 24(6):467–470

10. Safronova TV, Kazakova GK, Yevdokimov PV (2016) Ceramics based on calcium phosphate powder synthesized from calcium saccharate and ammonium hydrophosphate. Inorg Mater Appl Res 7(4):635–640
11. Manjubala I, Sivakumar M, Sureshkumar RV et al (2002) Bioactivity and osseointegration study of calcium ceramic of different chemical composition. J Biomed Mater Res 63(2):200–208
12. Kuznetsov VM, Sukhodub LB, Sukhodub LF (2014) Structural and substructural features of apatite-biopolymer composites: the comparison of data obtained using X-Ray diffraction and scanning electron microscopy with electron diffraction. J Nano Electr Phys 6(4) (Article ID 04039)
13. Nahirniak S, Dontsova TA, Astrelin I (2018) Directional synthesis of SnO_2-based nanostructures for use in gas sensors. Nanochem Biotechnol Nanomater Appl 214:233–245
14. Makarchuk O, Dontsova T, Perekos A et al (2017) Magnetic mineral nanocomposite sorbents for wastewater treatment. J Nanomater (Article ID 8579598)
15. Berzina-Cimdina L, Borodajenko N (2012) Research of calcium phosphates using fourier transform infrared spectroscopy. Riga Technical University, Institute of General Chemical Engineering Latvia, 510 pp.
16. Vidal BC, Mello MLS (2011) Collagen type I amide I band infrared spectroscopy. Micron 42:283–289
17. Nosenko VV, Yaremko AM, Dzhagan VM, Vorona IP, Romanyuk YA, Zatovsky IV (2016) Nature of some features in Raman spectra of hydroxyapatite-containing materials. J Raman Spectrosc 47(6):726–730
18. Lin S-Y, Li M-J, Cheng W-T (2007) FT-IR and Raman vibrational microspectroscopies used for spectral biodiagnosis of human tissues. Spectroscopy 21(1):1–30

TiO_2–ZnO Nanocomposites for Photodegradation of Dyes in Water Bodies

Olena Yanushevska, Tetiana Dontsova, Svitlana Nahirniak, and Vitalina Alisova

1 General

1.1 Introduction

The modern production of textiles, leather, plastics, rubber, medicines, paper, and other household goods generates a large number of highly carcinogenic wastewater containing persistent toxic organic compounds, including dyes [1–3]. The magnitude of the damage caused by coloring agents to the environment and human health induces the use of high-quality physical, chemical, and biological methods of water purification (adsorption, reagent chemical purification, flocculation, bioremediation, oxidation, ultraviolet treatment, etc.) [4, 5]. However, the formation of intermediate and secondary contaminants or the simple redistribution of pollutants between phases limit the use of conventional methods and this only leads to partial, temporary accumulation of toxicants, and does not guarantee their complete disposal. Therefore, recently a new direction of purification is being actively formed—the method of photocatalytic degradation of organic toxicants in the aquatic environment. This method is promising because of its high catalytic activity, the possibility of repeated

O. Yanushevska (✉) · T. Dontsova · S. Nahirniak · V. Alisova
Igor Sikorsky Kyiv Polytechnic Institute, Kyiv, Ukraine
e-mail: l_rrr@ukr.net

T. Dontsova
e-mail: dontsova@ua.fm

S. Nahirniak
e-mail: nagirnyak_sv@ukr.net

V. Alisova
e-mail: vitalinavitaliivna05@gmail.com

O. Fesenko and L. Yatsenko (eds.), *Nanomaterials and Nanocomposites, Nanostructure Surfaces, and Their Applications*, Springer Proceedings in Physics 246, https://doi.org/10.1007/978-3-030-51905-6_49

catalysts usage, and simultaneous purification of the pollutant mixtures, complete destruction of contaminants, low cost, and the ease of operation.

The photocatalytic degradation method is based on the generation of effective oxidizing hydroxyl radicals by metal oxides (semiconductors) under the action of UV radiation. Semiconductor materials (TiO_2, ZnO, Fe_2O_3, WO_3, CdS, ZnS, Cu_2O, CuO, SnO_2, and V_2O_5) are irradiated with UV waveband, which transmits additional energy to the electrons to overcome the bandgap for their transition into the conduction band. The generated hydroxyl and oxide radicals react with the pollutants, which leads to their complete degradation.

Given the large bandgap (3.2 eV) and the high rate of recombination of «electron–hole» processes that limit TiO_2 catalytic activity, studies to improve the efficiency of titanium (IV) oxide photocatalysts by its doping with metal oxides are conducted [6]. Among the metal oxides zinc oxide (ZnO), which has a similar bandgap (3.37 eV) and greater catalytic activity in corrosive environments, is distinguished. The TiO_2–ZnO nanocomposite exhibits better photocatalytic properties than both oxides separately by extending the «life» time of the photo-excited pairs—«hole–electron.»

Therefore, the use of binary nanocomposites increases the sorption and photocatalytic activity of nanomaterials, which opens the prospects for optimization of water treatment processes.

1.2 *Structural Modifications of Titanium (IV) and Zinc (II) Oxides*

TiO_2 can form many structural modifications, but only three of them are naturally occurring—rutile, anatase, and brookite (Fig. 1). They are characterized by tetragonal

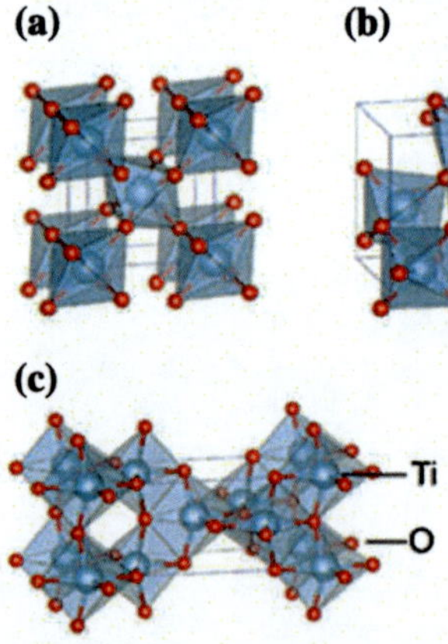

Fig. 1 Crystal structure of TiO_2 polymorphs: **a** rutile, **b** anatase, **c** brookite [7]

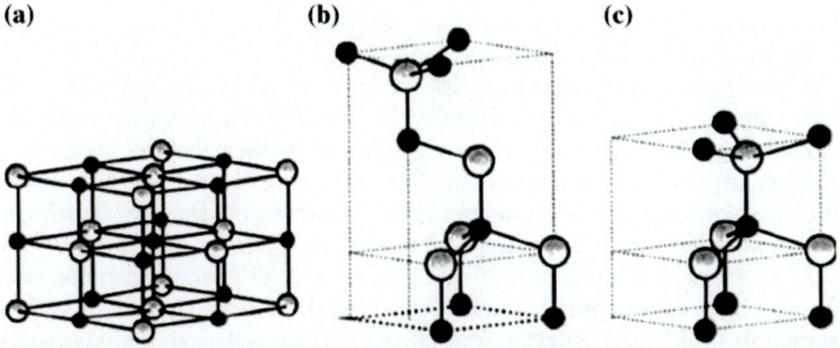

Fig. 2 Crystal structure of ZnO: **a** cubic rocksalt, **b** cubic zinc blende, **c** hexagonal wurtzite [8]

(rutile and anatase) and orthorhombic (brookite) structures. Rutile and anatase are the main types of TiO_2 structures that have prospects for use in photocatalytic processes.

ZnO_2 crystallizes in three modifications—cubic rocksalt, cubic zinc blende, and hexagonal wurtzite (Fig. 2).

2 Photocatalytic Activity of TiO_2–ZnO Nanocomposites

2.1 The Problem of Sewage Pollution by Organic Contaminants

Constantly growing part of the population in the world does not have access to safe water (water that meets current water supply standards), which has severe consequences, as failure to meet sanitary conditions for drinking water causes the diseases spread and increases the risk of environmental disasters. According to statistical studies, the growth of industry and consumer demand in both developing and developed countries, combined with the absence of rigid waste disposal standards, exacerbate the aforementioned problem, especially on municipal water resources. Heavy metals, microorganisms, pharmaceuticals, pesticides, dyes, etc., are among the most common pollutants of surface and groundwater. Undoubtedly, the scarcity of clean water is gradually becoming one of the most noticeable problems on the planet with unpredictable economic and social risks [9].

Currently, universal dyes are needed for the rapid development of modern industries such as textiles, leather, paper, printing, food, pharmacy, etc. Dyes are the main compounds that can give substances a bright color, and their structures usually contain unsaturated chromophore and auxochrome. For the preparation of dyes, auxochromes are attached to chromophores to achieve deep coloration of the product. That is, auxochromes are a group of atoms that, in combination with the corresponding chromophore, enhance color, and chromophores are constituents of molecules that absorb or reflect certain wavelengths when light is incident on them. They are used together for the manufacture of dyes [10].

Most organic dyes and pigments that are used in the industry, getting into the body in the form of dust, lead to a general toxic effect. They cause conjunctivitis, skin diseases of open body parts mainly, in severe cases spread throughout the body, sometimes because of contact with colored products. Numerous occupational diseases can arise not from the organic dyes and pigments themselves, but from the toxic impurities of intermediates or from dangerous metabolites formed from organics in the body. Wastewater treatment, which simultaneously contains more than one category of pollutants, i.e., pharmaceuticals and dyes, is a serious problem of modern pharmacological, food, and textile production. Preparation of photocatalysts that are effective for solving these problems is the goal of many scientific studies [9].

Such methods as adsorption, membrane filtration, coagulation/flocculation, ion exchange, microbial degradation, and advanced oxidation processes are successfully used in modern technologies to remove dyes from wastewater. In comparison with most of the above methods, adsorption has advantages because of its high efficiency, selectivity, ease of operation, and design of equipment. Instead, when using sorbents, there is a problem of regeneration of waste sorption materials, which only accumulate fluid or gas-phase pollutants but do not expose them to degradation, eliminating toxicity [10].

Methods based on oxidation reactions, such as heterogeneous photocatalysis, are considered to be the best among many methods of water purifying from organic compounds (petroleum products, dyes, bacteria, etc.). The variety of photocatalysts application brings this method to the level of a separate scientific direction in the processes of neutralization of most organic pollutants. Several semiconductors (e.g., TiO_2, ZnO, Fe_2O_3, CdS, GaP, and ZnS) have been successfully used as advanced oxidation photocatalysts. Nanostructured TiO_2 is the most studied material among the above-mentioned so far. In the recent researches, however, ZnO emerges as a strong competitor to TiO_2 in wastewater treatment technologies.

2.2 Prerequisites for Creating of TiO_2–ZnO Composites

Titanium oxide is mechanically strong, chemically resistant to dilute solutions of acids and alkalis, its own solubility is low, it does not give secondary contaminants (unlike the used natural minerals). The sorption activity of TiO_2 to various naturally occurring impurities is high due to the presence on its surface of both basic ($TiO^{\delta-}$) and acid ($Ti^{\delta+}$) centers.

Zinc oxide is a semiconductor and promising material due to its favorable piezoelectric and pyroelectric properties, which are used in optoelectronics, thin-film transistors, biosensors, gas sensors, bactericidal agents, and photocatalysis. Chemical stability, low toxicity, and resistance to oxidation contribute to these applications. It is noted that the morphology, crystal size, and orientation of ZnO nanocrystals play a key role in achieving the desired response [11]. This is due to the fact that, despite the broad bandgap in both semiconductors (absorbing only a small (~5%) fraction of solar radiation), ZnO exhibits mobility of photo-excited electrons at least 2 orders of magnitude higher than that of TiO_2 and it is characterized by stable photocatalytic properties in the degradation process of dye molecules in both acidic and alkaline environments [12].

The manifestation of similar properties of TiO_2 and ZnO under different conditions makes the use of these oxides to create photocatalysts composites logical and promising. Therefore, interest in the photocatalytic properties of mixed oxide nanomaterials, which can potentially combine the properties of individual oxides, leads to the search for new synthesis methods of composites. Increased photostability, accompanied by improved photoactivity, is one of the factors that justify the creation of mixed oxide nanocrystals, which are considered as promising photocatalysts for environmental applications, in particular for the decomposition of dyes. The interaction of TiO_2 and ZnO in nanostructures enhances the physical and chemical properties (such as electrical conductivity and charge carrier concentration) of devices used in lithium batteries, supercapacitors, and fuel cells.

Thus, the authors of [13] consider semiconductor oxides to be unique in nature due to the presence of partially filled d-orbitals. This feature of the transition metal structure allows changing the electronic properties of semiconductors, including by introducing impurities into their crystalline lattice, which can intensify the heterojunctions in the photolysis process. The effect of accelerating the generation and heterojunctions of photographic charges with the possibility of forming an additional energy level between the conduction band and the valence band leads in turn to a shift of the photo-irradiation to the visible light range [14, 15] and the bandgap reduction [16] (Fig. 3). The introduction of modifiers into the nodes of the crystal lattice, according to the authors [17], has a greater impact on the stability of the catalysts and the possibility of their repeated use in comparison with the introduction of alloying additives in the space between the nodes of the crystal lattice.

Due to the small particle size, semiconductor catalysts are distinguished by high specific surface area and the presence of active centers that stimulate photocatalytic efficiency. However, nanoscale photocatalysts tend to aggregate in aqueous solutions, which significantly reduce the use of light in the photocatalyst process. In addition, the

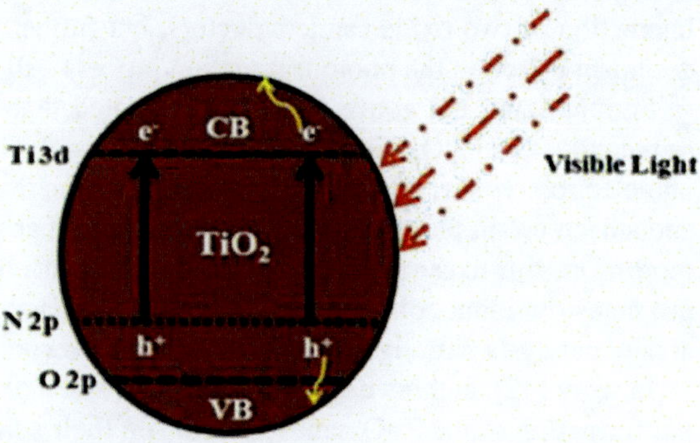

Fig. 3 Scheme of the additional energy level occurrence between the conduction band and the valence band after nitrogen doping

photocatalyst in nanoscale range complicates the further processing of suspensions, and suspended nanoparticles remaining in the water can have harmful toxic effects in the subsequent use of purified water.

Thus, titanium oxide (36 nm) was applied by using sol–gel method, adsorption, and catalytic activity were evaluated in the process of adsorption and photodegradation of the Methylene blue dye in the UV irradiation range [18]. Four samples with different TiO_2 content (12.3, 24.8, 40.7, and 48.3 wt%) were investigated, among which the highest photocatalytic activity showed the sample with 40.7 wt% TiO_2, which completely degraded the dye molecules within 60 min. It was also noted that the adsorption capacity of the nanocomposite on the microspherical carrier to the methylene blue dye was inversely proportional to the content of TiO_2 nanoparticles. Instead, the photocatalytic efficiency of the nanocomposites under study showed a direct correlation with the titanium (IV) oxide concentration. Surface electron microscope images of microspheres-based titanium-based photocatalysts indicate that an excessive increase in the content of TiO_2 nanoparticles may also lead to the aggregation of microspheres and a decrease in light energy access to the catalyst surface and, consequently, a decrease in their activity.

Similar studies were performed and presented in [19], in which synthesized TiO_2-based nanocomposites were investigated in the process of phenol photo-degradation during irradiation in the visible light range ($\lambda \geq 420$ nm). It was found that usage of the 95 wt% TiO_2 content composite leads to the 68% photodegradation degree of the pollutant.

Work [11] presented a simple and effective approach to increase the area available for light energy in photocatalyst based on dispersed TiO_2 particles and 1D ZnO structures. The formation of nanostructured zinc rods dispersed on polymer titanium nanoparticles, which serve as centers of crystallization and growth of ZnO nanostructures (21–24 nm), was realized. The efficiency of the obtained nanocomposite as a photocatalyst is explained by a significant increase in the surface area available for absorption of light energy, a high concentration of charge carriers, which minimizes the number of recombination of electrons and «holes» in photocatalytic processes. The obtained composite was studied in the degradation reaction of the organic dye Rhodamine B when irradiated with UV and visible light. The degree of photodegradation was 90% for 70 min with the maximum dye absorption in the UV region of the light spectrum. This indicates that there is a photo-synergistic effect due to the interaction of two oxide semiconductors, but further studies have revealed that the dominant effect on the photodegradation process still provides ZnO nanostructures.

Summarizing the above, it should be noted that the photocatalytic activity of semiconductors (TiO_2 and ZnO), if used separately, is significantly reduced due to one-electron recovery, a high degree of recombination of «electron–hole» transfers, and a decrease in the specific surface area of light absorption. Therefore, it is much more promising to use nanocomposite materials that combine photo-synergies, which promotes the induction of multi-electron transmissions, increases the total contact area of catalysts with light energy, and reduces recombinant effects.

In work [12], authors used photocatalysts based on TiO_2 nanotubes with sequential formation of the ZnO nanostructures on their surface by a chemical deposition method. The photocatalytic activity of the obtained catalysts was tested for the degradation reactions of Rhodamine B in UV and near-visible light irradiation (254 nm, 313 nm, 350 nm, and 420 nm, respectively). It has been shown that under UV irradiation, the decomposition of the dye occurs by 100% and gradually the degradation degree of the contaminant is reduced to 4% in the visible region of the irradiation. Studies have confirmed the presence of a synergistic effect in the reaction of Rodamin B photodestruction. The percentage of the dye solution degradation on TiO_2 nanotubes was 65%, whereas in the presence of ZnO nanoparticles, this percentage reached up to 72, 83, 91, and 94% according to the number of electrochemical deposition cycles—2, 4, 6, 8. The photocatalytic activity of the samples increased depending on the ZnO content, but had a certain limit, due to the fact that the deposition of more than eight cycles of zinc oxide was subjected to total agglomeration on the titanium (IV) oxide surface and screened the surface-active centers. That is, further deposition of zinc oxide leads to the formation of the dominant layer of ZnO nanoparticles on the surface of TiO_2 nanotubes and loss of synergistic effect between nanotubes and nanoparticles.

The obtained composites were titanium-containing nanotubes with deposited zinc oxide particles in the form of grains (Fig. 4). The effect of enhancing photocatalytic properties when using binary TiO_2–ZnO nanocomposites can be demonstrated as follows. Photo-irradiation generates the formation of «holes» (h^+) and electrons (e^-) that interact with surface hydroxyl and hydrate particles to form hydroxyl (OH·) and oxide (O_2·) radicals. Further, the active radical Oxygen forms cause degradation of organic dye molecules [12].

The confirmation of the synergistic effect of the nanostructured tandem TiO_2 and ZnO was obtained by the authors [20] for the photodegradation reaction of methylene orange and nitrophenol. Pure ZnO (NPs), pure TiO_2 (P25), urchin-like TiO_2 structure and ZnO–TiO_2 composite, which was synthesized by two-stage hydrothermal

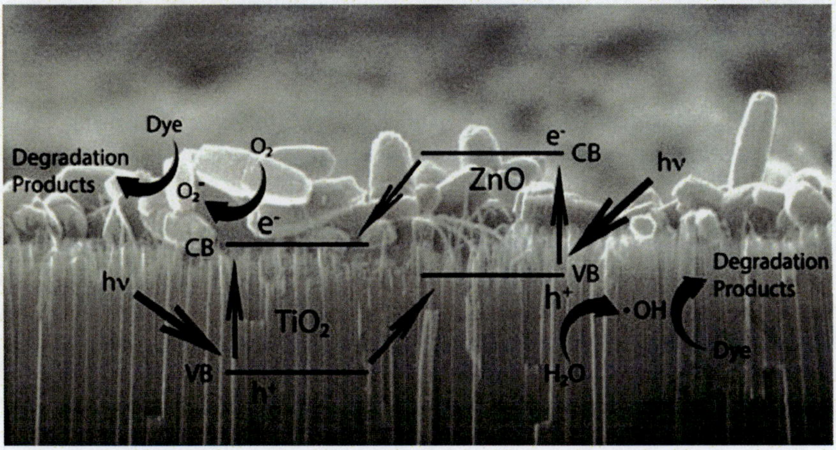

Fig. 4 Scheme of the photodegradation mechanism of organic pollutants [12]

synthesis, were investigated. In the first stage of this synthesis, 3D urchin-like TiO_2 nanospheres were grown, in the second stage 1D ZnO structures were synthesized. According to the BET, the surface area of the ZnO–TiO_2 composite was 187.3 m^2/g, a pore diameter of about 8 nm. The photocatalysis results showed a high catalytic activity of ZnO–TiO_2 compared to the pure ZnO (NPs), TiO_2 (P25) and urchin-like structure (TiO_2). Moreover, the photodegradation of both pollutants improves with increasing irradiation time, which indicates their complete decomposition. The authors explain this synergy by the fact that, unlike the short «life» time of the photo-induced charge carrier in a single semiconductor, rapid separation of photogenerated electrons and holes in the ZnO–TiO_2 composite occurs due to the high recombination rate of the «electron–hole» pairs. This caused by close phase interaction in the ZnO/TiO_2 surface heterostructure. Such heterostructure also contributes to the broadening of the incident light spectrum, which is able to intensify photocatalysis.

Thus, the above studies prove that the use of nanocomposite photocatalysts based on binary TiO_2–ZnO systems can be an effective method for removing hazardous organic toxicants from wastewater. This process is facilitated by the features of the surface structure of TiO_2–ZnO systems and the synergistic effect resulting from the activation of the adsorption process of organic molecules and increasing in the rate of charge transfer in «electron–hole» systems.

The synergistic effect in the photocatalytic decomposition of dyes on binary ZnO–TiO_2 systems observed in the above works is also confirmed by the study [21]. According to which, the highest photodegradation efficiency of methylene blue and naproxen (NPX) is inherent in zinc oxide. It was found that the photocatalytic activity of ZnO–TiO_2 depends on the type of zinc titanate molecule (Zn_2TiO_4) but is insignificantly higher than the photoactivity of TiO_2 nanoparticles.

In [22], commercial ZnO and TiO_2 nanopowders and a mixed ZnO–TiO_2 nanopowder (molar ratio of 2:1) were compared. Specific surface measurements showed that TiO_2 has a much larger surface area (9.3 m^2/g) than ZnO (5.1 m^2/g), but the structural features of the ZnO nanopowder have a dominant effect on the specific surface area of the ZnO–TiO_2 mixture (4.7 m^2/g). This fact confirms the absence of a clear correlation between the photocatalytic activity and the specific surface area of the metal oxides but leads to the conclusion about the significant influence of the ZnO nanoparticles geometry (nanorod structure) on the selective and active properties of the photocatalyst. The X-ray diffraction method showed the presence of the new phase of zinc titanate (Zn_2TiO_4) corresponds to the cubic spinel structure, in addition to the two initial phases of ZnO and TiO_2 metal oxides (hexagonal ZnO structure and tetragonal TiO_2 structure anatase). The calculated average values of crystallites are 83, 74, and 84 nm for ZnO, Zn_2TiO_4, and TiO_2, respectively. This fact also testifies to the dominant influence of the ZnO nanorod structure rather than its particle size.

According to a literature review of scientific articles [23–25], the authors observed a similar behavior of ZnO in comparison with TiO_2, which is explained by the higher efficiency of generation, mobility, and separation of photoinduced electrons and «holes» in ZnO. However, the authors also cited the facts [26] that such ZnO properties are not a general rule and TiO_2 was more active during the degradation of methylene blue along with the inactivation of certain bacteria. The authors attribute this effect to the higher concentration of active Oxygen forms induced on TiO_2.

In addition to the high photodegradation activity of methylene blue ZnO nanopowder also demonstrates greater efficiency to the destruction of naproxen among the tested samples, but with a significantly smaller advantage over TiO_2. It should be noted that no information on the ZnO usage as photocatalyst in the degradation reaction of naproxen has been found in the literature. At the same time, information on the use of TiO_2 for the destruction process of naproxen is present in the literature [27–29]. The analysis of the literature search shows that ZnO demonstrated greater efficiency in the reaction of degradation of naproxen (the rate of photocatalytic degradation is 0.00764 min compared with the rate of 0.006 min for TiO_2 in the similar reaction).

The authors of the following work [30] prefer the finding a scientific approach to explain the synergism and efficiency of photocatalytic activity of ZnO/TiO_2 catalysts precisely in the methods of catalyst synthesis (sol–gel method and deposition method) and in the mass ratio of ZnO to TiO_2 (ZnO:TiO_2 = 0.1…0.9). The authors argue that the crystalline structure, morphological, optical, catalytic properties of the composites are directly dependent on the mass ratio of the components and the synthesis method of ZnO–TiO_2 systems.

The study of the photocatalytic properties of Zn(SG)–Ti–R (sol–gel method) and Zn(Pr)–Ti–R (deposition method) was conducted using degradation of Methyl orange dye in aqueous solution under the action of UV irradiation (λ = 365 nm). In their previous studies [31], the authors reported the influence of methods and the nature of precursors on the morphology, structure, optical properties, and photocatalytic properties of synthesized ZnO. It is known [32–34] that the addition of ZnO to TiO_2 leads to the formation of the crystalline phase of zinc titanate ($ZnTiO_3$, Zn_2TiO_4). X-ray studies showed that crystalline anatase phase is present in pure TiO_2, but with an increase in the ZnO mass percentage in the sample, the intensity of the anatase peaks markedly decreases and, at mass ratio of 0.5, the Zn(SG)–Ti–R sample becomes amorphous. At a mass ratio of 0.67–0.9, the ZnO phase (wurtzite) appears. In contrast, the Zn(Pr)–Ti–R sample with ZnO content of 0.33% is characterized by the reflexes inherent the Zn_2TiO_4 titanate phase, and the wurtzite peaks are observed when the ZnO content is increased to 0.8…0.9%.

It should be noted that in no case crystalline phases of TiO_2 and ZnO–wurtzite are presented simultaneously in the samples obtained by different methods (sol–gel method and deposition method). This fact can be explained by the presence of these phases in the amorphous state, which is correlated with the results of studies [35]. The study of photocatalytic activity led to the conclusion that the activity of the Zn(SG)–Ti–R sample is better with a mass content of ZnO $\geq$ 0.67 (MO conversion ~ 97%). Whereas in the case of the Zn(Pr)–Ti–R sample, the methyl orange conversion decreases with a mass content of ZnO = 0.2 and gradually increases to 90% with an increase in the ratio of components to 0.9 (Fig. 5).

Increasing the initial concentration of the dye in all cases showed a decrease in photocatalytic activity in the process of its photodegradation. A similar effect has been claimed previously by many authors [36–38] and is explained by the fact that,

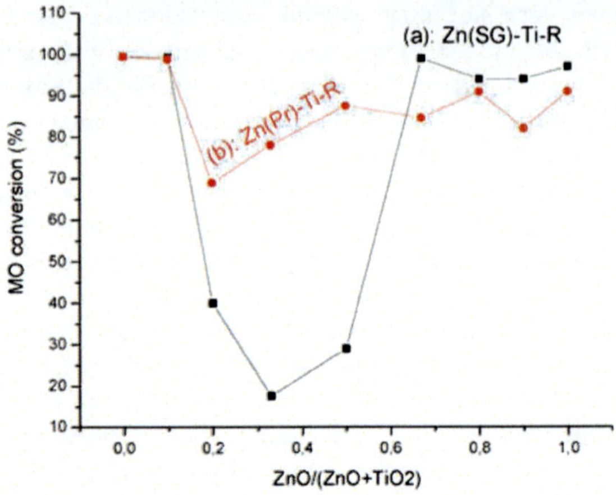

Fig. 5 Degradation of methyl orange (MO) depending on the synthesis method and the zinc oxide content [31]

at higher dye concentrations, the number of recombined photons increases before they reach particles of photocatalysts.

Interestingly, the authors' opinion that the efficiency of methyl orange degradation cannot be related only to the energy of the bandgap. This thesis, as stated above, correlates well with the conclusions of some authors, including in [39–43]. It is a well-known fact that the low bandgap energy of metal oxides has a positive effect on photocatalytic activity due to facilitated initiation of the photocatalytic reaction. However, as the experiments show, the morphology, crystalline structure and band gap have a significant influence on the photocatalysis, which only confirms the general thesis that new ways of ZnO and TiO_2 modification, optimal methods for synthesis of composites, conditions for photodegradation processes, etc., need to be found.

The analysis of the photodegradation efficiency of methyl orange, depending on the width of the forbidden zone of the catalysts, can serve as confirmation of this conclusion. Therefore, at the ZnO content in the range from 0.2 to 0.5, the Zn(SG)–Ti–R particles are amorphous and agglomerated as spheres up to 400 nm in size. In this case, the bandgap is maximal with zinc oxide content of 0.5, but the extraction of methyl orange at this value (0.5) on the contrary begins to increase. Based on this example, it can be concluded that the absence of the available correlation between bandwidth and photoreaction efficiency implies a much more complex mechanism for the interaction of different photocatalyst characteristics and the photolysis claimed above (Fig. 6).

Analysis and comparison of apparent degradation constants of methyl orange show that with increasing zinc oxide content up to 0.5, the velocity initially decreases and with subsequent increase in ZnO concentration it increases. These results do not confirm the findings of some researchers [41] but correlate very well with other authors [43].

Fig. 6 Dependence of band gap on zinc oxide content [33]

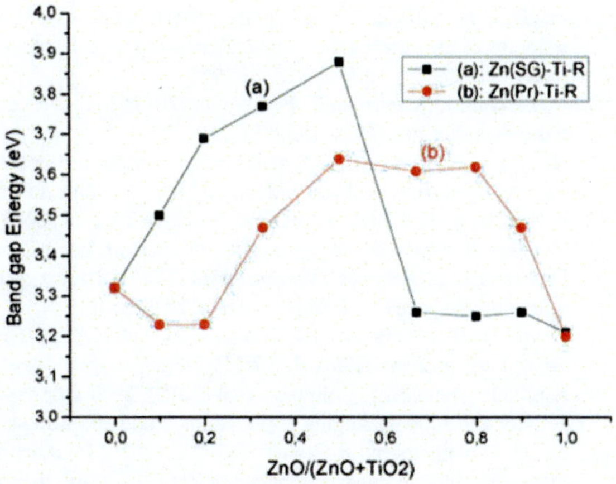

3 Conclusions

Therefore, the method of photocatalytic oxidation can effectively neutralize dyes and other organic compounds, decomposing them to low molecular weight nontoxic molecules, when used effective reusable photocatalysts, which makes their use economically feasible. The similarity of photocatalytic properties, including virtually the same bandwidth of TiO_2 and ZnO, makes their joint use a successful solution to the problems of low photocarriers concentration, high speed of recombination in tandem «electron–hole», complications of light absorption in the visible region, and therefore the photoreaction rate, which allows us to confirm the synergy of TiO_2–ZnO composites. It was shown that to create effective photocatalysts, researchers resort to finding different levers of influence on their activity, from which it is possible to distinguish: synthesis method, choice of carriers for oxide components immobilization, choice of metal oxides modifiers, synthesis method, particle morphology and surfaces of nanomaterials (including forming of 1D nanostructures), degree of crystallinity/amorphousness of photocatalysts, methods of increasing the photosensitivity of the surface of catalyst nanoparticles (sensitization), temperature regime of catalyst synthesis, mass ratios of components of ZnO–TiO_2 systems, etc.

Acknowledgments The authors thank students of the National Technical University of Ukraine «Igor Sikorsky Kyiv Polytechnic Institute» for actively participating in the discussion on the study topic.

References

1. Blachnio M, Budnyak TM, Derylo-Marczewska A et al (2018) Hybrid composites for removal of sulfonated azo dyes from aqueous solutions. Chitosan-Silica 34(6):2258–2273

2. Paździor K, Bilińska L, Ledakowicz S (2018) A review of the existing and emerging technologies in the combination of AOPs and biological processes in industrial textile wastewater treatment. Chem Eng J 696(15):133961
3. Makarchuk O, Dontsova T, Perekos A (2017) Magnetic nanocomposite sorbents on mineral base. Springer Proc Phys 195:705–719
4. Hassan MM, Carr CM (2018) A critical review on recent advancements of the removal of reactive dyes from dyehouse effluent by ion-exchange adsorbents. Chemosphere 209:201–219
5. Mykhailenko NO, Makarchuk OV, Dontsova TA et al (2015) Purification of aqueous media by magnetically operated saponite sorbents. Eastern Eur J Enterp Technol 10(76):13–20
6. Dontsova TA, Nahirniak SV, Astrelin IM (2019) Metaloxide nanomaterials and nanocomposites of ecological purpose. J Nanomaterials 2019:5942194
7. Haggerty JES, Schelhas LT, Kitchaev DA et al (2017) High-fraction brookite films from amorphous precursors. Sci Rep 7:15232
8. Espitia P, Soares NFF, Coimbra J et al (2012) Zinc oxide nanoparticles: synthesis, antimicrobial activity and food packaging applications. Food Bioprocess Technol 5(5):144–1464
9. Štrbac D, Aggelopoulos C, Štrbac G et al (2018) Photocatalytic degradation of naproxen and methylene blue: comparison between ZnO, TiO_2 and their mixture. Process Saf Environ Prot 113:174–183
10. Zhou X, Wang L, Liu X et al (2019) Organic/inorganic hybrid consisting of supportive poly (arylene ether nitrile) microspheres and photocatalytic titanium dioxide nanoparticles for the adsorption and photocatalysis of methylene blue. Compos B Eng 177:107414
11. Araújo ES, DaCosta BP, Oliveira RAP et al (2016) TiO_2/ZnO hierarchical heteronanostructures: synthesis, characterization and application as photocatalysts. J Environ Chem Eng 4:2820–2829
12. Bozkurt C, Caglar B, Kilinc T et al (2019) Synthesis and characterization of ZnO nanorice decorated TiO_2 nanotubes for enhanced photocatalytic activity. Mater Res Bull 106:160–167
13. Ani IJ, Akpan UG, Olutoye MA et al (2018) Photocatalytic degradation of pollutants in petroleum refinery wastewater by TiO_2- and ZnO-based photocatalysts: recent development. J Clean Prod 205:930–954
14. Lavand AB, Malghe YS (2015) Visible light photocatalytic degradation of 4-chlorophenol using C/ZnO/CdS nanocomposite. J Saudi Chem Soc 19(5):471–478
15. Wei Z, Liang F, Liu Y et al (2017) Photoelectrocatalytic degradation of phenol-containing wastewater by TiO_2/g-C_3N_4 hybrid heterostructure thin film. Appl Catal B 201:600–606
16. Liu F, Yu J, Tu G et al (2017) Carbon nitride coupled Ti-SBA15 catalyst for visible-light-driven photocatalytic reduction of Cr (VI) and the synergistic oxidation of phenol. Appl Catal B Environ 201:1–11
17. Dawson M, Soares GB, Ribeiro C (2014) Influence of calcination parameters on the synthesis of N-doped TiO_2 by the polymeric precursors method. J Solid State Chem 215:211–218
18. Zhou X, Wang L, Liu X et al (2019) Organic/inorganic hybrid consisting of supportive poly(arylene ether nitrile) microspheres and photocatalytic titanium dioxide nanoparticles for the adsorption and photocatalysis of methylene blue. Compos B 177:107414
19. Islam MdR, Kumar Chakraborty A, Gafur MA et al (2014) Easy preparation of recyclable thermally stable visible-light-active graphitic-C_3N_4/TiO_2 nanocomposite photocatalyst for efficient decomposition of hazardous organic industrial pollutants in aqueous medium. Res Chem Intermed 45(4):1753–1773
20. Cheng P, Wang Y, Xu L et al (2016) High specific surface area urchin-like hierarchical ZnO-TiO_2 architectures: hydrothermal synthesis and photocatalytic properties. Mater Lett 175:52–55
21. Strbac D, Aggelopoulos CA, Strbac G et al (2017) Photocatalytic degradation of naproxen and methylene blue: comparison between ZnO, TiO_2 and their mixture. Process Saf Environ Prot 113:174–183
22. Strbac D, Aggelopoulos CA, Strbac G et al (2017) Photocatalytic degradation of naproxen and metylene blue: comparison between ZnO, TiO_2 and their mixture. Process Saf Environ Prot 113:174–183

23. Li Y, Xie W, Hu X et al (2010) Comparison of dye photodegradation and its coupling with light-to-electricity conversion over TiO_2 and ZnO. Langmuir 26:591–597
24. Aggelopoulos CA, Dimitropoulos M, Govatsi K et al (2017) Influence of the surface-to-bulk defects ratio of ZnO and TiO_2 on their UV-mediated photocatalytic activity. Appl Catal B 205:292–301
25. Law M, Greene LE, Johnson JC et al (2005) Nanowire dyesensitized solar cells. Nat Mater 4:455–459
26. Barnes RJ, Molin R, Xu J et al (2013) Comparison of TiO_2 and ZnO nanoparticles for photocatalytic degradation of methylene blue and the correlated inactivation of gram-positive and gram-negative bacteria. J Nanopart Res 15:1432–1442
27. Houasa A, Lachheba H, Ksibia M et al (2001) Photocatalytic degradation pathway of methylene blue in water. Appl Catal B 31:145–157
28. Kanakaraju D, Motti CA, Glass BD et al (2015) TiO_2 photocatalysis of naproxen: effect of the water matrix, anions and diclofenac on degradation rates. Chemosphere 139:579–588
29. Méndez-Arriaga F, Esplugas S, Gimenez J (2008) Photocatalytic degradation of nonsteroidal anti-inflammatory drugs with TiO_2 and simulated solar irradiation. Water Res 42(3):585–594
30. Beegam A, Prasad P, Jose J et al (2016) Environmental fate of zinc oxide nanoparticles: risks and benefits. Toxicolology new aspects this scientific conundrum, p 33
31. Mragui A El, Daou I, Zegaoui O (2016) Influence of the preparation method and ZnO/(ZnO + TiO_2) weight ratio on the physicochemical and photocatalytic properties of ZnO-TiO_2 nanomaterials. Catal Today 321–322:41–55
32. Mullerova J, Sutta P, Medlin R et al (2016) Optical properties of zinc titanate perovskite prepared by reactive RF sputtering. J Electr Eng 68(7):10–16
33. García-Ramírez E, Mondragón-Chaparro M, Zelaya-Angel O (2012) Band gap coupling in photocatalytic activity in ZnO–TiO_2 thin films. Appl Phys A Mater Sci Process 108(2):291–297
34. Moradi S, Aberoomand-Azar P, Raeis-Farshid S, Abedini-Khorrami S (2016) The effect of different molar ratios of ZnO on characterization and photocatalytic activity of TiO_2/ZnO nanocomposite. J Saudi Chem Soc 20(4):373–378
35. Ramaseshan R, Ramakrishna S (2007) Zinc titanate nanofibers for the detoxification of chemical warfare simulants. J Am Ceram Soc 90:1836–1842
36. Kolen'ko YV, Kovnir KA, Gavrilov AI et al (2005) History of nanotechnology: from prehistoric to modern times. J Phys Chem B 109(6):20303–20309
37. Kumar R, Kumar G, Umar A (2013) ZnO nano-mushrooms for photocatalytic degradation of methyl orange. Mater Lett 97:100–103
38. Azam A, Ahmed F, Arshi N et al. (2010) Formation and characterization of ZnO nanopowder synthesized by sol–gel method. J Alloys Compd 496:399–402
39. Ge M, Guo Ch, Zhu X et al (2009) Characterization and high pollutant removal ability of buoyant (C, N)–TiO_2/PTFE flakes prepared by high-energy ball-milling. Front Environ Sci Eng China 3(3):271–280
40. Liao DL, Badour CA, Liao BQ (2008) Preparation of nanosized TiO_2/ZnO composite catalyst and its photocatalytic activity for degradation of methyl orange. J Photochem Photobiol A 194:11–19
41. Tian J, Chen L, Yin Y et al (2009) Innovations in engineered porous materials for energy generation and storage applications. Surf Coat Technol 204:205–214
42. Tian J, Wang J, Dai J et al (2009) N-doped TiO_2/ZnO composite powder and its photocatalytic performance for degradation of methyl orange. Surf Coat Technol 204:723–730
43. Štengl V, Bakardjieva S, Murafa N et al (2008) Visible-light photocatalytic activity of TiO_2/ZnS nanocomposites prepared by homogeneous hydrolysis. Microporous Mesoporous Mater 110:370–378

Nanostructured Electrolytic Composites Based on Cobalt Alloys with Refractory Metals: Composition and Functional Properties

Maryna Ved', T. A. Nenastina, N. D. Sakhnenko, Yu. I. Sachanova, and I. Yu. Yermolenko

The development of efficient power supply sources (PSS) has always been and remains to be one of the most promising areas that ensure the energy stability of any country. A wide range of PPSs also includes the fuel cells (FC) that are pegged as the most ecologically safe and promising energy sources because the reactions in them are reduced to the formation of water from the fuel, in particular, hydrogen and oxygen [1–3]. A sticking point of FC is the electrode materials that act as catalytic agents, and in the vast majority of cases, these are expensive materials related to the platinum and silver family [4–6]. In addition to their high electrocatalytic activity, such materials differ by their stability and chemical resistance when exposed to the media of different mineralization and corrosiveness. Earlier researches [7–11] show that transition metal-based alloys can be regarded as decent competitors to precious metals.

Lately, the research dealing with the study of new types of fuel for FC such as methanol, ethanol, formaldehyde, and hydrocarbons have found a wide application [12]. Liquid fuels such as low molecular alcohols have certain advantages in comparison with pure hydrogen because these are easy to store and transport. Moreover, their energy capacity is comparable with the energy capacity of usual liquid fuels (gasoline, kerosene, etc.), for example, it will be equal to 6 kW/kg for methanol.

The obtained research data show that different organic substances are reversibly oxidized on platinum family metals at relatively low anode potentials, and the electrocatalytic oxidation mechanism and the products depend on the electrode material [13]. Today, platinum is considered as one of the best catalysts of the electrooxidation of monoatomic alcohols; nevertheless, much attention is paid to the development of less expensive materials that can replace it [14]. The main requirements set to the

M. Ved' (✉) · T. A. Nenastina · N. D. Sakhnenko · Yu. I. Sachanova · I. Yu. Yermolenko
National Technical University "Kharkiv Polytechnic Institute", 2, Kyrpychova str., Kharkiv 61002, Ukraine
e-mail: kirilesha72@gmail.com

O. Fesenko and L. Yatsenko (eds.), *Nanomaterials and Nanocomposites, Nanostructure Surfaces, and Their Applications*, Springer Proceedings in Physics 246, https://doi.org/10.1007/978-3-030-51905-6_50

catalysts can be reduced just to three of them, in particular, a high catalytic activity, chemical and corrosion resistance, and nontoxicity. Undoubtedly, the activity of catalysts is defined by the material nature though it depends to a great extent on the surface condition. It is known that molybdenum compounds possess a certain catalytic activity with regard to many processes, such as electrochemical methanol oxidation, the reduction of iodates, oxygen, etc. [15]. Cobalt compounds, in particular, oxides and complexes also found a wide application in the electrocatalysis [16]. Evidently, it is related to the mobility of oxygen contained by the compounds of metals, their high coordination capacity, and ability to form nonstoichiometric oxides [17].

The use of alkaline electrolytes allows us to the extent the range of materials that can be used for the formation of catalytic matrixes. It is conditioned by the acceleration of the reduction reaction of oxygen, and simple alcohols in alkaline electrolytes in comparison with acid electrolytes, and a lower corrosiveness of the former with regard to many transition metals. The problems of the use of methanol as a fuel in the alkaline environment are related to gradual carbonization of electrolyte when alcohol is oxidized to CO_2 and the poisoning of precious metal-based electrodes. Transition metal alloys and manganese oxides turned out to be the efficient catalysts that reduce the poisoning effect [18, 19]. Hence, detection of the mechanism for the methanol oxidation on such alloys equally as the maintenance of phasic oxidation with no formation of carbon dioxide can become a highly promising approach to the replacement of precious metals by more economic catalytic systems.

Hence, the formation of the layer of electrolytic alloys based on cobalt with molybdenum (tungsten), and zirconium that has a branched globular surface can provide a preset route for the methanol electrooxidation process, and increase the FC operation efficiency using no precious metals. The subject of this paper deals with the studies of the influence exerted by the qualitative composition of ternary cobalt alloys with refractory metals on the anodic methanol oxidation rate in alkaline electrolytes, and it also deals with the analysis of the corrosion resistance of electrode materials.

1 Experimental

The electrolytic methanol oxidation mechanism, the kinetics, and the corrosion resistance were investigated using the electrodes with the electrolytic coatings of ternary Co–Mo–W(Zr), Co–Fe–Mo alloys of at least 5μm thick. The Co–Mo–W(Zr) coatings were applied onto the steel substrate from the biligand pyrophosphate–citrate or citrate electrolytes [9, 20]. The samples were subjected to the preliminary treatment using the methods given in [21]. The electrodeposition was conducted using the pulse current at a varied current density, the pulse frequency, and on/off time ratio [9, 20]. The electrolyte temperature at a level of 30–35 °C was maintained. The Co–Fe–Mo coatings were deposited onto the steel substrate from the citrate electrolytes based on the iron (III) by direct (DC) and pulse (PC) current [20].

The chemical composition of the coatings was determined by energy-dispersive X-ray spectroscopy by an Oxford INCA Energy 350 electron probe microanalysis integrated into the system of the SEM. The X-rays were excited by exposure of the samples to a beam of 15 keV electrons. The surface morphology of the deposits was studied with a Zeiss EVO 40XVP scanning electron microscope (SEM). Images were made registering secondary electrons (SEs) via scanning with an electron beam; this mode made it possible to study topography with a high resolution and contrast ratio.

The surface relief and morphology were evaluated by the contact method on 10 × 10 × 2 mm samples with an NT-206 scanning probe AFM microscope (lateral and vertical resolutions 2 and 0.2 nm, respectively; 1024 × 1024 scanning matrix, CSC cantilever B as a probe, probe tip radius being 10 nm) [22–24].

The structure of the coatings deposited onto the copper substrate was examined by X-ray diffraction analysis using a diffractometer (DRON-3.0) in the emission of cobalt anode and CuKα radiation. The compounds (phases) were identified by the method of comparison of interplane distances (d, A), and relative intensities (I/I_0) of the experimental curve with electron PCPDFWIN card index data.

The polarization dependences were obtained for Co–Mo–W(Zr) electrodes using the potentiostat PI-501.1, and the programmer PR-8 equipped with the special device intended for the digital data registration. Cyclic voltammetry data (CVA) were analyzed at the potential scanning rate of $s = 0.002$–0.05 V/s according to the algorithm given in [25]. The standard three-electrode cell was used for the measurements, the platinum spiral served as an auxiliary electrode, and the reference electrode was the saturated silver chloride connected to the cell via the salt bridge. The electrolytes of 0.25 M NaOH and 0.25 M NaOH + 1 M CH_3OH compositions were prepared using purely for analysis reagents and distilled water.

The coating corrosion resistance was studied using the method of electrode impedance in the model solution of 0.25 M NaOH. The electrochemical impedance spectra (EIS) of the electrodes with the coatings that have a working area of 1 cm^2 arranged at a distance of 1 cm from each other in the two-electrode cell were registered in the frequency range of 10^{-2}–10^6 Hz at the free corrosion potential using the electrochemical module Autolab-30 (type PGSTAT301N Metrohm Autolab) equipped with the frequency response analyzer module (type FRA-2). The measurements were performed at 18 ± 1 °C. The module was controlled using the Autolab 4.9 program according to the standard procedure with the subsequent processing by the Zview 2.0 package. The phase interface structure was simulated using the method of equivalent circuits. The EIS results are discussed according to known approaches [26, 27].

2 Theoretical Preconditions

A complete methanol oxidation is a multistage process that can be represented as a series of stages:

$$(CH_3OH)_L \rightarrow (CH_3OH)_S \rightarrow (HCHO)_S \rightarrow (HCOO^-)_S \rightarrow (CO)_S \rightarrow CO_2, \quad (1)$$

where $()_L$ are the particles in the solution, $()_S$ are intermediate products and the particles adsorbed by the electrode surface.

The given series fails to reflect the process mechanism in full because, in addition to the solution components, intermediate metal oxides that are formed on the catalyst electrode surface during the anode polarization can participate in it. In this case, the reversibility of the processes that occur on the electrodes is the precondition for the efficient operation of catalysts.

Let's consider the individual stages of the electrode process and detect the specific features of each stage. The first stage of the adsorption of methanol molecules on the electrode surface that corresponds to the $(CH_3OH)_L \rightarrow (CH_3OH)_S$ transformation can limit the electrode process only in the case of fast and completely reversible single-stage charge transfer in the electrochemical reaction. It is known that the oxidation of compound molecules is not of that type; therefore, we can conclude that the adsorption process will be reversible with regard to the charge transfer stage. The second stage $(CH_3OH)_S \rightarrow (HCHO)_S$ is directly related to the charge transfer, and it is electrochemical. The methanol oxidation that results in the formation of aldehyde is accompanied by the hydrogen atom and carbon σ-bond rupture, and the formation of the double bond between carbon and oxygen:

$$(CH_3OH)_S + 2OH^- \rightarrow (HCHO)_S + 2H_2O + 2e, \qquad (2)$$

Particularly this stage is considered to be limiting, and its rate can be controlled by the system cycling without carbon dioxide formation and solution carbonization. The conduction of the process in the interval of potentials that corresponds to the given stage is an essential and sufficient condition for the control of the entire oxidation process.

The next stage of methanol oxidation conducted to get methane acid is also electrochemical, and it is related to the substitution of hydrogen atom by oxygen one according to the gross scheme:

$$(HCHO)_S + 3OH^- \rightarrow (HCOO^-)_S + 2H_2O + 2e. \qquad (3)$$

A degree of the reversibility of this stage will depend on the lability of oxygen atoms whose donor can be the hydroxide ions that are present in the solution. However, it is more probable that the surface oxides of electrode material that are formed during the anode polarization will participate in this stage. For example, according to the transfer reaction of oxygen atom from the metal oxide MO_x to carbon:

$$(HCHO)_S + (MO_x)_S \rightarrow (HCOOH)_S + (MO_{x-1})_S, \qquad (4)$$

the reaction is transformed into the chemical one, and as a consequence, the process probability and the rate of it, on the whole, are increased. It is evident that the catalytic activity of the alloys of cobalt, molybdenum, and tungsten (zirconium) is explained particularly by a high donor acceptance capacity with regard to oxygen, and the formation of the oxides of a variable composition on the alloy surface during the anode polarization:

$$(MO_{x-1})_S + 2OH^- \rightarrow (MO_x)_S + H_2O + 2e, \quad (5)$$

oxygen in them is mobile and can be transferred to the carbon atom without additional energy inputs as in the case of the participation of hydroxide ions in the reaction (3).

Subsequent transitions in the system are related to the formation of the water molecule and intermediate carbon oxide (II) according to the chemical reaction:

$$(HCOOH)_S \rightarrow (CO)_S + H_2O, \quad (6)$$

and a further $(CO)_S \rightarrow CO_2$ transformation is of no interest for the functioning of fuel cells, and this link can be excluded from the general scheme if potential varying intervals will be limited.

Hence, the investigation of the process of methanol oxidation on the alloys of cobalt with molybdenum and tungsten (zirconium) is the foundation for the formation of efficient catalysts for FC.

3 Results and Discussion

The coatings formed by ternary Co–Mo–W alloys are characterized by the uniformly branched globular surface (Table 1, specimens 1, 2) whose relief differs considerably from that of the substrate onto which these are applied (Fig. 1a, b).

We can see the cracks and a number of pores on the surface of specimen 2 (Table 1) which is coated with Co–Mo–W at the current density 6 A/dm^2. Such a phenomenon is associated with two reasons. Firstly, a higher rate of nucleation and growth of crystallites at a higher current density leads to an increase in internal stresses in coatings and following cracking. Secondly, the hydrogen evolution reaction is intensified with increasing polarization and causes the porosity of the coating.

The analysis of ACM data is indicative of the nanoglobular character of the surface of Co–Mo–W coatings on which the cone-shaped associates with the diameter of 2–5 μm are formed due to the merging of fine globules with a diameter of 20–80 nm. The roughness parameters calculated for the section area of 20 × 20 μm are: R_a = 0.02 and R_q = 0.04.

Ternary coatings with Co–Mo–Zr alloy differ from Co–Mo–W ones by a more branched surface, and uniformly distributed cone-shaped associates with a diameter of 5–7 μm (Fig. 1c, d). The cobalt content increases with the current density and an extension of the pulse time. However, the percentage of alloying metals is decreased in comparison with that of Co–Mo–W coatings (Table 1, specimens 3–5). Despite this, the origination of cracks in Co–Mo–Zr coatings is independent of the content of refractory components as well as of current density. Many studies have paid attention to the cracking of binary and ternary cobalt and molybdenum alloys [28, 29]. A

Table 1 Deposition modes, composition, and morphology of the surface of ternary cobalt alloys

Specimen #	Mode	Composition (in the terms of metal)	SEM magnification ×5000
1	4 A/dm^2; t_{on}/t_{off} = 2 ms/10 ms	wt%: Co—59.94; Mo—23.28; W—16.77	
2	6 A/dm^2; t_{on}/t_{off} = 2 ms/10 ms	wt%: Co—61.06; Mo—20.58; W—18.36	
3	4 A/dm^2; t_{on}/t_{off} = 2 ms/10 ms	wt%: Co—62.13; M—33.29; Zr—4.58	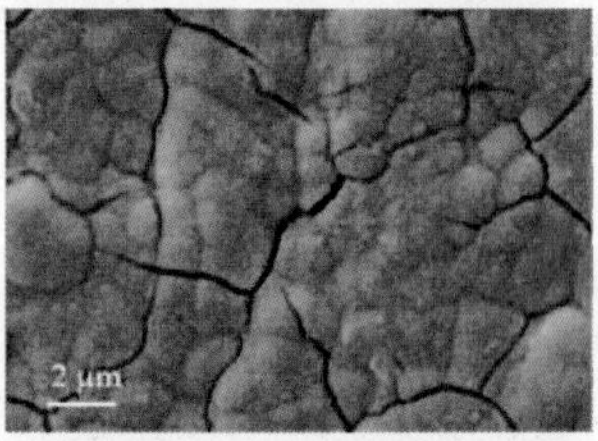
4	4 A/dm^2; t_{on}/t_{off} = 5 ms/10 ms	wt%: Co—63.61; Mo—32.60; Zr—3.79	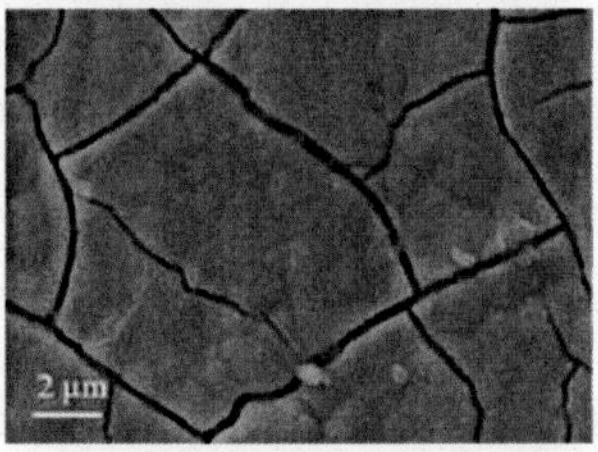
5	6 A/dm^2; t_{on}/t_{off} = 5 ms/10 ms	wt%: Co—68.96; Mo—26.68; Zr—4.36	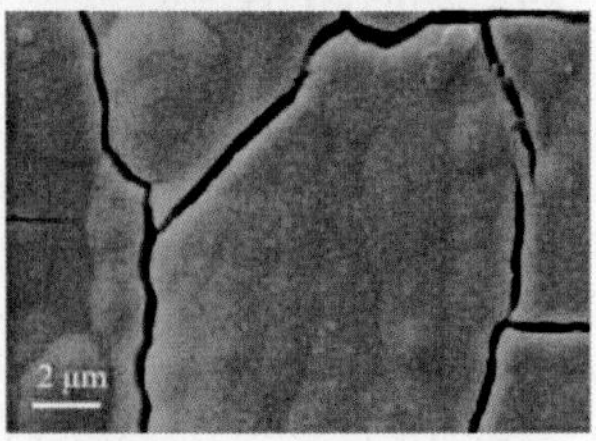

(continued)

Table 1 (continued)

Specimen #	Mode	Composition (in the terms of metal)	SEM magnification ×5000
6	4 A/dm^2; t_{on}/t_{off} = 2 ms/10 ms	wt%: Co—88.25; W—9.06; Zr—2.69	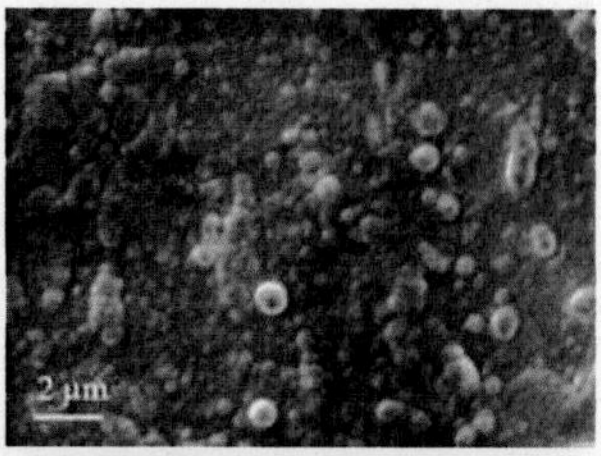
7	6 A/dm^2; t_{on}/t_{off} = 2 ms/10 ms	wt%: Co—73.87; W—24.35; Zr—1.78	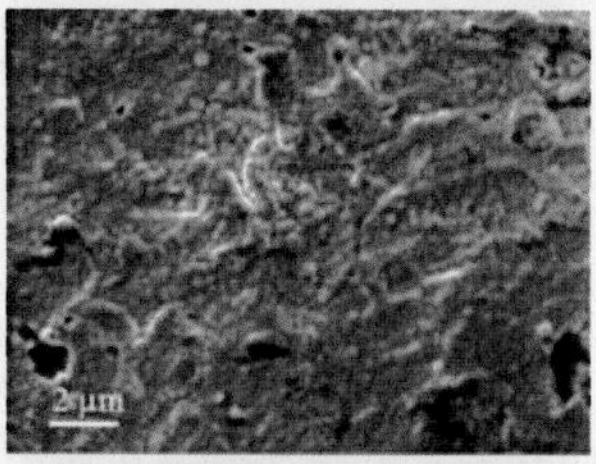
8	3 A/dm^2	at.%: Fe—51; Co—36; Mo—13	
9	3 A/dm^2; t_{on}/t_{off} = 2 ms/5 ms	at.%: Fe—50; Co—39; Mo—11	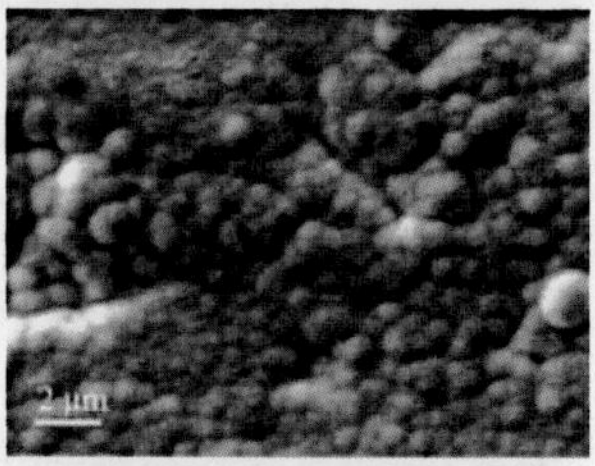
10	5 A/dm^2; t_{on}/t_{off} = 2 ms/5 ms	at.%: Fe—51; Co—40; Mo—9	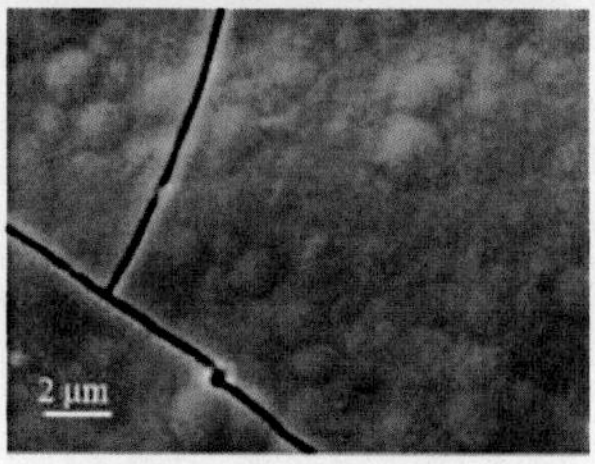

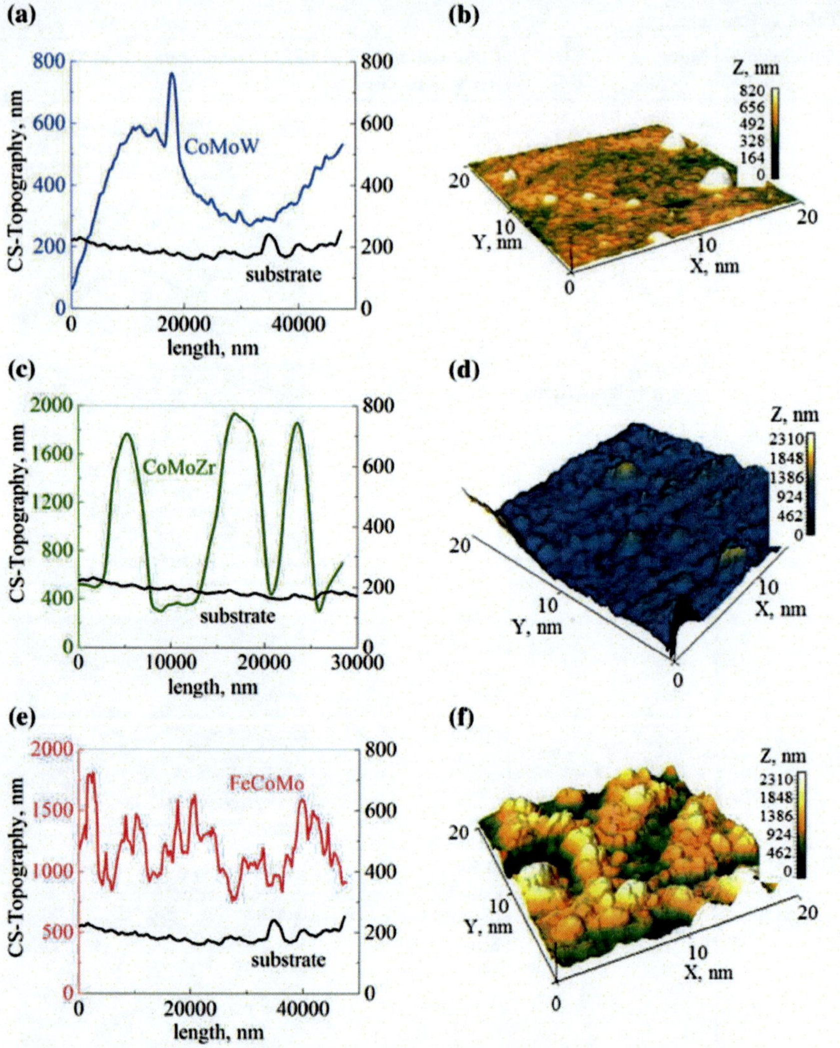

Fig. 1 Topography (**a**, **c**, **f**) and the relief (**b**, **d**, **g**) of the surface of Co–Mo–W (Specimen 1 Table 1a, b), Co–Mo–Zr (specimen 3, Table 1c, d), and Co–Fe–Mo (specimen 8, Table 1f, g) alloys

possible cause of the coatings fracturing can be considered the difference in the crystal lattice types of the alloying metals. It is known that cobalt and zirconium crystal structure is hexagonal close packed (hcp), and molybdenum, as well as tungsten, are characterized by is body-centered cubic (bcc) crystal lattice.

The analysis of the AFM data obtained for Co–Mo–Zr coatings on the section area of 20 × 20 μm allowed us to calculate the roughness parameters of R_a и R_q that are equal to 0.1 and 0.2, accordingly, and are one order of magnitude higher than those for Co–Mo–W coatings.

The Fe–Co–Mo coatings regardless of the deposition mode are characterized by a more developed surface compared to the Co–Mo–W and Co–Mo–Zr alloys. The AFM and SEM analyses show the Fe–Co–Mo coatings are micro-globular; and agglomerates on the surface of 1–2 μm are formed from smaller grains (Fig. 1f, g; Table 1).

The morphology and relief of Co–W–Zr coatings are similar to those of Co–Mo–W coatings (Table 1, specimens 6, 7), however, the tungsten content reduces in the alloys obtained for the same electrolysis conditions (Table 1, specimens 1 and 2). It should be noted the zirconium percentage in Co–W–Zr is also lower in comparison to that in Co–Mo–Zr coatings deposited under the same conditions (Table 1, specimens 3 and 5). A decrease in the size of associates (agglomerates) and unavailability of the cracks in the coatings can be related to the influence of tungsten on the electro-crystallization process. An increase in the current density contributes to the enrichment of the coating with tungsten, though the content of zirconium reduces to 1.78 wt% (Table 1, specimens 6 and 7).

Nevertheless, as it was shown in [21, 30–32] the relief and a degree of the surface development of all the coatings mentioned above are favorable for the catalytic processes that occur not only through the adsorption stage but are also realized in the diffusive mode.

Alongside with the formation of branched globular surface the phase composition of coatings can turn out to be an important factor that has an effect on their properties, and in particular on the catalytic activity because it predetermines the distribution of active accepting centers on the surface.

Figure 2 gives the X-ray patterns of the ternary coatings Co–Mo–W that are deposited onto the copper substrate in the pulse mode; the coating thickness is 20 μm. The X-ray diffraction patterns reflect the amorphous and crystalline alloy structure.

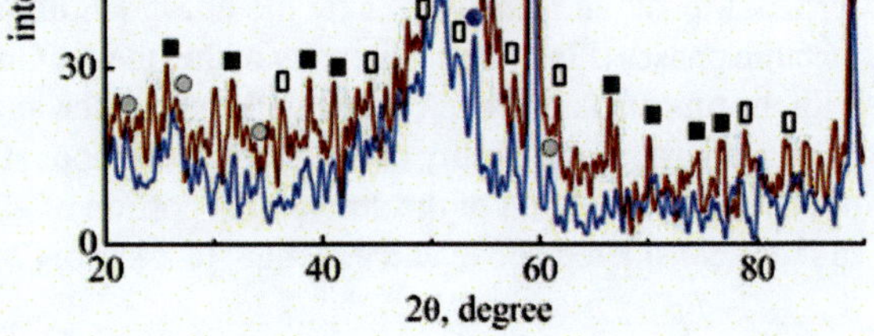

Fig. 2 The X-ray diffraction patterns for ternary Co–Mo–W coatings: 1 is for specimen 1, and 2 is for specimen 2 (Table 1)

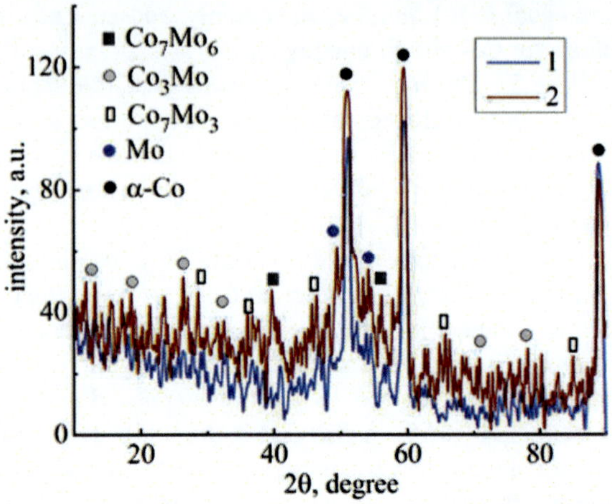

Fig. 3 X-ray diffraction patterns of ternary Co–Mo–Zr coatings: 1 is for specimen 5; 2 is for specimen 3

The patterns were detected that correspond to the α-Co phase, Co_7W_6, Co_7Mo_6, and Co_7Mo_3 intermetallic compounds. A rather broad halo is visible at the 2θ angles from 43° to 58° (Fig. 2) that reflects the X-ray amorphous structure of coatings. The most important event is the origination of the reflexes of metallic molybdenum and tungsten which are formed in compliance with the mechanism suggested in [8, 9, 20] during the reduction of intermediate oxides by hydrogen adatoms when the polarization current is interrupted. The sizes of the zone of coherent dissipation of the amorphous portion of alloy vary in the range of 2–8 nm [20].

The X-ray diffraction patterns (Fig. 3) show the difference in the phase composition of electrolytic Co–Mo–Zr alloys deposited in the pulse mode of a different current amplitude (see Table 1, specimens 3, 5). The figure shows the α-Co lines as well as the lines of intermetallic compounds Co_3Mo, Co_7Mo_3, Co_7Mo_6, and metallic molybdenum (specimen 3). A narrow halo is observed at the angles of 2θ from 48° to 58° reflecting thus the amorphous structure of materials. Unavailability of the lines that correspond to zirconium or its intermetallides is explained by a low metal percentage in the coatings. In addition, high intensity of the lines of intermetallides for specimen 3 is conditioned by the enrichment of alloy with a refractory component. The sizes of the zone of coherent dissipation of the amorphous portion of alloy vary in the range of 2 to 6 nm [20].

The X-ray diffraction patterns of Co–W–Zr alloys that are deposited in the pulse mode of different current amplitude (specimens 6 and 7, Table 1) differ from those for Co–Mo–W and Co–Mo–Zr by the unavailability of tungsten and molybdenum metallic phases (Fig. 4). The intensity of the lines of Co_7W_6 intermetallides increases with the tungsten content. A rather wide halo at the angles of 2θ in the range of 48°–65° confirms the availability of the X-ray amorphous structure. The sizes of the zones of coherent dissipation of the amorphous portion of alloy are smaller in comparison to those considered above and are equal to 2–5 nm [20].

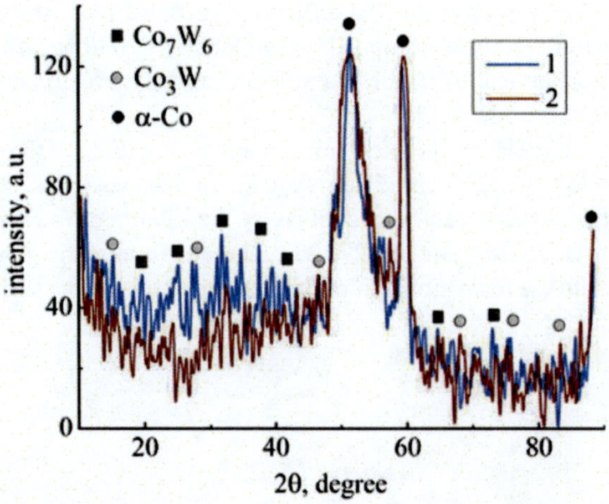

Fig. 4 X-ray diffraction patterns of ternary Co–W–Zr coatings: 1 is for specimen 7; 2 is for specimen 6

An analysis of the Fe–Co–Mo alloy diffraction patterns indicates the absence of Co_xMo_y intermetallides, in contrast to Co–Mo–W and Co–W–Zr coatings. Lines of intermetallic phases Fe_7Mo, Fe_7Co, and FeCo were found for Fe–Co–Mo coatings deposited by direct current (Fig. 5). The non-stationary electrolysis substantially effects on the Fe–Co–Mo coating phase composition (Fig. 5). The X-ray patterns visualize lines corresponding to metallic cobalt and molybdenum, but the lines of intermetallic compounds of iron with cobalt are disappearing. Considering that the co-deposition of iron and tungsten in the binary Fe–W alloy occurs with the formation of a solid solution of tungsten in α-Fe [23], we can assume the combination of the intermetallic compounds and a solid solution at the formation of the ternary Fe–Co–Mo alloy by pulse current.

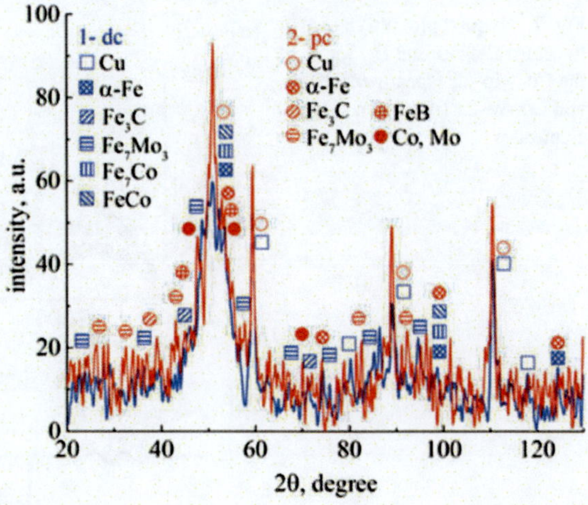

Fig. 5 X-ray diffraction patterns of ternary Fe–Co–Mo alloy, deposited by: 1—direct current (specimens 8, Table 1), 2—pulse current (specimens 9, Table 1)

It is evident that the difference in the surface structure and morphology, equally as in the phase composition of electrolytic alloys will have an effect on their electrochemical behavior, in particular on the corrosion resistance and catalytic activity in electrode reactions.

The electrochemical impedance spectra (EIS) of the electrodes coated with ternary alloys (Figs. 6 and 7) in Nyquist coordinates (Figs. 6a and 7a) reflect the fact that the systems can be described by the modified Voith equivalent circuit peculiar for multiphase systems [27, 33]. The above coatings corrosion process is associated with the formation oxides of alloying metals on the surface. These nonstoichiometric

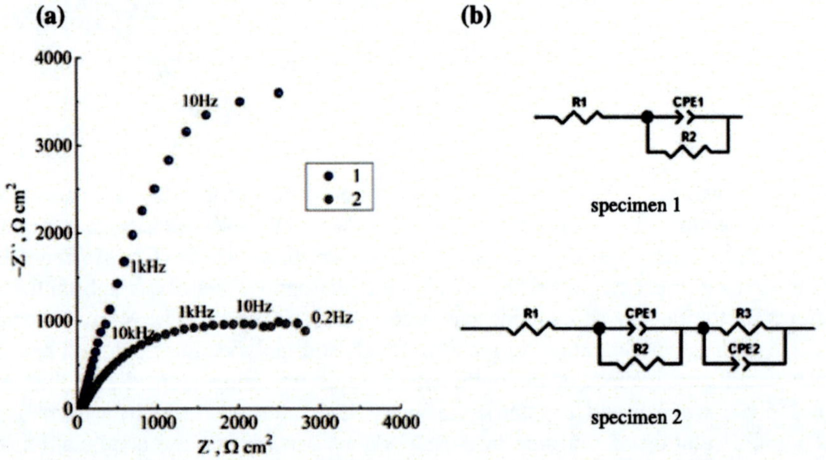

Fig. 6 Nyquist plots (a) and the equivalent circuits (b) for the Co–Mo–W electrode: 1 is for specimen 1; 2 is for specimen 2 (see Table 1)

Fig. 7 Nyquist plots (a) and the equivalent circuit (b) for the Co–Mo–Zr (specimen 3) and Co–W–Zr (specimen 6) electrodes

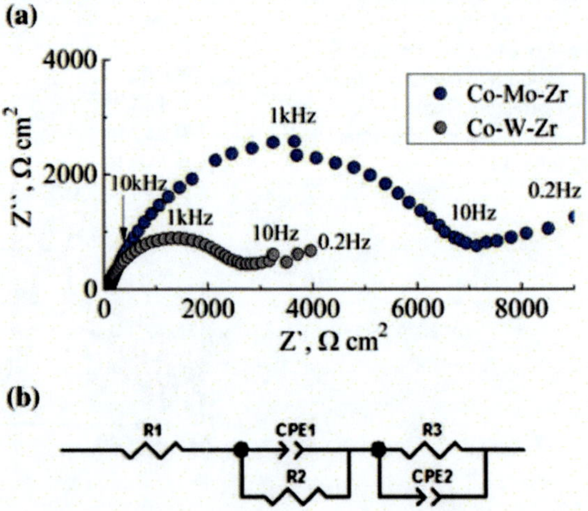

oxides differ in chemical nature, porosity and electrical conductivity. Therefore, the equivalent circuit includes at least two R–C chains, one of which corresponds to the oxide phase and the other describes the corrosion behavior of coating. Such a circuit may include the following elements: *R*1 is associated with the electrolyte resistance; CPE_i ($i = 1, 2$) is the constant phase element (the phase interface capacity); and R_i ($i = 2, 3$) is the resistance of oxide layer ($i = 2$) or corrosion one ($i = 3$). The CPE elements define the phase fractality that is peculiar for the metals (such as Co, Mo, W) of a variable valence that forms hydrated oxide films on the surface in alkali corrosive medium. The degree of deviation from the perfect capacitor (the degree of surface fractality) is determined by the parameter n ($0 < n < 1$), and for perfect capacitor $n = 1$ [27].

Figure 6 shows the different shape of Nyquist plots for specimens coated with Co–Mo–W alloy: specimen 1 demonstrates more capacitor behavior in comparison with specimen 2. The equivalent circuit for specimen 1 includes only three elements: the electrolyte resistance *R*1, CPE1 which is associated with oxide layer capacitance, and *R*2 which simulates the resistance of the non-porous oxide layer (Fig. 6b). The values of equivalent circuit parameters are: $R1 = 63 \pm 1\ \Omega\ cm^2$; $CP1 = 0.1 \pm 0.02\ \Omega\ cm^2$; $R2 = 9100 \pm 1100\ \Omega\ cm^2$.

The Nyquist plot for specimen 2 (Fig. 6a) is characterized by two flattened semicircles that reflects inhomogeneity of the surface namely some sections with pores and cracks (see Table 1 and as a result uneven distribution of corrosion process in the system. Therefore, the equivalent circuit for this specimen consists of two *R–C* chains (Fig. 6b), where the first one reflects the compact oxide layer properties as for specimen 1. The second chain is associated with corrosion behavior of porous oxide film, where CPE2 is the double-layer capacitance, and *R*3 simulates the corrosion resistance which is inversely proportional to the corrosion rate. The values of equivalent circuit parameters and the error (%) of parameters fitting to experimental data are given in Table 2. They characterize sample 2 as resistant to corrosion in an alkaline environment, but such that it has less ohmic resistance. Both facts are favorable for the implementation of electrocatalytic processes.

The Nyquist plots as well as the equivalent circuit for Co–Mo–Zr-coated electrodes (specimens 3–5) are similar to each other and to that of specimen 2, and to Co–W–Zr alloy (Fig. 2a). Since the difference of parameters does not exceed the

Table 2 Values of equivalent circuit parameters ($\Omega\ cm^2$) and error (%) of parameters fitting for coated specimens

Coating (specimen #)	R1	Error	R2	Error	CPE1	Error	CPE2	Error	R3	Error	n
Co–Mo–W (#2)	59	0.2	3700	24	0.74	1.4	0.53	2.2	1800	9	0.47
Co–Mo–Zr (#3)	61	1.2	7100	1.9	0.76	0.8	0.98	2.4	2500	4	0.48
Co–W–Zr (#6)	42	0.9	2800	2.1	0.76	0.7	0.9	1.7	1300	8	0.51

error of 5–7%, we present data only for specimen 3 (Co–Mo–Zr) and specimen 6 (Co–W–Zr) (Table 2) obtained at similar electrolysis conditions. The higher values R2 and R3 for specimen 3 in comparison with specimen 2 are due to the presence on the surface stoichiometric zirconium oxide ZrO_2 with high both electrical resistivity and chemical stability.

It should also be noted that the chemical stability of molybdenum oxides in an alkaline medium is higher than that of tungsten. Therefore, the parameters *R*2 and *R*3 for specimen 6 are lower compared to sample 3. The results for Co–Mo–Zr coatings are more reproducible than for Co–Mo–W ones, which makes it possible to consider Co–Mo–Zr alloy to be more corrosively stable in an alkaline environment. Moreover, we can conclude that coatings with metallic phases of Mo or W are characterized by higher corrosion resistance.

Electrocatalytic properties of Co–Mo–W and Co–Mo(W)–Zr-coated electrodes were studied using the model methanol oxidation reaction in the 1 M CH_3OH solution against the background of 0.25 M NaOH using the method of cyclic voltamometry (CVA), and the obtained data were compared with those for the electrochemical behavior of the platinum electrode in the analogous solution (Fig. 8) and the CVA of ternary alloys in the background solution of 0.25 M NaOH. The CVA for the platinum electrode shows that one peak of the methanol oxidation is displayed at the half-peak potential of $E_{pa/2}$ (from −0.05 to 0.0 V); the current peak is equal to 0.25 mA/cm^2, and the inverse peak is actually two times lower than the direct anode peak. The difference in the potentials of oxidation half-peaks $E_{pa/2}$, and the inverse peak of $E_{pc/2}$ is within 30 mV that is indicative of anode reaction to be quasi-reversibility, and it progresses with the participation of two electrons (stage 2).

The anodic cyclic voltamograms (CVA) obtained for ternary Co–Mo–W and Co–Mo–Zr alloys in the 0.25 M NaOH solution show the peak at the potentials of (0.35–0.4) V that can be related to the oxidation of alloy components to form the intermediate oxides. The peak current on the Co–Mo–W-coated electrodes is within

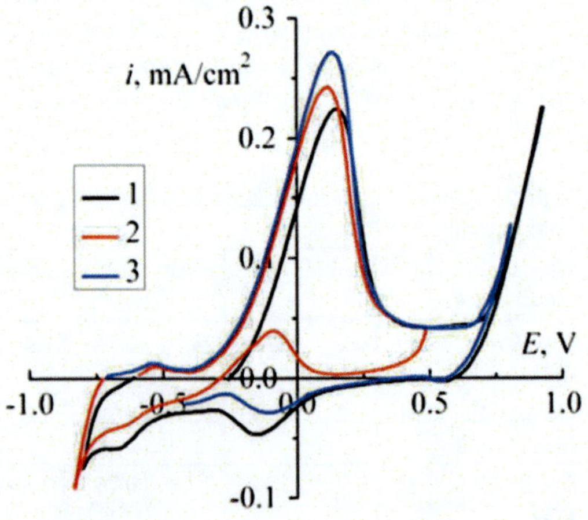

Fig. 8 Cyclic voltamograms of methanol oxidation on platinum nail electrode. Number of the cycle—1, 2, 3

7.0–7.5 mA/cm^2, and it is one order of magnitude higher than the current for Co–Mo–Zr electrode (Figs. 9 and 10a).

On the CVA for Co–Mo–W-coated electrodes in the presence of methanol (Fig. 9b, dependence 1), the anode current peak i_p of the first cycle is shifted to the domain of lower potentials ($E_{pa/2}$ = 0.22–0.24 V) in comparison with 0.25 M NaOH solution ($E_{pa/2}$ = 0.35–0.37 V), and the peak current is raised to 9.0–9.5 mA/cm^2 that is indicative of the surface oxides participation in the methanol oxidation reaction. Moreover, the difference of peak currents in the background solutions and in the presence of methanol is in the range of 2.0–2.2 mA/cm^2, and it exceeds the methanol oxidation current on the platinum electrode (see Fig. 8) that confirms the catalytic properties of the alloy. In the second cycle of the polarization the half-peak potential returns to the level peculiar for the background solution, and the peak current is increased twofold. In this case, the cycling results no change of the general pattern.

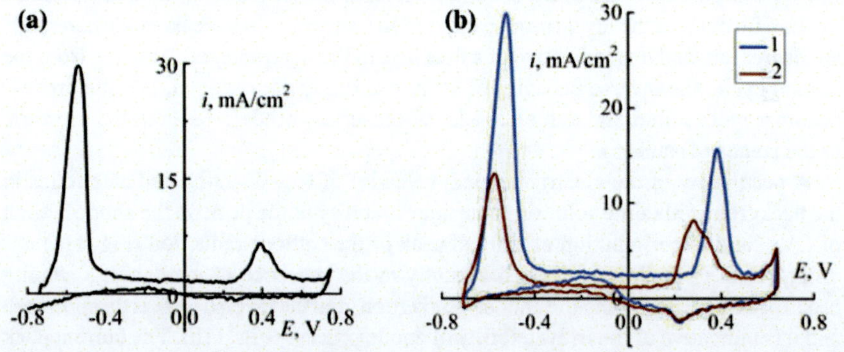

Fig. 9 Cyclic voltamograms of Co–Mo–W-coated electrodes (specimen 1) in the 0.25 M NaOH solution (**a**) and in the presence of methanol (**b**)

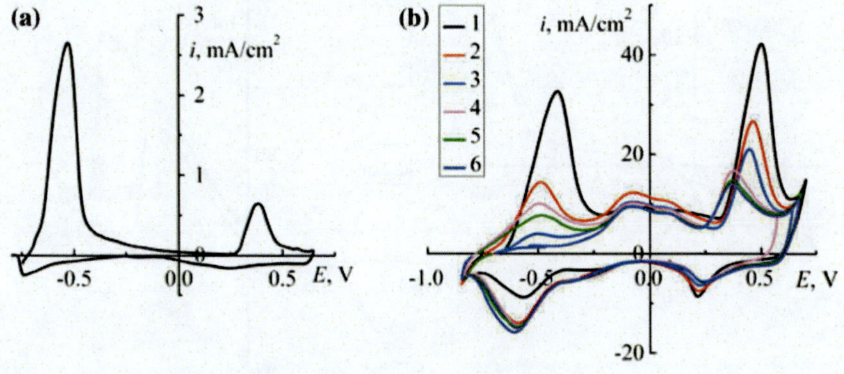

Fig. 10 Cyclic voltamograms for Co–Mo–Zr-coated electrodes (specimen 3) in the 0.25 M NaOH solution (**a**) and in the presence of methanol (**b**)

This behavior can be explained by the fact that particularly the intermediate oxides of alloying metals that are formed in the anodic process are reduced during the inverse polarization, i.e., the mechanism suggested in this paper that includes the stages (4) and (5) is realized.

A comparative analysis of the CVA shows that the activity of Co–Mo–Zr-coated electrodes in the methanol oxidation reaction is considerably higher than that of Co–Mo–W and platinum because the oxidation peak current is higher by a factor of 2.0–2.5 (Fig. 10). Such an increased efficiency of the electrodes can be explained by a higher degree of the surface branching (see Fig. 1), and the synergetic effect of the metals, in particular cobalt, molybdenum, and zirconium that are characterized by the different stable oxidation degrees and the affinity to oxygen [8, 9, 20].

It should be noted that methanol oxidation waves on the alloys are symmetric relatively i_p, and it confirms the availability of the adsorption stage and is indicative of the reaction quasi-reversibility. The oxidation peak current is increased in the range from the first cycle to the second cycle, and then it is decreased and stabilized by the fifth cycle at the level from 16 up to 18 mA/cm^2; the half-peak anode potentials are decreased, and it is indicative of a decrease in the overpotential. Starting from the second cycle, the inverse behavior of CVA is actually not changed, i.e., the process becomes more stable, and as for Co–Mo–W-coated electrode, it progresses according to the stages (4) and (5).

A peculiarity of the electrochemical behavior of Co–W–Zr-coated electrodes in the background alkaline solution is the unavailability of the peak on the anode branch of CVA, and the origination of a broad peak of the cathode reduction (Fig. 11a).

The type of CVA has changed abruptly in the presence of methanol; a sloping broad peak appears during the anode polarization, and the inverse peak is disappeared, and it is indicative of the irreversibility of anode process (Fig. 11b). The current peak is increased in the range from the first cycle to the fourth one, and afterward, it

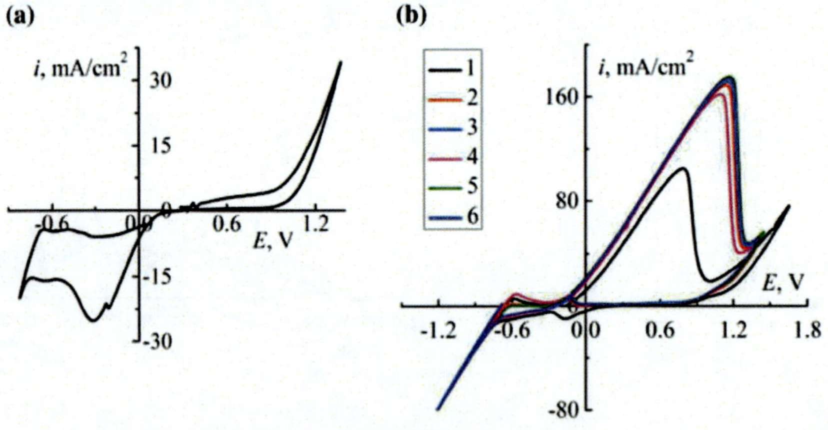

Fig. 11 Cyclic voltamograms of Co–W–Zr-coated electrodes (specimen 6) in the 0.25 M NaOH solution (**a**) and in the presence of methanol (**b**)

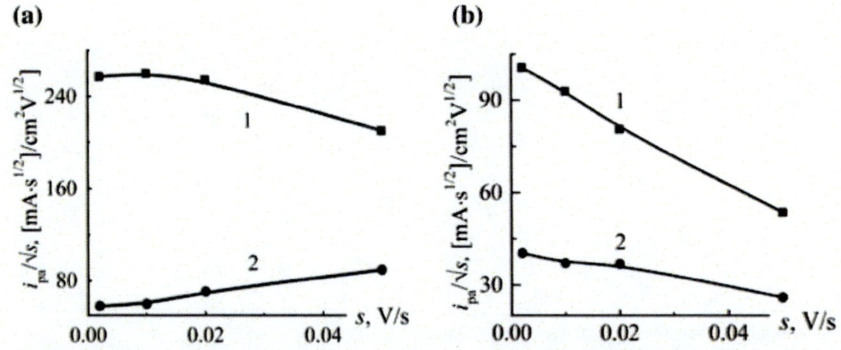

Fig. 12 Dependence of the square root of the peak current on the potential scanning rate for Co–Mo–Zr (**a**) and Co–Mo–W (**b**) coated electrodes for the cycles 1 (1) and 6 (2)

is slightly decreased, and it is stabilized at the level of 160–165 mA/cm^2 that is almost one order of magnitude higher than for Co–Mo–Zr-coated electrodes. Taking into consideration an increase in the slope of an ascending branch of the anodic polarization dependence and unavailability of the inverse peak, we can conclude that the process on the electrodes is irreversible, and a strong current is related to the simultaneous oxidation of coating components and methanol. In addition, it should be considered that Co–W–Zr coatings have a more branched surface in comparison with the series of the alloys in question.

Proceeding from the CVA analysis, we can make a conclusion that the electrolytic Co–Mo–Zr alloy is considered to be the most promising from the standpoint of catalytic activity in the reversible cyclic process, though Co–Mo–W and Co–W–Zr alloys can be used for noncyclic processes.

The kinetic peculiarities of the methanol oxidation on Co–Mo–Zr and Co–Mo–W-coated electrodes were studied by way of the analysis of obtained data according to the known algorithm [25]. A linear dependence of the characteristic criterion $i_{pa}/s^{1/2}$ on the potential scanning rate s (Fig. 12) confirms the quasi-reversible methanol oxidation. It should be noted that the potential scanning rate has an effect on the potentials of oxidation and reduction peaks, and in this case, these values approach each other as the scanning rate is increased.

The dependence of the logarithm difference in the peak current on the logarithm difference in the potential scanning rate at a constant methanol concentration, i.e., the linearized expression of the peak current of diffusion process (the Semerano criterion) is equal to $X_s = 0.60$ for the first cycle of scanning on the Co–Mo–Zr alloy, and it is increased to $X_s = 0.92$ for the sixth cycle, and it is indicative of the adsorption influence on the electrode process that is peculiar particularly for heterogeneous catalytic reactions. At the same time, the Semerano criterion for the Co–Mo–W alloy is not changed during the potential cycling, and it is equal to $X_s =$ 0.50 that is indicative of the limiting chemical or electrochemical stage. As a result, the catalytic effect of Co–Mo–W is lower in comparison to that of Co–Mo–Zr. A

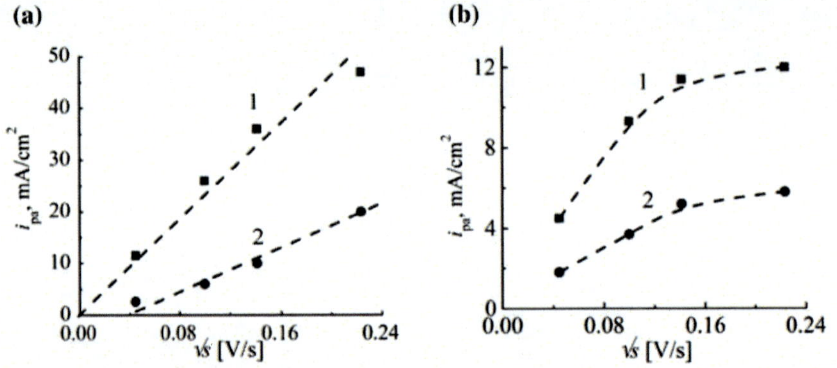

Fig. 13 Dependence of the peak of anode current on the potential scanning rate for Co–Mo–Zr (**a**) and Co–Mo–W (**b**)-coated electrodes for the cycles 1 (1) and 6 (2)

linear dependence of the current peak with an increase in the potential scanning rate for the Co–Mo–Zr alloy (Fig. 13a) has a positive angular coefficient, and for the first cycle of potential scanning it passes through the coordinate origin, while for the Co–Mo–W alloy this dependence is not linear (Fig. 13b). The potentials of the peaks E_p are shifted to the domain of positive values as the potential scanning rate is increased, and for Co–Mo–Zr-coated electrodes, these are linearized in the coordinates of $E - \lg s$.

The product of the transfer coefficient and the number of electrons (αz) calculated for the Co–Mo–Zr alloy using the Hochstein equation is equal to 0.3, and for the Co–Mo–W alloy, it is equal to 1. The polarization dependences of methanol oxidation on the Co–Mo–Zr alloy are linearized in the $E - \lg[i/i_d - i]$ coordinates and their slope is equal to 90 mV that is indicative of the irreversibility of the process. Nature of the stage that limits the electrode process can be judged from the type of the dependence of the i_{pa}/i_{pc} ratio of direct i_{pa} and inverse i_{pc} current peaks on the potential scanning rate s (Fig. 14).

The ratio of direct and inverse currents passes through a maximum with an increase in s remaining less than 1, and in this case, the direct peak is changed more explicitly than the indirect one (Table 3). It serves as a basis for the assertion that the methanol is adsorbed on the electrode surface. It was established that after the eighth cycle the values of potentials and peak currents are not changed for the test alloys.

Thus, electrodes coated with Co–Mo–Zr alloy can be considered as the most promising for electrocatalysis [34].

Two peaks at potentials −0.55 and 0.50 V are visualized at CVA anode part for the electrode coated with Fe–Co–Mo alloy (Fig. 15). These peaks are associated with oxidation/reduction reactions of alloying components namely molybdenum [35] which participates in the CH_3OH oxidation according to the scheme (4).

The CVA of the Fe–Co–Mo coatings deposited both by DC and PC exhibit ability to the pseudoreversible formation of intermediate oxides and their following reduction. As the number of polarization cycles increases, the current peaks increase during

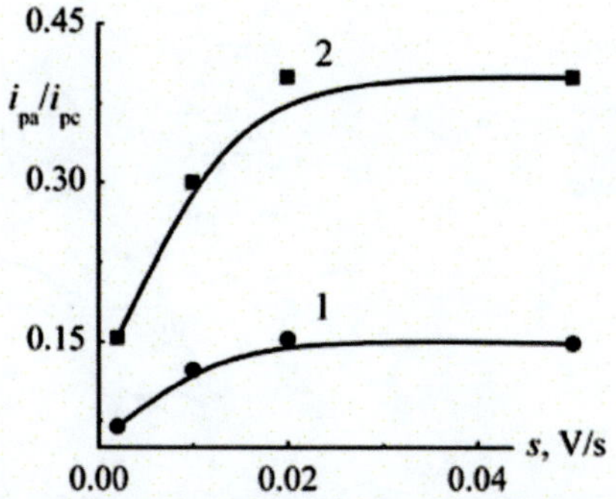

Fig. 14 The dependence of i_{pa}/i_{pc} on the potential scanning rate for the Co–Mo–Zr-coated electrode for the cycles 1 (1) and 6 (2)

Table 3 The influence of the number of potential scanning cycles on the methanol oxidation current peak

Current peak (mA)	# of the potential scanning cycle							
	1	2	3	4	5	6	7	8
i_{pa}	46.5	30.0	26.2	24.0	21.0	19.4	17.0	16.0
i_{pc}	5.2	5.3	5.6	5.9	6.2	6.5	6.8	7.1

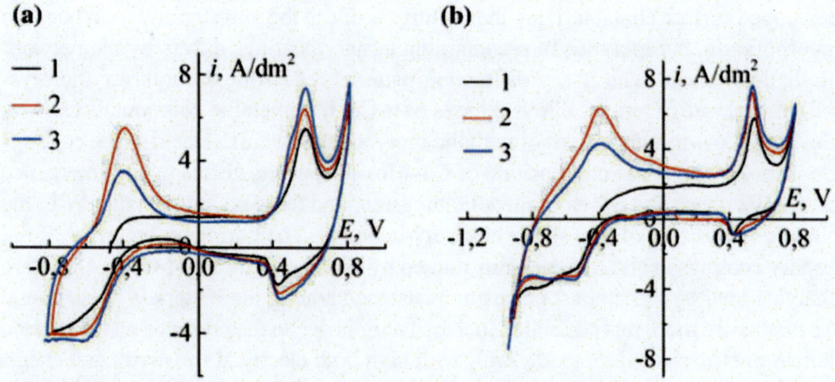

Fig. 15 Cyclic voltamograms of electrodes coated with Fe–Co–Mo deposited by DC (specimen 8) (**a**) and PC (specimen 9) (**b**) in methanol containing an alkaline solution

direct polarization, and the potentials of the peaks are shifted toward more positive values. The growth of current peaks in each subsequent cycle is due to the extension of the electrode material surface and the enrichment of the electrode surface by catalytically active centers associated with intermediate oxides.

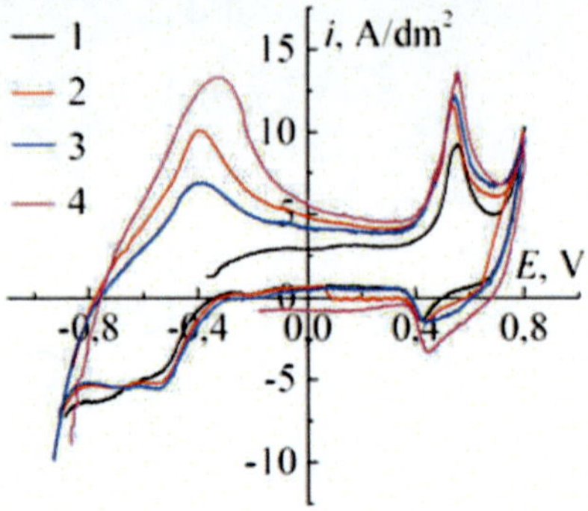

Fig. 16 Cyclic voltamograms of Fe–Co–Mo-coated electrodes (specimen 9) in methanol containing alkaline solution after 3 days of operation

Testing of the catalytic properties of the coatings was continued after 3 days (Fig. 16). The type of the dependences remains and the value of the current density continues to increase, which indicates the preservation of the catalytic properties of the coating after long-term operation.

Thus, the research results indicate that the Fe–Co–Mo galvanic coatings are competitive in the reactions of methanol electrochemical oxidation, which allows us to view them to be considered as promising anodic catalysts.

As was shown the ternary Co–Mo–W(Zr) coatings with the total content of refractory metals of 30–40 wt%, and Co–W–Zr alloys with that of 12–26 wt% are deposited from pyrophosphate–citrate electrolytes in pulse regime. A uniformly cone-shaped developed surface characterizing the coatings is due to the tungsten and molybdenum incorporation, but molybdenum-containing ternary coatings differs by the network of shallow cracks. The X-ray diffraction patterns reflect the amorphous and crystalline structure of ternary alloys. Phases of α-Co, intermetallic compounds Co_3Mo, Co_7Mo_3, Co_7Mo_6, and traces of metallic molybdenum were detected in the coatings Co–Mo–Zr. The phase composition of Co–Mo–W deposits differs by the emergence of Co_7W_6 phase and traces of metallic tungsten, and there no any metallic W in the phase composition of Co–W–Zr electrolytic alloys. The corrosion behavior of the ternary coatings in alkaline medium studied by EIS shows the Co–Mo–Zr alloys are characterized by the highest corrosion resistance among the materials studied, and the results are more reproducible. Such behavior is due to the presence on the surface stoichiometric zirconium oxide ZrO_2 with high both electrical resistivity and chemical stability. We can conclude that coatings Co–Mo–W and Co–Mo–Zr containing metallic phases of Mo or W are characterized by higher corrosion resistance. The mechanism and kinetics of electrolytic oxidation of methanol on electrodes with electrolytic ternary alloys coatings is determined. Based on the experiments carried out, we can conclude that the reaction of $(CH_3OH)_S \rightarrow (HCHO)_S$ is the limiting stage, and it allows us to provide the process cycling and prevent the formation of carbon dioxide. The obtained data are indicative of a rather high electrocatalytic activity of ternary alloys, in particular, Co–Mo–Zr, for the reaction of methanol oxidation in the alkaline medium. Such catalytic properties can be explained by a higher degree of the surface branching, and the synergetic effect of the metals, in particular cobalt, molybdenum, and zirconium that are characterized by the different stable oxidation states and the affinity to oxygen. Hence, a high catalytic activity and corrosion resistance of Co–Mo–Zr coatings allow us to view them as promising electrode materials for chemical current sources, in particular fuel cells.

Data Availability

The data used to support the findings of this study are available from the corresponding author upon request.

Acknowledgments This research was conducted with the support of the Ministry of Education and Science of Ukraine within the confines of the project (Registration Number 0118U002051).

Conflict of Interests

The authors certify that they have no conflicts of interest.

References

1. Averkov IS, Baykov AV, Yanovskiy LS, Volokhov VM (2017) Modeling of electrochemical processes in solid oxide fuel cells. Russ Chem Bull 65(10):2375–2380. https://doi.org/10.1007/s11172-016-1592-8
2. Korovin NV, Sedlov AS, Slavnov YA et al (2007) Calculating the efficiency of a hybrid power station employing a high-temperature fuel cell. Therm Eng 54:137. https://doi.org/10.1134/S0040601507020115
3. Mench MM (2008) Fuel cell engines. Wiley, Hoboken, New Jersey
4. Liu L, Corma A (2018) Metal catalysts for heterogeneous catalysis: from single atoms to nanoclusters and nanoparticles. Chem Rev 118(10):4981–5079. https://doi.org/10.1021/acs.chemrev.7b00776
5. Greeley J, Nørskov JK, Mavrikakis M (2002) Electronic structure and catalysis on metal surfaces. Annu Rev Phys Chem 53(1):319–348
6. Tarasevich MR, Bogdanovskaya VA (2017) Degradation mechanism of membrane fuel cells with monoplatinum and multicomponent cathode catalysts. In: Electrocatalysts for low temperature fuel cells: fundamentals and recent trends. https://doi.org/10.1002/9783527803873.ch6
7. Ved M, Glushkova M, Sakhnenko N (2013) Catalytic properties of binary and ternary alloys based on silver. Funct Mater 20(1):87–91. http://dx.doi.org/10.15407/fm20.01.087
8. Sakhnenko ND, Ved MV, Hapon YK, Nenastina TA (2015) Functional coatings of ternary alloys of cobalt with refractory metals. Russ J Appl Chem 88(12):1941. https://doi.org/10.1134/S1070427215012006X
9. Yar-Mukhamedova G, Ved' M, Sakhnenko N, Nenastina T (2018) Electrodeposition and properties of binary and ternary cobalt alloys with molybdenum and tungsten. Appl Surf Sci 445:298. http://doi.org/10.1016/j.apsusc.2018.03.171
10. Ved' MV, Koziar MA, Sakhnenko ND, Slavkova MA (2016) Functional properties of electrolytic alloys of cobalt with molybdenum and zirconium. Funct Mater 23(3):420. http://dx.doi.org/10.15407/fm23.03.420
11. Yapontseva YS, Dikusar AI, Kyblanovskii VS (2014) Study of the composition, corrosion, and catalytic properties of Co-W alloys electrodeposited from a citrate pyrophosphate electrolyte. Surf Eng Appl Electrochem 50:330. https://doi.org/10.3103/S1068375514040139

12. Poplavskii VV, Mishchenko TS, Matys VG (2010) Composition and electrocatalytic properties of the coatings formed by the ion-beam-assisted deposition of platinum from a pulsed arc-discharge plasma onto aluminum. Tech Phys 55:296. https://doi.org/10.1134/S1063784210020222
13. Wendlandt AE, Stahl SS (2015) Quinone-catalyzed selective oxidation of organic molecules. Angew Chem Int Ed Engl 54(49):14638–14658. https://doi.org/10.1002/anie.201505017
14. Zhang J, Shangguan L, Shuang S et al (2013) Electrocatalytic oxidation of formaldehyde and methanol on $Ni(OH)_2$/Ni electrode. Russ J Electrochem 49:888. https://doi.org/10.1134/S1023193512120166
15. Song C, Khanfar M, Pickup P (2006) Mo oxide modified catalysts for direct methanol, formaldehyde and formic acid fuel cells. J Appl Electrochem 36(3):339–345. https://doi.org/10.1007/s10800-005-9071-1
16. Pirskyy Y, Murafa N, Korduban OM et al (2014) Nanostructured catalysts for oxygen electroreduction based on bimetallic monoethanolamine complexes of Co(III) and Ni(II). J Appl Electrochem 44:1193. https://doi.org/10.1007/s10800-014-0732-9
17. Baklan V, Uminsky MV, Kolesnikova IP (2005) The state of fuel cells and its development in Ukraine. In: Sammes N, Smirnova A, Vasylyev O (eds) Fuel cell technologies: state and perspectives, vol 202. NATO Science Series (Mathematics, Physics and Chemistry). Springer, Dordrecht, pp 181–186
18. Zeng J, Lee JY (2005) Effects of preparation conditions on performance of carbon-supported nanosize Pt-Co catalysts for methanol electro-oxidation under acidic conditions. J Power Sources 140(2):268. https://doi.org/10.1016/j.jpowsour.2004.08.022
19. Fei J, Sun L, Zhou C et al (2017) Tuning the synthesis of manganese oxides nanoparticles for efficient oxidation of benzyl alcohol. Nanoscale Res Lett 12(1):23. https://doi.org/10.1186/s11671-016-1777-y
20. Ved' MV, Sakhnenko ND, Yermolenko IY, Nenastina TA (2018) Nanostructured functional coatings of iron family metals with refractory elements. Springer International Publishing AG, part of Springer Nature. https://doi.org/10.1007/978-3-319-92567-7_1
21. Ved' MV, Sakhnenko MD, Bohoyavlens'ka OV, Nenastina TO (2008) Modeling of the surface treatment of passive metals. Mater Sci 44(1):79. https://doi.org/10.1007/s11003-008-9046-6
22. Yar-Mukhamedova G, Sakhnenko N, Ved' M, Yermolenko I, Zyubanova S (2017) Surface analysis of Fe-Co-Mo electrolytic coatings. In: PCM 2017, IOP conference series: materials science and engineering, vol 213. https://doi.org/10.1088/1757-899x/213/1/012019
23. Ved' MV, Sakhnenko MD, Karakurkchi HV et al (2016) Functional properties of Fe-Mo and Fe-Mo-W galvanic alloys. Mater Sci 51(5):701. https://doi.org/10.1007/s11003-016-9893-5
24. Yermolenko IY, Ved' MV, Sakhnenko ND, Sachanova YI (2017) Composition, morphology, and topography of galvanic coatings Fe-Co-W and Fe-Co-Mo. Nanoscale Res Lett 12:352. http://doi.org/10.1186/s11671-017-2128-3
25. Thomas D, Rasheed Z, Jagan JS, Kumar KG (2015) Study of kinetic parameters and development of a voltammetric sensor for the determination of butylated hydroxyanisole (BHA) in oil samples. J Food Sci Technol 52(10):6719–6726. https://doi.org/10.1007/s13197-015-1796-1
26. Dutta G, Siddiqui Sh, Zeng H et al (2015) The effect of electrode size and surface heterogeneity on electrochemical properties of ultrananocrystalline diamond microelectrode. J Electroanal Chem 756:61–68. https://doi.org/10.1016/j.jelechem.2015.08.016
27. Cesiulis H, Tsyntsaru N, Ramanavicius A, Ragoisha G (2016) The study of thin films by electrochemical impedance spectroscopy. In: Tiginyanu I et al (eds) Nanostructures and thin films for multifunctional applications, nanoscience and technology. Springer International Publishing Switzerland, pp 3–42. https://doi.org/10.1007/978-3-319-30198-3_1
28. Casciano PNS, Ramon L, Benevides RL, Santana RAC (2016) Factorial design in the electrodeposition of Co-Mo coatings and their evaluations for hydrogen evolution reaction. J Alloy Compd 723:164–171. https://doi.org/10.1016/j.jallcom.2017.06.282
29. Kublanovsky VS, Yapontseva YS (2014) Electrocatalytic properties of Co-Mo alloys electrodeposited from a citrate-pyrophosphate electrolyte. Electrocatalysis 5:372–378. https://doi.org/10.1007/s12678-014-0197-y

30. Ved M, Sakhnenko N, Bairachnaya T, Tkachenko N (2008) Structure and properties of electrolytic cobalt-tungsten alloy coatings. Funct Mater 15(4):613–617
31. Yermolenko IY, Ved' MV, Sakhnenko ND, Fomina LP, Shipkova IG (2018) Galvanic ternary Fe-Co-W coatings: structure, composition and magnetic properties. Funct Mater 25(2):274. http://doi.org/10.15407/fm25.02.274
32. Yakushin R, Kuterbekov K, Grafov D, Kruchinina NE, Dyesyatov A, Nurakhmetov T (2015) Improving the efficiency and safety of operation of the hydrogen fuel cell. Saf Technosphere 3:40–43. https://doi.org/10.12737/11880
33. Chiu HW, Chuang JM, Lu CC, Lin WT, Lin CW, Lin ML (2013) In situ measurement of tissue impedance using an inductive coupling interface circuit. IEEE Trans Biomed Circuits Syst 7(3):225–235. https://doi.org/10.1109/TBCAS.2012.2199488
34. Shemet V, Piron-Abellan J, Quadakkers W, Singheiser L (2005) Metallic materials in solid oxide fuel cells. In: Sammes N, Smirnova A, Vasylyev O (eds) Fuel cell technologies: state and perspectives, vol 202. NATO Science Series (Mathematics, Physics and Chemistry). Springer, Dordrecht, pp 97–106
35. Ved' M, Sakhnenko N, Yermolenko I, Yar-Mukhamedova G, Atchibayev R (2018) Composition and corrosion behavior of iron-cobalt-tungsten. Eurasian Chem Technol J 20(2):145–152

Filtration Membranes Containing Nanoparticles of Hydrated Zirconium Oxide–Graphene Oxide

Ludmila Rozhdestvenska, Kateryna Kudelko, Vladimir Ogenko, Olexii Palchik, Tatiana Plisko, Alexander Bildyukevich, Vladimir Zakharov, Yurii Zmievskii, and Olexii Vishnevskii

1 Composite Materials for Baromembrane Separation

At the present time, membrane separation processes are the most widespread for solving environmental and industrial problems, which are directed to the principle of zero wastes or obtaining the concentrate of valuable components. Membrane separation methods are requested particularly by food, electronic, chemical industries, they are widely used for water treatment and wastewater processing. Among known membrane processes, filtration induced by the pressure gradient occupies a special position due to low energy consumptions [1] and high selectivity toward the components being recovered [2]. Baromembrane separation is applied to the concentration of proteins [3], polysaccharides [4], dyes [5], pharmaceutical drugs [6], and so on.

Polymer membranes are used most often for baromembrane separation [7]. For instance, polyacrylonitrile (PAN) is applied to the preparation of ultrafiltration membranes [8–11]. Asymmetric membranes, the active layer of which contains narrow pores, can be obtained from this polymer by means of a method of phase inversion.

L. Rozhdestvenska · K. Kudelko · V. Ogenko · O. Palchik
V.I. Vernadskii Institute of General and Inorganic Chemistry, National Academy of Science of Ukraine, Kiev, Ukraine

T. Plisko · A. Bildyukevich
Institute of Physical Organic Chemistry, National Academy of Science of Belarus, Minsk, Belarus

V. Zakharov (✉) · Y. Zmievskii
National University of Food Technologies, Kiev, Ukraine
e-mail: saharoff.911@gmail.com

O. Vishnevskii
M.P. Semenenko Institute of Geochemistry, Mineralogy and Ore Formation, National Academy of Science of Ukraine, Kiev, Ukraine

O. Fesenko and L. Yatsenko (eds.), *Nanomaterials and Nanocomposites, Nanostructure Surfaces, and Their Applications*, Springer Proceedings in Physics 246, https://doi.org/10.1007/978-3-030-51905-6_51

When solutions contain organic molecular and colloidal species as well as microorganisms, rapid clogging and fouling of membranes occur [8, 12–15]. Organic impurities are accumulated inside the membrane pores deteriorating the membrane elasticity [16]. In order to remove them, aggressive reagents are required for cleaning [15, 17–20]. Their usage reduces a lifetime of polymer membranes. Another fouling type is the adhesion of organic substances to the membrane surface, the cakes can be easily removed by backwashing with permeate enhanced by air flux [21]. Accumulation of organic impurities is explained by the hydrophobicity of polymer materials [22–24].

Modifying the outer surface of commercially available membranes with polymer hydrophilic constituents allows one to prevent fouling and, as a result, to reduce the operation costs of separation processes [25–28]. However, the protecting layer is not always mechanically stable. Another way is to insert nanoparticles of inorganic ion exchangers to the polymer matrix. Nanocomposite polysulfone membrane was designed with 5% of Al_2O_3 nanoparticles (20 nm, the modifier was in a form of hydrated oxide) [29]. The contact angle was found to reach 56°. The rejection degree toward BSA was 99%. In the case of larger particles (80 nm), the rejection was 96%. Considerable BSA rejection was found also for microfiltration membrane modified with hydrated zirconium dioxide (HZD) [30–32]. This ion exchanger forms smaller nanoparticles comparing with zirconium hydrophosphate [31, 32]. Thus, it is precipitated directly in the active layer of the membrane. HZD was also applied to modifying ceramic [33, 34] or polymer [35] membranes for electrodialysis. As opposed, zirconium hydrophosphate is precipitated in the macroporous support: the composite membrane shows mush lower rejection, particularly during filtration of corn distillery [31, 32, 36]. However, nanoparticles of this ion exchanger were obtained, when the deposition inside ion exchange polymer occurs [35, 37, 38]. Other inorganic modifiers for polymer filtration membranes are TiO_2 [39], silica [40]. Even Ag [41] and CuO coupled with carbon nitride [42] have been proposed that the composite membranes demonstrate antibacterial properties.

Now, carbon nanomaterials (one-dimensional nanotubes, CNT, and two-dimensional graphene oxide, GO) are considered as prospective modifiers. The fullerenes of CNTs provide a higher pore size on the membrane surface as well as empty space called porosity in membrane structures [43]. However, the hydrophobicity of CNTs deteriorates their distribution in polymers, since they tend to agglomerate with each other. This leads to a decrease in the dispersion ability in a polymer-containing solvent during membrane preparation, and, as a result, a decline in the compatibility of a membrane structure. The effect of synthesis conditions on nanotube aggregation was examined in [44]. Functionalization of CNTs is expected to solve those problems. The addition of 0.00084–0.0048% of nanotubes to polymers gives a possibility to vary the transport properties of the hollow fibrous membranes in a wide range [45]. Polyvinylpyrrolidone–polysulfone membrane modified with CNTs was tested to estimate the possibility of its usage in hemodialysis [46]. The liquid flux through the membrane was 0.72 dm m^{-2} h^{-1}, the rejection was found to achieve 56, 55, and 28% of urea, creatinine, and lysozyme, respectively.

The hydrophilicity of GO is mainly provided by functional groups: hydroxyl (C–OH), epoxy groups (C–O–C), carboxyl (–COOH). Oxygen-containing functional groups at the edges and in the plane of the graphene oxides are capable of both covalent and ionic interaction with different species [47]. Theoretical calculations of the specific surface area give a value of ≈2600 $m^2\ g^{-1}$. The experimental magnitudes, which were obtained with a method of standard contact porosimetry involving water or octane as working liquids [48–50], are 2200–2400 $m^2\ g^{-1}$ [51–54].

Filtration of GO suspension through macroporous support allows one to obtain a mechanically stable layer, which possesses ion sieving ability [55]. This layer is so-called "GO-membrane." Its stability is provided by the metal contaminants that are introduced during the synthesis and processing of GO nanosheets [56]. In fact, they are cross-linked with multivalent ions, particularly alumina (this metal is released from the porous filter during filtration of GO suspension in acid). Reduced GO and carbon quantum dots also support mechanical stability of "GO membrane" [57]. Titanium dioxide was purposefully inserted to GO suspension: the layer rejects dye molecules [58]. Another way is to fix GO in the volume [59, 60] or on the surface of polymer membranes [61]. However, embedding GO additions to the polymer matrix causes enlargement of its pores deteriorating rejection. Moreover, GO can be removed from the membrane pores during pressure-driven separation. The GO layer attached to the outer surface can be easily damaged.

The solution of these problems is embedding of GO-containing nanocomposite based on inorganic ion exchanger directly to the pores of the polymer. The inorganic ion exchanger will provide both binding and rejection functions. The secondary porosity of the membranes caused by the nanocomposite will ensure the retention of colloidal particles. The location of the nanocomposite in the pores of the active layer of membranes will provide mechanical properties and resistance against organic fouling due to the enhanced hydrophilicity.

2 Preparation of Composite Membranes

Commercial spectrally pure graphite was purchased from "Zaporizhskii graphit" Ltd.

Polymer ultrafiltration membranes were synthesized at the Institute of Physical Organic Chemistry of the National Academy of Sciences of the Republic of Belarus. The materials included the active PAN layer fixed at the macroporous substrate made of non-woven 'lavsan' (polyethylene terephthalate). This combination provides the necessary complex of membrane characteristics. The membranes were marked as PAN-2, PAN-3, and PAN-4.

The membrane characteristics are listed in Table 1, where J is the flux of deionized water at 1 bar, rejection corresponds to polyvinylpyrrolidone (40 kDa, the aqueous solution contained 3 g dm^{-3} of this substance).

Sol of insoluble zirconium hydroxocomplexes containing 1 M Zr(IV) was prepared similar to [33, 34]. Globular primary nanoparticles (6 nm) and their aggregates dominated in sol [34]. The sol was stable: the particle size did not change over

Table 1 Water flux through the membranes and their selectivity

Membrane	Thickness (μm)	J, dm^3 m^{-2} h^{-1}	Rejection (%)
PAN-2	55	20	99
PAN-3	60	130	75
PAN-4	60	70	90

time. GO was obtained by peeling off graphite layers using an advanced Hammer's method [62] similar to [63]. For this purpose, a mixture of concentrated H_2SO_4 (360 cm^3) and H_3PO_4 (40 cm^3) was prepared, and 3 g of graphite was added to the liquid with stirring. After stirring for 10 min, 18 g of $KMnO_4$ was gradually added to the reaction system, which was cooled on ice. The mixture was stirred for 36 h at 50 °C to form a dark viscous substance. The reaction system was then cooled to room temperature and slowly poured into 400 cm^3 of cold deionized water, to which 3 cm^3 of 50% H_2O_2 was previously added. The suspension was centrifuged and then washed with HCl, water and methanol to pH 6. The solid was collected after centrifugation and sprayed again in 1.2 dm^3 of water. The content of GO was determined after drying 1 cm^3 of the suspension. The solids content was found to be 2.8 mg cm^{-3}.

Zirconium-containing sol was diluted in four times, GO-suspension was added. The obtained suspension was sonicated for 10 min at 30 kHz using a Bandelin ultrasonic bath (Bandelin, Hungary). The resulting mixture was inserted to the active layer of polymer membranes, which was previously degassed in deionized water under vacuum at 343 K. The deposition of the composite modifier occurred directly in the polymer using a 0.1 M NH_4OH solution. The GO content in the composite was ≈2% relatively anhydrous ZrO_2.

The membranes were washed with deionized water, dried at room temperature to constant weight, treated with ultrasonic, and dried. The precipitate removed from the outer surface of the membrane was separated, dried, and investigated later with a method of transmission electron microscopy (TEM). For comparison, the membranes were modified only with HZD in a similar manner. In all cases, the deposition occurred both on the substrate fibres and in the pores of the active layer. It has been previously established that inorganic particles are deposited on the surface of the substrate fibres as aggregates, a size of which is up to several microns [31, 32]. They are much smaller than the pores between the fibres and do not affect the rejection ability of the membranes. At the same time, they slow down the rate of filtration. Therefore, the impregnation of the inorganic component was performed only on the side of the active layer of the polymer membrane.

The results of GO characterization are given in [63]. The membrane morphology was examined using scanning electron microscopy (SEM) by means of a JEOL JSM 6700 F microscope (Jeol, Japan). A thin layer of platinum was deposited onto the membrane surface at 3 Pa using a JEOL JFC-1600Auto fine coater (Jeol). TEM images of HZD–GO composite were obtained by means of a JEOL JEM 1230 transmission electron microscope (Jeol).

3 Membrane Test

The diffusion of ions was studied in a two-chamber cell divided with a tested membrane. The first compartment was filled with 0.1 M solutions of LiCl or NaCl, deionized water-filled other chambers. The concentration of Li^+ and Na^+ ions was determined by means of a PFM-U4 flame photometer (Analitpribor, Belarus) and a Pye Unicam 8800 UV/VIS atomic absorption spectrophotometer (Philips, the Netherlands).

Preliminarily the membranes were pressurized in deionized water at 1 bar and 298 K, the volume of permeate was monitored. When this volume was constant, tap water was filtered. Water contained 1 mol dm^{-3} Ca^{2+} and 0.5 mol dm^{-3} Mg^{2+}, as well as organic additions (COD was 15 mg O_2 m^{-3}) and colloidal additions of iron hydroxide, which were formed as a result of corrosion of water pipes. The permeate was analyzed with an atomic absorption method.

The experimental stack for filtration included a divided two-chamber cell of tangential type (the effective membrane area was 3.85×10^{-3} m^2), a fluid supply line, which provided the liquid circulation through the concentration compartment, a thermostat and measuring devices (manometer, rotameter) [31, 32]. The filtering of solutions of ovalbumin (OVA) and bovine serum albumin (BSA) was investigated (the protein concentration was 1 g dm^{-3}). The molecular mass of OVA and BSA is 40 and 69 kDa, respectively.

Filtration of protein solutions was performed at 1–2 bar. The protein concentration in the permeate was determined by the Bradford method using Coomassie brilliant blue G-250 [64]. The selectivity of the membranes (φ) was evaluated by the retention capacity according to:

$$f = \left(1 - \frac{C_p}{C_0}\right) \times 100\% \quad (1)$$

where C_p and C_0 are the concentrations of the permeate and initial solution, respectively.

4 Membrane Morphology

TEM images of the HZD–GO composite are given in Fig. 1a. GO nanosheets are aggregated through their edges forming a compact coating of primary HZD nanoparticles, which cannot be recognized.

In the case of composite membrane, the outer surface is a relief, protrusions (up to several tens of nanometers) are visible indicating a presence of the modifier inside the active layer. The size of these protrusions corresponds to the possible pore size of active layer. According to the data of Table 1, the pore size decreases within the order: PAN-2 < PAN-3 < PAN-4. Large particles of irregular shape, a size of which

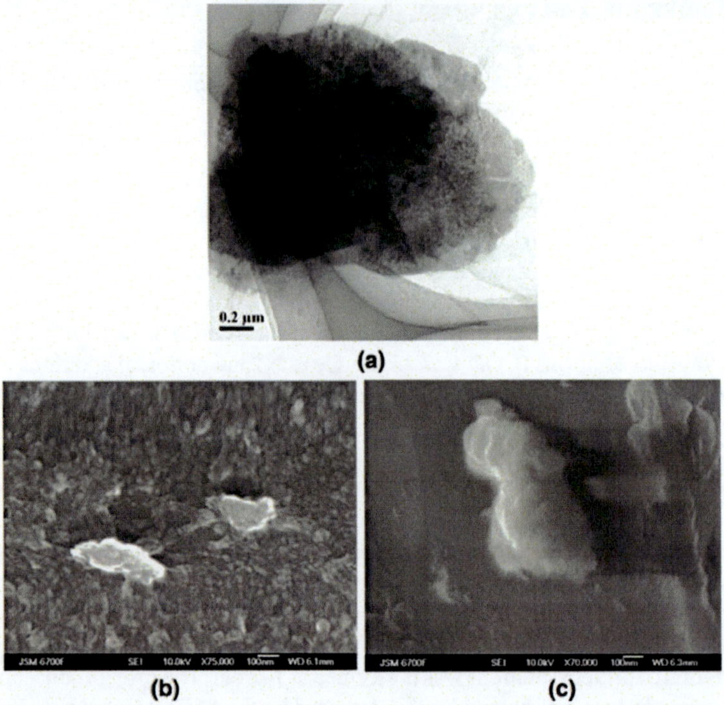

Fig. 1 TEM image of HZD–GO composite (**a**), SEM image of the outer surface of the modified composite membrane (**b, c**)

is several hundred nanometers and even up to 1 μm, are also visible (Fig. 1b, c). They evidently block structure defects of the polymer. Obviously, these formations are splices of the primary particles of the composite.

5 Diffusion of Li^+ and Na^+ Ions

The change in membrane permeability was estimated taking into consideration the diffusion of Li^+ and Na^+ ions since GO shows mainly cation exchange properties [65] enhancing cation exchangeability of HZD-based composite [63]. The flux of ions from the salt solution to deionized water is observed. The diffusion coefficient can be calculated after the steady state, which corresponds to the constant flux:

$$\frac{dC_i}{dt} = D_i \frac{d^2C}{dx^2}, \tag{2}$$

where x is the membrane thickness, C_i is the concentration of species, D_i is the diffusion coefficient. The interval of steady state is correlated with D_i. It can be determined by the second Fick's law (2), which would be expressed as a function of time.

This equation is solved graphically [66]. The concentration of diffused ions as a function of time is plotted; the linear region is extrapolated to the abscissa axis (Fig. 2). The intersection corresponds to the time of steady state (t_e). Under the steady-state conditions, the ion flux is constant. Thus, the t_e value is proportional to the square of the membrane thickness and inversely proportional to the diffusion coefficient:

$$t_e = \frac{x^2}{D_i} \tag{3}$$

In all cases, Li^+ ions show higher diffusion coefficients than Na^+ ions (Table 2). This is in agreement with their mobility in aqueous solutions. GO depresses diffusion

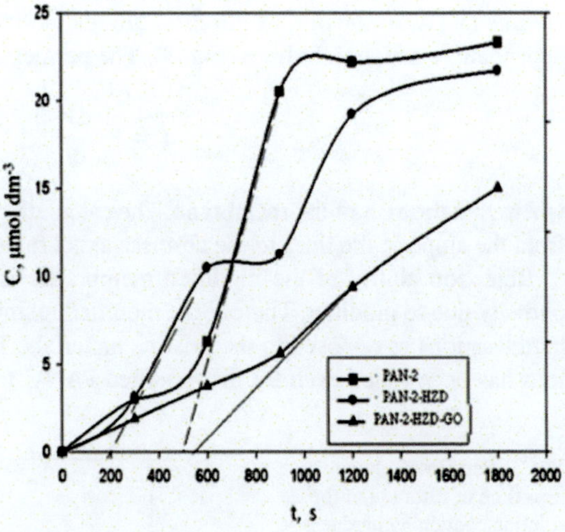

Fig. 2 Concentration of Li^+ ions in the solution, which is formed during diffusion from the chamber filled with the salt solution to deionized water, as a function of time

Table 2 Diffusion of Li^+ and Na^+ cations through the tested membranes

Membrane	Modifier	$D_i \times 10^{12}$ m^2 s^{-1}	
		Li^+	Na^+
PAN-2	HZD	1.187	0.914
	HZD–GO	0.996	0.847
PAN-3	HZD	9.524	0.755
	HZD–GO	12.562	12.441
PAN-4	HZD	2.992	1.573
	HZD–GO	3.814	5.235

through the PAN-2 membrane: lower values of diffusion coefficients have been found for the sample containing HZD–GO composite than for the PAN-2–HZD membrane. According to the data of Table 1, the active layer of this polymer is characterized by the smallest pore size: in this case, GO affects ion mobility. On the contrary, the PAN-3–HZD–GO and PAN-4–HZD–GO samples show higher values of diffusion coefficient comparing with the membranes modified with HZD. It means higher swelling of GO in larger pores despite the inorganic ion exchanger. Indeed, the effect of GO on the nobility of higher hydrated Na^+ ions is more sufficient than that of Li^+ ions.

Further, the membranes, for which GO slows down diffusion or changes it to the greatest degree (namely, the materials based on PAN-2 and PAN-4), were investigated in more detail.

6 Filtration of the Solution Containing Inorganic Ions

Filtration of the solution containing hardness ions was carried out. The permeate volume (V) as a function of filtration time is plotted in Fig. 3. As seen, the $V - t$ dependences are straight lines (Fig. 3). The permeate flux is defined as follows:

$$J = \frac{1}{A}\frac{dV}{dt} \tag{4}$$

where A is the area of the membrane. The calculation of the J values was performed from the slope of the lines to the abscissa axis (Table 3).

Rejection ability of the modified membranes is determined by the secondary porosity due to modifier. Thereof the modified membranes demonstrate rejection of hardness ions as opposed to the pristine materials. The rejection of Ca^{2+} and Mg^{2+} ions has been found even for the modified PAN-3 membrane despite GO swelling,

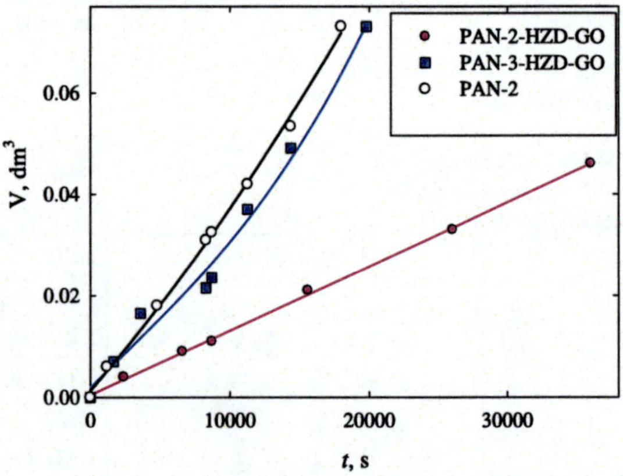

Fig. 3 Permeate volume over time of filtration of the solution containing hardness ions

Table 3 Filtration of liquids

Membrane	Solution containing					
	Ca^{2+}, Mg^{2+}		OVA		BSA	
	$J \times 10^7$ m^3 m^{-2} s^{-1}	φ (%)	$J \times 10^7$ m^3 m^{-2} s^{-1}	φ (%)	$J \times 10^7$ m^3 m^{-2} s^{-1}	φ (%)
PAN-2	12.9	–	10.9	92	6.8	97
PAN-2–HZD–GO	3.3	14.28	4.1	98	3.5	98
PAN-3	35.0	–	14.4	–	15.1	–
PAN-3–HZD–GO	9.6	9.74	7.3	95	4.9	98

which improves diffusion of alkaline ions. This indicates the effect of surface charge of the composite on the rejection of divalent ions. Besides dissociated and protonated –OH groups of HZD, functional groups of GO make sufficient contribution to the modifier surface charge.

The thickness of the secondary active layer caused by inorganic particles within the pores (l) can be calculated from the Kozeny–Carman equation [67]:

$$\frac{\Delta P}{1} = \frac{180\mu(1-\varepsilon^2)v}{\Phi^2 d^2 \varepsilon^2}, \tag{5}$$

where ε is the porosity (0.33 for compact packing of globules), Φ is the sphericity of the particles (we can assume $\Phi = 1$), d is the diameter of the particles, μ the dynamic viscosity (9×10^{-3} Pa at 298 K), and v is the liquid velocity. Since the effective membrane area was 21 cm^2, $v = 1.6 \times 10^{-4}$ (PAN-2–HZD–GO) and 4.8×10^{-4} (PAN-3–HZD–GO). Equation (5) can be rewritten as follows:

$$\frac{l}{d^2} = \frac{\Phi^2 \varepsilon^2 \Delta P}{180\mu(1-\varepsilon^2)v}, \tag{6}$$

Here, l/d^2 is the parameter, which characterizes the retarding ability of ion exchanger. This value is 1.66×10^7 (PAN-2–HZD–GO) and 5.2×10^8 m^{-1} (PAN-3–HZD–GO), or 166 and 52 nm^{-1}, respectively. Nanoparticles of insoluble zirconium hydroxocomplexes, a size of which is 6 nm, dominate in sol [34]. It is possible to assume no change of the size of primary nanoparticles after deposition in polymers. Thus, the thickness of the ion exchanger layer, which determines rejection, is 12 (PAN-2–HZD–GO) and 3% (PAN-3–HZD–GO) of the total membrane thickness. It means the formation of thicker ion exchanger layer in the membrane containing smaller pores.

The permeate flux is related to the hydrodynamic resistance of the membrane according to the Darcy equation [7, 67]:

$$J = \frac{\Delta P}{\eta R_m}, \tag{7}$$

Table 4 Hydrodynamic resistance of membranes and modelling of their fouling

Parameter	PAN-2	PAN-2–HZD–GO	PAN-3	PAN-3–HZD–GO
R_m, m^{-1}	8.55 × 10^{12}	6.73 × 10^{13}	3.13 × 10^{12}	6.5 × 10^{12}
n H_2O	–	0	–	0.5
n BSA	0.99	0	–	0.01
n OVA	0.67	0	–	1.5

where η is the dynamic viscosity of the liquid being filtered, R_m is the hydrodynamic resistance of a membrane. Comparing with pristine membranes, the hydrodynamic resistance of the composites is higher in two (PAN-3–HZD–GO) and eight times (PAN-3–HZD–GO), as shown in Table 4. It means blockage of membrane pores with modifier particles.

7 Filtration of the Protein Solutions

Despite GO swelling, which accelerates the diffusion of the alkaline metal cations, the modifier improves the membrane selectivity toward OVA and BSA (see Table 2). It means smaller pore size between the modifier particles comparing with protein macromolecules. For instance, the size of the elliptic BSA macromolecule is 14 × 4 × 4 nm. Regarding OVA, its size is about 3 nm [68]. It means, the size of pores between the modifier particles is ≈3 nm. Larger pore size has been found for HZD embedded to microfiltration membrane (14–18 nm) [31, 32].

The dependencies of the permeate volume on time are given in Fig. 4. As seen, the filtration rate gradually decreases (Table 2 shows the average permeate flux). Slower filtration was found for the BSA solution comparing with the OVA solution. Since the PAN-3 membrane shows no selectivity toward proteins, the data for this sample are not given.

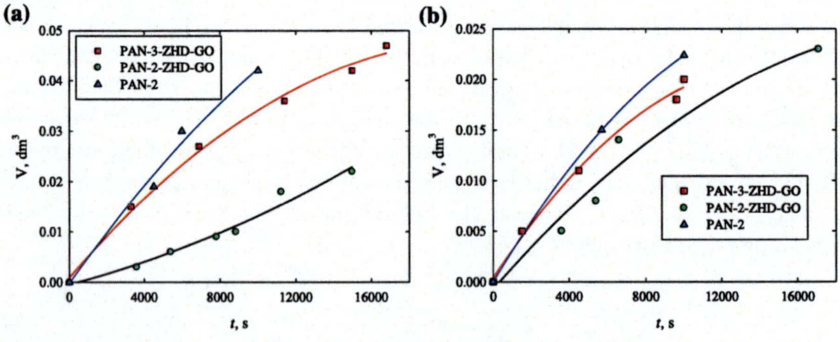

Fig. 4 The permeate volume over the time of OVA (**a**) and BSA (**b**) filtration

8 Mechanism of Fouling

According to theoretical models, the mechanism of fouling is divided into complete blocking of pores; pore narrowing (deposition on the walls of the pores); formation of a cake on the outer surface.

The mechanism of fouling was estimated according to the Hermia model [69], which takes into consideration a change of the permeate flux over time:

$$\frac{d^2t}{dV} = k\left(\frac{dt}{dV}\right)^n, \tag{8}$$

where k is the proportionality factor for describing the resistance of the protein layer. The exponent n characterizes the fouling model: $n = 0$ corresponds to the cake formation, $n = 1.5$ is related to partial pore blockage (a substance is adsorbed on the pore walls, the centre of pores remains free), $n = 1$ is attributed to the intermediate case, $n = 2$ means complete blockage of pores. Reducing the number of available pores leads to a significant resistance to filtration [69]. The exponent value can be determined from the logarithmic dependence of the second derivative of the $t - V$ function on the first derivative:

$$\log\frac{d^2t}{dV^2} = \log k + n\log\left(\frac{dt}{dV}\right) \tag{9}$$

The n value has been found as a slope of the curves to the abscissa axes (Fig. 5).

As shown from Table 4, the fouling mechanism for the pristine PAN-2 membrane is both adsorption of organic species in pores and cake formation on the outer surface. Insertion of hydrophilic modifier into pores leads to a change of the fouling mechanism: only cake is formed outside the membrane. This is evidently due to impossibility for the BSA and OVA macromolecules to penetrate into membranes,

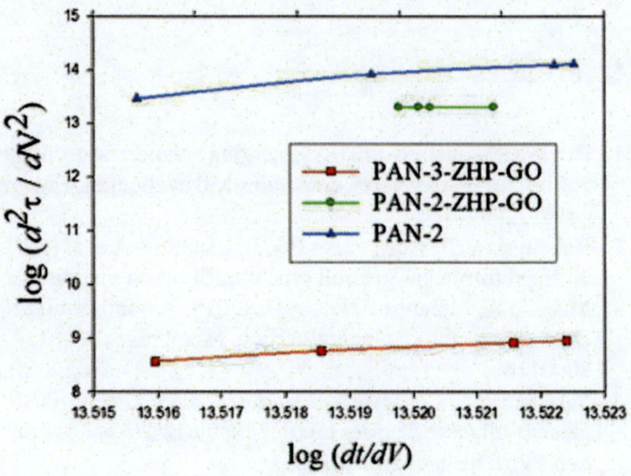

Fig. 5 Application of the fouling model to the filtration of the BSA solution

Regarding the PAN-3 sample, similar regularities are observed for the BSA filtration evidently due to impossibility of protein macromolecules to penetrate into membranes. However, smaller OVA species remain on the pore walls due to smaller size. They are adsorbed on the pore walls. It is similar for fouling of the composite membranes with organic species and inorganic colloidal particles, which are contained in tap water.

The reason for this is partially hydrophobic surface of the modifier. This is possible when aggregated GO nanosheets penetrate into the polymer pores during the membrane modifying. As known, the surface of aggregates is hydrophobic [52, 63].

9 Conclusions

Composite materials based on polyacrylonitrile polymers were synthesized by modifying commercial membranes with HZD–GO composite. The particles of the inorganic ion exchanger are covered with GO nanosheets. The composite forms a "secondary active layer" inside polymer membranes. Its thickness was estimated according to the Kozeny–Carman equation, these values are about 3–12% of the total membrane thickness.

The effect of polymer matrix on ion transport and fouling with organic substances, particularly with OVA and BSA proteins, has been found. Large pores of polymer matrix evidently provoke the deposition of GO aggregates inside the membrane. This provided hydrophobicity of pore walls. As a result, the fouling mechanism involves adsorption of organic species on pore walls similar to pristine polymer membranes. Moreover, GO swelling causes acceleration of diffusion of alkaline metal ions. The modified membrane with smaller pores shows slow diffusion of monovalent ions. The mechanism of fouling is the cake formation: the precipitate can be easily removed from the outer surface of the membranes. In all the cases, HZD–GO composite provides rejection of hardness ions (10–14%) as well as BSA and OVA (95–98%).

References

1. Roy S, Ragunath S (2018) Emerging membrane technologies for water and energy sustainability: future prospects, constrains and challenges. Energies 11:2997. https://doi.org/10.3390/en11112997
2. Babenyshev SP, Nesterenko PG, Bratsikhin AA et al (2018) Hydrodynamics and mass transfer with gel formation in a roll type ultrafiltration membrane. Foods Raw Mater 6(2):350–357
3. Macedo A, Monteiro J, Duarte E (2018) A contribution for the valorisation of sheep and goat cheese whey through nanofiltration. Membranes 8(4):114. https://doi.org/10.3390/membranes8040114
4. Santibáñez L, Córdova A, Astudillo-Castro C et al (2019) Effect of the lactose hydrolysis on galacto-oligosaccharides mixtures subjected to nanofiltration: a detailed fractionation analysis. Sep Purif Technol 222:342–351

5. Lin J, Ye W, Zen H et al (2015) Fractionation of direct dyes and salts in aqueous solution using loose nanofiltration membranes. J Membr Sci 477:183–193
6. Kamrani M, Akbari A (2018) Chitosan-modified acrylic nanofiltration membrane for efficient removal of pharmaceutical compounds. J Environ Chem Eng 6(1):583–587
7. Mulder M (1996) Basic principle of membrane technology. Springer, Dordrecht
8. Zahid M, Rashid A, Akram S et al (2018) A comprehensive review on polymeric nano-composite membranes for water treatment. J Membr Sci Technol 8:1–20
9. Marbelia L, Mulier M, Vandamme D et al (2016) Polyacrylonitrile membranes for microalgae filtration: influence of porosity, surface charge and microalgae species on membrane fouling. Algal Res 19:128–137
10. Chai X, Kobayashi T, Fujii N (1998) Ultrasound effect on cross-flow filtration of polyacrylonitrile ultrafiltration membranes. J Membr Sci 148(1):129–135
11. Scharnag N, Buschatz H (2001) Polyacrylonitrile (PAN) membranes for ultra- and microfiltration. Desalination 139(1–3):191–198
12. Chang IS, Lee CH (1998) Membrane filtration characteristics in membrane-coupled activated sludge system—the effect of physiological states of activated sludge on membrane fouling. Desalination 120(3):221–233
13. Lee NH, Amy G, Croue JP et al (2004) Identification and understanding of fouling in low-pressure membrane (MF/UF) filtration by natural organic matter (NOM). Water Res 38(20):4511–4523
14. Guo W, Ngo H-H, Li J (2012) A mini-review on membrane fouling. Bioresourse Technol 122:27–34
15. Shi X, Tal G, Hankin NP et al (2014) Fouling and cleaning of ultrafiltration membranes: a review. J Water Proc Eng 1:121–138
16. Herterich JC, Griffiths IM, Vella D (2019) Reproducing the pressure–time signature of membrane filtration: the interplay between fouling, caking, and elasticity. J Membr Sci 577:235–248
17. Zondervan E, Roffel B (2007) Evaluation of different cleaning agents used for cleaning ultra filtration membranes fouled by surface water. J Membr Sci 304(1–2):40–49
18. Tin MMM, Anioke G, Nakagoe O et al (2017) Membrane fouling, chemical cleaning and separation performance assessment of a chlorine-resistant nanofiltration membrane for water recycling applications. Separ Purif Technol 189:170–175
19. Kimura K, Uchida H (2019) Intensive membrane cleaning for MBRs equipped with flat-sheet ceramic membranes: controlling negative effects of chemical reagents used for membrane cleaning. Water Res 150:21–28
20. Kuzmenko D, Arkhangelsky E, Belfer S et al (2005) Chemical cleaning of UF membranes fouled by BSA. Desalination 179(1–3):323–333
21. Jiang T, Kennedy MD, van der Meer WGJ et al (2003) The role of blocking and cake filtration in MBR fouling. Desalination 157:335–343
22. Maximous N, Nakhla G, Wan W (2009) Comparative assessment of hydrophobic and hydrophilic membrane fouling in wastewater applications. J Membr Sci 339(1–2):93–99
23. Qu F, Liang H, Zhou J et al (2014) Ultrafiltration membrane fouling caused by extracellular organic matter (EOM) from *Microcystis aeruginosa*: effects of membrane pore size and surface hydrophobicity. J Membr Sci 449:58–66
24. Shen Y, Zhao W, Xiao K et al (2010) A systematic insight into fouling propensity of soluble microbial products in membrane bioreactors based on hydrophobic interaction and size exclusion. J Membr Sci 346(1):187–193
25. Gray S, Tsuru T, Cohen Y et al (2018) Advanced materials for membrane fabrication and modification. CRC Press, Boca Raton
26. Bildyukevich AV, Plisko TV, Liubimova AS et al (2017) Hydrophilization of polysulfone hollow fiber membranes via addition of polyvinylpyrrolidone to the bore fluid. J Membr Sci 524:537–549
27. Ulbricht M, Belfort G (1996) Surface modification of ultrafiltration membranes by low temperature plasma II. Graft polymerization onto polyacrylonitrile and polysulfone. J Membr Sci 111(2):193–215

28. Qin Y, Yang H, Xu Z et al (2018) Surface modification of polyacrylonitrile membrane by chemical reaction and physical coating: comparison between static and pore-flowing procedures. ACS Omega 3(4):4231–4241
29. Saki S, Uzal N, Ates N (2017) The size and concentration effects of Al_2O_3 nanoparticles on PSF membranes with enhanced structural stability and filtration performance. Desalination Water Treat 84:215–224
30. Pang R, Li X, Li J, Lu Z et al. (2014) Preparation and characterization of ZrO_2/PES hybrid ultrafiltration membrane with uniform ZrO_2 nanoparticles. Desalination 332:60–66
31. Myronchuk VG, Dzyazko YS, Zmievskii YG et al (2016) Organic-inorganic membranes for filtration of corn distillery. Acta Periodica Technologica 47:153–165
32. Zmievskii Y, Rozhdestvenska L, Dzyazko Y et al (2017) Organic-inorganic materials for baromembrane separation. Springer Proc Phys 195:675–686
33. Dzyaz'ko YS, Belyakov VN, Stefanyak NV et al (2006) Anion-exchange properties of composite ceramic membranes containing hydrated zirconium dioxide. Russ J Appl Chem 79(5):769–773
34. Dzyazko YS, Volfkovich YM, Sosenkin VE et al (2014) Composite inorganic membranes containing nanoparticles of hydrated zirconium dioxide for electrodialytic separation. Nanoscale Res Lett 9(1):271. https://doi.org/10.1186/1556-276X-9-27
35. Dzyazko YS, Rozhdestvenska LM, Vasilyuk SL, Kudelko KO, Belyakov VN (2017) Composite membranes containing nanoparticles of inorganic ion exchangers for electrodialytic desalination of glycerol. Nanoscale Res Lett 12:438. https://doi.org/10.1186/s11671-017-2208-4
36. Dzyazko YS, Rozhdestvenskaya LM, Zmievskii YG et al (2015) Organic-inorganic materials containing nanoparticles of zirconium hydrophosphate for baromembrane separation. Nanoscale Res Lett 10:64. https://doi.org/10.1186/s11671-015-0758-x
37. Dzyazko YS, Ponomareva LN, Volfkovich YM et al (2012) Effect of the porous structure of polymer on the kinetics of Ni^{2+} exchange on hybrid inorganic-organic ionites. Russ J Phys Chem 86(6):913–919
38. Dzyazko YS, Ponomaryova LN, Volfkovich YM et al (2014) Ion-exchange resin modified with aggregated nanoparticles of zirconium hydrophosphate. Morphology and functional properties. Microporous Mesoporous Mater 198:55–62
39. Said NN, Hamzah F, Ramlee NA et al (2018) The effect of TiO_2 particles addition on the characteristics of polysulfone membrane. Int J Adv Sci Eng Inform Technol 8(3):825–831
40. Tripathi BP, Dubey NC, Subair R (2016) Enhanced hydrophilic and antifouling polyacrylonitrile membrane with polydopamine modified silica nanoparticles. RSC Adv 6:4448–4457
41. Arumugham T, Amimodu RG, Kaleekkal NJ et al (2019) Nano CuO/g-C_3N_4 sheets-based ultrafiltration membrane with enhanced interfacial affinity, antifouling and protein separation performances for water treatment application. J Environ Sci 82:57–69
42. Kharaghani D, Kee JY, Qamar Khan Q et al (2018) Electrospun antibacterial polyacrylonitrile nanofiber membranes functionalized with silver nanoparticles by a facile wetting method. Eur Polym J 108:69–75
43. Sianipar M, Kim SH, Khoiruddin et al (2017) Functionalized carbon nanotube (CNT) membrane: progress and challenges. RSC Adv 7:51175–51198
44. Plisko TV, Bildyukevich AV (2014) Debundling of multiwalled carbon nanotubes in *N, N*-dimethylacetamide by polymers. Colloid Polym Sci 292(10):2571–2580
45. Plisko TV, Bildyukevich AV, Volkov VV et al (2015) Formation of hollow fiber membranes doped with multiwalled carbon nanotube dispersions. Pet Chem 55(4):318–332
46. Irfan M, Idris A, Yusof NM et al (2014) Surface modification and performance enhancement of nano-hybrid f-MWCNT/PVP90/PES hemodialysis membranes. J Membr Sci 467:73–84
47. Dreyer DR, Park S, Bielawski CW et al (2010) The chemistry of graphene oxide. Chem Soc Rev 39(1):228–240
48. Volfkovich YM, Filippov AN, Bagotsky VS (2014) Structural properties of different materials and powders used in different fields of science and technology. Springer, London
49. Volfkovich YM, Sosenkin VE (2012) Porous structure and wetting of fuel cell components as the factors determining their electrochemical characteristics. Russ Chem Rev 81(10):936–959

50. Volfkovich YM, Sakars AV, Volinsky AA (2005) Application of the standard porosimetry method for nanomaterials. Int J Nanotechnol 2(3):292–302
51. Volfkovich YM, Rychagov AY, Sosenkin VE (2014) Measuring the specific surface area of carbon nanomaterials by different methods. Russ J Electrochem 50(11):1099–1101
52. Volfkovich YM, Lobach AS, Spitsyna NG et al (2019) Hydrophilic and hydrophobic pores in reduced graphene oxide aerogel. J Porous Mat 26(4):1111–1119
53. Shulga YM, Baskakov SA, Baskakova YV et al (2017) Hybrid porous carbon materials derived from composite of humic acid and graphene oxide. Microporous Mesoporous Mater 245:24–30
54. Shulga YM, Baskakov SA, Baskakova YV et al (2018) Preparation of graphene oxide-humic acid composite-based ink for printing thin film electrodes for micro-supercapacitors. J Alloys Compd 730:88–95
55. Abraham J, Vasu KS, Williams CD et al (2017) Tunable sieving of ions using graphene oxide membranes. Nat Nanotechnol 12:546–550
56. Yeh CN, Raidongia K, Shao J et al (2015) On the origin of the stability of graphene oxide membranes in water. Nat Chem 7:166–170
57. Wang W, Eftekhari E, Zhu G et al (2014) Graphene oxide membranes with tunable permeability due to embedded carbon dots. Chem Commun 50(86):13089–13092
58. Xu C, Cui A, Xu Y et al (2013) Graphene oxide–TiO_2 composite filtration membranes and their potential application for water purification. Carbon 62:465–471
59. Zinadini S, Zinatizadeh A, Rahimi M et al (2014) Preparation of a novel antifouling mixed matrix PES membrane by embedding grapheme oxide nanoplates. J Membr Sci 453:292–301
60. Abdel-Karim A, Leaper S, Alberto M et al (2018) High flux and fouling resistant flat sheet polyethersulfone membranes incorporated with graphene oxide for ultrafiltration applications. Chem Eng J 334:789–799
61. Akbari A, Sheath P, Martin ST et al (2016) Large-area graphene-based nanofiltration membranes by shear alignment of discotic nematic liquid crystals of graphene oxide. Nat Commun 7:10891. https://doi.org/10.1038/ncomms10891
62. Marcano DC, Kosynkin DV, Berlin JM et al (2010) Improved synthesis of graphene oxide. ACS Nano 4(8):4806–4814
63. Dzyazko YS, Ogenko VM, Volfkovich YM et al (2018) Composite consisting of hydrated zirconium dioxide and grapheme oxide for removal of organic and inorganic components from water. Chem Phys Technol Surf 9(4):417–431
64. Bradford MM (1976) A rapid and sensitive method for the quantitation of microgram quantities of protein utilizing the principle of protein-dye binding. Anal Biochem 72(1–2):248–254
65. Szabo T, Tombacz E, Illes E et al (2018) Enhanced acidity and pH-dependent surface charge characterization of successively oxidized graphite oxides. Carbon 44:537–545
66. Helfferich F (1995) Ion exchange. Dover, New York
67. McCabe WL, Smith JC, Harriot P (2005) Unit operations of chemical engineering, 7th edn. McGraw-Hill, New York
68. Erickson HP (2009) Size and shape of protein molecules at the nanometer level determined by sedimentation, gel filtration, and electron microscopy. Biol Proced Online 11:32–51
69. Hermia J (1982) Constant pressure blocking filtration laws-application to power-law non-Newtonian fluids. Chem Eng Res Des 60a:183–187